Light Harvesting in Photosynthesis

FOUNDATIONS OF BIOCHEMISTRY AND BIOPHYSICS SERIES

Light Harvesting in Photosynthesis

Edited by

Roberta Croce
Department of Physics and Astronomy
Vrije Universiteit
Amsterdam, the Netherlands

Rienk van Grondelle
Department of Physics and Astronomy
Vrije Universiteit
Amsterdam, the Netherlands

Herbert van Amerongen
Laboratory of Biophysics
Wageningen University
Wageningen, the Netherlands

Ivo van Stokkum
Department of Physics and Astronomy
Vrije Universiteit
Amsterdam, the Netherlands

CRC Press
Taylor & Francis Group
Boca Raton London New York

CRC Press is an imprint of the
Taylor & Francis Group, an **informa** business

CRC Press
Taylor & Francis Group
6000 Broken Sound Parkway NW, Suite 300
Boca Raton, FL 33487-2742

First issued in paperback 2021

ISBN 13: 978-0-367-78149-1 (pbk)
ISBN 13: 978-1-4822-1835-0 (hbk)

Library of Congress Cataloging-in-Publication Data

Names: Croce, Roberta, editor. | Grondelle, Rienk van, editor. | Amerongen, Herbert van, editor. | Stokkum, Ivo van, editor.
Title: Light harvesting in photosynthesis / [edited by] Roberta Croce, Rienk van Grondelle, Herbert van Amerongen, Ivo van Stokkum.
Other titles: Foundations of biochemistry and biophysics.
Description: Boca Raton : Taylor & Francis/CRC Press, 2017. | Series: Foundations of biochemistry and biophysics | Includes bibliographical references.
Identifiers: LCCN 2017037100 | ISBN 9781482218350 (hardback : alk. paper)
Subjects: | MESH: Photosynthesis--physiology | Biophysical Phenomena | Light-Harvesting Protein Complexes--metabolism | Plants--metabolism
Classification: LCC QK882 .L47 2017 | NLM QK 882 | DDC 572/.45--dc23
LC record available at https://lccn.loc.gov/2017037100

Visit the Taylor & Francis Web site at
http://www.taylorandfrancis.com

and the CRC Press Web site at
http://www.crcpress.com

Contents

Preface

This book introduces the basic physical, chemical, and biological principles underlying the first steps in photosynthesis: light absorption, excitation energy transfer, and charge separation. In Part 1, we introduce pigments and their spectroscopic/redox properties. In Part 2, pigment-proteins as they occur in various natural systems (plants, algae, photosynthetic bacteria) are described, including the regulation of light harvesting. Part 3 deals with the physics underlying light harvesting: energy transfer and electron transport. Part 4 introduces basic and advanced spectroscopic methods, including data analysis. In Part 5, we discuss artificial and natural photosynthetic systems, how they are assembled, and what the energy transfer properties are.

Editors

Roberta Croce, PhD, (1968) studied chemistry at the University of Padova and earned her PhD at the University of Milano. Since 2011, she is a full professor in biophysics of photosynthesis/energy and head of the Biophysics Group at Vrije Universiteit Amsterdam, the Netherlands. She served as cochair of the International Congress of Photosynthesis Research held in Maastricht in 2016. She has published more than 130 scientific articles. Since 2013, she is a member of the Royal Holland Society of Sciences and Humanities (KHMW).

Rienk van Grondelle, PhD, (1949) studied physics at Vrije Universiteit Amsterdam (VU) and earned his PhD in Leiden. He joined the physics faculty of the VU in 1982, where he was appointed full professor in 1987. He built up a large research group and has made major contributions to elucidating the fundamental physical mechanisms that underlie light harvesting and charge separation. In 2001, he became a member of the Royal Netherlands Academy of Arts and Sciences (KNAW). In 2009, he was awarded an Academy Professorship for his outstanding contributions to the understanding of the first stage in photosynthesis. To date, he has published more than 600 scientific papers, which have attracted about 27K citations.

Herbert van Amerongen, PhD, (1959) studied physics at Vrije Universiteit Amsterdam and earned his PhD in biophysics at the same university. Since 2002, he is a full professor of biophysics at Wageningen University, where he is also director of the MicroSpectroscopy Research Facility. He published more than 170 publications in peer-refereed journals and, together with Leonas Valkunas and Rienk van Grondelle, he is author of the book *Photosynthetic Excitons*. In 2016, he served as cochair of the 17th International Congress of Photosynthesis Research in Maastricht.

Ivo van Stokkum, PhD, (1962) is associate professor of computational biophysics in the Department of Physics and Astronomy, Faculty of Science, at the Vrije Universiteit, Amsterdam. He earned his master's degree in experimental physics and his PhD in natural sciences from the Radboud University Nijmegen. He has coauthored 270 scientific publications, which have attracted more than 14K citations.

Contributors

Noam Adir
Schulich Faculty of Chemistry
The Technion
Haifa, Israel

Pascal van Alphen
Molecular Microbial Physiology Group
Swammerdam Institute for Life Sciences
University of Amsterdam
Amsterdam, the Netherlands

Herbert van Amerongen
Laboratory of Biophysics
Wageningen University
Wageningen, the Netherlands

Jessica M. Anna
Department of Chemistry
University of Pennsylvania
Philadelphia, Pennsylvania

Leeat Bar-Eyal
Department of Plant and Environmental
Sciences
The Hebrew University of Jerusalem
Jerusalem, Israel

Roberto Bassi
Department of Biotechnology
University of Verona
Verona, Italy

Robert E. Blankenship
Departments of Biology and Chemistry
Washington University in St. Louis
St. Louis, Missouri

Egbert J. Boekema
Groningen Biomolecular Sciences and
Biotechnology Institute
University of Groningen
Groningen, the Netherlands

Claudia Büchel
Institute of Molecular Biosciences
Goethe University Frankfurt
Frankfurt, Germany

Min Chen
School of Life and Environmental Sciences (A12)
University of Sydney
Sydney, New South Wales, Australia

Jevgenij Chmeliov
Faculty of Physics
Department of Theoretical Physics
Vilnius University
and
Department of Molecular Compound Physics
Center for Physical Sciences and Technology
Vilnius, Lithuania

Roberta Croce
Department of Physics and Astronomy
Vrije Universiteit
Amsterdam, the Netherlands

Carles Curutchet
Department of Pharmacy and Pharmaceutical
Technology and Physical Chemistry
and
Institute of Biomedicine
University of Barcelona
Barcelona, Spain

Jacob C. Dean
Department of Physical Science
Southern Utah University
Cedar City, Utah

Gregory S. Engel
Department of Chemistry
University of Chicago
Chicago, Illinois

David A. Farmer
Department of Molecular Biology and
Biotechnology
University of Sheffield
Sheffield, United Kingdom

Harry A. Frank
Department of Chemistry
University of Connecticut
Storrs, Connecticut

Arvi Freiberg
Institute of Physics
and
Institute of Molecular and Cell Biology
University of Tartu
Tartu, Estonia

Győző Garab
Institute of Plant Biology
Biological Research Centre
Hungarian Academy of Sciences
Szeged, Hungary

Rienk van Grondelle
Department of Biophysics
Vrije Universiteit
Amsterdam, the Netherlands

Devens Gust
School of Molecular Sciences
Arizona State University
Tempe, Arizona

Jeremy Harbinson
Department of Plant Sciences
Wageningen University
Wageningen, the Netherlands

Klaas J. Hellingwerf
Molecular Microbial Physiology Group
Swammerdam Institute for Life Sciences
University of Amsterdam
Amsterdam, the Netherlands

Andrew Hitchcock
Department of Molecular Biology and
Biotechnology
University of Sheffield
Sheffield, United Kingdom

Susana F. Huelga
Institute of Theoretical Physics
Ulm University
Ulm, Germany

C. Neil Hunter
Department of Molecular Biology and
Biotechnology
University of Sheffield
Sheffield, United Kingdom

Michael R. Jones
School of Biochemistry
University of Bristol
Bristol, United Kingdom

Nir Keren
Department of Plant and Environmental Sciences
The Hebrew University of Jerusalem
Jerusalem, Israel

Diana Kirilovsky
Institute for Integrative
Biology of the Cell (I2BC)
Université Paris-Saclay
Saclay, France

Gerdenis Kodis
School of Molecular Sciences
Arizona State University
Tempe, Arizona

Manuel J. Llansola-Portoles
Institute for Integrative Biology of the Cell
Université Paris-Saclay
Gif-sur-Yvette cedex, France

Tomáš Mančal
Faculty of Mathematics and Physics
Charles University
Prague, Czech Republic

Benedetta Mennucci
Department of Chemistry
University of Pisa
Pisa, Italy

Ana L. Moore
School of Molecular Sciences
Arizona State University
Tempe, Arizona

Thomas A. Moore
School of Molecular Sciences
Arizona State University
Tempe, Arizona

David J. Mothersole
Department of Molecular Biology and
Biotechnology
University of Sheffield
Sheffield, United Kingdom

Lauren Nicol
Department of Physics and Astronomy
Vrije Universiteit
Amsterdam, the Netherlands

Vladimir I. Novoderezhkin
A. N. Belozersky Institute of Physico-Chemical
Biology
Moscow State University
Moscow, Russia

Yossi Paltiel
Department of Applied Physics
The Hebrew University of Jerusalem
Jerusalem, Israel

Alberta Pinnola
Department of Biotechnology
University of Verona
Verona, Italy

Martin B. Plenio
Institute of Theoretical Physics
Ulm University
Ulm, Germany

Tomáš Polívka
Faculty of Science
University of South Bohemia
České Budějovice, Czech Republic

Jakub Pšenčík
Faculty of Mathematics and Physics
Charles University
Prague, Czech Republic

Tönu Pullerits
Division of Chemical Physics
Lund University
Lund, Sweden

Fabrice Rappaport (deceased)
Institut de Biologie Physico-Chimique
University Pierre and Marie CURIE
Paris, France

Thomas Renger
Institut für Theoretische Physik
Johannes Kepler Universität Linz
Linz, Austria

Bruno Robert
Institute of Integrative Biology of the Cell
University Paris South
Atomic Energy Commission
Saclay, France

Gregory D. Scholes
Department of Chemistry
Princeton University
Princeton, New Jersey

Dmitry A. Semchonok
Groningen Biomolecular Sciences and
Biotechnology Institute
University of Groningen
Groningen, the Netherlands

Anat Shperberg-Avni
Department of Biochemistry
Weizmann Institute of Science
Rehovot, Israel

Ivo van Stokkum
Department of Physics and Astronomy
Vrije Universiteit
Amsterdam, the Netherlands

Villy Sundström
Division of Chemical Physics
Lund University
Lund, Sweden

Leonas Valkunas
Faculty of Physics
Department of Theoretical Physics
Vilnius University
and
Department of Molecular Compound Physics
Center for Physical Sciences and Technology
Vilnius University
Vilnius, Lithuania

Katherine WongCarter
School of Molecular Sciences
Arizona State University
Tempe, Arizona

Donatas Zigmantas
Division of Chemical Physics
Lund University
Lund, Sweden

PART 1

Building the Light-Harvesting Apparatus: Pigments

Pigments: General properties and biosynthesis

MIN CHEN AND ROBERT E. BLANKENSHIP

1.1 PIGMENTS OVERVIEW

For the energy of sunlight to be converted and stored into biological systems, it must first be captured by the pigments present in the organisms. There are three major classes of pigments involved in photosynthesis: (bacterio)chlorophylls, carotenoids, and phycobilins. Each pigment interacts with light to absorb only a narrow range of the spectrum and reflects only certain wavelengths of light so as to produce their distinctive colors (Figure 1.1). The photosynthetic organisms broaden their light absorption region and increase their optical cross section by combining these various pigments, which have different maximum absorption peaks. Such a "rainbow" array of pigments permits photosynthetic organisms to capture maximally most of the available light; thus there is the opportunity for competitive advantage in any particular habitat by developing the most effective combination of pigments.

Chlorophylls are greenish molecules because they absorb mainly in the blue or near-UV region and red or near-infrared spectral region and leave a considerable gap in the green wavelength region. Carotenoids have their reddish, orange, or yellow colors due to their major absorption in the blue spectral region of 420–570 nm. Phycobilins are a group of molecules that absorb the light in the wavelength region of 500–650 nm, reflecting the blue and red photons and producing their blue or pink colors.

Chlorophylls are found in all oxygenic photosynthetic organisms, including plants, algae, and cyanobacteria, while bacteriochlorophylls occur in certain phototrophic bacteria. All chlorophylls have their common structural elements: a stable ring-shaped molecule with a magnesium atom in the center and an attached long carbon–hydrogen side chain (phytyl chain) (Figure 1.2). The electrons are free to migrate around the system of alternating single and double bonds (conjugated double bonds) in a chlorophyll macrocycle that provides the potential to

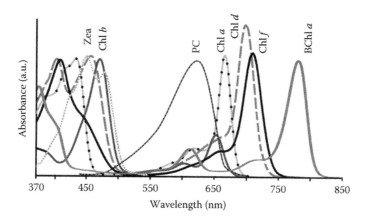

Figure 1.1 Absorption spectra of isolated pigments in 100% methanol (except isolated phycocyanin, which is recorded in 50 mM phosphate buffer, pH = 7.0). Chl, chlorophyll; PC, isolated phycocyanin; Zea, zeaxanthin.

gain or lose electrons easily. This plays a fundamental role for chlorophyll capturing the energy of light. There are five naturally occurring forms of chlorophylls: chlorophylls *a*, *b*, *c*, *d*, and *f*. Chlorophyll *a* (Chl *a*) is the most widespread pigment in nature (Björn et al. 2009). It is the only chlorophyll that is the primary electron donor in reaction centers of oxygen-evolving phototrophs, with the notable exception of the newly discovered prokaryotic oxygenic photosynthetic organism: *Acaryochloris marina*, where Chl *d* acts as the special pair P740 in photosystem I and almost entirely substitutes for Chl *a* (Hu et al. 1998).

There are six naturally occurring types of bacteriochlorophylls (BChl), all of which are found in the anoxygenic (non-oxygen-evolving) phototrophic bacteria: bacteriochlorophylls *a*, *b*, *c*, *d*, *e*, and *g* (Figure 1.3). The minor structural modifications on the periphery of the macrocycles of chlorophyll change the electron migration profile around the macrocycle structure and generate the different absorption features of chlorophylls (Blankenship 2014; Chen and Blankenship 2011).

Carotenoids are an important group of pigments in photosynthetic organisms. They function as accessory light-harvesting pigments to capture the energy of light in the blue-green light spectral region, where (bacterio)chlorophylls do not absorb efficiently. The absorbed energy then is transferred to the (bacterio)chlorophyll. Another important function of carotenoids is photoprotection. They protect the organisms from photodamage by quenching both singlet or triplet states of

(bacterio)chlorophylls under strong illumination (see Chapter 3). In chloroplasts, carotenoids also function as photosynthetic membrane stabilizers (Hayaux 1998). There are many hundreds of chemically distinct carotenoids, but they can be classified into two broad groups: xanthophylls and carotenes (Figure 1.4). Xanthophylls are carotenoids containing oxygen, such as zeaxanthin in cyanobacteria and plants, fucoxanthin in algae, and lutein in higher plants. Carotenes are oxygen-free carotenoids, which typically only contain carbon and hydrogen such as α-carotene and β-carotene (Figure 1.4). The most common backbone of carotenoid chains contains 40 carbons with 10–13 conjugated double bonds, which allows electrons to move freely across this area of the molecule.

There are a large number of various xanthophylls based on their oxygen-containing functional groups: carbonyl, epoxy, formyl, hydroxyl, methoxyl, or oxo groups. In green plants, the composition of carotenoids in chloroplasts is remarkably similar, and most chloroplast carotenoids are bound together with chlorophylls located in the photosynthetic membranes (thylakoid membranes) (Demming-Adams and Adams 1996; Ruiz-Sola and Rodriguez-Concepcion 2012). Both chlorophylls and carotenoids are commonly lipophilic due to the presence of long unsaturated aliphatic hydrocarbon chains, located in membrane-bound protein complexes. Xanthophylls are involved in the xanthophyll cycle in algae and higher plants, but there may be other uncharacterized roles for carotenoids (Chapter 3).

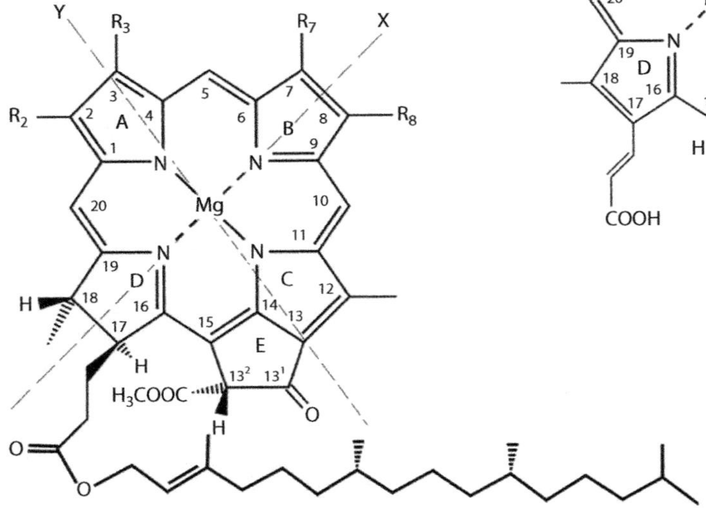

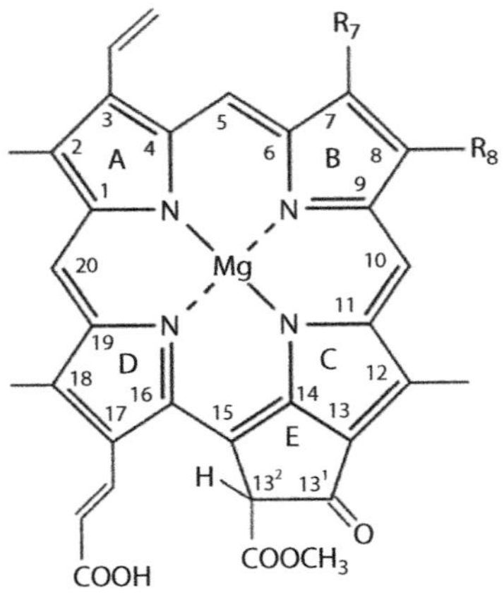

Porphyrin macrocycle

	R$_7$	R$_8$
Chl c1	CH$_3$	CH$_2$–CH$_3$
Chl c2	CH$_3$	CH=CH$_2$
Chl c3	COOCH$_3$	CH=CH$_2$

Chlorin macrocycle

	R$_2$	R$_3$	R$_7$	R$_8$
Chl a	CH$_3$	CH=CH$_2$	CH$_3$	CH$_2$–CH$_3$
Chl b	CH$_3$	CH=CH$_2$	CHO	CH$_2$–CH$_3$
Chl d	CH$_3$	CHO	CH$_3$	CH$_2$–CH$_3$
Chl f	CHO	CH=CH$_2$	CH$_3$	CH$_2$–CH$_3$
8-vinyl Chl a	CH$_3$	CH=CH$_2$	CH$_3$	CH=CH$_2$
8-vinyl Chl b	CH$_3$	CH=CH$_2$	CHO	CH=CH$_2$

Figure 1.2 Structures of chlorophylls. The numbering scheme is based on the current IUPAC standard system. The axis of x and y indicate the direction of Q$_x$ and Q$_y$ electronic transitions (see Chapter 2.). R$_\#$ refers to ring substitutions at the corresponding carbon positions.

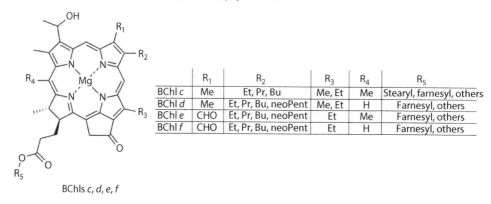

Figure 1.3 Structures of BChl *a* and of BChls *b* and *g* represented by the subunit changes replacing the respective BChl *a* substituents (as indicated). The structure of BChls *c*, *d*, *e*, and *f* represented with the R$_\#$ refers to ring substitutions at the corresponding carbon positions (as indicated).

Phycobilins are unique chromophores among the photopigments in that a thioether bond covalently links them to water-soluble proteins, phycobiliproteins. They are open-chain tetrapyrroles and are found in cyanobacteria as well as in some eukaryotic algal protists: glaucocystophytes, rhodophytes, and cryptophytes, but not in green algae and higher plants. The phycobiliproteins absorb in the red, orange, yellow, and green light regions; the wavelengths 500–650 nm are not well absorbed by chlorophylls (Figure 1.1). The three main classes of phycobiliproteins are allophycocyanin, phycocyanin, and phycoerythrin (MacColl 1998). Both phycobiliproteins and carotenoids function as accessory light-harvesting components and pass their absorbed light energy to chlorophyll for photosynthesis, but are not directly involved with photochemical reactions in the reaction center (see Chapter 5).

1.2 CHLOROPHYLLS AND BACTERIOCHLOROPHYLLS

1.2.1 Chemical structure and distribution

Several types of chlorophylls, Chls *a*, *b*, *c*, *d*, and *f*, serve in various oxygenic photosynthetic organisms (Table 1.1). All chlorophylls except the group of Chl *c* have a chlorin macrocycle and a phytyl chain as shown in Figure 1.2. Chl *c* is the common name for an increasing number of Mg-pheoporphyrins, which are found mainly in marine algal protists—in the chromophytes (with the exception of eustigmatophytes) (Larkum and Barrett 1983; Stauber and Jeffrey 1988). The molecule of Chl *c* has the fully unsaturated porphyrin macrocycle and does not carry a phytyl chain (Zapata et al. 2006).

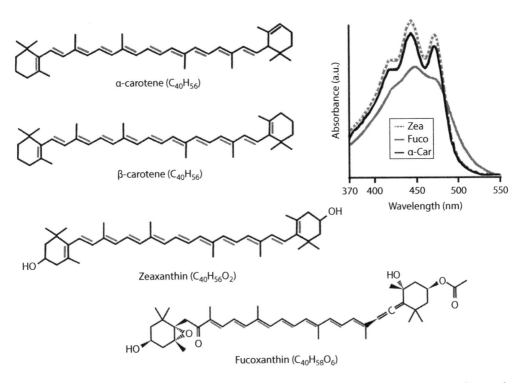

Figure 1.4 Structures of several carotenoids present in photosynthetic systems. Inset shows absorption spectra of several carotenoids in 100% methanol. The conjugated double bonds are indicated. α-Car, α-carotene; Fuco, fucoxanthin; Zea, zeaxanthin.

Table 1.1 Chlorophylls and phycobiliprotein functional distributions

		Chlorophyll					Phycobiliproteins
		a	b[a]	c[b]	d[c]	f[d]	
Cyanobacteria	LHC complexes	+	+		+	+	+
	RC complexes	+			+	?	
Algae	LHC complexes	+	+	+			+
	RC complexes	+					
Green plants	LHC complexes	+	+				
	RC complexes	+					

[a] Chl b is only found in prochlorophytes, a group of cyanobacteria containing Chl b (or 8-vinyl Chl b).
[b] Chl c represents a number of chlorophylls having Mg-pheoporphyrins.
[c] Chl d is only found in *Acaryochloris marina* spp.
[d] Chl f was reported in some cyanobacteria, but its function in photosynthesis is undefined yet (see text).

It had generally been accepted that Chl a exists in all known oxygenic photosynthetic organisms discovered to date (Scheer 2006). It occurs in both reaction center and all chlorophyll-bound light-harvesting complexes (Table 1.1). The importance of Chl a in the energy phase of oxygenic photosynthesis is made known as their special function in the reaction centers involving charge separation and passing of electrons on to the transport chain.

Chl b is the major accessory light-harvesting pigment in green algae and plants (see Chapter 4), and additionally, it has been found in the prochlorophytes, cyanobacteria containing Chl a and Chl b.

Up to date, no Chl *b* has been found in the reaction centers or the core antenna systems, such as CP43 and CP47 proteins in photosystem II (Chen and Scheer 2013).

Chl *d* differs from Chl *a* only in the C3 position at ring A, where a formyl group in Chl *d* replaces the vinyl group in Chl *a* (Figure 1.2). Chl *d* was first reported from a red algal pigment extract in 1943 (Manning and Strain 1943), and afterward it was considered as an artificial by-product of pigment extracts and not present in any organisms (Scheer 1991). However, this belief changed with the discovery of a cyanobacterium that contains over 95% Chl *d* as its major pigment in 1996 (Miyashita et al. 1996). Chl *d* is the only known chlorophyll to date that can replace the function of Chl *a* as a special pair in the reaction center in oxygenic photosynthetic systems (Hu et al. 1998; Itoh et al. 2007; Tomo et al. 2007). Interestingly, Chl *d* has been reported to occur widely around the world since then, although all cyanobacteria containing Chl *d* belong to one monophylogenetic clade, *Acaryochloris marina*, based on 16S rRNA classification (Kashiyama et al. 2008; Kühl et al. 2005; Loughlin et al. 2013; Miller et al. 2005).

Chl *f* is the most red-shifted chlorophyll found in oxygenic photosynthetic organisms to date (Chen et al. 2010; Kräutler 2011). It differs from Chl *a* only in the C2 position at ring A, where a formyl group in Chl *f* replaces the methyl group in Chl *a* (Figure 1.2) (Willows et al. 2013). It co-occurs with Chl *a* and other pigments in the cyanobacterium *Halomicronima hongdechloris*, isolated from stromatolite colonies, and also in the cyanobacterium CK1 collected from Biwa Lake, a freshwater lake in Japan (Akutsu et al. 2011; Chen et al. 2012). Several cyanobacteria containing Chl *f* belong to two unrelated genera and have different morphological characteristics (Gan et al. 2015). *H. hongdechloris* is a marine filamentous cyanobacterium having allophycocyanin and allophycocyanin (Chen et al. 2012); in contrast, cyanobacterium strain CK1 is a unicellular cyanobacterium containing phycocyanin, allophycocyanin, and phycoerythrin (Akutsu et al. 2011). The function of Chl *f* is uncharacterized yet; however, it is the first chlorophyll reported that is responsive to the changes of light spectrum. Chl *f* has been named as a red-light-induced chlorophyll (Chen et al. 2012; Gan et al. 2014).

Two additional chlorophyll derivatives are 3,8-divinyl Chl *a* (also named 8-vinyl Chl *a*) and 3,8-divinyl Chl *b* (also named 8-vinyl Chl *b*) (Figure 1.2). They are the major chlorophylls found in *Prochlorococcus* spp. and replace all functions of Chl *a* and Chl *b* in oxygenic photosynthesis in these organisms (Chisholm et al. 1992).

Pheophytins are a group of metal-free chlorophylls. Acidic conditions promote the displacement of the metal (Mg^{2+}) by two hydrogen ions. For each single chlorophyll (Chls *a*, *b*, *d*, and *f*), there is one corresponding pheophytin, which has a green color as it has a similar absorption spectral profile as their corresponding chlorophyll. Pheophytins are often regarded as primarily degradation products of chlorophylls from the loss of the central metal, Mg. However, the role of pheophytin *a* in photosystem II is an essential early electron acceptor in the sequence of electron carriers. *A. marina*, the Chl *d*–containing cyanobacterium, uses pheophytin *a* instead of pheophytin *d* as primary electron acceptor in photosystem II, although the percentage of Chl *a* is less than 5% of total chlorophyll (Tomo et al. 2007). Nevertheless, the biosynthetic pathway for making pheophytin *a* is not yet defined *in vivo*.

BChl *a* is widely distributed in many types of anoxygenic phototrophs. BChl *b* is found only in a few species of purple phototrophic bacteria, where it substitutes for BChl *a*. BChls *c*, *d*, and *e* are found in the green sulfur bacteria (GSB), where they are localized in chlorosome antenna complexes. In most cases, only one of these "chlorobium chlorophylls" is found in a given organism, depending on the light environment where the cells live. BChl *f* is not known from any naturally occurring organism, but has been produced in mutants of GSB. All GSB also contain BChl *a*, which is located in the chlorosome baseplate, the Fenna–Matthews–Olson (FMO) protein, and the reaction center. The filamentous anoxygenic phototrophs, formerly called the green nonsulfur bacteria, contain BChl *a*, and some species, such as the widely studied organism *Chloroflexus aurantiacus*, also contain BChl *c* localized in chlorosomes. BChl *g* is found only in the Gram-positive heliobacteria, where it is almost the only chlorophyll pigment found. Small quantities of a derivative of Chl *a* are found in the reaction centers of both the heliobacteria and the green sulfur bacteria, where it is proposed to function as an early electron acceptor.

1.2.2 Chlorophyll biosynthesis

1.2.2.1 FORMATION OF PROTOPORPHYRIN IX

Chlorophyll biosynthesis contains multiple enzymatic steps, which begin with the formation of δ-aminolevulinic acid (δ-ALA), the first common precursor in the biosynthesis of all tetrapyrroles including chlorophyll, heme, and phycobilins (Chen 2014). There are two different biosynthetic pathways of ALA: one is related to heme biosynthesis in mitochondria and another is related to the pigment biosynthesis in chloroplasts. In the chloroplasts of plants, ALA is synthesized from glutamic acid (Glu) via a two-step reaction: the reduction of Glu to glutamate 1-semialdehyde and the aminomutation of glutamate 1-semialdehyde to form ALA (Figure 1.5). The formation of ALA is one of a very small number of reactions known in biology in which a tRNA molecule is used in a molecular synthesis other than peptide synthesis. On the other hand, a completely different route for making ALA is defined in mitochondria of all eukaryotic cells, which is a one-step condensation from glycine and succinyl-CoA catalyzed by ALA synthase. Interestingly, the reaction is also found in some purple photosynthetic bacteria (Zappa and Bauer 2010) and chloroplasts of *Chromera velia* for chlorophyll biosynthesis (Kořený et al. 2011).

Then eight molecules of ALA form an open-chain tetrapyrrole hydroxymethylbilane via porphobilinogen (PBG) synthesis. The next four steps, including ring closure, oxidative decarboxylations, and the oxidation of protoporphyrinogen IX, lead to the formation of the symmetric metal-free porphyrin, protoporphyrin IX (Figure 1.5).

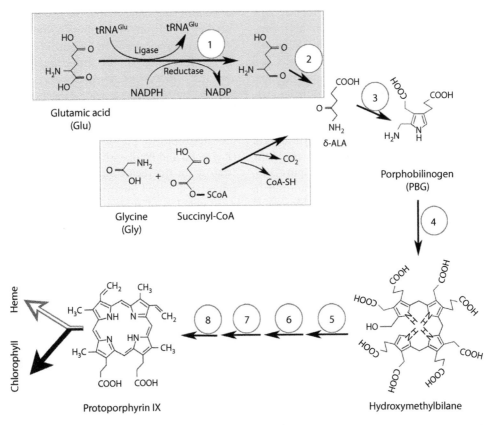

Figure 1.5 Outlined pathway of protoporphyrin IX biosynthesis from Glu. The enzymes that catalyze the individual reactions are (1) glutamyl-tRNA ligase and glutamyl-tRNA reductase, (2) glutamate 1-semialdehyde aminotransferase, (3) PBG synthase, (4) hydroxymethylbilane synthase, (5) uroporphyrinogen III synthase (ring closure), (6) uroporphyrinogen III decarboxylase, (7) coproporphyrinogen III oxidative decarboxylase, and (8) protoporphyrinogen IX oxidase. Two pathways of δ-ALA biosynthesis are highlighted.

Protoporphyrin IX is the first photosensitive colored intermediate and the last common precursor for chlorophyll and heme biosynthesis (Figure 1.5). Free protoporphyrin IX molecules can absorb photons; therefore, the concentration of free protoporphyrin and other photosensitive intermediates after this step is tightly regulated *in vivo* to avoid photodamage.

1.2.2.2 FORMATION OF PROTOCHLOROPHYLLIDE *a*

Inserting the magnesium into protoporphyrin IX and the formation of Mg-protoporphyrin IX is catalyzed by Mg-chelatase, a three-component enzyme (Figure 1.6). The three subunits, ChlD, ChlH, and ChlI, are conserved from cyanobacteria to higher plants and are commonly referred to as BchI, BchD, and BchH in bacteriochlorophyll biosynthesis (Chew and Bryant 2007). ChlH is the porphyrin and Mg^{2+} the binding subunit of Mg-chelatase (Willows et al. 1996). ChlI and ChlD require ATP and Mg^{2+} for the interaction of I-D complex formation (Jensen et al. 1999). The ChlI:ChlD oligomer subsequently interacts with the ChlH subunit and drives the ATP-dependent insertion of Mg^{2+} into protoporphyrin IX (Masuda 2008; Mochizuki et al. 2010). The insertion of magnesium needs a significant amount of energy, at least the hydrolysis of ~15 ATP (Reid and Hunter 2004). The synthesis of Mg-protoporphyrin IX is the first unique step in the biosynthesis of chlorophyll (step 9 in Figure 1.6). Therefore, Mg-chelatase plays an important role in channeling protoporphyrin IX into the chlorophyll branch in response to conditions suitable for photosynthetic growth, competitively separating the intermediates from the heme biosynthetic branch. The Mg insertion can be described as follows:

$$ATP + protoporphyrin\ IX + Mg^{2+} + H_2O \rightarrow ADP$$
$$+ phosphate + Mg\text{-}protoporphyrin\ IX + 2\,H^+$$

in which the four substrates are ATP, protoporphyrin IX, Mg^{2+}, and water. This reaction is catalyzed by Mg-chelatase.

Mg chelation is tightly coupled to the next step, the methyltransferase reaction for the transfer of a methyl group to the 13-propionate side chain of Mg-protoporphyrin IX and produces Mg-protoporphyrin IX monomethyl ester (step 10 in Figure 1.6). The next step of the biosynthesis is the oxidative cyclization of isocyclic ring E and generates the intermediate, divinyl protochlorophyllide. The formation of the fifth ring (ring E) structure represents the distinctive macrocycle feature of chlorophylls from all other tetrapyrroles (step 11 in Figure 1.6). This cyclization includes a complex of reactions that proceeds by first esterifying the carboxylic acid moiety and then undertakes a stereospecific oxidative cyclase reaction. There are two unrelated cyclization mechanisms based on the origin of the oxygen atom, the aerobic cyclase and the anaerobic cyclase (Ouchane et al. 2004; Raymond and Blankenship 2004). Most oxygenic photosynthetic organisms use aerobic cyclases carrying out the Mg-protoporphyrin IX monomethyl ester oxidative cyclic reaction in an oxygen-dependent manner. The anaerobic cyclase is a radical SAM enzyme and is widely distributed in anaerobic anoxygenic photosynthetic organisms such as green sulfur bacteria. However, the multistep formation of the fifth ring suggests that the reactions are catalyzed by a multisubunit enzymatic complex, which is still uncharacterized (Hollingshead et al. 2012).

The formation of the fifth ring is a decisive step for chlorophyll biosynthesis. The coexisting oxygen-dependent and oxygen-independent Mg-protoporphyrin IX methyl ester cyclases could represent the evolutionary transition from the anaerobic to the aerobic metabolic reaction (Raymond and Blankenship 2004), or an adaptation strategy for the regulation of reactions under continuously changed microenvironmental conditions. Accumulation of molecular oxygen in the atmosphere fundamentally changed the redox balance on Earth, permitted the development of aerobic metabolism, and led to development of advanced life forms (Samuilov 2005). By increasing the oxygen concentration in the atmosphere, ancient life forms, living in anaerobic or very limited accessible oxygen concentrations, had to adapt and evolve new strategies to deal with the increased oxygen concentration, including reactive oxygen species generated by excited porphyrin molecules in the cells. Nowadays, oxygen is directly involved in chlorophyll biosynthesis (Hohmann-Marriott and Blankenship 2011). It also has a close relationship with photoregulation of these processes (Kopp et al. 2005).

1.2.2.3 CHLOROPHYLL *a* BIOSYNTHESIS

From divinyl protochlorophyllide, two similar, yet separate, reactions on ring B and ring D are

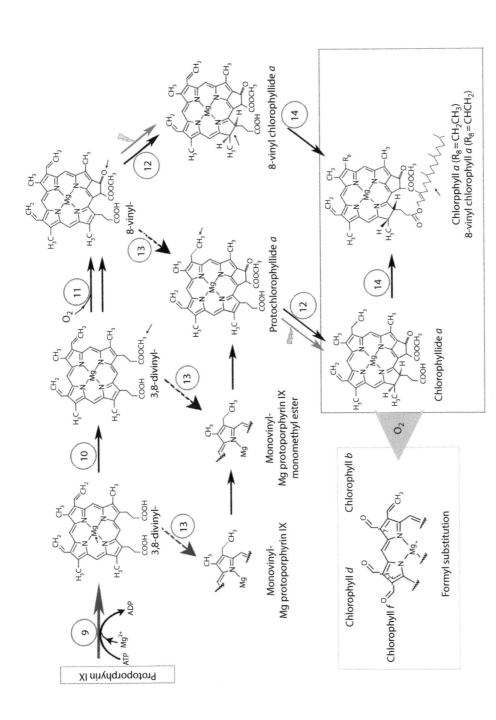

Figure 1.6 Outlined pathway of chlorophyll biosynthesis from protoporphyrin IX. The enzymes that catalyze the individual reactions are (9) protoporphyrin IX Mg-chelatase, (10) S-adenosyl-L-methionine:Mg-protoporphyrin IX methyltransferase, (11) Mg-protoporphyrin IX monomethyl ester oxidative cyclase (oxygen dependent or oxygen independent), (12) light-dependent NADPH:protochlorophyllide oxidoreductase or light-independent protochlorophyllide oxidoreductase, (13) 8-vinyl reductase, and (14) chlorophyll synthase. The arrows with dashed line relate the possible broad substrate specificity of 8-vinyl reductase. The triangle (with O_2) indicates the modification of Chls b, d, and f from either chlorophyll a or chlorophyllide a.

catalyzed by two enzymes, 8-vinyl reductase and protochlorophyllide oxidoreductase, respectively (steps 12 and 13 in Figure 1.6). Both enzymes are not sensitive to whether the other changes made at the opposite side of the macrocycle, that is, the 8-vinyl reductase, have taken place, so that either the reaction catalyzed by protochlorophyllide oxidoreductase or vice versa can take place (Figure 1.6). The 8-vinyl reductase step is the earliest step in chlorophyll biosynthesis that is responsible for the chemical diversity in this class of molecules. Marine *Prochlorococcus* spp. are exceptional because they lack the reductase for the 8-vinyl group (Nagata et al. 2005) and thus produce divinyl Chl *a* and divinyl Chl *b* (Chisholm et al. 1992). In higher plants and cyanobacteria, divinyl chlorophylls were observed in divinyl reductase (*dvr*) mutants. Those mutants grow photosynthetically under moderate and low-light conditions with reduced efficiency of energy transfer between light-harvesting components and reaction centers, but died within a day after being transferred to high-light conditions (Islam et al. 2008). One possible explanation is that the 8-vinyl group causes a higher degradation rate of chlorophyll-binding protein complexes than the recycle rate of the complexes, leading to photodamage under high-light conditions. This may be the reason for the common presence of 8-ethyl chlorophylls in photosynthetic organisms instead of 8-vinyl chlorophylls. However, the 8-vinyl chlorophylls in *Prochlorococcus* provide an advantage by absorbing wavelengths of light unused by other marine oxygenic phototrophs. The modified photosynthetic protein sequences are reported, and such distinct site modification plays an important role in increasing the stability of 8-vinyl chlorophyll-binding protein complexes (Ito and Tanaka 2011). Two types of 8-vinyl reductase are identified: one uses NADPH as reductant and another requires FAD and iron for activity. The potential application of unrelated enzymes to catalyze a single reaction is common in chlorophyll biosynthetic reactions, such as Mg-protoporphyrin IX methyl ester cyclization, or 8-vinyl reductase and protochlorophyllide reductase (Chen et al. 2016). The strategy for using unrelated enzymes to catalyze the same reaction provides the photosynthetic organisms with the capability to grow under particular environments. In higher plants, one 8-vinyl reductase is responsible for the multibranched chlorophyll biosynthesis

due to the broad substrate specificity of the enzyme (step 13 in Figure 1.6), although with significant differences in the efficiency of enzyme reactions (Wang et al. 2013).

The reduction of the carbon–carbon double bond between C17 and C18 in ring D is the step converting porphyrin-type macrocycle to chlorin-type macrocycle (step 12 in Figure 1.6). Chlorin-type macrocycle is the central structure for Chls *a*, *b*, *d*, and *f*, and porphyrin-type macrocycle forms the group of Chl *c*s. The reduction of the double band between C17 and C18 is involved directly with regulating plant development and the assembly of the photosynthetic apparatus. There are two unrelated enzymes that catalyze this reaction, the light-dependent protochlorophyllide oxidoreductase (LPOR) and the light-independent protochlorophyllide oxidoreductase (Tanaka and Tanaka 2007). The LPOR has an unconditional requirement for light to catalyze a hydride transfer reaction from NADPH to protochlorophyllide (Heyes et al. 2003). Upon illumination, light energy is captured into LPOR–enzyme complexes by protochlorophyllide, and a photochemical reaction occurs to yield chlorophyllide *a* (Gabruk and Mysliwa-Kurdziel 2015). This enzyme can be found in cyanobacteria, and all eukaryotic oxygenic photosynthetic organisms and some photosynthetic organisms contain more than one isoforms of LPOR. Importantly, LPOR catalytic site residues are highly conserved and essential for the catalytic function (Reinbothe et al. 2006).

Light-independent protochlorophyllide oxidoreductase is composed of three subunits, ChlL, ChlB, and ChlN, and uses ferredoxin as reductant instead of NADPH. It is ubiquitously distributed among prokaryotic phototrophs, including oxygenic photosynthetic prokaryotic organisms (cyanobacteria) and anoxygenic photosynthetic prokaryotic organisms (such as purple bacteria and green sulfur bacteria). However, it is not so universally distributed in eukaryotic photosynthetic organisms. It is only reported in some lineages of eukaryotic photosynthetic organisms: green algae (Choquet et al. 1992; Shi and Shi 2006), mosses (Kohchi et al. 1988), glaucophytes (Stirewalt et al. 1995), and red algae (Reith and Munholland 1995). Interestingly, cyanobacteria and green algae such as *Chlamydomonas reinhardtii* have both LPOR and light-independent protochlorophyllide oxidoreductase, making them model organisms for the study of the relationship between light-dependent and light-independent protochlorophyllide oxidoreductases.

The final step of chlorophyll a biosynthesis is the attachment of the phytol chain, catalyzed by "chlorophyll synthase." The phytol serves to make the pigment more hydrophobic and facilitates its binding to proteins, especially integral membrane proteins such as reaction centers and many antenna complexes. The reaction can be summarized as follows:

chlorophyllide a + phytol diphosphate chlorophyll a + diphosphate

Chlorophyll synthase is an enzyme belonging to the family of transferases, specifically those transferring aryl or alkyl groups other than methyl groups.

1.2.2.4 BACTERIOCHLOROPHYLL a BIOSYNTHESIS

The biosynthesis of BChl a is generally similar to that of chlorophyll a in the early steps but diverges in the later steps (Chew and Bryant 2007). The major differences are that pyrrole ring B is reduced by an enzyme complex related to the light-independent protochlorophyllide oxidoreductase discussed above and that the C3 vinyl group is converted to an acetyl by two enzymes called BchF and BchC.

1.2.2.5 CHLOROPHYLL MODIFICATION

All chlorophylls have very similar structures. The differences among them come at the end of the biosynthetic pathways (Figure 1.6). In general, Chlide a (or Chl a) represents the precursor of all other types of chlorophylls, Chl b, Chl d, and Chl f (113–115); however, the biosynthetic pathway leading to Chl c is unknown. Interestingly, Chl b, Chl d, and Chl f contain a formyl functional group at the C7, C3, or C2 positions, respectively, which determines their different spectral properties from Chl a. The formyl group contains a planar carbon that is connected by a double bond to oxygen (C=O) and a single bond to hydrogen (C–H), which is attributed to the electron-withdrawing quality of the electron-withdrawing formyl group and the more polar feature. The formyl group modification means that a conjugated carbonyl group (C=O) is introduced at the peripheral sites of the macrocycle. The formyl group substitution in Chl d and Chl f at the ring A, along the y-axis, leads to a red-shifted Q_Y absorbance, while the formyl group modification in Chl b at the ring B, along the x-axis, leads to a blue-shifted absorbance (Figure 1.2). All these fine-tuned absorption shifts are the results of changing π-electrons of the macrocycle by the orbital overlap of the C=O groups (see Chapter 2).

Chl b is made from Chl a or chlorophyllide a by oxidation of the methyl group in C7 to a formyl group. This reaction is catalyzed by the chlorophyll a oxygenase (CAO) (Espineda et al. 1999; Tanaka et al. 1998). The CAO enzyme is a Rieske-containing, nonheme–iron monooxygenase that uses molecular O_2 and NADPH to perform two successive hydroxylations at the $C7_1$ position of Chlide a (Qster et al. 2000). Chl b can be reduced back to Chl a through a chlorophyll cycle, so the relative amounts of the two pigments are subject to regulation (Rüdiger 2006; Tanaka and Tanaka 2011). The ratio of Chl a and Chl b regulated by the chlorophyll cycle interacts directly with the assembly of light-harvesting complexes in response to changing light environments. The mutual conversion between Chl a and Chl b may provide higher plants with the ability to optimize their adaptation to varying light conditions (see Chapter 4) and may also be important for chlorophyll degradation as needed.

Chl f biosynthesis has recently been proposed to be carried out by a divergent version of the D1 protein of photosystem II (Ho et al. 2016). Chl d biosynthesis must have a different mechanism from that of Chl b and Chl f, because it requires the transformation of a vinyl group to a formyl group. Small amounts of Chl d were detected from some cyanobacteria as the results of far-red light photoacclimation (Gan et al. 2014, 2015). Chl d synthase is proposed as a P-450-type enzyme due to the fact that Chl d biosynthesis is inhibited by bubbling CO-enriched air, and this reaction can be reversed by O_2-enriched air (Chen et al., unpublished data). Our current knowledge of the formyl substitutions in Chls b, d, and f is very limited, but one fact is common for the formyl formations in chlorophyll modification: molecular oxygen is required (Porra and Scheer 2000; Schliep et al. 2010). However, the oxygen level seems not to be a limiting element for the biosynthesis of Chls b, d, and f, because photosynthesis continuously provides it.

1.3 ANABOLIC PATHWAY FOR PHYCOBILINS IN PHOTOTROPHS

Bilins are a common name for open-chain tetra-pyrrole pigments. They are widely distributed in all kingdoms and have several distinct functions. In heterotrophic organisms, bilins can be considered as catabolic products mainly related to iron acquisition from heme. However, in phototrophs, the biosynthesis of bilins (phycobilins) is considered as an anabolic process because of their function as cofactors of light-harvesting protein complexes or light-sensing phytochromes (Dammeyer and Frankenberg-Dinkel 2008). Here we will only focus on the biosynthetic pathway of phycobilins, photosensitive linear tetrapyrroles, which act as chromophores in light-harvesting complexes (Chapter 5). The phycobilin pigment biosynthesis in photosynthetic

organisms starts with the cleavage of heme by heme oxygenase that produces biliverdin IXα (common name: biliverdin) (Kikuchi et al. 2005). Biliverdin is the first committed intermediate and is subsequently reduced by ferredoxin-dependent bilin reductases for the synthesis of the linear tetrapyrrole precursors of their phycobiliprotein light-harvesting antenna complexes (Figure 1.7). The differences in position of the carbon–carbon double bond in phycocya-nobilin (PC) and phycoerythrobilin (PE) result in a shifted spectral absorbance: purple-red-colored PE with its main absorbance peak at ~550 nm and blue-colored PC with its main peak at ~640 nm (Figure 1.7). As two major phycobiliprotein chro-mophore precursors, the biosynthesis of PC and PE requires different ferredoxin-dependent bilin reduc-tases and several double-bond isomerases in order to produce the specific double carbon-bond positions. The phycocyanobilin:ferredoxin oxidoreductase

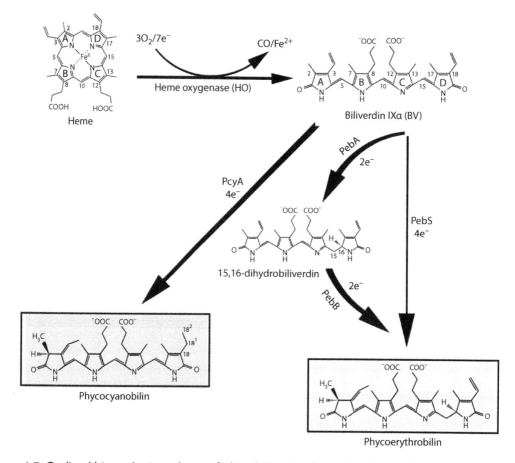

Figure 1.7 Outlined biosynthetic pathway of phycobilins from heme in photosynthetic organisms. PcyA, phycocyanin:ferredoxin oxidoreductase; PebA, 15,16-dihydrobiliverdin:ferredoxin oxidoreductase; PebB, phycoerythrobilin:ferredoxin oxidoreductase; PebS, phycoerythrobilin synthase.

(PcyA) is found in all cyanobacteria except *Prochlorococcus* spp., which catalyzes the two electron reduction reactions for formation of PC (Dammeyer et al. 2007). For example, PcyA from *Nostoc* sp. PCC 7120 is defined as a four-electron transferring enzyme reducing biliverdin via the semireduced intermediate 18^1, 18^2-dihydrobiliverdin (Figure 1.6) (Frankenberg and Lagarias 2003). PC is the chromophore of allophycocyanin and phycocyanin, and PE is the chromophore of phycoerythrin or phycoerythrocyanin (if present) (see Chapter 5). The biosynthesis of PE requires a four-electron reduction, which requires two ferredoxin-dependent bilin reduction enzymes functioning as a dual enzyme complex. The two enzymes designated as PebA and PebB are independently active in a consecutive reduction of biliverdin via 15,16-dihydrobiliverdin to PE (Busch et al. 2011). However, recently, a new enzyme, PebS, isolated from the cyanophage P-SSM2, was shown to catalyze the four-electron reduction of biliverdin to PE directly but has a different location of the pair of protons from PcyA, which reduces the biliverdin via the semireduced intermediate 15, 16-dihydrobiliverdin (instead of 18^1, 18^2-dihydrobiliverdin) (Figure 1.6) (Dammeyer et al. 2008).

1.4 CAROTENOID BIOSYNTHESIS

Carotenoids are abundant hydrophobic secondary metabolites derived from the isoprenoid pathway and are synthesized by all photosynthetic organisms and some nonphotosynthetic bacteria and fungi (Cunningham and Gantt 1998; Shumskaya and Wurtzel 2013). There are hundreds of structures with yellow, orange, and red colors, which function in flowers and fruits to attract pollinators and agents of seed dispersal. However, carotenoids are vital components of photosynthetic reactions and function in protection from photooxidation and light collection (see Chapter 3). The carotenoid biosynthesis pathway starts with the formation of a 40-carbon phytoene, which is generated by condensation of two 20-carbon geranylgeranyl pyrophosphate (GGPP). The biosynthesis of phytoene from GGPP is a two-step reaction catalyzed by one enzyme, the phytoene synthase (Figure 1.8). This reaction is considered as the gatekeeper in the carotenoid biosynthesis pathway (Ruiz-Sola and Rodriguez-Concepcion 2012). Then, a series of desaturation (dehydrogenation) and isomerization reactions increase the numbers of conjugated double bonds and transform the colorless phytoene into a pinkish-colored lycopene (Figure 1.7). In oxygenic photosynthetic organisms, four desaturation reactions are catalyzed by two enzymes: phytoene desaturase and ζ-carotene desaturase, which sequentially increase the number of conjugated double bonds to 5, 7, 9, and 11, respectively. Two isomerases are involved with converting 15-*cis*-phytoene to all-trans lycopene (Figure 1.8). However, the same series of four desaturation reactions and two steps of isomerization are catalyzed by one enzyme, which is a completely unrelated enzyme in bacteria (except cyanobacteria) and fungi (Paniagua-Michel et al. 2012).

The bicyclic carotenoids are the common forms found in photosynthetic organisms. Two types of cyclic end groups, β-ring and ε-ring, are produced by two different cyclases, the lycopene β-cyclase and the lycopene ε-cyclase, which make the first branch point in carotenoid biosynthesis pathway (Figure 1.8). The position of the double bond in the β-ring is in conjugation with the polyene chain, but the double-bond position in the ε-ring is not in conjugation, which may be the reason why the structure of the β-ring is more commonly found in photosynthetic organisms. The lycopene β-cyclase can catalyze the formation of the β-ring symmetrically at the two ends of the linear lycopene; however, the ε-cyclase adds only one ring to either linear lycopene or monocyclic γ-carotene. β-carotene and zeaxanthin have two β-rings, and α-carotene and lutein have one ε-ring and one β-ring. Interestingly, *Acaryochloris marina*, a Chl *d*–containing cyanobacterium, has only α-carotene and zeaxanthin. The biosynthesis pathways for the formation of α-carotene and zeaxanthin concurrently await further exploration, although there is a predicted pathway based on the genomic sequence comparison (Swingley et al. 2008).

The oxygenated carotenoids (xanthophylls) comprise most of the carotenoid pigments in photosynthetic organisms. They are generated by hydroxylation at the carbon number three of each ring of carotenes, such as zeaxanthin from β-carotene and lutein from α-carotene. Further epoxidation between carbons number five and six of each ring produces violaxanthin via antheraxanthin from zeaxanthin, in the so-called xanthophyll cycle (Jahns and Holzwarth 2012) (see Chapter 11). The xanthophyll cycle is driven reversely by epoxidase and de-epoxidase enzymes in response to the need to regulate the excited states levels of the

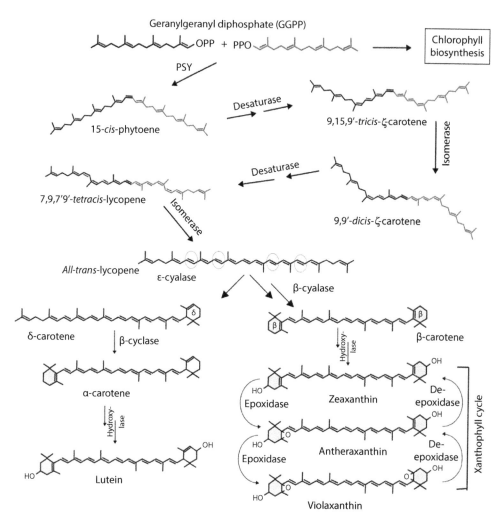

Figure 1.8 Outlined carotenoids biosynthetic pathway. Desaturase, either phytoene desaturase or ζ-carotene desaturase; hydroxylase, either carotenoid β-hydroxylase or carotenoid ε-hydroxylase; isomerase, either 15-cis-ζ-carotene isomerase or carotenoid isomerase; PSY, phytoene synthase.

chlorophylls in the membrane. During high light stress, the conversion from violaxanthin to zeaxanthin plays a direct photoprotective role as a lipid-protective antioxidant.

In contrast, algae are able to synthesize very diverse carotenoids for their photosynthetic apparatus, which are different from the common carotenoids of plants and cyanobacteria (Takaichi 2011). The three major unique carotenoids in algae are peridinin, fucoxanthin, and diadinoxanthin. Peridinin and fucoxanthin are involved in the assembly of light-harvesting complexes, fucoxanthin-chlorophyll-binding light-harvesting protein complexes, and peridinin-chlorophyll-binding

protein complexes. Diadinoxanthin forms a different xanthophyll cycle in diatoms (Lohr and Wilhelm 1999). The major carotenoids in cyanobacteria are β-carotene and zeaxanthin, although small amounts of echinenone, myxol pentosides (myxoxanthophyll) also exist in some cyanobacteria.

Pigment is the mediator molecule for photobiology: (1) Pigments are involved in light energy capture as the energy source for cellular metabolism; (2) Pigments play their light-sensing role to receive the environmental information. There are three main groups of pigments: chlorophylls, carotenoids, and tetrapyrroles (such as bilins) involved in sunlight energy absorption and conversion processes.

REFERENCES

Akutsu S, Fujinuma D, Furukawa H et al. 2011. Pigment analysis of a chlorophyll *f*-containing cyanobacterium strain KC1 isolated from Lake Biwa. *Photochem Photobiol* 33:35–40.

Björn L, Papageorgiou G, Blankenship R, Govindjee A. 2009. A viewpoint: Why chlorophyll *a*. *Photosynth Res* 99:85–98.

Blankenship RE 2014. *Molecular Mechanisms of Photosynthesis*, Wiley-Blackwell, Oxford, U.K.

Busch AWU, Rejierse EJ, Lubitz W et al. 2011. Structural and mechanistic insight into the ferredoxin-mediated two-electron reduction of bilins. *Brioche J* 439:257–264.

Chen GE, Hitchcock A, Jackson PJ et al. 2016. Two unrelated 8-vinyl reductases ensure production of mature chlorophylls in *Acaryochloris marina*. *J Bacteriol* 198:1393–1400.

Chen M. 2014. Chlorophyll modification and their spectral extension in oxygenic photosynthesis. *Annu Rev Biochem* 83:317–340.

Chen M and Blankenship RE. 2011. Expanding the solar spectrum used by photosynthesis. *Trends Plant Sci* 16:427–431.

Chen M, Li YQ, Birch D et al. 2012. A cyanobacterium that contains chlorophyll *f*—A red-absorbing photopigment. *FEBS Lett* 586:3249–3254.

Chen M and Scheer H. 2013. Extending the limits of natural photosynthesis and implications for technical light harvesting. *J Porphyrins Phthalocyanines* 17:1–15.

Chen M, Schliep M, Willows RD et al. 2010. A red-shifted chlorophyll. *Science* 329:1318–1319.

Chew AG and Bryant DA. 2007. Chlorophyll biosynthesis in bacteria: The origins of structural and functional diversity. *Annu Rev Microbiol* 61:113–129.

Chisholm SW, Frankel SL, Goericke R et al. 1992. *Prochlorococcus marinus* nov. gen. nov. sp.: An oxyphototrophic marine prokaryote containing divinyl chlorophyll *a* and *b*. *Arch Microbiol* 157:297–300.

Choquet Y, Rahire M, Girard-Bascou J et al. 1992. A chloroplast gene is required for the light- independent accumulation of chlorophyll in *Chlamydomonas reinhardtii*. *EMBO J* 11:1697–1704.

Cunningham Jr FX and Gantt E. 1998. Genes and enzymes of carotenoid biosynthesis in plants. *Annu Rev Plant Physiol Plant Mol Biol* 49:557–583.

Dammeyer T, Baggby SC, Sullivan MB et al. 2008. Efficient phage-mediated pigment biosynthesis in ocean cyanobacteria. *Curr Biol* 18:442–448.

Dammeyer T and Frankenberg-Dinkel N. 2008. Function and distribution of bilin biosynthesis enzymes in photosynthetic organisms. *Photochem Photobiol Sci* 7:1121–1131.

Dammeyer T, Michaelsen K and Frankenberg-Donkel N. 2007. Biosynthesis of open-chain tetrapyrroles in *Prochlorococcus marinus*. *FEMS Microbiol Lett* 271:251–257.

Demming-Adams B and Adams WW. 1996. The role of xanthophyll cycle carotenoids in the protection of photosynthesis. *Trends Plant Sci* 1:21–26.

Espineda CE, Linford AS, Devine D et al. 1999. The AtCAO gene, encoding chlorophyll a oxygenase, is required for chlorophyll b synthesis in Arabidopsis thaliana. *Proc Natl Acad Sci USA* 96:10507–10511.

Frankenberg N and Lagarias LC. 2003. Phycocyanobilin:ferredoxin oxidoreductase of *Anabaena* sp. PCC 7220. *J Biol Chem* 278:9219–9226.

Gabruk M and Mysliwa-Kurdziel B. 2015. Light-dependent protochlorophyllide oxidoreductase: Phylogeny, regulation, and catalytic properties. *Biochemistry* 54:5255–5262.

Gan F, Zhang S and Bryant DA. 2015. Occurrence of far-red light photoacclimation (FaRLiP) in diverse cyanobacteria. *elife* 5:4–24.

Gan F, Zhang S, Rockwell NC et al. 2014. Extensive remodeling of a cyanobacterial photosynthetic apparatus in far-red light. *Science* 345:1312–1317.

Hayaux M. 1998. Carotenoids as membrane stabilizers in chloroplast. *Trends Plant Sci* 3:147–151.

Heyes DJ, Ruban AV and Hunter CN. 2003. Protochlorophyllide oxidoreductase: 'dark' reactions of a light-driven enzyme. *Biochemistry* 42:523–528.

Ho MY, Shen GZ, Canniffe DP, Zhao C and Bryant DA. 2016. Light-dependent chlorophyll f synthase is a highly divergent paralog of PsbA of photosystem II. *Science* 353:886.

Hohmann-Marriott MF and Blankenship RE. 2011. Evolution of photosynthesis. *Annu Rev Plant Biol* 62:515–548.

Hollingshead S, Kopečná J, Jackson PJ et al. 2012. Conserved chloroplast open-reading frame ycf54 is required for activity of the magnesium protoporphyrin monomethylester oxidative cyclase in *Synechocystis* PCC 6803. *J Biol Chem* 287:27823–27833.

Hu Q, Miyashita H, Iwasaki I et al. 1998. A photosystem I reaction center driven by chlorophyll *d* in oxygenic photosynthesis. *Proc Natl Acad Sci USA* 95:13319–13323.

Islam MR, Aikawa S, Midorikawa T et al. 2008. Slr1923 of *Synechocystis* sp. PCC6803 is essential for conversion of 3,8-divinyl(proto) chlorophyll(ide) to 3-monovinyl(proto) chlorophyll(ide). *Plant Physiol* 148:1068–1081.

Ito H and Tanaka A. 2011. Evolution of divinyl chlorophyll-based photosystem in Prochlorococcus. *Proc Natl Acad Sci U S A* 108:18014–18019.

Itoh S, Mino H, Itoh K et al. 2007. Function of chlorophyll *d* in reaction centers of photosystems I and II of the oxygenic photosynthesis of *Acaryochloris marina*. *Biochemistry* 46:12473–12481.

Jahns P and Holzwarth AR. 2012. The role of the xanthophyll cycle and of lutein in photoprotection of photosystem II. *Biochim Biophys Acta* 1817:182–193.

Jensen PE, Gibson LC and Hunter CN. 1999. ATPase activity associated with the magnesium- protoporphyrin IX chelatase enzyme of *Synechocystis* PCC6803: Evidence for ATP hydrolysis during Mg^{2+} insertion, and the MgATP-dependent interaction of the ChlI and ChlD subunits. *Biochem J* 339:127–134.

Kashiyama Y, Miyashita H, Ohkubo S et al. 2008. Evidence of global chlorophyll *d*. *Science* 321:658–658.

Kikuchi G, Yoshida T and Noguchi M. 2005. Heme oxygenase and heme degradation. *Biochem Biophys Res Commun* 338:558–567.

Kohchi T, Shirai H and Fukuzawa H. 1988. Structure and organization of *Marchantia polymorpha* chloroplast genome: IV. Inverted repeat and small single copy regions. *J Mol Biol* 203:353–372.

Kopp RE, Kirschvink JL, Hilburn IA et al. 2005. The Paleoproterozoic Snowball Earth: A climate disaster triggered by the evolution of oxygenic photosynthesis. *Proc Natl Acad Sci USA* 102:11131–11136.

Kořený L, Sobotka R, Janouškovec J et al. 2011. Tetrapyrrole synthesis of photosynthetic chromerids is likely homologous to the unusual pathway of apicomplexan parasites. *Plant Cell* 23:3454–3462.

Kräutler B. 2011. A new factor in life's quest for energy. *Angew Chem Int Ed* 50:2439–2441.

Kühl M, Chen M, Ralph PJ et al. 2005. A niche for cyanobacteria containing chlorophyll *d*. *Nature* 433:820.

Larkum AWD and Barrett J. 1983. Light-harvesting processes in Algae. *Adv Bot Res* 10:1–219.

Lohr M and Wilhelm C. 1999. Algae displaying the diadinoxanthin cycle also possess the violaxanthin cycle. *Proc Natl Acad Sci USA* 96:8784–8789.

Loughlin PC, Lin Y and Chen M. 2013, Chlorophyll *d* and *Acaryochloris marina*: Current status. *Photosynth Res* 116:277–293.

MacColl R. 1998. Cyanobacterial phycobilisomes. *J Struct Biol* 124:311–334.

Manning WM and Strain HH. 1943. Chlorophyll *d*, a green pigment of red algae. *J Biol Chem* 151:1–19.

Masuda TT. 2008. Recent overview of the Mg branch of the tetrapyrrole biosynthesis leading to chlorophylls. *Photosynth Res* 96:121–143.

Miller SR, Augustine S, Olson TL et al. 2005. Discovery of a free-living chlorophyll *d*-producing cyanobacterium with a hybrid proteobacterial/cyanobacterial small-subunit rRNA gene. *Proc Natl Acad Sci USA* 102:850–855.

Miyashita H, Ikemoto H, Kurano N et al. 1996. Chlorophyll *d* as a major pigment. *Nature* 383:402.

Mochizuki N, Tanaka R, Grimm B et al. 2010. The cell biology of tetrapyrroles: A life and death struggle. *Trends Plant Sci* 15:488–498.

Nagata N, Tanaka R and Satoh S. 2005. Identification of a vinyl reductase gene for chlorophyll synthesis in *Arabidopsis thaliana* and implications for the evolution of *Prochlorococcus* species. *Plant Cell* 17:233–240.

Ouchane S, Steunou AS, Picaud M et al. 2004. Aerobic and anaerobic Mg-protoporphyrin monomethyl ester cyclases in purple bacteria: A strategy adopted to bypass the repressive oxygen control system. *J Biol Chem* 279:6385–6394.

Paniagua-Michel J, Olmos-Soto J and Ruiz MA. 2012. Pathways of carotenoid biosynthesis in bacteria and microalgae. In *Microbial Carotenoids from Bacteria and Microalgae: Methods and Protocols, Methods in Molecular Biology*, ed. J-L Barredo, vol. 892, pp. 1–12. Springer Science+Business Media, New York.

Porra RJ and Scheer H. 2000. O-18 and mass spectrometry in chlorophyll research: Derivation and loss of oxygen atoms at the periphery of the chlorophyll macrocycle during biosynthesis, degradation and adaptation. *Photosynth Res* 66:159–175.

Oster U, Tanaka R, Tanank A et al. 2000. Cloning and functional expression of the gene encoding the key enzyme for chlorophyll b biosynthesis (CAO) from *Arabidopsis thaliana*. *Plant J* 21:305–310.

Raymond J and Blankenship RE. 2004. Biosynthetic pathways, gene replacement and the antiquity of life. *Geobiology* 2:199–203.

Reid JD and Hunter CN. 2004. Magnesium dependent ATPase activity and cooperativity of magnesium chelatase from *Synechocystis* sp. PCC6803. *J Biol Chem* 279:26893–26899.

Reinbothe C, Buhr F, Bartsch S et al. 2006. In vitro-mutagenesis of NADPH:protochlorophyllide oxidoreductase B: Two distinctive protochlorophyllide binding sites participate in enzyme catalysis and assembly. *Mol Gen Genomics* 275:540–552.

Reith ME and Munholland J. 1995. Complete nucleotide sequence of the porphyra purpurea chloroplast genome. *Plant Mol Biol Report* 13:333–335.

Rüdiger W. 2006. Biosynthesis of chlorophylls a and b the last step. In *Chlorophylls and Bacteriochlorophylls: Biochemistry, Biophysics, Functions and Applications*, eds. B Grimm, RJ Porra, W Rudiger and H Scheer, vol 25, pp. 189–200. Springer, Dordrecht, the Netherlands.

Ruiz-Sola MA and Rodriguez-Concepcion M. 2012. Carotenoid biosynthesis in arabidopsis: A colorful pathway. *Arabidopsis Book* 10:e0158.

Samuilov VD. 2005 Energy problems in life evolution. *Biochemistry (Mosc)* 70:246–250.

Scheer H. 1991. Structure and occurrence of chlorophylls. In *Chlorophylls*, ed. H Scheer, pp. 3–30. CRC Press, Boca Raton, FL.

Scheer H. 2006. An overview of chlorophylls and bacteriochlorophylls: Biochemistry, biophysics, functions and applications. In *Chlorophylls and Bacteriochlorophylls: Biochemistry, Biophysics, Functions and Applications*, eds. B Grimm, RJ Porra, W Rudiger and H Scheer, vol 25, pp. 1–26. Springer, Dordrecht, the Netherlands.

Schliep M, Crossett B, Willows RD et al. 2010. ^{18}O labeling of chlorophyll d in *Acaryochloris marina* reveals that chlorophyll a and molecular oxygen are precursors. *J Biol Chem* 285: 28450–28456.

Shi C and Shi X. 2006. Characterization of three genes encoding the subunits of light- independent protochlorophyllide reductase in *Chlorella protothecoides* CS-41. *Biotechnol Prog* 22:1050–1055.

Shumskaya M and Wurtzel ET. 2013. The carotenoid biosynthetic pathway: Thinking in all dimensions. *Plant Sci* 208:58–63.

Stauber JL and Jeffrey SW. 1988. photosynthetic pigments in fifty-one species of marine diatoms. *J Phycol* 24:158–172.

Stirewalt VL, Michalowski CB, Loffelhardt W et al. 1995. Nucleotide sequence of the cyanelle genome from *Cyanophora paradoxa*. *Plant Mol Biol Report* 13:327–332.

Swingley WD, Chen M, Cheung PC et al. 2008. Niche adaptation and genome expansion in the chlorophyll d-producing cyanobacterium *Acaryochloris marina*. *Proc Natl Acad Sci USA* 105:9050–9055.

Takaichi S. 2011. Carotenoids in algae: Distribution, biosynthesis and functions. *Mar Drugs* 9:1101–1118.

Tanaka A, Ito H, Tanaka R et al. 1998. Chlorophyll a oxygenase (CAO) is involved in chlorophyll b formation from chlorophyll a. *Proc Natl Acad Sci USA* 95:12719–12723.

Tanaka R and Tanaka A. 2007. Tetrapyrrole biosynthesis in higher plants. *Annu Rev Plant Sci* 58:321–346.

Tanaka R and Tanaka A. 2011. Chlorophyll cycle regulates the construction and destruction of the light-harvesting complexes. *Biochim Biophys Acta* 1807:968–976.

Tomo T, Okubo T, Akimoto S et al. 2007. Identification of the special pair of photosystem II in a chlorophyll d-dominated cyanobacterium. *Proc Natl Acad Sci USA* 104:7283–7288.

Wang P, Wan C, Xu Z et al. 2013. One divinyl reductase reduces the 8-vinyl groups in various intermediate of chlorophyll biosynthesis in given higher plant species, but the isozyme differs between species. *Plant Physiol* 161:521–534.

Willows RD, Gibson LCD, Kanangara CG et al. 1996. Three separate proteins constitute the magnesium chelatase of *Rhodobacter sphaeroides*. *Eur J Biochem* 235:438–443.

Willows RD, Li Y, Scheer H et al. 2013. Structure of chlorophyll *f*. *Org Lett* 7:1588–1590.

Zapata M, Garrido JL and Jeffrey SW. 2006. Chlorophyll *c* pigment: Current. In *Chlorophylls and Bacteriochlorophylls: Biochemistry, Biophysics, Functions and Applications*, eds. B Grimm, RJ Porra, W Rudiger, H Scheer, vol 25, pp. 39–53. Springer, Dordrecht, the Netherlands.

Zappa S, Li K, and Bauer CE. 2010. The Tetrapyrrole biosynthetic pathway and its regulation in Rhodobacter capsulatus. *Adv Exp Med Biol* 675:229–250.

Chlorophylls in a protein environment: How to calculate their spectral and redox properties (from MO to DFT)

CARLES CURUTCHET AND BENEDETTA MENNUCCI

2.1 INTRODUCTION

Chlorophylls accomplish a key task in the overall machinery of photosynthesis. They are specialized light-absorbing molecules optimized in order to effectively capture sunlight and transfer and convert the absorbed energy into useful chemical energy that can be stored by photosynthetic organisms. There are three key aspects that chlorophylls must address in order for this process to be successful. They have to optimize both the spectral and spatial cross section for light absorption, but they also need to be able to funnel rapidly the absorbed sunlight to avoid energy losses in the form of heat (Mirkovic et al. 2017). The requirement of optimizing the spectral cross section by covering a significant range of the spectra of available sunlight is especially true for organisms living on low-light habitats, for instance, marine algae, where the water column at deep marine habitats, as well as the presence of overlaying organisms, can drastically reduce the amount of available sunlight. This is clearly reflected in nature. For instance, photosynthetic marine algae possess a richer variety of chlorophyll pigments (Chl a, Chl b, Chl c, Chl d), as well as other related light-absorbing molecules like bilins (open-chain tetrapyrrole chromophores), compared to land plants (Chl a and Chl b) (Collini et al. 2009). In Figure 2.1, we show the respective variation of the lowest-energy electronic transitions (known as Q_x, Q_y, and Soret) for different chlorophyll derivatives. This variety of chlorophyll types, each one possessing different spectral properties due to subtle chemical modifications on their structure, has allowed photosynthetic organisms to develop highly optimized antenna pigment–protein complexes adapted to their particular spectral needs.

Besides the chemical structure of the chlorophylls, however, a key aspect in the optimization of the light-harvesting (LH) abilities of pigment–protein complexes relies on the organization of the pigments in the protein scaffold supporting them. Increasing the number of pigment molecules in antenna complexes allows increasing the spatial cross section for light absorption. For instance, green bacteria possess highly specialized antenna structures called

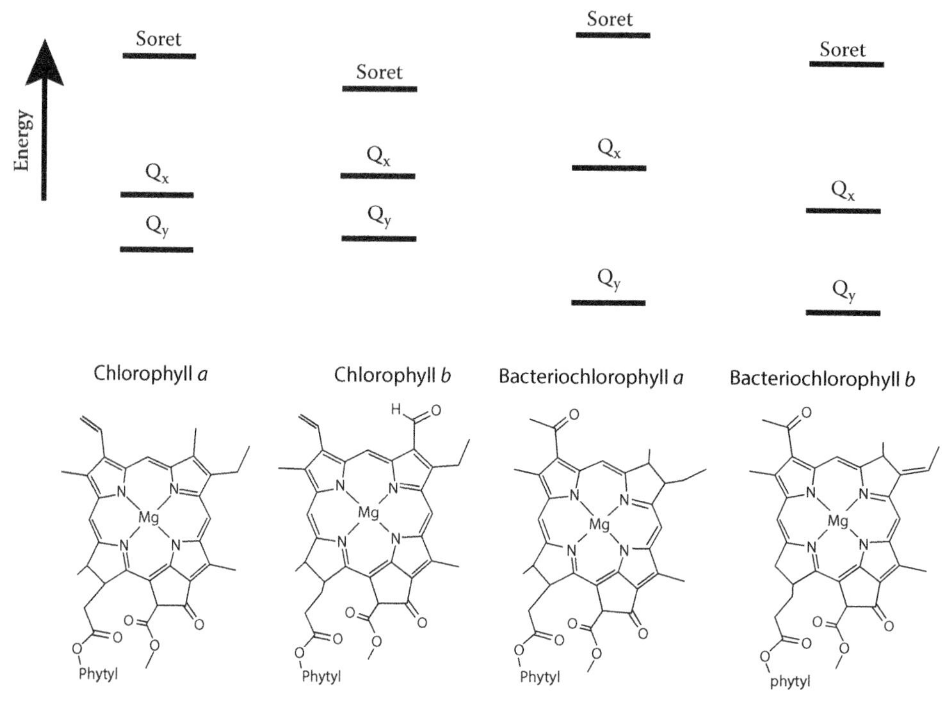

Figure 2.1 Energies of the Q_x, Q_y, and Soret bands of different chlorophylls as given by the absorption maxima in diethyl ether (Chl *a*, Chl *b*, and BChl *a*) and acetone (BChl *b*) solution. (From Blankenship, R.E., *Molecular Mechanisms of Photosynthesis*, Wiley Blackwell, Oxford, U.K., 2002.)

chlorosomes, carrying ~25,000 BChls, which allow them to live in environments with the lowest light intensity of any known photosynthetic organism (Blankenship et al. 1995; Blankenship and Matsuura 2003). The spatial cross section is also clearly increased by simply including a larger number of antenna complexes in the organism.

The protein environment, however, plays another delicate role in the LH process apart from binding a large amount of chlorophyll molecules of a particular type (Scholes et al. 2011). The local protein environment surrounding chlorophylls can significantly tune their spectral and redox properties through specific pigment–protein interactions, for example, by shifting their maximum absorption wavelength via hydrogen bonds between the pigment and the protein amino acids, and such tuning can have key consequences for the mechanism and overall efficiency of the LH process (Milne et al. 2015). For instance, specific pigment–protein interactions shift the energies of the eight BChl *a* chromophores in the Fenna–Matthews–Olson (FMO) complex of green sulfur bacteria in slightly different ways leading to an overall range of ~500 cm^{-1} covered by the energies of the different sites, as illustrated in

Figure 2.2 (Milder et al. 2010; Schmidt am Busch et al. 2011). Electrostatic pigment–protein interactions are considered to be the leading mechanism for shifting the energy levels—the site energies—of the chromophores, and theoretical approaches used to predict such shifts are typically focused on such interactions (König and Neugebauer 2012; Renger and Müh 2013; Jurinovich et al. 2015). However, also short-range dispersion and repulsion interactions dictate the actual properties of a chromophore surrounded by a particular environment and can be eventually included in theoretical models through more sophisticated approaches (König and Neugebauer 2011). In addition, the protein can also constrain the configurational space of the chlorophylls depending on the characteristics of the binding pocket holding the chromophore, and this can also lead to modulation of the energy levels responsible for its spectral and redox properties. For example, the slightly different conformations induced by the protein on the pigments' structures have been suggested to be the leading mechanism for site energy tuning in the phycoerythrin 545 (PE545) antenna of cryptophyte algae, probably owing to the increased conformational

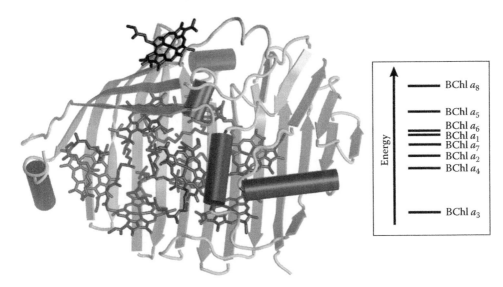

Figure 2.2 Specific pigment–protein interactions shift the energies of the eight BChl *a* chromophores in the FMO complex of green sulfur bacteria leading to an overall range of ~500 cm^{-1} covered by the energies of the different sites. (From Schmidt am Busch, M. et al., *J. Phys. Chem. Lett.*, 2, 93, 2011.)

flexibility of the bilin chromophores it contains, characterized by an open disposition of the tetrapyrrole ring structure characteristic of chlorophylls (Curutchet et al. 2013).

By tuning the energy levels of the chromophores, the protein environment plays an important role in the optimization of the spectral cross section for the capture of sunlight. In addition, the presence of site energy disorder is of fundamental importance in order to appropriately direct the captured energy toward the reaction centers. In other words, the protein can modulate the energy and charge migration pathways in order to deliver the energy in the right place by manipulating the chlorophylls' energy levels. For instance, such tuning should avoid the presence of undesired low-energy states—trapping sites for the absorbed energy—far from the reaction center.

However, the introduction of disorder among the site energies not only affects *where* the energy will travel. Such modulation can also affect *how* it travels, and therefore changes the overall mechanism and rate of the LH process. This is because the proximity of energy levels among chlorophylls promotes the appearance of so-called excitonic states, which can be delocalized over several pigments that are electronically coupled. In this case, the whole mechanism of energy transfer among chlorophylls is modified, because the energy-transferring states are now coherently shared among several molecules.

This effect leads to significant shifts in the transition energies of such states, as well as in a redistribution of dipole strength among transitions, thereby allowing for potential manipulation of the probability for light absorption associated to a particular wavelength. In turn, also the mechanisms of energy migration can be affected by exciton delocalization. A paradigmatic example is illustrated by the properties of the peripheral light-harvesting complex II (LHCII) from purple photosynthetic bacteria. This complex is characterized by two strong absorption bands associated to the B800 and B850 rings of BChl *a* molecules in their structure. Because the chlorophylls in the B850 ring are closely packed, they are strongly electronically coupled, thus promoting exciton delocalization along the ring. Such delocalization effects have been shown to be key in the energy migration mechanism by providing better B850 energy acceptor states for excitations in the B800 ring (Scholes and Fleming 2000).

From the earlier discussion, it is clear that understanding how the protein environment modulates the chlorophylls' energy levels is key in order to understand the molecular mechanisms and overall efficiency of photosynthetic LH. A common strategy to determine the site energies of a given pigment–protein complex consists of empirically fitting their values through spectral and kinetic modeling of multiple steady-state and time-resolved spectra measured for the system, for instance, absorption, circular dichroism, fluorescence, or transient absorption kinetics (van Grondelle and Novoderezhkin 2006; Novoderezhkin and van Grondelle 2010). This strategy has been very useful in understanding the mechanisms of LH in a variety of systems. A drawback from this approach is that it can lead to multiple sets of energies able to describe the experiments, although such sets are usually not very different (Novoderezhkin et al. 2010; Jankowiak et al. 2016). In other cases, when the number of pigment energies to fit is too large, often several pigments have to be assigned a common value (Novoderezhkin et al. 2005). On the other hand, the empirical fits do not provide insights into the molecular mechanism leading to site energy tuning. An attractive alternative is naturally given by theoretical calculations rooted in quantum chemistry, which allow the determination of such electronic states from a powerful and varied pool of methodologies, from simple models based on consideration of frontier orbitals of the pigment to advanced methodologies based on density functional theory (DFT). If the protein's three-dimensional structure has been solved by X-ray crystallography, theory also allows investigating how the protein and solvent environment surrounding the chlorophylls shifts their energies. Because in this case the number of atoms to be included in the theoretical calculations grows drastically, and so does the computational cost associated, theoretical strategies in this case typically rely on a dual description of the system, where the pigment molecule is fully described using quantum chemistry, whereas the rest of the system—the protein and solvent environment—is approximated based on a simpler classical description (König and Neugebauer 2012; Renger and Müh 2013; Jurinovich et al. 2015). This description can either implicitly or explicitly account for the environment, based on continuum solvation models or relying on an atomistic classical force field, respectively.

In the last decade, there has been an increased effort toward the accurate theoretical determination of site energies in photosynthetic pigment–protein complexes. In the following section, we will first describe the methods provided by quantum chemistry that can be used to calculate the spectral and redox properties of chlorophylls. Then, we will describe how these quantum chemistry methods can be coupled to a simpler classical description of the environment, based on either classical force fields or continuum solvation models, in order to account for the effect of the protein surroundings.

2.2 SPECTRAL AND REDOX PROPERTIES FROM QUANTUM CHEMISTRY METHODS

Quantum chemistry methods currently represent the only valid alternative to spectroscopic techniques to study structural and electronic properties of molecular systems. Obviously, during the years these methods have enormously changed following the progress in the development of new and more accurate quantum mechanical descriptions. Since the first applications generally using molecular orbital (MO)-based models (such as Hartree–Fock, HF), many alternative methods have been proposed to achieve better and better descriptions as well as to treat larger and larger systems. In particular, large efforts have been focused on obtaining methods accounting for electronic correlations that in HF and similar old quantum chemistry methods were neglected. It is convenient to recognize two types of electron correlation, the first called dynamical correlation and the second called nondynamical (near degeneracy) correlation. While the latter is generally limited to specific systems, dynamical correlation, being the short-range effect by which electrons avoid one another to reduce electron repulsion, is present in all finite systems containing two or more electrons.

DFT has shown to be a reliable and inexpensive approach for the description of dynamical correlation, and during the last years, it has become the most used approach to study properties and processes of molecular systems of medium/large dimensions including the various types of LH pigments. Indeed, DFT calculations on systems comprising several hundreds of atoms can be done in a few days on a standard personal computer, whereas expensive MO-based models including electron correlation involve much longer computational times. DFT describes the electronic states of atoms, molecules, and materials in terms of the three-dimensional many-electron spin density of the system, which is a great simplification over MO-based theories, which involve a 3N-dimensional antisymmetric wave function for a system with N electrons. In DFT, the key aspect to obtain reliable descriptions is the choice of the functional representing the exchange and correlation effects (xc) in terms of the spin density. The oldest approximation to a density functional is the so-called local spin density approximation (LSDA) since it depends only on spin densities (not their derivatives). The next level of complexity in density functionals is to add a dependence on the gradients of the spin densities. Such functionals are called generalized gradient approximations (GGAs). DFT with LSDA or GGA functionals includes self-exchange and self-correlation, both of which are unphysical, because the self-interaction in the Coulomb and exchange parts do not cancel each other exactly. As a consequence, such functionals tend to predict too small highest occupied molecular orbital (HOMO)–lowest unoccupied molecular orbital (LUMO) gaps and to underestimate the relative stability of high-spin states. An important consequence of the error in LSDA and GGA exchange functionals is that an electron interacts with its own charge density; this artificially raises the energy of localized states and causes DFT to produce excessively delocalized charge distributions. By including partial HF exchange, one can decrease the self-exchange problem, and in fact, nowadays, hybrid functionals incorporating a contribution of exact orbital exchange represent very effective approaches, and they have been largely used to study pigments' structures and properties. More recently, alternative strategies based on the generalization of GGA (meta-GGA) and its combination with HF exchange (hybrid meta-GGA) have shown to give very accurate descriptions and are becoming quite popular also for systems of biological interest.

Despite their general success, DFT methods may encounter difficulties for some specific systems and/or processes. Among them, the redox processes surely represent a challenging task especially when transition metal complexes are considered. General-purpose density functionals have intrinsic errors of 100–300 mV for gas-phase ionization potentials and electron affinities; this provides a rough upper limit for the accuracy of calculated standard redox potentials.

When moving to the study of electronic excited-state properties, things become even more delicate, and until now there is no optimal quantum mechanics (QM) approach that gives the same high accuracy for different systems and/or different types of excited states. Indeed, quantum chemistry has developed many alternative theoretical methods and computational approaches to describe excited states for a long time.

Some of them belong to the family of wave function–based ab initio, and they can be divided into single reference and multireference methods on the one hand and into configuration interaction (CI) and coupled-cluster (CC) methods on the other. In contrast to single reference methods, multireference methods describe the ground state as a combination of several reference electronic configurations, thereby accounting for nondynamical correlation effects.

In CI-type calculations, the electronic wave function is constructed as a linear combination of the ground-state HF determinant and "excited" determinants, which are obtained by replacing occupied orbitals with virtual ones. Within this framework, the exact numerical solution of the Schrödinger equation within the chosen atomic basis set would correspond to the inclusion of all possible "excited" determinants (full CI). Such calculations are extremely expensive, and one has necessarily to truncate the CI expansion; in particular, if one stops after the "singly" excited determinants, the popular method called configuration interaction singles (CIS) is obtained (Foresman et al. 1992). The main problem of the CIS method is the basic lack of correlation energy. As a result, excitation energies computed with the CIS method are usually overestimated by 0.5–2 eV compared to their experimental values. Despite these limitations, the CIS method possesses some useful properties. It is cheap and allows the description of singlet and triplet excited states in large systems comprising several hundreds of atoms. In addition, the excited-state energies are analytically differentiable with respect to external parameters as, for example, nuclear displacements and external fields, which make possible the application of analytic gradient techniques for the calculation of excited-state properties such as equilibrium geometries and vibrational frequencies.

To make CI calculations computationally feasible, a different approximation with respect to the CI truncation is the definition of an active space of occupied and virtual orbitals in which all possible "excited" determinants are constructed. To keep the flexibility of the CI wave function, in general, one has to reoptimize the MOs during the minimization procedure to improve the accuracy. The resulting approach is known as complete active space self-consistent field (CASSCF) (Roos et al. 1980). In analogy to CIS, CASSCF analytical derivatives are available allowing for efficient optimization of excited-state geometries and localization of conical intersections; in addition, effects of nondynamical correlation are automatically included so that the CASSCF method is one of the most suited approaches to treat systems and processes presenting a multiconfigurational character. Unfortunately, from a computational point of view, CASSCF calculations become quickly very expensive, since the computational effort increases exponentially with the size of the active space. In addition, the choice of the active space is not unique, but a careful analysis of both the chemical nature of the system and the physical nature of the orbitals is required. Moreover, too small active spaces and the concomitant neglect of large parts of dynamical electron correlation can lead to significant errors and an unbalanced treatment of electronic states of different nature. A very successful (but also expensive) approach to include dynamic electron correlation applies second-order perturbation theory on wave functions obtained from a state-averaged CASSCF calculation: this approach is known as CASPT2 (Andersson et al. 1990).

As an alternative to multireference approaches, correlation can also be included in single-reference wave function methods, for instance, through CC theory (Bartlett and Purvis 1978). CC theory has been also extended to calculation of excited states (Stanton and Bartlett 1993); however, the computational cost is generally very high, limiting its application to small-sized molecules. Only the most approximate formulations of the method (CC singles and the second-order approximate CC singles and doubles, CC2 model) are currently applicable to large molecules such as LH pigments.

Alternative to ab initio methods, semiempirical approaches have been also successfully applied to the calculation of photoinitiated processes in molecular systems; the most common approach is the intermediate neglect of differential overlap (INDO/S) method in combination with a CIS formalism developed by Zerner (the method is also known as ZINDO)(Zerner 1991) or the more recent PM3, PM6, and PM7 family of methods proposed by Stewart and collaborators (Stewart 2007, 2013). These semiempirical methods neglect the overlap of specific pairs of atomic orbitals, and the surviving two-electron integrals as well as the one-electron ones are empirically parameterized. In the case of ZINDO, for example, the parameterization is designed for spectroscopic applications. The simple structure of the one- and two-electron integrals allows for an efficient and fast computation of excited states of very large molecular systems. The accuracy of these methods, however, is unpredictable, and careful comparison with experiment and/or higher-level computations must be made. In other words, the quantitative aspect of the obtained results has been found to be highly system dependent, and thus practically unpredictable.

MO-based methods, in summary, remain either very expensive or extremely approximated, and thus they do not represent a very simple strategy for systems of real interest for biological applications. Once again, DFT has represented the real discontinuity with respect to the past and its generalization to excited states, known as time-dependent DFT (TDDFT) (Runge and Gross 1984; Casida 1995), is now the most widely used approach to study excited-state properties and processes for systems of increasing complexity and dimension. The success of TDDFT with respect to previous, even more accurate, approaches is due to its unique characteristics combining computational efficiency with ease of use. TDDFT performs usually very well for valence excited states, but it is now well known that it has severe problems with the correct description of Rydberg and charge-transfer (CT) excited states (Dreuw and Head-Gordon 2005). In the case of CT excited states, which play an important role in photosynthesis (Reimers et al. 2016), the excitation energies are much too low (by up to 1 eV) and the potential energy curves do not exhibit the correct $1/R$ asymptote when R corresponds to a distance coordinate between the positive and negative charges of the CT state. The $1/R$ failure of TDDFT employing pure standard xc functionals arises from the self-interaction error discussed before. This electron-transfer self-interaction effect is cancelled in the parallel HF-based approach by the response of the HF exchange term. At present, several different pathways have been proposed to address this substantial failure of TDDFT for CT states and to correct for it. A very effective way to improve the TDDFT performances for these difficult cases is to split the Coulomb operator of the Hamiltonian into two parts, a short-range and a long-range part, which are treated primarily using a local functional and an exact orbital exchange, respectively.

TDDFT presents also another problem due to the single determinant ansatz of DFT; as a result, excited states characterized by a significant double excitation character cannot be properly described. To overcome this limit that prevents, for example, the use of TDDFT methods to investigate excitations of carotenoids, a combination of DFT and a multireference CI ansatz has been proposed (Grimme and Waletzke 1999). The resulting DFT/MRCI method has shown to be very effective in describing energies and properties of excited states of large systems.

In the last years, many of these different methods have been applied to the study of excited states of LH pigments. In particular, chlorophylls have been extensively studied (see the reviews by König and Neugebauer [2012], by Linnanto and Korppi-Tommola [2006], and by us [Curutchet and Mennucci 2017] for a more complete list of references), and in light of this large literature, some general conclusions can be drawn.

Traditionally, the analysis of chlorophyll electronic spectra is based on the four-orbital model proposed by Gouterman (1961) for porphyrin derivatives, illustrated in Figure 2.3. Due to the D_{4h} symmetry of porphyrin, the LUMO is twofold degenerate.

The two HOMOs (HOMO and HOMO − 1) are almost degenerate by accident. The configurations arising from excitation of one electron from the HOMO or HOMO − 1, respectively, to the LUMO are therefore also almost degenerate and interact strongly. This CI gives rise to two states and therefore to two excitation bands, the so-called Q and B (or Soret) bands. In lower-symmetry porphyrin derivatives like chlorophylls, the degeneracy of the LUMO is removed. Thus, the Q band is split into two bands, which have transition dipole vectors lying in the plane of the macrocycle (Q_x and Q_y).

This effect of lowering the symmetry was reproduced by Hasegawa et al. (1998) for model structures from free-base porphyrin to a truncated model of Chl a by applying an accurate wave function–based approach. It was concluded that a reduced π conjugation destabilizes the HOMO and the LUMO + 1, which leads to a smaller HOMO–LUMO gap. This trend is continued if the symmetry is further lowered by an axial ligand to the central Mg. More recent calculations based on DFT/MRCI have also shown that the ground state of the bacteriochlorophylls in FMO is near multiconfigurational in nature (Holmgaard List et al. 2013). The TDDFT methods when combined to hybrid (and long-range corrected) functionals, however, seem relatively robust to the multireference character

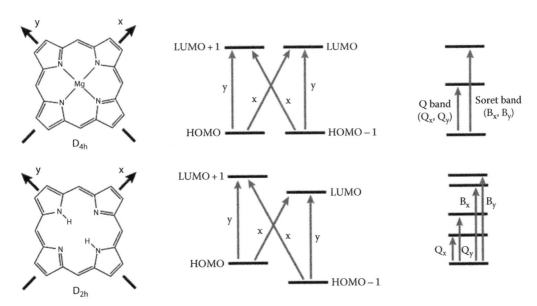

Figure 2.3 Scheme describing the Soret and Q bands of porphyrin derivatives according to the Gouterman four-orbital model.

and constitute the best choice among the considered methods for studying the variations among the Q transition energies and moments of the chlorophylls. Semiempirical approaches or other cheap alternatives such as CIS can also lead to good results, for instance, when applied to the determination of the relative energies of the chlorophylls in a complex; however, the good performances can be due to some unpredictable cancellation of errors, and a systematic improvement is difficult to introduce.

From this brief summary, it appears evident that the accurate QM modeling of both spectral and redox properties of molecular systems is not an easy task even if TDDFT has represented a very effective step forward in the search of accurate but still computationally feasible methods. Nowadays, it surely represents the best option to achieve a reliable simulation of the systems of interest in LH, energy-transferring, and charge-separating processes in photosynthesis. The identification of the best available QM approach however is not the only preliminary aspect that needs to be solved as the calculations should not only give us accurate results but also help us to understand the mechanisms beyond the photophysical processes.

For example, to really understand the different absorption peaks in an electronic spectrum of chlorophyll, we need to investigate the origin of each peak, the nature of the corresponding excitation, the possible coupling of the excitation with vibrational effects induced by geometrical fluctuations, and the effects that the surrounding environment have on all these aspects.

Also the calculation of redox properties involves many aspects in addition to the specific one related to the QM method. As the redox potentials provide a direct measure of the free-energy cost of adding or removing electrons from a species and thus of their relative stability, a proper selection of the QM method is surely fundamental to obtain reliable results (in particular, the QM method should equally describe the neutral and the radical species). However, the presence of the environment will also play a crucial role in differently affecting the reduced and oxidized species. The role of the environment and its influence on the rate of electron transfer was made particularly clear by Marcus in his treatment of electron-transfer reactions.

All the aspects related to the effects of the environment on spectral and redox properties will be reviewed in the following section.

2.3 ACCOUNTING FOR ENVIRONMENT EFFECTS IN CHLOROPHYLL PROPERTIES

In the introduction of this chapter, we discussed the importance of the tuning effect exerted by the protein environment on the site energies of chlorophylls. Clearly, it would be desirable to accurately account for the environment describing the entire pigment–protein system using quantum chemistry. There are promising attempts in this direction (König and Neugebauer 2011). However, the computational cost associated to such calculations is still high, especially considering the need to statistically average the computed properties over the structural fluctuations of the system, which are usually accounted for through classical molecular dynamics (MD) simulations.

A very promising alternative strategy is to couple the quantum chemistry methods described in Section 2.2 to classical descriptions of the protein and solvent environment in order to estimate the energy shifts experienced by chlorophylls due to their surrounding environment. This strategy benefits from the notion that the relevant property (the energy of the excited/oxidized/reduced state) is localized on the chlorophyll molecules. Thus, the protein amino acids do not participate directly in the property, but rather they perturb it through pigment–protein interactions, which can be described at a classical level. This dual strategy is very popular in the quantum chemistry community, giving rise to so-called multiscale modeling techniques, which combine, for example, a high-level quantum-chemical description of the molecule of interest with a simpler classical description of the solvent or surrounding protein matrix.

In the following Sections 2.3.1 and 2.3.2, we will give a description of the two main multiscale approaches used to model the effect of the protein and solvent environment on quantum chemical calculations of redox or spectral properties of chlorophylls. These two strategies rely on either an implicit or an explicit description of the environment in the calculations.

2.3.1 Implicit QM/classical models: The continuum solvation models

Continuum solvation models are a mature technique that provides a cost-effective yet accurate description of solvation effects in the estimation of a variety of properties of molecular systems in a condensed phase, including the prediction of chemical equilibria, reactivity, and spectroscopic properties (Tomasi et al. 2005). Their origins are rooted on the first versions proposed by Onsager and Kirkwood in the 1930s, and nowadays continuum solvation models have become one of the most used computational techniques in the theoretical chemistry community. This success arises from their ability to account for the environment effects in a QM calculation at almost the same cost of an analogous calculation for the system in gas phase. This is possible by resorting to a classical continuum description of the environment in terms of its macroscopic dielectric properties. In particular, in modern continuum solvation models, the solute is placed in a proper molecular-shaped cavity inside the dielectric medium representing the environment, and the polarization response of the dielectric medium to the solute charge distribution is obtained by solving the Poisson equation of classical electrostatics. Such polarization response is termed the solvent reaction field, and it polarizes back the solute charge distribution. Thus, mutual polarization effects among the solute and the solvent are taken into account, leading to the self-consistent reaction field term also used to denote quantum mechanical continuum solvation models.

Different flavors of continuum solvation models resort to different strategies to solve the electrostatic problem of a charge distribution hosted by a cavity immersed in a dielectric medium. In this chapter, we will focus on one of the most extended approaches, the polarizable continuum model (PCM) (Mennucci 2012). Other classes of continuum models are described in different excellent reviews on the subject (Cramer and Truhlar 1999; Tomasi et al. 2005). In PCM, the solvent polarization is described in terms of an apparent surface charge spreading over the cavity surface. Moreover, this surface charge is discretized into point charges placed over small surface elements, called tesserae. In Figure 2.4, we show the cavity corresponding to a Chl a molecule as described in the PCM model, where the small surface elements hosting the point charges representing the solvent polarization can be visualized. In recent developments, the point charges describing the environment polarization have been replaced by spherical Gaussian functions, which avoid some numerical problems, for example, in geometry optimizations, by ensuring continuous and robust derivatives of the reaction field with respect to atomic positions (Scalmani and Frisch 2010).

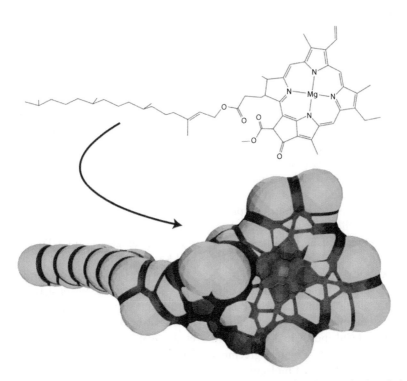

Figure 2.4 Graphical representation of the cavity hosting a Chl a molecule inside the dielectric medium that represents the protein environment in a PCM calculation.

Continuum solvation models can be coupled to a given quantum chemistry method by introducing the solute–solvent interaction as an additional term in the usual Hamiltonian of the isolated molecule. The resulting effective Hamiltonian can then be used to solve the Schrödinger equation taking into account the environment's effect:

$$\hat{H}_{eff} = \hat{H}_M^0 + \hat{V}_{int} \qquad (2.1)$$

where the solute–solvent interaction is given by

$$\hat{V}_{int} = \sum_{i=1}^{N} \frac{q_i}{|\mathbf{r}_i - \mathbf{r}|} \qquad (2.2)$$

This effective Hamiltonian can be used to extend the quantum chemical approaches described in Section 2.2 to estimate the spectral or redox properties of chlorophylls in a condensed phase. The definition of the operator in Equation 2.2 describes the electrostatic interactions between the chromophore and the dielectric environment, which is considered to be the main contribution shifting the redox and spectral properties of pigments (König and Neugebauer 2012; Curutchet and Mennucci 2017). Other contributions can arise from the more complex interactions of the chromophore with its first neighbors. These often-called first-solvation-shell effects include dispersion and exchange repulsion, and changes associated to the perturbation of the environment structure due to the presence of the chromophore. Interesting formulations have been developed to couple the estimation of dispersion effects on solvatochromic shifts (Weijo et al. 2010; Marenich et al. 2013). Usually, however, these are obtained from empirical expressions parameterized to reproduce experimental solvation free energies in a variety of solvents. Because such empirical terms do not depend explicitly on the chromophore electron density, they cannot be used to estimate solvatochromic shifts. A suitable strategy in this case can be to explicitly include the amino acids directly interacting with the chlorophyll in the definition of the QM region (Gudowska-Nowak et al. 1990). For instance, continuum solvent calculations performed on different Chl and BChl derivatives in solution were found to provide a better agreement with experiments if explicit solvent molecules were included in the calculation (Linnanto and Korppi-Tommola 2004).

An important consideration in the description of environment effects on ultrafast electronic processes, such as light absorption promoting a molecule from its ground to an electronic excited state, are the timescales of the environmental polarization. For example, the ground-state equilibrium properties of a neutral or oxidized/reduced form of a pigment can be described through the static dielectric constant of the medium, which describes the complete environment polarization response, including rotational and libration motions of environment molecules, as well as atomic (vibrations) and electronic motions. Photophysical processes like light absorption, fluorescence, or energy and electron transfer reactions, however, occur in a nonequilibrium solvation regime. In this case, the electronic polarization of the environment is able to adapt almost instantaneously to the new electronic state, whereas nuclear motions relax on a slower timescale, leading, for instance, to the time-dependent fluorescence Stokes shifts, associated to the reorganization of the environment. Continuum solvation models can handle these nonequilibrium solvation effects in excited states by partitioning the environment response in two contributions, which can be described from the static and the optical dielectric constant of the medium (Tomasi et al. 2005).

The ability to describe nonequilibrium solvation effects on excited-states, and the possibility to obtain thermodynamically meaningful properties from a single QM calculation, thus avoiding the need to statistically average the calculated properties over the environment degrees of freedom, is a powerful attribute of continuum solvation models. This is especially true for the calculation of the properties of chromophores in homogeneous solutions. When the environment is heterogeneous, for example, the protein surrounding the chlorophylls in a photosynthetic complex, this strategy has however some limitations. Time-dependent fluorescence Stokes shifts (Cohen et al. 2002) and MD simulations (Golosov and Karplus 2007) show that polar solvation dynamics in proteins are position dependent and highly heterogeneous. Predictions of static dielectric constants for different proteins or protein binding sites predict ε values ranging from 4 to 40 (King et al. 1991; Smith et al. 1993; Simonson and Brooks 1996; Pitera et al. 2001). On the other hand, optical dielectric constants in a photosynthetic pigment–protein complex have been estimated to vary between 1.5 and 2.5, based on the simulation of dielectric screening effects in energy transfer from combined MD–QM/MM calculations, although the average protein value estimated was close to the value $\varepsilon_{opt} = 2$ usually assumed in studies of photosynthetic LH (Curutchet et al. 2011). It is thus difficult to assign a particular value to the dielectric constants used to describe the protein polarization response in continuum solvation models.

If the fine-tuning induced in the chlorophylls energies from different local pigment–protein interactions needs to be accounted for, it is better to resort to an explicit description of the environment, although at the price of a more complex and costly calculation. This strategy is described in the next section, focused on combined QM/MM approaches.

2.3.2 Explicit QM/classical approaches: The hybrid QM/MM models

Continuum solvation models provide a robust and efficient way to calculate how the properties of chlorophylls are shifted in a condensed-phase environment. The assumption of a continuum dielectric environment, however, precludes the possibility to describe individual pigment–protein interactions that can be important to understand, for example, the different energies of the eight BChl a pigments in the FMO photosynthetic complex illustrated in Figure 2.2, which span a

range of ~60 meV (Milder et al. 2010; Schmidt am Busch et al. 2011). Because this fine-tuning of the energies of chlorophylls by the protein has important consequences on the efficiency and mechanisms of LH, an attractive alternative adopted in several theoretical studies in this area resorts to so-called combined quantum-mechanics molecular-mechanics (QM/MM) models (Senn and Thiel 2009), first introduced in the seminal paper by Warshel and Levitt (1976). In QM/MM models, the chlorophyll pigment, and eventually some neighboring amino acids or solvent molecules, can be accurately described through the QM methods described in Section 2.2, whereas the environment is described using a classical force field, that is, a collection of empirical energy functions widely used in the simulation of biomolecules. In Figure 2.5, we show a scheme of a QM/MM model of the FMO complex, where the BChl pigments are treated with quantum chemistry, whereas the protein amino acids are described using classical molecular mechanics.

Classical force fields typically involve bonded and nonbonded terms. Bonded (intramolecular) interactions are described through bond, angle, and torsional energy terms, whereas nonbonded (intra- and intermolecular) interactions are associated to pairwise dispersion–repulsion and electrostatic contributions.

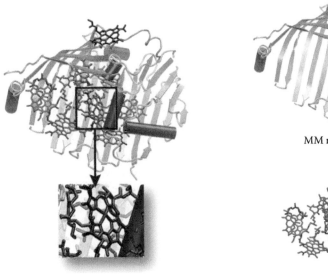

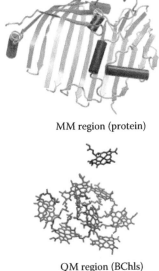

MM region (protein)

QM region (BChls)

Figure 2.5 Scheme of a QM/MM model of the FMO complex, where the BChl pigments are treated with quantum chemistry, whereas the protein amino acids are described using classical molecular mechanics.

A variety of force fields have been parameterized to describe biological systems (Cieplak et al. 2009). A popular example is the Amber force field (Cornell et al. 1995), which describes the energy using the following expression:

$$E_{\text{total}} = E_{\text{bonded}} + E_{\text{nonbonded}} \qquad (2.3)$$

$$E_{\text{bonded}} = \sum_{\text{bonds}} K_r \left(r - r_{\text{eq}} \right)^2 + \sum_{\text{angles}} K_\theta \left(\theta - \theta_{\text{eq}} \right)^2$$

$$+ \sum_{\text{dihedrals}} \frac{V_n}{2} \left[1 + \cos \left(n\phi - \gamma \right) \right] \qquad (2.4)$$

$$E_{\text{nonbonded}} = \sum_{\text{van der Waals}}^{i<j} \left[\frac{A_{ij}}{R_{ij}^{12}} - \frac{B_{ij}}{R_{ij}^{6}} \right] + \sum_{\text{electrostatic}}^{i<j} \frac{q_i q_j}{\varepsilon R_{ij}}$$

$$(2.5)$$

Development of a classical force field involves calibration of the different parameters in the energy function, such as the r_{eq} and θ_{eq} equilibrium bond lengths and angles; n and γ dihedral multiplicity and angle phases; K_r, K_θ, and V_n force constants; A and B van der Waals coefficients; and q partial charges located on the atomic positions. Standard additive force fields like the one given by Equations 2.3 through 2.5 take into account electronic polarization effects implicitly by assigning partial charges to the atoms that describe their charge distribution in a condensed phase. Important efforts, however, have been devoted to develop polarizable force fields, which add a further term to the total energy associated to many-body polarization effects. Several models to include polarization effects have been developed, including the fluctuating charge, the Drude oscillator, and the induced dipole model (Cieplak et al. 2009). Here we will focus on the induced dipole model, adopted in the context of the Amber force field. In this case, an additional energy term is added to the total energy:

$$E_{\text{pol}} = -\frac{1}{2} \sum_i \mu_i E_i = -\frac{1}{2} \sum_i \alpha_i E_i^{(0)} E_i \qquad (2.6)$$

where

α_i denotes the isotropic point polarizability of atom i

$E_i^{(0)}$ is the electrostatic field on atom i due to partial charges

E_i is the electrostatic field on atom i due to charges and induced dipoles

Polarizable force fields based on the induced dipole model, therefore, assign isotropic polarizability parameters in addition to partial charges in order to describe electrostatic interactions among the atoms.

Because of the empirical description of bonded and van der Waals terms, in practice, only electrostatic and polarization terms from a force field are explicitly coupled to the QM determination of the redox or excited states of a chlorophyll in QM/MM models. In a similar way as previously described for continuum solvation models, the QM/MM approach defines an effective Hamiltonian for the chromophore in the presence of the protein environment:

$$\hat{H}_{\text{eff}} = \hat{H}_{\text{M}}^0 + \hat{H}_{\text{QM/MM}} + \hat{H}_{\text{MM}} \qquad (2.7)$$

$$\hat{H}_{\text{QM/MM}} + \hat{H}_{\text{QM/MM}}^{\text{ele}} + \hat{H}_{\text{QM/MM}}^{\text{pol}} \qquad (2.8)$$

$$\hat{H}_{\text{MM}} = \hat{H}_{\text{MM}}^{\text{ele}} = \hat{H}_{\text{MM}}^{\text{pol}} \qquad (2.9)$$

where

$\hat{H}_{\text{M}}^0$ is the usual Hamiltonian of the isolated chromophore

$\hat{H}_{\text{QM/MM}}$ term describes the interaction between the QM and MM regions of the system (the chromophore and the protein + solvent environment, respectively)

$\hat{H}_{\text{MM}}$ accounts for the interactions among the atoms of the MM environment

Each term $\hat{H}_{\text{QM/MM}}$ and $\hat{H}_{\text{MM}}$ includes electrostatic and polarization contributions, associated to the interactions involving either the point charges or the induced dipoles in the MM environment. We note that the $\hat{H}_{\text{MM}}$ term has to be included in the QM/MM calculation because the induced dipoles in the MM region explicitly depend on the electron density of the QM system.

If a nonpolarizable MM description is used that only involves assignment of partial charges to the atoms of the protein and solvent, the effective Hamiltonian (2.7) simplifies as follows:

$$\hat{H}_{\text{eff}} = \hat{H}_{\text{M}}^0 + \hat{H}_{\text{QM/MM}}^{\text{ele}} = \hat{H}_{\text{M}}^0 + \sum_i \frac{q_i}{|\mathbf{r}_i - \mathbf{r}|} \qquad (2.10)$$

By using either the polarizable or the nonpolarizable effective Hamiltonian, classical MM models can be coupled to the QM methods described in

Section 2.2 to determine the spectral and redox properties of chlorophylls in a photosynthetic protein. A strong advantage of these methods is the ability to explicitly account for both local short-range as well as long-range electrostatic interactions. Indeed, the solvatochromic shift in a chlorophyll energy can arise from hydrogen bonds among the chlorophylls and the surrounding amino acids, or by the presence of waters or different amino acids, especially aromatic, charged, or polar ones, at their ligation site, the central Mg atom.

In addition, these methods can be coupled to MD simulations, which simulate the time evolution of the system, to explore how the motions of the pigment–protein complex modulate the energies of the pigments (Vrandecic et al. 2015). This combined MD–QM/MM strategy has been applied, for example, to the prediction of the spectral density of pigment–protein coupling in photosynthetic proteins (Damjanović et al. 2002; Olbrich et al. 2011; Jing et al. 2012; Shim et al. 2012), a key quantity in the modeling of photosynthetic LH.

A disadvantage of QM/MM methods, compared to continuum solvation models, is however the need to be careful in the selection of the many parameters used to describe the MM region through the force field, which can significantly affect the obtained results (Curutchet et al. 2013). Also, continuum models provide a much simpler and fast approach, because QM/MM calculations performed on the three-dimensional crystal structure of the protein offer a limited static picture of the protein environment. In general, QM/MM calculations have to be performed over a significant number of structures of the system, extracted typically from a classical MD trajectory. This allows estimation of statistically averaged properties that account for the dynamic fluctuations of the environment. A problem of this strategy, however, is the mismatch of the geometries obtained from the classical simulation as compared to a more accurate QM description. This problem can be circumvented by related quantum chemical/electrostatic two-step (QC/E2) approaches (Renger and Müh 2013). In this case, a first step involves QM calculations of the pigments in gas phase and fitting a set of atomic partial charges to the corresponding electrostatic potentials. Such charges can then be used in subsequent classical electrostatic calculations to derive the protein-induced energy shifts. These methods provide an efficient

route for the determination of the site energies of a photosynthetic complex. However, they neglect changes in the pigment properties (charges) arising from pigment–protein interactions and ignore the impact of steric effects leading to deformation of the chlorophyll conformation in the site energies (MacGowan and Senge 2016), which are otherwise accounted for in QM/MM models. A promising alternative is given by subsystem QM/QM models or novel linear-scaling DFT algorithms, in which the entire system is described at a QM level (König and Neugebauer 2011; Cole et al. 2013). These strategies, however, are still very expensive in terms of computational cost, especially if the fluctuations of the environment need to be taken into account.

Despite the increased effort in terms of both computational resources and definition of the parameters involved in the model compared to a continuum dielectric approach, QM/MM or related QC/E2 models represent nowadays probably the best-suited techniques to study how the protein environment modulates the properties of chlorophylls in photosynthesis. Many studies based on multiscale techniques have been useful to characterize the congested spectral features of photosynthetic systems like the FMO complex of green sulfur bacteria (Muh et al. 2007), the LHCII of higher plants (Müh et al. 2010), or the phycoerythrin 545 (PE545) antenna of cryptophyte algae (Curutchet et al. 2013).

Inclusion of explicit environment polarization in these models allows accounting for nonequilibrium solvation effects arising from the different relaxation timescales of the electronic and nuclear protein degrees of freedom, as discussed in Section 2.3.1 (Curutchet et al. 2013). Polarizable QM/MM models, in addition, allow accounting for the impact of the heterogeneous polarizable environment of the protein on the dielectric screening effects that modulate exciton interactions (Curutchet et al. 2011).

Finally, QM/MM calculations have also been used to investigate the electron transfer processes that take place in photosynthetic complexes, for instance, in the photosystem II reaction center of plants and cyanobacteria (Kitagawa et al. 2011).

Despite the partial success of the models described in this section in the calculation of chlorophyll properties in protein environments, we remark that such calculations are still a considerable challenge. This is because exciton delocalization, optical spectra, and energy migration mechanisms and pathways in LH complexes are exquisitely sensitive to small variations of the BChls' relative site energies (tens of meV). Considering, however, the continuous improvement of both theoretical approaches and computer resources available to theoretical chemists, the impact and success of multiscale methods for the prediction of chlorophyll properties in protein environments is expected to be even better in the near future.

REFERENCES

Andersson, K., P. A. Malmqvist, B. O. Roos, A. J. Sadlej, and K. Wolinski. 1990. Second-order perturbation-theory with a CASSCF reference function. *J. Phys. Chem.* 94:5483–5488.

Bartlett, R. J. and G. D. Purvis. 1978. Many-body perturbation theory, coupled-pair many-electron theory, and the importance of quadruple excitations for the correlation problem. *Int. J. Quantum Chem.* 14:561–581.

Blankenship, R. E. 2002. *Molecular Mechanisms of Photosynthesis.* Oxford, U.K.: Wiley-Blackwell.

Blankenship, R. E., M. T. Madigan, and C. E. Bauer. 1995. *Anoxygenic Photosynthetic Bacteria.* Dordrecht, the Netherlands: Kluwer Academic Press.

Blankenship, R. E. and K. Matsuura. 2003. Antenna complexes from green photosynthetic bacteria. In *Light-Harvesting Antennas*, eds. B. R. Green and W. W. Parson, pp. 195–217. Dordrecht, the Netherlands: Kluwer Academic Publishers.

Casida, M. E. 1995. Time-dependent density functional response theory for molecules. In *Recent Advances in Density Functional Methods*, ed. D. P. Chong, pp. 155–192. Singapore: World Scientific.

Cieplak, P., F.-Y. Y. Dupradeau, Y. Duan, and J. M. Wang. 2009. Polarization effects in molecular mechanical force fields. *J. Phys. Condens. Matter* 21:333102.

Cohen, B. E., T. B. McAnaney, E. S. Park, Y. N. Jan, S. G. Boxer, and L. Y. Jan. January 2002. Probing protein electrostatics with a synthetic fluorescent amino acid. *Science* 296:1700–1703.

Cole, D. J., A. W. Chin, N. D. M. Hine, P. D. Haynes, and M. C. Payne. 2013. Toward Ab initio optical spectroscopy of the Fenna–Matthews–Olson complex. *J. Phys. Chem. Lett.* 4:4206–4212.

Collini, E., C. Curutchet, T. Mirkovic, and G. D. Scholes. 2009. Electronic energy transfer in photosynthetic antenna systems. In *Energy Transfer Dynamics in Biomaterial Systems*, eds. I. Burghardt, V. May, D. A. Micha, and E. R. Bittner, pp. 3–34. Berlin, Germany: Springer.

Cornell, W. D., P. Cieplak, C. I. Bayly et al. 1995. A second generation force field for the simulation of proteins, nucleic acids, and organic molecules. *J. Am. Chem. Soc.* 117:5179–5197.

Cramer, C. J. and D. G. Truhlar. 1999. Implicit solvation models: Equilibria, structure, spectra, and dynamics. *Chem. Rev.* 99:2161–2200.

Curutchet, C., J. Kongsted, A. Muñoz-Losa, H. Hossein-Nejad, G. D. Scholes, and B. Mennucci. 2011. Photosynthetic light-harvesting is tuned by the heterogeneous polarizable environment of the protein. *J. Am. Chem. Soc.* 133:3078–3084.

Curutchet, C. and B. Mennucci. 2017. Quantum chemical studies of light harvesting. *Chem. Rev.* 117:294–343.

Curutchet, C., V. I. Novoderezhkin, J. Kongsted et al. 2013. Energy flow in the cryptophyte PE545 antenna is directed by bilin pigment conformation. *J. Phys. Chem. B* 117:4263–4273.

Damjanović, A., I. Kosztin, U. Kleinekathöfer, and K. Schulten. 2002. Excitons in a photosynthetic light-harvesting system: A combined molecular dynamics, quantum chemistry, and polaron model study. *Phys. Rev. E* 65:31919.

Dreuw, A. and M. Head-Gordon. 2005. Single-reference Ab initio methods for the calculation of excited states of large molecules. *Chem. Rev.* 105:4009–4037.

Foresman, J. B., M. Head-Gordon, J. A. Pople, and M. J. Frisch. 1992. Toward a systematic molecular-orbital theory for excited-states. *J. Phys. Chem.* 96:135–149.

Golosov, A. A. and M. Karplus. 2007. Probing polar solvation dynamics in proteins: A molecular dynamics simulation analysis. *J. Phys. Chem. B* 111:1482–1490.

Gouterman, M. 1961. Spectra of porphyrins. *J. Mol. Spectrosc.* 6:138.

Grimme, S. and M. Waletzke. 1999. A combination of Kohn–Sham density functional theory and multi-reference configuration interaction methods. *J. Chem. Phys.* 111:5645.

Gudowska-Nowak, E., M. D. Newton, and J. Fajer. 1990. Conformational and environmental effects on bacteriochlorophyll optical spectra: Correlations of calculated spectra with structural results. *J. Phys. Chem.* 94:5795–5801.

Hasegawa, J., K. Ohkawa, and H. Nakatsuji. 1998. Excited states of the photosynthetic reaction center of *Rhodopseudomonas viridis*: SAC-CI study. *J. Phys. Chem. B* 102:10410–10419.

Holmgaard List, N., C. Curutchet, S. Knecht, B. Mennucci, and J. Kongsted. 2013. Toward reliable prediction of the energy ladder in multichromophoric systems: A benchmark study on the FMO light-harvesting complex. *J. Chem. Theory Comput.* 9:4928–4938.

Jankowiak, R., M. Jassas, J. Chen et al. 2016. On the conflicting estimations of pigment site energies in photosynthetic complexes: A case study of the CP47 complex. *Anal. Chem. Insights* 11:35–48.

Jing, Y. Y., R. H. Zheng, H. X. Li, and Q. Shi. 2012. Theoretical study of the electronic-vibrational coupling in the Q(y) states of the photosynthetic reaction center in purple bacteria. *J. Phys. Chem. B* 116:1164–1171.

Jurinovich, S., L. Viani, C. Curutchet, and B. Mennucci. 2015. Limits and potentials of quantum chemical methods in modelling photosynthetic antennae. *Phys. Chem. Chem. Phys.* 17:30783–30792.

King, G., F. S. Lee, and A. Warshel. 1991. Microscopic simulations of macroscopic dielectric constants of solvated proteins. *J. Chem. Phys.* 95:4366–4377.

Kitagawa, Y., K. Matsuda, and J. Hasegawa. 2011. Theoretical study of the excited states of the photosynthetic reaction center in photosystem II: Electronic structure, interactions, and their origin. *Biophys. Chem.* 159:227–236.

König, C. and J. Neugebauer. 2011. First-principles calculation of electronic spectra of light-harvesting complex II. *Phys. Chem. Chem. Phys.* 13:10475–10490.

König, C. and J. Neugebauer. 2012. Quantum chemical description of absorption properties and excited-state processes in photosynthetic systems. *ChemPhysChem* 13:386–425.

Linnanto, J. and J. Korppi-Tommola. 2004. Semiempirical PM5 molecular orbital study on chlorophylls and bacteriochlorophylls: Comparison of semiempirical, Ab initio, and density functional results. *J. Comput. Chem.* 25:123–137.

Linnanto, J. and J. Korppi-Tommola. 2006. Quantum chemical simulation of excited states of chlorophylls, bacteriochlorophylls and their complexes. *Phys. Chem. Chem. Phys.* 8:663–687.

MacGowan, S. A. and M. O. Senge. 2016. Contribution of bacteriochlorophyll conformation to the distribution of site-energies in the FMO protein. *BBA Bioenergetics* 1857:427–442.

Marenich, A. V., C. J. Cramer, and D. G. Truhlar. 2013. Uniform treatment of solute–solvent dispersion in the ground and excited electronic states of the solute based on a solvation model with state-specific polarizability. *J. Chem. Theory Comput.* 9:3649–3659.

Mennucci, B. 2012. Polarizable continuum model. *WIREs Comput. Mol. Sci.* 2:386–404.

Milder, M. T. W., B. Bruggemann, R. van Grondelle, and J. L. Herek. 2010. Revisiting the optical properties of the FMO protein. *Photosynth. Res.* 104:257–274.

Milne, B. F., Y. Toker, A. Rubio, and S. B. Nielsen. 2015. Unraveling the intrinsic color of chlorophyll. *Angew. Chem. Int. Ed.* 54:2170–2173.

Mirkovic, T., E. E. Ostroumov, J. M. Anna, R. van Grondelle, Govindjee, and G. D. Scholes. 2017. Light absorption and energy transfer in the antenna complexes of photosynthetic organisms. *Chem. Rev.* 117:249–293.

Muh, F., M. E.-A. Madjet, J. Adolphs et al. 2007. α-Helices direct excitation energy flow in the fenna Matthews Olson protein. *Proc. Natl. Acad. Sci. U. S. A.* 104:16862–16867.

Müh, F., M. E.-A. Madjet, and T. Renger. 2010. Structure-based identification of energy sinks in plant light-harvesting complex II. *J. Phys. Chem. B* 114:13517–13535.

Novoderezhkin, V. I., A. B. Doust, C. Curutchet, G. D. Scholes, and R. van Grondelle. 2010. Excitation dynamics in phycoerythrin 545: Modeling of steady-state spectra and transient absorption with modified redfield theory. *Biophys. J.* 99:344–352.

Novoderezhkin, V. I., M. A. Palacios, H. van Amerongen, and R. van Grondelle. 2005. Excitation dynamics in the LHCII complex of higher plants: Modeling based on the 2.72 Angstrom crystal structure. *J. Phys. Chem. B* 109:10493–10504.

Novoderezhkin, V. I. and R. van Grondelle. 2010. Physical origins and models of energy transfer in photosynthetic light-harvesting. *Phys. Chem. Chem. Phys.* 12:7352–7365.

Olbrich, C., J. Strumpfer, K. Schulten, and U. Kleinekathofer. 2011. Theory and simulation of the environmental effects on FMO electronic transitions. *J. Phys. Chem. Lett.* 2:1771–1776.

Pitera, J. W., M. Falta, and W. F. van Gunsteren. 2001. Dielectric properties of proteins from simulation: The effects of solvent, ligands, pH, and temperature. *Biophys. J.* 80:2546–2555.

Reimers, J. R., M. Biczysko, D. Bruce et al. 2016. Challenges facing an understanding of the nature of low-energy excited states in photosynthesis. *BBA Bioenergetics* 1857:1627–1640.

Renger, T. and F. Müh. 2013. Understanding photosynthetic light-harvesting: A bottom up theoretical approach. *Phys. Chem. Chem. Phys.* 15:3348–3371.

Roos, B. O., P. R. Taylor, and P. E. M. Siegbahn. 1980. A Complete Active Space SCF Method (CASSCF) using a density-matrix formulated super-CI approach. *Chem. Phys.* 48:157–173.

Runge, E. and E. K. U. Gross. 1984. Density-functional theory for time-dependent systems. *Phys. Rev. Lett.* 52:997–1000.

Scalmani, G. and M. J. Frisch. 2010. Continuous surface charge polarizable continuum models of solvation I. general formalism. *J. Chem. Phys.* 132:1–15.

Schmidt am Busch, M., F. Müh, M. El-Amine Madjet, and T. Renger. 2011. The eighth bacteriochlorophyll completes the excitation energy funnel in the FMO protein. *J. Phys. Chem. Lett.* 2:93–98.

Scholes, G. D. and G. R. Fleming. 2000. On the mechanism of light harvesting in photosynthetic purple bacteria: B800 to B850 energy transfer. *J. Phys. Chem. B* 104:1854–1868.

Scholes, G. D., G. R. Fleming, A. Olaya-Castro, and R. van Grondelle. 2011. Lessons from nature about solar light harvesting. *Nat. Chem.* 3:763–774.

Senn, H. M. and W. Thiel. 2009. QM/MM methods for biomolecular systems. *Angew. Chem. Int. Ed.* 48:1198–1229.

Shim, S., P. Rebentrost, S. Valleau, and A. Aspuru-Guzik. 2012. Atomistic study of the long-lived quantum coherences in the Fenna–Matthews–Olson complex. *Biophys. J.* 102:649–660.

Simonson, T. and C. L. Brooks. 1996. Charge screening and the dielectric constant of proteins: Insights from molecular dynamics. *J. Am. Chem. Soc.* 118:8452–8458.

Smith, P. E., R. M. Brunne, A. E. Mark, and W. F. van Gunsteren. 1993. Dielectric properties of trypsin inhibitor and lysozyme calculated from molecular dynamics simulations. *J. Phys. Chem.* 97:2009–2014.

Stanton, J. F. and R. J. Bartlett. 1993. The equation of motion coupled-cluster method. A systematic biorthogonal approach to molecular excitation energies, transition probabilities, and excited state properties. *J. Chem. Phys.* 98:7029–7039.

Stewart, J. J. P. 2007. Optimization of parameters for semiempirical methods V: Modification of NDDO approximations and application to 70 elements. *J. Mol. Model.* 13:1173–1213.

Stewart, J. J. P. 2013. Optimization of parameters for semiempirical methods VI: More modifications to the NDDO approximations and re-optimization of parameters. *J. Mol. Model.* 19:1–32.

Tomasi, J., B. Mennucci, and R. Cammi. 2005. Quantum mechanical continuum solvation models. *Chem. Rev.* 105:2999–3093.

van Grondelle, R and V. I. Novoderezhkin. 2006. Energy transfer in photosynthesis: Experimental insights and quantitative models. *Phys. Chem. Chem. Phys.* 8:793–807.

Vrandecic, K., M. Rätsep, L. Wilk et al. 2015. Protein dynamics tunes excited state positions in light-harvesting complex II. *J. Phys. Chem. B* 119:3920–3930.

Warshel, A. and M. Levitt. 1976. Theoretical studies of enzymic reactions: Dielectric, electrostatic and steric stabilization of the carbonium ion in the reaction of lysozyme. *J. Mol. Biol.* 103:227–249.

Weijo, V., B. Mennucci, and L. Frediani. 2010. Toward a general formulation of dispersion effects for solvation continuum models. *J. Chem. Theory Comput.* 6:3358–3364.

Zerner, M. C. 1991. Semiempirical molecular orbital methods. In *Reviews of Computational Chemistry*, vol. 2, eds. K. B. Lipkowitz and D. B. Boyd, pp. 313–365. New York: VCH Publishers Inc.

Carotenoids: Electronic states and biological functions

HARRY A. FRANK AND BRUNO ROBERT

ABBREVIATIONS

Chl	Chlorophyll
HPLC	High-performance liquid chromatography
HOMO	Highest occupied molecular orbital
ICT	Intramolecular charge transfer
ISRS	Impulsive stimulated Raman scattering
LUMO	Lowest unoccupied molecular orbital
NPQ	Nonphotochemical quenching
rR	Resonance Raman

3.1 INTRODUCTION

Carotenoids are synthesized by photosynthetic organisms and provide vibrant red, orange, and yellow coloration in nature (Isler 1971). They are found in many fruits, vegetables, and flowers as well as in a wide variety of bacterial and algal species. Upon ingestion of these organisms by birds, crustaceans, and fish, carotenoids can produce striking displays of color (Figure 3.1).

Moreover, dietary intake of carotenoids by mammals is thought to be associated with reduced risks of several chronic health disorders including heart disease, age-related macular degeneration, and certain cancers (Cooper et al. 1999). It has been postulated that these actions are related to the ability of carotenoids to inhibit reactions involving reactive oxygen species (Krinsky 1971).

It is also well known that carotenoids act as protective devices against the photodestruction of photosynthetic organisms either by quenching chlorophyll (Chl) triplet states, which prevents the Chl-sensitized formation of singlet-state oxygen, a major oxidizing agent of Chl (Foote 1968; Krinsky 1968; Goodwin 1976; Boucher et al. 1977; Renger and Wolff 1977; Cogdell and Frank 1987; Frank and Cogdell 1993; Young and Britton 1993; Frank et al. 1999), by scavenging singlet oxygen directly (Foote et al. 1970; Krinsky 1971), or by dissipating excess excitation energy (Demmig-Adams 1990;

Figure 3.1 Carotenoid pigmentation in birds, crustaceans, and fish. (Flamingo and lobster photos by H. A. Frank. Goldfish photo provided by Creative Commons: k_millo.)

Niyogi et al. 1997; Müller et al. 2001; Holt et al. 2004; Ruban et al. 2007; Staleva et al. 2015). The molecular mechanisms carotenoids use to accomplish photoprotection are the subject of intense debate. Carotenoids also perform a vital role as light-harvesting pigments by absorbing light in regions of the visible spectrum where Chl is not a very efficient absorber (Goedheer 1969; Govindjee 1975; Cogdell 1978; Frank and Christensen 1995). Carotenoids transfer the absorbed excited state energy to Chl where it is trapped by a reaction-center pigment–protein complex and converted into electrical potential (Kirmaier and Holten 1987; Blankenship et al. 1995). Carotenoids also stabilize protein structures. Many photosynthetic pigment–protein complexes will not assemble properly without these molecules (Lang and Hunter 1994; Yamamoto and Bassi 1996; Sandonà et al. 1998).

be reached by the absorption of a photon that promoted a single electron from its highest occupied molecular orbital (HOMO) to its lowest unoccupied molecular orbital (LUMO) (two left-hand columns in Figure 3.2).

The allowedness of this transition is consistent with quantum mechanical selection rules, which require a change in the symmetry of the electronic state upon light absorption. Thus, a one-electron HOMO → LUMO transition, which originates from the ground (gerade, even, g) electronic state, denoted S_0, and terminates in an excited (ungerade, odd, u) state, satisfies this requirement. While examining the optical spectroscopic properties of α,ω-diphenyloctatetraene in an n-alkane, mixed crystalline Shpolski'i matrix at cryogenic

3.2 ENERGY LEVELS OF CAROTENOIDS

In order to understand how carotenoids function in nature, it is important to be familiar with their excited states from which many of their photochemical properties originate. There is a rich history of scientific literature pertaining to carotenoids that dates back over a century, but for the present purposes, it will suffice to begin with the landmark report by Hudson and Kohler (1972) on the electronic absorption spectra of polyenes. The spectroscopic experiments performed by these investigators changed forever our view of the excited state complexion of linear π-electron conjugated systems. Prior to that time, molecular orbital theory predicted that the lowest-energy excited state in all π-electron conjugated molecules could

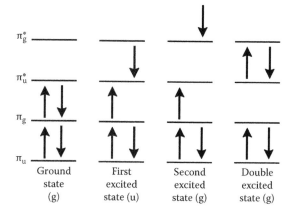

Figure 3.2 Molecular orbital diagram of a short π-electron conjugated system. π and π* are bonding and antibonding orbitals, respectively. g and u represent the symmetry of the configurations (gerade and ungerade).

temperatures, Hudson and Kohler (1972) observed a narrow absorption feature that was significantly lower in energy than that corresponding to the λ_{max} of the absorption spectrum of molecule. The narrow band corresponded precisely with the Stokes-shifted spectral origin of the fluorescence spectrum of the molecule, which led the authors to postulate that the state from which fluorescence occurs is different from the state into which absorption is strongly allowed. Schulten and Karplus (1972) and subsequently Tavan and Schulten and others (Schulten et al. 1976; Ohmine et al. 1978; Tavan and Schulten 1979, 1986, 1987; Ohmine and Morokuma 1980) provided a theoretical basis for this interpretation by invoking configuration interaction between higher-lying g states that included both single and double excited configurations (two right-hand columns of Figure 3.2). The configurational mixing of these states gave rise to new g states formed from linear combinations of the original single and double excited g states, with one of these, denoted S_1, dropping below that of the lowest-lying u state. The implication of this discovery was immediately clear: The state into which absorption is strongly allowed for polyenes (and possibly also carotenoids and other molecules having a linear π-electron conjugated chain of double bonds) is not the lowest-lying excited state, S_1. Rather, it is the second exited state, S_2, which upon photoexcitation can then decay by internal conversion to the low-lying S_1 state from which fluorescence occurs. The forbidden $S_1 \rightarrow S_0$ fluorescence transition gains allowedness by borrowing intensity from the S_2 state via Herzberg–Teller coupling. This interpretation rationalized the large Stokes shift between the maxima in the absorption and fluorescence spectra and also reconciled the different dependencies of the spectra on changes in solvent polarizability. These findings also explained the discrepancy between the anomalously long fluorescence lifetimes of polyene excited states measured experimentally and those predicted from integrating the absorption bands according to the Strickler–Berg relationship (Hudson and Kohler 1973; Hudson et al. 1982; Hudson and Kohler 1984). Assignments of appropriate group theoretical irreducible representations for these linear π-electron conjugated molecules having approximate C_{2h} symmetry, and subsequent higher-order computational treatments that include pseudoparity restrictions, settled on the state designations S_0 $(1^1A_g^-)$, S_1 $(2^1A_g^-)$, and S_2 $(1^1B_u^+)$ for both carotenoids and polyenes (Figure 3.3). The + and − signs accompanying the group theoretical symmetry representations are pseudoparity elements and derive from π-electron molecular orbital pairing relationships when configuration interaction among singly excited configurations is included (Pariser 1955; Callis et al. 1983; Birge 1986). One-photon transitions from the ground S_0 state are allowed to S_2 but forbidden to S_1 on the basis of both symmetry (g/u) and pseudoparity (+/−) selection rules. Hence, the S_1 $(2^1A_g^-)$ state is sometimes referred to as a "forbidden" or "dark" state.

The notion that a low-lying state of A_g symmetry exists for carotenoids had profound implications for how the molecules function as light-harvesting pigments in photosynthetic systems. The possibility now existed for more than one excited electronic state to transfer energy to Chl (Figure 3.3), which suggested the potential for improved efficiency of

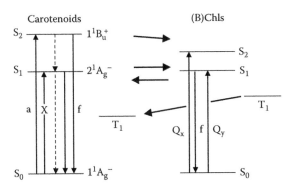

Figure 3.3 Energy state diagram of carotenoids and potential routes of energy transfer to and from the singlet and triplet states of (bacterio)chlorophylls, (B)Chl. The X denotes a one-photon-forbidden transition; a, one-photon-allowed transition; f, fluorescence; Q_X and Q_Y are absorption transitions of (B)Chls.

carotenoid-to-Chl energy transfer (Cogdell and Frank 1987). Clearly, a knowledge of the precise energies of the S_1 ($2^1A_g^-$) states of various naturally occurring carotenoids having different π-electron conjugated chain lengths and attached functional groups would be important in understanding the precise pathway(s) by which the molecules transfer energy to (B)Chl. Therefore, many research groups turned their attention to achieving this goal (Koyama 1991; Andersson et al. 1992; Andersson and Gillbro 1992; DeCoster et al. 1992). Moreover, a very low-lying S_1 ($2^1A_g^-$) state could potentially act as a quencher of excess Chl excited states by the reverse process of Chl-to-carotenoid energy transfer (Figure 3.3), thereby suggestive of a mechanism of photoprotection (Demmig-Adams 1990; Frank et al. 1994).

However, many technical issues stood in the way of characterizing the S_1 ($2^1A_g^-$) states of carotenoids. First and foremost was the fact that the S_0 ($1^1A_g^-$) → S_1 ($2^1A_g^-$) transition is forbidden by one-photon absorption. This makes the lower-energy S_0 ($1^1A_g^-$) → S_1 ($2^1A_g^-$) absorption bands in polyenes and carotenoids extremely weak and difficult to detect on top of the tailing, long-wavelength edge of the much stronger S_0 ($1^1A_g^-$) → S_2 ($1^1B_u^+$) absorption bands. Deducing the energies of the S_1 ($2^1A_g^-$) states of carotenoids by direct absorption spectroscopy from the ground state remains to this day a very challenging undertaking.

3.2.1 Fluorescence spectroscopy of carotenoids

As demonstrated by Hudson and Kohler (1972), fluorescence spectroscopy offered a promising approach to determining the energy of the low-lying S_1 ($2^1A_g^-$) states of carotenoids. The fluorescence quantum yields of short polyenes, for example, octatetraene, which has 4 conjugated carbon–carbon double bonds, N, are ~0.6 at low temperatures (Gavin et al. 1978), but unfortunately, longer carotenoid molecules, for example, β-carotene, have fluorescence yields on the order of 10^{-4} or less (Bondarev et al. 1988, 1989; Gillbro and Cogdell 1989; Cosgrove et al. 1990; Petek et al. 1992), and therefore presented an additional challenge to characterizing the lowest excited singlet S_1 ($2^1A_g^-$) states of the biologically important carotenoids.

The earliest reports of fluorescence spectra from β-carotene showed that the weak emissions

coincided with the spectral origin of the strongly allowed absorption spectra, suggesting that absorption and fluorescence occurred to and from the same excited state (Cherry et al. 1968; Haley and Koningstein 1983; van Riel et al. 1983; Watanabe et al. 1986; Bondarev et al. 1988, 1989). In some investigations (van Riel et al. 1983), good agreement was found between the absorption and fluorescence excitation spectra, and this confirmed the existence of fluorescence from what were thought previously to be nonfluorescent molecules. Subsequent work (Gillbro and Cogdell 1989; Cosgrove et al. 1990) confirmed the reports of fluorescence emission from β-carotene and related molecules. Yet, it was peculiar that carotenoids were not exhibiting detectable emission from the lowest-lying S_1 ($2^1A_g^-$) state, which was known to be present in the manifold of excited states based on the previous detailed studies on polyenes described earlier. Extrapolations of the S_1 ($2^1A_g^-$) energies from the more highly fluorescent polyenes to carotenoids were suggestive of the presence of such a low-lying state, but the location of the S_1 ($2^1A_g^-$) states of carotenoids was still unclear.

A few years after the pioneering discovery of the low-lying S_1 ($2^1A_g^-$) state in polyenes by Hudson and Kohler (1972), an attempt to locate the state in β-carotene was made by Thrash et al. (1977, 1979) using resonance Raman (rR) excitation spectroscopy. The idea behind this approach was to monitor a strong rR band of the carotenoid and scan the Raman excitation spectrometer through the preresonance region. A change in the spectral intensity is expected to be seen when the excitation wavelength corresponds to the energy of the S_0 ($1^1A_g^-$) → S_1 ($2^1A_g^-$) absorption. The data suggested that the S_1 ($2^1A_g^-$) state was 17,230 cm^{-1} above the ground state, which placed it ~3500 cm^{-1} below the strongly allowed S_2 ($1^1B_u^+$) state. This assignment would then position the S_1 ($2^1A_g^-$) state well above that of the S_1 states of Chls a and b and therefore was suggestive of a possible role for the S_1 ($2^1A_g^-$) state of carotenoids in photosynthetic light harvesting (Figure 3.3). However, the relatively small 3500 cm^{-1} energy difference between S_2 ($1^1B_u^+$) and S_1 ($2^1A_g^-$) for β-carotene was not consistent with trends observed for shorter polyenes (Snyder et al. 1985; Cosgrove et al. 1990), and another group (Watanabe et al. 1987) questioned the conclusions of the study by Thrash et al. (1977, 1979).

In the early 1990s, the technical obstacles for reliably detecting the S_1 ($2^1A_g^-$) emission from carotenoids were largely overcome by Cosgrove et al. (1990) who carried out a careful, systematic study comparing the fluorescence properties of shorter polyenes having relatively strong emissions from the S_1 ($2^1A_g^-$) state with those of longer carotenoids having relatively weak emissions from the S_2 ($1^1B_u^+$) state. These authors used state-of-the-art high-performance liquid chromatography (HPLC) to obtain ultrapure samples free of fluorescing impurities and recorded absorption, fluorescence, and, very importantly so as to unequivocally assign the nature of the fluorescing molecules, fluorescence excitation spectra from a systematic series of carotenols having N = 7–11. They found that the fluorescence of the molecules having N > 8 showed dominant S_2 ($1^1B_u^+$) → S_0 ($1^1A_g^-$) emission as originally observed for β-carotene (van Riel et al. 1983; Watanabe et al. 1986; Bondarev et al. 1988; Gillbro and Cogdell 1989). However, molecules having N < 8, for example, β-apo-12′-carotenol, showed fluorescence spectra that were dominated by the Stokes-shifted S_1 ($2^1A_g^-$) → S_0 ($1^1A_g^-$) emission similar to that observed from shorter polyenes such as hexadecaheptaene (N = 7).

Decoster et al. (1992) expanded on the work by Cosgrove et al. (1990) and recorded absorption fluorescence and fluorescence excitation spectra at room temperature of a naturally occurring carotenoid, spheroidene (N = 10), and three analogs having N = 7–9 synthesized in the laboratory of J. Lugtenburg (Gebhard et al. 1991) (Figure 3.4). Interestingly, the fluorescence spectrum from the analog with N = 7 was dominated by broad emission from the S_1 ($2^1A_g^-$) state. The fluorescence spectrum from the molecule with N = 8 displayed emission of comparable intensity from both the S_1 ($2^1A_g^-$) and S_2 ($1^1B_u^+$) states. The fluorescence from the spheroidene analog having N = 9 and spheroidene (N = 10) was dominated by S_2 ($1^1B_u^+$) emission and displayed approximate mirror-image symmetry with their corresponding absorption spectra. The crossover from dominant S_1 ($2^1A_g^-$) → S_0 ($1^1A_g^-$) emission for the molecules having N < 8 to dominant S_2 ($1^1B_u^+$) → S_0 ($1^1A_g^-$) emission for those having N > 8 can be attributed at least in part to the fact that the decay of the S_1 ($2^1A_g^-$) state to S_0 ($1^1A_g^-$) accelerates with increasing N. This has the effect of quenching the fluorescence from the S_1 ($2^1A_g^-$) state making the weak, residual S_2 ($1^1B_u^+$) → S_0 ($1^1A_g^-$) emission more noticeable. Most importantly, however, this study indicated that the S_1 ($2^1A_g^-$) energy of the biologically important

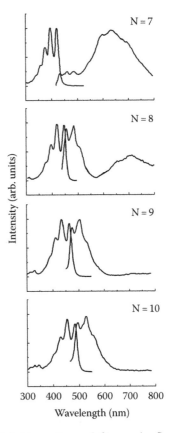

Figure 3.4 Absorption (left) and fluorescence (right) spectra of spheroidene analogs having N = 7–10. Spectra were recorded at room temperature in methanol. (Adapted from DeCoster, B. et al., *Biochim. Biophys. Acta*, 1102(1), 107, 1992.)

carotenoid, spheroidene (N = 10), was located at ~14,250 cm^{-1}, which was well below the 17,230 cm^{-1} value for the even longer β-carotene (N = 11) reported by Thrash et al. (1977, 1979). This work revealed very clearly that the S_1 ($2^1A_g^-$) states of carotenoids were much lower in energy than anyone had previously thought.

Subsequently, many groups reported S_1 ($2^1A_g^-$) and S_2 ($1^1B_u^+$) fluorescence spectra of carotenoids (Andersson et al. 1995; Koyama et al. 1996; Frank et al. 1997) and gradually improved on the technical aspects and data analysis. Spectrometers equipped with double grating excitation and emission monochromators to reduce stray light and improve the ability to detect weak emissions were used, as was laser excitation for efficient and stable optical pumping. Cryogenic temperatures increased the vibronic resolution of the spectral features, and Gaussian deconvolution of the emission bands revealed the spectral origins of the electronic transitions (Josue and Frank 2002). Now, fluorescence spectra from many different carotenoids having a wide range of π-electron conjugated chain lengths have been reported (Frank and Christensen 1995; Christensen 1999; Frank et al. 2000, 2002; Frank 2001; Josue and Frank 2002). Although there is still some debate regarding the precise energies of the S_1 ($2^1A_g^-$) states of some carotenoids, due primarily to the broad emission lineshapes and interference from the long-wavelength tail of the strong S_2 ($1^1B_u^+$) emission, both of which make the assignment of the spectral origins of some of the S_1 ($2^1A_g^-$) fluorescence bands difficult, much progress has been made in assigning the S_1 ($2^1A_g^-$) energies of carotenoids by fluorescence methods.

3.2.2 Time-resolved optical spectroscopic determinations of S_1 ($2^1A_g^-$) energies of carotenoids

Optical transitions from the S_1 ($2^1A_g^-$) state of carotenoids to higher singlet excited states have been reported to be very strong even for polyenes (Bachilo and Bondarev 1988). Thus, an alternative approach to fluorescence spectroscopy for directly determining the S_1 ($2^1A_g^-$) state energies of carotenoids was pioneered by Polívka et al. (1999) who employed an ultrafast time-resolved laser pulse to initially excite the molecule into its S_2 ($1^1B_u^+$) state. This state relaxes in 100–200 fs to the S_1 ($2^1A_g^-$) state whose excited state absorption to the S_2 ($1^1B_u^+$) state can then be detected using a probe laser pulse tune to the near-infrared (NIR) spectral region. Subtracting the energy of the spectral origin of the S_1 ($2^1A_g^-$) → S_2 ($1^1B_u^+$) transition (in the NIR region) from the energy of the spectral origin of the S_0 ($1^1A_g^-$) → S_2 ($1^1B_u^+$) transition (in the visible region), obtained from steady-state absorption or fluorescence spectroscopy, allowed these researchers to compute the position of the S_1 ($2^1A_g^-$) levels of the carotenoids. This method was applied to spheroidene, violaxanthin, and zeaxanthin (Figure 3.5), the latter two of which are xanthophylls implicated in carrying out photoprotection (Polívka et al. 1999, 2001). Interestingly, the values of the S_1 ($2^1A_g^-$) energy levels obtained in this manner turned out to be lower by 400–800 cm^{-1} than those obtained by fluorescence (Fujii et al. 1998; Frank et al. 2000). The authors explained this discrepancy by suggesting that the fluorescence and transient absorption methods are probing different

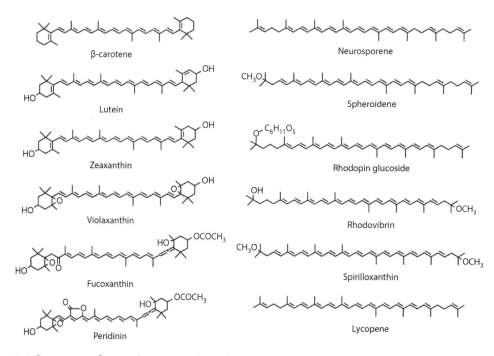

Figure 3.5 Structures of several carotenoids and xanthophylls.

conformations of the carotenoids in their S_1 ($2^1A_g^-$) excited states, which can lead to systematically different energy values (Polívka et al. 2001). This interpretation is borne out by subsequent studies on other carotenoids and polyenes (Niedzwiedzki et al. 2009; Christensen et al. 2013).

3.2.3 Two-photon spectroscopy

Whereas one-photon transitions between states of the same symmetry are forbidden, quantum mechanical selection rules allow transitions between such states via the simultaneous absorption of two photons (Birge 1986). Thus, two-photon spectroscopy has been used to measure the energy of the S_0 ($1^1A_g^-$) $\rightarrow$ S_1 ($2^1A_g^-$) transition for a number of carotenoids including fucoxanthin (Shreve et al. 1990), spheroidene (Krueger et al. 1999), lutein (Walla et al. 2000), and peridinin (Shima et al. 2003) (Figure 3.5). The technique works particularly well when applied to light-harvesting pigment–protein complexes in which carotenoids reside in proximity to Chl (Shreve et al. 1990; Walla et al. 2000) or BChl (Krueger et al. 1999) and are able to transfer energy efficiently to these energy acceptors (Figure 3.3). The experiment is typically done by monitoring the strong Chl fluorescence while scanning the wavelength region of the excitation spectrum for two-photon absorption of the carotenoid. In addition to providing excited state energies, these experiments provide direct experimental verification of energy transfer from the S_1 ($2^1A_g^-$) state of the carotenoid-to-(B)Chl and provide critical information on how the state energies of carotenoids may be modulated upon binding to different proteins.

3.2.4 Resonance Raman excitation spectroscopy and quantum computations

As suggested by the work of Thrash et al. (1977, 1979) and despite the uncertainty in the results, rR excitation spectroscopy offered the potential to determine the S_1 ($2^1A_g^-$) energies of carotenoids. Koyama and coworkers (Sashima et al. 1998, 1999, 2000) adopted this approach and reported determinations of the energies of the S_1 ($2^1A_g^-$) states of several carotenoids using crystalline samples to mitigate the effects of self-absorption of the rR lines due to the strongly allowed S_0 ($1^1A_g^-$) $\rightarrow$ S_2 ($1^1B_u^+$) absorption in the background. These authors not

only reported rR excitation profiles of the S_1 ($2^1A_g^-$) state but also made assignments to the $1^1B_u^-$ state, which also is associated with a forbidden transition from the ground state. The state assigned as $1^1B_u^-$ was reported to have an energy between the S_1 ($2^1A_g^-$) and S_2 ($1^1B_u^+$). Therefore, if more excited states reside between S_1 and S_2, which were thought to be the two lowest excited singlet states, then the potential exists for additional routes of energy transfer to Chl. Quantum computations on model polyenes suggested additional low-lying forbidden states between S_1 ($2^1A_g^-$) and S_2 ($1^1B_u^+$) especially for the longer molecules (Tavan and Schulten 1979, 1986), and significant progress in computational methods is aiding in the modeling of the electronic and vibrational levels of carotenoid molecules in complex systems (see, e.g., Wirtz et al. 2007; Duffy et al. 2013; Macernis et al. 2014). It is important to point out that, to avoid confusion in the sequential numbering of the excited states, most authors retain the traditional notation of S_1 ($2^1A_g^-$) and S_2 ($1^1B_u^+$) for the lowest-lying forbidden state and the state into which absorption is strongly allowed, respectively (Figure 3.3), and refer to any additional states either by their group theoretical representations (e.g., $1^1B_u^-$) if they emerge from quantum computations or by less specific designations (e.g., S_X or S^*; see following text) if they are derived from ultrafast time-resolved spectroscopic observations.

3.3 OTHER DARK EXCITED STATES OF CAROTENOIDS

Although quantum computations on model polyenes and carotenoids have suggested several low-lying forbidden states in the vicinity of S_1 ($2^1A_g^-$) and S_2 ($1^1B_u^+$) (Tavan and Schulten 1979, 1986), confirmation of their existence has been contentious. Work in the group of Koyama (Zhang et al. 2000; Fujii et al. 2003, 2004) proposed that transient absorption bands recorded in an ultrafast experiment carried out on carotenoids having different π-electron conjugated chain lengths originated from a $1^1B_u^-$ dark state. They argued for a sequential deactivation pathway whereby after photoexcitation, the S_2 ($1^1B_u^+$) state decays in less than 70 fs to an intermediate $1^1B_u^-$ state and then subsequently to the S_1 ($2^1A_g^-$) state. Cerullo and coworkers (Cerullo et al. 2002; Polli et al. 2004; Maiuri et al. 2012) used sub-10 fs laser excitation to record the transient absorption spectra of several different carotenoids including β-carotene and spheroidene, and concurred with the conclusion that an intermediate state, which they denoted S_X, resides between S_1 ($2^1A_g^-$) and S_2 ($1^1B_u^+$). However, another group suggested that the observations may be due to coherent laser artifacts (Kosumi et al. 2005). In an important review paper, Polívka and Sundstrom (2009) summarized very well the problematic issues that arise if an intermediate state does in fact exist. These include the fact that if a $1^1B_u^-$ state exists between S_1 ($2^1A_g^-$) and S_2 ($1^1B_u^+$), the origin of the "anti-Kasha" fluorescence, that is, emission from a state higher than the lowest-lying excited state, would not be the S_2 ($1^1B_u^+$) state, but rather the intermediate state. If this were the case, the line-shape of the emission would be expected to deviate substantially from the approximate mirror-image symmetry that it exhibits with its corresponding absorption spectrum. This should become more obvious as the π-electron conjugation chain length becomes longer, and the S_2 ($1^1B_u^+$) → $1^1B_u^-$ energy gap becomes larger. Experiments on a series of linear carotenoids having N= 9–13 show that this does not appear to be the case (Fujii et al. 1998, 2001). Another issue is that if the strong, rapidly decaying, transient absorption signal that appears after photoexcitation of the S_2 ($1^1B_u^+$) state can be attributed to the $1^1B_u^-$ state, then the final state for this transition must have $1^1A_g^+$ symmetry according to the quantum mechanical selection rules. The energy of at least one of these $1^1A_g^+$ states can be readily discerned from the steady-state absorption spectra of cis-geometric isomers of carotenoids where the S_0 ($1^1A_g^-$) → $1^1A_g^+$ transition is allowed. Unfortunately, the expected trend in the energy of the $1^1B_u^-$ → $1^1A_g^+$ transition does not appear to be borne out by experiment (Polívka and Sundström 2009). However, uncertainty exists in precisely how to assign the spectral origin of this transition, so the failure of the spectroscopic observations to adhere to expectations cannot be taken as a strong evidence against an intermediate state. In fact, Ostroumov et al. (2013) have presented a broadband 2D electronic spectroscopic investigation of spheroidene (N = 10) and rhodopin glucoside (N = 11) in solution and in LH2 complexes from photosynthetic bacteria that offer compelling new evidence for an intermediate state. The authors argue that the state is indeed the $1^1B_u^-$ state but suggest that additional theoretical and experimental studies will be needed to incorporate the role of this state in a comprehensive model for carotenoid-to-Chl energy transfer in light-harvesting complexes.

Ultrafast time-resolved optical spectroscopic experiments carried out by Van Grondelle and coworkers (Gradinaru et al. 2001) on the bacterial carotenoid, spirilloxanthin (N = 13), suggested yet another excited singlet state, which they denoted S*, existing between S_1 and S_2, was formed from a branched decay pathway from S_2 and was involved in photoprotection. The observation was manifested as a shoulder on the short-wavelength side of the strong S_1 → S_n transition and which exhibited a longer lifetime than the S_1 state. Subsequent work supported this interpretation and emphasized the possibility that S* is actively involved in the process of energy transfer to BChl (Papagiannakis et al. 2002, 2003; Wohlleben et al. 2003; Christensson et al. 2009). Frank and coworkers explored the nature of the S* state using a series of open-chain carotenoids having systematically increasing N values and tested several different global fitting target models to see which best accounted for the ultrafast dynamics of the excited states and spectral changes associated with their decay (Niedzwiedzki et al. 2007). The data were found to be consistent with a model whereby S* is populated from S_2 but is identified with a twisted conformational structure of the molecule in the S_1 state, the yield of which increases in molecules having long π-electron configurations. In particular, for the longest molecule analyzed, spirilloxanthin (N = 13), the experiments and a detailed quantum computational analysis revealed the presence of two different S* states associated with relaxed S_1 conformations involving nearly planar 6-s-cis and 6-s-trans geometries. It was proposed that in polar solvents, the ground state of spirilloxanthin (Figure 3.5) takes on a corkscrew conformation that generates a net solute dipole moment while decreasing the cavity formation energy. Upon excitation and relaxation into the S_1 state, the polyene unravels and flattens into a more planar geometry with comparable populations of 6-s-trans and 6-s-cis conformations (Niedzwiedzki et al. 2007). Recent ultrafast experiments probing the energetics and dynamics of the low-lying electronic states of constrained polyenes have shown that the features associated with S* are observed only in the longest of the molecules suggesting a role of conformational twisting in the formation of the state (Christensen et al. 2013).

However, the assignment of S* to an excited state has been met with significant skepticism by other researchers who argue that the optical spectroscopic observations are not associated with an excited state transition, but instead should be assigned to a vibrationally hot ground state formed either by impulsive stimulated Raman scattering (ISRS) or direct internal conversion from the S_1 state (Wohlleben et al. 2004; Buckup et al. 2006; Jailaubekov et al. 2010; Lenzer et al. 2010). Larsen and coworkers performed the definitive experiment to resolve this dispute (Jailaubekov et al. 2010). They rationalized that since the ISRS mechanism requires an appreciable bandwidth of the excitation pulse to generate a vibrationally excited hot ground, femtosecond time-resolved experiments comparing results from broadband and narrowband excitation conditions would answer the question. Using both types of excitation pulses, they found that the resulting signals were nearly identical and hence bandwidth independent, thereby disproving ISRS as the mechanism for generating S*. Yet, controversy regarding the nature of S* and its mechanism of formation remains (Lenzer et al. 2010; Ostroumov et al. 2011).

The forbiddenness of the S_0 ($1^1A_g^-$) → S_1 ($2^1A_g^-$) transition prevents the transition from having a significant transition dipole. Hence, the energy of this transition and the kinetics of the S_1 ($2^1A_g^-$) state show very little dependence on solvent environment (Hudson et al. 1982). An exception to this rule was discovered in 1999 for peridinin (Figure 3.5), which was found to have an S_1 ($2^1A_g^-$) lifetime of ~165 ps in nonpolar solvents and ~10 ps in very polar solvents such as methanol and acetonitrile (Bautista et al. 1999). However, the solvent effect on the excited state kinetics was absent unless a carbonyl group was present in the conjugated π-electron polyene chain of the molecule. This was subsequently verified for many other carbonyl-containing carotenoids and polyenes (Bautista et al. 1999; Frank et al. 2000; Polívka and Sundström 2004; Wild et al. 2006; Ehlers et al. 2007; Kopczynski et al. 2007; Stalke et al. 2008; Niedzwiedzki et al. 2009; Enriquez et al. 2010, 2012). The strong correlation of the S_1 ($2^1A_g^-$) excited state kinetics with solvent polarity was ascribed to the presence of an intramolecular charge transfer (ICT) state, and this idea has been supported by theoretical computations (Vaswani et al. 2003; Wagner et al. 2013) and further experimentation by several groups (Zigmantas et al. 2001, 2002, 2003, 2004; Herek

et al. 2004; Papagiannakis 2004; Papagiannakis et al. 2004; Van Tassle et al. 2007; Enriquez et al. 2010). The energy of these types of states is very sensitive to solvent polarity and thereby can induce a change in the excited state dynamics. The precise electronic nature of the ICT state is still an active area of investigation, but a recent detailed combined experimental/theoretical analysis of peridinin (Wagner et al. 2013) is strongly suggestive of the fact that the ICT state is an evolved state formed via excited state bond-order reversal, quantum mechanical mixing of the S_2 ($1^1B_u^+$) ionic state and the lowest-lying S_1 ($2^1A_g^-$) covalent state, and solvent reorganization in polar media. All of these factors contribute to the shifting of electron density from the allenic side of the molecule into the lactone ring region (Figure 3.5) resulting in the formation of the ICT state. The ability of the ICT state to influence the dynamics of the S_1 ($2^1A_g^-$) state of carbonyl-containing carotenoids suggests that it may play a role in regulating flow of energy to Chl in photosynthetic pigment–protein complexes. However, such a role has not yet been established.

3.4 TRIPLET EXCITED STATES OF CAROTENOIDS

Carotenoid molecules also possess triplet states, that is, electronic states in which two electrons in different molecular orbitals have parallel spins. In carotenoids, triplet states are not populated by direct excitation, and the first characterization of these states was performed by flash photolysis of the molecules in solution (Chessin et al. 1966), and very shortly thereafter from experiments on photosynthetic systems (Mathis 1969; Mathis and Kleo 1973). The transition from the lowest-lying triplet state T_1 (Figure 3.3) to higher energy triplet states is strongly allowed and results in an intense absorption band around 540 nm, the exact position of which depends on the length of the conjugated polyene chain (see, e.g., Angerhofer et al. 1995). Although the energy of this transition is readily observed, the absence of an unambiguous measurement of emission from T_1, that is, carotenoid phosphorescence, despite an early report on β-carotene (Cherry et al. 1968) has not provided reliable information on the precise energy of the triplet states of carotenoids. However, it should lie below that of the (B)Chl triplet state, since in photosynthetic systems, it can be populated by fast (B)Chl-to-carotenoid triplet–triplet energy transfer (see, e.g., Angerhofer et al. 1995). Moreover, the triplet states of carotenoids should lie well below the energy level of the $^1\Delta_g{}^*$ excited state of molecular oxygen (7882 cm^{-1}), because one of the main biological functions of carotenoids is to quench this reactive oxygen species (Foote and Denny 1968). Many experimental investigations have focused on triplet energy transfer from Chl (Van der Vos et al. 1994; Peterman et al. 1995; Mozzo et al. 2008) and have also examined the effect of varying the extent of π-electron conjugation on the efficiency of triplet energy transfer to carotenoids (Farhoosh et al. 1994). Carotenoids with shorter π-electron chain lengths have higher triplet state energies. This undoubtedly accounts for the fact that the vast majority of carotenoids found in nature have N values >9. Any value shorter than this would have too high a triplet energy rendering it ineffective at quenching the triplet state of Chl (Figure 3.3) and unable to protect against singlet oxygen formation. Typically, the lifetime of carotenoid triplet states resides in the range of 2–10 ms (Fuciman et al. 2011). Despite their important role in quenching (B)Chl triplet states, carotenoid triplet states are not very well characterized. This is due largely to the fact that they are extremely difficult to model theoretically, even with advanced computational methods (see, e.g., Takahashi et al. 1998; Ma et al. 2004).

3.5 STRUCTURES OF CAROTENOIDS

It is clear that from the linear nature of their structures (Figure 3.5), carotenoid molecules may adopt different conformations (rotations about C–C single bonds) and configurations (rotations about C=C double bonds), the spectral properties of which are expected to be slightly different. For example, several different configurational isomers of β-carotene can be obtained after the illumination of this molecule in the presence of I_2 (Hashimoto et al. 1991). The isomers include mono- and di-cis isomers, which were reported to have different band positions and intensities (Vetter et al. 1971; Koyama et al. 1983). In addition, carotenoid molecules may exhibit distorted conformations, that is, torsions and/or s-cis isomerization about their C–C single bonds. For instance, in β-carotene, the conjugated terminal rings have been reported in crystalline samples to be in an s-cis conformation

(Sterling 1964). However, it was recently proposed that, in solution, the conjugated rings from carotenoid molecules may adopt conformations that are either out-of-plane or s-cis, which would explain their apparent shorter effective conjugation length (Mendes-Pinto et al. 2013). There are not many methods available to determine the precise structures of carotenoids bound to proteins. X-ray diffraction patterns obtained from single crystals can help determine the structures of protein-bound carotenoids. When it is possible to extract carotenoids from the proteins without perturbing their configurations, the structures may be determined by NMR spectroscopy (Lutz et al. 1987). However, since the barriers to rotation about C–C single bonds are low, information on conformations of the molecules is usually lost upon extraction. The method of choice to characterize the structures of carotenoids in situ is rR spectroscopy. This method is highly sensitive to changes in the structure of carotenoids (Koyama et al. 1983). Moreover, some specific rR bands exhibit an intensity dependence due to out-of-plane deformations of the molecule. Therefore, the spectra are helpful in revealing conformation distortions as well (Lutz et al. 1987).

An interest in the configuration of protein-bound carotenoids was initially raised by the discovery that the carotenoid bound in photosynthetic bacterial reaction centers was in a cis configuration (Lutz et al. 1976, 1978; Koyama et al. 1982, 1983, 1988), which ultimately was found to be a distorted 15,15′ (central) cis configuration (Lutz et al. 1987; Koyama et al. 1988). However, in light-harvesting proteins, carotenoids are generally found to be in all-trans configurations (Lutz et al. 1976), also sometimes with distorted conformations (Iwata et al. 1985). It is not yet clear whether the variations in carotenoid structures, induced primarily by the protein-binding sites in which these molecules reside, have a functional significance. However, recent work has shown that the distortions of lutein in LHCII and β-carotene in photosystem II (PSII) reaction centers from higher plants could be responsible for the tuning of the absorption spectra of these pigments (Mendes-Pinto et al. 2013). Similarly, distortion of a carbonyl-containing carotenoid was recently proposed to enhance the charge transfer state character, thereby affecting the efficiency of energy transfer (Šlouf et al. 2013).

3.6 ENERGY TRANSFER BY CAROTENOIDS

The complex nature of the excited singlet-state manifold of carotenoids presents an unprecedented challenge for understanding the mechanism by which carotenoids act as energy donors to Chl in the photosynthetic apparatus (Figure 3.3). Depending on whether the donor state is the strongly allowed S_2 ($1^1B_u^+$) state, the quantum mechanically forbidden S_1 ($2^1A_g^-$) state, or any of the other states (S_X, S^* or ICT) described earlier, the electronic coupling that controls energy transfer as well as the dependence on the distance between pigments will be different. The Förster dipole–dipole mechanism (Förster 1948, 1968) is frequently invoked to explain energy transfer between Chls where both donor and acceptor molecules have appreciable transition dipole moments. This model does not perform well at short donor–acceptor distances and is wholly inadequate when a quantum mechanically forbidden state is the donor or acceptor state. In the latter case, it was previously thought that the Dexter electron exchange mechanism was appropriate (Dexter 1953). However, it is now generally accepted that a Coulomb mechanism that accounts for strong coupling between closely spaced molecules (Davydov 1962), possibly involving higher-order multipole interactions (Nagae et al. 1993) when a forbidden state is the donor state, is the most suitable model to explain the rates and efficiencies of carotenoid-to-Chl energy transfer in photosynthetic pigment–protein complexes (Scholes 2003).

There are two low-lying electronic states of Chl that can act as energy acceptors. These are the states associated with the Q_X and Q_Y transitions of Chl (Figure 3.3). The relative contributions of these states to the overall energy transfer efficiency will depend on the energies of the S_1 and S_2 states of the carotenoid that are modulated by their extent of π-electron conjugation and environmental factors in the case of S_2. The energy transfer route involving the S_2 state of the carotenoid as the donor, and the state associated with the Q_x transition of Chl as the acceptor, is active for a large number of carotenoids (Shreve et al. 1991; Polivka and Frank 2010). The S_1 state of a carotenoid that is populated by internal conversion following photoexcitation of the S_2 state can act as an energy donor only if the acceptor state of (B)Chl has an energy below S_1. This is the case for many carotenoid–(B)Chl donor–acceptor pairs in several light-harvesting complexes. In light harvesting from purple bacteria, the tuning of this route by the number of conjugated double bonds can be clearly observed by experiment. When the carotenoids have <10 effective π-electron conjugated double bonds, the S_1 state lies above the state associated with the Q_Y transition of Chl (Figure 3.3).

3.7 PHOTOPROTECTION BY CAROTENOIDS

Carotenoids are involved either directly or indirectly in protecting the photosynthetic apparatus from photoinduced damage following light absorption. Hypothetical mechanisms for protection include direct transfer of excess excited state energy of Chl to carotenoids (Snyder et al. 1985; Frank et al. 1994), quenching of this energy by charge transfer states (Beddard et al. 1977; Holt et al. 2004), aggregation of protein complexes that create quenching centers, or some combination of these processes, to diminish the population of Chl excited states (Demmig-Adams 1990; Gust et al. 1992; Rees et al. 1992; Mullineaux et al. 1993; Cardoso et al. 1996; Demmig-Adams and Adams 1996; Horton et al. 1996; Phillip et al. 1996; Yamamoto and Bassi 1996; Connelly et al. 1997; Walters et al. 1997; Young et al. 1997; Crimi et al. 1998; Gradinaru et al. 1998; Ruban et al. 1998; Pogson et al. 2005). An important mechanism involving the dissipation of excess absorbed light energy is referred to as nonphotochemical quenching (NPQ) (Horton and Boyer 1990; Demmig-Adams and Adams 1992; Krause and Weis 1992; Yamamoto and Bassi 1996; Niyogi 1999; Bassi and Caffarri 2000; Müller et al. 2001). NPQ is a major regulation mechanism for the photoprotection of photosynthetic organisms from exposure to excess light and can be measured in plants and algae by the extent to which Chl fluorescence from the light-harvesting pigment–protein complexes associated with PSII is quenched under different conditions (Demmig-Adams and Adams 2002; Kulheim et al. 2002; Horton and Ruban 2005). Many mechanisms have been proposed to account for the quenching of excess excitation energy during NPQ, including quenching by charge transfer states between carotenoid and

Chl molecules (Holt et al. 2004), between Chls (Holzwarth et al. 2009), or quenching of the Chl excited states by the S_1 state of a lutein (Ruban et al. 2007), governed by a protein conformational change (Pascal et al. 2005). NPQ will be dealt with in detail in Chapter 11 by Pinnola et al. and will not be discussed further here.

As mentioned earlier, carotenoids can protect photosynthetic systems by quenching Chl triplet states, which otherwise have the potential to sensitize the formation of the $^1\Delta_g^*$ excited singlet state of oxygen. This excited form of molecular oxygen has the capacity to damage many of the components of the photosynthetic apparatus. In the light-harvesting proteins from oxygenic photosynthetic organisms studied so far, no population of Chl molecules in their triplet states can be detected, meaning that this excited state is quenched (by carotenoids) at a higher rate than it is produced (Gall et al. 2011). This was recently proposed to be the cause of the adaptation of the photosynthetic apparatus of these organisms to an oxygen-rich earth environment (Gall et al. 2011). The molecular mechanisms underlying this ultrafast triplet–triplet transfer are not fully characterized, but it was proposed to involve partial delocalization of the triplet state over Chl and carotenoid molecules.

3.8 CONCLUSIONS

This chapter has attempted to provide a brief history and background on carotenoids from which many of the spectroscopic and energy transfer properties of these molecules have been revealed. Much remains to be elucidated regarding the nature of their so-called dark excited states, which have been a formidable challenge to examine. Technological advances in spectroscopic and computational methods are developing rapidly and promise to reveal important new information regarding the photophysics and biological function of carotenoids.

ACKNOWLEDGMENTS

The authors wish to thank Professors Robert Birge and Ronald Christensen for helpful discussions and Ms. Amy LaFountain for her assistance in preparing this manuscript. Work in the laboratory of H.A.F. was supported by grants from the National Science Foundation (MCB-1243565) and the University of Connecticut Research Foundation.

REFERENCES

Andersson PO, Bachilo SM, Chen R-L, Gillbro T (1995) Solvent and temperature effects on dual fluorescence in a series of carotenes. Energy gap dependence of the internal conversion rate. J Phys Chem 99 (44):16199–16209.

Andersson PO, Gillbro T (1992) Ultrafast radiationless relaxation in macro-carotenes and application of the energy gap law. Laser Spectrosc Biomol 1921:48–56.

Andersson PO, Gillbro T, Asato AE, Liu RSH (1992) Dual singlet state emission in a series of mini-carotenes. J Lumin 51 (1–3):11–20.

Angerhofer A, Bornhäuser F, Gall A, Cogdell RJ (1995) Optical and optically detected magnetic resonance investigation on purple photosynthetic bacterial antenna complexes. Chem Phys 194:259–274.

Bachilo SM, Bondarev SL (1988) Absorption from the diphenyl-polyene S_1 state in the 580-1650-nm range. Opt Spectrosc 65:295–300.

Bassi R, Caffarri S (2000) LHC proteins and the regulation of photosynthetic light harvesting function by xanthophylls. Photosynth Res 64 (2–3):243–256.

Bautista JA, Connors RE, Raju BB, Hiller RG, Sharples FP, Gosztola D, Wasielewski MR, Frank HA (1999) Excited state properties of peridinin: Observation of a solvent dependence of the lowest excited singlet state lifetime and spectral behavior unique among carotenoids. J Phys Chem B 103 (41):8751–8758.

Beddard GS, Davidson RS, Tretheway KR (1977) Quenching of chlorophyll fluorescence by β-carotene. Nature 267:373–374.

Birge RR (1986) Two photon spectroscopy of protein-bound chromophores. Acc Chem Res 19:138–146.

Blankenship RE, Madigan MT, Bauer CE (1995) Anoxygenic photosynthetic bacteria. Advances in Photosynthesis, vol 2. Kluwer Academic Publishers, Dordrecht, the Netherlands.

Bondarev SL, Bachilo SM, Dvornikov SS, Tikhomirov SA (1989) $S_2 \rightarrow S_0$ fluorescence and transient $S_n \rightarrow S_1$ absorption of all-trans-β-carotene in solid and liquid solutions. J Photochem Photobiol A Chem 46:315–322.

Bondarev SL, Dvornikov SS, Bachilo SM (1988) Fluorescence of β-carotene at 77 and 4.2 K. Opt Spectrosc (USSR) 64:268–270.

Boucher F, van der Rest M, Gingras G (1977) Structure and function of carotenoids in the photoreaction center from *Rhodospirillum rubrum*. *Biochim Biophys Acta* 461:339–357.

Buckup T, Savolainen J, Wohlleben W, Herek JL, Hashimoto H, Correia RRB, Motzkus M (2006) Pump-probe and pump-deplete-probe spectroscopies on carotenoids with N = 9–15 conjugated bonds. *J Chem Phys* 125:194505.

Callis PR, Scott TW, Albrecht AC (1983) Perturbation selection rules for multiphoton electronic spectroscopy of neutral alternant hydrocarbons. *J Chem Phys* 78 (1):16–22.

Cardoso SL, Nicodem DE, Moore TA, Moore AL, Gust D (1996) Synthesis and fluorescence quenching studies of a series of carotenoporphyrins with carotenoids of various lengths. *J Braz Chem Soc* 7 (1):19–30.

Cerullo G, Polli D, Lanzani G, De Silvestri S, Hashimoto H, Cogdell RJ (2002) Photosynthetic light harvesting by carotenoids: Detection of an intermediate excited state. *Science* 298 (5602):2395–2398.

Cherry RJ, Chapman D, Langelaar J (1968) Fluorescence and phosphorescence of b-carotene. *Trans Faraday Soc* 64:2304–2307.

Chessin M, Livingston R, Truscott TG (1966) Direct evidence for the sensitized formation of a metastable state of β-carotene. *Trans Faraday Soc* 62 (6):1519–1524.

Christensen RL (1999) The electronic states of carotenoids. In: Frank HA, Young AJ, Britton G, Cogdell RJ (eds). *The Photochemistry of Carotenoids*, vol 8. Advances in Photosynthesis. Kluwer Academic Publishers, Dordrecht, the Netherlands, pp. 137–159.

Christensen RL, Enriquez MM, Wagner NL, Peacock-Villada AY, Scriban C, Schrock RR, Polívka T, Frank HA, Birge RR (2013) Energetics and dynamics of the low-lying electronic states of constrained polyenes: Implications for infinite polyenes. *J Phys Chem A* 117:1449–1465.

Christensson N, Milota F, Nemeth A, Sperling J, Kauffmann HF, Pullerits T, Hauer J (2009) Two-dimensional electronic spectroscopy of β-carotene. *J Phys Chem B* 113:16409–16419.

Cogdell RJ (1978) Carotenoids in photosynthesis. *Philos Trans R Soc Lond Ser B* 284 (1002):569–579.

Cogdell RJ, Frank HA (1987) How carotenoids function in photosynthetic bacteria. *Biochim Biophys Acta* 895 (2):63–79.

Connelly JP, Müller MG, Hucke M, Gatzen G, Mullineaux CW, Ruban AV, Horton P, Holzwarth AR (1997) Ultrafast spectroscopy of trimeric light-harvesting complex II from higher plants. *J Phys Chem* 101B:1902–1909.

Cooper DA, Eldridge AL, Peters JC (1999) Dietary carotenoids and certain cancers, heart disease, and age-related macular degeneration: A review of recent research. *Nutr Rev* 57 (7):201–214.

Cosgrove SA, Guite MA, Burnell TB, Christensen RL (1990) Electronic relaxation in long polyenes. *J Phys Chem* 94:8118–8124.

Crimi M, Croce R, Sandona D, Varotto C, Simonetto R, Bassi R (1998) Mutation analysis of either protein or chromophore moieties in Higher Plant Light Harvesting Proteins. In: Garab G (ed). *Photosynthesis: Mechanisms and Effects, Proceedings of the XI International Congress on Photosynthesis, Budapest, Hungary*, vol 1. Kluwer Academic Publishers, Dordrecht, the Netherlands, pp. 253–258.

Davydov AS (1962) *Theory of Molecular Excitons*, trans: Kasha M, M. Oppenheimer J. McGraw-Hill, New York.

DeCoster B, Christensen RL, Gebhard R, Lugtenburg J, Farhoosh R, Frank HA (1992) Low-lying electronic states of carotenoids. *Biochim Biophys Acta* 1102 (1):107–114.

Demmig-Adams B (1990) Carotenoids and photoprotection in plants: A role for the xanthophyll zeaxanthin. *Biochim Biophys Acta* 1020:1–24.

Demmig-Adams B, Adams WW (1996) The role of xanthophyll cycle carotenoids in the protection of photosynthesis. *Trends Plant Sci* 1 (1):21–26.

Demmig-Adams B, Adams WW (2002) Antioxidants in photosynthesis and human nutrition. *Science* 298:2149–2153.

Demmig-Adams B, Adams WWI (1992) Photoprotection and other responses of plants to high light stress. *Annu Rev Plant Physiol Mol Biol* 43:599–626.

Dexter DL (1953) A theory of sensitized luminescence in solids. *J Chem Phys* 21:836–860.

Duffy CDP, Chmeliov J, Macernis M, Sulskus J, Valkunas L, Ruban AV (2013) Modeling of fluorescence quenching by lutein in the plant light-harvesting complex LHCII. *J Phys Chem B* 117:10974–10986.

Ehlers F, Wild DA, Lenzer T, Oum K (2007) Investigation of the S_1/ICT -> S_0 internal conversion lifetime of 4'-apo-β-caroten-4'-al and 8'-apo-β-caroten-8'-al: Dependence on conjugation length and solvent polarity. *J Phys Chem A* 111 (12):2257–2265.

Enriquez MM, Fuciman M, LaFountain AM, Wagner NL, Birge RR, Frank HA (2010) The intramolecular charge transfer state in carbonyl-containing polyenes and carotenoids. *J Phys Chem B* 114:12416–12426.

Enriquez MM, Hananoki S, Hasegawa S, Kajikawa T, Katsumura S, Wagner NL, Birge RR, Frank HA (2012) Effect of molecular symmetry on the spectra and dynamics of the intramolecular charge transfer (ICT) state of peridinin. *J Phys Chem B* 116 (35):10748–10756.

Farhoosh R, Chynwat V, Gebhard R, Lugtenburg J, Frank HA (1994) Triplet energy transfer between bacteriochlorophyll and carotenoids in B850 light-harvesting complexes of *Rhodobacter sphaeroides* R-26.1. *Photosynth Res* 42 (2):157–166.

Foote CS (1968) Mechanisms of photosensitized oxidation. *Science* 162:963–970.

Foote CS, Chang YC, Denny RW (1970) Chemistry of singlet oxygen. X. Carotenoid quenching parallels biological protection. *J Am Chem Soc* 92:5216–5218.

Foote CS, Denny RW (1968) Chemistry of singlet oxygen. VII. Quenching by β-carotene. *J Am Chem Soc* 90:6233–6235.

Förster T (1948) Zwischenmolekulare energiewanderung und fluoreszenz. *Ann Phys* 437:55.

Förster T (1968) Intermolecular energy transfer and fluorescence. *Ann Phys* 2:55–75.

Frank HA (2001) Spectroscopic studies of the low-lying singlet excited electronic states and photochemical properties of carotenoids. *Arch Biochem Biophys* 385:53–60.

Frank HA, Bautista JA, Josue J, Pendon Z, Hiller RG, Sharples FP, Gosztola D, Wasielewski MR (2000) Effect of the solvent environment on the spectroscopic properties and dynamics of the lowest excited states of carotenoids. *J Phys Chem B* 104 (18):4569–4577.

Frank HA, Bautista JA, Josue JS, Young AJ (2000) Mechanism of nonphotochemical quenching in green plants: Energies of the lowest excited singlet states of violaxanthin and zeaxanthin. *Biochemist* 39:2831–2837.

Frank HA, Christensen RL (1995) Singlet energy transfer from carotenoids to bacteriochlorophylls. In: Blankenship RE, Madigan MT, Bauer CE (eds). *Anoxygenic Photosynthetic Bacteria*, vol 2. Advance in Photosynthesis. Kluwer Academic Publishers, Dordrecht, the Netherlands, pp. 373–384.

Frank HA, Cogdell RJ (1993) The photochemistry and function of carotenoids in photosynthesis. In: Young AJ, Britton G (eds). *Carotenoids in Photosynthesis*. Chapman & Hall, London, U.K., pp. 252–326.

Frank HA, Cua A, Chynwat V, Young A, Gosztola D, Wasielewski MR (1994) Photophysics of the carotenoids associated with the xanthophyll cycle in photosynthesis. *Photosynth Res* 41 (3):389–395.

Frank HA, Desamero RZB, Chynwat V, Gebhard R, van der Hoef I, Jansen FJ, Lugtenburg J, Gosztola D, Wasielewski MR (1997) Spectroscopic properties of spheroidene analogs having different extents of π-electron conjugation. *J Phys Chem A* 101 (2):149–157.

Frank HA, Josue JS, Bautista JA, van der Hoef I, Jansen FJ, Lugtenburg J, Wiederrecht G, Christensen RL (2002) Spectroscopic and photochemical properties of open-chain carotenoids. *J Phys Chem B* 106 (8):2083–2092.

Frank HA, Young AJ, Britton G, Cogdell RJ (1999) The photochemistry of carotenoids. In: Govindjee (ed). *Advances in Photosynthesis*, vol 8. Kluwer Academic Publishers, Dordrecht, the Netherlands.

Fuciman M, Enriquez MM, Kaligotla S, Niedzwiedzki DM, Kajikawa T, Aoki K, Katsumura S, Frank HA (2011) Singlet and triplet state spectra and dynamics of structurally modified peridinins. *J Phys Chem B* 115:4436–4445.

Fujii R, Fujino T, Inaba T, Nagae H, Koyama Y (2004) Internal conversion of $1B_u^+$ -> $1B_u^-$ -> $2A_g^-$ and fluorescence from the $1B_u^-$ state in all-*trans*-neurosporene as probed by up-conversion spectroscopy. *Chem Phys Lett* 384 (1–3):9–15.

Fujii R, Inaba T, Watanabe Y, Koyama Y, Zhang JP (2003) Two different pathways of internal conversion in carotenoids depending on the length of the conjugated chain. *Chem Phys Lett* 369:165–172.

Fujii R, Onaka K, Kuki M, Koyama Y, Watanabe Y (1998) The $2A_g^-$ energies of all-*trans*-neurosporene and spheroidene as determined by fluorescence spectroscopy. *Chem Phys Lett* 288 (5,6):847–853.

Fujii R, Onaka K, Nagae H, Koyama Y, Watanabe Y (2001) Fluorescence spectroscopy of all-*trans*-lycopene: Comparison of the energy and the potential displacements of its $2A_g^-$ state with those of neurosporene and spheroidene. *J. Lumin* 92:213–222.

Gall A, Berera R, Alexandre MTA, Pascal AA, Bordes L, Mendes-Pinto MM, Andrianambinintsoa S et al. (2011) Molecular adaptation of photoprotection: Triplet states in light-harvesting proteins. *Biophys J* 101:934–942.

Gavin RM, Weisman C, McVey JK, Rice SA (1978) Spectroscopic properties of polyenes. III. 1,3,5,7-Octatetraene. *J Chem Phys* 68:522–529.

Gebhard R, Van Dijk JTM, Van Ouwerkerk E, Boza MVTJ, Lugtenburg J (1991) Synthesis and spectroscopy of chemically modified spheroidenes. *Recl Trav Chim Pays-Bas* 110 (11):459–469.

Gillbro T, Cogdell RJ (1989) Carotenoid fluorescence. *Chem Phys Lett* 158 (3–4):312–316.

Goedheer JC (1969) Energy transfer from carotenoids to chlorophyll in blue-green, red and green algae and greening bean leaves. *Biochim Biophys Acta* 172:252–265.

Goodwin TW (1976) *Chemistry and Biochemistry of Plant Pigments*, vol 1. Academic Press, New York.

Govindjee, R (1975) Bioenergetics of photosynthesis. In: Govindjee (ed). *Cell Biology*. Academic Press, New York, pp. 2–50.

Gradinaru CC, Kennis JTM, Papagiannakis E, van Stokkum IHM, Cogdell RJ, Fleming GR, Niederman RA, van Grondelle R (2001) An unusual pathway of excitation energy deactivation in carotenoids: Singlet-to-triplet conversion on an ultrafast timescale in a photosynthetic antenna. *Proc Natl Acad Sci USA* 98:2364–2369.

Gradinaru CC, Pascal AA, van Mourik F, Robert B, Horton P, van Grondelle R, van Amerongen H (1998) Ultrafast evolution of the excited states in the chlorophyll a/b complex CP29 from green plants studied by energy-selective pump-probe spectroscopy. *Biochemistry* 37 (4):1143–1149.

Gust D, Moore TA, Moore AL, Devadoss C, Liddell PA, Heman R, Nieman RA, Demanche LJ, Degraziano JM, Gouni I (1992) Triplet and singlet energy transfer in carotene-porphyrin dyads: Role of the linkage bonds. *J Am Chem Soc* 114:3591–3603.

Haley LV, Koningstein JA (1983) Space and time-resolved resonance-enhanced vibrational Raman spectroscopy from femtosecond-lived singlet excited state of b-carotene. *Chem Phys* 77:1–9.

Hashimoto H, Koyama Y, Hirata Y, Mataga N (1991) S1 and T1 species of b-carotene generated by direct photoexcitation from the all-trans, 9-*cis*, 13-*cis*, and 15-*cis* isomers as revealed by picosecond transient absorption and transient Raman spectroscopies. *J Phys Chem* 95 (8):3072–3076.

Herek JL, Wendling M, He Z, Polivka T, Garcia-Asua G, Cogdell RJ, Hunter CN, van Grondelle R, Sundström V, Pullerits T (2004) Ultrafast carotenoid band shifts: Experiment and theory. *J Phys Chem B* 108 (29):10398–10403.

Holt NE, Fleming GR, Niyogi KK (2004) Toward an understanding of the mechanism of nonphotochemical quenching in green plants. *Biochemist* 43 (26):8281–8289.

Holzwarth AR, Miloslavina Y, Nilkens M, Jahns P (2009) Identification of two quenching sites active in the regulation of photosynthetic light-harvesting studied by time-resolved fluorescence. *Chem Phys Lett* 483:262–267.

Horton P, Boyer JR (1990) Chlorophyll fluorescence transients. In: Harwood JL, Boyer JR (eds). *Methods in Plant Biochemistry*, vol 4. Academic Press, London, U.K., pp. 259–296.

Horton P, Ruban A (2005) Molecular design of the photosystem II light-harvesting antenna: Photosynthesis and photoprotection. *J Exp Bot* 56:365–373.

Horton P, Ruban AV, Walters RG (1996) Regulation of light harvesting in green plants. *Annu Rev Plant Physiol Mol Biol* 47:655–684.

Hudson BS, Kohler BE (1972) A low lying weak transition in the polyene α,ω-diphenyloctatetraene. *Chem Phys Lett* 14:299–304.

Hudson BS, Kohler BE (1973) Polyene spectroscopy: The lowest energy excited single state of diphenyloctatetraene and other linear polyenes. *J Chem Phys* 59:4984–5002.

Hudson BS, Kohler BE (1984) Electronic structure and spectra of finite linear polyenes. *Synth Met* 9:241–253.

Hudson BS, Kohler BE, Schulten K (1982) Linear polyene electronic structure and potential surfaces. In: Lim ED (ed). *Excited States*, vol 6. Academic Press, New York, pp. 1–95.

Isler O (1971) *Carotenoids*. Birkhäuser, Basel, Switzerland.

Iwata K, Hayashi H, Tasumi M (1985) Resonance Raman studies of the conformations of all-*trans* carotenoids in light-harvesting systems of photosynthetic bacteria. *Biochim Biophys Acta* 810:269–273.

Jailaubekov AE, Song S-H, Vengris M, Cogdell RJ, Larsen DS (2010) Using narrowband excitation to confirm that the S* state in carotenoids is not a vibrationally-excited ground state species. *Chem Phys Lett* 487:101–107.

Josue JS, Frank HA (2002) Direct determination of the S₁ excited-state energies of xanthophylls by low-temperature fluorescence spectroscopy. *J Phys Chem A* 106 (19):4815–4824.

Kirmaier C, Holten D (1987) Primary photochemistry of reaction centers from the photosynthetic purple bacteria. *Photosynth Res* 13:225–260.

Kopczynski M, Ehlers F, Lenzer T, Oum K (2007) Evidence for an intramolecular charge transfer state in 12′-apo-β-caroten-12′-al and 8′-apo-β-caroten-8′-al: Influence of solvent polarity and temperature. *J Phys Chem A* 111 (25):5370–5381.

Kosumi D, Komukai M, Hashimoto H, Yoshizawa M (2005) Ultrafast dynamics of all-*trans*-β-carotene explored by resonant and non-resonant photoexcitations. *Phys Rev Lett* 95:213601–213604.

Koyama Y (1991) Structures and functions of carotenoids in photosynthetic systems. *J Photochem Photobiol B* 9 (3–4):265–280.

Koyama Y, Kanaji M, Shimamura T (1988) Configurations of neurosporene isomers isolated from the reaction center and the light-harvesting complex of Rhodobacter spheroides G1C. A resonance Raman, electronic absorption, and proton-NMR study. *Photochem Photobiol* 48 (1):107–114.

Koyama Y, Kito M, Takii T, Saiki K, Tsukida K, Yamashita J (1982) Configuration of the carotenoid in the reaction centers of photosynthetic bacteria. Comparison of the resonance Raman spectrum of the reaction centers of *Rhodopseudomonas*

sphaeroides G1C with those of cis-trans isomers of β-carotene. *Biochim Biophys Acta* 680 (2):109–118.

Koyama Y, Kuki M, Andersson PO, Gillbro T (1996) Singlet excited states and the light-harvesting function of carotenoids in bacterial photosynthesis. *Photochem Photobiol* 63 (3):243–256.

Koyama Y, Takii T, Saiki K, Tsukida K (1983) Configuration of the carotenoid in the reaction centers of photosynthetic bacteria. 2. Comparison of the resonance Raman lines of the reaction centers with those of the 14 different cis-trans isomers of β-carotene. *Photobiochem Photobiophys* 5 (3):139–150.

Krause GH, Weis E (1992) Chlorophyll fluorescence and photosynthesis: The basics. *Annu Rev Plant Physiol Mol Biol* 42:313–349.

Krinsky NI (1968) The protective function of carotenoid pigments. In: Giese AC (ed). *Photophysiology III*. Academic Press, New York, pp. 123–195.

Krinsky NI (1971) Function. In: Isler O, Guttman G, Solms U (eds). *Carotenoids*, Birkhauser Verlag, Basel, Switzerland, pp. 669–716.

Krueger BP, Yom J, Walla PJ, Fleming GR (1999) Observation of the S₁ state of spheroidene in LH2 by two- photon fluorescence excitation. *Chem Phys Lett* 310 (1–2):57–64.

Kulheim C, Agren J, Jansson S (2002) Rapid regulation of light harvesting and plant fitness in the field. *Science* 297:91–93.

Lang HP, Hunter CN (1994) The relationship between carotenoid biosynthesis and the assembly of the light-harvesting LH2 complex in *Rhodobacter sphaeroides*. *Biochem J* 298:197–205.

Lenzer T, Ehlers F, Scholz M, Oswald R, Oum K (2010) Assignment of carotene S* state features to the vibrationally hot ground electronic state. *Phys Chem Chem Phys* 12:8832–8839.

Lutz M, Agalidis I, Hervo G, Cogdell RJ, Reiss-Husson F (1978) On the state of carotenoids bound to reaction centers of photosynthetic bacteria: A resonance Raman study. *Biochim Biophys Acta* 503 (2):287–303.

Lutz M, Kleo J, Reiss-Husson F (1976) Resonance Raman scattering of bacteriochlorophyll, bacteriophaeophytin and spheroidene in reaction centers from *Rhodopseudomonas spheroides*. *Biochem Biophys Res Commun* 69:711–717.

Lutz M, Szponarski W, Berger G, Robert B, Neumann J-M (1987) The stereoisomerization of bacterial, reaction-center-bound carotenoids revisited: An electronic absorption, resonance Raman and NMR study. *Biochim Biophys Acta* 894:423–433.

Ma H, Liu C, Yuansheng J (2004) "Triplet-excited region" in polyene oligomers revisited: Pariser-Parr-Pople model studied with the density matrix renormalization group method. *J Chem Phys* 120:9316–9320.

Macernis M, Sulskus J, Malickaja S, Robert B, Valkunas L (2014) Resonance Raman spectra and electronic transitions in carotenoids: A density functional theory study. *J Phys Chem A* 118:1817–1825.

Maiuri M, Polli D, Brida D, Lüer L, LaFountain AM, Fuciman M, Cogdell RJ, Frank HA, Cerullo G (2012) Solvent-dependent activation of intermediate excited states in the energy relaxation pathways of spheroidene. *Phys Chem Chem Phys* 14 (18):6312–6319.

Mathis P (1969) Flash photolysis of chlorophyll-carotenoid energy transfer. *Photochem Photobiol* 9:55–63.

Mathis P, Kleo J (1973) Triplet state of β-carotene and of analog polyenes of different length. *Photochem Photobiol* 18:343–346.

Mendes-Pinto MM, Galzerano D, Telfer A, Pascal AA, Robert B, Ilioaia C (2013) Mechanisms underlying carotenoid absorption in oxygenic photosynthetic proteins. *J Biol Chem* 288:18758–18765.

Mendes-Pinto MM, Sansiaume E, Hashimoto H, Pascal AA, Gall A, Robert B (2013) Electronic absorption and ground state structure of carotenoid molecules. *J Phys Chem B* 117:11015–11021.

Mozzo M, Dall'Osto L, Hienerwadel R, Bassi R, Croce R (2008) Photoprotection in the antenna complexes of photosystem II: Role of individual xanthophylls in chlorophyll triplet quenching. *J Biol Chem* 283 (10):6184–6192.

Müller P, Li X-P, Niyogi KK (2001) Non-photochemical quenching. A response to excess light energy. *Plant Physiol* 125:1558–1566.

Mullineaux CW, Pascal AA, Horton P, Holzwarth AR (1993) Excitation-energy quenching in aggregates of the LHC II chlorophyll-protein complex: A time-resolved fluorescence study. *Biochim Biophys Acta* 1141:23–28.

Nagae H, Kikitani T, Katoh T, Mimuro M (1993) Calculation of the excitation transfer matrix elements between the S_1 or S_2 state of carotenoid and the S_2 or S_1 state of bacteriochlorophyll. *J Chem Phys* 98:8012–8023.

Niedzwiedzki D, Koscielecki JF, Cong H, Sullivan JO, Gibson GN, Birge RR, Frank HA (2007) Ultrafast dynamics and excited state spectra of open-chain carotenoids at room and low temperatures. *J Phys Chem B* 111 (21):5984–5998.

Niedzwiedzki DM, Chatterjee N, Enriquez MM, Kajikawa T, Hasegawa S, Katsumura S, Frank HA (2009) Spectroscopic investigation of peridinin analogues having different π-electron conjugated chain lengths: Exploring the nature of the intramolecular charge transfer state. *J Phys Chem B* 113:13604–13612.

Niedzwiedzki DM, Sandberg DJ, Cong H, Sandberg MN, Gibson GN, Birge RR, Frank HA (2009) Ultrafast time-resolved absorption spectroscopy of geometric isomers of carotenoids. *Chem Phys* 357:4–16.

Niyogi KK (1999) Photoprotection revisited: Genetic and molecular approaches. *Annu Rev Plant Physiol Plant Mol Biol* 50:333–359.

Niyogi KK, Björkman O, Grossman AR (1997) The roles of specific xanthophylls in photoprotection. *Proc Natl Acad Sci USA* 94:14162–14167.

Ohmine I, Karplus M, Schulten K (1978) Renormalized configuration interaction method for electron correlation in the excited states of polyenes. *J Chem Phys* 68:2298–2318.

Ohmine I, Morokuma K (1980) Photoisomerization of polyenes: Potential energy surfaces and normal mode analysis. *J Chem Phys* 73 (4):1907–1917.

Ostroumov EE, Müller MG, Reus M, Holzwarth AR (2011) On the nature of the "dark S*" excited state of β-carotene. *J Phys Chem A* 115 (16):3698–3712.

Ostroumov EE, Mulvaney RM, Cogdell RJ, Scholes GD (2013) Broadband 2D electronic spectroscopy reveals a carotenoid dark state in purple bacteria. *Science* 340:52–56.

Papagiannakis E (2004) Shedding light on the dark states of carotenoids. PhD dissertation, Free University of Amsterdam, Amsterdam, the Netherlands.

Papagiannakis E, Das SK, Gall A, Stokkum IHM, Robert B, van Grondelle R, Frank HA, Kennis JTM (2003) Light harvesting by carotenoids incorporated into the B850 light-harvesting complex from *Rhodobacter sphaeroides* R-26.1: Excited-state relaxation, ultrafast triplet formation, and energy transfer to bacteriochlorophyll. *J Phys Chem B* 107:5642–5649.

Papagiannakis E, Kennis JTM, van Stokkum IHM, Cogdell RJ, van Grondelle R (2002) An alternative carotenoid-to-bacteriochlorophyll energy transfer pathway in photosynthetic light harvesting. *Proc Natl Acad Sci USA* 99:6017–6022.

Papagiannakis E, Larsen DS, van Stokkum IHM, Vengris M, Hiller RG, van Grondelle R (2004) Resolving the excited state equilibrium of peridinin in solution. *Biochemist* 43 (49):15303–15309.

Pariser R (1955) Theory of electronic spectra and structure of the polyacenes and of alternant hydrocarbons. *J Chem Phys* 24:250–268.

Pascal AA, Liu Z, Broess K, van Oort B, van Amerongen H, Wang C, Horton P, Robert B, Chang W, Ruban AV (2005) Molecular basis of photoprotection and control of photosynthetic light-harvesting. *Nature* 436:134–137.

Petek H, Bell AJ, Kandori H, Yoshihara K, Christensen RL (1992) Spectroscopy and dynamics of a model polyene decatetraene: A study of non-radiative pathways in S_1 and S_2 states under isolated conditions. In: Takahashi H (ed). *Time Resolved Vibrational Spectroscopy*, vol 5. Springer, Berlin, Germany, pp. 198–199.

Peterman EJG, Dukker FM, van Grondelle R, van Amerongen H (1995) Chlorophyll a and carotenoid triplet states in light-harvesting complex II of higher plants. *Biophys J* 69 (6):2670–2678.

Phillip D, Ruban AV, Horton P, Asato A, Young AJ (1996) Quenching of chlorophyll fluorescence in the major light-harvesting complex of photosystem II: A systematic study of the effect of carotenoid structure. *Proc Natl Acad Sci USA* 93:1492–1497.

Pogson BJ, Rissler HM, Frank HA (2005) The role of carotenoids in energy quenching. In: Wydrzynski T, Satoh K (eds). *Photosystem II: The Water/Plasoquinone Oxido-Reductase in Photosynthesis*, vol 21. Springer, Dordrecht, the Netherlands, pp. 515–537.

Polivka T, Frank HA (2010) Molecular factors controlling photosynthetic light harvesting by carotenoids. *Acc Chem Res* 43 (8):1125–1134.

Polívka T, Herek JL, Zigmantas D, Akerlund HE, Sundström V (1999) Direct observation of the (forbidden) S_1 state in carotenoids. *Proc Natl Acad Sci USA* 96 (9):4914–4917.

Polívka T, Sundström V (2004) Ultrafast dynamics of carotenoids excited states: From solution to natural and artificial systems. *Chem Rev* 104 (4):2021–2071.

Polivka T, Sundström V (2009) Dark excited states of carotenoids: Consensus and controversy. *Chem Phys Lett* 477:1–11.

Polívka T, Zigmantas D, Frank HA, Bautista JA, Herek JL, Koyama Y, Fujii R, Sundström V (2001) Near-infrared time-resolved study of the S1 state dynamics of the carotenoid spheroidene. *J Phys Chem B* 105 (5):1072–1080.

Polli D, Cerullo G, Lanzani G, De Silvestri S, Yanagi K, Hashimoto H, Cogdell RJ (2004) Conjugation length dependence of internal conversion in carotenoids: Role of the intermediate state. *Phys Rev Lett* 93:163002.

Rees D, Noctor G, Ruban AV, Crofts J, Young AJ, Horton P (1992) pH dependent chlorophyll fluorescence in spinach thylakoids from light treated or dark adapted leaves. *Photosynth Res* 31:11–19.

Renger G, Wolff C (1977) Further evidence for dissipative energy migration via triplet states in photosynthesis. The protective mechanism of carotenoids in *Rhodopseudomonas sphaeroides* chromatophores. *Biochim Biophys Acta* 460:47–57.

Ruban AV, Berera R, Ilioaia C, van Stokkum IHM, Kennis JTM, Pascal AA, van Amerongen H, Robert B, Horton P, van Grondelle R (2007) Identification of a mechanism of photoprotective energy dissipation in higher plants. *Nature* 450 (7169):575–578.

Ruban AV, Phillip D, Young AJ, Horton P (1998) Excited-state energy level does not determine the differential effect of violaxanthin and zeaxanthin on chlorophyll fluorescence quenching in the isolated light-harvesting complex of photosystem II. *Photochem Photobiol* 68:829–834.

Sandonà D, Croce R, Pagano A, Crimi M, Bassi R (1998) Higher plants light harvesting proteins. Structure and function as revealed by mutation analysis of either protein or chromophore moieties. *Biochim Biophys Acta* 1365:207–214.

Sashima T, Koyama Y, Yamada T, Hashimoto H (2000) The $1B_u{}^+$, $1B_u{}^-$, and $2A_g{}^-$ energies of crystalline lycopene, β-carotene, and mini-9-β-carotene as determined by resonance-Raman excitation profiles: Dependence of the $1B_u{}^-$ state energy on the conjugation length. *J Phys Chem B* 104 (20):5011–5019.

Sashima T, Nagae H, Kuki M, Koyama Y (1999) A new singlet-excited state of all-*trans*-spheroidene as detected by resonance-Raman excitation profiles. *Chem Phys Lett* 299 (2):187–194.

Sashima T, Shiba M, Hashimoto H, Nagae H, Koyama Y (1998) The $2^1A_g{}^-$ energy of crystalline all-*trans*-spheroidene as determined by resonance-Raman excitation profiles. *Chem Phys Lett* 290 (1–3):36–42.

Scholes GD (2003) Long-range resonance energy transfer in molecular systems. *Annu Rev Phys Chem* 54:57–87.

Schulten K, Karplus M (1972) On the origin of a low-lying forbidden transition in polyenes and related molecules. *Chem Phys Lett* 14:305–309.

Schulten K, Ohmine I, Karplus M (1976) Correlation effects in the spectra of polyenes. *J Chem Phys* 64:4422–4441.

Shima S, Ilagan RP, Gillespie N, Sommer BJ, Hiller RG, Sharples FP, Frank HA, Birge RR (2003) Two-photon and fluorescence spectroscopy and the effect of environment on the photochemical properties of peridinin in solution and in the peridinin-chlorophyll-protein from *Amphidinium carterae. J Phys Chem A* 107:8052–8066.

Shreve AP, Trautman JK, Frank HA, Owens TG, Albrecht AC (1991) Femtosecond energy-transfer processes in the B800-850 light-harvesting complex of *Rhodobacter sphaeroides* 2.4.1. *Biochim Biophys Acta* 1058 (2):280–288.

Shreve AP, Trautman JK, Owens TG, Albrecht AC (1990) Two-photon excitation spectroscopy of thylakoid membranes from *Phaeodactylum tricornutum*: Evidence for an in vivo two-photon-allowed carotenoid state. *Chem Phys Lett* 170:51–56.

Šlouf V, Fuciman M, Dulebo A, Kaftan D, Koblížek M, Frank HA, Polívka T (2013) Carotenoid charge transfer states and their role in energy transfer processes in LH1-RC complexes from aerobic anoxygenic phototrophs. *J Phys Chem B* 117:10987–10999.

Snyder R, Arvidson E, Foote C, Harrigan L, Christensen RL (1985) Electronic energy levels in long polyenes: S_2 -> S_0 emission in all-*trans*-1,2,5,7,9,11,13-tetradecaheptaene. *J Am Chem Soc* 107:4117–4122.

Staleva H, Komenda J, Shukla MK, Šlouf V, Kaňa R, Polívka T, Sobotka R (2015) Mechanism of photoprotection in the cyanobacterial ancestor of plant antenna proteins. *Nat Chem Biol* 11 (4):287–291.

Stalke S, Wild DA, Lenzer T, Kopczynski M, Lohse PW, Oum K (2008) Solvent-dependent ultrafast internal conversion dynamics of n'-apo-β-carotenoic-n'-acids (n = 8, 10, 12). *Phys Chem Chem Phys* 10 (16):2180–2188.

Sterling C (1964) Crystal structure analysis of β-carotene. *Acta Crystallogr* 17:1224–1228.

Takahashi O, Watanabe M, Kikuchi O (1998) Structure of the T1-state wave function of linear polyenes. *Int J Quantum Chem* 67:101–106.

Tavan P, Schulten K (1979) The 2^1A_g-1^1B_u energy gap in the polyenes: An extended configuration interaction study. *J Chem Phys* 70:5407–5413.

Tavan P, Schulten K (1986) The low-lying electronic excitations in long polyenes: A PPP-MRD-CI study. *J Chem Phys* 85:6602–6609.

Tavan P, Schulten K (1987) Electronic excitations in finite and infinite polyenes. *Phys Rev B Condens Matter* 36 (8):4337–4358.

Thrash RJ, Fang H, Leroi GE (1979) On the role of forbidden low-lying excited state of light-harvesting carotenoids in energy transfer in photosynthesis. *Photochem Photobiol* 29:1049–1050.

Thrash RJ, Fang HLB, Leroi GE (1977) The Raman excitation profile spectrum of b-carotene in the preresonance region: Evidence for a low-lying singlet state. *J Chem Phys* 67:5930–5933.

Van der Vos R, Franken EM, Hoff AJ (1994) ADMR studies of oligomerisation on the carotenoid triplets and on triplet-triplet transfer in light harvesting complex II (LHCII) of spinach. *Biochim Biophys Acta* 1188:243–250.

van Riel M, Kleinen-Hammans J, van de Ven M, Verwer W, Levine Y (1983) Fluorescence excitation profiles of β-carotene in solution and in lipid/water mixtures. *Biochem Biophys Res Commun* 113:102–107.

Van Tassle AJ, Prantil MA, Hiller RG, Fleming GR (2007) Excited state structural dynamics of the charge transfer state of peridinin. *Is J Chem* 47 (1):17–24.

Vaswani HM, Hsu CP, Head-Gordon M, Fleming GR (2003) Quantum chemical evidence for an intramolecular charge-transfer state in the carotenoid peridinin of peridinin-chlorophyll-protein. *J Phys Chem B* 107 (31):7940–7946.

Vetter W, Englert G, Rigassi N, Schwieter U (1971) Spectroscopic methods. In: Isler O, Guttman G, Solms U (eds). *Carotenoids*. Birkhauser Verlag, Basel, Germany, pp. 189–266.

Wagner NL, Greco JA, Enriquez MM, Frank HA, Birge RR (2013) The nature of the intramolecular charge transfer (ICT) state in peridinin. *Biophys J* 104:1314–1325.

Walla PJ, Yom J, Krueger BP, Fleming GR (2000) Two-photon excitation spectrum of light-harvesting complex II and fluorescence upconversion after one- and two-photon excitation of the carotenoids. *J Phys Chem B* 104 (19):4799–4806.

Walters RG, Ruban AV, Horton P (1997) Identification of proton-active residues in a higher plant light-harvesting complex. *Proc Natl Acad Sci USA* 93:14204–14209.

Watanabe J, Kinoshita S, Kushida T (1986) Non-motional narrowing effect in excitation profiles of second-order optical processes: Comparison between Stochastic Theory and experiments in b-carotene. *Chem Phys Lett* 126:197–200.

Watanabe J, Kinoshita S, Kushida T (1987) Effects of nonzero correlation time of system-reservoir interaction on the excitation profiles of second-order optical processes in b-carotene. *J Phys Chem* 87:4471–4477.

Wild DA, Winkler K, Stalke S, Oum K, Lenzer T (2006) Extremely strong solvent dependence of the S1 -> S0 internal conversion lifetime of 12′-apo-β-caroten-12′-al. *Phys Chem Chem Phys* 8 (21):2499–2505.

Wirtz AC, van Hemert MC, Lugtenburg J, Frank HA, Groenen EJJ (2007) Two stereoisomers of spheroidene in the *Rhodobacter sphaeroides* R26 reaction center: A DFT analysis of resonance Raman spectra. *Biophys J* 93:981–991.

Wohlleben W, Buckup T, Hashimoto H, Cogdell RJ, Herek JL, Motzkus M (2004) Pump-deplete-probe spectroscopy and the puzzle of carotenoid dark states. *J Phys Chem B* 108 (10):3320–3325.

Wohlleben W, Buckup T, Herek JL, Cogdell RJ, Motzkus M (2003) Multichannel carotenoid deactivation in photosynthetic light harvesting as identified by an evolutionary target analysis. *Biophys J* 85 (1):442–450.

Yamamoto H, Bassi R (1996) Carotenoids: Localization and function. In: Ort DR, Yocum CF (eds). *Oxygenic Photosynthesis: The Light Reactions.* Kluwer Academic Publishers, Dordrecht, the Netherlands, pp. 539–563.

Young A, Britton G (1993) *Carotenoids in Photosynthesis.* Kluwer Academic, London, U.K..

Young AJ, Phillip D, Ruban AV, Horton P, Frank HA (1997) The xanthophyll cycle and carotenoid-mediated dissipation of excess excitation energy in photosynthesis. *Pure Appl Chem* 69 (10):2125–2130.

Zhang JP, Inaba T, Watanabe Y, Koyama Y (2000) Excited-state dynamics among the $1B_u^+$, $1B_u^-$ and $2A_g^-$ states of all-trans-neurosporene as revealed by near-infrared time-resolved absorption spectroscopy. *Chem Phys Lett* 332 (3–4):351–358.

Zigmantas D, Hiller RG, Sharples FP, Frank HA, Sundström V, Polivka T (2004) Effect of a conjugated carbonyl group on the photophysical properties of carotenoids. *Phys Chem Chem Phys* 6 (11):3009–3016.

Zigmantas D, Hiller RG, Sundström V, Polivka T (2002) Carotenoid to chlorophyll energy transfer in the peridinin chlorophyll-a protein complex involves an intramolecular charge transfer state. *Proc Natl Acad Sci USA* 99:16760–16765.

Zigmantas D, Hiller RG, Yartsev A, Sundström V, Polivka T (2003) Dynamics of excited states of the carotenoid peridinin in polar solvents: Dependence on excitation wavelength, viscosity, and temperature. *J Phys Chem B* 107:5339–5348.

Zigmantas D, Polivka T, Hiller RG, Yartsev A, Sundström V (2001) Spectroscopic and dynamic properties of the peridinin lowest singlet excited states. *J Phys Chem A* 105 (45):10296–10306.

PART 2

Building the Light-Harvesting Apparatus: Proteins

4

Light harvesting in higher plants and green algae

LAUREN NICOL AND ROBERTA CROCE

4.1 INTRODUCTION

At the heart of photosynthesis lies the photochemical reaction center (RC). Solar radiation provides the energy required for the RC pigments to donate an electron to an acceptor molecule. This charge separation sets in motion a series of downhill redox reactions that ultimately produce the chemical potential energy required for CO_2 fixation. However, in isolation, this system is inefficient and energetically costly. An RC chlorophyll exposed to the light of the midday sun (1800 $\mu E/m^2/s$) would absorb approximately 10 photons per second. Considering the electron transport chain can process ~200 electrons per second, the photosynthetic machinery would sit idle the majority of the time.

In light of this, nature has coevolved light-harvesting antenna systems made up of pigment–protein complexes coordinating a large number of chlorophyll and other accessory pigments. The antenna increases the effective absorption cross section of the RC in two ways. First, by increasing the total number of Chls

contributing to absorption, and second, by broadening the spectral range over which absorption can occur. The antenna is then able to efficiently transfer the absorbed energy to the RC. This process is largely dictated by the protein scaffold, which imparts specific properties on the pigments and fixes them in specific orientations and distances with respect to one another, facilitating efficient excitation energy transfer (EET) and minimizing losses in the competing pathways of intersystem crossing, internal conversion, and fluorescence.

The antenna surrounding a single RC can be divided into two broad groups: the core and the peripheral antenna. The core antenna is highly conserved in all oxygenic photosynthetic organisms and exists in a fixed stoichiometry. In contrast, a number of peripheral antenna systems have evolved independently and diversified to suit the light intensities and spectral qualities available in a particular environment. Furthermore, the size and composition of the peripheral antenna can vary in response to changing environmental conditions.

In higher plants and green algae, the peripheral antenna is made up of pigment–protein complexes from the light-harvesting complex (LHC) family. These complexes are integral thylakoid membrane proteins, which coordinate chlorophylls and carotenoids. Together they constitute 40% of the proteins found in the thylakoid membrane, making them the most abundant membrane proteins on Earth.

4.2 LHC FAMILY OF HIGHER PLANTS AND GREEN ALGAE

The original light-harvesting pigment–protein complexes of oxygenic photosynthesis were the membrane-extrinsic phycobilisomes of cyanobacteria. Following the primary endosymbiotic event, in which a cyanobacterium was engulfed by a non-photosynthetic eukaryotic cell, light-harvesting antenna diversified to also include the LHCs, a family of integral membrane proteins binding both chlorophylls and carotenoids (Koziol et al. 2007). The separation of green algae was characterized by the evolution of Chl *a/b*–binding LHCs, distinct to the Chl *a*–binding proteins of red algae and the Chl *a/c*–binding proteins of chromalveolates (see Chapter 8). The separation was also accompanied by a complete loss of phycobilisomes, resulting in a massive expansion of the Chl *a/b*–binding family and specialization into distinct photosystem I (PSI) and photosystem II (PSII) antennae. This basic antenna structure diversified throughout the evolution of the green lineage as organisms adapted to different environments and light conditions.

The genes encoding LHCs are located in the nucleus, and their products are imported to the chloroplast upon recognition of a signal peptide (Schuemann 2004). Once in the stroma, the signal peptide is removed by the stromal processing peptidase (SPP) and the mature polypeptide is targeted to the thylakoid membrane. The only known targeting pathway is SRP-dependent; however, there is evidence to suggest that another currently unknown pathway also exists (Reinbothe et al. 2006; Hoober et al. 2007). The final steps of pigment incorporation and protein folding occur via a largely unknown mechanism. While this process occurs spontaneously *in vitro*, it requires pigment concentrations not typically found *in vivo*, thus postulating the question of

pigment delivery mediated by a specific carrier (Paulsen et al. 1993; Dall'Osto et al. 2015). Once fully folded, LHCs adopt the "LHCII-general fold" of three transmembrane helices and two short amphipathic helices located on the lumenal surface (Kühlbrandt 1994). Two of the transmembrane helices intertwine to form a left-handed supercoil, while the other, located further from the supercoil, breaks this twofold pseudosymmetry. With a number of conserved pigment-binding sites, LHCs have a very high pigment-to-protein ratio, that is, 15 kDa of pigment coordinated by 25 kDa of protein.

In *Arabidopsis thaliana*, PSII is associated with both major (LHCII) and minor (CP29, CP26, CP24) antenna proteins. Major LHCII is encoded by three classes of genes (Lhcb1, Lhcb2, and Lhcb3), and the minor antenna proteins, CP29, CP26 and CP24, are encoded by Lhcb4, Lhcb5, and Lhcb6, respectively. PSI is typically associated with LHCI proteins encoded by Lhca1–Lhca4.

In the green alga *Chlamydomonas reinhardtii*, a distinct set of genes (Lhcbm1–Lhcbm9) encode the major antenna of PSII, and no homologue for CP24 exists. There are also approximately twice as many genes encoding the PSI antenna protein, LHCI.

4.3 SYSTEMS TO CHARACTERIZE LHCs

There are three main systems to study individual LHCs: (1) purification of native complexes from the thylakoid membrane, (2) *in vitro* reconstitution, and (3) reverse genetics. The following section will discuss the pros and cons of each of these methods for the purpose of analyzing biochemical, spectroscopic, and structural characteristics.

1. *Purification of native complexes from the thylakoid membrane*: While this method offers the obvious advantage that the complex is assembled into the native state with the correct pigment complement and required cofactors, it is complicated by many reasons. First, the need to solubilize the thylakoid membrane with detergents can damage the protein and remove essential cofactors. However, this problem has largely been mitigated with the use of mild nonionic detergents such as n-dodecyl α-D-maltoside (Caffarri et al. 2001).

The second problem lies in the separation of the solubilized thylakoid membrane proteins. Mild separation techniques such as sucrose density gradient centrifugation have been useful in providing isolated LHCII, but many LHCs have very similar physical and chemical properties or tight associations with one another, making homogeneous preparations difficult. This has been the case for the LHCII isoforms, minor antenna complexes, and PSI antenna complexes. For these complexes, multiple purification steps are required, often including isoelectric focusing (IEF) or high detergent concentrations, resulting in pigment loss from peripheral sites (Dainese et al. 1990; Pascal et al. 1999a; Ruban et al. 1999). Even LHCII, which is far more stable than the minor antenna complexes, loses both a carotenoid and a Chl when subject to IEF (Caffarri et al. 2001; Natali et al. 2015). This problem has recently been circumvented with the addition of affinity tags to individual complexes, allowing a mild one-step purification from solubilized membranes (Passarini et al. 2014).

2. *In vitro reconstitution*: This approach, first introduced in 1990, involves overexpression of LHC apoproteins in *Escherichia coli* followed by reconstitution *in vitro* with the addition of purified pigments (Paulsen et al. 1990). The resulting complexes have been shown to possess characteristics almost identical to those isolated in the native state (Natali et al. 2014). The advantage of this technique is the ease in which one can obtain high levels of yield and purity. It also confers the additional advantage of being a system that is easy to manipulate. By removing or substituting specific residues of the protein using site-directed mutagenesis, one can elucidate the properties of individual pigments and the factors important for oligomerization, pigment binding specificity, protein stability or protein–protein interactions (Cammarata and Schmidt 1992; Hobe et al. 1995; Bassi et al. 1999). Pigment composition of the reconstitution mix can also be manipulated to assess pigment contribution to structure and function. This has been particularly important for the study of different xanthophyll species (Croce et al. 1999; Hobe et al. 2000).

However, reconstituted complexes are not always accepted as viable substitutes for the native ones. For example, it is known that the reconstituted LHCII is unable to coordinate violaxanthin in the peripheral V1 site, even in its trimeric state (Paulsen and Hobe 1992; Peterman et al. 1997; Croce et al. 1999; Rogl and Kühlbrandt 1999). In addition, the Chl a/b ratio of the recombinant complex strongly depends on the Chl a/b ratio of the reconstitution mix (Sandonà et al. 1998). This is due to the fact that most of the pigment-binding sites are not completely selective for Chl a or Chl b but can accommodate both with differing affinities (Hobe et al. 1995; Giuffra et al. 1996; Bassi et al. 1999; Kleima et al. 1999). This is also the case for most carotenoid-binding sites (Croce et al. 1999; Hobe et al. 2000). Although, pigment preparations with a Chl a/b ratio of 3:1 and a Chl/car ratio of 2.9:1 typically produce a reconstituted protein with the same properties as the native one (Natali et al. 2014).

3. *Reverse genetics*: This technique is particularly important for studying the role of the protein complex *in vivo*. Thanks to global efforts in generating T-DNA insertion mutant collections in *Arabidopsis thaliana*, there are knockout lines for most LHCs. The two major exceptions are Lhcb1 and Lhcb2, as they are encoded by multiple closely linked genes, making T-DNA KO mutants almost impossible to generate. Instead, artificial microantisense RNA (amiRNA) has been introduced to silence their expression, although Lhcb1 is able to partially escape this silencing (Pietrzykowska et al. 2014). Minor antenna knockouts have been crossed to produce all combinations of double mutants, the Lhcb4/Lhcb5 knockout lacking all three minor antenna due to hampered accumulation of CP24 (Dall'Osto et al. 2014). There are knockouts of individual CP29 isoforms and all combinations of double mutants (de Bianchi et al. 2011). There are also knockouts available for individual Lhcas, including a triple mutant of Lhca2–Lhca4 which is also devoid of Lhcb1 (Benson et al. 2015). A T-DNA insertional mutant for Lhca1 exists; however, the insertion is found in the promoter region, and as a consequence, the plant still contains a small amount of Lhca (less than 10% of wild-type levels) (Ganeteg et al. 2004).

4.4 LHCs OF HIGHER PLANTS

4.4.1 LHCII

LHCII is the so-called major antenna of PSII as it coordinates approximately 70% of total PSII chlorophyll. However, it is also acts as an antenna of PSI in practically all light conditions (Wientjes et al. 2013a). In its native state, LHCII exists mainly as a trimer, consisting of varying combinations of three similar but distinct Lhcb isoforms Lhcb1, Lhcb2, and Lhcb3. In *A. thaliana*, Lhcb1 is encoded by five genes and is the largest (230–233 residues) and most abundant isoform. Lhcb2 is encoded by three genes and is slightly shorter with 228 residues. Lhcb3 is encoded by a single gene and is the shortest isoform (223 residues), as it has a truncated N-terminus. It is also one of the newest members of the LHC family, appearing in plants shortly after land colonization. Overall, the three isoforms share over 77% sequence similarity, and it has been shown that the pigment organization and spectroscopic properties are very similar (Caffarri et al. 2004; Standfuss and Kuhlbrandt 2004; Palacios et al. 2006).

In *A. thaliana*, under normal light conditions, the isoforms have a ratio of about 6:2:1 for Lhcb1, Lhcb2, and Lhcb3, respectively, although this ratio varies depending on growth conditions (Bailey et al. 2001; Jackowski et al. 2001; Kouřil et al. 2013).

In vitro, LHCII trimers exist in all possible combinations of the three isoforms, with the exception of an Lhcb3 homotrimer (Caffarri et al. 2004). *In vivo*, however, only Lhcb1 homotrimers and Lhcb1/Lhcb2, Lhcb1/Lhcb3, and Lhcb1/Lhcb2/Lhcb3 heterotrimers are observed (Jackowski et al. 2001; Standfuss and Kuhlbrandt 2004). Although it has been suggested that Lhcb2 homotrimers and Lhcb2/Lhcb3 heterotrimers exist in the absence of Lhcb1 (Pietrzykowska et al. 2014). Trimerization of the isoforms is dependent upon several factors: (1) the sequence motif WYGPDR at the N-terminal region (Hobe et al. 1995), (2) the presence of phosphatidyl glycerol (PG) (Hobe et al. 1994), and (3) and the presence of lutein in L2 (Lokstein et al. 2002).

The molecular structure of the LHCII monomer, which has been obtained at 2.5–2.72 Å (Liu et al. 2004; Standfuss et al. 2005), is presented below in Figure 4.1. Each monomer binds 14 Chl molecules (8 Chls *a* and 6 Chls *b*), four xanthophylls (2 luteins, 1 neoxanthin, and 1 violaxanthin), and one structurally important phospholipid. The chlorophylls are arranged in two distinct layers on both the stromal and lumenal sides of the protein. On the stromal side, there are two Chl *a* clusters (a610–a611–a612 and a602–a603) and one Chl *b* cluster (b601'-b609-b608, where b601' belongs to a neighboring monomer). On the lumenal side,

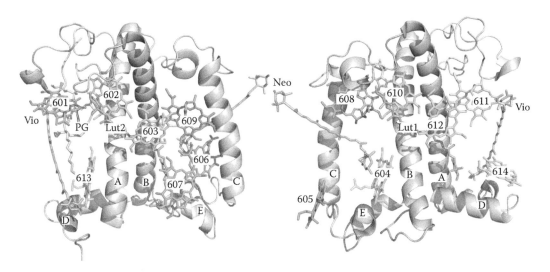

Figure 4.1 Molecular model of the LHCII monomer based on the crystal structure of the LHCII trimer. The front and back faces are shown with selected chromophores. For clarity, Chl phytyl chains are not shown. Gray, polypeptide; dark green, Chl *b*; light green, Chl *a*; orange, xanthophylls; pink, PG. The structure was obtained from the PDB code 1RWT.

there is a a613-a614 dimer and b605–b606–b607–a604 tetramer. The coordination of chlorophylls is achieved by interaction of the central magnesium ion with nucleophilic amino acid side chains, backbone carbonyls, water, and, in one case, the phosphodiester group of a PG (Table 4.1). While the crystallographic data suggests a fixed Chl a/b ratio, it has been shown both *in vitro* (Remelli et al. 1999; Hobe et al. 2003) and *in vivo* (Pattanayak et al. 2005) that some sites have low selectivity depending on the ratio of available pigments. In terms of the xanthophylls, the two luteins are bound with high affinity to the center of the protein in sites L1 and L2. Neoxanthin is stably bound in site N1 near helix C, and violaxanthin and lutein are bound with much lower affinity to the periphery of the protein in site V1 (Croce et al. 1999; Caffarri et al. 2001). In contrast to the chlorophylls, coordination of the xanthophylls cannot be narrowed down to specific residues, instead it is achieved by varying combinations of hydrogen bonds, hydrophobic interactions, and van der Waals forces. It also appears that these interactions do not confer a significant amount of specificity. Reconstitution studies have shown that while the L1 and L2 sites have the highest affinity for lutein, they can also bind violaxanthin, zeaxanthin, and even carotenoids not present in native Lhcb proteins (Croce et al. 1999; Phillip et al. 2002).

As can be seen in Figure 4.1, pigments within LHCII are tightly packed. An equivalent concentration of chlorophyll in solution would result in almost all excitation energy being lost as heat (Beddard and Porter 1976). However, due to the influence of the protein matrix, this quenching is avoided and the excitation energy can be efficiently transferred to the core. How does the protein matrix achieve this?

1. Directly interacting with pigments to alter their transition energies
2. Specifically arranging chromophores of different transition energies
3. Determining the relative orientation and distance between pigments

The importance of these three factors in influencing the rate of energy transfer is partly described by the Förster equation (Förster 1948). Here the rate is extremely sensitive to the center-to-center distance between the interacting chromophores, the relative orientations of the pigments, and the overlap of energy levels. However, Förster only applies when coupling between pigments is weak. If pigments of similar energy levels are very close to each other, which is the case for many pigments in LHCII, they become excitonically coupled, that is, they act as a "supermolecule" with delocalized transitions, rather than individual molecules with localized transitions (Fassioli et al. 2014). This leads to new excitonic energy levels and fast energy transfer dynamics. A number of different theoretical approaches such as the combined Redfield–Förster have been developed to describe the energy transfer pathways within antenna complexes, but this will be discussed in detail in Chapter 13. Very briefly, within LHCII, EET first occurs from Chls b to Chls a; this is followed by fast exciton relaxation within the Chl a clusters and then transfer from clusters on the lumenal side to clusters on the stromal side (Novoderezhkin et al. 2004, 2011). The lowest excited energy state is the Chls 610–611–612 cluster, and from here excitation energy can be transferred to other antenna proteins.

The xanthophylls, which absorb in the blue-green region (350–550 nm), also partake in this energy transfer pathway, although they are often neglected in modeling due to their complex photophysical properties (discussed in Chapter 3). Due to their close proximity to Chls, the average efficiency of energy transfer from Cars to Chls in LHCII is high, approximately 80% (Peterman et al. 1997; Caffarri et al. 2001). This efficiency varies depending on the specific xanthophyll, for example, the luteins in L1 and L2 are more efficient than neoxanthin, whereas violaxanthin does not participate in any energy transfer to Chls (Caffarri et al. 2001; Croce et al. 2001). It was also determined that energy transfer occurs almost exclusively from the initially excited S_2 state (Croce et al. 2001). Transfer of energy from the lower-lying S_1 state is less favorable with an efficiency well below 20% in LHCII (Gradinaru et al. 2000). The preise ratio of Chl a or Chl b molecules that act as primary acceptors of this energy transfer is unknown; however, studies on reconstituted LHCII indicate that the luteins transfer preferentially to Chls a and neoxanthin to Chls b (Gradinaru et al. 2000; Croce et al. 2001), which is in agreement with the structure of the complex (Liu et al. 2004).

Table 4.1 Pigment-binding sites of spinach LHCII, CP29, and CP26 and Pisum sativum (pea) Lhca1–Lhca4 according to Su et al. 2017 and Qin et al. 2015

Pigment	LHCII	CP29	CP26	CP24	Lhca1	Lhca2	Lhca3	Lhca4
601	Chl b (Tyr 24)	Chl a (Trp 14)	Chl b (Phe 34)	Chl b (Trp 9)	Chl b (Trp 37)	Chl b (Trp 57)	—	Chl a (Trp 56)
602	Chl a (Glu 65)	Chl a (Glu 96)	Chl a (Glu 78)	Chl a (Glu 54)	Chl a (Glu 76)	Chl a (Glu 96)	Chl a (Glu 96)	Chl a (Glu 95)
603	Chl a (His 68)	Chl a (His 99)	Chl a (His 81)	Chl a (His 57)	Chl a (His79)	Chl a (His99)	Chl a (Asn99)	Chl a (Asn 98)
604	Chl a (H_2O)	Chl a (H_2O)	Chl a (H_2O)	Chl a (H_2O)	Chl a (H_2O)	Chl a (H_2O)	Chl a (H_2O)	Chl a (H_2O)
605	Chl b (Val 119)	—	—	—	—	—	—	—
606	Chl b (H_2O)	Chl b (H_2O)	Chl b (H_2O)	Chl b (H_2O)	Chl a (H_2O)	Chl b (H_2O)	Chl a (H_2O)	Chl b (H_2O)
607	Chl b (H_2O)	Chl b (H_2O)	Chl b (H_2O)	Chl b (H_2O)	Chl b (Gln 105)	Chl b (H_2O)	Chl a (Val 135)	Chl b (H_2O)
608	Chl b (H_2O)	Chl b (H_2O)	Chl b (H_2O)	Chl b (H_2O)	Chl a (H_2O)	Chl b (H_2O)	Chl b (H_2O)	Chl b (H_2O)
609	Chl b (Glu 139)	Chl a (Glu 159)	Chl a (Glu 150)	Chl b (Glu 109)	Chl a (Glu 142)	Chl a (Glu 154)	Chl a (Glu 163)	Chl a (Glu 153)
610	Chl a (Glu 180)	Chl a (Glu 198)	Chl a (Glu 189)	Chl a (Glu 180)	Chl a (Glu 180)	Chl a (Glu 211)	Chl a (Glu 221)	Chl a (Glu 204)
611	Chl a (PG)	Chl a (PG)	Chl a (PG)	Chl a (PG)	Chl a (PG)	Chl a (PG)	Chl a (PG)	Chl a (H_2O)
612	Chl a (Asn 183)	Chl a (His 201)	Chl a (Asn 192)	Chl a (His 183)	Chl a (Asn 183)	Chl a (Asn 214)	Chl a (Asn 224)	Chl a (Asn 207)
613	Chl a (Gln 197)	Chl a (Gln 215)	Chl a (Gln 206)	—	Chl a (Gln197)	Chl a (Gln228)	Chl a (Gln238)	Chl a (Gln 221)
614	Chl a (His 212)	Chl b (His 230)	Chl a (His 221)	—	Chl a (His 213)	Chl a (His 243)	Chl a (His 253)	Chl a (His 236)
616	—	Chl a (Phe 80)	—	—	Chl a (Leu226)	—	Chl a (His 68*PsaK)	—
617	—	—	—	—	—	—	Chl a (His 164)	Chl a (His 150)
618	—	—	—	—	—	Chl b (Asp 170)	—	Chl b (Asp 169)
619	—	—	—	—	—	—	Chl a (unknown)	—
L1	Lutein	Lutein	Lutein	Lutein	Lutein	Lutein	Lutein	Lutein
L2	Lutein	Violaxanthin	Lutein	Violaxanthin	Violaxanthin	Violaxanthin	Violaxanthin	Violaxanthin
N1	Neoxanthin	Neoxanthin	Neoxanthin	β-carotene	β-carotene	β-carotene	β-carotene	β-carotene
V1	Violaxanthin	—	—	—	β-carotene	β-carotene	—	β-carotene
L3	—	—	—	—	—	—	—	Lutein

Note: Central ligands of chlorophylls are shown in parentheses. Pigments in CP24 are tentatively assigned based on biochemcial data obtained by Passarini et al. 2009.

Aside from the role of carotenoids in light harvesting and EET, they are also required for structure stabilization and folding. LHCII is unable to be reconstituted unless the L1 site is occupied (Croce et al. 1999). The occupancy of both the N and the L2 site has only a minor effect on structure stabilization, whereas violaxanthin can easily be removed from site V1 without affecting the stability of the trimers (Formaggio et al. 2001). Carotenoids are also important for chlorophyll triplet quenching, singlet oxygen scavenging, and excess energy dissipation; however, this is discussed in detail in other chapters (see Chapters 3 and 11).

LHCII is also involved in maintaining the excitation balance between PSI and PSII, a process known as state transitions (for a recent review, see Goldschmidt-Clermont and Bassi 2015). This process is regulated by the reversible phosphorylation of the N-terminal threonine residues of Lhcb1 and Lhcb2. Lhcb3, which has a truncated N-terminus, does not have these threonines and thus is not phosphorylated. Reversible phosphorylation is carried out by the STN7 kinase and the constitutively active phosphatase PPH1 (also known as TAP38) (Bellafiore et al. 2005; Shapiguzov et al. 2010).

4.4.2 Minor antenna complexes

The minor antenna complexes of PSII are CP29, CP26, and CP24. They are designated as minor due to their relatively low prevalence in the membrane, accounting for only ~15% of the total PSII chlorophyll. Despite their intrinsically low light-harvesting capacity, they are highly conserved in all plant species, thus pointing to other important roles in the membrane such as the regulation of energy transfer. Overall, they have high sequence and structural homology to LHCII; however, small differences in primary structure (Figure 4.2) result in their monomeric state and unique pigment-binding characteristics (Figure 4.3).

4.4.2.1 CP29 (Lhcb4)

CP29 is the most well characterized of the three minor light-harvesting proteins. It is encoded by at least two highly conserved genes, lhcb4.1 and lhcb4.2, and in *A. thaliana* there is a third and less abundant isoform encoded by lhcb4.3 (Jansson 1999; Koziol et al. 2007). This isoform lacks a large part of the C-terminal domain, has a distinct expression pattern, and appears to be present only in dicots (Klimmek et al. 2006).

The overall sequence and structure of CP29 is highly homologous to LHCII and other LHCs, except that it contains a 42–amino acid insert at the N-terminus (Figure 4.2). The first full structure of CP29 became available as part of the C2S2 PSII–LHCII supercomplex (see Chapter 10), and it shows the N-terminal insert forming two structured hairpin loops, a longer one extending over neighboring core subunits and a shorter one extending into the interface below (Wei et al. 2016). Compared to LHCII, the 605 site is empty and a unique Chl-binding site, a616, is observed in the short hairpin loop of the N-terminal insert. The 614 site is also empty in this structure; however, it is likely that it was lost during the purification process as it is present in other structures (Pan et al. 2011; van Bezouwen et al. 2017). This gives CP29 a total of 14 Chls (10 Chls *a* and 4 Chls *b*), 6 more than predicted by previous studies of the purified protein (Bassi et al. 1999; Pascal et al. 1999b). The three carotenoids are assigned as lutein, violaxanthin, and neoxanthin in the L1, L2, and N1 sites, respectively (Caffarri et al. 2007).

In *A. thaliana*, CP29 is the only reversibly phosphorylated minor antenna protein. The phosphorylation sites in Lhcb4.1 and Lhcb4.2 are Thr109 and Thr111. Lhcb4.2 has an additional site, Thr37. Lhcb4.3 is phosphorylated at a single site, Ser34. (Hansson and Vener 2003; Mikko et al. 2006; Reiland et al. 2009). The physiological significance of this reversible phosphorylation is not well understood, although it has been suggested to be involved in the disassembly of PSII supercomplexes under high-light stress (Fristedt et al. 2011).

4.4.2.2 CP26 (Lhcb5)

CP26 evolved prior to green algal diversification and is encoded by the single gene, lhcb5. The first structure of CP26 was also solved as part of the C2S2 PSII–LHCII supercomplex, and it is very similar to that of LHCII (Wei et al. 2016). Thirteen Chl-binding sites are observed (nine Chls *a* and four Chls *b*), four more than predicted by previous studies of both the native protein purified by IEF and the reconstituted protein (Croce et al. 2002a; Ballottari et al. 2009). The four new Chls, b601, a604, b607, and b608, are only tentatively assigned (as Chl *a* or Chl *b*) based on the corresponding sites in LHCII. This gives the

601

Lhcb1	1	... RKTVAKPKGPSGS... SDRVKYLGPF ... SGE PSYLTGEFPGDYGWDTAGLSAD ...	53
Lhcb2	1	... RKTVKS...TPQ. AVGPDRPKYLGPF ... SFNTPSYLTGEYPGDYGWDTAGLSAD ...	50
Lhcb3	1	... GNI...U PDRVKYLGPF ... SVQTPSYLTGEFPGDYGWDTAGLSAD ...	43
Lhcb4.1	1	VFGFGKKKAAPKKSAI... TDRPLWYPGA ... ISPDWLDGSLVGDYGFDPFGLGKP...	61
Lhcb4.2	1	RFGFGTKKASPKKAI... SDRPLWFPGA ... KSPEYLDGSLVGDYGFDPFGLGKP...	59
Lhcb4.3	1	...GTRLVWFPGA ... NPPEWLDGSMIGDRGFDPFGLGKP	65
Lhcb5	1	...PCRRIFLPDGL LDRSE IPEYLNGEVAGDGYDPFGLGKK...	63
Lhcb6	1	...PKSWIPAVGGGNL PEWLDGSLPGDFGFDPLGLGKD	44
Lhca1	1	...AAHWMPGI ... PRPAYLDGSAPGDFGFDPLGLGEV ...	32
Lhca2	1	...PCRPIWFPGI ... TPPEWLDGSLPGDFGFDPLGLSSD ...	38
Lhca3	1	...ANRPIWFASS ... QSLSYLDGSLPGDFGFDPLGLSDP ...	49
Lhca4	1	...KGEWLPGI ... ASPDYLTGSLAGDNGFDPLGLAED ...	33
Lhca5	1	...YIRATWLPGI ... NPPPYLDGNLAGDYGFDPLGLGED ...	45
Lhca6	1	...PCRPIWFPGI ... SPPEWLDGSLPGDFGFDPLGLGSD ...	45

Helix B

602 603

Lhcb1	54	PETFARNRELEVIHSRWAMLGALGCVFPELLARN	87
Lhcb2	51	PETFAKNRELEVIHSRWAMLGALGCTFPEILSKN	84
Lhcb3	44	PEAFAKNRALEVIHGRWAMLGAFGCITPEVLQK	77
Lhcb4.1	62	VFGIQRFRECELIHGRWAMLATGALSVEWLTGV	131
Lhcb4.2	60	VFGLQRFRECELIHGRWAMLATLGAITVEWLTGV	129
Lhcb4.3	66	VFGIQRFRECELIHGRWAMLGTLGAIAVEALTGI	135
Lhcb5	64	PENFAKYQAFELIHARWAMLGAAGFIIPEALNKY	97
Lhcb6	45	PAFLKWYREAELIHGRWAMAAVLGIFVGQAWSGV	78
Lhca1	33	PANLERYKESELIHCRWAMLAVPGILVPEALGYG	66
Lhca2	39	PDSLKWNVQAEIVHCRWAMLGAAGFIPEFLTKI	72
Lhca3	50	FIFPRWLAYGEIINGRFAMLGAAGAIAPEILGKA	83
Lhca4	34	PENLKWFVQAELVNGRWAMLGVAGMLLPEVFTKI	67
Lhca5	46	PLSLKWYVQAELVHSRFAMLGVAGILFTDLLRTT	79
Lhca6	46	PDTLKWFAQAELIHSRWAMLAVTGIIIPECLERL	79

Helix E   Helix C

605   606   609

Lhcb1	88	...V.....VWFKAGSQIF ...LDYLGNPSIVHA SILAIWATQVILMGAVEGYRVAGNGPIGE...	150
Lhcb2	85	...V.....VWFKAGSQIF ...LDYLGNPNLIHA SILAIWAVQVVLMGFIEGYRIGG GPLGE	146
Lhcb3	78	...RVDF VVWFKAGSQIF ...LDYLGNPNLVHA SILAVLGFQVILMGLVEGFRINGLDGVGE	141
Lhcb4.1	132	TWQDAGKVEL ...SSYIGQP IP SISTLIWIEVLVIGYIEFQRNAELD	179
Lhcb4.2	130	TWQDAGKVELVDG SSYIGQP IP SISTLIWIEVLVVGYIEFQRNAELD	177
Lhcb4.3	136	AWQDAGKVELVFG SSYIGQP LP SITTLIWIEVLVVGYIEFQRNSELD	183
Lhcb5	98	CA NCGPEAVWFKTGALLLDGNT LNYEGKN NLVLAVVAEVVLLGGAEYYRITNGLD	155
Lhcb6	79	AWFEAGAQPDAIA PF SFGSLLGTQLLLMGWVESKRWVDEFNPDSQSVE	126
Lhca1	67	NWVKAQEWAAIPGGQATYLGNP VPW TLFTILAIEFLAIAFVEHQRSME KD	117
Lhca2	73	GII NTPSWYTAGEQEY F DKTTLFVVELILIGWAEGRRWADILKPGSVNID	123
Lhca3	84	GLIPAETALEWFQTGVIEF AGTYIY W DNYTLFVLEMALMGFAEHRRLQDAYNPGSMGKQ	143
Lhca4	68	NVPEWYDAGKFQY SSSTLFVIEFILFHYVEIRRWQDIKNPGSVNQD	118
Lhca5	80	GIF NLFSWYEAGAVKID F STKTLIVVQFLLMGFAETKRYMDFVSPGSQAKE	131
Lhca6	80	GFL ENFSWYDAGSRFF F DSTTLFVAQMVLMGWAEGRRWADILKPGSVDIE	130

Helix A

610 612

Lhcb1	151	AEDLLYPGG S FDPLGLAT	DPEAFAELKVKELKNGRLAMFSMFGFF	195
Lhcb2	147	GLDPLYPGGA FDPLNLAE	DPEAFSELKVKELKNGRLAMFSMFGFF	191
Lhcb3	142	GNDLYPGG FDPLGLAD	DPVTFAELKVKELKNGRLAMFSMFGFF	186
Lhcb4.1	180	SEKRLYPGGKFFDPLGLAA	DPEKTAQLQLAEIKHARLAMVGFLGFA	225
Lhcb4.2	178	SEKRLYPGGKFFDPLGLAS	DPLKKAQLQLAEIKHARLAMVGFLGFA	223
Lhcb4.3	184	TEKRIYPGGY FDPLGLAA	DPEKLDTLKLEIKHSRLAMVAFLIFA	228
Lhcb5	156	FEDKLHPGGP FDPLGLAK	LPEOGALLKVELKNGRLAMFAMLGFF	200
Lhcb6	127	WSTPWSKTAENEANYTGDQGYPGGRFFDPLGLAGKNPDGVYF PDFLKLERLKLAEIKHSRLAMVAMLIFY	196
Lhca1	118	PEKKKYPGGA FDPLGYSK	DPKKLEELKVKEIKNGRLALLAVVGA	162
Lhca2	124	PVEPNNKLT GTDVGYPGGLWFDPLGWGSG	SPAKLRLRTSEIKNGRLAMLAVMGA	179
Lhca3	144	VELGLA DGNPAYPGGPFFNPLGFGK	DEKSLKEELKLEEVKNGRLAMLAIILGYF	200
Lhca4	119	PIFKQ YSLFFGEVGYPGG FNPLNFAP	TQEAKEKELANGRLAMLAFLGFV	168
Lhca5	132	GSKFT GLEAALEGLEPGYPGGPLLNPLGLAK	DVKNAHDWKLEIKNGRLAMMAMLGFF	189
Lhca6	131	PSYR HKVNPKPDVGYPGGIWFDFMMWGRG	SPEPVMVLRTSEIKNGRLAMLAFLGFC	186

Helix D

613   614

Lhcb1	196	VQAIVT GKGPIENLADHLADPVNNNAWAFATNFVPGK	232
Lhcb2	192	VQAIVT GKGPIENLFDHLADPVANNAWSYATNFVPGK	228
Lhcb3	187	VQAIVT GKGPLENLLDHLDNPVANAWAFATKFAPGA	223
Lhcb4.1	226	VQAAAT GKGPLNNWATHLSDPLHTTIDTFSSS	258
Lhcb4.2	224	VQAAAT GKGPLNNWATHLSDPLHTTIDTFSSS	256
Lhcb4.3	229	LQAAT T GKGPVSFLATFNN	247
Lhcb5	201	IQAYVT GEGPVENLAKHLSDPFHNLLTVIAGTAEPAPTL	240
Lhcb6	197	FEAG Q GKTPLGALGL	211
Lhca1	163	VQQSAYPGTGPLENLATHLADPWHNNIGDIVIPEN	197
Lhca2	180	FQHIYT GIGPIDNLFAHLADPGHTIFAALIPK	212
Lhca3	201	IQGLVT GVGPYQNLLDHLADPVHNNVLTSLKFH	233
Lhca4	169	VQHNVT GKGPFENLLQHLSDPWHNTIVQTEN	199
Lhca5	190	VQASVT HTGPIDNLVEHLSNPWHTTLIQTLETSTS	224
Lhca6	187	FQATYT SQDPIENLMAHLADPGHNNVFSAFTSH	219

Figure 4.2 Sequence alignment of mature LHC polypeptides from *A. thaliana*. Pigment-binding sites are indicated. Green background, transmembrane helix; pink background, amphipathic lumenal helix; yellow, N-terminal trimerization motif. The alignment was done using MUSCLE then manually adjusted. (From Edgar, R.C., *Nucl. Acids Res.*, 32, 1792, 2004.)

complex a Chl *a/b* ratio of 2.25, which is similar to values previously reported (Croce et al. 2002a; Ballottari et al. 2009). The carotenoids are assigned as two luteins (L1 and L2) and one neoxanthin (N1), although it has been shown that the L2 site is also able to accommodate violaxanthin (Croce et al. 2002a; Ballottari et al. 2009).

CP26 is the only minor antenna protein that contains the N-proximal trimerization motif (Figure 4.2). In antisense plants lacking Lhcb1 and Lhcb2, CP26 is overexpressed and suggested to form trimers that can replace LHCII (Andersson et al. 2003; Ruban et al. 2003).

4.4.2.3 CP24 (Lhcb6)

CP24, appearing during land colonization, is, together with Lhcb3, the most recent member of the LHC family (Koziol et al. 2007). It is encoded by the single gene lhcb6. Overall, the sequence is similar to LHCII except that it lacks the amphipathic helix D at the C-terminal domain and has two small inserts in the AC loop. The structure has recently become available in the C2S2M2 PSII–LHCII supercomplex (Su et al. 2017). Eleven Chls were detected, six Chls *a* and five Chls *b*. However, the identities as either Chls *a* or Chls *b* are only tentatively assigned based on previous biochemical characterization (Passarini et al. 2009). This gives CP24 the lowest Chl *a/b* ratio of all Lhcb complexes and explains why it does not accumulate in the Chl *b*–less mutant (Bossmann et al. 1999). The carotenoids in the structure were assigned as lutein, violaxanthin, and β-carotene in the L1, L2, and N1 sites, respectively. The β-carotene was a tentative assignment because it had previously only been found in PSI and the PSII core. However, its presence in the PSII antenna has since been confirmed biochemically, with one β-carotene molecule identified per one CP29-LHCII-CP24 complex (Xu et al. 2017). A unique feature of CP24 compared to other Lhcbs is that carotenoid binding in the L1 site is not required for protein stability (Passarini et al. 2009).

4.4.3 LHCI

In Arabidopsis, LHCI is composed of four nuclear gene products (Lhca1–Lhca4) that are 20–24 kDa polypeptides. Two more Lhca gene products have been identified (Lhca5–Lhca6), but they are transcribed only at a very low level and their gene products exist in substoichiometric amounts (Klimmek et al. 2006).

The Lhca proteins share a relatively high degree of sequence and structural homology to each other and to LHCII; the most notable differences exist in the loop regions AC and BC and the N-terminus (Figure 4.3) (Ben-Shem et al. 2003; Amunts et al. 2007; Mazor et al. 2015, 2017; Qin et al. 2015). Perhaps the most striking structural difference compared to LHCII is the oligomerization state. In its native state, LHCI exists in the form of two heterodimers, the Lhca1–Lhca4 and Lhca2–Lhca3 pairs. The dimerization is mediated via the N-terminus of one Lhca coming into contact with the extended helix C of the adjacent Lhca. Hydrophobic interactions of Chls and carotenoids in this interface further strengthen the interaction, providing an explanation for the greater stability of the Lhca1–Lhca4 dimer, which contains two bridging carotenoids (BCR623 and Lut624) compared to the one (BCR623) of Lhca2–Lhca3.

While the majority of the Chl-binding sites are conserved between LHCI and LHCII, there are a few marked differences. First, LHCI contains a number of unique Chls, although these are typically located at the interface with the PSI core or in connecting regions between adjacent Lhcas. Second, LHCI has a significantly lower affinity for Chl *b*. Together the four Lhca subunits bind 45 Chls *a* and 12 Chls *b*, giving a Chl *a/b* ratio of 3.75, consistent with biochemical data that suggests a ratio of 3.7 (Wientjes and Croce 2011; Qin et al. 2015). However, the distribution of Chl *b* is uneven, with Lhca2 and Lhca4 binding much more Chl *b* than Lhca1 and Lhca3. In terms of the carotenoids, each Lhca binds one lutein, one violaxanthin, and one β-carotene in the L1, L2, and N1 sites, respectively. Another lutein (Lut624) is bound between Lhca1 and Lhca4, giving rise to a total of 13 carotenoids in the four Lhcas.

Spectrally, LHCI differs from LHCII due to the presence of low-energy chlorophylls also known as "red forms". These red forms are responsible for fluorescence emission that is 50 nm redshifted compared to the emission of Lhcb complexes (Lam et al. 1984). The properties of the individual Lhcas have been studied by *in vitro* reconstitution as it is still

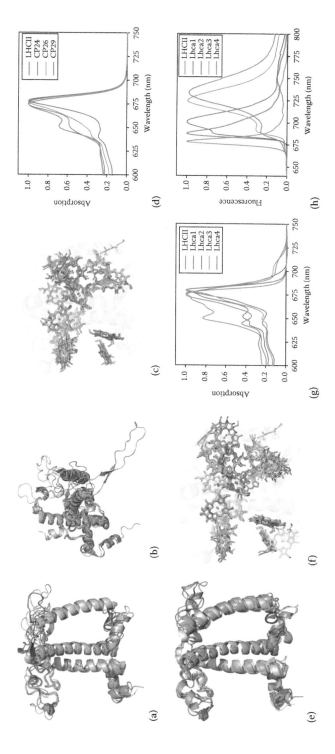

Figure 4.3 Comparison of higher plant LHCs. (a–d) LHCII (green), CP24 (orange), CP26 (pink) and CP29 (blue). (a) Structural superposition of all Lhcb minor antenna proteins onto the LHCII protein backbone (PDB code: 5XNL). (b) View from the stromal side to highlight structural differences in the loop regions. (c) Superposition of the corresponding Chl molecules on a LHCII protein backbone. (d) Absorption spectra taken at 77k. (e–h) LHCII (grey), Lhca1 (green), Lhca2 (blue), Lhca3 (pink) and Lhca4 (orange). (e) Structural superposition of Lhca1-4 (PDB code 4XK8) onto a LHCII protein backbone. (f) Superposition of all corresponding Chl molecules on a LHCII protein backbone. (g) Absorption spectra taken at 77k. (h) Fluorescence emission spectra taken at 77k.

not possible to obtain the native versions; in fact, they were only recently purified as two functional heterodimers (Schmid et al. 1997, 2002; Croce et al. 2002b; Castelletti et al. 2003; Wientjes and Croce 2011). At 4K, Lhca1, Lhca2, Lhca3, and Lhca4 have emissions at 690, 702, 725, and 733 nm, respectively. Lhca5 does not contain red forms and emits at 684 nm, and so far nothing is known about Lhca6 (Storf et al. 2005).

The origin of the red forms was studied by mutation analysis of the reconstituted complexes. It was shown that the Chls 603–609 dimer was responsible for the low-energy absorption in all Lhcas (Morosinotto et al. 2002, 2005b; Croce et al. 2004; Mozzo et al. 2006). In Lhca3 and Lhca4 that harbor the reddest forms, the ligand for Chl 603 is an Asn, whereas in Lhca1 and Lhca 2 it is a His. The phytol tails of the red Chls extend into the gap region between LHCI and the PSI core. The connection with the core potentially influences their conformation as it has been shown that binding to the core enhances the red-shifted emission (Morosinotto et al. 2005a). The biological function of the red forms is still under debate although they are generally thought to expand the light-harvesting capacity of PSI in the lower levels of a plant canopy, where far-red light is abundant (Rivadossi et al. 1999).

Excitation energy transfer in LHCI complexes appears to be very similar to that observed within Lhcb complexes. The only difference is represented by the energy transfer component in the red forms, which takes place in approximately 5 ps. Energy transfer between the monomers is extremely efficient as a result of the chlorophylls located in connecting regions between adjacent Lhcas. For a review, see Croce and Van Amerongen (2013).

4.5 LHCs in *Chlamydomonas reinhardtii*

In *C. reinhardtii*, nine genes (LhcbM1–LhcbM9) encode trimeric LHCII. However, none of these can be specifically associated with a single *A. thaliana* isoform, indicating that its diversification between the LhcbM isoforms occurred after the divergence of *Chlamydomonas* from the green lineage (Ballottari et al. 2012). The gene products have been divided into four groups based

on their sequence similarity: Type I (LhcbM3, LhcbM4, LhcbM6, LhcbM8, and LhcbM9), Type II (LhcbM5), Type III (LhcbM2 and LhcbM7), and Type IV (LhcbM1). Overall, they share sequence homology of 75–80%.

It appears that all isoforms are comparable in terms of pigment content, organization, and spectroscopic properties (Natali et al. 2015). They also show many similarities to plant LHCII, including conservation of all chlorophyll-binding sites and the presence of lutein, neoxanthin, and violaxanthin. They also have a similar Chl *a/b* ratio (1.28 in *C. reinhardtii* vs. 1.33 in plants) indicating that at most there is a change in affinity of one binding site (Drop et al. 2014a). There are however some important differences, the first being the presence of loroxanthin as opposed to a lutein in one of the two internal xanthophyll binding sites (Drop et al. 2014a). In addition, the absorption spectra of all LhcbMs are 2 nm blue-shifted compared to that of LHCII of plants. This is proposed to be a result of loroxanthin influencing the lowest-energy chlorophylls 610, 611, and 612 (Natali et al. 2015). Several of the isoforms have been suggested to have distinct functional roles in the thylakoid membrane (Elrad et al. 2002; Ferrante et al. 2012; Grewe et al. 2014); however, the highly conserved biochemical spectroscopic properties suggest these differences are not intrinsic but instead related to interactions with other proteins or cofactors, for example, LhcSR (see Chapter 11).

Of the minor antenna proteins, CP26 and CP29 are present in *C. reinhardtii*, whereas CP24 is not. CP29 is a phosphoprotein, although all six CP29 phosphorylation sites (Thr7, Thr11, Thr17, Thr18, Thr33, and Ser103) are different to those found in higher plants (Turkina et al. 2006). The phosphorylation of CP29 is involved in state transitions and is crucial for the association of LHCII to PSI (Tokutsu et al. 2009).

LHCI of *C. reinhardtii* is encoded by nine Lhca genes (Lhca1–Lhca9), all of which are expressed under normal conditions. Functional analysis of the reconstituted proteins revealed that despite their relatively low sequence identity, the pigment binding and the spectroscopic properties of the subunits are very similar. The main difference between the complexes is the emission maxima of the red forms, where Lhca2, Lhca4, and Lhca9 display the most redshifted emission with maxima between 707 and 715 nm (Mozzo et al. 2010).

4.6 ASSEMBLY OF THE ANTENNA COMPLEXES IN SUPERCOMPLEXES

In vivo, the antenna complexes are associated with the core complexes of PSI and PSII and form supercomplexes (Figure 4.4); the structures of which are discussed in detail in Chapter 10. In these functional units, light energy absorbed by the antennas is efficiently transferred to the RC, where it promotes charge separation. Effective energy transfer requires that the pigments of the LHCs and those of the core are in close proximity, and this is achieved in the supercomplexes.

PSII is normally present as a dimer in the thylakoid membranes of plants and green algae. Each monomeric core (C) is associated with one monomer of CP29 and CP26 and one trimer of LHCII, which is strongly (S) bound to it. Together they form the basic unit of the PSII supercomplex, called C2S2, which contains more than 100 Chls per RC. The addition of one monomer of CP24 and one LHCII trimer, moderately (M) bound to

the supercomplex, gives rise to the C2S2M2 supercomplex. Additional trimers (called L or "extra") are loosely associated with the supercomplex and therefore cannot be purified with it. The amount of extra LHCII varies depending on the growth conditions, increasing in low light to augment the absorption cross section of the core and decreasing in high light, where the harvested energy exceeds the capacity of the electron transport chain (Ballottari et al. 2007; Wientjes et al. 2013b). Due to the difficulty in purifying PSII supercomplexes to homogeneity, there are only a limited number of studies investigating EET and trapping in these systems. It was shown that the overall trapping time (from the absorption of a photon by a pigment in the supercomplex to the use of the energy in the RC) is around 150 ps in the C2S2M2 supercomplex (Caffarri et al. 2011). This corresponds to a quantum yield of charge separation of 0.92, indicating that the LHCs are very efficient in transferring excitation energy to the core. The trapping time increases to ~310 ps (and the quantum yield decreases to 0.84) in low-light conditions when 2–3 extra LHCII are present per monomeric core (Wientjes et al. 2013b). This indicates that the extra LHCII trimers are less efficient in EET and have weak functional connectivity to the core. Modeling of energy transfer pathways in these large complexes and of those in the membrane is still in its infancy, and several

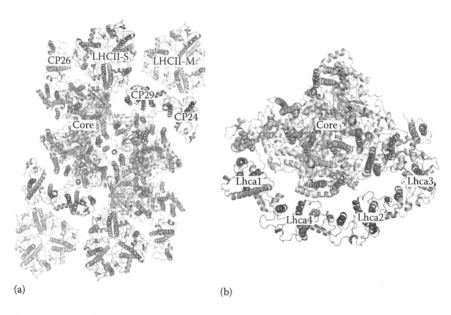

(a) (b)

Figure 4.4 Organization of the PSII–LHCII supercomplex (a) and PSI–LHCI supercomplex (b). Structures were obtained from PDB codes 5MDX and 5L8R, respectively.

approaches are being developed (Broess et al. 2008; Caffarri et al. 2011; Bennett et al. 2013; Amarnath et al. 2016).

Due to its high stability, the PSI–LHCI supercomplex has been studied in much more detail than the PSII–LHCII supercomplexes. In plants, the core, which contains around 100 Chls a, is complemented by four LHCIs to give a total of 156 Chls per RC. Despite the presence of the red forms, which slow down EET, the overall trapping time of this complex is around 50 ps, and the quantum yield of charge separation is close to 1.0. Even an increase of the antenna size by five additional LHCIs per core, as observed in the PSI–LHCI complex of *C. reinhardtii*, only leads to a small increase of the trapping time, indicating that the functional connectivity of the antenna with the core is excellent (e.g., Slavov et al. 2008; Wientjes et al. 2011; Le Quiniou et al. 2015).

In addition to the four LHCI antennae, in *A. thaliana* one trimer of LHCII is stably associated with PSI in all light conditions, forming the PSI–LHCI–LHCII supercomplex (Wientjes et al. 2013a). This trimer is associated with the opposite side of the core to LHCI and can easily be dissociated *in vitro* by detergent and *in vivo* by dephosphorylation. Despite this, this LHCII is still very efficient in transferring excitation energy to the PSI RC (Galka et al. 2012; Wientjes et al. 2013a). A larger supercomplex, containing one additional LHCII trimer and one monomer, was purified from *C. reinhardtii* (Drop et al. 2014b) and also showed a very fast trapping time. Membrane nanodisks in which up to 5 LHCII per PSI RC are present also show efficient EET from the antenna to the core (Bell et al. 2015, 2017). These results indicate that PSI has several docking sites for LHCs, and all permit good functional connectivity between the antenna and the core, supporting fast and thus efficient EET to the RC. This is at variance with PSII, for which an enlargement of the antenna size leads to a decrease in trapping efficiency.

REFERENCES

Amarnath, K., D. I. G. Bennett, A. R. Schneider, and G. R. Fleming. 2016. Multiscale model of light harvesting by photosystem II in plants. *Proc. Natl. Acad. Sci. U. S. A.* 113: 1156–1161.

Amunts, A., O. Drory, and N. Nelson. 2007. The structure of a plant photosystem I supercomplex at 3.4 Å resolution. *Nature* 447: 58–63.

Andersson, J., M. Wentworth, R. G. Walters et al. 2003. Absence of the Lhcb1 and Lhcb2 proteins of the light-harvesting complex of photosystem II: Effects on photosynthesis, grana stacking and fitness. *Plant J.* 35: 350–361.

Bailey, S., R. G. Walters, S. Jansson, and P. Horton. 2001. Acclimation of *Arabidopsis thaliana* to the light environment: The existence of separate low light and high light responses. *Planta* 213: 794–801.

Ballottari, M., L. Dall'Osto, T. Morosinotto, and R. Bassi. 2007. Contrasting behavior of higher plant photosystem I and II antenna systems during acclimation. *J. Biol. Chem.* 282: 8947–8958.

Ballottari, M., J. Girardon, L. Dall'Osto, and R. Bassi. 2012. Evolution and functional properties of photosystem II light harvesting complexes in eukaryotes. *Biochim. Biophys. Acta Bioenerg.* 1817: 143–157.

Ballottari, M., M. Mozzo, R. Croce, T. Morosinotto, and R. Bassi. 2009. Occupancy and functional architecture of the pigment binding sites of photosystem II antenna complex Lhcb5. *J. Biol. Chem.* 284: 8103–8113.

Bassi, R., R. Croce, D. Cugini, and D. Sandonà. 1999. Mutational analysis of a higher plant antenna protein provides identification of chromophores bound into multiple sites. *Proc. Natl. Acad. Sci. U. S. A.* 96: 10056–10061.

Beddard, G. S. and G. Porter. 1976. Concentration quenching in chlorophyll. *Nature* 260: 366–367.

Bell, A. J., L. K. Frankel, and T. M. Bricker. 2015. High yield non-detergent isolation of photosystem I-light-harvesting chlorophyll II membranes from spinach thylakoids: Implications for the organization of the ps i antennae in higher plants. *J. Biol. Chem.* 290: 18429–18437.

Bellafiore, S., F. Barneche, G. Peltier, and J.-D. Rochaix. 2005. State transitions and light adaptation require chloroplast thylakoid protein kinase STN7. *Nature* 433: 892–895.

Ben-Shem, A., F. Frolow, and N. Nelson. 2003. Crystal structure of plant photosystem I. *Nature* 426: 630–635.

Bennett, D. I. G., K. Amarnath, and G. R. Fleming. 2013. A structure-based model of energy transfer reveals the principles of light harvesting in photosystem II supercomplexes. *J. Am. Chem. Soc.* 135: 9164–9173.

Benson, S. L., P. Maheswaran, M. A. Ware et al. 2015. An intact light harvesting complex I antenna system is required for complete state transitions in Arabidopsis. *Nat. Plants* 1: 15176.

van Bezouwen, L. S., S. Caffarri, R. S. Kale et al. 2017. Subunit and chlorophyll organization of the plant photosystem II supercomplex. *Nat. Plants* 3: 17080.

de Bianchi, S., N. Betterle, R. Kouril et al. 2011. Arabidopsis mutants deleted in the light-harvesting protein Lhcb4 have a disrupted photosystem II macrostructure and are defective in photoprotection. *Plant Cell* 23: 2659–2679.

Bos, I., K. M. Bland, L. Tian et al. 2017. Multiple LHCII antennae can transfer energy efficiently to a single Photosystem I. *Biochim. Biophys. Acta Bioenerg.* 1858: 371–378.

Bossmann, B., L. H. Grimme, and J. Knoetzel. 1999. Protease-stable integration of Lhcb1 into thylakoid membranes is dependent on chlorophyll b in allelic chlorina-f 2 mutants of barley (*Hordeum vulgare* L.). *Planta* 207: 551–558.

Broess, K., G. Trinkunas, A. van Hoek, R. Croce, and H. van Amerongen. 2008. Determination of the excitation migration time in photosystem II. *Biochim. Biophys. Acta Bioenerg.* 1777: 404–409.

Caffarri, S., K. Broess, R. Croce, and H. van Amerongen. 2011. Excitation energy transfer and trapping in higher plant photosystem II complexes with different antenna sizes. *Biophys. J.* 100: 2094–2103.

Caffarri, S., R. Croce, J. Breton, and R. Bassi. 2001. The major antenna complex of photosystem II has a xanthophyll binding site not involved in light harvesting. *J. Biol. Chem.* 276: 35924–35933.

Caffarri, S., R. Croce, L. Cattivelli, and R. Bassi. 2004. A Look within LHCII: Differential analysis of the Lhcb1-3 complexes building the major trimeric antenna complex of higher-plant photosynthesis. *Biochemistry* 43: 9467–9476.

Caffarri, S., F. Passarini, R. Bassi, and R. Croce. 2007. A specific binding site for neoxanthin in the monomeric antenna proteins CP26 and CP29 of photosystem II. *FEBS Lett.* 581: 4704–4710.

Cammarata, K. V. and G. W. Schmidt. 1992. *In vitro* reconstitution of a light-harvesting gene product: Deletion mutagenesis and analyses of pigment binding. *Biochemistry* 31: 2779–2789.

Castelletti, S., T. Morosinotto, B. Robert et al. 2003. Recombinant Lhca2 and Lhca3 subunits of the photosystem I antenna system. *Biochemistry* 42: 4226–4234.

Croce, R. and H. Van Amerongen. 2013. Light-harvesting in photosystem I. *Photosynth. Res.* 116: 153–166.

Croce, R., G. Canino, F. Ros, and R. Bassi. 2002a. Chromophore organization in the higher-plant photosystem II antenna protein CP26. *Biochemistry* 41: 7334–7343.

Croce, R., T. Morosinotto, S. Castelletti, J. Breton, and R. Bassi. 2002b. The Lhca antenna complexes of higher plants photosystem I. *Biochim. Biophys. Acta Bioenerg.* 1556: 29–40.

Croce, R., T. Morosinotto, J. A. Ihalainen et al. 2004. Origin of the 701-nm fluorescence emission of the Lhca2 subunit of higher plant photosystem I. *J. Biol. Chem.* 279: 48543–48549.

Croce, R., M. G. Müller, R. Bassi, and A. R. Holzwarth. 2001. Carotenoid-to-chlorophyll energy transfer in recombinant major light-harvesting complex (LHCII) of higher plants. I. Femtosecond transient absorption measurements. *Biophys. J.* 80: 901–915.

Croce, R., S. Weiss, and R. Bassi. 1999. Carotenoid-binding sites of the major light-harvesting complex II of higher plants. *J. Biol. Chem.* 274: 29613–29623.

Dainese, P., G. Hoyer Hansen, and R. Bassi. 1990. The resolution of chlorophyll a b binding proteins by a preparative method based on flat bed isoelectric focusing. *Photochem. Photobiol.* 51: 693–794.

Dall'Osto, L., C. Ünlü, S. Cazzaniga, and H. van Amerongen. 2014. Disturbed excitation energy transfer in *Arabidopsis thaliana* mutants lacking minor antenna complexes of photosystem II. *Biochim. Biophys. Acta Bioenerg.* 1837: 1981–1988.

Dall'Osto, L., M. Bressan, and R. Bassi. 2015. Biogenesis of light harvesting proteins. *BBA Bioenergetics* 1847: 861–871.

Drop, B., M. Webber-Birungi, S. K. N. Yadav et al. 2014a. Light-harvesting complex II (LHCII) and its supramolecular organization in *Chlamydomonas reinhardtii*. *Biochim. Biophys. Acta Bioenerg.* 1837: 63–72.

Drop, B., K. N. S. Yadav, E. J. Boekema, and R. Croce. 2014b. Consequences of state transitions on the structural and functional organization of photosystem I in the green alga *Chlamydomonas reinhardtii*. *Plant J.* 78: 181–191.

Edgar, R. C. 2004. MUSCLE: Multiple sequence alignment with high accuracy and high throughput. *Nucl. Acids Res.* 32: 1792–1797.

Elrad, D., K. K. Niyogi, and A. R. Grossman. 2002. A major light-harvesting polypeptide of photosystem II functions in thermal dissipation. *Plant Cell* 14: 1801–1816.

Fassioli, F., R. Dinshaw, P. C. Arpin, and G. D. Scholes. 2014. Photosynthetic light harvesting: Excitons and coherence. *J. R. Soc. Interface* 11: 20130901.

Ferrante, P., M. Ballottari, G. Bonente, G. Giuliano, and R. Bassi. 2012. LHCBM1 and LHCBM2/7 polypeptides, components of major LHCII complex, have distinct functional roles in photosynthetic antenna system of *Chlamydomonas reinhardtii*. *J. Biol. Chem.* 287: 16276–16288.

Formaggio, E., G. Cinque, and R. Bassi. 2001. Functional architecture of the major light-harvesting complex from higher plants. *J. Mol. Biol.* 314: 1157–1166.

Förster, T. 1948. Zwischenmolekulare Energiewanderung und Fluoreszenz. *Ann. Phys.* 437: 55–75.

Fristedt, R., A. V. Vener, A. Shevchenko, M. Wilm, and M. Mann. 2011. High light induced disassembly of photosystem II supercomplexes in Arabidopsis requires STN7-dependent phosphorylation of CP29. *PLoS One* 6: e24565.

Galka, P., S. Santabarbara, T. T. H. Khuong et al. 2012. Functional analyses of the plant photosystem I–light-harvesting complex II supercomplex reveal that light-harvesting complex II loosely bound to photosystem II Is a very efficient antenna for photosystem I in state II. *Plant Cell* 24: 2963–2978.

Ganeteg, U., C. Külheim, J. Andersson, and S. Jansson. 2004. Is each light-harvesting complex protein important for plant fitness? *Plant Physiol.* 134: 502–509.

Giuffra, E., D. Cugini, R. Croce, and R. Bassi. 1996. Reconstitution and pigment-binding properties of recombinant CP29. *Eur. J. Biochem.* 238: 112–120.

Goldschmidt-Clermont, M. and R. Bassi. 2015. Sharing light between two photosystems: Mechanism of state transitions. *Curr. Opin. Plant Biol.* 25: 71–78.

Gradinaru, C. C., I. H. M. van Stokkum, A. A. Pascal, R. Van Grondelle, and H. Van Amerongen. 2000. Identifying the pathways of energy transfer between carotenoids and chlorophylls in LHCII and CP29. A multicolor, femtosecond pump–probe study. *J. Phys. Chem. B* 104: 9330–9342.

Grewe, S., M. Ballottari, M. Alcocer et al. 2014. Light-harvesting complex protein LHCBM9 Is critical for photosystem II activity and hydrogen production in *Chlamydomonas reinhardtii*. *Plant Cell* 26: 1598–1611.

Hansson, M. and A. V. Vener. 2003. Identification of three previously unknown *in vivo* protein phosphorylation sites in thylakoid membranes of *Arabidopsis thaliana*. *Mol. Cell. Proteomics* 2: 550–559.

Hobe, S., H. Fey, H. Rogl, and H. Paulsen. 2003. Determination of relative chlorophyll binding affinities in the major light-harvesting chlorophyll a/b complex. *J. Biol. Chem.* 278: 5912–5919.

Hobe, S., R. Förster, J. Klingler, and H. Paulsen. 1995. N-proximal sequence motif in light-harvesting chlorophyll a/b-binding protein is essential for the trimerization of light-harvesting chlorophyll a/b complex. *Biochemistry* 34: 10224–10228.

Hobe, S., H. Niemeier, A. Bender, and H. Paulsen. 2000. Carotenoid binding sites in LHCIIb. Relative affinities towards major xanthophylls of higher plants. *Eur. J. Biochem.* 267: 616–624.

Hobe, S., S. Prytulla, W. Kuhlbrandt, and H. Paulsen. 1994. Trimerization and crystallization of reconstituted light-harvesting chlorophyll a/b complex. *EMBO J.* 13: 3423–3429.

Hoober, J. K., L. L. Eggink, and M. Chen. 2007. Chlorophylls, ligands and assembly of light-harvesting complexes in chloroplasts. *Photosynth. Res.* 94: 387–400.

Jackowski, G., K. Kacprzak, and S. Jansson. 2001. Identification of Lhcb1/Lhcb2/Lhcb3 heterotrimers of the main light-harvesting chlorophyll a/b–protein complex of photosystem II (LHC II). *Biochim. Biophys. Acta Bioenerg.* 1504: 340–345.

Jansson, S. 1999. A guide to the Lhc genes and their relatives in Arabidopsis. *Trends Plant Sci.* 4: 236–240.

Kleima, F. J., S. Hobe, F. Calkoen et al. 1999. Decreasing the chlorophyll a/b ratio in reconstituted LHCII: Structural and functional consequences. *Biochemistry* 38: 6587–6596.

Klimmek, F., A. Sjödin, C. Noutsos, D. Leister, and S. Jansson. 2006. Abundantly and rarely expressed Lhc protein genes exhibit distinct regulation patterns in plants. *Plant Physiol.* 140: 793–804.

Kouřil, R., E. Wientjes, J. B. Bultema, R. Croce, and E. J. Boekema. 2013. High-light vs. low-light: Effect of light acclimation on photosystem II composition and organization in *Arabidopsis thaliana*. *Biochim. Biophys. Acta Bioenerg.* 1827: 411–419.

Koziol, A. G., T. Borza, K.-I. Ishida et al. 2007. Tracing the evolution of the light-harvesting antennae in chlorophyll a/b-containing organisms. *Plant Physiol.* 143: 1802–1816.

Kühlbrandt, W. 1994. Structure and function of the plant light-harvesting complex, LHC-II. *Curr. Opin. Struct. Biol.* 4: 519–528.

Lam, E., W. Ortiz, and R. Malkin. 1984. Chlorophyll *a/b* proteins of photosystem I. *FEBS Lett.* 168: 10–14.

Le Quiniou, C., B. van Oort, B. Drop, I. H. M. van Stokkum, and R. Croce. 2015. The high efficiency of photosystem I in the green alga *Chlamydomonas reinhardtii* is maintained after the antenna size is substantially increased by the association of light-harvesting complexes II. *J. Biol. Chem.* 290: 30587–30595.

Liu, Z., H. Yan, K. Wang et al. 2004. Crystal structure of spinach major light-harvesting complex at 2.72 Å resolution. *Nature* 428: 287–292.

Lokstein, H., L. Tian, J. E. W. Polle, and D. DellaPenna. 2002. Xanthophyll biosynthetic mutants of *Arabidopsis thaliana*: Altered non-photochemical quenching of chlorophyll fluorescence is due to changes in photosystem II antenna size and stability. *Biochim. Biophys. Acta Bioenerg.* 1553: 309–319.

Mazor, Y., A. Borovikova, I. Caspy, and N. Nelson. 2017. Structure of the plant photosystem I supercomplex at 2.6 Å resolution. *Nat. Plants* 3: 17014.

Mazor, Y., A. Borovikova, and N. Nelson. 2015. The structure of plant photosystem I super-complex at 2.8 Å resolution. *elife* 4: e07433.

Mikko, T., P. Mirva, S. Marjaana et al. 2006. State transitions revisited—A buffering system for dynamic low light acclimation of Arabidopsis. *Plant Mol. Biol.* 62: 779–793.

Morosinotto, T., M. Ballottari, F. Klimmek, S. Jansson, and R. Bassi. 2005a. The association of the antenna system to photosystem I in higher plants. Cooperative interactions stabilize the supramolecular complex and enhance red-shifted spectral form. *J. Biol. Chem.* 280: 31050–31058.

Morosinotto, T., S. Castelletti, J. Breton, R. Bassi, and R. Croce. 2002. Mutation analysis of Lhca1 antenna complex. Low energy absorption forms originate from pigment-pigment interactions. *J. Biol. Chem.* 277: 36253–36261.

Morosinotto, T., M. Mozzo, R. Bassi, and R. Croce. 2005b. Pigment-pigment interactions in Lhca4 antenna complex of higher plants photosystem I. *J. Biol. Chem.* 280: 20612–20619.

Mozzo, M., M. Mantelli, F. Passarini et al. 2010. Functional analysis of photosystem I light-harvesting complexes (Lhca) gene products of *Chlamydomonas reinhardtii*. *Biochim. Biophys. Acta Bioenerg.* 1797: 212–221.

Mozzo, M., T. Morosinotto, R. Bassi, and R. Croce. 2006. Probing the structure of Lhca3 by mutation analysis. *Biochim. Biophys. Acta Bioenerg.* 1757: 1607–1613.

Natali, A., R. Croce, E. Ostendorf et al. 2015. Characterization of the major light-harvesting complexes (LHCBM) of the green alga *Chlamydomonas reinhardtii*. *PLoS One* 10: e0119211.

Natali, A., L. M. Roy, and R. Croce. 2014. *In vitro* reconstitution of light-harvesting complexes of plants and green algae. *J. Vis. Exp.* 92: e51852.

Novoderezhkin, V. I., M. A. Palacios, H. Van Amerongen, and R. Van Grondelle. 2004. Energy-transfer dynamics in the LHCII complex of higher plants: Modified redfield approach. *J. Phys. Chem. B* 108: 10363–10375.

Novoderezhkin, V., A. Marin, and R. Van Grondelle. 2011. Intra- and inter-monomeric transfers in the light harvesting LHCII complex: The Redfield–Foster picture. *Phys. Chem.* 13: 17093–17103.

Palacios, M. A., J. Standfuss, M. Vengris et al. 2006. A comparison of the three isoforms of the light-harvesting complex II using transient absorption and time-resolved fluorescence measurements. *Photosynth. Res.* 88: 269–285.

Pan, X., M. Li, T. Wan et al. 2011. Structural insights into energy regulation of light-harvesting complex CP29 from spinach. *Nat. Struct. Mol. Biol.* 18: 309–315.

Pascal, A., C. Gradinaru, U. Wacker et al. 1999a. Spectroscopic characterization of the spinach Lhcb4 protein (CP29), a minor light-harvesting complex of photosystem II. *Eur. J. Biochem.* 262: 817–823.

Pascal, A., C. Gradinaru, U. Wacker et al. 1999b. Spectroscopic characterization of the spinach Lhcb4 protein (CP29), a minor light-harvesting complex of photosystem II. *Eur. J. Biochem.* 262: 817–823.

Passarini, F., E. Wientjes, R. Hienerwadel, and R. Croce. 2009. Molecular basis of light harvesting and photoprotection in CP24: Unique features of the most recent antenna complex. *J. Biol. Chem.* 284: 29536–29546.

Passarini, F., P. Xu, S. Caffarri, J. Hille, and R. Croce. 2014. Towards *in vivo* mutation analysis: Knock-out of specific chlorophylls bound to the light-harvesting complexes of *Arabidopsis thaliana*—The case of CP24 (Lhcb6). *Biochim. Biophys. Acta Bioenerg.* 1837: 1500–1506.

Pattanayak, G. K., A. K. Biswal, V. S. Reddy, and B. C. Tripathy. 2005. Light-dependent regulation of chlorophyll b biosynthesis in chlorophyllide a oxygenase overexpressing tobacco plants. *Biochem. Biophys. Res. Commun.* 326: 466–471.

Paulsen, H., B. Finkenzeller, and N. Kühlein. 1993. Pigments induce folding of light harvesting chlorophyll a/b binding protein. *Eur. J. Biochem.* 215: 809–816.

Paulsen, H. and S. Hobe. 1992. Pigment-binding properties of mutant light-harvesting chlorophyll-a/b-binding protein. *Eur. J. Biochem.* 205: 71–76.

Paulsen, H., U. Rümler, and W. Rüdiger. 1990. Reconstitution of pigment-containing complexes from light-harvesting chlorophyll a/b-binding protein overexpressed in *Escherichia coli*. *Planta* 181: 204–211.

Peterman, E. J. G., R. Monshouwer, I. H. M. van Stokkum, R. van Grondelle, and H. van Amerongen. 1997. Ultrafast singlet excitation transfer from carotenoids to chlorophylls via different pathways in light-harvesting complex II of higher plants. *Chem. Phys. Lett.* 264: 279–284.

Phillip, D., S. Hobe, H. Paulsen et al. 2002. The binding of Xanthophylls to the bulk light-harvesting complex of photosystem II of higher plants. A specific requirement for carotenoids with a 3-hydroxy-beta-end group. *J. Biol. Chem.* 277: 25160–25169.

Pietrzykowska, M., M. Suorsa, D. A. Semchonok et al. 2014. The light-harvesting chlorophyll a/b binding proteins Lhcb1 and Lhcb2 play complementary roles during state transitions in Arabidopsis. *Plant Cell* 26: 3646–3660.

Qin, X., M. Suga, T. Kuang, and J.-R. Shen. 2015. Structural basis for energy transfer pathways in the plant PSI-LHCI supercomplex. *Science* 348: 989–995.

Reiland, S., G. Messerli, K. Baerenfaller et al. 2009. Large-scale Arabidopsis phosphoproteome profiling reveals novel chloroplast kinase substrates and phosphorylation networks. *Plant Physiol.* 150: 889–903.

Reinbothe, C., S. Bartsch, L. L. Eggink et al. 2006. A role for chlorophyllide a oxygenase in the regulated import and stabilization of light-harvesting chlorophyll ab proteins. *PNAS* 103: 4777–4782.

Remelli, R., C. Varotto, D. Sandonà, R. Croce, and R. Bassi. 1999. Chlorophyll binding to monomeric light-harvesting complex A mutation analysis of chromophore-binding residues. *J. Biol. Chem.* 274: 33510–33521.

Rivadossi, A., G. Zucchelli, F. M. Garlaschi, and R. C. Jennings. 1999. The importance of PS I chlorophyll red forms in light-harvesting by leaves. *Photosynth. Res.* 60: 209–215.

Rogl, H. and W. Kühlbrandt. 1999. Mutant trimers of light-harvesting complex II exhibit altered pigment content and spectroscopic features. *Biochemistry* 38: 16214–16222.

Ruban, A. V., M. Wentworth, A. E. Yakushevska et al. 2003. Plants lacking the main light-harvesting complex retain photosystem II macroorganization. *Nature* 421: 648–652.

Ruban, A. V., P. J. Lee, M. Wentworth, A. J. Young, and P. Horton. 1999. Determination of the stoichiometry and strength of binding of xanthophylls to the photosystem II light harvesting complexes. *J. Biol. Chem.* 274: 10458–10465.

Sandonà, D., R. Croce, A. Pagano, M. Crimi, and R. Bassi. 1998. Higher plants light harvesting proteins. Structure and function as revealed by mutation analysis of either protein or chromophore moieties. *Biochim. Biophys. Acta Bioenerg.* 1365: 207–214.

Schmid, V. H. R., K. V. Cammarata, B. U. Bruns, and G. W. Schmidt. 1997. *In vitro* reconstitution of the photosystem I light-harvesting complex LHCI-730: Heterodimerization is required for antenna pigment organization. *Plant Biol.* 94: 7667–7672.

Schmid, V. H. R., S. Potthast, M. Weiner et al. 2002. Pigment binding of photosystem I light-harvesting proteins. *J. Biol. Chem.* 277: 37307–37314.

Schuemann, D. 2004. Structure and function of the chloroplast signal recognition particle. *Curr. Genet.* 44: 295–304.

Shapiguzov, A., B. Ingelsson, I. Samol et al. 2010. The PPH1 phosphatase is specifically involved in LHCII dephosphorylation and state transitions in Arabidopsis. *Proc. Natl. Acad. Sci. U. S. A.* 107: 4782–4787.

Slavov, C., M. Ballottari, T. Morosinotto, R. Bassi, and A. R. Holzwarth. 2008. Trap-limited charge separation kinetics in higher plant photosystem I complexes. *Biophys. J.* 94: 3601–3612.

Standfuss, J. and W. Kuhlbrandt. 2004. The three isoforms of the light-harvesting complex II. *J. Biol. Chem.* 279: 36884–36891.

Standfuss, J., A. C. Terwisscha van Scheltinga, M. Lamborghini, and W. Kühlbrandt. 2005. Mechanisms of photoprotection and nonphotochemical quenching in pea light- harvesting complex at 2.5 Å resolution. *EMBO J.* 24: 919–928.

Storf, S., S. Jansson, and V. H. R. Schmid. 2005. Pigment binding, fluorescence properties, and oligomerization behavior of Lhca5, a novel light-harvesting protein. *J. Biol. Chem.* 280: 5163–5168.

Su, X., J. Ma, X. Wei et al. 2017. Structure and assembly mechanism of plant C2S2M2-type PSII-LHCII supercomplex. *Science* 357: 815–820.

Tokutsu, R., M. Iwai, and J. Minagawa. 2009. CP29, a monomeric light-harvesting complex II protein, is essential for state transitions in *Chlamydomonas reinhardtii. J. Biol. Chem.* 284: 7777–7782.

Turkina, M. V., J. Kargul, A. Blanco-Rivero et al. 2006. Environmentally modulated phosphoproteome of photosynthetic membranes in the green alga *Chlamydomonas reinhardtii. Mol. Cell. Proteomics* 5: 1412–1425.

Wei, X., X. Su, P. Cao et al. 2016. Structure of spinach photosystem II–LHCII supercomplex at 3.2 Å resolution. *Nature* 534: 69–74.

Wientjes, E., I. H. M. van Stokkum, H. Van Amerongen, and R. Croce. 2011. Excitation-energy transfer dynamics of higher plant photosystem I light-harvesting complexes. *Biophys. J.* 100: 1372–1380.

Wientjes, E., H. van Amerongen, and R. Croce. 2013a. LHCII is an antenna of both photosystems after long-term acclimation. *Biochim. Biophys. Acta Bioenerg.* 1827: 420–426.

Wientjes, E., H. van Amerongen, and R. Croce. 2013b. Quantum yield of charge separation in photosystem II: Functional effect of changes in the antenna size upon light acclimation. *J. Phys. Chem. B* 117: 11200–11208.

Wientjes, E. and R. Croce. 2011. The light-harvesting complexes of higher-plant photosystem I: Lhca1/4 and Lhca2/3 form two red-emitting heterodimers. *Biochem. J.* 433: 477–485.

Xu, P., L. M. Roy, and R. Croce. 2017. Functional organization of photosystem II antenna complexes: CP29 under the spotlight. *Biochim. Biophys. Acta Bioenerg.* 1858: 815–822.

Light harvesting in cyanobacteria: The phycobilisomes

LEEAT BAR-EYAL, ANAT SHPERBERG-AVNI, YOSSI PALTIEL,
NIR KEREN, AND NOAM ADIR

The phycobilisome (PBS) is the major photosynthetic light-harvesting complex (LHC) in cyanobacteria and red algae [1–5]. These species inhabit the entire span of aquatic and terrestrial habitats, from hot springs to glaciers, from deserts with extreme light fluencies to ocean niches with depleted visible light [6]. As with all photosynthetic systems, the balance between energy absorption and photochemistry must be maintained at all times. Therefore, for species exposed to variable environmental conditions, the balance must be maintained in a dynamic fashion.

LHCs in plants, green algae, or purple nonsulfur bacteria have evolved to spatially exist within the same confined intracellular compartment as the photochemical reaction centers (RC)—the thylakoid membranes [7–11]. The membranes' physicochemical characteristics, along with the mechanisms of energy transfer, limit the dimensions of transmembrane LHCs. Having both LHCs and RCs in the membrane facilitates intercomplex interactions leading to efficient intercomplex energy transfer.

The fluid nature of the membrane affords the potential for control over the flow of energy from LHCs to RCs. In contrast to other LHCs, the PBS assembles on the outside of the membrane, removing the dimensional constraints of the membrane. And indeed, the PBS can assemble into huge complexes with changeable dimensions that can reach $80 \times 50 \times 12$ nm. Electron microscopy images of various PBS-containing strains have revealed that the thylakoid membranes are almost completely covered with PBSs [6]. Thus in essence, the PBS serves as a physical barrier between adjacent membranes in the cytoplasm. PBSs can represent up to 60% of the soluble protein content of these organisms. The sheer size of each PBS could present a difficult task for efficient light harvesting, since energy transfer may have to overcome dozens if not hundreds of events, each with only a finite possibility of success as opposed to other pathways of relaxation. In this chapter, we describe the latest understanding on how the unique structural features and configuration change of the PBS afford best functionalities.

5.1 STRUCTURE OF THE PBS

All PBSs are assembled by the association of a class of proteins called phycobiliproteins (PBPs). PBPs have been suggested to be evolutionary products of the globin protein family. The smallest PBP unit (typically called a subunit) has many similarities to the basic fold found in globin-type proteins [12,13]; however, the overall sequence homology is quite low. The globins are one of the major heme-binding protein families, while PBP subunits bind 1–3 bilin chromophores, which are heme oxidation and cleavage products. By opening the heme ring into a linear tetrapyrrole, the bilin chromophores are free to adopt protein-induced conformational changes and chemical modifications (especially the addition or subtraction of conjugated double bonds) as well as form hydrogen bonds with surrounding residues and solvent molecules. The result is that PBPs have the ability to strongly absorb light with a wavelength in the range of 500–680 nm and therefore can serve as improved light-harvesting molecules when compared to the parent heme or other cyclical derivatives. As we describe later, the PBS takes advantage of bilin chromophore flexibility to carefully cover the absorption of available light as well as ensure that the excited states are efficiently transferred down into the RCs. Like often in biology, evolutionary change that brings about benefits can also produce new problems that need to be overcome. In the case of the PBPs, the flexibility of the bilin molecule could also result in a lack of binding stability. It appears that this problem was evolutionarily solved by covalently attaching the bilin to the protein backbone via a thioether bond to cysteine residues. This solution prevents the loss of the bilin chromophore; however, it requires the activity of an additional class of enzymes, the bilin lyases, to assure proper covalent attachment of the correct bilin to the correct cysteine site [14].

All PBPs contain two types of subunits, α and β. These two subunits are preferably chromophorylated prior to assembly into the basic PBP unit, the (αβ) heterodimer known in the literature as the PBP monomer (Figure 5.1a). Monomers self-assemble rapidly into disk-form (αβ)₃ trimers, which are typically quite stable and in most cases serve as the material used for structure determination by X-ray crystallography (Figure 5.1b). The trimers have a diameter of 11.5 nm with a triangular aperture of 25–50 Å in the center. Upon trimerization,

chromophores found at the ends of the monomer are brought into relative proximity—2 nm center to center. All other distances between chromophores, in any of the different levels of assembly, are significantly larger. Bringing these chromophores into proximity, coupled with the surrounding protein environment, has been proposed to have a major effect on the absorption and emission characteristics of certain PBP trimers but may have additional critical functional importance for all PBPs in the higher levels of assembly. It is interesting to note that a 1–2 nm distance marks an intermediate coupling regime weaker than in LH2

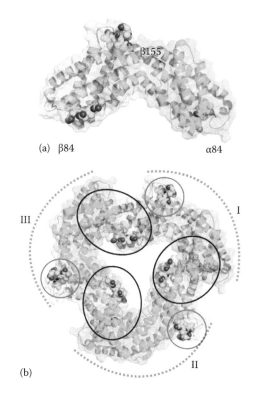

Figure 5.1 Steps in phycobilisome assembly. **(a)** The monomer. Shown is phycocyanin (PC), with the α subunit in green and the β subunit in cyan. The three phycocyanobilin (PCB) chromophores are depicted in spheres, with nitrogens and oxygens colored in blue and red, respectively. The numbering of the PCBs is according to the positions of the conserved cysteine residues. **(b)** The PC trimer. The three individual monomers are marked by the dashed arcs. The coupled PCBs are indicated by the black ovals, while the β155 PCBs situated on the circumference are indicated by the blue circles. These outer PCBs may enable energy transfer between rods.

complexes. The assembly process continues with trimers assembling into $(\alpha\beta)_6$ hexamers, mostly through interaction between α subunits. This step has been proposed to require the presence of a class of proteins called linker proteins (LPs) [15,16]. These proteins usually lack chromophores and are sequestered within the trimeric/hexameric aperture. Hexamers continue the assembly procedure forming the two main PBS substructures, the core and the rods, via interactions between β subunits. Cores contain 2–5 cylinders made up of 1–2 hexamers of the lowest energy-absorbing PBP variant called allophycocyanin (APC; λ_{max} = 652 nm). Two core cylinders typically assemble onto the membrane surface (the basal cylinders) while up to three additional cylinders assemble onto the basal layer. Each core cylinder has dimensions similar to an entire Photosystem II (PSII) complex, and thus the basal layer can overlay onto the PSII dimer, proposed to be the common form in cyanobacteria and red algae. The basal core cylinders contain two copies each of APC variants ApcE and ApcD that are considered to be terminal emitters, with superior overlap with the absorption characteristics of chlorophyll *a*. These cylinders are considered to be spatially situated close to the RC and thus are the most likely conduit of energy transfer between the PBS and the RCs.

According to the most widely accepted model, the full PBS structure is composed of 6–8 rods that extend from the core assembly, with each rod containing 2–4 hexamers (Figure 5.2). The PBPs found in the rods absorb at higher energy than APC, and thus it is assumed that energy is funneled down through the rod components into the core. Rods contain 1–2 PBP types: phycocyanin (PC; λ_{max} = 620 nm) that is always found adjacent to the core and in some species, either phycoerythrin (PE; λ_{max} = 560 nm) or phycoerythrocyanin (PEC; λ_{max} = 575 nm) found distal to the core. The fashion of arrangement of the rods onto the cores will be discussed in detail in succeeding text.

Transmission electron microscopy (TEM) of whole cells in thin slices often shows the thylakoid membranes arranged as uneven concentric circles within the cytoplasm [17]. Between every two membrane assemblies (composed of a double membrane enclosing the lumen), PBSs can be seen to nearly cover the entire membrane surface and almost completely fill the intermembrane space (the stroma). Indeed, it has been suggested that the PBS lining the cytoplasmic surface of two adjacent membranes can serve as spacers between the thylakoid membranes [18]. Mutants of the cyanobacterium *Synechocystis* sp. PCC 6803 that are deficient in rod assembly (and thus contain smaller PBSs) exhibit much smaller spacing between adjacent thylakoid membranes. Tight packing of the PBS on the thylakoid membranes has the possibility of bringing each PBS complex to be in close proximity to 4–6 other PBS complexes on a single membrane if arranged in a crystalline type array. This arrangement can also bring any one PBS to be spatially close to additional PBS complexes on the adjacent

PC rods

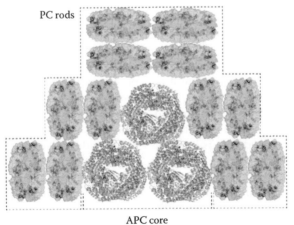

APC core

Figure 5.2 Model of the PBS. The depicted complex is a tricylindrical type. PC rods are shown as in Figure 5.1, while the APC α and β subunits in the cores are in pink and wheat. The small linker protein ApcC is in orange.

membrane [19]. In essence, the highly dense formation described here could afford absorbed energy transfer not only within a single PBS but also between adjacent complexes before arriving at an RC. A mechanism of this type would, however, require efficient energy transfer between rods that is nearly as efficient as energy transfer along the rod, toward the core. All directions of transfer would have to be kinetically more efficient than quenching processes (heat or fluorescence). A possible mechanism for this energy transfer efficiency will be discussed later in further detail.

Understanding the overall structure of the PBS has been addressed by a variety of methodologies. The earliest, most prevalent method is imaging of single molecules by negatively stained TEM [20–23]. In the earliest studies, each complex was visualized separately, resulting in the presentation of what was proposed to represent the most likely PBS structure. In later studies, particle averaging and reconstruction methods were used to obtain higher-resolution images [19,24]. It should be mentioned here that a severe physicochemical drawback exists in all studies of the PBS, and this is its lack of stability in solution upon isolation from the cellular milieu [21,25]. It was shown quite early in the study of the PBS that preservation of the most critical aspect of PBS function, energy transfer from high-energy PBPs to lower-energy PBPs, is dependent on isolation of the complex in high concentrations (>0.75 M) of phosphate buffer (in some cases, phosphate could be replaced by citrate but not simply by high–ionic strength buffers). The need for such a high concentration of phosphate buffer (HPB) was suggested to mimic the crowded natural environment of the PBS in the stroma [26]. The presence of such high concentrations of salts is typically deleterious in EM staining procedures, and thus sample preparation protocols usually include removal of the HPB prior to the application of the staining material. The possibility that both the washing and the staining procedure might have an effect on the resulting image cannot be excluded. It is however quite reasonable that the model of stacked cylinders in the core and the fashion of elongation of the rods by assembled hexamers is indeed correct. The two main structural issues that are unclear are the mode of attachment of the rods onto the core assembly and the fashion by which the unique, large LP, known in the literature as ApcE or L_{CM} [16,27,28], organizes and stabilizes the entire core substructure.

The nomenclature of the LPs is not consistent in the literature, with the name obtained by its source gene (such as *apcE*) or using a subscript representing what is believed to be the two objects linked by the LP (such as core-membrane, in this case). The L_{CM} has molecular weights of between ≈70–120 kDa (in bicylindrical to pentacylindrical cores, respectively) and has been proposed to loop in and out of the core assembly, serving as a template or scaffold for the subcomplex and also to assist in the association of the PBS with the membrane protein complexes (especially PSII). However, it should be reiterated that the stabilization effect of the L_{CM} is completely lost in low–ionic strength buffers, with the core disintegrating rapidly into its representative trimeric rings. The connection between the trimeric rings and all of the LPs is also quickly lost, and in chromatographic separations, the LPs are typically absent. This would appear to indicate that the LPs are hydrophobic, and upon release from the assembly (in the absence of HPB), the LPs precipitate irreversibly. The lack of solubility has hampered the study of the LPs, inhibiting their expression in heterologous systems (such as *Escherichia coli*). For these reasons, structural determination (by X-ray crystallography or NMR) has only been successful for small domains of the larger (>25 kDa) LPs [29]. A single X-ray structure of the complex between an APC trimer and the small core linker (L_c) exists [30]. Other assemblies of PBPs with LPs present have been crystallized; however, due to the highly symmetric nature of the PBP assemblies, the contribution of the LPs to the diffraction is averaged out and the LPs do not appear in the electron density maps [31–34].

While in most cases, the EM images of isolated negatively stained PBS complexes are visually pleasing [2,35], with the rods emanating out in a hemi- or semicircular fan fashion, other experiments have shown much less ordered aggregations. Only a few TEM or AFM studies on the PBS attached to the thylakoid membrane have been performed [36]. The results indicated very tightly arranged complexes. Negative staining on thin slices of cells appears to show ordered two-dimensional arrays covering the membrane surface, while AFM does not indicate the existence of such a high degree of order. We have recently developed a methodology to preserve the functional intactness of the PBS by glutaraldehyde cross-linking [32]. By carefully using the minimal concentration of both cross-linker and PBS (in HPB), we could obtain complexes that showed clear energy transfer from PC to the terminal core emitters, ApcD and L_{CM}, without the presence of HPB. The protocol could be used for PBS from a number of cyanobacterial species, the thermophile *Thermosynechococcus vulcanus (T. vulcanus)*, the mesophile *Synechocystis* sp. PCC 6803, and the chlorophyll *d*–containing *Acaryochloris marina (A. marina)* [37,38]. These stabilized PBS complexes could be utilized for crystallization and X-ray diffraction, negatively stained TEM, and nonstained cryo-TEM. Both negatively stained TEM and single particle reconstruction of 50,000 single particles by cryo-TEM of the stabilized PBS from *T. vulcanus* did not visualize a thin fan-type complex. Rather, a more globular arrangement could be identified (Figure 5.3). Crystals were also obtained from the same material, and diffraction to

2.3 Å was collected. The crystals were shown to contain all components, as identified by mass spectrometry, and confocal fluorescence microscopy showed clear energy transfer from PC to APC. However, the asymmetric unit was quite small, containing a single PC monomer. Reconstruction of the unit cell shows that the PBS rearranged into infinitely long rods, with the APC components probably interspersed between the ordered PC rods. The conclusion of these experiments is quite remarkable—although the PBS appeared in different forms of assembly, energy transfer from PC to APC was remarkably efficient. What could be the source of this apparent robustness?

In the past, it has been shown that the APC trimer absorbs maximally with a bathochromic shift of about 30 nm in comparison to PC trimers [39].

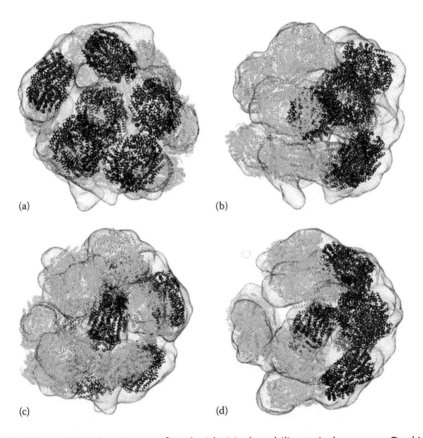

(a)

(b)

(c)

(d)

Figure 5.3 Refined cryo-TEM density map fitted with 14 phycobiliprotein hexamers. Docking into the EM map was performed using the fit module in Chimera. Five hexamers that appear to form a dome-like structure are in blue, eight hexamers that were docked into six elongated protrusions connected to the dome are in green, and a single hexamer docked into a central body is in black. (a) View of the particle from above the dome. (b) View from below the particle, looking onto the aperture formed by the six appendages. (c) and (d) Two side views of the particle. (Adapted from David, L. et al., *Biochim. Biophys. Acta*, 1837, 385, 2014.)

Both APC and PC have identical phycocyanobilin (PCB) chromophores bound at identical positions within the protein. The two PCBs bound at position 84 of both the α and β subunits are also held by the protein in a very similar configuration. A bathochromic shift of this nature is typically indicative of coupled excitonic states between the two chromophores [39–41]. However, as already mentioned earlier, with a center-to-center distance of 2 nm, the coupling should be far too weak to have any major effect on the absorption characteristics of the chromophores. Indeed, in PC trimers that have similar structures and bilin conformations, there is no apparent coupling. It was suggested that the unique layered surroundings of the chromophore pair in APC enable stronger coupling [42]. APC chromophores are tightly held within a rather hydrophobic pocket. This first shell is surrounded by a second, extremely polar/charged shell. In this fashion, the effect of the charges on the energy levels of the excitonic couple is increased. Other shifts in the absorption/emission spectra of PBS components can also be attributed to the effect of the protein environment on the chromophores' excited state levels. The surprising functional robustness of the PBPs might be such that the rod–core interface allows collective effects. All PBPs can be strongly coupled with the threefold symmetry in more than a single association face. Recently, cross-linking/mass spectrometry was used to identify potential nearest-neighbor residues in the PC/APC interface [43]. The results of this analysis were employed to model the interface, taking into account the limitations imposed by the length of the cross-linker. The most prevalent interaction shows that the rods sit flush on the circumference of the core cylinder. The distance between the PC chromophores closest to the interface (β84-PCB) and the nearest APC chromophores is about 3 nm. While rather distant, this energy jump is quite efficient, with minimal energy loss. During the editorial production of this review, a structure of an entire PBS from the Rhodophyta (red alga) *Griffithsia pacifica* was determined by Sui and coworkers at 3.5 Å [44] by cryo-TEM. The PBS structure obtained from this species is much larger than the PBS typically found in cyanobacteria and reached 16.8 MDa, with dimensions of 68 × 39 × 45 nm. The structure includes 862 protein subunits and 2048 bilin chromophores. This truly remarkable study shows the entire assembly of rod and core, with all of the associated linker proteins. A number of facets of

this PBS stand out as different than the "classical" PBS already mentioned. The core contains three APC cylinders, with only three trimers in the basal cylinders and only two trimers on the top cylinder. There are 14 rods (with between 2 and 5 hexamers) surrounding the core, with additional hexamers attached to the rods. Many of the rods exhibit the parallel interactions seen in Figure 5.2, while others extend out at other angles from the central core. Not all of the elements of the rods attach to the core directly, indicating cross-rod energy transfer pathways. The intricate fashion by which the linkers attach to the PBP components shows a remarkable diversity in interaction types. Measurements on intact PBS complexes have shown that energy transfer from the rods to the cores occurs in tens of picoseconds [45], although a recent report claims that it occurs in a few ps at most [46], which is faster than typically measured for long-distance transfer. We believe that the PBPs form an interface that is very powerful in its effect on this critical step in photosynthesis, suggesting strong excitonic coupling. The physical requirements for long-range excitonic coupling will be described in succeeding text.

The final step in the function of the PBS is energy transfer from the terminal emitters of the core (ApcE and ApcD), both of which absorb and emit at about 10–15 nm further to the red than bulk APC [47–49]. The emission of these two components overlaps well with the absorption cross section of chlorophyll *a*, the major chromophores in cyanobacterial RCs [39]. The distance between the terminal PBS chromophores and the nearest chlorophyll molecules is most likely on the order of 4 nm, even further than the separation between PBS components. The remarkable ability of the PBS to tune chromophores was shown by a number of groups [14,37]. An extreme example was studied by Bryant and coworkers [50] in describing the PBS absorption in a unique cyanobacterium, *Leptolyngbya* sp. strain JSC-1, that contains the far-red absorbing chlorophylls *d* and *f*. When grown in red light (710 nm), the PBS of this organism exhibits an absorption feature at 708 nm, a bathochromic shift of 40 nm when compared to the absorption of typical ApcD or ApcE. When grown in "normal" (645 nm) light, this feature is absent. Analysis of the complex indicates that the appearance of the farred absorbing species is due to the expression of a special ApcE gene (ApcE2), which reduces the size of the core (from pentacylindrical to tricylindrical), but more importantly does not contain the conserved cysteine residue needed for covalent attachment of the PCB. The researchers hypothesize that the result is a noncovalently linked PCB with one additional conjugated double bond, leading to a significant redshift in absorption.

5.2 ENERGY TRANSFER TO REACTION CENTERS

Where does the PBS transfer the absorbed energy to? It should be noted that in both cyanobacteria and red algae, the ratio of PSI:PSII is not 1:1 as it is in plants and green algae. The ratio can change, depending on the light intensity and quality, but it has been typically found to be between 2:1 and 10:1 [51–53]. The amount of PBSs appears to be rather invariant, covering the stromal surface of the thylakoid membrane. Since the major pigment of the accessory internal antennas of both PSII and PSI is chlorophyll *a*, it is quite reasonable to assume that energy transfer to both photosystems is possible. It has been shown that the PBS can associate with PSII functionally through interactions with ApcE, and thus it has typically been assumed that the PBS mostly serves as an antenna to PSII. This can also be rationalized by the need to ensure the flow of enough electrons from PSII (which are fewer and have only 35 chlorophyll *a* molecules per RC) to PSI (which are in excess and have nearly 100 chlorophyll

a molecules per RC). Using the said considerations, it is apparent that in most species, the PBS will be found in close contact with both PSII and PSI. A rod-like PBS was isolated in close contact with a unique tetrameric PSI complex in *Anabaena* sp. PCC 7120 [54]. Containing only PC and the rod–core linker CpcG3, it was suggested that this is a "supercomplex" (Figure 5.4). The requirement for a special LP for such associations was also previously suggested by Kondo and coworkers [55]. Liu and coworkers described an *in vivo* cross-link-stabilized complex between the PBS, PSII, and PSI from *Synechocystis* [56]. *In vivo* cross linking, under certain conditions, can allow retention of PBS structures transferring energy to the RC [32,57]. The complex isolated by Liu and coworkers was denoted a "megacomplex" in *Synechocystis* sp. PCC 6803. The complex contained a PSII dimer, a PSI trimer, and an entire PBS, with a molecular mass of over 7.5 MDa (Figure 5.4). The isolated complex was functional in energy transfer to both PSII and PSI, although the kinetic rise of the fluorescence component from PSII was within the minimal time resolution of the measurement (and thus denoted instantaneous), while the fluorescence from PSI was considerably delayed. These authors thus suggested that the connection of the PBS to PSII is tighter and functionally optimized. The slower rate of transfer to PSI could be the result of a looser interaction interface, spillover from PSII, or due to the stabilizing cross-linking reagent.

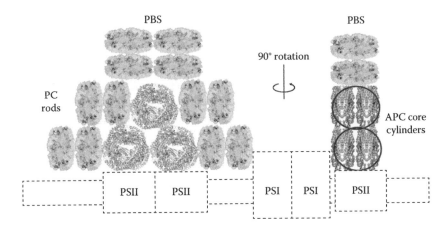

Figure 5.4 Schematic representation of possible associations between the PBS and the photosystems. The major interaction shows the PBS core above a PSII dimer (green). The rods can interact with their distal ends with a trimeric PSI complex (light blue, one PSI removed for clarity). A second PBS complex is shown on the right, rotated 90° with respect to the PBS on the left. Only one rod is shown on top, while the positions of two additional rods that would point toward the viewer are denoted by the red circles. The core cylinders contact both a PSII dimer (green, one PSII shown) and the PSI trimer.

5.3 PHYCOBILIPROTEINS *IN VITRO*: STUDIES OF PC NANOWIRES IN BIO-HYBRID DEVICES

PBSs share structural properties with artificial molecular aggregates. Molecular aggregates are defined as self-assembling clusters of small molecules with intermolecular separations typically close to the dimensions of the individual molecule. Coulomb interactions stabilize their configuration, determine their optical properties, and enable neutral electronic excitation energy to migrate through them. PBSs are, essentially, natural molecular aggregates [58]. The advent of new lithographic techniques [59,60] has opened up the possibilities of controlling the surface arrangements of groups of photosynthetic pigment-protein complexes in order to examine their collective properties for energy propagation [61].

Cyanobacterial PBS components were utilized to construct nanometric wires and bundles [62]. These nanowire bundles transmitted energy through hundreds of pigment–protein complexes over micron distances, enabling the realization of quantum interference components. When *T. vulcanus* PC [34] were dried under appropriate conditions, PC trimers formed bundles of nanowires. Patterning the structure of the dendrites is possible by filling microtrenches created using optical lithography with PBS solution (Figure 5.5a). In these very large–scale PC assemblies, a redshift of the organized sample was detected, as well as exciton lifetime shortening (Figure 5.5b and c). Two-probe near-field scanning optical microscopy (NSOM) measurements provide evidence for energy transfer over micron distances (Figure 5.5d). Such long-distance energy transfer and lifetime shortening indicate strong coupling between the PC units. Coupling can occur within a wire or between adjacent wires. The long-range energy transfer might be due to quantum coherence at room temperature. In any case, this long-range transfer comes at an energetic price. A recent study isolated PC and APC and measured their optical properties in a solution and when aggregated by drying on a glass surface. Absorption measurements showed inhomogeneous spectral broadening. Luminescence measurements exhibited a redshift at dry phase. Most interestingly, while in the wet phase, the PC peak is higher in energy in comparison to APC, while in the dry phase, the order is reversed and the APC peak is higher in energy. These effects were coupled with shortening of the fluorescence lifetime and loss of fluorescence intensity [63]. These results imply a changed energy level structure in the aggregate. Changes in aggregation state may provide the PBS system with yet another mechanism for switching between efficient photosynthesis and energy quenching [63].

5.4 PHYCOBILISOME DYNAMICS

In natural environments, the abiotic environment changes constantly. To sustain photosynthetic life, the structure and function of the PBS antenna must react to these changes. In the following section, we will take a look at "short-term" acclimation processes like state transition and nonphotochemical quenching (NPQ) as well as at "long-term" processes like PBS degradation and chromatic acclimation.

5.4.1 State transition

State transitions enable redistribution of absorbed light between photosystems in response to changes of light quality and intensity. This process provides photosynthetic organisms with the ability to acclimate to changing environmental conditions. The phenomenon was first described in green and red algae ([64,65] respectively) in the 1960s. The state transition allows excitation energy reallocation between PSI and PSII to support efficient energy flow through the photosynthetic apparatus. Moreover, differential energy transfer to either PSI or PSII by PBSs has also been shown to be induced by changes in osmolality [66].

The extent of the state transition differs between photosynthetic organisms. Much of the work in this field was performed using the green alga *Chlamydomonas reinhardtii*, an organism in which the state transition plays a prominent role in photosynthesis [67]. Although components of the photosynthetic apparatus differ between green algae and cyanobacteria [11], the concept of state transition remains similar. When illumination is such that it leads to excess excitation of PSII, a transition from State 1 to State 2, a state favoring excitation of PSI, occurs. When PSI is excited in excess, a transition from State 2 to State 1, favoring excitation of PSII, occurs.

In cyanobacteria, state transition is often measured using dark vs. low light conditions as opposed to the red vs. far-red light protocol used commonly in algae and plants (Figure 5.6). There are a number

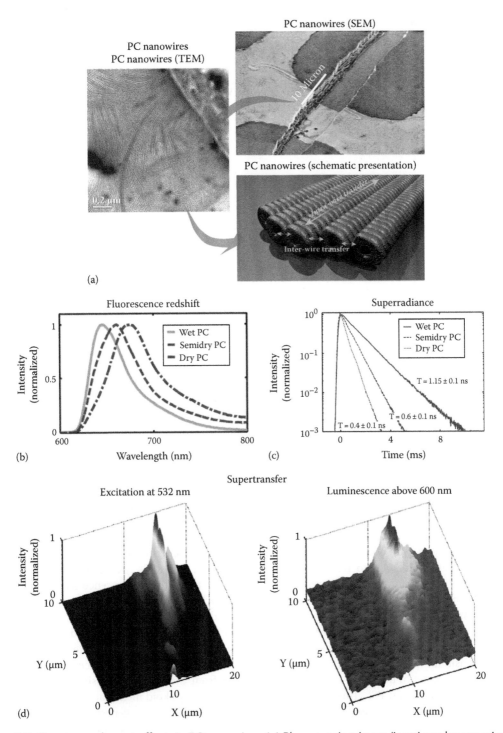

Figure 5.5 Quantum coherent effects in PC nanowires. (a) Phycocyanin trimers (inset) can be organized in "nanowire" structures by self-assembly processes. The figure presents a PC nanowire grown on an etched silicone surface. (b) The fluorescence peak of organized structures of PC is redshifted when compared to solution (wet vs. dry). (c) The fluorescence lifetime in such structures, measured by photon counting in the ns range, is shortened by a factor of ≈3. (d) Two-probe NSOM measurements indicate that the luminescence map of these nanowire (right) structures is broadened by ≈1 μm as compared to the excitation map (left). (Redrawn from Eisenberg, I. et al., *Phys. Chem. Chem. Phys.*, 16, 11245, 2014.)

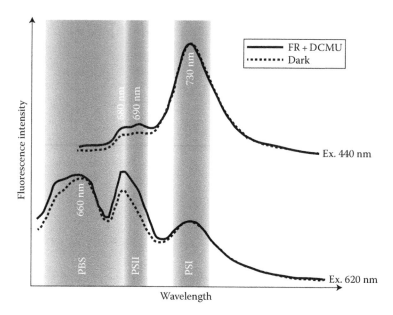

Figure 5.6 Measurements of state transition and of NPQ in cyanobacteria. 77 K fluorescence spectroscopy. In a typical state transition measurement, samples are "State I adapted" by 10 min of far-red illumination followed by the addition of 10 μM 3-(3,4-dichlorophenyl)-1,1-dimethylurea (FR+DCMU; solid line) or State II adapted by dark adaptation (dark, dashed line). In this example, excitations were applied at 440 nm (top, mainly exciting Chl) and 620 nm (bottom, mainly exciting PBSs) wavelengths. Spectra were normalized to the PSI peak emission at 730 nm. In State I, a larger fraction of the energy is distributed to PSII than in State II. The figure schematically represents typical datasets.

of chlorophyll fluorescence–based tools for detecting state transition in cyanobacteria. The state transition was shown to effect the slow (tens of seconds) "S-to-M" transition in fluorescence rise kinetics, for example [68]. The major tool for detecting state transition is measurement of 77 K chlorophyll fluorescence spectra (Figure 5.6). The samples are exposed to a specific light state prior to freezing in liquid nitrogen. Low temperature detection resolves the PSI fluorescence component that cannot be observed at room temperature. Shifts in the relative intensity of the PSI and PSII peaks record the distribution of excitation energy between the photosystems.

State transitions in cyanobacteria are likely to be prompted by changes in the redox state of electron carriers between PSII and PSI. This was shown by Mullineaux and Allen, who, in their experiment, used various chemical and illumination treatments to modify the redox state of the electron transport chain. They found that state transitions occur in response to a change in the redox state of the plastoquinone pool and therefore concluded that they are controlled by the redox state of plastoquinone or a closely associated electron carrier [69].

Two mechanisms have been proposed for the state transition, the first involving PBSs, in which case PBSs transfer energy differentially to either photosystem according to light conditions [70]; the second mechanism involves energy transfer between RC chlorophylls and is described as the spillover of energy from PSII directly to PSI [71]. Diffusion of PBSs between photosystems was shown, using the fluorescence recovery after photobleaching (FRAP) method, to be required for PBS-dependent state transitions [72]. However, this matter is still debated, and it has also been reported that illumination intensity used in the FRAP method may cause excitonic decoupling of PBS from the RCs and possibly even detachment from the thylakoid membrane [73,74].

A random mutagenesis approach used by Emlyn-Jones and coworkers identified genes essential for state transition in cyanobacteria. State transition–deficient mutants all appeared to have one defective gene—*rpaC*. These mutants are locked in State 1 [75]. The only documented phenotype of *rpaC* deletion mutants is the lack of state transitions, a phenotype expressed uniquely under low light intensities and wavelengths relevant to PBSs. This led Mullineaux and Emlyn-Jones to conclude that state transitions in cyanobacteria are only physiologically important under low light, allowing optimization of light absorption by the photosystems [76]. Dong et al. [77] described a PBS terminal core emitter protein, ApcD, as being required for both the state transition and protection against photodamage under high-light conditions. ApcD is required for direct energy transfer from PBSs to PSI [77]. Mutants lacking ApcD were unable to perform state transitions and were more sensitive to photoinhibition when grown under green light, which is preferentially absorbed by the PBSs. This effect is likely due to an inability to redirect the light to PSI via a transition to State 2 [77].

Evidence for the structural basis of state transition can be drawn from the analysis of *in vivo* cross-linking and LC-MS results. A megacomplex formed in cyanobacteria by PBSs, PSII, and PSI has been described [56]. Beyond the structural implications discussed earlier, time-resolved fluorescence spectroscopy measurements indicate that energy transfer occurs from the attached PBS core to both photosystems. Modulations of this structure may provide a basis for the state transition phenomenon [78].

5.4.2 Nonphotochemical quenching

Fluorescence quenching terminology was developed for higher plants where it is defined as a decline in the maximal fluorescence (Fm) following a transition from dark to light [79]. Fluorescence quenching is composed of a number of components, the state transition discussed previously being one of them.

While state transition is a dark/low-light phenomenon, NPQ takes place under excess light conditions. Under these conditions, the risk of photoinhibitory damage increases [80,81]. With NPQ, the risk is mitigated by releasing a larger fraction of the absorbed light energy as heat [82,83]. Different mechanisms were proposed for the NPQ phenomenon in plant and algae, including the effect of pH, carotenoid de-epoxidation, and the involvement of the PsbS protein [83,84]. In cyanobacteria, where PsbS and the LHCII antenna are missing, NPQ is driven by different mechanisms. An in-depth discussion of these mechanisms is included in Chapter 11 by Pinnola and coauthors.

5.4.3 PBS degradation

Under extreme conditions, cyanobacteria degrade their PBS antenna. The most studied condition in this respect is nitrogen limitation. The initial observation was that nondiazotrophic cyanobacteria lose their typical blue-green color under nitrogen limitation [85]. This effect was termed nitrogen bleaching and was found to be caused by the breakdown of PC. The bleaching could be reversed by the addition of nitrogen salts to the growth media [85]. The process of PBS breakdown and recovery following nitrogen supplementation, its kinetic properties, and molecular control mechanisms have been subjected to intensive research conducted with different cyanobacterial strains [86–95]. The common feature reported in these studies was a massive, regulated breakdown of PBS on a time scale of a few hours to 2 days following a nitrogen step-down. Degradation of the PBS is catalyzed by the NblA protein. Activation of the NblA degradation pathway is under the control of the response regulator NblR [95]. It is a small polypeptide that mediates the complete degradation of PBS complexes under nitrogen limitation [86,88]. NblA has no proteolytic activity by itself. The NblA monomer has a helix-loop-helix motif, which dimerizes into an open, four-helical bundle [96]. It was suggested that this structure allows NblA to interact with the PBS via "structural mimicry" due to similarity in structural motifs found in all PBPs. NblA was found to be present in close association with PBS complexes *in vivo*, and the current hypothesis is that its function is in tagging with the PBS complex for degradation [97], possibly through the action of the Clp protease complex [98].

Degradation of PBS was suggested to allow for the reallocation of nitrogen from this highly abundant protein to other critical pathways in the cell [99]. At the same time, nitrogen limitation grinds biosynthetic mechanisms to a halt, resulting in overreduction of photosynthetic electron carrier pools [100]. Under nitrogen limitation, metabolic processes slow down and shift from an anabolic to a catabolic metabolism, represented by the degradation of proteins, chlorophyll, and other cellular components [95]. These conditions expose the photosynthetic apparatus to damage from reactive oxygen species. Decreasing the PBS absorption cross section alleviates this stress by reducing excitation pressure. Reduced PBS degradation under a combination of nitrogen limitation and a low-light condition provides evidence for this hypothesis [101].

5.4.4 Chromatic acclimation

An additional acclimation mechanism used by many cyanobacterial species is the synthesis of different PBPs or their associated chromophores for specific wavelengths. This phenomenon is referred to as complementary chromatic acclimation. Cyanobacteria capable of chromatic acclimation must perceive the wavelengths of absorbed light and respond by induction of signal transduction pathways, which lead to the synthesis of the relevant PBPs and linker polypeptides [102,103]. Regulation of the mechanisms of chromatic acclimation has recently been further elucidated, with the identification of genes involved in the process as well as their mode of action [104–106].

The prominent PBPs involved in the process are PC and two types of phycoerythrin (PE), PEI and PEII. PC accumulates in cultures grown under red light, while PE accumulates in cultures grown under green light. Not all cyanobacterial species expressing both PC and PE genes are capable of chromatic acclimation, and the mechanism also requires expression of structural and regulatory genes for this purpose [102]. Initially, three groups of responses to shifts between green and red light were identified in cyanobacteria expressing both PC and PE. In Group I, PC and PE content do not change in response to the light wavelength; Group II exhibits a partial response to changing light wavelength, regulating only synthesis of PE; and Group III, to which most cyanobacteria expressing both PC and PE belong, exhibits a more comprehensive response, showing differential accumulation of either PC or PE in accordance with the light wavelength [107]. More recent work on chromatic acclimation has shown that it includes a response to a wider range of wavelengths. Response to blue and green light, common among the marine cyanobacterial species *Synechococcus*, does not involve changes in the PBP composition of the PBS rods but rather of the chromophores associated with PEII [108]. The PBS structure of cyanobacteria from hot spring bacterial mats was shown to be able to adjust to far-red light [50].

Chromatic acclimation is controlled by photoreceptors rather than by photosynthesis. Group III chromatic acclimation is regulated by two light-responsive pathways: Rca and Cgi. The Rca system regulates transcription under red-light conditions (transcription of PC-related genes and repression of PE-related genes). It is controlled by the phytochrome-class photoreceptor RcaE that has been suggested to act as a kinase in red light and phosphatase in green light. The Cgi system also operates under red-light conditions by posttranscriptionally repressing PE-related genes, apparently via transcription attenuation, a process in which the translation initiation factor IF3 has been suggested to take part. Group II chromatic acclimation is controlled by a phytochrome-class photoreceptor-based two-component system, CcaS/R, which is capable of sensing both red and green light [105,106,108]. Type 4 chromatic acclimation is still only poorly understood; however, progress has been made, beginning with the identification of MpeZ, an enzyme involved in ligation and isomerization of one of the PEII-linked chromophores [104].

REFERENCES

1. Adir, N. 2005. Elucidation of the molecular structures of components of the phycobilisome: Reconstructing a giant. *Photosynth. Res.* 85: 15–32.
2. Glazer, A. 1989. Light guides directional energy transfer in a photosynthetic antenna. *J. Biol. Chem.* 264: 1–4.
3. Grossman, A.R., D. Bhaya, and Q. He. 2001. Tracking the light environment by cyanobacteria and the dynamic nature of light harvesting. *J. Biol. Chem.* 276: 11449–11452.
4. Marx, A., L. David, and N. Adir. 2014. Piecing together the phycobilisome. In: *The Structural Basis of Biological Energy Generation.* Hohmann-Marriott MF, ed. Springer, Dordrect, the Netherlands. pp. 59–76.
5. Watanabe, M. and M. Ikeuchi. 2013. Phycobilisome: Architecture of a light-harvesting supercomplex. *Photosynth. Res.* 116: 265–276.
6. Samsonoff, W.A. and R. MacColl. 2001. Biliproteins and phycobilisomes from cyanobacteria and red algae at the extremes of habitat. *Arch. Microbiol.* 176: 400–405.
7. Cheng, Y.-C. and G.R. Fleming. 2009. Dynamics of light harvesting in photosynthesis. *Annu. Rev. Phys. Chem.* 60: 241–262.

8. Cogdell, R.J., A.T. Gardiner, A.W. Roszak, C.J. Law, J. Southall, and N.W. Isaacs. 2004. Rings, ellipses and horseshoes: How purple bacteria harvest solar energy. *Photosynth. Res.* 81: 207–214.

9. Croce, R. and H. van Amerongen. 2013. Light-harvesting in photosystem I. *Photosynth. Res.* 116: 153–166.

10. van Amerongen, H. and R. Croce. 2013. Light harvesting in photosystem II. *Photosynth. Res.* 116: 251–263.

11. Hohmann-Marriott, M.F. and R.E. Blankenship. 2011. Evolution of photosynthesis. *Annu. Rev. Plant Biol.* 62: 515–548.

12. Pastore, A. and A.M. Lesk. 1990. Comparison of the structures of globins and phycocyanins: Evidence for evolutionary relationship. *Proteins* 8: 133–155.

13. Schirmer, T., W. Bode, R. Huber, W. Sidler, and H. Zuber. 1985. X-ray crystallographic structure of the light-harvesting biliprotein C-phycocyanin from the thermophilic cyanobacterium *Mastigocladus laminosus* and its resemblance to globin structures. *J. Mol. Biol.* 184: 257–277.

14. Scheer, H. and K.-H. Zhao. 2008. Biliprotein maturation: The chromophore attachment. *Mol. Microbiol.* 68: 263–276.

15. Marsac, N.T., L. De, K. Bogorad, D. Boresch, G. Bryant, N. Cohen-bazire, Y. Esenbeck et al. 2003. Phycobiliproteins and phycobilisomes: The early observations. *Photosynth. Res.* 76: 197–205.

16. Liu, L.-N., X.-L. Chen, Y.-Z. Zhang, and B.-C. Zhou. 2005. Characterization, structure and function of linker polypeptides in phycobilisomes of cyanobacteria and red algae: An overview. *Biochim. Biophys. Acta* 1708: 133–142.

17. Edwards, M. and E. Gantt. 1971. Phycobilisomes of the thermophilic blue-green alga *Synechococcus lividus. J. Cell Biol.* 50: 896–900.

18. Liberton, M., L.E. Page, W.B. O'Dell, H. O'Neill, E. Mamontov, V.S. Urban, and H.B. Pakrasi. 2013. Organization and flexibility of cyanobacterial thylakoid membranes examined by neutron scattering. *J. Biol. Chem.* 288: 3632–3640.

19. Arteni, A.A., L.-N. Liu, T.J. Aartsma, Y.-Z. Zhang, B.-C. Zhou, and E.J. Boekema. 2008. Structure and organization of phycobilisomes on membranes of the red alga *Porphyridium cruentum. Photosynth. Res.* 95: 169–174.

20. Gantt, E. and S. Conti. 1966. Granules associated with the chloroplast lamellae of *Porphyridium cruentum. J. Cell Biol.* 29: 423–434.

21. Gantt, E. and C. Lipschultz. 1972. Phycobilisomes of *Porphyridium cruentum* I. Isolation. *J. Cell Biol.* 54: 313–324.

22. Bryant, D., A. Glazer, and F. Eiserling. 1976. Characterization and structural properties of the major biliproteins of *Anabaena* sp. *Arch. Microbiol.* 75: 61–75.

23. Marsac, N. De and G. Cohen-Bazire. 1977. Molecular composition of cyanobacterial phycobilisomes. *Proc. Natl. Acad. Sci. USA* 74: 1635–1639.

24. Arteni, A.A., G. Ajlani, and E.J. Boekema. 2009. Structural organisation of phycobilisomes from *Synechocystis* sp. strain PCC6803 and their interaction with the membrane. *Biochim. Biophys. Acta* 1787: 272–279.

25. Zilinskas, B. and R. Glick. 1981. Noncovalent intermolecular forces in phycobilisomes of Porphyridium cruentum. *Plant Physiol.* 68: 447–452.

26. Stagg, L., S.-Q. Zhang, M.S. Cheung, and P. Wittung-Stafshede. 2007. Molecular crowding enhances native structure and stability of alpha/beta protein flavodoxin. *Proc. Natl. Acad. Sci. USA* 104: 18976–18981.

27. Zhao, K.-H., P. Su, S. Böhm, B. Song, M. Zhou, C. Bubenzer, and H. Scheer. 2005. Reconstitution of phycobilisome core-membrane linker, LCM, by autocatalytic chromophore binding to ApcE. *Biochim. Biophys. Acta* 1706: 81–87.

28. Lundell, D., R. Williams, and A. Glazer. 1981. Molecular architecture of a light-harvesting antenna. *In vitro* assembly of the rod substructures of *Synechococcus* 6301 phycobilisomes. *J. Biol. Chem.* 256: 3580–3592.

29. Gao, X., N. Zhang, T.-D. Wei, H.-N. Su, B.-B. Xie, C.-C. Dong, X.-Y. Zhang et al. 2011. Crystal structure of the N-terminal domain of linker L(R) and the assembly of cyanobacterial phycobilisome rods. *Mol. Microbiol.* 82: 698–705.

30. W. Reuter, G. Wiegand, R. Huber, and M.E. Than. 1999. Structural analysis at 2.2 Å of orthorhombic crystals presents the asymmetry of the allophycocyanin–linker complex, AP·LC7.8, from phycobilisomes of *Mastigocladus laminosus. Proc. Natl. Acad. Sci. USA* 96: 1363–1368.

31. Chang, W., T. Jiang, Z. Wan, and J. Zhang. 1996. Crystal structure of R-phycoerythrin from *Polysiphonia urceolata* at 2.8 Å resolution. *J. Mol. Biol.* 262: 721–731.

32. David, L., M. Prado, A. Arteni, D. Elmlund, R.E. Blankenship, and N. Adir. 2014. Structural studies show energy transfer within stabilized phycobilisomes independent of the mode of rod-core assembly. *Biochim. Biophys. Acta* 1837: 385–395.

33. Contreras-Martel, C., J. Martinez-Oyanedel, M. Bunster, P. Legrand, C. Piras, X. Vernede, and J.C. Fontecilla-Camps. 2001. Crystallization and 2.2 A resolution structure of R-phycoerythrin from *Gracilaria chilensis*: A case of perfect hemihedral twinning. *Acta Crystallogr. D Biol Crystallogr.* 57: 52–60.

34. David, L., A. Marx, and N. Adir. 2011. High-resolution crystal structures of trimeric and rod phycocyanin. *J. Mol. Biol.* 405: 201–213.

35. Anderson, L. and C. Toole. 1998. A model for early events in the assembly pathway of cyanobacterial phycobilisomes. *Mol. Microbiol.* 30: 467–474.

36. Liu, L.-N., T.J. Aartsma, J.-C. Thomas, G.E.M. Lamers, B.-C. Zhou, and Y.-Z. Zhang. 2008. Watching the native supramolecular architecture of photosynthetic membrane in red algae: Topography of phycobilisomes and their crowding, diverse distribution patterns. *J. Biol. Chem.* 283: 34946–34953.

37. Chen, M., M. Floetenmeyer, and T.S. Bibby. 2009. Supramolecular organization of phycobiliproteins in the chlorophyll d-containing cyanobacterium *Acaryochloris marina*. *FEBS Lett.* 583: 2535–2539.

38. Hu, Q., I. Iwasaki, H. Miyashita, N. Kurano, E. Mo, and S. Miyachi. 1999. Molecular structure, localization and function of biliproteins in the chlorophyll a/d containing oxygenic photosynthetic prokaryote *Acaryochloris marina*. *Biochim. Biophys. Acta* 1412: 250–261.

39. MacColl, R. 2004. Allophycocyanin and energy transfer. *Biochim. Biophys. Acta* 1657: 73–81.

40. Holzwarth, A. and E. Bittersmann. 1990. Studies on chromophore coupling in isolated phycobiliproteins: III. Picosecond excited state kinetics and time-resolved fluorescence spectra of different. *Biophys. J.* 57: 133–145.

41. Csatorday, K., R. MacColl, and V. Csizmadia. 1984. Exciton interaction in allophycocyanin. *Biochemistry* 23: 6466–6470.

42. McGregor, A., M. Klartag, L. David, and N. Adir. 2008. Allophycocyanin trimer stability and functionality are primarily due to polar enhanced hydrophobicity of the phycocyanobilin binding pocket. *J. Mol. Biol.* 384: 406–421.

43. Tal, O., B. Trabelcy, Y. Gerchman, and N. Adir. 2014. Investigation of phycobilisome subunit interaction interfaces by coupled cross-linking and mass spectrometry. *J. Biol. Chem.* 289: 33084–33097.

44. Zhang, J., J. Ma, D. Liu, S. Qin, S. Sun, J. Zhao, and S.-F. Sui. 2017. Structure of phycobilisome from the red alga *Griffithsia pacifica*. *Nature* 551(7678): 57–63.

45. Tian, L., I.H.M. Van Stokkum, R.B.M. Koehorst, A. Jongerius, D. Kirilovsky, and H. Van Amerongen. 2011. Site, rate, and mechanism of photoprotective quenching in cyanobacteria. *J. Am. Chem. Soc.* 133: 18304–18311.

46. Nganou, C., L. David, N. Adir, and M. Mkandawire. 2015. Linker proteins enable ultrafast excitation energy transfer in the phycobilisome antenna system of *Thermosynechococcus vulcanus*. *Photochem. Photobiol. Sci.* 15: 31–44.

47. Glazer, A., C. Chan, R. Williams, S. Yeh, and J. Clark. 1985. Kinetics of energy flow in the phycobilisome core. *Science* 230: 16–18.

48. Capuano, V. and A. Braux. 1991. The" anchor polypeptide" of cyanobacterial phycobilisomes. Molecular characterization of the *Synechococcus* sp. PCC 6301 apce gene. *J. Biol. Chem.* 266: 7239–7247.

49. Glazer, A. and D. Bryant. 1975. Allophycocyanin B (lambdamax 671, 618 nm): A new cyanobacterial phycobiliprotein. *Arch. Microbiol.* 104: 15–22.

50. Gan, F., S. Zhang, N.C. Rockwell, S.S. Martin, J.C. Lagarias, and D.A. Bryant. 2014. Extensive remodeling of a cyanobacterial photosynthetic apparatus in far-red light. *Science* 345: 1312–1317.

51. Fraser, J.M., S.E. Tulk, J. Jeans, D. Campbell, T.S. Bibby, and A.M. Cockshutt. 2013. Photophysiological and photosynthetic complex changes during iron starvation in *Synechocystis* sp. PCC 6803 and *Synechococcus elongatus* PCC 7942. *PLoS One* 8: e59861.

52. Sonoike, K., Y. Hihara, and M. Ikeuchi. 2001. Physiological significance of the regulation of photosystem stoichiometry upon high light acclimation of *Synechocystis* sp. PCC 6803. *Plant Cell Physiol.* 42: 379–384.

53. Kawamura, M., M. Mimuro, and Y. Fujita. 1979. Quantitative relationship between two reaction centers in the photosynthetic system of blue-green algae. *Plant Cell Physiol.* 20: 697–705.

54. Watanabe, M., D.A. Semchonok, M.T. Webber-Birungi, S. Ehira, K. Kondo, R. Narikawa, M. Ohmori, E.J. Boekema, and M. Ikeuchi. 2014. Attachment of phycobilisomes in an antenna–photosystem I supercomplex of cyanobacteria. *Proc. Natl. Acad. Sci. USA* 111: 2512–2517.

55. Kondo, K., Y. Ochiai, M. Katayama, and M. Ikeuchi. 2007. The membrane-associated CpcG2-phycobilisome in *Synechocystis*: A new photosystem I antenna. *Plant Physiol.* 144: 1200–1210.

56. Liu, H., H. Zhang, D.M. Niedzwiedzki, M. Prado, G. He, M.L. Gross, and R.E. Blankenship. 2013. Phycobilisomes supply excitations to both photosystems in a megacomplex in cyanobacteria. *Science* 342: 1104–1107.

57. Papageorgiou, G. 1977. Photosynthetic activity of diimidoester-modified cells, permeaplasts, and cell-free membrane fragments of the blue-green alga *Anacystis nidulans*. *Biochim. Biophys. Acta* 461: 379–391.

58. van Grondelle, R. and V.I. Novoderezhkin. 2006. Energy transfer in photosynthesis: Experimental insights and quantitative models. *Phys. Chem. Chem. Phys.* 8: 793–807.

59. Falconnet, D., D. Pasqui, S. Park, R. Eckert, H. Schift, J. Gobrecht, R. Barbucci, and M. Textor. 2004. A novel approach to produce protein nanopatterns by combining nanoimprint lithography and molecular self-assembly. *Nano Lett.* 4: 1909–1914.

60. Escalante, M., P. Maury, C.M. Bruinink, K. van der Werf, J.D. Olsen, J. A. Timney, J. Huskens, C. Neil Hunter, V. Subramaniam, and C. Otto. 2008. Directed assembly of functional light harvesting antenna complexes onto chemically patterned surfaces. *Nanotechnology* 19: 25101.

61. Escalante, M., A. Lenferink, Y. Zhao, N. Tas, J. Huskens, C.N. Hunter, V. Subramaniam, and C. Otto. 2010. Long-range energy propagation in nanometer arrays of light harvesting antenna complexes. *Nano Lett.* 10: 1450–1457.

62. Eisenberg, I., S. Yochelis, R. Ben-Harosh, L. David, A. Faust, N. Even-Dar, H. Taha et al. 2014. Room temperature biological quantum random walk in phycocyanin nanowires. *Phys. Chem. Chem. Phys.* 16: 11245–11250.

63. Eisenberg, I., F. Caycedo-Soler, D. Harris, S. Yochelis, S.F. Huelga, M.B. Plenio, N. Adir, N. Keren, and Y. Paltiel. 2017. Regulating the energy flow in a cyanobacterial light harvesting antenna complex. *J. Phys. Chem. B* 121: 1240–1247.

64. Murata, N. 1969. Control of excitation transfer in photosynthesis I. Light-induced change of chlorophyll a fluorescence in *Porphyridium cruentum*. *Biochim. Biophys. Acta* 172: 242–251.

65. Bonaventura, C. and J. Myers. 1969. Fluorescence and oxygen evolution from *Chlorella pyrenoidosa*. *Biochim. Biophys. Acta* 172: 366–383.

66. Stamatakis, K. and G. Papageorgiou. 2001. The osmolality of the cell suspension regulates phycobilisome-to-photosystem I excitation transfers in cyanobacteria. *Biochim. Biophys. Acta* 1506: 172–181.

67. Rochaix, J.-D., S. Lemeille, A. Shapiguzov, I. Samol, G. Fucile, A. Willig, and M. Goldschmidt-Clermont. 2012. Protein kinases and phosphatases involved in the acclimation of the photosynthetic apparatus to a changing light environment. *Philos. Trans. R. Soc. Lond. Ser. B Biol. Sci.* 367: 3466–3474.

68. Kaňa, R., E. Kotabová, O. Komárek, B. Sedivá, G.C. Papageorgiou, Govindjee, and O. Prášil. 2012. The slow S to M fluorescence rise in cyanobacteria is due to a state 2 to state 1 transition. *Biochim. Biophys. Acta* 1817: 1237–1247.

69. Mullineaux, C. and J. Allen. 1990. State 1-state 2 transitions in the cyanobacterium *Synechococcus* 6301 are controlled by the redox state of electron carriers between photosystems I and II. *Photosynth. Res.* 23: 297–311.

70. van Thor, J. and C. Mullineaux. 1998. Light harvesting and state transitions in cyanobacteria. *Bot. Acta* 111: 430–443.

71. Federman, S., S. Malkin, and A. Scherz. 2000. Excitation energy transfer in aggregates of Photosystem I and Photosystem II of the cyanobacterium *Synechocystis* sp. PCC 6803: Can assembly of the pigment-protein complexes control the extent of spillover? *Photosynth. Res.* 64: 199–207.

72. Joshua, S. and C. Mullineaux. 2004. Phycobilisome diffusion is required for light-state transitions in cyanobacteria. *Plant Physiol.* 135: 2112–2119.

73. Tamary, E., V. Kiss, R. Nevo, Z. Adam, G. Bernat, S. Rexroth, M. Rogner, and Z. Reich. 2012. Structural and functional alterations of cyanobacterial phycobilisomes induced by high-light stress. *Biochim. Biophys. Acta* 1817: 319–327.

74. Chukhutsina, V., L. Bersanini, E.-M. Aro, and H. van Amerongen. 2015. Cyanobacterial light-harvesting phycobilisomes uncouple from Photosystem I during dark-to-light transitions. *Sci. Rep.* 5: 14193.

75. Emlyn-Jones, D., M.K. Ashby, and C.W. Mullineaux. 1999. A gene required for the regulation of photosynthetic light harvesting in the cyanobacterium *Synechocystis* 6803. *Mol. Microbiol.* 33: 1050–1058.

76. Mullineaux, C. and D. Emlyn-Jones. 2005. State transitions: An example of acclimation to low-light stress. *J. Exp. Bot.* 56: 389–393.

77. Dong, C., A. Tang, J. Zhao, C.W. Mullineaux, G. Shen, and D.A. Bryant. 2009. ApcD is necessary for efficient energy transfer from phycobilisomes to photosystem I and helps to prevent photoinhibition in the cyanobacterium *Synechococcus* sp. PCC 7002. *Biochim. Biophys. Acta* 1787: 1122–1128.

78. Acuña, A.M., J.J. Snellenburg, M. Gwizdala, D. Kirilovsky, R. Van Grondelle, and I.H.M. Van Stokkum. 2016. Resolving the contribution of the uncoupled phycobilisomes to cyanobacterial pulse-amplitude modulated (PAM) fluorometry signals. *Photosynth. Res.* 127: 91–102.

79. Krause, G. and E. Weis. 1991. Chlorophyll fluorescence and photosynthesis: The basics. *Annu. Rev. Plant Biol.* 42: 313–349.

80. Keren, N. and A. Krieger-Liszkay. 2011. Photoinhibition: Molecular mechanisms and physiological significance. *Physiol. Plant.* 142: 1–5.

81. Tikkanen, M., N.R. Mekala, and E.-M. Aro. 2014. Photosystem II photoinhibition-repair cycle protects Photosystem I from irreversible damage. *Biochim. Biophys. Acta* 1837: 210–215.

82. Papageorgiou, G.C. and Govindjee. 2011. Photosystem II fluorescence: Slow changes—Scaling from the past. *J. Photochem. Photobiol. B* 104: 258–270.

83. Horton, P. 2012. Optimization of light harvesting and photoprotection: Molecular mechanisms and physiological consequences. *Philos. Trans. R. Soc. Lond. Ser. B Biol. Sci.* 367: 3455–3465.

84. Li, Z., S. Wakao, B.B. Fischer, and K.K. Niyogi. 2009. Sensing and responding to excess light. *Annu. Rev. Plant Biol.* 60: 239–260.

85. Allen, M.M. and A.J. Smith. 1969. Nitrogen chlorosis in blue-green algae. *Arch. Mikrobiol.* 69: 114–120.

86. Baier, K., S. Nicklisch, C. Grundner, J. Reinecke, and W. Lockau. 2001. Expression of two nblA-homologous genes is required for phycobilisome degradation in nitrogen-starved *Synechocystis* sp. PCC6803. *FEMS Microbiol. Lett.* 195: 35–39.

87. Collier, J. and A. Grossman. 1992. Chlorosis induced by nutrient deprivation in *Synechococcus* sp. strain PCC 7942: Not all bleaching is the same. *J. Bacteriol.* 174: 4718–4726.

88. Collier, J. and A. Grossman. 1994. A small polypeptide triggers complete degradation of light-harvesting phycobiliproteins in nutrient-deprived cyanobacteria. *EMBO J.* 13: 1039–1047.

89. Elmorjani, K. and M. Herdman. 1987. Metabolic control of phycocyanin degradation in the cyanobacterium *Synechocystis* PCC 6803: A glucose effect. *J. Gen. Microbiol.* 133: 1685–1694.

90. Gilbert, S. and G. Allison. 1996. Expression of genes involved in phycocyanin biosynthesis following recovery of *Synechococcus* PCC 6301 from nitrogen starvation, and the effect of gabaculine on cpcBa transcript levels. *FEMS Microbiol.* 140: 93–98.

91. Görl, M., J. Sauer, T. Baier, and K. Forchhammer. 1998. Nitrogen-starvation-induced chlorosis in *Synechococcus* PCC 7942: Adaptation to long-term survival. *Microbiology* 144: 2449–2458.

92. Krasikov, V., E. Aguirre von Wobeser, H.L. Dekker, J. Huisman, and H.C.P. Matthijs. 2012. Time-series resolution of gradual nitrogen starvation and its impact on photosynthesis in the cyanobacterium *Synechocystis* PCC 6803. *Physiol. Plant.* 145: 426–439.

93. Lau, R., M. MacKenzie, and W. Doolittle. 1977. Phycocyanin synthesis and degradation in the blue-green bacterium *Anacystis nidulans*. *J. Bacteriol.* 132: 771–778.

94. Sauer, J., U. Schreiber, and R. Schmid. 2001. Nitrogen starvation-induced chlorosis in *Synechococcus* PCC 7942. Low-level photosynthesis as a mechanism of long-term survival. *Plant Physiol.* 126: 233–243.

95. Schwarz, R. and A. Grossman. 1998. A response regulator of cyanobacteria integrates diverse environmental signals and is critical for survival under extreme conditions. *Proc. Natl. Acad. Sci. USA* 95: 11008–11013.

96. Dines, M., E. Sendersky, L. David, R. Schwarz, and N. Adir. 2008. Structural, functional, and mutational analysis of the NblA protein provides insight into possible modes of interaction with the phycobilisome. *J. Biol. Chem.* 283: 30330–30340.

97. Sendersky, E., N. Kozer, M. Levi, Y. Garini, Y. Shav-Tal, and R. Schwarz. 2014. The proteolysis adaptor, NblA, initiates protein pigment degradation by interacting with the cyanobacterial light-harvesting complexes. *Plant J.* 79: 118–126.

98. Baier, A., W. Winkler, T. Korte, W. Lockau, and A. Karradt. 2014. Degradation of phycobilisomes in *Synechocystis* sp. PCC6803: Evidence for essential formation of an NblA1/NblA2 heterodimer and its codegradation by A Clp protease complex. *J. Biol. Chem.* 289: 11755–11766.

99. Boussiba, S. and A. Richmond. 1980. C-phycocyanin as a storage protein in the blue-green alga Spirulina platensis. *Arch. Microbiol.* 147: 143–147.

100. Schwarz, R. and K. Forchhammer. 2005. Acclimation of unicellular cyanobacteria to macronutrient deficiency: Emergence of a complex network of cellular responses. *Microbiology* 151: 2503–2514.

101. Salomon, E., L. Bar-Eyal, S. Sharon, and N. Keren. 2013. Balancing photosynthetic electron flow is critical for cyanobacterial acclimation to nitrogen limitation. *Biochim. Biophys. Acta* 1827: 340–347.

102. Kehoe, D.M. and A. Gutu. 2006. Responding to color: The regulation of complementary chromatic adaptation. *Annu. Rev. Plant Biol.* 57: 127–150.

103. Grossman, A. 1990. Chromatic adaptation and the events involved in phycobilisome biosynthesis. *Plant Cell Environ.* 13: 651–666.

104. Shukla, A., A. Biswas, N. Blot, F. Partensky, J.A. Karty, L.A. Hammad, L. Garczarek, A. Gutu, W.M. Schulchter, and D.M. Kehoe. 2012. Phycoerythrin-specific bilin lyase–isomerase controls blue-green chromatic acclimation in marine *Synechococcus*. *Proc. Natl. Acad. Sci. USA* 109: 20136–20141.

105. Bezy, R.P., L. Wiltbank, and D.M. Kehoe. 2011. Light-dependent attenuation of phycoerythrin gene expression reveals convergent evolution of green light sensing in cyanobacteria. *Proc. Natl. Acad. Sci. USA* 108: 18542–18547.

106. Gutu, A., A.D. Nesbit, A.J. Alverson, J.D. Palmer, and D.M. Kehoe. 2013. Unique role for translation initiation factor 3 in the light color regulation of photosynthetic gene expression. *Proc. Natl. Acad. Sci. USA* 110: 16253–16258.

107. Marsac, N.D. and J. Houmard. 1988. Complementary chromatic adaptation: Physiological conditions and action spectra. *Methods Enzymol.* 167: 318–328.

108. Gutu, A. and D.M. Kehoe. 2012. Emerging perspectives on the mechanisms, regulation, and distribution of light color acclimation in cyanobacteria. *Mol. Plant* 5: 1–13.

6

Photosynthetic apparatus in purple bacteria

DAVID J. MOTHERSOLE, DAVID A. FARMER, ANDREW HITCHCOCK, AND C. NEIL HUNTER

6.1 PURPLE PHOTOTROPHIC BACTERIA

Purple phototrophs are the most metabolically versatile organisms on Earth and display a huge diversity of energy modes and metabolic capabilities (Hunter et al. 2009). They can dispense with photosynthesis and grow as heterotrophs in the dark; they can use organic acids, amino acids, fatty acids, alcohols, carbohydrates, C1, or aromatic compounds; they can ferment or perform anaerobic respiration; and they can fix nitrogen and CO_2 and produce hydrogen. Purple phototrophs have been invaluable models for investigating the assembly, structure, and function of photosynthetic membranes because of their amenability to genetic manipulation, the level of biochemical, structural, and spectroscopic characterization, and the relative simplicity of their assembly pathways and photosynthetic apparatus. This chapter will focus on the most heavily studied purple phototroph, *Rhodobacter (Rba.) sphaeroides*.

6.1.1 *Rhodobacter sphaeroides*

Rba. sphaeroides is a Gram-negative, rod-shaped bacterium, which has been isolated from anoxic zones of water at the bottom of deep lakes, soil, mud, sludge, sewage (Siefert et al. 1978), and waste lagoons (Cooper et al. 1975). *Rba. sphaeroides* requires no unusual conditions for growth and is capable of growing rapidly in liquid media in the laboratory under both anaerobic photoheterotrophic and aerobic chemotrophic conditions. It has a small, fully sequenced, and well-annotated genome (Choudhary et al. 2007; Kontur et al. 2012), and rapid and specific genomic manipulations can be performed using various molecular genetic techniques. Most of the open reading frames within the photosynthesis gene cluster (PGC) have been functionally assigned (Coomber et al. 1990; Naylor et al. 1999) and the carotenoid and bacteriochlorophyll

(BChl) biosynthesis pathways have been well studied, as well as the genes encoding apoproteins of the photosynthetic complexes. The ability to dispense with photosynthesis and grow chemoheterotrophically makes *Rba. sphaeroides* a valuable model for studies of photosynthesis. Genes essential for phototrophic growth can be deleted or altered, allowing in-depth studies of photosystem assembly and function that are not possible with obligate phototrophs.

6.1.2 The photosynthetic apparatus in purple phototrophs utilizes light to generate ATP

BChl and carotenoid pigments attached to light-harvesting apoproteins absorb photons, and the excitation energy is transferred between pigment molecules in the peripheral light-harvesting complexes, called LH2, until it reaches the belt of LH1 units that surrounds a specialized pigment–protein complex, the reaction center (RC). Here, the excitation energy is transduced to a photochemical charge separation (Figure 6.1),

Figure 6.1 Basic concept of energy transfer from pigmented light-harvesting complexes to reaction centers. Photons (yellow arrow) are absorbed by pigment molecules in the peripheral light-harvesting complexes (green) creating excited states that migrate from pigment to pigment and complex to complex until a reaction center (blue) is reached via a closely attached antenna complex (red). The reaction center conserves the excitation energy as a charge separation.

and an electron then traverses the membrane within the RC, arriving at an exchangeable quinone, Q_B. Two successive excitations within the RC, coupled with the arrival of two protons from the cytoplasmic side of the membrane, convert the Q_B quinone to a quinol, which migrates to the cytochrome bc_1 complex, where a proton-motive force is generated. The adenosine triphosphate (ATP) synthase consumes the proton gradient generating ATP, the energy currency of the cell (Figure 6.2). Repeated turnovers of this cyclic system are made possible by the membrane-extrinsic cytochrome c_2 and the membrane-intrinsic quinone/quinol (Q/QH$_2$) molecules, each shuttling between the RC–LH1 and cytochrome bc_1 complexes (Lavergne et al. 2009). The whole "photon to ATP" process has been summarized in a movie of bacterial photosynthesis (https://www.youtube.com/watch?v=cUHxfpPkN6E), which arises from *in silico* modeling studies of bacterial membrane vesicles (see Section 6.4.3).

6.2 COMPONENTS OF THE BACTERIAL PHOTOSYNTHETIC APPARATUS

These complexes are termed either LH2 for the peripheral antenna or LH1 for the complexes that receive energy from LH2. They were purified for the first time many years ago (Clayton and Clayton 1972; Broglie et al. 1980), and the sequences of their polypeptides were determined by Brunisholz and Zuber (1992), who showed that the antenna complexes of purple bacteria comprise a 1:1 ratio of two membrane-spanning hydrophobic polypeptides known as α and β, each consisting of 50–60 amino acids (Brunisholz and Zuber 1992). The function of these polypeptides is to determine the position, orientation, and environment for the light-harvesting BChl and carotenoid pigments. Carotenoids are found in abundance in photosynthetic organisms and have two major functions in photosynthesis. First, as accessory light-harvesting pigments they absorb light energy in

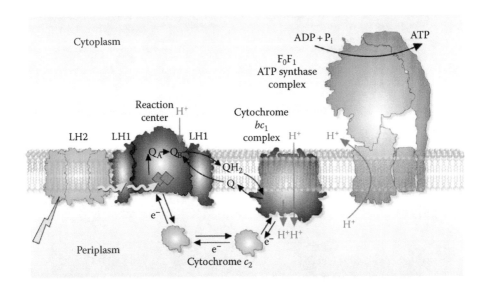

Figure 6.2 Diagram of the transfers of excitation energy, electrons, and protons in photosynthetic membranes of *Rba. sphaeroides*. The diagram shows an LH2 antenna complex (green) absorbing light and transferring the excitation energy to a RC–LH1–PufX complex (blue RC, red LH1). Subsequent electron and quinone transfers (black arrows) reduce a cytochrome bc_1 complex (purple), and a proton-motive force is generated (orange arrows). The ATP synthase (orange) consumes the proton-motive force and produces ATP. The mobile electron carrier cytochrome c_2 (cyan) accepts an electron from the cytochrome bc_1 complex, migrates to the RC, and transfers an electron to reduce the oxidized primary donor, resetting the cyclic system so the RC can begin another charge separation.

the 450–600 nm range and transfer it to neighboring BChl molecules, thereby increasing the spectral range over which light energy can be absorbed beyond the limitations of BChl absorption. Second, as photoprotective agents carotenoids play a vital role in preventing photooxidative damage by quenching singlet oxygen directly, or by quenching the triplet excited BChl (^{3}BChl a^*) sensitizer, preventing the production of singlet oxygen (Frank and Cogdell 1996).

6.2.1 Peripheral light-harvesting LH2 complex

The structure of the LH2 complex from *Rhodoblastus* (*Rbl.*) *acidophilus* (formerly *Rhodopseudomonas acidophila*) is shown in Figure 6.3. A detailed discussion of the properties of this complex is beyond the scope of this chapter, and the reader is referred to some reviews (Cogdell et al. 2006; Gabrielsen et al. 2009; Saer and Blankenship 2017). The LH2 structure beautifully illustrates how nature has complied with the need to pack different types of pigment, both BChl and carotenoid, within a single complex in order to absorb light from 400 to 900 nm while also redshifting and tuning the absorption of the BChls in order to create absorption bands centered at both 800 and 850 nm, known as B800 and B850, respectively. Moreover, the spacings and orientations of the transition dipoles of the pigments allow for fast and efficient couplings between the B800 and B850 pigments and especially between the 18 BChls that comprise the B850 ring of pigments (Freer et al. 1996).

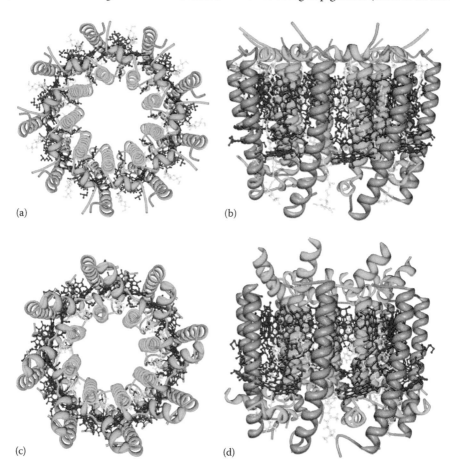

(a) (b) (c) (d)

Figure 6.3 The LH2 complexes of *Rhodoblastus acidophilus* and *Phaeospirillum molischianum*. In these models, the components have been colored as follows: LH2 α-polypeptides in green, LH2 β polypeptides in cyan, B850 BChl in pink, B800 BChl in red, and carotenoids in yellow. (**a,b**) LH2 from *Rbl. acidophilus* (previously *Rps. acidophila*) (Prince et al. 1997, PDB ID 1KZU). (**c,d**) LH2 from *Phs. molischianum* (previously *Rsp. molischianum*) (Koepke et al. 1996, PDB ID 1LGH). (**a,c**) Projection views of the periplasmic side of each complex. (**b,d**) Views of each complex in the plane of the membrane.

The BChls are bound to the α and β transmembrane polypeptides. The B850 BChls form an 18-membered ring of overlapping molecules coordinated to a circular array of α- and β-polypeptides via histidine residues; hydrogen bonding networks also stabilize BChl–protein interactions and modify the light-absorbing properties of the pigments (Fowler et al. 1992, 1994). The B850 BChls are positioned vertically with respect to the membrane plane and are situated toward the periplasmic face of the membrane. In contrast, the BChls that comprise the ring of nine B800 BChls are relatively well separated and are positioned between the helices of the β-polypeptides in the outer ring, lying toward the cytoplasmic face of the membrane and almost parallel to it. Nine membrane-spanning carotenoids traverse the membranes, making close contacts with both the B800 and B850 BChls (Freer et al. 1996).

Nine αβ heterodimers comprise the ring structures of the LH2 complexes of *Rbl. acidophilus* (McDermott et al. 1995; Prince et al. 1997), *Rba. sphaeroides* (Walz et al. 1998), *Rubrivivax gelatinosus* (Scheuring et al. 2001), and *Rhodospirillum (Rsp.) photometricum* (Scheuring et al. 2004), while eight αβ heterodimers are found in the LH2 complexes of *Phaeospirillum (Phs.) molischianum* (Koepke et al. 1996), *Rhodovulum sulfidophilum* (Savage et al. 1996), and *Rhodopseudomonas (Rps.) palustris* (Hartigan et al. 2002). There are complicating factors such as the presence of both octameric and nonameric LH2 complexes in *Rps. palustris* and *Rsp. photometricum* (Scheuring 2006a; Scheuring et al. 2004, 2005) as well as mixed LH2 rings containing more than one type of α and β polypeptide (Brotosudarmo et al. 2009; Mascle-Allemand et al. 2010). However, in all cases, the αβ heterodimers form a hollow cylinder that spans the membrane.

It is noteworthy that purple bacteria often encode multiple LH2 or LH2-like polypeptides (LH3 and LH4), resulting in variations of the peripheral antenna complexes with distinct spectral properties (Cogdell et al. 1983; Gardiner et al. 1993; McLuskey et al. 2001; Hartigan et al. 2002; Gabrielsen et al. 2009; Fixen et al. 2016; Magdaong et al. 2016).

6.2.2 RC–LH1 core complexes

The topic of bacterial "core" complexes has been reviewed (Bullough et al. 2009), and it will not be covered in detail here. Briefly, core complexes can be monomeric, with an RC surrounded by a continuous ring of 16 LH1 αβBChl$_2$ units or with an LH1 ring of 15 LH1 αβBChl$_2$ units interrupted by a PufX polypeptide. Sometimes, two such interrupted rings form a dimeric complex. Complexes with a continuous LH1 ring include the RC–LH1 complex of *Rsp. rubrum* (Jamieson et al. 2002), *Rps.* (now *Blastochloris*) *viridis* (Scheuring et al. 2003; Fotiadis et al. 2004), *Rsp. photometricum* (Scheuring and Sturgis 2005), and *Phs. molischianum* (Goncalves et al. 2005). Monomeric complexes with an RC surrounded by an LH1 (αβ)$_{15}$ ring plus a PufX homolog are found in *Rps. palustris* (Roszak et al. 2003) and *Rba. veldkampii* (Busselez et al. 2007). Finally, dimeric complexes with two RCs intertwined with an (αβ)$_{28}$ S-shaped LH1 complex and including two PufX polypeptides are found in *Rba. sphaeroides* and *Rba. blasticus* (Qian et al. 2005; Scheuring et al. 2005).

The level of structural information available for the RC–LH1 cores from *Rsp. rubrum* and *Rps. palustris* has only been sufficient to reveal the overall architecture of the complexes rather than the positions of amino acid sidechains (Jamieson et al. 2002; Roszak et al. 2003). There are likely to be difficulties in obtaining highly diffracting crystals when one of the major components, in this case LH1, is known to be flexible and able to adopt a number of conformations (Bahatyrova et al. 2004b). The 4.8 Å structure from X-ray crystallography of the *Rps. palustris* complex (Figure 6.4) shows a monomeric structure with the RC surrounded by an LH1 ring comprising 15 αβ heterodimers (Roszak et al. 2003). This LH1 ring is interrupted by a polypeptide termed W, proposed to be a homolog of the PufX polypeptide found in the dimeric RC–LH1–PufX core complex of *Rba. sphaeroides* (Qian et al. 2005). The structure of this dimeric complex (Qian et al. 2013) was based on a number of inputs, including an 8 Å electron density map from X-ray crystallography, site-directed mutagenesis (Olsen et al. 1994, 1997; Sturgis et al. 1997), mass spectrometry analysis, the solution structures of LH1 β and PufX (Conroy et al. 2000; Tunnicliffe et al. 2006), the 2.65 Å structure of the RC (Ermler et al. 1994), the 8.5 Å electron microscopy (EM) projection map of the complex (Qian et al. 2005), and the structures of the LH2 complex (McDermott et al. 1995; Koepke et al. 1996). Figure 6.5 shows that each RC is surrounded by an inner ring of 14 LH1 α polypeptides, an outer ring of 14 LH1 β polypeptides, and two BChls sandwiched between each αβ pair of transmembrane helices. The LH1 αβ units form

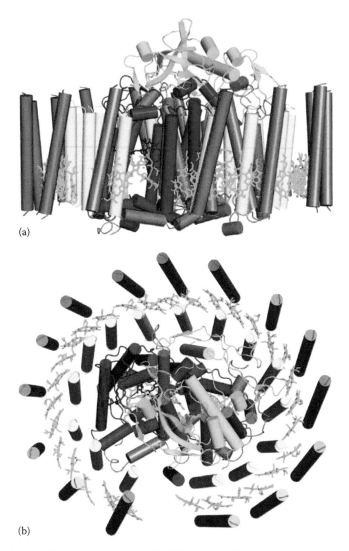

Figure 6.4 The *Rhodopseudomonas palustris* RC–LH1–protein W complex. The LH1α polypeptides are shown in yellow, the LH1β polypeptides in blue, the helix W polypeptide in red, and BChl a molecules in green. The RC-L, M, and H subunits are in orange, magenta, and cyan, respectively. **(a)** View of the complex in the plane of the membrane with the cytoplasmic face uppermost. **(b)** Projection view of the complex, perpendicular to the membrane. Model generated using PDB ID 1PYH. (From Roszak, A.W. et al., *Science*, 302, 1969, 2003.)

an arc that opens out slightly near to the RC Q_B site, allowing room for several quinones. Extractions of purified core dimer complexes showed the presence of 10–15 quinones per RC (Dezi et al. 2007). PufX is positioned to prevent the LH1 αβBChl$_2$ units from surrounding the RC, which otherwise could hinder the traffic of the quinones and quinols that shuttle between the Q_B site, the external quinone pool, and the cytochrome bc_1 complex.

Figure 6.5 also shows how LH1 BChls surround and interconnect the two RCs in the dimeric core complex. Excitation energy transfer from LH1 BChls to the RC "special pair" has a time constant of 37 ps at 77 K (Visscher et al. 1989) because of the distance from LH1 BChls to the RC special pair, which is 47 ± 2 Å. For comparison, energy transfer from LH2 to LH1 complexes is approximately 3 ps (Hess et al. 1995b). The two halves of the complexes bring the two arcs of LH1 BChls sufficiently close that migration of excited states round all 56 BChls is likely. A high degree of energy transfer connectivity between the two halves of the complex was

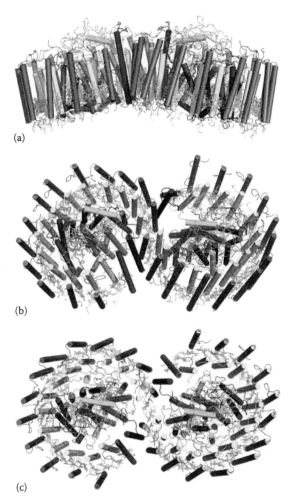

(a)

(b)

(c)

Figure 6.5 The dimeric RC–LH1–PufX complex from *Rhodobacter sphaeroides*. Color coded as in Figure 6.4. The PufX polypeptides are in red. **(a)** View of the complex in the plane of the membrane with the cytoplasmic face uppermost. **(b)** Tilted view of the cytoplasmic face. **(c)** Projection view of the cytoplasmic face of the complex. Model generated using PDB ID 4V9G. (From Qian, P. et al., *Biochemistry*, 52, 7575, 2013.)

shown by measuring the fluorescence yield of isolated dimeric complexes (Comayras et al. 2005). It was suggested that excitation sharing between the RCs might maintain efficient charge separation under high light by allowing energy to migrate from the half of the dimer with an already photo-oxidized ("closed") RC to the side with an "open" RC (Sener et al. 2009).

The thermostable RC–LH1 complex from *Thermochromatium* (*Tch.*) *tepidum* allowed the production of high-quality 3D crystals, and the

resulting 3.0 Å structure showed a detailed model of a core complex for the first time, revealing a closed LH1 ring of 16 αβ heterodimers enclosing a heterodimeric RC (Figure 6.6). Interestingly, this complex required no PufX/protein W to provide a portal for exporting quinols across the LH1 barrier surrounding the RC. Instead, the lack of carotenoids in the *Tch. tepidum* LH1 complex relative to the complex in *Rba. sphaeroides*, for example, was proposed to provide a series of small channels for quinol/quinone exchange (Niwa et al. 2014).

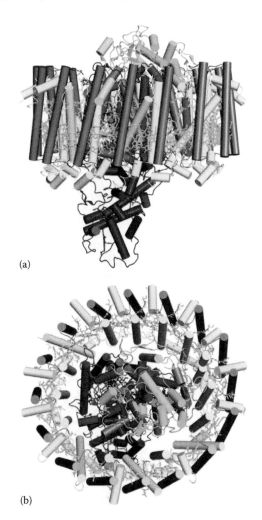

(a)

(b)

Figure 6.6 The *Thermochromatium tepidum* RC–LH1 complex. Color coded as in Figure 6.4. The cytochrome subunit is shown in dark red. **(a)** View of the complex in the plane of the membrane with the cytoplasmic face uppermost. **(b)** Projection view of the complex, perpendicular to the membrane. Model generated using PDB ID 4V8K. (From Niwa, S. et al., *Nature*, 508, 228, 2014.)

6.2.3 The reaction center

RCs are the site of energy transduction, converting excitation energy to electron transfers. RCs in *Rba. sphaeroides* consist of three subunits known as L (light), M (medium), and H (heavy), historically named according to their apparent molecular weights as determined by SDS-PAGE (Clayton and Haselkorn 1972). Their true masses are significantly different with H having the lowest mass, followed by L and M, respectively (Williams et al. 1986). The L and M subunits both contain five transmembrane helices and are related by pseudotwofold symmetry; the H subunit is significantly less hydrophobic and only has one membrane-spanning helix with the bulk of its mass forming a globular domain on the cytoplasmic side of the membrane.

High-resolution structures have been determined for RCs from two species of purple bacteria. The first structure was solved for the *Blastochloris* (*Blc.*) *viridis* RC, which consists of four subunits (Deisenhofer et al. 1985, 1995; Deisenhofer and Michel 1989); the second is that of *Rba. sphaeroides* (Figure 6.7) obtained and refined by numerous

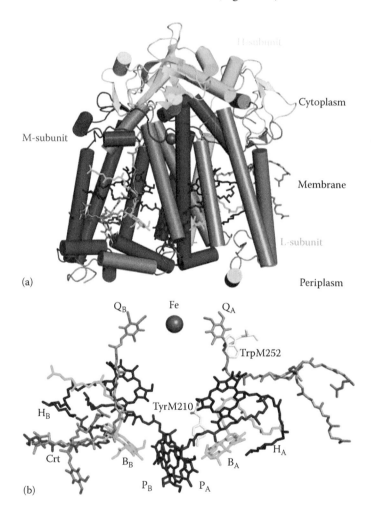

Figure 6.7 The reaction center of *Rhodobacter sphaeroides*. (a) Model of the reaction center showing the L, M, and H subunits in orange, magenta, and cyan, respectively. Pigments are shown in the same colors as (b). (b) RC cofactors. P_A and P_B (magenta) are the "special pair" BChls; the two accessory BChls B_A and B_B are in green and bacteriopheophytins H_A and H_B are in dark blue. The quinones Q_A and Q_B are in orange, and the carotenoid (spheroidene) is in pink. The TyrM210 and TrpM252 aromatic residues (yellow) and the Fe atom are also shown. Model generated using PDB ID 3I4D.

groups (Chang et al. 1986, 1991; Allen et al. 1987a,b; Ermler et al. 1994; McAuley et al. 2000; Katona et al. 2003) with 2.01 Å as the highest-resolution structure available at present (PDB entry 3I4D). The fourth subunit of the *Blc. viridis* RC, cytochrome *c*, "hardwires" electron transfer to the photooxidized complex; in *Rba. sphaeroides* and many other RCs, this function is performed by the membrane-extrinsic, periplasmic protein cytochrome c_2. Noncovalently bound cofactors associated with the complex (Figure 6.7) include two molecules of BChl *a* known as the special pair, two accessory molecules of BChl, two molecules of bacteriopheophytin, one Fe^{2+} ion, two quinones (ubiquinone), and one carotenoid. A detailed account of RCs can be found in Chapter 9.

6.3 FUNCTIONS OF THE PHOTOSYSTEM, CYTOCHROME, AND ATP SYNTHASE COMPONENTS IN BACTERIAL PHOTOSYNTHETIC MEMBRANES

6.3.1 Excitation energy transfer within LH1 and LH2

Energy transfer dynamics within LH1 and LH2 after the absorption of photons by pigment molecules within the complexes are ultrafast (van Grondelle et al. 1994). Energy transfer between B800 and B850 in LH2 from both *Rba. sphaeroides* and *Rbl. acidophilus* takes approximately 650–800 fs at room temperature (Jimenez et al. 1996; Joo et al. 1996; Ma et al. 1997); this is reduced to a transfer time of approximately 1.2 ps at 77 K (Monshouwer et al. 1995). Polarizing pump-probe spectroscopy measurements concluded that transfer time between two neighboring B800 molecules in LH2 complexes was approximately 0.7–1.25 ps (Monshouwer et al. 1995; Jimenez et al. 1996). The transfer time between B850 molecules in LH2 is approximately 110 fs (Jimenez et al. 1996). Energy transfer between B875 molecules in LH1 from *Rba. sphaeroides*, identified by ultrafast fluorescence depolarization and annihilation studies, is reported to have a hopping time of approximately 80–100 fs (Bradforth et al. 1995). A detailed summary of excitation energy transfers within and between these complexes can be found in Chapters 10 through 16 (theory) and 17 through 20 (spectroscopy).

6.3.2 Excitation energy transfer to the reaction center

Excitation energy is transferred from LH2 to LH2, LH2 to LH1, and finally from LH1 to the RC. The time constant for energy transfer between LH2 complexes is approximately 5 ps; transfer between LH2 and LH1 complexes is approximately 3.3–4.6 ps (Hess et al. 1995a,b; Nagarajan and Parson 1997; Agarwal et al. 2002). The rate-limiting step in excitation trapping is the transfer of excitation energy from LH1 BChls to the RC $(BChl)_2$ dimer, which is approximately 35 ps at 77 K (Beekman et al. 1994; Hunter et al. 1988; Visscher et al. 1989). This relatively long timescale for excitation energy transfer reflects the distance between the LH1 and RC special pair, which is approximately 4.7 nm (Qian et al. 2008). An advantage of this rate-limiting step is the reduction in the probability of back transfer from the RC to LH1, with the early events in charge separation occurring within 1 ps at room temperature.

6.3.3 Transduction of excitation energy to electron flow in the reaction center

Electron flow in the RC has been the subject of many reviews including (Deisenhofer and Michel 1991; Hoff and Deisenhofer 1997; Bixon and Jortner 1999;) and is covered in detail in Chapter 9. Briefly, upon excitation of the RC $(BChl)_2$ special pair $(P_A P_B)$, an electron is promoted to an excited state $(P_A P_B^*)$, then transferred via the accessory BChl (B_A) to reduce a bacteriopheophytin (H_A). This process takes approximately 2.8 ps and produces the charge-separated state $P_A P_B^+ H_A^-$ (Martin et al. 1986). The adjacent quinone Q_A is reduced approximately 200 ps later, forming the Q_A^- semiquinone (Holzapfel et al. 1990), and within 200 µs, the more stable Q_B^- semiquinone has formed.

Meanwhile, the soluble periplasmic electron carrier cytochrome c_2, already reduced by receiving an electron from the cytochrome bc_1 complex, has diffused to the RC. X-ray structures show that the cytochrome c_2 docks onto the periplasmic surface of this complex (Axelrod et al. 2002). The reduced cytochrome c_2 donates an electron to the charge-separated state $P_A P_B^+$, returning it to the ground state $P_A P_B$, a process that takes approximately 0.9 µs. The $(BChl)_2$ special pair is then ready to be excited by transfer of excitation energy from LH1 once again. A second photon initiates transfer of a second electron to the exchangeable Q_B; this quinone is then reduced to its quinol form by the simultaneous uptake of two protons from the cytoplasm to form Q_BH_2. Q_BH_2 dissociates from the RC, and the Q_B site is refilled with a quinone from the pool that is sequestered close to the RC by the encircling LH1 ring (Dezi et al. 2007; Qian et al. 2008, 2013).

6.3.4 Completion of electron flow through the cytochrome bc_1 complex and generation of a proton-motive force

Quinols produced by RC photochemistry diffuse to a membrane complex, ubihydroquinone: cytochrome c oxidoreductase, or the cytochrome bc_1 complex (see Figure 6.2). This multisubunit membrane protein complex is composed of cytochrome b, cytochrome c_1, and the Rieske protein, which contains a 2Fe–2S cluster (Hunte et al. 2007). X-ray crystal structures of the cytochrome bc_1 from *Rba. sphaeroides* and *Rba. capsulatus* show that the complex forms an intertwined homodimer (Berry et al. 2009).

The cycling of electrons and protons through the cytochrome bc_1 complex is known as the Q-cycle (Mitchell 1976). On entering the catalytic site, Q_O, of the cytochrome bc_1 complex QH_2 releases its electrons into two different electron-acceptor chains; the first is released to the Rieske iron–sulfur cluster and the heme of cytochrome c_1; the second electron is delivered via heme b_L and heme b_H to the Q_i site where it reduces a quinone to a stable semiquinone. Oxidation of a second molecule of QH_2 to Q via the Q_O site reduces the semiquinone located in the Q_i site to QH_2, which is then released back into the membrane quinone pool. The net oxidation of one QH_2 results in the release of four protons into the periplasm, and generates reduced cytochrome c_2 (Berry et al. 2004; Esser et al. 2006, 2007). Each photooxidation event at the RC requires an electron from reduced cytochrome c_2 to reset the RC to its "dark," reduced state. Thus, repeated shuttling of cytochrome c_2 and Q/QH_2 between the cytochrome bc_1 and RC

complexes sustains the cyclic electron transfer process and generates a proton-motive force. Solar energy has been conserved in a stable and versatile form that can be used either to drive the production of ATP (see Section 6.3.5) or directly for endergonic process such as flagellar rotation or for reversed electron transport to generate NADH/NADPH for biosynthesis.

6.3.5 Formation of ATP

ATP is the major currency of energy in living cells. ATP synthase catalyzes the synthesis of ATP from adenosine diphosphate (ADP) and inorganic phosphate (P_i). The majority of ATP in *Rba. sphaeroides* is synthesized by the F_1F_0 ATP synthase utilizing the proton gradient generated by the cytochrome bc_1 complex.

The F_1 component of the ATP synthase is a water-soluble protein complex consisting of the subunit composition $\alpha_3\beta_3\gamma\delta\varepsilon$. The F_0 component is a membrane-embedded complex composed of the subunits ab_2c_{10-15}. The components are physically connected by a central stalk composed of the γ and ε subunits and a peripheral stalk composed of the δ and b subunits. The F_1 component catalyzes ATP synthesis using a rotary mechanism, while the F_0 component facilitates proton translocation across the membrane (Feniouk and Junge 2009).

6.4 ORGANIZATION OF THE BACTERIAL PHOTOSYNTHETIC APPARATUS

6.4.1 Organization of light harvesting and reaction center complexes

Diffusion processes are fundamental for every level of photosynthesis, from migration of excitations in the light-harvesting antenna to the percolation of quinones and quinols in the membrane bilayer between the RC and the cytochrome bc_1 complexes. These processes are assisted by the supramolecular organization of photosynthetic complexes, and the light-harvesting antenna has evolved to optimize the absorption of solar energy by packing a number of repeating light-absorbing complexes closely together in the membrane to ensure their connectivity for excitation transfer. Various forms of microscopy have revealed the organization of large assemblies of membrane proteins. EM is suitable for imaging whole membranes but has been most valuable when there is a degree of order in the arrangement of complexes (Siebert et al. 2004). Atomic force microscopy (AFM) of membranes under liquid has the advantage of avoiding staining or freezing of the sample, and it can be applied to membranes with no discernible order. AFM studies of bacterial photosynthetic membranes has been reviewed extensively (Scheuring 2006b; Sturgis et al. 2009; Liu and Scheuring 2013), and only a very brief summary is presented here.

The application of AFM has been profoundly important for studies of photosynthetic membrane organization in purple phototrophs; topographs of native membranes have revealed the architecture of the photosynthetic apparatus in different photosynthetic bacteria, such as *Blc. viridis* (Scheuring et al. 2003), *Rsp. photometricum* (Scheuring and Sturgis 2005), *Rba. sphaeroides* (Bahatyrova et al. 2004a; Olsen et al. 2008; Tucker et al. 2010; Adams and Hunter 2012), and *Rba. blasticus* (Scheuring et al. 2005). In some cases, it has been possible to augment AFM data with spectroscopic data on intact membranes by exploiting the ability of polarized light to interrogate the organization of the BChls contained within the LH and RC complexes; for example, Frese and coworkers used linear dichroism to demonstrate the long-range organization of pigment–protein complexes in the membranes from *Rba. sphaeroides* (Frese et al. 2000, 2004). Membranes containing only monomeric core complexes such as those of *Blc. viridis*, adopt a hexagonal packing (Scheuring et al. 2003), while those with monomeric cores and LH2 do not have a fixed stoichiometry; analysis of AFM topographs recorded on membranes from *Rps. photometricum* showed that there were LH2–LH2, core–core, and core–LH2 contacts with varying stoichiometries (Scheuring et al. 2004). The intracytoplasmic membranes (ICMs) of *Rba. sphaeroides*, however, showed a degree of organization with rows of 4–5 dimeric RC–LH1–PufX core complexes associated with LH2-enriched domains (Bahatyrova et al. 2004a). The AFM topographs in Figure 6.8

show examples of partitioning into LH2-enriched and RC–LH1–PufX-enriched domains, the origin of which lies in the membrane curvature of these complexes. The 146° bend in the RC–LH1–PufX dimer imposes curvature on the membrane to the extent of promoting the assembly of tubular membranes in the absence of LH2 (Hunter et al. 1988; Jungas et al. 1999; Siebert et al. 2004; Qian et al. 2008; Hsin et al. 2009). Monte Carlo simulations showed that the size and curvature mismatch between the LH2 and core dimer complexes was a likely driver of domain formation and membrane curvature in the ICM of *Rba. sphaeroides* (Frese et al. 2008).

6.4.2 Organization of the cytochrome bc_1 complex

Cyclic electron flow in purple phototrophs is sustained by quinols migrating from the RC to the cytochrome bc_1 complexes, carrying their cargo of protons and electrons. Oxidized quinones then make the return journey to the RC. This process can be monitored following excitation of RCs by short flashes of light and recording the consequent optical changes in the b and c_1-type cytochromes. Such experiments, conducted over many years, produced various models of the association between RC–LH1 and cytochrome bc_1 complexes, with some suggesting a close physical linkage with fixed stoichiometry and others, a looser and less constrained relationship. The coupling between RC–LH1 and cytochrome bc_1 complexes is reviewed in Lavergne et al. (2009).

In order to investigate this problem in a direct way, EM and AFM were used to image *Rba. sphaeroides* membranes where the cytochrome bc_1 complexes had been labelled with nanogold beads (Cartron et al. 2014). This work showed that the majority of the cytochrome bc_1 complexes are positioned adjacent to RC–LH1–PufX complexes. Biochemical fractionation of these membranes ruled out a fixed stoichiometric association between cytochrome bc_1 and RC–LH1–PufX complexes, however. The proximity between these complexes suggests that quinol/quinone shuttling is a short-range process confined to cytochrome bc_1–RC–LH1–PufX nanodomains within the bacterial photosynthetic membrane, as originally proposed by Comayras et al. (2005) on the basis of their kinetic experiments.

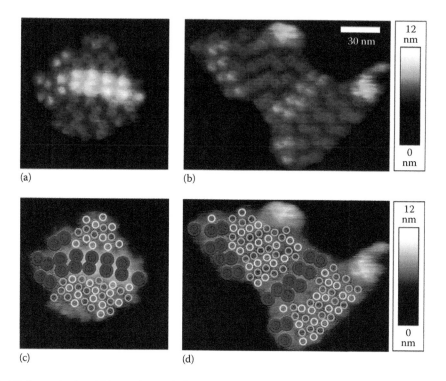

Figure 6.8 High-resolution AFM analysis of photosynthetic membranes from *Rhodobacter sphaeroides*. **(a,b)** AFM topographs of two patches of membranes prepared from cells grown under low-light conditions. The Z-range is shown to the right. The small ring features are LH2 complexes, observed in LH2-rich domains. Alternating complexes of high and low topology originate from the flattening of natively curved membranes onto the flat mica surface (Olsen et al. 2008). RC–LH1–PufX core complexes with the cytoplasmic side uppermost can be identified by their protruding RC-H subunits, which appear as bright spots. **(c,d)** Schematic analyses corresponding to **(a,b)** respectively. LH2 complexes of high and low topology are denoted by dark and light green rings, respectively. Dimeric RC–LH1–PufX complexes are shown with RCs (blue) surrounded by LH1 (red). Occasionally, monomeric complexes can also be seen.

6.4.3 *In silico* modeling studies of bacterial membrane vesicles

The availability of structural information for the major components of the bacterial photosynthetic apparatus, as well as AFM-derived images of membranes, provided the opportunity to construct models of complete membranes in order to understand the behavior of complex sequential processes such as energy migration and trapping, which potentially involve many hundreds of LH2 and RC–LH1 complexes. The model of a spherical ICM from *Rba. sphaeroides* (Sener et al. 2007; Sener and Schulten 2009) was generated using structural (NMR, EM, X-ray crystallography, and AFM) inputs and data from spectroscopy and molecular genetics studies. The original atomic-level membrane model

consists of 18 dimeric RC–LH1 complexes and 101 LH2 complexes, representing a total of 4464 BChls (Sener et al. 2007). This *in silico* membrane model was used to calculate a 50 ps excitation lifetime and a 95% quantum efficiency, in agreement with experimental data. Computational studies have also been applied to energy migration in membranes of *Rsp. photometricum* (Fassioli et al. 2009), building on earlier AFM membrane mapping studies (Scheuring and Sturgis 2005).

The awareness that capturing light to generate ATP involves a concerted series of energy and electron transfer steps, and the generation of a proton gradient and its controlled dissipation to power the ATP synthase, prompted a systems treatment of the ICM vesicle (Geyer and Helms 2006). Such a systems approach was recently combined with the EM and AFM data on the location of the cytochrome bc_1 complex to produce a 1.9 million-atom structural model of an ICM vesicle (chromatophore) of *Rba. sphaeroides*. This model (Figure 6.9) shows the arrangement of the major membrane complexes involved in conversion of solar energy into ATP. The stoichiometries in the model were established from spectroscopic measurements and quantitative mass spectrometry; the vesicle comprises 67 LH2 complexes, 11 LH1–RC–PufX dimers, 2 RC–LH1–PufX monomers, 4 cytochrome bc_1 dimers, and 2 ATP synthases. This model allows computation of the processes of energy harvesting and transfer, electron transfer, formation of the proton-motive force, and ATP synthesis; it shows that the architecture of the chromatophore vesicle is optimized for growth of *Rba. sphaeroides* at low light intensities with half-maximal ATP turnover at only 1% of bright sunlight (Cartron et al. 2014; Sener et al. 2016). The conversion of harvested solar energy to ATP is depicted in a movie of bacterial photosynthesis (https://www.youtube.com/watch?v=cUHxfpPkN6E). The gradual incorporation of other components such as lipids and other proteins into the model will yield an invaluable *in silico* testbed with predictive and design capabilities for fabricated artificial assemblies.

6.5 ASSEMBLY OF PHOTOSYSTEM COMPONENTS OF THE BACTERIAL PHOTOSYNTHETIC APPARATUS

6.5.1 Genetic aspects: The photosynthesis gene cluster in purple phototrophic bacteria

Gene organization and regulation in many species of purple bacteria has been well studied and is relatively well understood. Genome sequences for multiple species including *Rba. sphaeroides*,

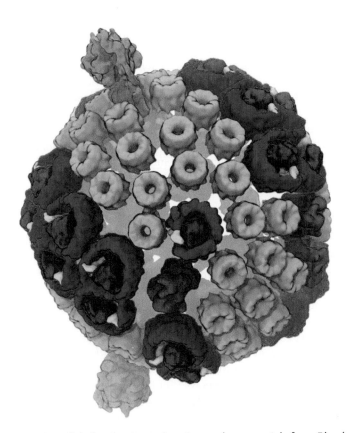

Figure 6.9 Atomic structural model of an intracytoplasmic membrane vesicle from *Rhodobacter sphaeroides*. The vesicle comprises 67 LH2 complexes (green), 11 LH1–RC–PufX dimers and 2 RC–LH1–PufX monomers (blue/red), 4 cytochrome bc_1 dimers (magenta), and 2 ATP synthases (orange).

Rba. capsulatus, Rsp. rubrum, Rps. palustris, Phs. molischianum, and *Blc. viridis* are publicly available from databases such as those found at NCBI (http://www.ncbi.nlm.nih.gov/).

In the majority of purple bacteria, most of the photosynthesis-related genes are located in a 40–50 kb DNA region of the genome, known as the photosynthesis gene cluster (PGC). The *Rba. sphaeroides* PGC (Figure 6.10) contains at least 38 open reading frames with a further five genes downstream of *puhA* with potential functions related to photosynthesis (Naylor et al. 1999). Fifty percent of the ORFs in the cluster are dedicated to the later stages of BChl biosynthesis, 18% are for carotenoid biosynthesis, 8% are for structural genes for the antennae and RC complexes, and 10% are genes essential for photosynthetic growth. The remaining DNA accounts for genes of unknown function and non-protein-coding regions.

Genes required for photosynthesis that are shared with other metabolic pathways are not located within the PGC. For example, the early stages of BChl biosynthesis are shared with the heme biosynthesis pathway (up to protoporphyrin IX); these genes are located elsewhere in the genome. Similarly, the genes associated with electron transport components such as the cytochrome bc_1 complex and ATP synthase are not located within the cluster. Notable proteins with important functions in photosynthesis that are also not encoded in the PGC include the *puc* operons that encode the LH2 antenna complex proteins and the genes encoding the enzymes of the Calvin cycle (Naylor et al. 1999).

6.5.1.1 THE *puf* OPERON, *puhA*, AND THE *lhaA* GENE

The *puf* operon, located within the PGC, encodes the LH1 α and β polypeptides (*pufA* and *pufB*), the RC-L and M subunits (*pufL* and *pufM*), the PufX polypeptide (*pufX*), a regulator of BChl biosynthesis (*pufQ*), and a transcription regulator (*pufK*) (Kiley et al. 1987; Lee et al. 1989a; Hunter et al. 1991; Naylor et al. 1999; Chidgey et al. 2017).

The *puf* operon is transcribed as two mRNAs; the 0.5 kb transcript encodes the LH1 α and β polypeptides, and the 2.6 kb transcript encodes the LH1 α and β polypeptides, the RC-L and M subunits, and the PufX polypeptide. The *pufLMX* transcript is relatively unstable (Belasco et al. 1985; Zhu and Kaplan 1985), resulting in a 10–20-fold excess of

pufBA mRNA and sufficient LH1 α and β polypeptides to encircle the RC. Transcription of the *puf* operon is upregulated in low-oxygen/low-light conditions (Zhu and Hearst 1986).

At the other end of the PGC, *puhA* encodes the RC-H subunit (Donohue et al. 1986). The *lhaA* open reading frame is located just upstream of *puhA* (Zsebo and Hearst 1984); *lhaA* is highly conserved among purple bacteria both in terms of sequence and location within the photosynthesis gene cluster. Insertion mutations of *lhaA* showed that this gene encodes an assembly factor for the LH1 complex (Young et al. 1998).

6.5.1.2 THE *pucBAC* AND *puc2BA* OPERONS

The *puc1B* and *puc1A* genes, which encode the LH2 β and α polypeptides (Ashby et al. 1987; Kiley and Kaplan 1987), are found in a *pucBAC* operon located approximately 20 kb downstream of the PGC. The *puc* genes are transcribed as a shorter 0.5 kb *pucBA* mRNA, the levels of which are regulated by both light and oxygen, and as a 2.3 kb *pucBAC* transcript. The half-life of the smaller transcript was determined to be around 20.5 min in both photosynthetic and aerobic conditions, whereas the 2.3 kb transcript half-life is less than 5 min. The biosynthesis of LH2 complexes was found to be dependent on the expression of the 2.3 kb transcript (Lee et al. 1989b). Similar to LH1 and the RC, levels of the LH2 complex are strongly influenced by variations in the light or oxygen intensity of the surrounding environment. In comparison to high-oxygen conditions, LH2 *puc1BA* transcripts are approximately 37 times more abundant in cultures grown under high light and 60-fold more abundant under low-light conditions (Zhu and Hearst 1986; Kiley and Kaplan 1987). A small lag in the rise of *pucBA* mRNA transcripts is observed in cells of *Rba. sphaeroides* induced to synthesize photosynthetic complexes, in comparison with the rise in RC and LH1 gene mRNA levels (Klug et al. 1985; Hunter et al. 1987).

The *pucC* gene encodes a protein essential for LH2 assembly in *Rba. capsulatus* (Tichy et al. 1989) and in *Rba. sphaeroides* (Gibson et al. 1992); interruption of the gene abolished LH2 complex formation and resulted in a sixfold reduction in *pucBA* mRNA. The *pucBAC-hemF* region of the *Rba. sphaeroides* chromosome was mapped by insertional mutagenesis and complementation analysis, and the sequence of the gene revealed that PucC is a hydrophobic protein.

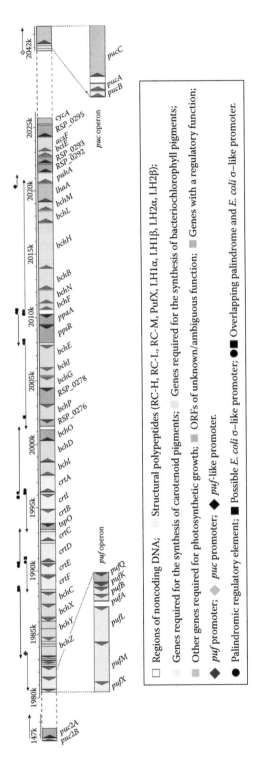

Figure 6.10 Overview of the photosynthetic gene cluster of *Rhodobacter sphaeroides*. Genes with no assigned name are labeled with their NCBI genome sequence number starting RSP_ (Chromosome 1, reference NC_007493.1). Gene lengths are approximately to scale. (Modified from Mothersole, D.J. et al., *Mol. Microbiol.*, 99, 307, 2016.)

□ Regions of noncoding DNA; ▨ Structural polypeptides (RC-H, RC-L, RC-M, PufX, LH1α, LH1β, LH2α, LH2β);

▨ Genes required for the synthesis of carotenoid pigments; ▨ Genes required for the synthesis of bacteriochlorophyll pigments;

▨ Other genes required for photosynthetic growth; ▨ ORFs of unknown/ambiguous function; ▨ Genes with a regulatory function;

◆ *puf* promoter; ◇ *puc* promoter; ◆ *puf*-like promoter.

■ Palindromic regulatory element; ● Possible *E. coli* σ–like promoter; ●■ Overlapping palindrome and *E. coli* σ–like promoter.

Both Tichy et al. (1989) and Gibson et al. (1992) postulated a role for PucC in membrane assembly of the LH2 complex on the basis of its high degree of hydrophobicity. Topology studies of *Rba. capsulatus* using *pho'A and lac'Z* fusions demonstrated that PucC has 12 membrane-spanning segments and that both the N and C termini are located in the cytoplasm (LeBlanc and Beatty 1996).

A second *puc2BA* operon is located approximately 1.36 Mb apart from its paralog in the *Rba. sphaeroides* genome (Zeng et al. 2003). Similar to the *puc1BAC* operon, the *puc2BA* operon is tightly controlled by both oxygen and light with PpsR and FnrL repressor binding sites located upstream. These genes were designated as *puc2B* and *puc2A*; an additional *pucC* gene was not found. The polypeptide encoded by the *puc2B* gene was found to be in approximately 30% of LH2 complexes; despite this, deletion of *puc1B* abolishes the biosynthesis of LH2 completely. The *puc2A* gene encodes a protein significantly different to that of *puc1A* with only 58% identity in the N-terminal 48 amino acid residues; the remaining 215 amino acids bear no resemblance to the *puc1A* polypeptide or to any other sequence in the genome. The protein encoded by the *puc2A* gene is expressed but does not appear to be located within LH2 complexes (Zeng et al. 2003).

6.5.2 Assembly of photosynthetic complexes

6.5.2.1 ASSEMBLY OF THE RC–LH1 CORE COMPLEX

Studies by Chory et al. (1984), which involved light-based induction of the photosystems of *Rba. sphaeroides*, established that the RC-H subunit was constitutively synthesized and was present at low levels in the membranes of aerobically grown cells that otherwise had very low levels of pigment. Further studies on the RC-H subunit in the assembly of RCs in *Rps. rubrum* described it as a foundation for assembly of the RC-M and RC-L subunits (Cheng et al. 2000).

A time-based comparison of mRNA transcripts and protein components of the core complex (Pugh et al. 1998) proposed the formation of a compact RC–PufX–LH1$\alpha_1\beta_1$ complex prior to encirclement by LH1 α and β polypeptides. PufX was envisaged as an important docking protein early in the assembly process that precedes full encirclement of the

RC by LH1$\alpha_1\beta_1$BChl$_2$ (B820) units. These minimal B820 building blocks of the LH1 antenna have also been used for *in vitro* reconstitution studies of LH1 assembly (Todd et al. 1998, 1999). Davis et al. (1995) further disassembled the components of LH1 for reassociation assays. Using LH1 α and β polypeptides, BChl *a* and the carotenoid spheroidene, a functional LH1 ring was produced. However, as noted earlier, LhaA appears to be essential for efficient assembly of the LH1 complex *in vivo*. An *lhaA* deletion mutant still makes some LH1$\alpha_1\beta_1$BChl$_2$ subunits that, instead of being shared among the RCs, form a few normal RC–LH1–PufX core complexes and leave a population of RCs with no LH1. This observation shows that once initiated, LH1 assembly round the RC is cooperative and proceeds to completion. This study also showed that LhaA is found at sites where membranes grow and that it associates with RCs, the bacteriochlorophyll synthase enzyme, and the protein machinery for membrane protein assembly (Mothersole et al. 2016).

6.5.2.2 ASSEMBLY OF THE LH2 COMPLEX

Lang and Hunter (1994) studied the effects of altering the carotenoid biosynthesis pathway on the synthesis of the LH2 complex of *Rba. sphaeroides*. They observed that in the absence of colored carotenoids, LH2 α and β polypeptides were synthesized but rapidly turned over, failing to become integrated into a stable membrane complex. Biosynthesis of carotenoids from neurosporene onward in the pathway yielded stable LH2 complexes; carotenoid integration is therefore essential for LH2 formation in *Rba. sphaeroides*. LH2 from the purple sulfur bacterium *Allochromatium minutissimum* has been shown to form complexes without the presence of carotenoid (Makhneva et al. 2008). Todd et al. (1998, 1999) observed that while the α and β polypeptides of *Rba. sphaeroides* LH2 were unable to form a complex *in vitro*, the α and β polypeptides of *Phs. molischianum* were capable of successfully reconstituting an LH2 complex.

As noted in Section 6.5.1.2, the PucC protein is essential for efficient LH2 assembly in *Rba. sphaeroides*, in the same way that LhaA is essential for LH1. Like LhaA, PucC is predicted to consist of 12 membrane-spanning segments and it forms oligomers at sites of initiation of membrane invagination (Mothersole et al. 2016). However, the mechanistic role of PucC in LH2 biogenesis has not been determined.

6.6 ULTRASTRUCTURE AND ASSEMBLY OF PHOTOSYNTHETIC MEMBRANES

6.6.1 Intracytoplasmic membrane

In purple bacteria, the photosynthetic apparatus is located within invaginations of the cytoplasmic membrane known as the ICM. In *Rba. sphaeroides* and *Rsp. rubrum*, these ICMs are vesicular, 40–80 nm diameter structures packed tightly throughout the cell (Vatter and Wolfe 1958; Cohen-Bazire and Kunisawa 1963; Peters and Cellarius 1972). In some species of purple bacteria such as *Phs. molischianum* and *Blc. viridis*, the ICMs are composed of stacked lamellar sheets (Drews 1960; Miller 1979; Konorty et al. 2008). This section will concentrate on the vesicular ICMs of *Rba. sphaeroides*. A structural model of a vesicle is displayed in Figure 6.9.

6.6.2 Changes in membrane organization in response to alterations in light intensity and oxygen

6.6.2.1 EFFECTS OF LIGHT INTENSITY ON MEMBRANES OF *Rba. sphaeroides*

The general response of phototrophs with LH2 complexes to reduced light intensity is to increase the size of the LH2-rich domains in relation to the RC–LH1 complexes. An early study of *Rba. sphaeroides* grown at different light intensities showed that this bacterium responds to lowered light by increasing the overall BChl content of the cell and by increasing the level of LH2 complexes relative to LH1 and RCs (Aagaard and Sistrom 1972; Adams and Hunter 2012).

Adams and Hunter (2012) examined the effects of a transition from high- to low-light growth conditions in *Rba. sphaeroides* and found that a four-fold increase in cellular BChl levels from 1.6×10^6 to 6.5×10^6 molecules per cell was reflected in the increased number of ICM vesicles per cell from approximately 274 to 1468. The effect of this internal proliferation of membrane is to increase the surface area for harvesting and utilizing solar energy from 3×10^6 nm^2 to 16×10^6 nm^2. AFM showed that high-light ICM consisted mainly of monomeric and dimeric core complexes with only small clusters of LH2, whereas extensive LH2-only domains form

during adaptation to low light (see Figure 6.8). The more random organization of *Rhodospirillum* (*Rsp.*) *photometricum* membranes was unaffected by the incorporation of more LH2 complexes under low-light conditions (Scheuring and Sturgis 2005).

6.6.2.2 EFFECTS OF OXYGEN ON MEMBRANES OF *Rba. sphaeroides*

The response of some purple phototrophs to oxygen is more marked than the responses to light intensity, and high levels of oxygen can repress the synthesis of photosynthetic membranes almost completely, leaving cells with very low levels of pigment that can grow rapidly in the dark and generate their energy by aerobic respiration (Bauer et al. 2009). The underlying rationale for this oxygen response is not clear, but some purple phototrophs are found in stratified layers in lakes where they balance the conflicting requirements of harvesting light from the surface and using organic carbon produced by the reducing environment nearer the bottom of the water column. Thus, high-oxygen/high-light and low-oxygen/low-light go hand in hand, and the genetic triggers for this transition are oxygen based.

It was realized many years ago that it was possible to grow cells under high aeration in the dark so that nearly all pigment biosynthesis is repressed, then to relieve this repression by transfer to oxygen-limited growth conditions. This convenient experimental ploy was exploited by Koblizek and coworkers to show that assembly of the photosynthetic apparatus proceeds in a stepwise manner over a 24 h timecourse, with RC–LH1 dominating early on and LH2 the latter stages of assembly (Koblizek et al. 2005).

Further experimental benefits of respiratory growth in the dark enabled researchers to create mutations in genes essential for photosynthesis without killing the cell. This formed the basis for the discovery and mutagenesis of the genes encoding enzymes for the BChl and carotenoid biosynthetic pathways and the LH2, LH1, and RC apoproteins (see also Section 5.1).

6.6.3 Maturation of intracytoplasmic membranes from precursor membranes

Disruption of photosynthetically grown *Rba. sphaeroides* cells by passage through the French press and subsequent separation of membrane fractions on continuous, rate-zonal sucrose density gradients by ultracentrifugation results in the formation of two discrete bands. The more slowly migrating and less abundant fraction is termed the upper pigmented band (UPB) (Niederman et al. 1979); the faster moving fraction arises from the ICM.

Several pulse-chase radio-labelling studies with [S³⁵] methionine demonstrated that the UPB and ICM bands have a biosynthetic relationship, originating from developing immature membranes and developed mature membranes, respectively. After a short pulse, the radio-labelled [S³⁵] methionine is incorporated almost instantaneously into the UPB membrane and declines over the timecourse of the chase; the ICM initially has a significantly lower level of radiolabel incorporated, which increases during the chase. When the protein inhibitor chloramphenicol is added along with the chase, the radiolabel is maintained in the UPB fraction; ongoing protein synthesis is therefore necessary for the development of the ICM from the immature UPB fraction (Niederman et al. 1979; Inamine et al. 1984; Reilly and Niederman 1986; Tucker et al. 2010).

Functional studies established the importance of UPB membranes in the early development of primary photochemistry and energy transfer in functioning photosynthetic membranes. In particular, the photosynthetic units are assembled sequentially; LH1 and RCs appear first, followed by activation of functional electron transfer and then accumulation of the peripheral LH2 antenna (Hunter et al. 1979a,b; Bowyer et al. 1985; Koblizek et al. 2005). Further studies of membrane development (Hunter et al. 2005) demonstrated that the increase in the LH2 antenna size is primarily restricted to the ICM fraction with the antenna composition of UPB fraction remaining constant. Some LH2 complexes are synthesized and assembled in the UPB, packing between rows of core complex dimers creating new LH2–LH1 interactions. The LH2 complexes that assemble in the ICM in response to lowered light intensity tend to aggregate, forming an LH2-only antenna pool.

Mature ICM and precursor UPB membranes have been purified and compared using EM and AFM. EM revealed that in contrast to the spherical ICM, purified UPB membranes consisted of 50 nm discs with a convex cytoplasmic surface (Tucker et al. 2010). AFM detected RC–LH1–PufX dimers in

the UPB, which would account in part for the curvature of this membrane (Qian et al. 2008), and single membrane fluorescence spectroscopy showed that each UPB membrane examined had a significant impairment in energy transfer from LH2 to LH1 complexes in line with a previous measurement on bulk samples (Hunter et al. 1979b). Cryo-electron tomography showed that some ICM vesicles are isolated structures within the cell and others form interconnected structures that retain a connection to the peripheral cytoplasmic membrane (Tucker et al. 2010). Small indents of this membrane were also observed, which were proposed to give rise to the UPB precursor membranes upon cell disruption. A subsequent quantitative proteomics study of UPB and ICMs showed that the UPB proteome is significantly more complex than the specialized ICM; the UPB is enriched in proteins representing a wide range of metabolic and biosynthetic functions, whereas the ICM is enriched in proteins involved in photosynthesis (Jackson et al. 2012).

The number of UPB precursor sites in an *Rba. sphaeroides* cell was estimated between 700 and 800, with the UPB accounting for about a third of the intracellular membrane in cells grown at high light intensity. The abundance of precursor sites in high-light cells, the relatively inefficient LH2-to-LH1 energy transfer seen in the UPB, and the suppressed levels of ICM synthesis make it likely that high-light grown cells are relatively inefficient at energy trapping. It has been speculated that by avoiding maturation of its photosynthetic membranes, *Rba. sphaeroides* gains some protection against the damaging effects of high light intensity (Adams and Hunter 2012).

REFERENCES

Aagaard, J. and Sistrom, W. R. 1972. Control of synthesis of reaction centre bacteriochlorophyll in photosynthetic bacteria. *Photochem. Photobiol.* 15: 209–225.

Adams, P. G. and Hunter, C. N. 2012. Adaptation of intracytoplasmic membranes to altered light intensity in *Rhodobacter sphaeroides. Biochim. Biophys. Acta* 1817: 1616–1627.

Agarwal, G., Kovac, L., Radziejewski, C., and Samuelsson, S. J. 2002. Binding of discoidin domain receptor 2 to collagen I: An atomic force microscopy investigation. *Biochemistry* 41: 11091–11098.

Allen, J. P., Feher, G., Yeates, T. O., Komiya, H., and Rees, D. C. 1987a. Structure of the reaction center from *Rhodobacter sphaeroides* R-26: The protein subunits. *Proc. Natl. Acad. Sci. USA* 84: 6162–6166.

Allen, J. P., Feher, G., Yeates, T. O., Komiya, H., and Rees, D. C. 1987b. Structure of the reaction center from *Rhodobacter sphaeroides* R26: The cofactors. *Proc. Natl. Acad. Sci. USA* 84: 5730–5734.

Ashby, M. K., Coomber, S. A., and Hunter, C. N. 1987. Cloning, nucleotide sequence and transfer of genes for the B800-850 light harvesting complex of *Rhodobacter sphaeroides*. *FEBS Lett.* 213: 245–248.

Axelrod, H. L., Abresch, E. C., Okamura, M. Y., Yeh, A. P., Rees, D. C., and Feher, G. 2002. X-ray structure determination of the cytochrome c_2: Reaction center electron transfer complex from *Rhodobacter sphaeroides*. *J. Mol. Biol.* 319: 501–515.

Bahatyrova, S., Frese, R. N., Siebert, C. A. et al. 2004a. The native architecture of a photosynthetic membrane. *Nature* 430: 1058–1062.

Bahatyrova, S., Frese, R. N., van der Werf, K. O., Otto, C., Hunter, C. N., and Olsen, J. D. 2004b. Flexibility and size heterogeneity of the LH1 light harvesting complex revealed by atomic force microscopy: Functional significance for bacterial photosynthesis. *J. Biol. Chem.* 279: 21327–21333.

Bauer, C. E., Setterdahl, A., Wu, J., and Robinson, B. R. 2009. Regulation of gene expression in response to oxygen tension. In *The Purple Phototrophic Bacteria*, eds. C. N. Hunter et al., pp. 707–725. Springer, Dordrecht, the Netherlands.

Beekman, L. M., van Mourik, F., Jones, M. R., Visser, H. M., Hunter, C. N., and van Grondelle, R. 1994. Trapping kinetics in mutants of the photosynthetic purple bacterium *Rhodobacter sphaeroides*: Influence of the charge separation rate and consequences for the rate-limiting step in the light-harvesting process. *Biochemistry* 33: 3143–3147.

Belasco, J. G., Beatty, J. T., Adams, C. W., Vongabain, A., and Cohen, S. N. 1985. Differential expression of photosynthesis genes in *Rhodopseudomonas capsulata* results from segmental differences in stability within the polycistronic *rxcA* transcript. *Cell* 40: 171–181.

Berry, E. A., Huang, L., Saechao, L. K., Pon, N. G., Valkova-Valchanova, M., and Daldal, F. 2004. X-ray structure of *Rhodobacter capsulatus* cytochrome bc_1: Comparison with its mitochondrial and chloroplast counterparts. *Photosynth. Res.* 81: 251–275.

Berry, E. A., Lee, D. W., Huang, L. S., and Daldal, F. 2009. Structural and mutational studies of the cytochrome bc_1 complex. In *The Purple Phototrophic Bacteria*, eds. C. N. Hunter et al., pp. 425–450. Springer, Dordrecht, the Netherlands.

Bixon, M. and Jortner, J. 1999. Electron transfer: From isolated molecules to biomolecules. *Adv. Chem. Phys.* 106: 35–202.

Bowyer, J. R., Hunter, C. N., Ohnishi, T., and Niederman, R. A. 1985. Photosynthetic membrane development in *Rhodopseudomonas sphaeroides* spectral and kinetic characterization of redox components of light-driven electron flow in apparent photosynthetic membrane growth initiation sites. *J. Biol. Chem.* 260: 3295–3304.

Bradforth, S. E., Jinenez, R., Vanmourik, F., Vangrondelle, R., and Fleming, G. R. 1995. Excitation transfer in the core light-harvesting complex (LH-1) of *Rhodobacter-sphaeroides*: An ultrafast fluorescence depolarization and annihilation study. *J. Phys. Chem.* 99: 16179–16191.

Broglie, R. M., Hunter, C. N., Delepelaire, P., Niederman, R. A., Chua, N. H., and Clayton, R. K. 1980, Isolation and characterization of the pigment-protein complexes of *Rhodopseudomonas sphaeroides* by lithium dodecyl sulfate/polyacrylamide gel electrophoresis. *Proc. Natl. Acad. Sci. USA* 77: 87–91.

Brotosudarmo, T. H. P., Kunz, R., Böhm, P. et al., 2009. Single-molecule spectroscopy reveals that individual low-light LH2 complexes from Rhodopseudomonas palustris 2.1.6. Have a heterogeneous polypeptide composition. *Biophys. J.* 97: 1491–1500.

Brunisholz, R. A. and Zuber, H. 1992. Structure, function and organization of antenna polypeptides and antenna complexes from the three families of *Rhodospirillaceae*. *J. Photochem. Photobiol. B Biol.* 15: 113–140.

Bullough, P. A., Qian, P., and Hunter, C. N. 2009. Reaction center-light-harvesting core complexes of purple bacteria. In *The Purple Phototrophic Bacteria*, eds. C. N. Hunter et al., pp. 155–179. Springer, Dordrecht, the Netherlands.

Busselez, J., Cottevieille, M., Cuniasse, P., Gubellini, F., Boisset, N., and Lévy, D. 2007. Structural basis for the PufX-mediated dimerization of bacterial photosynthetic core complexes. *Structure* 15: 1674–1683.

Cartron, M. L., Olsen, J. D., Sener, M. et al. 2014. Integration of energy and electron transfer processes in the photosynthetic membrane of *Rhodobacter sphaeroides*. *Biochim. Biophys. Acta* 1837: 1769–1780.

Chang, C. H., El-Kabbani, O., Tiede, D., Norris, J., and Schiffer, M. 1991. Structure of the membrane-bound protein photosynthetic reaction center from *Rhodobacter sphaeroides*. *Biochemistry* 30: 5352–5360.

Chang, C. H., Tiede, D., Tang, J., Smith, U., Norris, J., and Schiffer, M. 1986. Structure of *Rhodopseudomonas sphaeroides* R-26 reaction center. *FEBS Lett.* 205: 82–86.

Cheng, Y. S., Brantner, C. A., Tsapin, A., and Collins, M. L. 2000. Role of the H protein in assembly of the photochemical reaction center and intracytoplasmic membrane in *Rhodospirillum rubrum*. *J. Bacteriol.* 182: 1200–1207.

Chidgey, J.W., Jackson, P.J., Dickman, M. J., and Hunter, C.N. 2017. PufQ regulates porphyrin flux at the haem/bacteriochlorophyll branch-point of tetrapyrrole biosynthesis via interactions with ferrochelatase. *Mol. Microbiol.* DOI: 10.1111/mmi.13861 [Epub ahead of print].

Chory, J., Donohue, T. J., Varga, A. R., Staehelin, L. A., and Kaplan, S. 1984. Induction of the photosynthetic membranes of *Rhodopseudomonas sphaeroides*: Biochemical and morphological studies. *J. Bacteriol.* 159: 540–554.

Choudhary, M., Zanhua, X., Fu, Y. X., and Kaplan, S. 2007. Genome analyses of three strains of *Rhodobacter sphaeroides*: Evidence of rapid evolution of chromosome II. *J. Bacteriol.* 189: 1914–1921.

Clayton, R. K. and Clayton, B. J. 1972. Relations between pigments and proteins in the photosynthetic membranes of *Rhodopseudomonas sphaeroides*. *Biochim. Biophys. Acta* 283: 492–504.

Clayton, R. K. and Haselkorn, R. 1972. Protein components of bacterial photosynthetic membranes. *J. Mol. Biol.* 68: 97–105.

Cogdell, R. J., Durant, I., Valentine, J., Lindsay, J. G., and Schimidt, K. 1983. The isolation and partial characterisation of the light-harvesting pigment-protein complement of *Rhodopseudomonas acidophila*. *Biochim. Biophys. Acta* 722: 427–435.

Cogdell, R. J., Gall, A., and Köhler, J. 2006. The architecture and function of the light-harvesting apparatus of purple bacteria: From single molecules to *in vivo* membranes. *Q. Rev. Biophys.* 39: 227–324.

Cohen-Bazire, G. and Kunisawa, R. 1963. The fine structure of *Rhodospirillum rubrum*. *J. Cell Biol.* 16: 401–419.

Comayras, R., Jungas, C., and Lavergne, J. 2005. Functional consequences of the organization of the photosynthetic apparatus in *Rhodobacter sphaeroides*. I. Quinone domains and excitation transfer in chromatophores and reaction center antenna complexes. *J. Biol. Chem.* 280: 11203–11213.

Conroy, M. J., Westerhuis, W. H., Parkes-Loach, P. S., Loach, P. A., Hunter, C. N., and Williamson, M. P. 2000. The solution structure of *Rhodobacter sphaeroides* LH1β reveals two helical domains separated by a more flexible region: Structural consequences for the LH1 complex. *J. Mol. Biol.* 298: 83–94.

Coomber, S. A., Chaudhri, M., Connor, A., Britton, G., and Hunter, C. N. 1990. Localized transposon Tn5 mutagenesis of the photosynthetic gene cluster of *Rhodobacter sphaeroides*. *Mol. Microbiol.* 4: 977–989.

Cooper, D. E., Rands, M. B., and Woo, C. P. 1975. Sulfide reduction in fellmongery effluent by red sulfur bacteria. *J. Water Pollut. Control Fed.* 47: 2088–2100.

Davis, C. M., Bustamante, P. L., and Loach, P. A. 1995. Reconstitution of the bacterial core light-harvesting complexes of *Rhodobacter sphaeroides* and *Rhodospirillum rubrum* with isolated α- and β-polypeptides, bacteriochlorophyll *a*, and carotenoid. *J. Biol. Chem.* 270: 5793–5804.

Deisenhofer, J., Epp, O., Miki, K., Huber, R., and Michel, H. 1985. Structure of the protein subunits in the photosynthetic reaction centre of *Rhodopseudomonas viridis* at 3 Å resolution. *Nature* 318: 618–624.

Deisenhofer, J., Epp, O., Sinning, I., and Michel, H. 1995. Crystallographic refinement at 2.3 Å resolution and refined model of the photosynthetic reaction centre from *Rhodopseudomonas viridis*. *J. Mol. Biol.* 246: 429–457.

Deisenhofer, J. and Michel, H. 1989. The photosynthetic reaction centre from the purple bacterium *Rhodospeudomonas viridis*. *EMBO J.* 8: 2149–2170.

Deisenhofer, J. and Michel, H. 1991. High-resolution structures of photosynthetic reaction centers. *Annu. Rev. Biophys. Biomol. Struct.* 20: 247–266.

Dezi, M., Francia, F., Mallardi, A., Colafemmina, G., Palazzo, G., and Venturoli, G. 2007. Stabilization of charge separation and cardiolipin confinement in antenna-reaction center complexes purified from *Rhodobacter sphaeroides*. *Biochim. Biophys. Acta* 1767: 1041–1056.

Donohue, T. J., McEwan, A. G., and Kaplan, S. 1986. Cloning, DNA-sequence, and expression of the *Rhodobacter sphaeroides* cytochrome c_2 gene. *J. Bacteriol.* 168: 962–972.

Drews, G. 1960. Untersuchungen zur Substruktur der Chromatophoren von *Rhodospirillum rubrum* und *Rhodospirillum molischianum*. *Arch. Mikrobiol.* 36: 99–108.

Ermler, U., Fritzsch, G., Buchanan, S. K., and Michel, H. 1994. Structure of the photosynthetic reaction centre from *Rhodobacter sphaeroides* at 2.65 Å resolution: Cofactors and protein-cofactor interactions. *Structure* 2: 925–936.

Esser, L., Elberry, M., Zhou, F., Yu, C.-A., Yu, L., and Xia, D. 2007. Inhibitor-complexed structures of the cytochrome bc_1 from the photosynthetic bacterium *Rhodobacter sphaeroides*. *J. Biol. Chem.* 283: 2846–2857.

Esser, L., Gong, X., Yang, S., Yu, L., Yu, C. A., and Xia, D. 2006. Surface-modulated motion switch: Capture and release of iron-sulfur protein in the cytochrome bc_1 complex. *Proc. Natl. Acad. Sci. USA* 103: 13045–13050.

Fassioli, F., Olaya-Castro, A., Scheuring, S., Sturgis, J. N., and Johnson, N. F. 2009. Energy transfer in light-adapted photosynthetic membranes: From active to saturated photosynthesis. *Biophys. J.* 97: 2464–2473.

Feniouk, B. A. and Junge, W. 2009. Proton translocation and ATP synthesis by the F_oF_1-ATPase of purple bacteria. In *The Purple Phototrophic Bacteria*, eds. C. N. Hunter et al., pp. 475–493. Springer, Dordrecht, the Netherlands.

Fixen, K. R., Oda, Y., and Harwood, C. S. 2016. Clades of photosynthetic bacteria belonging to the genus rhodopseudomonas show marked diversity in light-harvesting antenna complex gene composition and expression. *mSystems* 1: e00006-15.

Fotiadis, D., Qian, P., Pilippsen, A., Bullough, P. A., Engel, A., and Hunter, C. N. 2004. Structural analysis of the RC-LH1 photosynthetic core complex of *Rhodospirillum rubrum* using atomic force microscopy. *J. Biol. Chem.* 279: 2063–2068.

Fowler, G. J. S., Sockalingum, G. D., Robert, B., and Hunter, C. N. 1994. Blue shifts in bacteriochlorophyll absorbance correlate with changed hydrogen bonding patterns in light-harvesting 2 mutants of *Rhodobacter sphaeroides* with alterations at α-Tyr-44 and α-Tyr-45. *Biochem. J.* 299: 695–700.

Fowler, G. J. S., Visschers, R. W., Grief, G. G., van Grondelle, R., and Hunter, C. N. 1992. Genetically modified photosynthetic antenna complexes with blueshifted absorbance bands. *Nature* 35563: 848–850.

Frank, H. A. and Cogdell, R. J. 1996. Carotenoids in photosynthesis. *Photochem. Photobiol.* 63: 257–264.

Freer, A. A., Prince, S., Sauer, K. et al. 1996. Pigment-pigment interactions and energy transfer in the antenna complex of the photosynthetic bacterium *Rhodopseudomonas acidophila*. *Structure* 4: 449–462.

Frese, R. N., Olsen, J. D., Branvall, R., Westerhuis, W. H., Hunter, C. N., and van Grondelle, R. 2000. The long-range supraorganization of the bacterial photosynthetic unit: A key role for PufX. *Proc. Natl. Acad. Sci. USA* 97: 5197–5202.

Frese, R. N., Pámies, J. C., Olsen, J. D. et al. 2008. Protein shape and crowding drive domain formation and curvature in biological membranes. *Biophys. J.* 94: 640–647.

Frese, R. N., Siebert, C. A., Niederman, R. A., Hunter, C. N., Otto, C., and van Grondelle, R. 2004. The long-range organization of a native photosynthetic membrane. *Proc. Natl. Acad. Sci. USA* 101: 17994–17999.

Gabrielsen, M., Gardiner, A. T., and Cogdell, R. J. 2009. Peripheral complexes of purple bacteria. In *The Purple Phototrophic Bacteria*, eds. C. N. Hunter et al., pp. 135–153. Springer, Dordrecht, The Netherlands.

Gardiner, A. T., Cogdell, R. J., and Takaichi, S. 1993. The effect of growth conditions on the light-harvesting apparatus in *Rhodopseudomonas acidophila*. *Photosynth. Res.* 38: 159–167.

Geyer, T. and Helms, V. 2006. Reconstruction of a kinetic model of the chromatophore vesicles from *Rhodobacter sphaeroides*. *Biophys. J.* 91: 927–937.

Gibson, L. C. D., McGlynn, P., Chaudhri, M., and Hunter, C. N. 1992. A putative anaerobic coproporphyrinogen III oxidase in *Rhodobacter sphaeroides*. II. Analysis of a region of the genome encoding *hemF* and the *puc* operon. *Mol. Microbiol.* 6: 3171–3186.

Goncalves, R. P., Bernadac, A., Sturgis, J. N., and Scheuring, S. 2005. Architecture of the native photosynthetic apparatus of *Phaeospirillum molischianum*. *J. Struct. Biol.* 152: 221–228.

Hartigan, N., Tharia, H. A., Sweeney, F., Lawless, A. M., and Papiz, M. Z. 2002. The 7.5-Å electron density and spectroscopic properties of a novel low-light B800 LH2 from *Rhodopseudomonas palustris*. *Biophys. J.* 82: 963–977.

Hess, S., Akesson, E., Cogdell, R. J., Pullerits, T., and Sundström, V. 1995a. Energy transfer in spectrally inhomogeneous light-harvesting pigment-protein complexes of purple bacteria. *Biophys. J.* 69: 2211–2225.

Hess, S., Chachisvilis, M., Timpmann, K. et al. 1995b. Temporally and spectrally resolved subpicosecond energy transfer within the peripheral antenna complex (LH2) and from LH2 to the core antenna complex in photosynthetic purple bacteria. *Proc. Natl. Acad. Sci. USA* 92: 12333–12337.

Hoff, A. J. and Deisenhofer, J. 1997. Photophysics of photosynthesis. Structure and spectroscopy of reaction centers of purple bacteria. *Phys. Rep.* 287: 1–247.

Holzapfel, W., Finkele, U., Kaiser, W. et al., 1990. Initial electron-transfer in the reaction center from *Rhodobacter sphaeroides*. *Proc. Natl. Acad. Sci. USA* 87: 5168–5172.

Hsin, J., Gumbart, J., Trabuco, L. G. et al. 2009. Protein-induced membrane curvature investigated through molecular dynamics flexible fitting. *Biophys. J.* 97: 321–329.

Hunte, C., Solmaz, S., Palsdottir, H., and Wenz, T. 2007. A structural perspective on mechanism and function of the cytochrome bc_1 complex. In *Bioenergetics*, eds. G. Schafer and H. S. Penefsky, pp. 253–278. Springer, Berlin, Germany.

Hunter, C. N., Ashby, M. K., and Coomber, S. A. 1987. Effect of oxygen on levels of mRNA coding for reaction-center and light-harvesting polypeptides of *Rhodobacter sphaeroides*. *Biochem. J.* 247: 489–492.

Hunter, C. N., Daldal, F., Thurnauer, M. C., and Beatty, J. T. 2009. *The Purple Phototrophic Bacteria*. Springer, Dordrecht, the Netherlands.

Hunter, C. N., McGlynn, P., Ashby, M. K., Burgess, J. G., and Olsen, J. D. 1991. DNA sequencing and complementation/deletion analysis of the *bchA-puf* operon region of *Rhodobacter sphaeroides*: In vivo mapping of the oxygen-regulated *puf* promoter. *Mol. Microbiol.* 5: 2649–2661.

Hunter, C. N., Pennoyer, J. D., Sturgis, J. N., Farrelly, D., and Niederman, R. A. 1988. Oligomerization states and associations of light-harvesting pigment protein complexes of *Rhodobacter sphaeroides* as analyzed by lithium dodecyl-sulfate polyacrylamide-gel electrophoresis. *Biochemistry* 27: 3459–3467.

Hunter, C. N., Tucker, J. D., and Niederman, R. A. 2005. The assembly and organisation of photosynthetic membranes in *Rhodobacter sphaeroides*. *Photochem. Photobiol. Sci.* 4: 1023–1027.

Hunter, C. N., van Grondelle, R., Holmes, N. G., and Jones, O. T. G. 1979a. The reconstitution of energy transfer in membranes from a bacteriochlorophyll-less mutant of *Rhodopseudomonas sphaeroides* by addition of light-harvesting and reaction centre pigment-protein complexes. *Biochim. Biophys. Acta* 548: 458–470.

Hunter, C. N., van Grondelle, R., Holmes, N. G., Jones, O. T. G., and Niederman, R. A. 1979b. Fluorescence yield properties of a fraction enriched in newly synthesized bacteriochlorophyll *a* protein complexes from *Rhodopseudomonas sphaeroides*. *Photochem. Photobiol.* 30: 313–316.

Inamine, G. S., Van Houten, J., and Niederman, R. A. 1984. Intracellular localization of photosynthetic membrane growth initiation sites in *Rhodopseudomonas sphaeroides*. *J. Bacteriol.* 158: 425–429.

Jackson, P. J., Lewis, H. J., Tucker, J. D., Hunter, C. N., and Dickman, M. J. 2012. Quantitative proteomic analysis of intracytoplasmic membrane development in *Rhodobacter sphaeroides*. *Mol. Microbiol.* 84: 1062–1078.

Jamieson, S. J., Wang, P., Qian, P. et al. 2002. Projection structure of the photosynthetic reaction centre-antenna complex of *Rhodospirillum rubrum* at 8.5 Å resolution. *EMBO J.* 21: 3927–3935.

Jimenez, R., Dikshit, S. N., Bradforth, S. E., and Fleming, G. R. 1996. Electronic excitation transfer in the LH2 complex of *Rhodobacter sphaeroides*. *J. Phys. Chem.* 100: 6825–6834.

Joo, T. H., Jia, Y. W., Yu, J. Y., Jonas, D. M., and Fleming, G. R. 1996. Dynamics in isolated bacterial light harvesting antenna (LH2) of *Rhodobacter sphaeroides* at room temperature. *J. Phys. Chem.* 100: 2399–2409.

Jungas, C., Ranck, J. L., Rigaud, J. L., Joliot, P., and Verméglio, A. 1999. Supramolecular organization of the photosynthetic apparatus of *Rhodobacter sphaeroides*. *EMBO J.* 18: 534–542.

Katona, G., Andreasson, U., Landau, E. M., Andreasson, L. E., and Neutze, R. 2003. Lipidic cubic phase crystal structure of the photosynthetic reaction centre from *Rhodobacter sphaeroides* at 2.35 Å resolution. *J. Mol. Biol.* 331: 681–692.

Kiley, P. J., Donohue, T. J., Havelka, W. A., and Kaplan, S. 1987. DNA sequence and *in vitro* expression of the B875 light-harvesting polypeptides of *Rhodobacter sphaeroides*. *J. Bacteriol.* 169: 742–750.

Kiley, P. J. and Kaplan, S. 1987. Cloning, DNA sequence, and expression of the *Rhodobacter sphaeroides* light-harvesting B800-850a and B800-850b genes. *J. Bacteriol.* 169: 3268–3275.

Klug, G., Kaufmann, N., and Drews, G. 1985. Gene expression of pigment-binding proteins of the bacterial photosynthetic apparatus: Transcription and assembly in the membrane of *Rhodopseudomonas capsulata*. *Proc. Natl. Acad. Sci. USA* 82: 6485–6489.

Koblizek, M., Shih, J. D., Breitbart, S. I. et al. 2005. Sequential assembly of photosynthetic units in *Rhodobacter sphaeroides* as revealed by fast repetition rate analysis of variable bacteriochlorophyll a fluorescence. *Biochim. Biophys. Acta* 1706: 220–231.

Koepke, J., Hu, X. C., Muenke, C., Schulten, K., and Michel, H. 1996. The crystal structure of the light-harvesting complex II (B800-B850) from *Rhodospirillum molischanum*. *Structure* 4: 581–597.

Konorty, M., Kahana, N., Linaroudis, A., Minsky, A., and Medalia, O. 2008. Structural analysis of photosynthetic membranes by cryo-electron tomography of intact *Rhodopseudomonas viridis* cells. *J. Struct. Biol.* 161: 393–400.

Kontur, W. S., Schwackwitz, W. S., Ivanova, N. et al. 2012. Revised sequence and annotation of the *Rhodobacter sphaeroides* 2.4.1 genome. *J. Bacteriol.* 192: 7016–7017.

Lang, H. P. and Hunter, C. N. 1994. The relationship between carotenoid biosynthesis and the assembly of the light-harvesting LH2 complex in *Rhodobacter sphaeroides*. *Biochem. J.* 298: 197–205.

Lavergne, J., Verméglio, A., and Joliot, P. 2009. Functional coupling between reaction centers and cytochrome bc_1 complexes. In *The Purple Phototrophic Bacteria*, eds. C. N. Hunter et al., pp. 509–536. Springer, Dordrecht, the Netherlands.

LeBlanc, H. N. and Beatty, J. T. 1996. Topological analysis of the *Rhodobacter capsulatus* PucC protein and effects of C-terminal deletions on light-harvesting complex II. *J. Bacteriol.* 178: 4801–4806.

Lee, J. K., DeHoff, B. S., Donohue, T. J., Gumport, R. I., and Kaplan, S. 1989a. Transcriptional analysis of *puf* operon expression in *Rhodobacter sphaeroides* 2.4.1. and an intercistronic transcription terminator mutant. *J. Biol. Chem.* 264: 19354–19365.

Lee, J. K., Kiley, P. J., and Kaplan, S. 1989b. Posttranscriptional control of *puc* operon expression of B800-850 light-harvesting complex formation in *Rhodobacter sphaeroides*. *J. Bacteriol.* 171: 3391–3405.

Liu, L. N. and Scheuring, S. 2013. Investigation of photosynthetic membrane structure using atomic force microscopy. *Trends Plant Sci.* 18: 277–286.

Ma, Y. Z., Cogdell, R. J., and Gillbro, T. 1997. Energy transfer and exciton annihilation in the B800-850 antenna complex of the photosynthetic purple bacterium *Rhodopseudomonas acidophila* (Strain 10050). A femtosecond transient absorption study. *J. Phys. Chem. B* 101: 1087–1095.

Magdaong, N. M., LaFountain, A. M., Hacking, K., Niedzwiedzki, D. M., Gibson, G. N., Cogdell, R. J., and Frank, H. A. 2016. Spectral heterogeneity and carotenoid-to-bacteriochlorophyll energy transfer in LH2 light-harvesting complexes from *Allochromatium vinosum*. *Photosynth. Res.* 127: 171–187.

Makhneva, Z., Bolshakov, M., and Moskalenko, A. A. 2008. Heterogeneity of carotenoid content and composition in LH2 of the purple sulphur bacterium Allochromatium minutissimum grown under carotenoid-biosynthesis inhibition. *Photosynth. Res.* 98(1–3): 633–641.

Martin, J. L., Breton, J., Hoff, A. J., Migus, A., and Antonetti, A. 1986. Femtosecond spectroscopy of electron transfer in the reaction center of the photosynthetic bacterium *Rhodopseudomonas sphaeroides* R-26: Direct electron transfer from the dimeric bacteriochlorophyll primary donor to the bacteriopheophytin acceptor with a time constant of 2.8 ± 0.2 psec. *Proc. Natl. Acad. Sci. USA* 83: 957–961.

Mascle-Allemand, C., Duquesne, K., Lebrun, R., Scheuring, S., and Sturgis, J. N. 2010. Antenna mixing in photosynthetic membranes from *Phaeospirillum molischianum. Proc. Natl. Acad. Sci. USA* 107: 5357–5362.

McAuley, K. E., Fyfe, P. K., Cogdell, R. J., Isaacs, N. W., and Jones, M. R. 2000. X-ray crystal structure of the YM210W mutant reaction centre from *Rhodobacter sphaeroides. FEBS Lett.* 467(2–3): 285–290.

McDermott, G., Prince, S. M., Freer, A. A. et al., 1995. Crystal structure of an integral membrane light-harvesting complex from photosynthetic bacteria. *Nature* 374: 517–521.

McLuskey, K., Prince, S. M., Cogdell, R. J., and Isaacs, N. W. 2001. The crystallographic structure of the B800-820 LH3 light-harvesting complex from the purple bacteria *Rhodopseudomonas Acidophila* strain 7050. *Biochemistry* 40: 8783–8789.

Miller, K. R. 1979. Structure of a bacterial photosynthetic membrane. *Proc. Natl. Acad. Sci. USA* 76: 6415–6419.

Mitchell, P. 1976. Possible molecular mechanisms of the protonmotive function of cytochrome systems. *J. Theor. Biol.* 62: 327–367.

Monshouwer, R., Visschers, R. W., Vanmourik, F., Freiberg, A., and van Grondelle, R. 1995. Low-temperature absorption and site-selected fluorescence of the light-harvesting antenna of *Rhodopseudomonas viridis*—Evidence for heterogeneity. *Biochim. Biophys. Acta* 1229: 373–380.

Mothersole, D. J., Jackson, P. J., Vasilev, C. et al. 2016. PucC and LhaA direct efficient assembly of the light-harvesting complexes in *Rhodobacter sphaeroides. Mol. Microbiol.* 99: 307–327.

Nagarajan, V. and Parson, W. W. 1997. Excitation energy transfer between the B850 and B875 antenna complexes of *Rhodobacter sphaeroides. Biochemistry* 36: 2300–2306.

Naylor, G. W., Addlesee, H. A., Gibson, L. C. D., and Hunter, C. N. 1999. The photosynthesis gene cluster of *Rhodobacter sphaeroides. Photosynth. Res.* 62: 121–139.

Niederman, R. A., Mallon, D. E., and Parks, L. C. 1979. Membranes of *Rhodopseudomonas sphaeroides.* VI. Isolation of a fraction enriched in newly synthesized bacteriochlorophyll a-protein complexes. *Biochim. Biophys. Acta* 555: 210–220.

Niwa, S., Yu, T. J., Takeda, K. et al. 2014. Structure of the LH1-RC complex from *Thermochromatium tepidum* at 3.0 Å. *Nature* 508: 228–232.

Olsen, J. D., Sockalingum, G. D., Robert, B., and Hunter, C. N. 1994. Modification of a hydrogen bond to a bacteriochlorophyll *a* molecule in the light harvesting 1 antenna of *Rhodobacter sphaeroides. Proc. Natl. Acad. Sci. USA* 91: 7124–7128.

Olsen, J. D., Sturgis, J. N., Westerhuis, W. H., Fowler, G. J. S., Hunter, C. N., and Robert, B. 1997. Site-directed modification of the ligands to the bacteriochlorophylls of the light-harvesting LH1 and LH2 complexes of *Rhodobacter sphaeroides. Biochemistry* 36: 12625–12632.

Olsen, J. D., Tucker, J. D., Timney, J. A., Qian, P., Vassilev, C., and Hunter, C. N. 2008. The organization of LH2 complexes in membranes from *Rhodobacter sphaeroides. J. Biol. Chem.* 283: 30772–30779.

Peters, G. A. and Cellarius, R. A. 1972. Photosynthetic membrane development in *Rhodopseudomonas spheroides.* II. Correlation of pigment incorporation with morphological aspects of thylakoid formation. *J. Bioenerg.* 3: 345–359.

Prince, S. M., Papiz, M. Z., Freer, A. A. et al. 1997. Apoprotein structure in the LH2 complex from *Rhodopseudomonas acidophila* strain 10050: Modular assembly and protein pigment interactions. *J. Mol. Biol.* 268: 412–423.

Pugh, R. J., McGlynn, P., Jones, M. R., and Hunter, C. N. 1998. The LH1-RC core complex of *Rhodobacter sphaeroides*: Interaction between components, time-dependent assembly, and topology of the PufX protein. *Biochim. Biophys. Acta* 1366: 301–316.

Qian, P., Bullough, P., and Hunter, C. N. 2008. Three-dimensional reconstruction of a membrane-bending complex. *J. Biol. Chem.* 283: 14002–14011.

Qian, P., Hunter, C. N., and Bullough, P. A. 2005. The 8.5 Å projection structure of the core RC-LH1-PufX dimer of *Rhodobacter sphaeroides*. *J. Mol. Biol.* 349: 948–960.

Qian, P., Papiz, M. Z., Jackson, P. J. et al. 2013. Three-dimensional structure of the *Rhodobacter sphaeroides* RC-LH1-PufX complex: Dimerization and quinone channels promoted by PufX. *Biochemistry* 52: 7575–7585.

Reilly, P. A. and Niederman, R. A. 1986. Role of apparent membrane growth initiation sites during photosynthetic membrane development in synchronously dividing *Rhodopseudomonas sphaeroides*. *J. Bacteriol.* 167: 153–159.

Roszak, A. W., Howard, T. D., Southall, J. et al., 2003. Crystal structure of the RC-LH1 core complex from *Rhodopseudomonas palustris*. *Science* 302: 1969–1972.

Saer, R. G. and Blankenship, R. E. 2017. Light harvesting in phototrophic bacteria: Structure and function. *Biochem. J.* 474: 2107–2131.

Savage, H., Cyrklaff, M., Montoya, G., Kühlbrandt, W., and Sinning, I. 1996. Two-dimensional structure of light harvesting complex II (LHII) from the purple bacterium *Rhodovulum sulfidophilum* and comparison with LHII from *Rhodopseudomonas acidophila*. *Structure* 4: 243–252.

Scheuring, S. 2006a. AFM studies of the supramolecular assembly of bacterial photosynthetic core-complexes. *Curr. Opin. Chem. Biol.* 10: 387–393.

Scheuring, S. 2006b. AFM studies of the supramolecular assembly of bacterial photosynthetic core-complexes. *Curr. Opin. Chem. Biol.* 10: 387–393.

Scheuring, S., Levy, D., and Rigaud, J. L. 2005. Watching the components of photosynthetic bacterial membranes and their in situ organisation by atomic force microscopy. *Biochim. Biophys. Acta* 1712: 109–127.

Scheuring, S., Reiss-Husson, F., Engel, A., Rigaud, J. L., and Ranck, J. L. 2001. High-resolution AFM topographs of *Rubrivivax gelatinosus* light- harvesting complex LH2. *EMBO J.* 20: 3029–3035.

Scheuring, S., Rigaud, J. L., and Sturgis, J. 2004. Variable LH2 stoichiometry and core clustering in native membranes of *Rhodspirillum photometricum*. *EMBO J.* 23: 4127–4133.

Scheuring, S., Seguin, J., Marco, S., Levy, D., Robert, B., and Rigaud, J. L. 2003. Nanodissection and high-resolution imaging of the *Rhodopseudomonas viridis* photosynthetic core complex in native membranes by AFM. *Proc. Natl. Acad. Sci. USA* 100: 1690–1693.

Scheuring, S. and Sturgis, J. N. 2005. Chromatic adaption of photosynthetic membranes. *Science*. 309: 484–487.

Sener, M., Hsin, J., Trabuco, L. G. et al. 2009. Structural model and excitonic properties of the dimeric RC-LH1-PufX complex from *Rhodobacter sphaeroides*. *Chem. Phys.* 357: 188–197.

Sener, M. K., Olsen, J. D., Hunter, C. N., and Schulten, K. 2007. Atomic-level structural and functional model of a bacterial photosynthetic membrane vesicle. *Proc. Natl. Acad. Sci. USA* 104: 15723–15728.

Sener, M. K. and Schulten, K. 2009. From atomic-level structure to supramolecular organization in the photosynthetic unit of purple bacteria. In *The Purple Phototrophic Bacteria*, eds. C. N. Hunter et al., pp. 275–294. Springer, Dordrecht, The Netherlands.

Sener, M. K., Strumpfer, J., Singharoy, A., Hunter, C. N., and Schulten, K. 2016. Overall energy conversion efficiency of a photosynthetic vesicle. *elife* 5: e09541.

Siebert, C. A., Qian, P., Fotiadis, D., Engel, A., Hunter, C. N., and Bullough, P. A. 2004. Molecular architecture of photosynthetic membranes in *Rhodobacter sphaeroides*: The role of PufX. *EMBO J.* 23: 690–700.

Siefert, E., Irgens, R. L., and Pfennig, N. 1978. Phototrophic purple and green bacteria in a sewage treatment plant. *Appl. Environ. Microbiol.* 35: 38–44.

Sturgis, J. N., Olsen, J. D., Robert, B., and Hunter, C. N. 1997. Functions of conserved tryptophan residues of the core light-harvesting complex of *Rhodobacter sphaeroides*. *Biochemistry* 36: 2772–2778.

Sturgis, J. N., Tucker, J. D., Olsen, J. D., Hunter, C. N., and Niederman, R. A. 2009. Atomic force microscopy studies of native photosynthetic membranes. *Biochemistry* 48: 3679–3698.

Tichy, H. V., Oberlé, B., Stiehle, H., Schiltz, E., and Drews, G. 1989. Genes downstream from *pucB* and *pucA* are essential for formation of the B800-850 complex of *Rhodobacter capsulatus*. *J. Bacteriol.* 171: 4914–4922.

Todd, J. B., Parkes-Loach, P. S., Leykam, J. F., and Loach, P. A. 1998. *In vitro* reconstitution of the core and peripheral light-harvesting complexes of *Rhodospirillum molischianum* from separately isolated components. *Biochemistry* 37: 17458–17468.

Todd, J. B., Recchia, P. A., Parkes-Loach, P. S. et al. 1999. Minimal requirements for *in vitro* reconstitution of the structural subunit of light-harvesting complexes of photosynthetic bacteria. *Photosynth. Res.* 62: 85–98.

Tucker, J. D., Siebert, C. A., Escalante, M. et al. 2010. Membrane invagination in *Rhodobacter sphaeroides* is initiated at curved regions of the cytoplasmic membrane, then forms both budded and fully detached spherical vesicles. *Mol. Microbiol.* 76: 833–847.

Tunnicliffe, R. B., Ratcliffe, E. C., Hunter, C. N., and Williamson, M. P. 2006. The solution structure of the PufX polypeptide from *Rhodobacter sphaeroides. FEBS Lett.* 580: 6967–6971.

van Grondelle, R., Dekker, J. P., Gillbro, T., and Sundström, V. 1994. Energy-transfer and trapping in photosynthesis. *Biochim. Biophys. Acta* 1187: 1–65.

Vatter, A. E. and Wolfe, R. S. 1958. The structure of photosynthetic bacteria. *J. Bacteriol.* 75: 480–488.

Visscher, K. J., Bergström, H., Sundström, V., Hunter, C. N., and van Grondelle, R. 1989. Temperature dependence of energy-transfer from the long wavelength antenna BChl-896 to the reaction center in *Rhodospirillum rubrum, Rhodobacter sphaeroides* (w.t. and M2 mutant) from 77 to 177K, studied by picosecond absorption spectroscopy. *Photosynth. Res.* 22: 211–217.

Walz, T., Jamieson, S. J., Bowers, C. M., Bullough, P. A., and Hunter, C. N. 1998. Projection structures of three photosynthetic complexes from *Rhodobacter sphaeroides*: LH2 at 6 Å, LH1 and RC-LH1 at 25 Å. *J. Mol. Biol.* 282: 833–845.

Williams, J. C., Steiner, L. A., and Feher, G. 1986. Primary structure of the reaction center from *Rhodopseudomonas sphaeroides. Proteins* 1: 312–325.

Young, C. S., Reyes, R. C., and Beatty, J. T. 1998. Genetic complementation and kinetic analyses of *Rhodobacter capsulatus* ORF1696 mutants indicates that the ORF1696 protein enhances assembly of the light-harvesting I complex. *J. Bacteriol.* 180: 1759–1765.

Zeng, X., Choudhary, M., and Kaplan, S. 2003. A second and unusual *pucBA* operon of *Rhodobacter sphaeroides* 2.4.1: Genetics and function of the encoded polypeptides. *J. Bacteriol.* 185: 6171–6184.

Zhu, Y. S. and Hearst, J. E. 1986. Regulation of expression of genes for light-harvesting antenna proteins LH-I and LH-II; reaction center polypeptides RC-L, RC-M and RC-H; and enzymes of bacteriochlorophyll and carotenoid biosynthesis in *Rhodobacter capsulatus* by light and oxygen. *Proc. Natl. Acad. Sci. USA* 83: 7613–7617.

Zhu, Y. S. and Kaplan, S. 1985. Effects of light, oxygen and substrates on steady-state levels of mRNA coding for ribulose 1,5-bisphosphate carboxylase and light-harvesting and reaction center polypeptides in *Rhodopseudomonas sphaeroides. J. Bacteriol.* 162: 925–932.

Zsebo, K. M. and Hearst, J. E. 1984. Genetic-physical mapping of a photosynthetic gene cluster from *Rhodopseudomonas capsulata. Cell* 37: 937–947.

Light harvesting in green bacteria

JAKUB PŠENČÍK AND TOMÁŠ MANČAL

ABBREVIATIONS

2DES	2D electronic spectroscopy
BChl	Bacteriochlorophyll
Cab.	*Chloracidobacterium*
Cba.	*Chlorobaculum*
Cfl.	*Chloroflexus*
Chl.	*Chlorobium*
cryo-EM	Electron cryomicroscopy
FAP	Filamentous anoxygenic phototrophs
FMO	Fenna–Matthews–Olson
GSB	Green sulfur bacteria
LH	Light-harvesting
RC	Reaction center
Rfl.	*Roseiflexus*

7.1 INTRODUCTION

The term green bacteria comprises photosynthetic organisms containing large light-harvesting (LH) complexes called chlorosomes. It includes most of the members of the phylum *Chlorobi* (green sulfur bacteria, or GSB), some of the *Chloroflexi* (filamentous anoxygenic phototrophs, or FAP), and one known member of *Acidobacteria*, *Chloracidobacterium* (*Cab.*) *thermophilum* (Tank and Bryant 2015). The most studied representative of GSB is *Chlorobaculum* (*Cba.*) *tepidum*, while for FAP it is *Chloroflexus* (*Cfl.*) *aurantiacus*. The green bacteria can be divided into two groups according to the organization of their photosynthetic apparatus: GSB and phototrophic acidobacterium *Cab. thermophilum* exhibit similar compositions (Figure 7.1a) including the presence of the so-called Fenna–Matthews–Olson (FMO) complex and a type I reaction center (RC), while FAP contain circular membrane antennas similar to both LH1 and LH2 of purple bacteria, and a type II RC (Figure 7.1b). FAP can be further divided into two major groups: one includes bacteria containing chlorosomes (green FAP) and the second includes species without chlorosomes (red FAP) (Klappenbach and Pierson 2004; Le Olson et al. 2007). According to

the definition mentioned earlier, the red FAP do not belong to green bacteria, but except for the missing chlorosomes, their photosynthetic apparatus is very similar to the one of the green FAP.

The chlorosome is thus the unifying LH element of green bacteria. It is a peripheral membrane complex located on the inner side of the cytoplasmic membrane. Typically, it has an ellipsoidal shape and dimensions of about $150 \times 50 \times 25$ nm, which makes the chlorosome the largest natural photosynthetic antenna complex. The chlorosome in fact represents two antenna complexes. The interior of the chlorosome does not contain any protein, and it is instead composed of $\sim 10^5$ specialized bacteriochlorophyll (BChl) molecules arranged in aggregates in addition to carotenoids and quinones. The photons absorbed by the BChl aggregates and carotenoids in the chlorosome interior are converted to excitations that are transferred to a

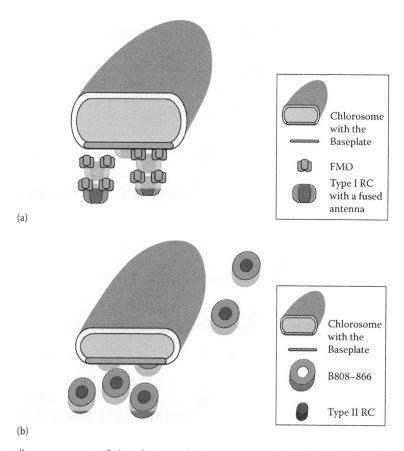

(a)

(b)

Figure 7.1 Overall arrangement of the photosynthetic apparatus in *Chlorobi* and *Acidobacteria* **(a)** and in *Chloroflexi* **(b)**. Chlorosomes are attached to the inner side of a cytoplasmic membrane, which is represented by the light yellow layer. The chlorosomes are shown in cross section to depict the position of the baseplate and to facilitate the observation of the complexes located below the chlorosomes.

second antenna system represented by the baseplate of the chlorosome, a pigment–protein complex located on the side of the chlorosome facing the cytoplasmic membrane. In the case of *Chlorobi* and *Acidobacteria*, the chlorosome is attached to the cytoplasmic membrane via another peripheral antenna complex, the FMO protein complex. FMO is in contact with a type I RC buried in the cytoplasmic membrane. This type of RC contains additional core LH pigments, and for *Chlorobi* the primary electron donor is called P840. Energy is thus transferred from the chlorosome baseplate to the RC through the FMO complex. In the case of FAP, the chlorosome is attached directly to the membrane containing circular antenna complexes called B808–866 for the green FAP and B800–880 for the red FAP. They surround a type II RC with an electron donor denoted as P865 or P870. Excitations flow from the baseplate presumably to B808 (B800) and then to B866 (B880) pigments of the membrane antenna and finally to the RC.

Green bacteria can be found in microbial mats or lower layers of aquatic ecosystems (Castenholz and Pierson 1995; van Gemerden and Mas 1995; Overmann and Garcia-Pichel 2006), often under the habitat of other photosynthetic organisms, and thus at low-light conditions. As an example, BChl *e*–containing GSB were found in the Black Sea at the depths of 100 m living at light intensities of ~0.001–0.002 $\mu E \times m^{-2} \times s^{-1}$ (Manske et al. 2005), which is about 10^6 times less than the maximal irradiance by the Sun at the Earth's surface during the day. The doubling time of this bacterium was estimated to be between 3 and 26 years, being the slowest-growing phototroph known to date (Manske et al. 2005). Another extreme example of a low-light-adapted *Chlorobi* species is a bacterium designated GSB1, which was found in the Pacific Ocean at the depth of 2500 m (Beatty et al. 2005). There is no sunlight penetrating at this depth, and the bacterium lives phototrophically using the near-infrared black-body radiation emitted by the nearby hydrothermal vents. It is the only known organism performing photosynthesis using a source of light other than the sun.

The organization and properties of the LH apparatus are responsible for the light-gathering capability of green bacteria. However, the exact structure of photosynthetic complexes from green bacteria is often not known. Except for the FMO complex, which was the first chlorophyll-containing protein

that was crystallized (Fenna and Matthews 1975), there is no X-ray structure available for any of the LH and RC complexes employed by green bacteria. Also the overall arrangement of the photosynthetic apparatus is much less well characterized as compared to that of, for example, purple bacteria or higher plants. Despite that, a large amount of information about the structure and function of the LH complexes of green bacteria has been obtained by other methods. This information is reviewed in this chapter. Particular attention is dedicated to two of the LH complexes: the chlorosome and the FMO protein. The chlorosome is the main LH complex of green bacteria and the only photosynthetic antenna where pigments are self-assembled into aggregates without a protein support. Its size and pigment organization is the key for understanding the ability of green bacteria to survive at low-light conditions. The FMO protein became one of the most important model LH complexes, especially for theoretical studies. It is mostly due to its favorably small size, known structure, the relatively strong exciton coupling between the BChl *a* molecules, and its well-studied spectroscopic properties. Although it attracted a lot of attention from theoreticians, it is probably not the most typical photosynthetic antenna. It is, for example, one of the few LH complexes that do not contain carotenoids, and it acts as a wire connecting the baseplate with the RC complex rather than a light harvester.

7.2 CHLOROSOME

The chlorosome is a remarkable LH complex differing from other known photosynthetic antennas by its large size and the organization of its pigments. While in other LH complexes pigments are held in appropriate positions by a protein scaffold, the chlorosome interior is composed of ~10^5 strongly interacting BChl *c*, *d*, or *e* molecules (Montano et al. 2003a; Saga et al. 2007; Psencik et al. 2014), self-assembled into aggregates without any direct involvement of proteins. The aggregation determines the spectral, light-gathering, and energy-transferring properties of the pigment complex as summarized in the following text. More detailed information can be found in recent reviews on chlorosomes (Oostergetel et al. 2010; Orf and Blankenship 2013; Psencik et al. 2014) and references therein. The baseplate of the chlorosome is described in a separate section.

7.2.1 Organization of the chlorosome interior

As already mentioned, chlorosomes are the largest known photosynthetic antennas. A typical chlorosome (Figure 7.2) is an ellipsoidal body that is 100–200 nm in length, 30–70 nm in width, and 10–40 nm in height, but exhibiting a large variability between species and growing conditions (Martinez-Planells et al. 2002; Psencik et al. 2006; Adams et al. 2013). The interior of the chlorosome contains mainly aggregates of BChl *c*, *d*, or *e*, with a minor contribution from carotenoids and quinones. Proteins are confined to the periphery of the chlorosome where they form two structural elements. On the side where the chlorosome is facing the cytoplasmic membrane, the proteins form the baseplate. The baseplate contains BChl *a* and carotenoids and

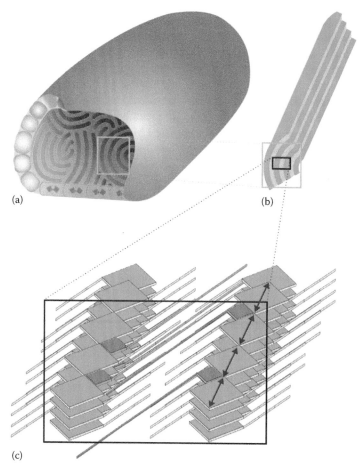

Figure 7.2 Schematic representation of a typical chlorosome. **(a)** The chlorosome is an ellipsoidal body separated from the cytoplasm by an envelope consisting of proteins (gray blobs) and lipids (cyan background). On the side facing the cytoplasmic membrane, there is a baseplate (orange) composed of the CsmA protein containing BChl *a* (green squares). The orange color of the baseplate refers to the presence of carotenoids. The interior of the chlorosomes is composed of curved lamellar layers of BChl aggregates (green) with carotenoids and quinones filling the space between them (orange). **(b)** Lamellar quarter-cylinder as a basic structural motif of both open curved lamellar structures prevailing in wild-type chlorosomes and multilayered cylinders observed mostly in a bchQRU mutant of *Cba. tepidum*. **(c)** Scheme of the pigment organization. The arrangement of BChl molecules (green) shown here is compatible with both an antiparallel and syn-anti BChl dimer as the building block of the aggregate. The orientation of the Q_y transition dipole moments is depicted by red arrows. Carotenoids and quinones (orange) are localized in the hydrophobic space between the BChl layers, and their close contact with BChls is ensured via π–π or CH–π interactions.

represents a distinct antenna system that functionally connects the chlorosome to the rest of the photosynthetic apparatus (Section 7.3). The remaining surface layer of the chlorosome facing the cytoplasm is denoted as an envelope here. It is composed mainly of proteins with minor contribution from lipids (Sorensen et al. 2008), but it does not contain any pigments, and thus has no LH function. The proteins found in the chlorosome can be clustered into four structural groups (Frigaard and Bryant 2006). CsmA is the most abundant protein and (possibly together with CsmE) forms the baseplate of the chlorosome. The remaining proteins are found in the envelope, and their functions are discussed in Section 7.2.2.2.

Chlorosomes from *Chlorobi* contain predominantly either BChl *c*, *d*, or *e*, while only BChl *c* was found in chlorosomes from *Chloroflexi* and acidobacterium *Cab. thermophilum*. BChl *c*, *d*, and *e* differ from each other by the substituents at positions C7 and C20 (Figure 7.3). The fourth possible

combination of the substituents leads to BChl *f*, which was not found to occur in nature, but was recently prepared by chemical synthesis (Tamiaki et al. 2011) and then by mutagenesis of BChl *e*–synthesizing *Chlorobi* species (Vogl et al. 2012). As discussed in Section 7.2.2.2, BChl *f* is not well suited for excitation energy transfer (EET) to the BChl *a*–containing baseplate, which is probably the reason why no wild-type bacterium producing BChl *f* was found in nature so far (Orf et al. 2013). One of the interesting properties of BChl *c*, *d*, and *e* is that they are found in chlorosomes as a mixture of several homologs and diastereoisomers differing in substituent at C8, C12, and C17[3] and in stereochemistry at C3[1] (Figure 7.3). The aggregation of BChl *c*, *d*, and *e* in the chlorosome interior is enabled by another unique property of their molecular structure. All these BChls contain a hydroxy group at C3[1] and a keto group at C13[1] (Figure 7.3). BChl molecules are held together by intermolecular coordination of the central Mg ion of one BChl molecule to the hydroxy group of a second BChl, forming a stack of BChls. The second interaction is a formation of a hydrogen bond between the hydroxy group at C3[1] of the second BChl and the keto group at C13[1] of a third BChl molecule, connecting the stacks into a layer and keeping the transition dipole moments in a direction allowing a strong exciton coupling between them (see following text). The third interaction involved in the aggregate formation is the π–π stacking between the chlorin rings, which contributes most to the stabilization energy (Alster et al. 2012). These three interactions allow several different possibilities for building the aggregate. Currently, the most frequent models are based on the so-called antiparallel dimer (Nozawa et al. 1994; Egawa et al. 2007) and syn-anti dimer (Ganapathy et al. 2009) as the basic unit. In both of these models, the esterifying alcohols of the two BChls forming the dimer extend to opposite sides of the layer. This allows the hydrophobic interaction between the esterifying alcohols from the neighboring layers to combine the layers into lamellar structures (Figure 7.2) as observed in electron cryomicroscopy (cryo-EM) projections and X-ray scattering experiments (Psencik et al. 2004; Oostergetel et al. 2007, 2010; Psencik et al. 2010). The hydrophobic interaction thus represents the fourth interaction involved in the aggregate formation. The lamellar layers are usually curved and may form closed cylinders in the case of a minimal disorder.

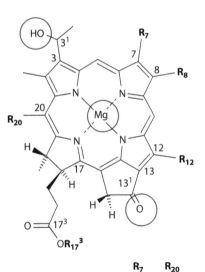

	R$_7$	R$_{20}$
Bacteriochlorophyll *c*	CH$_3$	CH$_3$
Bacteriochlorophyll *d*	CH$_3$	H
Bacteriochlorophyll *e*	CHO	CH$_3$
Bacteriochlorophyll *f*	CHO	H

Figure 7.3 Molecular structure of BChl *c*, *d*, *e*, and *f*, which differ in the substituents at R$_7$ and R$_{20}$. R$_8$ substituents are mainly ethyl, propyl, or isobutyl; R$_{12}$ can be methyl or ethyl. R$_{17}$^{3} stands for the alkyl group of the esterifying alcohol. A representative esterifying alcohol found in GSB is farnesol and stearol for FAP. The red circles label the groups involved in an aggregate formation.

Such lamellar cylinders were observed mostly in a triple mutant of *Cba. tepidum* producing a single BChl *c* homolog (Oostergetel et al. 2007). In contrast, the wild-type *Cba. tepidum* synthesizes four homologs of BChl *c*, and the internal arrangement is more disordered; nevertheless, the tubular structures still can be observed (Oostergetel et al. 2007). In other structurally characterized chlorosomes, a much more disordered organization was observed (for a review, see Psencik et al. 2014). In BChl *e*–containing *Chlorobium (Chl.) phaeovibrioides* and *Chl. phaeobacteroides*, the aggregates form lamellar domains with varying orientations. In BChl *c*–containing *Cfl. aurantiacus*, the lamellar system is parallel to the long axis of the chlorosome as in *Cba. tepidum*, but it is more disordered and the chlorosomes are often too thin to contain even a double-layered cylinder. In recently characterized acidobacterium, *Cab. thermophilum* domains of curved lamellae were observed (Costas et al. 2011). *Cba. tepidum* thus remains the only species in which the chlorosomes with the multilayered cylinders were observed, but even there they probably do not represent the prevailing type of arrangement. The lack of cylindrical symmetry in wild-type *Cba. tepidum* chlorosomes was confirmed in single-molecule spectroscopy studies (Shibata et al. 2009; Furumaki et al. 2011). In another single-molecule study, cylindrical symmetry was proposed for the wild-type chlorosomes from *Cba. tepidum* (Tian et al. 2011); however, this conclusion was based on downplaying the fact that chlorosomes orient preferentially with their baseplate toward the glass surface, as confirmed in other single-molecule spectroscopy and atomic force microscopy studies (Shibata et al. 2006; Adams et al. 2013).

Due to the self-assembly, the overall organization of BChl aggregates in each chlorosome is unique and exhibits a significant variability even inside a single chlorosome. This prevents the determination of the chlorosome structure by X-ray crystallography. The unifying feature of all the observed arrangements is that they consist of repeating layers, so they have a lamellar character (Figure 7.2). The alternating layers of relatively polar chlorin rings and layers of nonpolar esterifying alcohols resemble the structure of block copolymers. In the hydrophobic space formed by interdigitated esterifying alcohols, carotenoids and quinones are located (Psencik et al. 2006). Carotenoids serve light-harvesting (Section 7.2.2.1) and protective (Section 7.2.2.3) functions in

the chlorosomes and were shown to strengthen the hydrophobic interaction between esterifying alcohols leading to aggregate formation (Klinger et al. 2004). Quinones are involved in protective excitation quenching, which occurs in chlorosomes from species with type I RCs in the presence of oxygen (Section 7.2.2.3). As there is no protein in the chlorosome interior, there must be some other way how to keep a close contact of both carotenoids and quinones with BChls. This contact is required for their function in the chlorosomes. GSB contain mainly carotenoids with a conjugated ϕ-ring (e.g., chlorobactene and isorenieratene), while *Cfl. aurantiacus* accommodates mainly β- and γ-carotene, which possess a β-ring. The conjugated ϕ-ring is well suited for the π–π interaction with the conjugated system of BChls, while β-rings may interact with it via CH–π interactions (Psencik et al. 2013, 2014). Similarly, the quinones found in chlorosomes possess a conjugated bicyclic dione structure, which is likely to be involved in a π–π interaction with a conjugated system of BChls. The proposed arrangement (Figure 7.2c) allows BChls to functionally interact with other molecules located in their vicinity without disrupting the excitonic coupling between BChls.

The excitonic coupling between BChls is the result of close intermolecular distances between the molecules within one lamellar layer of the aggregate and suitable orientations of their transition dipole moments (Figure 7.2c). It manifests itself as a red shift of the Q_y absorption band. This band is found around 650–670 nm for the monomeric form of the pigments in organic solvents, while it shifts to ~745 nm for BChl *c*, ~730 nm for BChl *d*, ~715 nm for BChl *e*, and 705 nm for BChl *f* in chlorosomes (Orf et al. 2013; Psencik et al. 2014). Because of the red shift, the BChl aggregates are often denoted as J-aggregates. However, it should be stressed that except for the red shift, other properties of BChl aggregates are very different compared to J-aggregates. In pure J-aggregates, all the oscillator strength is concentrated into the lowest exciton level, which results in a high fluorescence quantum yield. On the contrary, in chlorosomal aggregates, the states with the largest oscillator strength (corresponding to the absorption maximum) lie ~500 cm^{-1} above the lowest exciton levels (corresponding to the fluorescence maximum), which have much lower oscillator strengths. This leads to an exceptionally large shift between the fluorescence and absorption maxima. In any excitonically coupled systems, absorption into the higher levels is followed by fast exciton relaxation to the red-shifted lowest exciton levels. As these states of BChl aggregates have smaller oscillator strengths, also the probability of excitation loss by fluorescence is lower. While the large oscillator strength is required for a strong fluorescence, it is not necessary for an efficient energy transfer between molecules with short distances between them (Fleming and Scholes 2004), as in the chlorosome. It should be also mentioned that a discrepancy exists between the oscillator strengths and lifetimes of the lowest exciton states. Smaller oscillator strengths of these states should imply a longer fluorescence lifetime. However, the fluorescence lifetime of the aggregates was reported to be in the order of tens of picoseconds (Causgrove et al. 1990a; van Noort et al. 1997), which is much shorter than, for an example, the lifetimes of BChl monomers, which were determined to be in the order of nanoseconds (Niedzwiedzki and Blankenship 2010). This may be caused by a presence of an additional relaxation channel shortening the fluorescence decay. One possibility is a presence of low-lying dark states, similar to what has been observed for aggregates of chlorophyll and pseudoisocyanine dyes (Oksanen et al. 1997; Vacha et al. 1998). These dark states may have a charge-transfer character (Frese et al. 1997) and are likely to be involved in energy transfer.

Another difference between the BChl- and J-aggregates is that exchange narrowing, which is typical for J-aggregates, is not observed for BChl aggregates (Oostergetel et al. 2010). Instead, broadening of the absorption band is observed upon aggregation, even if a single BChl homolog or diastereoisomer is used (Uehara et al. 1994; Steensgaard et al. 2000a). All wild-type bacteria produce more than one BChl homolog, which do not differ from each other spectroscopically in the monomeric form, but they do differ in the aggregated form (Uehara et al. 1994; Steensgaard et al. 2000a). It seems that the main reason why bacteria invest their energy into producing more homologs is that the aggregation of molecules with different substituents at the chlorin ring and with different stereochemistry at $C3^1$ increases the structural disorder. The increased disorder is translated into variable site energies and couplings between the monomers that further broaden the

absorption band to improve spectral coverage, and it is thus beneficial for light harvesting. The presence of substantial disorder is thus another feature that is typical for BChl aggregates from wild-type bacteria and manifests itself not only spectroscopically but also in X-ray scattering (Psencik et al. 2010) and cryo-EM (Oostergetel et al. 2007).

7.2.2 Function of the chlorosome interior

The main function of the chlorosomes is to absorb photons and transfer the excitation energy to the baseplate and then toward the RC. The EET steps following excitation are summarized in Sections 7.2.2.1 and 7.2.2.2. Due to the very high pigment density in the chlorosome interior, the EET processes may become easily affected by exciton annihilation if high excitation intensities are used in an experiment. In addition, efficient EET in GSB and phototrophic acidobacteria occurs only at anaerobic conditions. In the presence of oxygen, a protective excitation quenching occurs. The redox-dependent excitation quenching is an example of regulative and protective mechanisms, which are summarized in Section 7.2.2.3.

7.2.2.1 EET WITHIN THE CHLOROSOME INTERIOR

The BChl aggregates inside the chlorosome can be excited either directly or via EET from carotenoids. The transfer from carotenoids was studied mainly by measuring fluorescence excitation spectra, which provided direct information about the efficiency of the process. Carotenoids in the chlorosomes are absorbing mostly in the spectral region between ~425 and 525 nm and transfer excitation energy to BChls with a quantum efficiency of 50%–80% (van Dorssen et al. 1986; Melo et al. 2000; Psencik et al. 2002). As the absorption into the S_1 state of carotenoids is dipole forbidden, it is the S_2 state that is responsible for most of the carotenoid absorption in the visible spectral region. Absorption is followed by a rapid relaxation into the S_1 state, resulting in an extremely short S_2 state lifetime. For example, β-carotene has an S_2 state lifetime of ~150 fs (Macpherson and Gillbro 1998). Despite the very short lifetime, rather efficient EET from the carotenoid S_2 state to BChl e was observed in chlorosomes from GSB *Chl. phaeobacteroides*. The efficiency of EET from carotenoids to BChls was determined to be between 60% and 70% in this particular case, and the carotenoid S_2 state lifetime shortened to ~50 fs. This shortening corresponds to a transfer time of 65–100 fs (Psencik et al. 2002). The S_1 state lifetime of the carotenoids in these chlorosomes was observed to be about 10 ps, which is similar to the lifetime in a solvent (Fuciman et al. 2010). This result indicates that the S_1 state is not significantly involved in the EET. The prevailing EET from the carotenoid S_2 state was observed also in an artificial antenna consisting of BChl c aggregates with β-carotene (Alster et al. 2010). Since BChl c and β-carotene are the most abundant pigments in *Cfl. aurantiacus*, these results suggest that the S_2 channel plays a dominant role also in green FAP.

Other possibilities how to populate the exciton manifold within the Q_y band of BChl aggregates are direct excitation and internal relaxation after excitation of the Soret band. Internal relaxation occurs on a sub-100 fs timescale (Psencik et al. 2002). In a strongly coupled system, like one lamellar layer of a BChl aggregate, the excitation cannot be considered localized on a single pigment, but it is delocalized over a certain part of the aggregate. However, a significant amount of the structural disorder present in the chlorosomes prevents the delocalization over a larger scale and splits the lamellar layer into domains of strongly coupled pigments (Dostal et al. 2012). While the excitation is transferred coherently within the domain, the EET between the domains can be described as incoherent hopping. The domains thus act as "supermolecules," and this organization contributes to the high overall EET efficiency, as discussed in Section 7.6.2. To be consistent with the observed red shift, the number of BChl molecules within a single domain should be in the order of 10–100 molecules. The EET between the domains occurs on a sub-100 fs timescale and manifests itself as a diffusion through a disordered energy landscape of the chlorosome (Dostal et al. 2012). The disorder induces a rapid loss of coherence during the migration between the domains and prevents the chlorosome from acting as a coherent light harvester. On the other hand, the disorder improves the spectral coverage of the chlorosome by broadening the Q_y absorption band to a bandwidth of ~500–1000 cm^{-1} (Psencik et al. 2014). Such a large bandwidth imposes that each of the coherent domains exhibits several allowed transitions and

that there is significant spectral variability between the domains. Convolution of the inhomogeneous broadening of the lowest exciton level distribution (100–250 cm^{-1} [Psencik et al. 2014]) with the homogeneous linewidth of a single upper exciton level (≤350 cm^{-1} [Dostal et al. 2012]) alone would not explain the observed bandwidth.

Exciton relaxation within the domains occurs simultaneously with the excitation diffusion, and therefore the excitation migration between the domains is energetically downhill biased. While the diffusion between the domains was not resolved before the use of coherent 2D electronic spectroscopy (2DES) with chlorosomes, the exciton relaxation was observed already by using the pump–probe method. It manifests itself as EET from the blue part of the Q_y absorption band of BChl aggregates to their red-shifted states and was first observed in chlorosomes from *Cfl. aurantiacus* at low temperature (Savikhin et al. 1996). Owing to the low excitation intensities used, the process could also be resolved at room temperature. The transfer times were observed to increase with the energetic separation between the pump and probe wavelengths from ~200 to 1000 fs for BChl e–containing *Chl. phaeobacteroides* (Psencik et al. 2003) and from 150 to 250 fs in BChl c–containing *Cfl. aurantiacus* (Martiskainen et al. 2009). The smaller span of values for the latter most likely reflects a narrower exciton manifold in *Cfl. aurantiacus* compared to *Chlorobi* species. However, for BChl c–containing *Chlorobi* members, *Cba. tepidum* and *Prosthecochloris aestuarii,* this process was observed to be similarly fast (100–250 fs) (Martiskainen et al. 2012). This is perhaps due to the better ordered BChl c aggregates in *Cba. tepidum* as compared to other bacteria species or due to the presence of an additional fast quenching process (Section 7.2.2).

The relaxation within the chlorosomes on a slower timescale was studied by several groups, and many different decay components have been identified on the 1–100 ps timescale depending on the species, growth, and experimental conditions (for reviews, see Oostergetel et al. 2010; Orf and Blankenship 2013; Psencik et al. 2014). These decays were often observed without a corresponding rise in the spectral region of the excitation acceptor, making a reliable assignment to EET steps difficult. They probably reflect EET within the chlorosome interior and from the chlorosome to the baseplate. Since the organization of aggregates in the chlorosome interior is determined by self-assembly, it is hard to imagine that the domains and the lamellar layers are organized in a way that would ensure an energy funnel toward the baseplate. Most likely, the exciton relaxation is followed by a random incoherent transfer of the partially localized excitations within the whole chlorosome until it is close to the baseplate and can be transferred to BChl *a* therein (Psencik et al. 2003). The random walk may not lead to a decrease of the EET efficiency since the excitations are found in low-energy states with smaller oscillator strengths after exciton relaxation, and the probability of excitation loss by fluorescence is low. Thermal energy may be needed to overcome differences between lowest exciton levels of different domains (Psencik et al. 2014; Jun et al. 2015), which may explain the observed strong dependence of the EET efficiency on temperature (Section 7.2.2.2).

Nevertheless, the existence of an energy gradient within the chlorosome interior cannot be excluded. In many studies, the presence of different spectral forms of aggregated BChls was observed, and it was often proposed that the red forms are located closer to the baseplate in the chlorosome to increase the efficiency of EET (for a review, see Holzwarth et al. 1990; Blankenship and Matsuura 2003). It is, however, not clear by which mechanism such an arrangement could be achieved. A regulation by a different degree of BChl side group alkylation has been suggested (Otte et al. 1991). Perhaps the bands attributed to the long-wavelength form of aggregated BChls in fact reflect the distribution of the lowest exciton levels. In such a case, they would be distributed over the entire chlorosome interior. Chlorosomes from some species contain two different types of aggregated BChl in one chlorosome, BChl *c* and *d*. It was proposed that the two BChls form two separated pools within the chlorosome (Steensgaard et al. 1999) and the pool of BChl *c*, which has a red-shifted absorption compared to BChl *d*, is located closer to the baseplate (Steensgaard et al. 2000b). Again, the mechanism explaining how this could be achieved is unknown.

7.2.2.2 EET FROM THE CHLOROSOME INTERIOR TO THE BASEPLATE

EET from the chlorosome interior to the baseplate may occur either after excitation energy equilibration within the chlorosome or directly after excitation of the part of the BChl aggregate, which is in close contact with the baseplate. In the former case, the transfer time will depend on the size of the chlorosome. The two pathways explain why a bi-exponential rise in the baseplate is observed, consisting of one slower (on the order of 10–100 ps) and one faster (~1 ps or less) component (Psencik et al. 2003; Martiskainen et al. 2009). The main portion of energy from BChl aggregate is transferred to BChl *a* on a timescale slower than that of exciton relaxation and therefore likely via the Förster-type excitation hopping from the localized sites within the aggregates. This transfer step strongly depends on the overlap between the absorption spectrum of BChl *a* and the emission spectrum of the aggregated BChl. The BChl *a* in the baseplate is absorbing around 795 nm, and the spectral overlap is therefore the highest for BChl *c* with its fluorescence maximum at ~775 nm, followed by BChl *d* (~760 nm), BChl *e* (~745 nm), and BChl *f* (~730 nm) having the lowest overlap (Orf et al. 2013). Consequently, the transfer rates between the aggregates and BChl *a* decrease in the same order, at least between BChl *c* and BChl *e* (Causgrove et al. 1992). Among the BChl *c*–containing chlorosomes, the fastest EET has been observed for *Cfl. aurantiacus* chlorosomes (~10 ps) (for a review, see Blankenship et al. 1995; Martiskainen et al. 2009), while somewhat longer lifetimes (~15 ps) were resolved when the same process was observed in whole cells (Causgrove et al. 1990b; Muller et al. 1993). The chlorosomes from *Cfl. aurantiacus* are usually thinner than those of BChl *c*–containing *Chlorobi* members, where the main transfer component was most often determined to be between 30 and 40 ps (Causgrove et al. 1992; van Noort et al. 1997; Steensgaard et al. 2000b) though also longer times have been reported (van Walree et al. 1999). This confirms that the size of the chlorosome contributes to the transfer rates as well (Fetisova et al. 1996). The rate constants mentioned earlier are well established, as they are based on measurements where corresponding decay and rise were observed for the donor and acceptor. Some of the faster transfer times reported in the literature, which were determined without observing the corresponding rise and without the previously resolved slower component, are probably caused by some quenching processes (Section 7.2.2). Slower transfer times were observed for BChl *d*–containing chlorosomes (~65 ps), which exhibit a smaller spectral overlap (Causgrove et al. 1990b; Blankenship et al. 1993). Finally, the longest dominant transfer times (~120 ps) were observed for BChl *e*–containing chlorosomes (Causgrove et al.

1992; Psencik et al. 2003), apparently due to the smallest spectral overlap between BChl e emission and BChl a absorption, and the large size of these chlorosomes. Surprisingly, the relaxation kinetics in chlorosomes containing BChl f was recently shown to be only slightly slower than that obtained for BChl e (Niedzwiedzki et al. 2014).

The spectral overlap determines not only the transfer time but also the quantum efficiency of the EET from the BChl aggregates into the baseplate. This efficiency was studied mainly by the use of fluorescence excitation spectra. In contrast to the kinetics, the excitation spectra obtained for BChl f-containing chlorosomes clearly indicate a lower suitability of BChl f for efficient EET due to the low overlap of its fluorescence spectrum with absorption of BChl a. The efficiency of EET to BChl a in the baseplate was lower (~40%) than for BChl e (65%) in intact photosynthetic membranes at room temperature (Orf et al. 2013). Similar spectral coverage as BChl e but smaller efficiency in EET is probably the reason why BChl f was so far not found in any naturally occurring bacteria. For BChl c and d, the efficiencies are higher, as expected based on the higher spectral overlap, but a wide range of values were reported depending on the sample type and measuring conditions. For instance, EET from BChl c to the membrane-bound LH B866 pigments (Section 7.4.2) in whole cells of the thermophilic $Cfl.$ $aurantiacus$ was observed to occur with an efficiency close to 100% at temperatures above 40°C, while the efficiency was about 60% at 10°C and 15% at 4 K (van Dorssen and Amesz 1988). The efficiency of EET from BChl c to the baseplate has to be the same or higher, since it is one of the steps of the EET to the membrane. The same reasoning is applicable also to the results of the studies discussed in the rest of this paragraph where the entire photosynthetic apparatus was studied and the transfer efficiency to the red-most antenna was determined. An efficiency of EET close to 100% for whole cells of $Cfl.$ $aurantiacus$ was observed also in other studies (Brune et al. 1987; Wang et al. 1990; Taisova et al. 2014); however, only 60% efficiency was determined in membranes with attached chlorosomes (Brune et al. 1987). These results indicate that the EET efficiency may be very susceptible to environment variations. Very low efficiencies are obtained for GSB at aerobic conditions, which are caused by the protective excitation quenching (Section 7.2.2.3).

In the whole cells of BChl c– and d–containing bacteria, the efficiency of EET from aggregates to BChl a was found to drop from nearly 100% at anaerobic conditions to about 10% at aerobic conditions (Wang et al. 1990). The whole cells of $Cfl.$ $aurantiacus$ are not sensitive to redox conditions (Wang et al. 1990).

It is interesting to note that the EET efficiency is not significantly decreased by the backward EET from BChl a in the baseplate to the BChl aggregates. The probability of this process increases with temperature and decreases with the energetic separation between BChl aggregates and BChl a. The backward transfer manifests itself as a changing ratio between the intensity of the emission band from BChl aggregates and BChl a. The ratio is the largest for BChl c–containing bacteria as a consequence of the smallest energy separation (Causgrove et al. 1992). Despite the fact that the portion of photons emitted by aggregates is the largest for BChl c–containing bacteria, they exhibit a higher efficiency of EET from aggregates to BChl a than bacteria containing BChl e, where the backward transfer is much less prominent. This is probably caused by the low quantum yield of fluorescence from the aggregates. As a consequence, the losses by fluorescence do not notably affect the efficiency of EET as determined from the fluorescence excitation spectra.

7.2.2.3 REGULATION AND PROTECTION

Although there is no protein in the chlorosome interior, several mechanisms explaining how green bacteria may regulate the excitation flow within and from the chlorosomes exist. Like all other photosynthetic complexes, also chlorosomes have to deal with the production of harmful singlet oxygen that can be generated under aerobic conditions via the triplet states of BChls. Photosynthetic organisms deal with this threat by employing carotenoids, which are able to both directly scavenge singlet oxygen and quench the triplet states of chlorophylls. In chlorosomes, efficient quenching of BChl triplet states by carotenoids was also reported (Arellano et al. 2000; Melo et al. 2000) at room temperature. The triplet states from BChl aggregates could only be observed at low temperatures (Psencik et al. 1994; Carbonera et al. 2001) as a result of a decrease of the quenching efficiency.

Besides the protective role of carotenoids, two additional processes preventing singlet oxygen formation were proposed, which are based on the unique properties of the BChl aggregates. Upon aggregation, not only singlet but also triplet excited-state lifetimes shorten (Krasnovsky 1982). This property is at the basis of a first self-protecting mechanism: the shorter S_1 lifetime decreases the probability of intersystem crossing and thus reduces the population of the triplet states. The shorter triplet state lifetime further decreases the probability of singlet oxygen formation (Arellano et al. 2002). The second proposed mechanism is based on the existence of a low-lying triplet exciton state of BChl aggregates with an energy below that of singlet oxygen (Kim et al. 2007). However, such a state was not yet spectroscopically observed.

Another protective mechanism, which is specific to green bacteria possessing the type I RC, is present in GSB and the acidobacterium *Cba. thermophilum*. In these bacteria, efficient EET from the chlorosome to the baseplate and further toward the RC occurs only in the absence of oxygen. At aerobic conditions, the excitation is rapidly quenched in a process that prevents photooxidative damage of a type I RC containing low potential electron acceptors (Blankenship and Matsuura 2003). Interestingly, the excitation quenching is not triggered by the presence of oxygen itself, but by the change of the redox potential (Blankenship et al. 1993). The excitation quenching is mediated by chlorosomal quinones (Frigaard et al. 1997). CsmI and CsmJ proteins from the chlorosome envelope were shown to be involved in restoring EET upon return to anaerobic conditions in *Cba. tepidum*. These two proteins, together with CsmX, are iron–sulfur proteins with sequence similarity in their amino-terminal domains to [2Fe-2S] ferredoxins, and they probably serve as the carriers of electrons to and from the quinones in the chlorosome interior through the chlorosome envelope (Li et al. 2013).

The last mechanism, which can be considered to regulate light harvesting and which will be mentioned here, is the variable size of the chlorosome. It has been shown that chlorosomes from cells grown under low-light conditions contain more pigments, both in *Chlorobi* (light intensities above 1 μE × m^{-2} × s^{-1}) (Borrego and Garcia-Gil 1995; Borrego et al. 1999a) and *Cfl. aurantiacus* (Schmidt et al. 1980; Taisova et al. 2014). Changes in morphology between the high- and low-light chlorosomes observed in cryo-EM of *Cfl. aurantiacus* show that it is mainly the thickness of the chlorosome that increases at low-light conditions (Psencik et al. 2013) due to additionally formed layers of BChl aggregates. These additional layers are often arranged into domains of lamellar aggregates with slightly different orientations with regard to the long axis of the chlorosome and may assist in light harvesting (Psencik et al. 2006). The larger thickness and consequently longer distance for the excitation to travel by a random walk contribute to the longer transfer times observed in larger chlorosomes (Section 7.2.2.2) (Fetisova et al. 1996). Two groups of chlorosome proteins found in *Cba. tepidum*, CsmB/CsmF and CsmC/CsmD, were shown to affect the size and shape of the chlorosome (Li and Bryant 2009). These proteins (and similar proteins found in other species) may be involved in regulating the chlorosome size in response to the variable light conditions.

7.3 BASEPLATE

The baseplate and the chlorosome interior form an integrated unit, and the chlorosome cannot be isolated without the baseplate. However, the two subunits represent independent LH complexes, which differ from each other in their composition and architecture. The baseplate is attached to the side of the chlorosome, which is oriented toward the cytoplasmic membrane, ensuring a close contact with the subsequent antenna complexes. It contains BChl *a* and carotenoid molecules and mediates EET from the chlorosome interior toward the RC.

7.3.1 Organization of the baseplate

The baseplate is composed of many copies of a single protein, referred to as CsmA. CsmA is the only protein vital for chlorosome biogenesis (Frigaard and Bryant 2006), and its structure (Figure 7.4a) was determined by nuclear magnetic resonance (NMR) (Pedersen et al. 2008). A conserved histidine in CsmA from different species suggests binding of one BChl *a* per CsmA (Pedersen et al. 2010), while the experimentally obtained values are between one and three BChl *a* molecules (Sakuragi et al. 1999; Bryant et al. 2002; Montano et al. 2003b). The orientation of the transition dipole moments of BChl *a* molecules in the baseplate was shown to be more or less perpendicular to the plane of the baseplate and to the long axis of the chlorosome, while the main absorbing states of the BChl aggregates are oriented more parallel

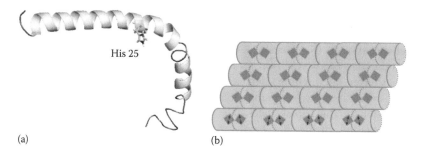

His 25

(a)   (b)

Figure 7.4 Baseplate of the chlorosome. **(a)** Structure of the CsmA protein from *Cba. tepidum* as determined by NMR spectroscopy (Pedersen et al. 2008, PDB 2K37). The conserved histidine-25 is shown, which is the most probable binding site of BChl *a*. The FMO binding site (hydrophilic C-terminus) is at the bottom of the figure, the part embedded in the chlorosome (hydrophobic N-terminal helix is on top). **(b)** Tentative arrangement of the CsmA proteins based on cryo-EM projections. BChl *a* molecules are shown as green squares forming a dimer, and the orange background represents the presence of carotenoids. Red arrows indicate the orientations of the Q_y transition dipole moments of BChl *a* molecules. Recent results indicate the presence of a BChl *a* tetramer in the baseplate, but the structural information is missing to confirm this proposal.

(van Amerongen et al. 1988). This might lower the efficiency of EET, but experimental data also show that the transition dipole moments of the low-energy states of the aggregates exhibit sufficient component parallel to that of BChl *a* (for a review, see Psencik et al. 2014). Compared to the number of aggregated BChls inside the chlorosome, the content of BChl *a* in the baseplate is relatively low. It represents typically 1% of the total BChl content in GSB and 5% in green FAP. Apart from BChl *a*, also carotenoids are found in the chlorosome baseplate (Montano et al. 2003b).

In contrast to the disordered arrangement of BChl aggregates in the chlorosome interior, the baseplate is very well ordered. CsmA proteins are arranged into rows, which in turn form a two-dimensional crystalline lattice with a spacing of 3.2–3.3 nm between the rows. The rows are oriented approximately perpendicular to the long axis of the chlorosome in *Cfl. aurantiacus*, while they make an angle of ~40° in *Cba. tepidum* (Staehelin et al. 1978; Psencik et al. 2009; Oostergetel et al. 2010). Although a crystal structure of the CsmA protein was not yet reported, a highly sophisticated model based on electron cryomicroscopy, NMR, and optical spectroscopy was recently reported (Nielsen et al. 2016). Based on the CsmA structure and observation of an exciton circular dichroism signal, it has been proposed that the baseplate building block is a CsmA dimer (Frigaard et al. 2005; Pedersen et al. 2010; Nielsen et al. 2016), as depicted in Figure 7.4b. However, results of 2DES indicate

the presence of four excitonically coupled BChl *a* molecules (Dostal et al. 2014). This would require two BChl *a* molecules per one CsmA in a dimer of the proteins. Although the baseplate composed of CsmA proteins is found in chlorosomes of all green bacteria, important differences in the baseplate organization may exist between different groups. For instance, the exciton CD signal was not observed for the isolated baseplate from *Cfl. aurantiacus* (Montano et al. 2003b).

7.3.2 Function of the baseplate

BChl *a* in the baseplate has a relatively broad absorption spectrum around 795 nm at room temperature, which is red-shifted compared to the absorption spectrum of the BChl aggregates in the chlorosome interior. This facilitates the collection of energy absorbed in the chlorosome interior and its transport toward the RC and increases the efficiency of light harvesting (Section 7.6.2) (Oostergetel et al. 2010). Anisotropy of the baseplate BChl *a* decays with lifetimes around 1 ps after excitation of BChl aggregates (Psencik et al. 2003; Martiskainen et al. 2012), faster than the decays observed for BChl aggregates (Lin et al. 1991). The fast anisotropy decay of baseplate BChl *a* is probably caused by the EET within the baseplate. Alternatively, it may reflect depolarization that occurs during the EET from low-lying states of BChl aggregates to the baseplate. Experimental data suggest that a small portion of the excitation inside the chlorosome interior can be found close enough to the baseplate, so it can be transferred directly to the BChl a with a ~1 ps transfer time (Psencik et al. 2003), that is, substantially faster than the main part of the excitation that arrives to the baseplate after being largely equilibrated over the chlorosome.

In *Chloroflexi*, the transfer time from BChl a in the baseplate to the membrane LH complexes is about 40 ps, as characterized by measuring fluorescence decays in whole cells of *Cfl. aurantiacus*. The decay of BChl a in the baseplate is multiexponential, and the component of ~40 ps matches the rise component of the B866 fluorescence in the B808–866 complex, which also contains BChl a (Mimuro et al. 1989; Causgrove et al. 1990b; Muller et al. 1993). Although the energy is transferred from the baseplate presumably via the B800 molecules, only B866 emission is observed due to the fast B800 to B866 EET (Section 7.4.2.2).

In *Chlorobi* and *Acidobacteria*, the energy is transferred from the baseplate to the FMO complex, which again contains BChl a. However, in contrast to the situation encountered in *Chloroflexi*, this time the assessment of transfer times is complicated by a spectral overlap between the contributions from the baseplate and the FMO. The same lifetimes observed for EET from BChl d aggregates to BChl a in isolated chlorosomes and whole cells (~65 ps) suggest fast equilibration between the baseplate and FMO in whole cells (Causgrove et al. 1990b). On the other hand, the observation of relatively fast relaxation within the four exciton levels of the BChl a manifold of the baseplate (Dostal et al. 2014) seems to be inconsistent with this suggestion, since the relaxation competes with the EET from the higher baseplate energy levels to higher states of FMO. This controversy has led to the proposal of "lateral" EET from all four baseplate levels to spectrally overlapping FMO states (Dostal et al. 2014). This proposal requires the spatial vicinity of the pigments in both complexes. Due to the lack of structural information, this is difficult to assess. Another possibility is that the EET from the lowest baseplate states occurs with the help of thermal energy. In any case, the energy is transferred from the chlorosome to the FMO, as directly observed in a recent 2DES study on whole cells of *Cba. tepidum* (Dostal et al. 2016). The EET from BChl c aggregates to FMO occurred with a transfer time of 70 ps, while the transfer step to the baseplate

could not be resolved due to its weak absorption and spectral overlap with FMO.

Energy absorbed by the carotenoids in the baseplate is transferred to BChl a with an efficiency of about 30% for *Cfl. aurantiacus* (Montano et al. 2003b), that is, lower than what is observed for carotenoids inside the chlorosome interior. Probably, the main function of the baseplate carotenoids is protective. Efficient quenching of the BChl a triplet states by carotenoids was observed at room temperature (Melo et al. 2000), and evidence of quenching was obtained even at low temperatures, when carotenoids are not effective in quenching of the triplet states formed in BChl aggregates (Carbonera et al. 2001).

7.4 INTERMEDIATE ANTENNAS

Although the structures of the chlorosome interior and baseplate are similar for all green bacteria, the LH complexes subsequent in the EET chain are rather different for each group of organisms. The chlorosome is connected to the membrane-bound RC complexes via the FMO complex in GSB and acidobacterium *Cab. thermophilum*, while circular membrane antennas similar to LH complexes of purple bacteria serve the same function in green FAP.

7.4.1 FMO complex of *Chlorobi* and *Acidobacteria*

The first photosynthetic pigment–protein complex to have its structure resolved with an accuracy that allowed the identification of the positions and mutual orientations of its pigment molecules was the FMO complex of *Prosthecochloris aestuarii* (Fenna and Matthews 1975). In 1975, FMO has thus started the era of the X-ray crystallographic structural insights into the inner working of photosynthetic antennas. Later, the structure of the FMO complex of *Cba. tepidum* was resolved (Li et al. 1997) showing a strong structural similarity between the two species. The absorption spectra of the FMOs of the two species are nevertheless distinct (Tronrud et al. 2009). Thirty years after its structure was first determined, FMO has confirmed its role as a model system to introduce pioneering experimental techniques into the studies of photosynthesis by being the first photosynthetic antenna subject to a study by coherent 2DES (Brixner et al. 2005).

7.4.1.1 ORGANIZATION OF THE FMO COMPLEX

FMO is present in *Chlorobi* and phototrophic acidobacteria where it is situated between the baseplate of the chlorosome and the membrane, and it thus connects the chlorosome to the RCs. The only pigments it contains are BChls *a* with no involvement of carotenoids. It has a homotrimeric structure with C3 symmetry (Figure 7.5a), and probably two copies of the FMO protein are associated with a RC (Hauska et al. 2001). Recently, it has been proposed that two RCs with four FMOs form a single supercomplex (Bina et al. 2016). The orientation of the trimer was revealed by chemical labeling and mass spectroscopy (Wen et al. 2009; Huang et al. 2012). Interestingly, the orientation was also predicted theoretically based on the BChl *a* transition energy calculations that take into account the interaction between the pigments and their protein surroundings, especially the charged protein residues (Adolphs and Renger 2006) and even earlier by site energy assignment based on fitting of optical spectra (Louwe et al. 1997; Vulto et al. 1998). The assumption that the role of the FMO complex is to funnel energy to the RC, meaning that the transition energies should decrease toward the RC, leads to the orientation of the FMO consistent with its independent experimental determination.

The three monomers of FMO were until recently believed to contain seven BChl *a* each. Using this structural information, transition energies of individual BChls in the complex were determined by fitting optical spectra and assuming dipole–dipole interaction between the transitions (Wendling et al. 2002) (see Milder et al. [2010] for an exhausting review of the various approaches). Later, quantum chemical calculations confirmed and refined the general assignment of the transition energies (Adolphs and Renger 2006; Muh et al. 2007; Adolphs et al. 2008). The lowest energy point of the excitation transfer through FMO was determined to be BChl 3 (see Figure 7.5b for the labeling). In the recent structures, an eighth BChl *a* molecule has been resolved on the side of FMO, which is oriented toward the chlorosomal baseplate (Tronrud et al. 2009). In accord with the excitation energy funneling concept and the position of this BChl *a* near the baseplate, calculations of the transition energy of this eight BChl *a* based on the composition of its protein environment assign it to the top of the energy funnel (Busch et al. 2011).

7.4.1.2 FUNCTION OF THE FMO COMPLEX

The main function of the FMO complex seems to be conducting excitation energy from the chlorosomal baseplate to the RC. This conclusion was mainly inferred from the spatial organization and spectral properties of the involved complexes. Due to the spectral overlap between these complexes, the transfer processes are difficult to study when all complexes are present. Recently, most of the transfer steps were resolved in a 2DES study on

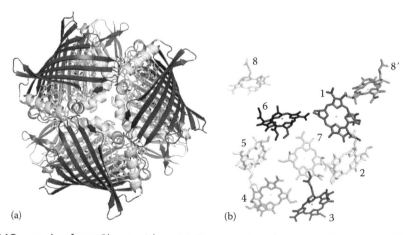

(a) (b)

Figure 7.5 FMO complex from *Cba. tepidum*. **(a)** Top view (i.e., from the chlorosome side) of the FMO complex as determined by X-ray crystallography (Tronrud et al. 2009, PDB 3ENI, biological assembly generated by PISA software). **(b)** A side view (chlorosome side on top) of the pigments in a monomer unit of the same structure. Note that the closest BChl *a* no. 8 is the one from the neighboring unit (shown as well and labeled as 8′). The phytyl chains have been truncated for clarity.

whole cells of *Cba. tepidum* (Dostal et al. 2016). The efficiency and transfer times of the FMO to the RC transfer step are discussed in more detail in Section 7.5.1. Another possible role of FMO is to simultaneously allow the access of ferredoxin to the RC. This function is discussed in Section 7.6.1. Here we will first concentrate on the FMO's energy conducting capacity.

Theoretical modeling of the optical spectra and quantum chemical calculations of site energies established a fairly clear picture of the energy landscape of the FMO complex. The BChl 8', belonging to the neighboring FMO monomer (Figure 7.5b), provides the most obvious entry point for the excitation that is then transferred through two branches of the FMO, one formed by the BChls 4–7 and one consisting of BChl 2. BChl 2 seems to be the bottleneck of the EET through FMO. The EET through the other branch is estimated to take 0.5 ps, and the combined transfer time through the FMO complex is roughly 1.5 ps. The energy gap between the BChl 8' (the point of entry) and BChl 3 (the presumed exit point) is on the order of 400 cm^{-1} (Busch et al. 2011), that is, much higher than the thermal energy at physiological temperatures. The EET through FMO is therefore most likely irreversible with only a small probability to return the excitation to the chlorosome.

When discussing EET within FMO, it is worth noting that the presence of oxygen induces a redox-dependent excitation quenching in the FMO complex (Zhou et al. 1994) similar to what was observed within the chlorosomes from GSB and phototrophic acidobacteria (Section 7.2.2.3). It has been suggested that tyrosine residues adjacent to BChl *a* molecules within the FMO complex are involved in this process (Zhou et al. 1994), but recently cysteine residues located near two low-energy BChls were identified as quenchers (Orf et al. 2016). This quenching mechanism probably protects the RCs from accepting energy at aerobic conditions (Orf et al. 2014), similar to the case of quenching in chlorosomes. The quenching also prevents the formation of triplet states at aerobic conditions. However, it might not be a part of the protection against singlet oxygen formation, since the BChl *a* triplet levels in FMO may lie below the energy of singlet oxygen (Orf et al. 2014). This would explain why FMO does not contain any carotenoids, which are important scavengers of singlet oxygen in all other LH complexes. It would be interesting to explore how this quenching affects EET within FMO, as most of the experiments on isolated FMO were done at aerobic conditions.

7.4.1.3 FMO AS A MODEL ANTENNA FOR THEORETICAL AND EXPERIMENTAL STUDIES

Although FMO is not the most typical nor abundant antenna, it has nevertheless played an extremely important role in developing an atomistic picture of photosynthesis. Apart from the early determination of its structure, it has also served as an important model system for developing theoretical concepts. For instance, it enabled the theoreticians to pinpoint the importance of the site energy variability in establishing the energy funneling effect both by fitting experimental optical spectra and by quantum chemical calculations. The small number of pigments in the complex, an absorption spectrum rich in relatively sharp features, and a favorable central absorption wavelength of around 800 nm (i.e., accessible by Ti:sapphire lasers) made it an ideal system for testing an application of new spectroscopic techniques in the field of photosynthesis. As the coherent multipulse spectroscopic techniques developed in NMR made their way to the infrared and visible regions (Jonas 2003; Brixner et al. 2004), the FMO complex from *Cba. tepidum* became the first photosynthetic antenna to be studied by this new technique (Brixner et al. 2005). The 2DES was originally hoped to have the potential to directly reveal couplings between the pigments in a fashion similar to the coherent NMR techniques. Although the relation between the excitonic coupling and the spectral features of the 2D techniques turned out to be less straightforward than in NMR, the added dimension of the spectrum, as compared to more conventional pump–probe techniques, enabled some refinement of the general understanding of pathways the excitation takes through the FMO complex (Cho et al. 2005). Theoretical studies of the 2DES made clear that the time evolution of the 2D spectral features not only holds information on the EET between excitonic levels of the system but also about the degree of coherence of these processes (Kjellberg et al. 2006; Pisliakov et al. 2006). Frequencies and lifetimes of these time-dependent features (coherent oscillations) were predicted for the FMO complex (Pisliakov et al. 2006). The presence of coherent oscillations was later confirmed in the FMO from

Cba. tepidum (Engel et al. 2007), but the measured lifetime of the oscillations significantly exceeded the predicted lifetime of only several hundreds of femtoseconds. The predicted lifetime of electronic coherence stated by Pisliakov et al. (2006) was 100–300 fs. The calculations were based on the EET rates calculated by modified Redfield method by Cho et al. (2005). The measurements found the lifetime of the coherence in FMO to be longer than 1 ps at 77 K, and the oscillations were clearly visible even at room temperature (Panitchayangkoon et al. 2010). This discrepancy led to the suspicion that some important part of the theoretical description of the EET phenomena was missing and to the suggestion that a coherent mode of excitation EET is fundamental to the high efficiency of photosynthetic antenna function (Engel et al. 2007).

To resolve the conundrum of the electronic coherence lifetime has become an important issue for theoretical research. The presence of long-lived electronic coherence was taken as a signature of the quantum mechanical nature of the EET process. Mechanisms to explain the extended lifetime of the electronic coherence were suggested based on the correlation between protein fluctuations on different pigment molecules in the antenna (Lee et al. 2007; Wolynes 2009), which were however, in the case of FMO, not confirmed by molecular dynamics studies (Olbrich et al. 2011a). Alternatively, it was claimed that the correct treatment of the electron–phonon (pigment–protein) coupling already leads to long-lived electronic coherences (Ishizaki et al. 2010; Pachon and Brumer 2011). Nevertheless, the achieved lifetimes are still in the realm of hundreds of femtosecond. Lifetimes of oscillatory features in 2DES consistent with experimental results were obtained by assuming a direct vibrational involvement in the EET. Vibrational modes resonant with electronic energy gaps were found to lead to time-dependent effects with the lifetime and magnitude in the 2D signal originating from the ground state (Tiwari et al. 2013) and the electronically excited states (Christensson et al. 2012; Chenu et al. 2013; Chin et al. 2013). It is important to note that the suggested effects are neither of purely electronic nor of purely vibrational origin. Thanks to the intensive studies of FMO, a new research avenue into the synergetic effects of electronic and vibrational transitions and their role in excitation EET in photosynthesis was thus recently opened.

FMO as a model antenna and the experiments performed on it, with both new and standard spectroscopic techniques, have played an important role in stimulating the development of new theoretical methods simulating EET (Mohseni et al. 2008; Plenio and Huelga 2008; Ishizaki and Fleming 2009), as well as the implementation of computational methods on massively parallel graphics processing unit architectures (Kreisbeck et al. 2011). FMO is one of the systems on which experimental determination of electron–phonon coupling spectral density was performed (Wendling et al. 2000). As a consequence, it serves as a model system for the efforts for atomistic determination of the spectral density (Olbrich et al. 2011a,b; Shim et al. 2012; Renger et al. 2012). The debate stimulated by the 2DES experiment on FMO involves fundamental questions of the role of quantum mechanics in the initial steps of photosynthesis. Although such fundamental inquiries often criticized the idea that dynamic electronic quantum coherence, in particular, can be responsible for the efficiency of photosynthetic process (Mancal and Valkunas 2010; Brumer and Shapiro 2012; Kassal et al. 2013; Chenu et al. 2014), they have nevertheless always stressed that the proper treatment of photosynthetic antennas is only possible with the use of quantum mechanics. Such treatments, however, should not be limited to the pigments only but have to involve also the protein environment, that is, the electron–phonon interaction (Mancal 2013). FMO and other small photosynthetic antennas of bacteria and plants have thus become the prime showcases of quantum biology.

7.4.2 Membrane LH complexes of *Chloroflexi*

Not all the FAP contain chlorosomes, but they all contain membrane complexes, which are denoted B808–866 in the case of *Cfl. aurantiacus* (Feick and Fuller 1984), a member of the green FAP, and B800–880 for the *Roseiflexus* (*Rfl.*) *castenholzii*, a member of the red FAP (Collins et al. 2009). They form circular structures, with their outer diameter being approximately 15 nm (Collins et al. 2009; Bina et al. 2014), and contain BChl *a* and carotenoid molecules.

7.4.2.1 ORGANIZATION OF THE MEMBRANE LH COMPLEXES

Membrane LH complexes in FAP resemble circular LH1 and LH2 antennas of purple bacteria (originally called B875 and B800–B850) and are similarly named for their respective BChl *a* Q_y absorbance maxima with the letter "B" referring to "BChl *a*" as the main pigment. The main difference with the LH complexes of purple bacteria is that the membrane-bound LH complexes of FAP combine the properties of LH1 and LH2 in one complex: they contain a RC in the center of the LH complex (Figure 7.1b), as in the case of LH1, and, at the same time, exhibit two absorption maxima in the near infrared, as in the case of LH2.

Similar to purple bacteria, the membrane antenna of FAP consists of subunits each containing a pair of αβ apoproteins (Wechsler et al. 1987). The number of subunits per one complex is 15–16 (Collins et al. 2009; Bina et al. 2014), that is, close to what was observed for the monomeric LH1 complexes from purple bacteria. Most likely, there are two BChl *a* molecules that comprise B866 and one BChl *a* that constitutes B808. The FAP membrane antennas contain also carotenoids (γ-carotene in the B808–866 complex and predominantly methoxy-keto-myxocoxanthin in the B800–880 complex [Niedzwiedzki et al. 2010]). The BChl *a* to carotenoid ratio was reported to be 3:2 (Xin et al. 2005; Collins et al. 2009), which represents a major difference from both LH2 and LH1 complexes, where only one carotenoid per subunit is found.

In analogy with the LH2 complex, it is assumed that B808/B800 BChl *a* molecules can be considered monomeric, forming the outer ring, while the red shift of the B866/880 pigments is caused by strong

exciton interaction between closely packed BChls in an aggregate (Figure 7.6) (Bordignon et al. 2003). The orientation of the Q_y transition dipole moments of the B866 BChl *a* molecules is close to parallel with the plane of the ring and therefore also with the cytoplasmic membrane (Vasmel et al. 1986; Novoderezhkin and Fetisova 1999). The Q_x transitions are close to perpendicular to the membrane (Collins et al. 2009), and so are the entire B880 molecules (Figure 7.6). A similar arrangement is found in the B850 BChl *a* pigments in LH2. In contrast, the orientation of the Q_y band of B808 seems to be significantly different from that of the B800 molecules in LH2, with an angle of about 45° (Vasmel et al. 1986; Novoderezhkin and Fetisova 1999). This difference is probably related to the fact that B808 molecules accept energy from the baseplate of the chlorosome located above. If the Q_y transition dipoles were parallel as in B800 of purple bacteria, they would be nearly perpendicular to the Q_y dipoles of BChl *a* in the baseplate (3.1), which would not be favorable for EET by Förster mechanism. Carotenoids also make a large angle with the membrane plane (Collins et al. 2009), which was earlier estimated to be about 50° (Vasmel et al. 1986).

7.4.2.2 FUNCTION OF THE MEMBRANE LH COMPLEXES

The membrane LH system of FAP was proposed to be oriented in a way that the B800/B808 pigments are located close to the cytoplasmic side, while the

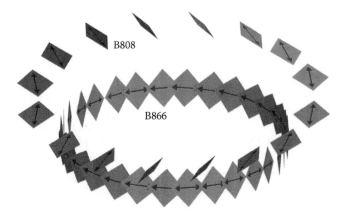

Figure 7.6 Scheme of a plausible arrangement of BChl *a* molecules in the B808–866 complex. The complex is shown oriented in such a way that the chlorosome (cytoplasmic side) is on top. Red arrows depict the orientation of the Q_y transition dipole moments of the BChl *a* molecules, which are close to parallel with the plane of the circle for the B866 molecules (inner circle) and make an angle of about 45° with the plane of the circle for the B808 molecules. The complex also contains carotenoids, which are not shown due to the lack of structural information.

B880/B866 pigments are closer to the periplasmic side and to the primary donor in the RC (Xin et al. 2012) (Figure 7.6). This arrangement is again based on analogy with the LH2 complexes and significantly increases the efficiency of EET from the chlorosome baseplate (in the case of green FAP) through the B808 molecules to B866 pigments and to the special pair. The proposed orientation ensures the presence of an energetic gradient, and also the optimal distances between BChl a in the baseplate, and the B808 and B866 complexes (Xin et al. 2012).

The EET from B808/B800 to B866/B880 occurs with a transfer time of about 2 ps in membranes (Novoderezhkin et al. 1998; Xin et al. 2012), about twice slower than the corresponding transfer in LH2 of purple bacteria. In previous studies, even slower transfer times were determined (5–6 ps) with close to 100% efficiency (Vasmel et al. 1986; Griebenow et al. 1991). No EET between B808 molecules prior to the transfer to B866 has been observed, whereas the redistribution of the excitation within the B866 pool occurs within ~0.5 ps (Novoderezhkin et al. 1998).

EET from the carotenoid to B866 was observed to occur with ~15% efficiency with a dominant pathway from the S_1 state and a transfer time of 4.3 ps in $Cfl.$ $aurantiacus$ (Montano et al. 2004), while an overall 35%–40% efficiency was determined for the red FAP $Rfl.$ $castenholzii$ (Vasmel et al. 1986; Collins et al. 2009). The increased efficiency observed for the red FAP is caused by an additional and prevailing EET channel from the S_2 state of the carotenoid to B880 molecules (Xin et al. 2012). Some evidence for carotenoid triplet formation in the B808–866 complex by quenching of the BChl a triplet states was observed by optically detected magnetic resonance (Bordignon et al. 2002). Based on the structural similarity with LH2, it is likely that this protective quenching is highly efficient (Limantara et al. 1998 and references therein).

7.5 EET TO THE REACTION CENTERS

The ultimate goal of the LH machinery is to deliver the excitations to the RCs, where they are converted into chemical energy. The last step of this process is EET from the intermediate antennas to the RCs located in the cytoplasmic membrane. The RCs of $Chlorobi$ and $Acidobacteria$ contain their own core antenna pigments, which represent the fourth LH system of these bacteria.

7.5.1 Type I reaction centers of $Chlorobi$ and $Acidobacteria$

In GSB and $Acidobacteria$ member $Cab.$ $thermophilum$, energy is transferred to the RC via the FMO complex. The involvement and efficiency of the EET from the FMO was a subject of long-lasting controversy. Although clear evidence of the association between FMO trimers and the RC complex existed (Remigy et al. 1999), efficient EET from the FMO to the RC has not been observed. Both in isolated FMO–RC complexes at 275–300 K (Neerken et al. 1998; Oh-Oka et al. 1998, He et al. 2015) and in membrane preparations at 6 K or below (Kramer et al. 1982; Francke et al. 1996), the highest determined EET efficiency was 40%. As there is general agreement that FMO has to be involved in efficient EET from the baseplate to the RC, these results were explained by unspecified damage during sample preparation. FMO is bound rather loosely to the RC and can be easily lost during RC isolation. In those preparations, where FMO was certainly still bound, a critical alternation of the interaction between FMO and RC was proposed to take place (Hauska et al. 2001). Another explanation of the observed low efficiency of the EET might be the presence of oxygen during the measurements, and the redox-dependent regulation of the EET through FMO, which was discussed in Section 7.4.1.2. Recently, excitation energy flow from the FMO complex to the RC was observed in whole cells from $Cba.$ $tepidum$ at 77 K and anaerobic conditions by 2DES (Dostal et al. 2016) with a transfer time estimated to be around 15 ps and with an efficiency of ~75%. The determined efficiency might be an underestimate of the real value due to the possibility that some of the RCs were closed during the measurements at low temperatures.

GSB and $Acidobacteria$ contain a homodimeric type I RC, which is similar to the photosystem I RC from oxygenic photosynthetic organisms, including the fact that it contains a fused intrinsic antenna. However, although the size of the two pigment-binding proteins in the RC core is very similar (164 vs. 165 kDa) and the number of transmembrane helices appears to be the same as in photosystem I (11), the RCs of $Chlorobi$ and $Acidobacteria$ contain a much smaller number of pigments. As an example, the RC core from $Cba.$ $tepidum$ was estimated to contain 16 BChl a molecules, 4 Chl a molecules, and only 2 carotenoids bound to two copies of PscA proteins, as compared to 96 Chl *a* molecules and 22 carotenoids in photosystem I RC core (Hauska et al. 2001; Jordan et al. 2001). Two of the four Chl *a* molecules found in the RC from *Cba. tepidum* act as accessory molecules and the other two as primary acceptors A_0 (Permentier et al. 2000; Ohashi et al. 2010). Two of the BChl *a* molecules form the special pair, which serves as a primary electron donor and is denoted P840. The remaining 14 BChl *a* molecules form the core antenna. Each of the seven pairs of BChl *a* molecules exhibits a different energy of the Q_y transition in the spectral region between ~775 and 840 nm, reflecting their different environments (Permentier et al. 2000; Hauska et al. 2001). Excitation of antenna molecules leads to a fast equilibration (~1 ps) among BChl *a* molecules of the core complex irrespective of the excitation wavelength, which was explained by strong interactions between them, and rapid exciton relaxation to the lowest state at ~835 nm (Neerken et al. 1998). The charge separation connected with the photooxidation of the primary donor in the RC occurs with a trapping time of 25–35 ps (Neerken et al. 1998; Oh-Oka et al. 1998; He et al. 2015). Based on the comparison with the related system of heliobacteria, it seems likely that the EET to the RC is trap limited (Amesz and Neerken 2002). After charge separation occurs in the RC, the electron is transferred through a sequence of intermediate acceptors, eventually reducing a water-soluble ferredoxin protein. This fact is important for the following discussion of the overall arrangement of the LH apparatus.

The RC from *Cab. thermophilum* has a somewhat different composition, and interestingly, it contains a pair of Zn-BChl *a*, which may act as the primary electron donor or an electron acceptor (Tsukatani et al. 2012). Its primary electron donor is denoted P840 as in *Chlorobi*. It is the first homodimeric RC found in an aerobic phototroph.

7.5.2 Type II reaction centers of *Chloroflexi*

In green/red FAP, excitation is transferred to the RC from the B866/B880 molecules of the B808–866/B800–880 membrane complex. The kinetics of this process depends on the state of the RC. In *Cfl. aurantiacus*, as a representative of green FAP, the transfer times were reported to be between 40 and 90 ps for the open RC (i.e., photochemically active, with a reduced primary donor) (Causgrove et al. 1990b; Muller et al. 1993), while much longer times (180–250 ps) were measured for closed RCs (with an oxidized primary donor) (Nuijs et al. 1986; Mimuro et al. 1989; Muller et al. 1993). In the red FAP *Rfl. castenholzii*, similar transfer times of 60 ps (open RC) and 210 ps (closed RC) were observed (Xin et al. 2012).

The RC of FAP is of type II, like the one of purple bacteria, or photosystem II of oxygenic photosynthetic organisms, and correspondingly it is not associated with any intrinsic antenna (Blankenship 2002). The primary electron donor is formed by two BChl *a* molecules and is referred to as P865 or P870. In contrast to the RC of purple bacteria, the two accessory pigments in FAP are not both BChl molecules, but one of them is replaced by a BPh *a* molecule, making the ratio BChl *a*/BPheo *a* 3:3 in the RC from FAP (Pierson and Thornber 1983; Yamada et al. 2005). The second main difference as compared to purple bacteria is that the RC of FAP does not contain carotenoids (for a review, see Feick et al. 1995). This is the only known RC without carotenoids.

7.6 COMPLETE LH MACHINERY

Thanks to the development of isolation protocols, numerous important pieces of information on EET within well-characterized isolated complexes were obtained in recent years. However, to properly understand the function of the entire LH machinery, it is also important to put together information about the functional relations between the complexes, their stoichiometry, and spectral properties.

7.6.1 Organization, stoichiometry, and their implications

The LH apparatus of green FAP is a textbook example of the so-called funnel concept. Photons with the highest energy are harvested by the most peripheral antenna of the photosynthetic system (the chlorosome), which contains the largest amount of pigments. The resulting excitations are transferred via several intermediate complexes (the baseplate and the B808–B866 membrane complex) with their excited-state energies gradually decreasing. This allows efficient EET to the RC. The transfer is efficient in terms of the quantum yield, while part of the energy is sacrificed for the sake of directionality and irreversibility of the process (Blankenship 2002). The funnel concept is applicable also to the LH apparatus of GSB and *Cab. thermophilum*. The excited-state energies of the baseplate, FMO complex, and RC overlap significantly (Dostal et al. 2014); nevertheless, the fluorescence maxima observed for the BChl *a* in the baseplate (~810 nm), FMO (~825 nm), and RC (~835 nm) indicate the presence of states ensuring the energy gradient also in GSB. It is interesting to note that except for the core antenna of the RC, none of the LH complexes found in GSB (chlorosome, baseplate, FMO complex) are located in the cytoplasmic membrane.

The stoichiometry between all the elements of the LH machinery of green bacteria is a matter of a long-standing debate. Thanks to a recent electron microscopy study, it became better understood for green FAP (Bina et al. 2014). By observing the cytoplasmic membranes from *Cfl. aurantiacus*, it was estimated that there are about 11 membrane LH complexes (B808–866) and thus also 11 RCs per chlorosome. This is a much lower number than proposed previously. A typical distance between the centers of two neighboring complexes is 18 nm, which is much larger than that for densely packed LH complexes of purple bacteria. It seems that the distribution of the B808–866 complexes in *Cfl. aurantiacus* is optimized for a "vertical" EET from the above lying chlorosomes, and the "horizontal" EET between the complexes does not play an important role, since each of them contains their own RC. In red FAP, the membrane LH complexes are found without attached chlorosomes, and it is therefore possible that even in green FAP, like *Cfl. aurantiacus*, not all membrane LH complexes are associated with chlorosomes, as depicted in Figure 7.1. It has been suggested that even the complexes of FMO and RC of GSB may not be always coupled with the chlorosomes (Wen et al. 2009).

Until recently, estimates for *Chlorobi* were based on rather indirect observations and pigment stoichiometry. For instance, Frigaard et al. (2003) based their estimates on the assumption of 200,000 BChl *c* molecules per chlorosome as reported in Montano et al. (2003a) for *Cba. tepidum*. Here we will use a similar discussion with the following differences: based on the known density and volume of the chlorosome, we came to a maximal number of 100,000 BChl *c* per typical chlorosome (Psencik et al. 2014). In addition, we assume that the RCs can be located under the whole area of the chlorosome,

not only at its edges (see following text), and that all the RCs and FMOs are functionally connected to a chlorosome. BChl *a* constitutes about 3% of the total BChl content in *Cba. tepidum* (Frigaard et al. 2003 and references therein) and about 1% of the BChl content of the chlorosome, where it is found in the baseplate. This represents about 1000 BChl *a* molecules per baseplate and 2000 BChl *a* molecules in all FMOs and RCs connected to one chlorosome. One FMO trimer contains 24 BChl *a* molecules; one RC complex contains 16 BChl *a* molecules. Preparations with 3, 1, and 0 FMO trimers per RC core complex were originally reported (Francke et al. 1997), but later it was concluded that the naturally occurring ratio is two FMO trimers per RC (Hauska et al. 2001; Bina et al. 2016). Using this assumption, one gets ~30 RCs per chlorosome, which agrees very well with the results of a recent electron microscopy study (Bina et al. 2016). This number is almost three times larger than what was observed for *Cfl. aurantiacus* (Bina et al. 2014), but since also the areas of the studied chlorosomes from *Cba. tepidum* were almost three times larger, the spatial density of the RCs is similar in both GSB and FAP. These assumptions lead to ~60 FMO trimers per chlorosome, which means that approximately one third of the baseplate area (a value of ~9000 nm² was determined in the work by Bina et al. [2016]) is covered by FMOs (~50 nm² for a single FMO trimer approximated by a circle with 8 nm diameter).

These estimates are important for a discussion of the role of the FMO complex in the photosynthetic machinery of *Chlorobi* and *Acidobacteria*. The type I RCs of these bacteria reduce ferredoxin. These water-soluble molecules have to travel to the cytoplasmic side of the RC, which is also the binding site for FMO and the attached chlorosome (Figure 7.1). To rationalize the interaction of the RC with a ferredoxin molecule, the RCs are located around the edges of the baseplate in some models (Frigaard et al. 2003). However, such an arrangement seems to be unfavorable for efficient EET from the chlorosomes and contradicts the recent observations (Bina et al. 2016). Therefore, we consider the model in which the RCs are located below the chlorosomes. In such a case, a spacer would be necessary allowing the ferredoxin to travel between the chlorosome and the cytoplasmic side of the RCs (Hauska et al. 2001; Wen et al. 2009). The distribution of the FMOs suggested in the previous paragraph seems to be compatible with the diffusion of ferredoxin molecules between them. This interpretation is supported by the recent observation of the four baseplate peaks (~795, 805, 815, 825 nm; Dostal et al. 2014) in low-temperature 2DES experiment, which spectrally overlap with the four main absorbing states of FMO resolved at low temperatures (~800, 805, 815, 825 nm; Otte et al. 1991). As the absorption spectrum of the type I RC of these bacteria (775–840 nm) overlaps with spectra of both complexes, the presence of FMO does not seem to be necessary from an energetic point of view. However, due to this large spectral overlap, FMO may have an important dual role as a spacer, which allows the ferredoxin molecule to access the RC, simultaneously ensuring efficient EET from the baseplate to the RCs. Such a double role may explain why FMO is not present in green filamentous bacteria.

Another conclusion, which can be drawn from the considerations mentioned earlier, is that horizontal EET within the baseplate is required to reach the closest underlying membrane LH complex in the case of FAP, or the closest FMO complex for GSB and *Cab. thermophilum*.

7.6.2 Efficiency of overall EET

Some of the green bacteria are able to survive at extremely low-light conditions, and this fact is often used as an argument supporting the claim about a very high LH efficiency of the chlorosome. However, it is not clear whether this ability is a consequence of extraordinary properties of the chlorosome itself or if it is caused by the overall organization of the LH apparatus. In the following text, we argue that both contributions are important. Some of the unique properties of chlorosomes were mentioned already in Section 7.2.1 (e.g., the decreased probability of the excitation loss due to the lower oscillator strength of the lowest exciton levels). As discussed already in Section 7.2.2, the quantum efficiency of the EET from the chlorosome to the baseplate and subsequent LH complexes depends mainly on the type of BChl found in the chlorosome interior and the temperature but may also depend on other factors like the redox conditions. In some cases, the efficiency reaches values close to 100%. However, quantum efficiencies approaching 100% are not exceptional among other photosynthetic antennas (Wraight and Clayton 1974; Ruban et al. 2011).

It may seem also striking that some species living at extremely low-light conditions do not use the chlorosomes with the highest observed quantum efficiency of EET. For instance, the green sulfur bacterium *Chlorobium phaeobacteroides* lives often at large water depths and contains BChl *e* as the main pigment (Borrego et al. 1999b; Garcia-Gil et al. 1999). Due to the light scattering and absorption by the water column and phytoplankton, the light reaching larger depths is usually enriched in the spectral region between 500 and 600 nm (Vila and Abella 1994). This is the region where carotenoids absorb. Also aggregates of BChl *e* are very suitable for light harvesting at these conditions due to the unusual splitting of their Soret absorption band (Shibata et al. 2010). However, the efficiency of EET from BChl *e* to BChl *a* (65% [Orf and Blankenship 2013]) is lower than that of BChl *c* and *d* (close to 100%; see Section 7.2.2.2). Clearly, the quantum efficiency of EET is not the only important factor for the survival of the bacterium, and, for example, spectral coverage is also essential.

The main function of all LH antennas is to increase the cross section for light absorption. Therefore, the key property of the chlorosome for photosynthesis in dim light is that it is the largest known photosynthetic antenna. As there is no protein in the chlorosome interior, the density of pigments is also very high. Using the above-introduced average number of 10^5 molecules per chlorosome (ellipsoid of a volume of $\sim10^{-22}$ m^3), a concentration of 1.7 M can be calculated, which is much higher than in other LH complexes (Scholes et al. 2011). A direct consequence of the large size and the high pigment concentration is that the chlorosome-containing bacteria exhibit the highest number of antenna pigments per RC among all characterized photosynthetic organisms. This ratio is largest even when growing the bacteria at moderate-light conditions and can be further increased at low-light conditions in both GSB and green FAP (Section 7.2.2.3). The experimental data suggest that the size of the baseplate and the number of RCs remain the same, but the thickness of the chlorosome increases in low light (Psencik et al. 2013). The large size of the chlorosome and large ratio between the LH pigments and RCs seem to be the most crucial properties ensuring survival at low-light conditions. As mentioned earlier, the BChl *e*–containing chlorosomes do not exhibit a superior efficiency, but these chlorosomes are in general much larger than those containing BChl c or d (Martinez-Planells et al. 2002; Psencik et al. 2006).

Increasing the size of the LH complex, the number of pigments inside, and their spectral coverage is an efficient strategy to increase the number of absorbed photons. However, with an increasing size of the system, it becomes more demanding to ensure a reasonable efficiency of EET. In smaller LH complexes, it is the protein scaffold that controls the optimal organization of the pigments, but this strategy is not applicable to a large system based on self-assembly. As pointed out in a recent review on chlorosomes (Oostergetel et al. 2010), the presence of the baseplate is crucial to achieve a reasonable efficiency for the EET from the chlorosome toward the RCs. This was demonstrated by using an example where it was considered that only 10 out of 10^5 BChls in the chlorosomes were in a close contact to the RC ($N_{CRC} = 10$, $N_{TOT} = 10^5$). Oostergetel et al. (2010) have shown that if the transfer time from any of these 10 pigments to the RC was 1 ps, then even if the transfer time between the BChl molecules was infinitely fast, the overall transfer time would be N_{TOT}/N_{CRC} ps = 10 ns. This is because the probability of the excitation being on the BChl molecule in contact to the RC was always $N_{CRC}/N_{TOT} = 10^{-4}$. At 10 ns transfer times, the losses by competitive processes would be too high. The baseplate contains BChl a, the absorption of which is red-shifted as compared to that of BChl aggregates and therefore acts as an efficient trap for the excitations from the chlorosome (Section 7.2.2.2). The typical ratio between the number of aggregated BChls in the chlorosome interior and BChl a molecules in the baseplate is 100:1 for GSB. This means that the transfer from the baseplate is 100× faster than it would be directly from BChl aggregates (Oostergetel et al. 2010). In green FAP, the molar ratio between BChl c in the chlorosome interior and BChl a in the baseplate is smaller (~20:1), but the quantum efficiency is increased by a larger energy gradient.

Here it is worthwhile to note that the presence of coherent domains recently described in the chlorosome interior (Section 7.2.2.1) (Dostal et al. 2012) further increases the efficiency of EET. While the excitation is delocalized within each domain (coherent transfer), it is transferred by incoherent hopping between the domains. The domains thus effectively act as "supermolecules" effectively

decreasing the ratio between the number of pigments within the chlorosome and the baseplate and increasing the efficiency of EET from the aggregates to the baseplate. To reach the observed red shift, one domain should consist of 10–100 strongly coupled molecules. If the chlorosome contains 10^3–10^4 coherent domains, this might lead to a one to two orders of magnitude increase of the EET efficiency when compared to 10^5 monomeric pigments. This means that the aggregation of the pigments employed in chlorosomes is an effective strategy to increase the efficiency of EET (Yakovlev et al. 2002; Taisova et al. 2014). The existence of domains is a direct consequence of the substantial disorder present in the chlorosome. Now the question arises whether a less disordered arrangement leading to a smaller number of coherent domains (and thus higher ratio between the number of "supermolecules" inside the chlorosome and BChl a in the baseplate) could further increase the efficiency of light harvesting in dim light. The most apparent disadvantage of such an arrangement would be a narrower spectral coverage.

In conclusion, it is the large size of the chlorosomes, the large number of LH pigments (and their ratio to RC), unique properties given by the strong coupling between the pigments, and the overall architecture of the complete LH machinery that allow the chlorosome-possessing bacteria to survive at extremely low-light conditions. However, the limited potential for regulation due to the lack of protein in the chlorosome interior makes chlorosomes unsuitable for high-light conditions. This is probably the reason why higher plants and other photosynthetic bacteria developed LH systems based completely on pigment–protein complexes.

7.6.3 EET in whole cells

The EET processes through the entire photosynthetic apparatus of green bacteria are summarized in Figure 7.7. EET from the chlorosome to the baseplate is well documented for isolated chlorosomes from both GSB and green FAP. To obtain information about the functional connectivity between the individual LH complexes, and most complete information on the overall function of the entire LH machinery, one has to study the intact photosynthetic apparatus, either in whole cells or membranes with attached chlorosomes. Most of the measurements on whole cells were

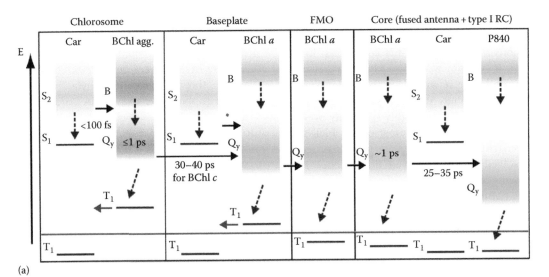

(a)

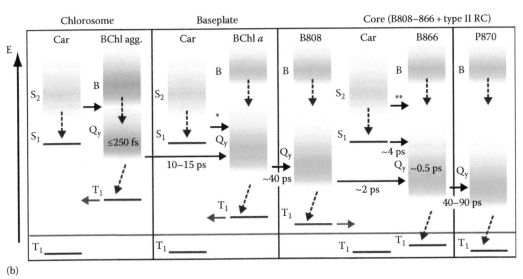

(b)

Figure 7.7 Scheme of the main absorbing states and EET processes in the photosynthetic apparatus of a typical member of GSB and presumably also in phototrophic acidobacteria (a) and green FAP (b). Ground states S_0 are not depicted. States into which absorption is allowed are shown in color, and the intensity approximately corresponds to the amplitude of the absorption spectrum. Optically forbidden states are shown by black horizontal lines, and their position is tentative. Horizontal arrows indicate EET between singlet states (blue) and triplet states (red). Black arrows represent the conversion between the states of the same multiplicity (vertical arrows) and of different multiplicity (oblique arrows). The violet line indicates the position of singlet oxygen energy. The vertical energy scale is arbitrary. Energies of the carotenoid triplet states were assumed to lie below the energy of singlet oxygen. The same assumption was used for the primary donors, since they are both composed of BChl a (Takiff and Boxer 1988). Energies of BChl triplet states were approximated to lie 4430 cm^{-1} below the energy of the fluorescence maxima (Orf et al. 2014). Typical relaxation and transfer times are indicated when available, and the references are given in the main text.* It is not clear whether the process occurs from the S_1 or S_2 state.** This process was observed only in red FAP.

carried out several decades ago due to the lack of isolation protocols. Such measurements are, however, experimentally challenging due to intense light scattering and spectrally overlapping contributions from individual LH complexes. In this respect, it is easier to study FAP. Their LH complexes are better separated in absorption spectra than for GSB, where the absorptions of the baseplate, FMO, and RC are highly spectrally overlapping. In *Cfl. aurantiacus*, most of the EET steps between chlorosome, baseplate, and B866 could be resolved in time-resolved fluorescence measurements on whole cells or photosynthetic membranes (Mimuro et al. 1989; Muller et al. 1993), although the time resolution was not high enough to resolve the fast B808 to B866 EET step within the membrane complexes (for transfer times, see Sections 7.2.2.2, 7.3.2, and 7.4.2.2, and Figure 7.7). The problem of spectrally overlapping contributions in GSB was largely overcome by the use of 2DES (Dostal et al. 2016). The method enabled the observation of previously unresolved EET channels between the chlorosome and FMO, and from FMO to the RC. The results also showed that EET between the individual complexes is slower than transfer within the complexes.

EET in the complete photosynthetic apparatus of green bacteria was studied also theoretically (Linnanto and Korppi-Tommola 2013; Huh et al. 2014). These studies provide insight into the mechanisms of EET, but they also suffer from the lack of experimentally determined structural information, which had to be replaced by estimates. The use of idealized structures for BChl aggregates did not allow an explanation of the role of the significant structural disorder experimentally observed in chlorosomes. It is thus obvious that further research on both structural and functional aspects of light harvesting in green bacteria is needed. Modern spectroscopy techniques, including 2DES, increasing amount of structural information, improving theoretical approaches, and rising computational capabilities have the potential to improve our understanding of primary processes of photosynthesis in the coming years.

ACKNOWLEDGMENTS

We are grateful to Prof. Herbert van Amerongen and Dr. Aaron Collins for their critical reading of the manuscript and their valuable comments.

REFERENCES

Adams, P. G., A. J. Cadby, B. Robinson, Y. Tsukatani, M. Tank, J. Z. Wen, R. E. Blankenship, D. A. Bryant, and C. N. Hunter. 2013. Comparison of the physical characteristics of chlorosomes from three different phyla of green phototrophic bacteria. *BBA Bioenergetics* 1827:1235–1244.

Adolphs, J., F. Muh, M. E. A. Madjet, and T. Renger. 2008. Calculation of pigment transition energies in the FMO protein. *Photosynth. Res.* 95:197–209.

Adolphs, J. and T. Renger. 2006. How proteins trigger excitation energy transfer in the FMO complex of green sulfur bacteria. *Biophys. J.* 91:2778–2797.

Alster, J., M. Kabelac, R. Tuma, J. Psencik, and J. V. Burda. 2012. Computational study of short-range interactions in bacteriochlorophyll aggregates. *Comput. Theor. Chem.* 998:87–97.

Alster, J., T. Polivka, J. B. Arellano, P. Chabera, F. Vacha, and J. Psencik. 2010. beta-Carotene to bacteriochlorophyll c energy transfer in self-assembled aggregates mimicking chlorosomes. *Chem. Phys.* 373:90–97.

Amesz, J. and S. Neerken. 2002. Excitation energy trapping in anoxygenic photosynthetic bacteria. *Photosynth. Res.* 73:73–81.

Arellano, J. B., T. B. Melo, C. M. Borrego, J. Garcia-Gil, and K. R. Naqvi. 2000. Nanosecond laser photolysis studies of chlorosomes and artificial aggregates containing bacteriochlorophyll e: Evidence for the proximity of carotenoids and bacteriochlorophyll a in chlorosomes from *Chlorobium phaeobacteroides* strain CL1401. *Photochem. Photobiol.* 72:669–675.

Arellano, J. B., T. B. Melo, C. M. Borrego, and K. R. Naqvi. 2002. Bacteriochlorophyll e monomers, but not aggregates, sensitize singlet oxygen: Implications for a self-protection mechanism in chlorosomes. *Photochem. Photobiol.* 76:373–380.

Beatty, J. T., J. Overmann, M. T. Lince, A. K. Manske, A. S. Lang, R. E. Blankenship, C. L. Van Dover, T. A. Martinson, and F. G. Plumley. 2005. An obligately photosynthetic bacterial anaerobe from a deep-sea hydrothermal vent. *Proc. Natl. Acad. Sci. U. S. A.* 102:9306–9310.

Bina, D., Z. Gardian, F. Vacha, and R. Litvin. 2014. Supramolecular organization of photosynthetic membrane proteins in the chlorosome-containing bacterium *Chloroflexus aurantiacus*. *Photosynth. Res.* 122: 13–21.

Bina, D., Z. Gardian, F. Vacha, and R. Litvin. 2016. Native FMO-reaction center supercomplex in green sulfur bacteria: An electron microscopy study. *Photosynth. Res.* 128:93–102.

Blankenship, R. E. 2002. *Molecular Mechanisms of Photosynthesis*. Blackwell Science, Oxford, U.K.

Blankenship, R. E., P. Cheng, T. P. Causgrove, D. C. Brune, S. H. Wang, J. Chon, and J. Wang. 1993. Redox regulation of energy transfer efficiency in antennas of green photosynthetic bacteria. *Photochem. Photobiol.* 57:103–107.

Blankenship, R. E. and K. Matsuura. 2003. Antenna complexes from green photosynthetic bacteria. *In Light-Harvesting Antennas in Photosynthesis*. B. R. Green and W. W. Parson, eds. Kluwer Academic Publishers, Dordrecht, the Netherlands, pp. 195–217.

Blankenship, R. E., J. M. Olson, and M. Miller. 1995. Antenna complexes from green photosynthetic bacteria. *In Anoxygenic Photosynthetic Bacteria*. R. E. Blankenship, M. T. Madigan, and C. E. Bauer, eds. Kluwer Academic Publisher, Dordrecht, the Netherlands, pp. 399–435.

Bordignon, E., G. Giacometti, and D. Carbonera. 2003. A structural model for the assembly of the reaction centre and the B808-866 complex in the membranes of *Chloroflexus aurantiacus* based on the calculation of the triplet minus singlet spectrum of the primary donor. *Chem. Phys.* 294:267–275.

Bordignon, E., M. Scarzello, G. Agostini, G. Giacometti, A. Vianelli, C. Vannini, and D. Carbonera. 2002. Optically detected magnetic resonance of intact membranes from *Chloroflexas aurantiacus*. Evidence for exciton interaction between the RC and the B808-866 complex. *Photosynth. Res.* 71:45–57.

Borrego, C. M., L. Baneras, and J. Garcia-Gil. 1999a. Temporal variability of *Chlorobium phaeobacteroides* antenna pigments in a meromictic karstic lake. *Aquat. Microb. Ecol.* 17:121–129.

Borrego, C. M. and L. J. Garcia-Gil. 1995. Rearrangement of light harvesting bacteriochlorophyll homologues as a response of green sulfur bacteria to low light intensities. *Photosynth. Res.* 45:21–30.

Borrego, C. M., P. D. Gerola, M. Miller, and R. P. Cox. 1999b. Light intensity effects on pigment composition and organisation in the green sulfur bacterium *Chlorobium tepidum*. *Photosynth. Res.* 59:159–166.

Brixner, T., T. Mancal, I. V. Stiopkin, and G. R. Fleming. 2004. Phase-stabilized two-dimensional electronic spectroscopy. *J. Chem. Phys.* 121:4221–4236.

Brixner, T., J. Stenger, H. M. Vaswani, M. Cho, R. E. Blankenship, and G. R. Fleming. 2005. Two-dimensional spectroscopy of electronic couplings in photosynthesis. *Nature* 434:625–628.

Brumer, P. and M. Shapiro. 2012. Molecular response in one-photon absorption via natural thermal light vs. pulsed laser excitation. *Proc. Natl. Acad. Sci. U. S. A.* 109:19575–19578.

Brune, D. C., G. H. King, A. Infosino, T. Steiner, M. L. Thewalt, and R. E. Blankenship. 1987. Antenna organization in green photosynthetic bacteria. 2. Excitation transfer in detached and membrane-bound chlorosomes from *Chloroflexus aurantiacus*. *Biochemistry* 26:8652–8658.

Bryant, D. A., E. V. Vassilieva, N. U. Frigaard, and H. Li. 2002. Selective protein extraction from *Chlorobium tepidum* chlorosomes using detergents. Evidence that CsmA forms multimers and binds bacteriochlorophyll *a*. *Biochemistry* 41:14403–14411.

Busch, M. S. A., F. Muh, M. E. Madjet, and T. Renger. 2011. The eighth bacteriochlorophyll completes the excitation energy funnel in the FMO protein. *J. Phys. Chem. Lett.* 2:93–98.

Carbonera, D., E. Bordignon, G. Giacometti, G. Agostini, A. Vianelli, and C. Vannini. 2001. Fluorescence and absorption detected magnetic resonance of chlorosomes from green bacteria *Chlorobium tepidum* and *Chloroflexus aurantiacus*. A comparative study. *J. Phys. Chem. B* 105:246–255.

Castenholz, R. W. and B. K. Pierson. 1995. Ecology of thermophilic anoxygenic phototrophs. *In Anoxygenic Photosynthetic Bacteria*. R. E. Blankenship, M. T. Madigan, and C. E. Bauer, eds. Kluwer Academic Publisher, Dordrecht, the Netherlands, pp. 87–103.

Causgrove, T. P., D. C. Brune, and R. E. Blankenship. 1992. Förster energy transfer in chlorosomes of green photosynthetic bacteria. *J. Photochem. Photobiol. B* 15:171–179.

Causgrove, T. P., D. C. Brune, R. E. Blankenship, and J. M. Olson. 1990a. Fluorescence lifetimes of dimers and higher oligomers of bacteriochlorophyll *c* from *Chlorobium limicola*. *Photosynth. Res.* 25:1–10.

Causgrove, T. P., D. C. Brune, J. Wang, B. P. Wittmershaus, and R. E. Blankenship. 1990b. Energy transfer in whole cells and isolated chlorosomes of green photosynthetic bacteria. *Photosynth. Res.* 26:39–48.

Chenu, A., N. Christensson, H. F. Kauffmann, and T. Mancal. 2013. Enhancement of vibronic and ground-state vibrational coherences in 2D spectra of photosynthetic complexes. *Sci. Rep.* 3:2029.

Chenu, A., P. Maly, and T. Mancal. 2014. Dynamic coherence in excitonic molecular complexes under various excitation conditions. *Chem. Phys.* 439:100–110.

Chin, A. W., J. Prior, R. Rosenbach, F. Caycedo-Soler, S. F. Huelga, and M. B. Plenio. 2013. The role of non-equilibrium vibrational structures in electronic coherence and recoherence in pigment-protein complexes. *Nat. Phys.* 9:113–118.

Cho, M. H., H. M. Vaswani, T. Brixner, J. Stenger, and G. R. Fleming. 2005. Exciton analysis in 2D electronic spectroscopy. *J. Phys. Chem. B* 109:10542–10556.

Christensson, N., H. F. Kauffmann, T. Pullerits, and T. Mancal. 2012. Origin of long-lived coherences in light-harvesting complexes. *J. Phys. Chem. B* 116:7449–7454.

Collins, A. M., Y. Y. Xin, and R. E. Blankenship. 2009. Pigment organization in the photosynthetic apparatus of *Roseiflexus castenholzii*. *BBA Bioenergetics* 1787:1050–1056.

Costas, A. M. G., Y. Tsukatani, S. P. Romberger, G. T. Oostergetel, E. J. Boekema, J. H. Golbeck, and D. A. Bryant. 2011. Ultrastructural analysis and identification of envelope proteins of "*Candidatus Chloracidobacterium thermophilum*" chlorosomes. *J. Bacteriol.* 193:6701–6711.

Dostal, J., T. Mancal, R. Augulis, F. Vacha, J. Psencik, and D. Zigmantas. 2012. Two-dimensional electronic spectroscopy reveals fast excitation energy redistribution in a chlorosome. *J. Am. Chem. Soc.* 134:11611–11617.

Dostal, J., J. Psencik, and D. Zigmantas. 2016. In situ mapping of the energy flow through the entire photosynthetic apparatus. *Nat. Chem.* 8:705–710.

Dostal, J., F. Vacha, J. Psencik, and D. Zigmantas. 2014. 2D electronic spectroscopy reveals excitonic structure in the baseplate of a chlorosome. *J. Phys. Chem. Lett.* 5:1743–1747.

Egawa, A., T. Fujiwara, T. Mizoguchi, Y. Kakitani, Y. Koyama, and H. Akutsu. 2007. Structure of the light-harvesting bacteriochlorophyll *c* assembly in chlorosomes from *Chlorobium limicola* determined by solid-state NMR. *Proc. Natl. Acad. Sci. U. S. A.* 104:790–795.

Engel, G. S., T. R. Calhoun, E. L. Read, T. K. Ahn, T. Mancal, Y. C. Cheng, R. E. Blankenship, and G. R. Fleming. 2007. Evidence for wavelike energy transfer through quantum coherence in photosynthetic systems. *Nature* 446:782–786.

Feick, R., J. A. Shiozawa, and A. Ertmaier. 1995. Biochemical and spectroscopic properties of the reaction center of the green filamentous bacterium, *Chloroflexus aurantiacus*. In *Anoxygenic Photosynthetic Bacteria*. R. E. Blankenship, M. T. Madigan, and C. E. Bauer, eds. Kluwer Academic Publisher, Dordrecht, the Netherlands, pp. 699–708.

Feick, R. G. and R. C. Fuller. 1984. Topography of the photosynthetic apparatus of *Chloroflexus aurantiacus*. *Biochemistry* 23:3693–3700.

Fenna, R. E. and B. W. Matthews. 1975. Chlorophyll arrangement in a bacteriochlorophyll protein from *Chlorobium limicola*. *Nature* 258:573–577.

Fetisova, Z., A. Freiberg, V. Novoderezhkin, A. Taisova, and K. Timpmann. 1996. Antenna size dependent exciton dynamics in the chlorosomal antenna of the green bacterium *Chloroflexus aurantiacus*. *FEBS Lett.* 383:233–236.

Fleming, G. R. and G. D. Scholes. 2004. Physical chemistry: Quantum mechanics for plants. *Nature* 431:256–257.

Francke, C., S. C. M. Otte, M. Miller, J. Amesz, and J. M. Olson. 1996. Energy transfer from carotenoid and FMO-protein in subcellular preparations from green sulfur bacteria. Spectroscopic characterization of an FMO-reaction center core complex at low temperature. *Photosynth. Res.* 50:71–77.

Francke, C., H. P. Permentier, E. M. Franken, S. Neerken, and J. Amesz. 1997. Isolation and properties of photochemically active reaction center complexes from the green sulfur bacterium *Prosthecochloris aestuarii*. *Biochemistry* 36:14167–14172.

Frese, R., U. Oberheide, I. H. M. van Stokkum, R. van Grondelle, M. Foidl, J. Oelze, and H. van Amerongen. 1997. The organization of bacteriochlorophyll *c* in chlorosomes from *Chloroflexus aurantiacus* and the structural role of carotenoids and protein: An absorption, linear dichroism, circular dichroism and Stark spectroscopy study. *Photosynth. Res.* 54:115–126.

Frigaard, N. U. and D. A. Bryant. 2006. Chlorosomes: Antenna organelles in photosynthetic green bacteria. In *Complex Intracellular Structures in Prokaryotes*. Microbiology Monographs, vol. 2. J. M. Shively, ed. Springer, Berlin, Germany, pp. 79–114.

Frigaard, N. U., A. G. M. Chew, H. Li, J. A. Maresca, and D. A. Bryant. 2003. *Chlorobium tepidum*: Insights into the structure, physiology, and metabolism of a green sulfur bacterium derived from a complete genome sequence. *Photosynth. Res.* 78:93–117.

Frigaard, N. U., H. Li, P. Martinsson, S. K. Das, H. A. Frank, T. J. Aartsma, and D. A. Bryant. 2005. Isolation and characterization of carotenosomes from a bacteriochlorophyll *c*-less mutant of *Chlorobium tepidum*. *Photosynth. Res.* 86:101–111.

Frigaard, N. U., S. Takaichi, M. Hirota, K. Shimada, and K. Matsuura. 1997. Quinones in chlorosomes of green sulfur bacteria and their role in the redox-dependent fluorescence studied in chlorosome-like bacteriochlorophyll *c* aggregates. *Arch. Microbiol.* 167:343–349.

Fuciman, M., P. Chabera, A. Zupcanova, P. Hribek, J. B. Arellano, F. Vacha, J. Psencik, and T. Polivka. 2010. Excited state properties of aryl carotenoids. *Phys. Chem. Chem. Phys.* 12:3112–3120.

Furumaki, S., F. Vacha, S. Habuchi, Y. Tsukatani, D. A. Bryant, and M. Vacha. 2011. Absorption linear dichroism measured directly on a single light-harvesting system: The role of disorder in chlorosomes of green photosynthetic bacteria. *J. Am. Chem. Soc.* 133:6703–6710.

Ganapathy, S., G. T. Oostergetel, P. K. Wawrzyniak, M. Reus, A. G. M. Chew, F. Buda, E. J. Boekema, D. A. Bryant, A. R. Holzwarth, and H. J. M. de Groot. 2009. Alternating syn-anti bacteriochlorophylls form concentric helical nanotubes in chlorosomes. *Proc. Natl. Acad. Sci. U. S. A.* 106:8525–8530.

Garcia-Gil, L. J., E. Vicente, A. Camacho, C. M. Borrego, X. Vila, X. P. Cristina, and J. Rodriguez-Gonzalez. 1999. Vertical distribution of photosynthetic sulphur bacteria linked to saline gradients in Lake 'El Tobar' (Cuenca, Spain). *Aquat. Microb. Ecol.* 20:299–303.

Griebenow, K., A. R. Holzwarth, F. van Mourik, and R. van Grondelle. 1991. Pigment organization and energy transfer in green bacteria. 2. Circular and linear dichroism spectra of protein-containing and protein-free chlorosomes isolated from *Chloroflexus aurantiacus* strain Ok-70-fl. *Biochim. Biophys. Acta* 1058:194–202.

Hauska, G., T. Schoedl, H. Remigy, and G. Tsiotis. 2001. The reaction center of green sulfur bacteria. *BBA Bioenergetics* 1507:260–277.

He, G. N., D. M. Niedzwiedzki, G. S. Orf, H. Zhang, and R. E. Blankenship. 2015. Dynamics of energy and electron transfer in the FMO-reaction center core complex from the phototrophic green sulfur bacterium *Chlorobaculum tepidum*. *J. Phys. Chem. B* 119:8321–8329.

Holzwarth, A. R., M. G. Müller, and K. Griebenow. 1990. Picosecond energy transfer kinetics between pools in different preparations of chlorosomes from the green bacterium *Chloroflexus aurantiacus* Ok-70-fl. *J. Photochem. Photobiol. B* 5:457–465.

Huang, R. Y. C., J. Z. Wen, R. E. Blankenship, and M. L. Gross. 2012. Hydrogen-deuterium exchange mass spectrometry reveals the interaction of Fenna-Matthews-Olson protein and chlorosome CsnnA protein. *Biochemistry* 51:187–193.

Huh, J., S. K. Saikin, J. C. Brookes, S. Valleau, T. Fujita, and A. Aspuru-Guzik. 2014. Atomistic study of energy funneling in the light-harvesting complex of green sulfur bacteria. *J. Am. Chem. Soc.* 136:2048–2057.

Ishizaki, A., T. R. Calhoun, G. S. Schlau-Cohen, and G. R. Fleming. 2010. Quantum coherence and its interplay with protein environments in photosynthetic electronic energy transfer. *Phys. Chem. Chem. Phys.* 12:7319–7337.

Ishizaki, A. and G. R. Fleming. 2009. Unified treatment of quantum coherent and incoherent hopping dynamics in electronic energy transfer: Reduced hierarchy equation approach. *J. Chem. Phys.* 130:234111.

Jonas, D. M. 2003. Two-dimensional femtosecond spectroscopy. *Annu. Rev. Phys. Chem.* 54:425–463.

Jordan, P., P. Fromme, H. T. Witt, O. Klukas, W. Saenger, and N. Krauss. 2001. Three-dimensional structure of cyanobacterial photosystem I at 2.5 angstrom resolution. *Nature* 411:909–917.

Jun, S., C. Yang, T. W. Kim, M. Isaji, H. Tamiaki, H. Ihee, and J. Kim. 2015. Role of thermal excitation in ultrafast energy transfer in chlorosomes revealed by two-dimensional electronic spectroscopy. *Phys. Chem. Chem. Phys.* 17:17872–17879.

Kassal, I., J. Yuen-Zhou, and S. Rahimi-Keshari. 2013. Does coherence enhance transport in photosynthesis? *J. Phys. Chem. Lett.* 4:362–367.

Kim, H., H. Li, J. A. Maresca, D. A. Bryant, and S. Savikhin. 2007. Triplet exciton formation as a novel photoprotection mechanism in chlorosomes of *Chlorobium tepidum*. *Biophys. J.* 93:192–201.

Kjellberg, P., B. Bruggemann, and T. Pullerits. 2006. Two-dimensional electronic spectroscopy of an excitonically coupled dimer. *Phys. Rev. B* 74:024303.

Klappenbach, J. A. and B. K. Pierson. 2004. Phylogenetic and physiological characterization of a filamentous anoxygenic photoautotrophic bacterium 'Candidatus Chlorothrix halophila' gen. nov., sp nov., recovered from hypersaline microbial mats. *Arch. Microbiol.* 181:17–25.

Klinger, P., J. B. Arellano, F. E. Vacha, J. Hala, and J. Psencik. 2004. Effect of carotenoids and monogalactosyl diglyceride on bacteriochlorophyll *c* aggregates in aqueous buffer: Implications for the self-assembly of chlorosomes. *Photochem. Photobiol.* 80:572–578.

Kramer, H. M., H. Kingma, T. Swarthoff, and J. Amesz. 1982. Prompt and delayed fluorescence in pigment-protein complexes of a green photosynthetic bacterium. *Biochim. Biophys. Acta* 681:359–364.

Krasnovsky, A. A. 1982. Delayed fluorescence and phosphorescence of plant pigments. *Photochem. Photobiol.* 36:733–741.

Kreisbeck, C., T. Kramer, M. Rodriguez, and B. Hein. 2011. High-performance solution of hierarchical equations of motion for studying energy transfer in light-harvesting complexes. *J. Chem. Theory Comput.* 7:2166–2174.

Le Olson, T., A. M. L. van de Meene, J. N. Francis, B. K. Pierson, and R. E. Blankenship. 2007. Pigment analysis of "Candidatus Chlorothrix halophila," a green filamentous anoxygenic phototrophic bacterium. *J. Bacteriol.* 189:4187–4195.

Lee, H., Y. C. Cheng, and G. R. Fleming. 2007. Coherence dynamics in photosynthesis: Protein protection of excitonic coherence. *Science* 316:1462–1465.

Li, H. and D. A. Bryant. 2009. Envelope proteins of the CsmB/CsmF and CsmC/CsmD Motif families influence the size, shape, and composition of chlorosomes in *Chlorobaculum tepidum*. *J. Bacteriol.* 191:7109–7120.

Li, H., N. U. Frigaard, and D. A. Bryant. 2013. [2Fe-2S] proteins in chlorosomes: CsmI and CsmJ participate in light-dependent control of energy transfer in chlorosomes of *Chlorobaculum tepidum*. *Biochemistry* 52:1321–1330.

Li, Y. F., W. L. Zhou, R. E. Blankenship, and J. P. Allen. 1997. Crystal structure of the bacteriochlorophyll *a* protein from *Chlorobium tepidum*. *J. Mol. Biol.* 271:456–471.

Limantara, L., R. Fujii, J. P. Zhang, T. Kakuno, H. Hara, A. Kawamori, T. Yagura, R. J. Cogdell, and Y. Koyama. 1998. Generation of triplet and cation-radical bacteriochlorophyll *a* in carotenoidless LH1 and LH2 antenna complexes from *Rhodobacter sphaeroides*. *Biochemistry* 37:17469–17486.

Lin, S., H. van Amerongen, and W. S. Struve. 1991. Ultrafast pump-probe spectroscopy of bacteriochlorophyll *c* antenna in bacteriochlorophyll *a*-containing chlorosomes from the green photosynthetic bacterium *Chloroflexus aurantiacus*. *Biochim. Biophys. Acta* 1060:13–24.

Linnanto, J. M. and J. E. L. Korppi-Tommola. 2013. Exciton description of chlorosome to baseplate excitation energy transfer in filamentous anoxygenic phototrophs and green sulfur bacteria. *J. Phys. Chem. B* 117:11144–11161.

Louwe, R. J. W., J. Vrieze, A. J. Hoff, and T. J. Aartsma. 1997. Toward an integral interpretation of the optical steady-state spectra of the FMO-complex of *Prosthecochloris aestuarii*. 2. Exciton simulations. *J. Phys. Chem. B* 101:11280–11287.

Macpherson, A. N. and T. Gillbro. 1998. Solvent dependence of the ultrafast S_2-S_1 internal conversion rate of b-carotene. *J. Phys. Chem. A* 102:5049–5058.

Mancal, T. 2013. Excitation energy transfer in a classical analogue of photosynthetic antennae. *J. Phys. Chem. B* 117:11282–11291.

Mancal, T. and L. Valkunas. 2010. Exciton dynamics in photosynthetic complexes: Excitation by coherent and incoherent light. *New J. Phys.* 12:065044.

Manske, A. K., J. Glaeser, M. A. M. Kuypers, and J. Overmann. 2005. Physiology and phylogeny of green sulfur bacteria forming a monospecific phototrophic assemblage at a depth of 100 meters in the Black Sea. *Appl. Environ. Microbiol.* 71:8049–8060.

Martinez-Planells, A., J. B. Arellano, C. M. Borrego, C. Lopez-Iglesias, F. Gich, and J. S. Garcia-Gil. 2002. Determination of the topography and biometry of chlorosomes by atomic force microscopy. *Photosynth. Res.* 71:83–90.

Martiskainen, J., J. Linnanto, V. Aumanen, P. Myllyperkio, and J. Korppi-Tommola. 2012. Excitation energy transfer in isolated chlorosomes from *Chlorobaculum tepidum* and *Prosthecochloris aestuarii*. *Photochem. Photobiol.* 88:675–683.

Martiskainen, J., J. Linnanto, R. Kananavicius, V. Lehtovuori, and J. Korppi-Tommola. 2009. Excitation energy transfer in isolated chlorosomes from *Chloroflexus aurantiacus*. *Chem. Phys. Lett.* 477:216–220.

Melo, T. B., N. U. Frigaard, K. Matsuura, and K. R. Naqvi. 2000. Electronic energy transfer involving carotenoid pigments in chlorosomes of two green bacteria: *Chlorobium tepidum* and *Chloroflexus aurantiacus*. *Spectrochim Acta A Mol. Biol. Spectrosc.* 56:2001–2010.

Milder, M. T. W., B. Bruggemann, R. van Grondelle, and J. L. Herek. 2010. Revisiting the optical properties of the FMO protein. *Photosynth. Res.* 104:257–274.

Mimuro, M., T. Nozawa, N. Tamai, K. Shimada, I. Yamazaki, S. Lin, R. S. Knox, B. P. Wittmershaus, D. C. Brune, and R. E. Blankenship. 1989. Excitation energy flow in chlorosomes of green photosynthetic bacteria. *J. Phys. Chem.* 93:7503–7509.

Mohseni, M., P. Rebentrost, S. Lloyd, and A. Aspuru-Guzik. 2008. Environment-assisted quantum walks in photosynthetic energy transfer. *J. Chem. Phys.* 129:174106.

Montano, G. A., B. P. Bowen, J. T. LaBelle, N. W. Woodbury, V. B. Pizziconi, and R. E. Blankenship. 2003a. Characterization of *Chlorobium tepidum* chlorosomes: A calculation of bacteriochlorophyll *c* per chlorosome and oligomer modeling. *Biophys. J.* 85:2560–2565.

Montano, G. A., H. M. Wu, S. Lin, D. C. Brune, and R. E. Blankenship. 2003b. Isolation and characterization of the B798 light-harvesting baseplate from the chlorosomes of *Chloroflexus aurantiacus*. *Biochemistry* 42:10246–10251.

Montano, G. A., Y. Y. Xin, S. Lin, and R. E. Blankenship. 2004. Carotenoid and bacteriochlorophyll energy transfer in the b808-866 complex from *Chloroflexus aurantiacus*. *J. Phys. Chem. B* 108:10607–10611.

Muh, F., M. E. A. Madjet, J. Adolphs, and T. Renger. 2007. Structure-based calculation of pigment transition energies in light-harvesting antennae. *Photosynth. Res.* 91:165.

Muller, M. G., K. Griebenow, and A. R. Holzwarth. 1993. Picosecond energy transfer and trapping kinetics in living cells of the green bacterium *Chloroflexus aurantiacus*. *Biochim. Biophys. Acta* 1144:161–169.

Neerken, S., H. P. Permentier, C. Francke, T. J. Aartsma, and J. Amesz. 1998. Excited states and trapping in reaction center complexes of the green sulfur bacterium *Prosthecochloris aestuarii*. *Biochemistry* 37:10792–10797.

Niedzwiedzki, D. M. and R. E. Blankenship. 2010. Singlet and triplet excited state properties of natural chlorophylls and bacteriochlorophylls. *Photosynth. Res.* 106:227–238.

Niedzwiedzki, D. M., A. M. Collins, A. M. LaFountain, M. M. Enriquez, H. A. Frank, and R. E. Blankenship. 2010. Spectroscopic studies of carotenoid-to-bacteriochlorophyll energy transfer in LHRC photosynthetic complex from *Roseiflexus castenholzii. J. Phys. Chem. B* 114:8723–8734.

Niedzwiedzki, D. M., G. S. Orf, M. Tank, K. Vogl, D. A. Bryant, and R. E. Blankenship. 2014. Photophysical properties of the excited states of Bacteriochlorophyll *f* in solvents and in chlorosomes. *J. Phys. Chem. B* 118:2295–2305.

Nielsen, J. T., N. V. Kulminskaya, M. Bjerring, J. M. Linnanto, M. Ratsep, M. O. Pedersen, P. H. Lambrev et al. 2016. In situ high-resolution structure of the baseplate antenna complex in *Chlorobaculum tepidum. Nat. Commun.* 7:12454.

Novoderezhkin, V. and Z. Fetisova. 1999. Exciton delocalization in the B808–866 antenna of the green bacterium *Chloroflexus aurantiacus* as revealed by ultrafast pump-probe spectroscopy. *Biophys. J.* 77:424–430.

Novoderezhkin, V. I., A. S. Taisova, Z. Fetisova, R. E. Blankenship, S. Savikhin, D. R. Buck, and W. S. Struve. 1998. Energy transfers in the B808-866 antenna from the green bacterium *Chloroflexus aurantiacus. Biophys. J.* 74:2069–2075.

Nozawa, T., K. Ohtomo, M. Suzuki, H. Nakagawa, Y. Shikama, H. Konami, and Z. Y. Wang. 1994. Structures of chlorosomes and aggregated BChl *c* in *Chlorobium tepidum* from solid state high resolution CP/MAS ^{13}C NMR. *Photosynth. Res.* 41:211–223.

Nuijs, A. M., H. Vasmel, L. N. M. Duysens, and J. Amesz. 1986. Antenna and reaction-center processes upon picosecond-flash excitation of membranes of the green photosynthetic bacterium *Chloroflexus aurantiacus. Biochim. Biophys. Acta* 849:316–324.

Ohashi, S., T. Iemura, N. Okada, S. Itoh, H. Furukawa, M. Okuda, M. Ohnishi-Kameyama et al. 2010. An overview on chlorophylls and quinones in the photosystem I-type reaction centers. *Photosynth. Res.* 104:305–319.

Oh-Oka, H., S. Kamei, H. Matsubara, S. Lin, P. I. van Noort, and R. E. Blankenship. 1998. Transient absorption spectroscopy of energy-transfer and trapping processes in the reaction center complex of *Chlorobium tepidum. J. Phys. Chem. B* 102:8190–8195.

Oksanen, J. A. I., E. I. Zenkevich, V. N. Knyukshto, S. Pakalnis, P. H. Hynninen, and J. E. I. KorppiTommola. 1997. Investigations

of Chl a aggregates cross-linked by dioxane in 3-methylpentane. *BBA Bioenergetics* 1321:165–178.

Olbrich, C., T. L. C. Jansen, J. Liebers, M. Aghtar, J. Strumpfer, K. Schulten, J. Knoester, and U. Kleinekathofer. 2011a. From atomistic modeling to excitation transfer and two-dimensional spectra of the FMO light-harvesting complex. *J. Phys. Chem. B* 115:8609–8621.

Olbrich, C., J. Strumpfer, K. Schulten, and U. Kleinekathofer. 2011b. Quest for spatially correlated fluctuations in the FMO light-harvesting complex. *J. Phys. Chem. B* 115:758–764.

Oostergetel, G. T., M. Reus, A. Gomez Maqueo Chew, D. A. Bryant, E. J. Boekema, and A. R. Holzwarth. 2007. Long-range organization of bacteriochlorophyll in chlorosomes of *Chlorobium tepidum* investigated by cryo-electron microscopy. *FEBS Lett.* 581:5435–5439.

Oostergetel, G. T., H. van Amerongen, and E. J. Boekema. 2010. The chlorosome: A prototype for efficient light harvesting in photosynthesis. *Photosynth. Res.* 104:245–255.

Orf, G. S. and R. E. Blankenship. 2013. Chlorosome antenna complexes from green photosynthetic bacteria. *Photosynth. Res.* 116:315–331.

Orf, G. S., D. M. Niedzwiedzki, and R. E. Blankenship. 2014. Intensity dependence of the excited state lifetimes and triplet conversion yield in the Fenna-Matthews-Olson antenna protein. *J. Phys. Chem. B* 118:2058–2069.

Orf, G. S., R. G. Saer, D. M. Niedzwiedzki, H. Zhang, C. L. McIntosh, J. W. Schultz, L. M. Mirica, and R. E. Blankenship. 2016. Evidence for a cysteine-mediated mechanism of excitation energy regulation in a photosynthetic antenna complex. *Proc. Natl. Acad. Sci. U. S. A.* 113:E4486–E4493.

Orf, G. S., M. Tank, K. Vogl, D. M. Niedzwiedzki, D. A. Bryant, and R. E. Blankenship. 2013. Spectroscopic insights into the decreased efficiency of chlorosomes containing bacteriochlorophyll *f. BBA Bioenergetics* 1827:493–501.

Otte, S. C. M., J. C. van der Heiden, N. Pfennig, and J. Amesz. 1991. A comparative study of the optical characteristics of intact cells of photosynthetic green sulfur bacteria containing bacteriochlorophyll *c*, *d* or *e*. *Photosynth. Res.* 28:77–87.

Overmann, J. and F. Garcia-Pichel. 2006. The phototrophic way of life. In *The Prokaryotes*. M. Dworkin, S. Falkow, E. Rosenberg, K. H. Schleifer, and E. Stackebrandt, eds. Springer, New York, pp. 32–85.

Pachon, L. A. and P. Brumer. 2011. Physical basis for long-lived electronic coherence in photosynthetic light-harvesting systems. *J. Phys. Chem. Lett.* 2:2728–2732.

Panitchayangkoon, G., D. Hayes, K. A. Fransted, J. R. Caram, E. Harel, J. Z. Wen, R. E. Blankenship, and G. S. Engel. 2010. Long-lived quantum coherence in photosynthetic complexes at physiological temperature. *Proc. Natl. Acad. Sci. U. S. A.* 107:12766–12770.

Pedersen, M. O., J. Linnanto, N. U. Frigaard, N. C. Nielsen, and M. Miller. 2010. A model of the protein-pigment baseplate complex in chlorosomes of photosynthetic green bacteria. *Photosynth. Res.* 104:233–243.

Pedersen, M. O., J. Underhaug, J. Dittmer, M. Miller, and N. C. Nielsen. 2008. The three-dimensional structure of CsmA: A small antenna protein from the green sulfur bacterium *Chlorobium tepidum*. *FEBS Lett.* 582:2869–2874.

Permentier, H. P., K. A. Schmidt, M. Kobayashi, M. Akiyama, C. Hager-Braun, S. Neerken, M. Miller, and J. Amesz. 2000. Composition and optical properties of reaction centre core complexes from the green sulfur bacteria *Prosthecochloris aestuarii* and *Chlorobium tepidum*. *Photosynth. Res.* 64:27–39.

Pierson, B. K. and J. P. Thornber. 1983. Isolation and spectral characterization of photochemical-reaction centers from the thermophilic green bacterium *Chloroflexus aurantiacus* strain J-10-F1. *Proc. Natl. Acad. Sci. U. S. A.* 80:80–84.

Pisliakov, A. V., T. Mancal, and G. R. Fleming. 2006. Two-dimensional optical three-pulse photon echo spectroscopy II. Signatures of coherent electronic motion and exciton population transfer in dimer two-dimensional spectra. *J. Chem. Phys.* 124:234505.

Plenio, M. B. and S. F. Huelga. 2008. Dephasing-assisted transport: Quantum networks and biomolecules. *New J. Phys.* 10:113019.

Psencik, J., J. B. Arellano, A. M. Collins, P. Laurinmaki, M. Torkkeli, B. Loflund, R. E. Serimaa, R. E. Blankenship, R. Tuma, and S. J. Butcher. 2013. Structural and functional roles of carotenoids in chlorosomes. *J. Bacteriol.* 195:1727–1734.

Psencik, J., J. B. Arellano, T. P. Ikonen, C. M. Borrego, P. A. Laurinmaki, S. J. Butcher, R. E. Serimaa, and R. Tuma. 2006. Internal structure of chlorosomes from brown-colored *Chlorobium* species and the role of carotenoids in their assembly. *Biophys. J.* 91:1433–1440.

Psencik, J., S. J. Butcher, and R. Tuma. 2014. Chlorosomes: Structure, function and assembly. In *The Structural Basis of Biological Energy Generation*. M. F. Hohmann-Marriott, ed. Springer, Dordrecht, the Netherlands, pp. 77–109.

Psencik, J., A. M. Collins, L. Liljeroos, M. Torkkeli, P. Laurinmaki, H. M. Ansink, T. P. Ikonen et al. 2009. Structure of chlorosomes from the green filamentous bacterium *Chloroflexus aurantiacus*. *J. Bacteriol.* 191:6701–6708.

Psencik, J., T. P. Ikonen, P. Laurinmäki, M. C. Merckel, S. J. Butcher, R. E. Serimaa, and R. Tuma. 2004. Lamellar organization of pigments in chlorosomes, the light harvesting complexes of green photosynthetic bacteria. *Biophys. J.* 87:1165–1172.

Psencik, J., Y. Z. Ma, J. B. Arellano, J. Garcia-Gil, A. R. Holzwarth, and T. Gillbro. 2002. Excitation energy transfer in chlorosomes of *Chlorobium phaeobacteroides* strain CL1401: The role of carotenoids. *Photosynth. Res.* 71:5–18.

Psencik, J., Y. Z. Ma, J. B. Arellano, J. Hala, and T. Gillbro. 2003. Excitation energy transfer dynamics and excited-state structure in chlorosomes of *Chlorobium phaeobacteroides*. *Biophys. J.* 84:1161–1179.

Psencik, J., G. F. W. Searle, J. Hala, and T. J. Schaafsma. 1994. Fluorescence-detected magnetic-resonance (FDMR) of green sulfur photosynthetic bacteria *Chlorobium* sp. *Photosynth. Res.* 40:1–10.

Psencik, J., M. Torkkeli, A. Zupcanova, F. Vacha, R. E. Serimaa, and R. Tuma. 2010. The lamellar spacing in self-assembling bacteriochlorophyll aggregates is proportional to the length of the esterifying alcohol. *Photosynth. Res.* 104:211–219.

Remigy, H. W., H. Stahlberg, D. Fotiadis, S. A. Muller, B. Wolpensinger, A. Engel, G. Hauska, and G. Tsiotis. 1999. The reaction center complex from the green sulfur bacterium *Chlorobium tepidum*: A structural analysis by scanning transmission electron microscopy. *J. Mol. Biol.* 290:851–858.

Renger, T., A. Klinger, F. Steinecker, M. S. A. Busch, J. Numata, and F. Muh. 2012. Normal mode analysis of the spectral density of the Fenna-Matthews-Olson light-harvesting protein: How the protein dissipates the excess energy of excitons. *J. Phys. Chem. B* 116:14565–14580.

Ruban, A. V., M. P. Johnson, and C. D. P. Duffy. 2011. Natural light harvesting: Principles and environmental trends. *Energy Environ. Sci.* 4:1643–1650.

Saga, Y., Y. Shibata, S. Ltoh, and H. Tamiaki. 2007. Direct counting of submicrometer-sized photosynthetic apparatus dispersed in medium at cryogenic temperature by confocal laser fluorescence microscopy: Estimation of the number of bacteriochlorophyll *c* in single light-harvesting antenna complexes chlorosomes of green photosynthetic bacteria. *J. Phys. Chem. B* 111:12605–12609.

Sakuragi, Y., N. U. Frigaard, K. Shimada, and K. Matsuura. 1999. Association of bacteriochlorophyll *a* with the CsmA protein in chlorosomes of the photosynthetic green filamentous bacterium *Chloroflexus aurantiacus*. *BBA Bioenergetics* 1413:172–180.

Savikhin, S., Y. Zhu, R. E. Blankenship, and W. S. Struve. 1996. Intraband energy transfers in the BChl *c* antenna of chlorosomes from the green photosynthetic bacterium *Chloroflexus aurantiacus*. *J. Phys. Chem.* 100:17978–17980.

Schmidt, K., M. Maarzahl, and F. Mayer. 1980. Development and pigmentation of chlorosomes in *Chloroflexus aurantiacus* strain Ok-70-fl. *Arch. Microbiol.* 127:87–97.

Scholes, G. D., G. R. Fleming, A. Olaya-Castro, and R. van Grondelle. 2011. Lessons from nature about solar light harvesting. *Nat. Chem.* 3:763–774.

Shibata, Y., Y. Saga, H. Tamiaki, and S. Itoh. 2006. Low-temperature fluorescence from single chlorosomes, photosynthetic antenna complexes of green filamentous and sulfur bacteria. *Biophys. J.* 91:3787–3796.

Shibata, Y., Y. Saga, H. Tamiaki, and S. Itoh. 2009. Anisotropic distribution of emitting transition dipoles in chlorosome from *Chlorobium tepidum*: Fluorescence polarization anisotropy study of single chlorosomes. *Photosynth. Res.* 100:67–78.

Shibata, Y., S. Tateishi, S. Nakabayashi, S. Itoh, and H. Tamiaki. 2010. Intensity borrowing via excitonic couplings among soret and Q(y) transitions of bacteriochlorophylls in the pigment aggregates of chlorosomes, the light-harvesting antennae of green sulfur bacteria. *Biochemistry* 49:7504–7515.

Shim, S., P. Rebentrost, S. Valleau, and A. Aspuru-Guzik. 2012. Atomistic study of the long-lived quantum coherences in the Fenna-Matthews-Olson complex. *Biophys. J.* 102:649–660.

Sorensen, P. G., R. P. Cox, and M. Miller. 2008. Chlorosome lipids from *Chlorobium tepidum*: Characterization and quantification of polar lipids and wax esters. *Photosynth. Res.* 95:191–196.

Staehelin, L. A., J. R. Golecki, R. C. Fuller, and G. Drews. 1978. Visualization of the supramolecular architecture of chlorosome (*Chlorobium* type vesicles) in freeze-fractured cells of *Chloroflexus aurantiacus*. *Arch. Microbiol.* 119:269–277.

Steensgaard, D. B., C. A. van Walree, L. Baneras, C. M. Borrego, J. Garcia-Gil, and A. R. Holzwarth. 1999. Evidence for spatially separate bacteriochlorophyll *c* and bacteriochlorophyll *d* pools within the chlorosomal aggregate of the green sulfur bacterium *Chlorobium limicola*. *Photosynth. Res.* 59:231–241.

Steensgaard, D. B., C. A. van Walree, H. Permentier, L. Baneras, C. M. Borrego, J. Garcia-Gil, T. J. Aartsma, J. Amesz, and A. R. Holzwarth. 2000a. Fast energy transfer between BChl *d* and BChl *c* in chlorosomes of the green sulfur bacterium *Chlorobium limicola*. *BBA Bioenergetics* 1457:71–80.

Steensgaard, D. B., H. Wackerbarth, P. Hildebrandt, and A. R. Holzwarth. 2000b. Diastereoselective control of bacterio-chlorophyll e aggregation. 3(1)-S-BChl e is essential for the formation of chlorosome-like aggregates. *J. Phys. Chem. B* 104:10379–10386.

Taisova, A. S., A. G. Yakovlev, and Z. G. Fetisova. 2014. Size variability of the unit building block of peripheral light-harvesting antennas as a strategy for effective functioning of antennas of variable size that is controlled in vivo by light intensity. *Biochemistry (Mosc.)* 79:251–259.

Takiff, L. and S. G. Boxer. 1988. Phosphorescence from the primary electron donor in *Rhodobacter sphaeroides* and *Rhodopseudomonas viridis* reaction centers. *Biochim. Biophys. Acta* 932:325–334.

Tamiaki, H., J. Komada, M. Kunieda, K. Fukai, T. Yoshitomi, J. Harada, and T. Mizoguchi. 2011. In vitro synthesis and characterization of bacterio-chlorophyll-f and its absence in bacteriochloro-phyll-e producing organisms. *Photosynth. Res.* 107:133–138.

Tank, M. and D. A. Bryant. 2015. *Chloracidobacterium thermophilum* gen. nov., sp nov.: An anoxygenic microaerophilic chlo-rophotoheterotrophic acidobacterium. *Int. J. Syst. Evol. Microbiol.* 65:1426–1430.

Tian, Y. X., R. Camacho, D. Thomsson, M. Reus, A. R. Holzwarth, and I. G. Scheblykin. 2011. Organization of Bacteriochlorophylls in individual chlorosomes from *Chlorobaculum tepidum* studied by 2-dimensional polarization fluorescence microscopy. *J. Am. Chem. Soc.* 133:17192–17199.

Tiwari, V., W. K. Peters, and D. M. Jonas. 2013. Electronic resonance with anticorrelated pigment vibrations drives photosynthetic energy transfer outside the adiabatic framework. *Proc. Natl. Acad. Sci. U. S. A.* 110:1203–1208.

Tronrud, D. E., J. Z. Wen, L. Gay, and R. E. Blankenship. 2009. The structural basis for the difference in absorbance spectra for the FMO antenna protein from various green sulfur bacteria. *Photosynth. Res.* 100:79–87.

Tsukatani, Y., S. P. Romberger, J. H. Golbeck, and D. A. Bryant. 2012. Isolation and characterization of homodimeric type-I reaction center complex from *Candidatus Chloracidobacterium thermophilum*, an aerobic chlorophototroph. *J. Biol. Chem.* 287:5720–5732.

Uehara, K., M. Mimuro, Y. Ozaki, and J. M. Olson. 1994. The formation and characterization of the in-vitro polymeric aggregates of bacterio-chlorophyll c homologs from *Chlorobium limicola* in aqueous suspension in the presence of monogalactosyl diglyceride. *Photosynth. Res.* 41:235–243.

Vacha, M., M. Furuki, and T. Tani. 1998. Origin of the long wavelength fluorescence band in some preparations of J-aggregates: Low-temperature fluorescence and hole burning study. *J. Phys. Chem. B* 102:1916–1919.

van Dorssen, R. J. and J. Amesz. 1988. Pigment organization and energy transfer in the green photosynthetic bacterium *Chloroflexus aurantiacus*. III. Energy transfer in whole cells. *Photosynth. Res.* 15:177–189.

van Dorssen, R. J., H. Vasmel, and J. Amesz. 1986. Pigment organization and energy transfer in the green photosynthetic bacterium *Chloroflexus aurantiacus*. II. The chlorosome. *Photosynth. Res.* 9:33–45.

van Gemerden, H. and J. Mas. 1995. Ecology of phototrophic sulfur bacteria. In *Anoxygenic Photosynthetic Bacteria*. R. E. Blankenship, M. T. Madigan, and C. E. Bauer, eds. Kluwer Academic Publisher, Dordrecht, the Netherlands.

van Amerongen, H., H. Vasmel, and R. van Grondelle. 1988. Linear dichroism of chloro-somes from *Chloroflexus aurantiacus* in compressed gels and electric fields. *Biophys. J.* 54:65–76.

van Noort, P. I., Y. Zhu, R. LoBrutto, and R. E. Blankenship. 1997. Redox effects on the excited-state lifetime in chlorosomes and bacteriochlorophyll c oligomers. *Biophys. J.* 72:316–325.

van Walree, C. A., Y. Sakuragi, D. B. Steensgaard, N. U. Frigaard, R. P. Cox, A. R. Holzwarth, and M. Miller. 1999. Effect of alkaline treatment on bacteriochlorophyll a, quinones and energy transfer in chlorosomes from *Chlorobium tepidum* and *Chlorobium phaeobacteroides*. *Photochem. Photobiol.* 69:322–328.

Vasmel, H., R. J. Vandorssen, G. J. Devos, and J. Amesz. 1986. Pigment organization and energy-transfer in the green photosynthetic bacterium *Chloroflexus aurantiacus* .1. The cytoplasmic membrane. *Photosynth. Res.* 7:281–294.

Vila, X. and C. A. Abella. 1994. Effects of light quality on the physiology and the ecology of planktonic green sulfur bacteria in lakes. *Photosynth. Res.* 41:53–65.

Vogl, K., M. Tank, G. S. Orf, R. E. Blankenship, and D. A. Bryant. 2012. Bacteriochlorophyll *f*: Properties of chlorosomes containing the "forbidden chlorophyll". *Front. Microbiol.* 3: 298.

Vulto, S. E., M. A. de Baat, R. J. W. Louwe, H. P. Permentier, T. Neef, M. Miller, H. van Amerongen, and T. J. Aartsma. 1998. Exciton simulations of optical spectra of the FMO complex from the green sulfur bacterium *Chlorobium tepidum* at 6 K. *J. Phys. Chem. B* 102:9577–9582.

Wang, J., D. C. Brune, and R. E. Blankenship. 1990. Effects of oxidants and reductants on the efficiency of excitation transfer in green photosynthetic bacteria. *Biochim. Biophys. Acta* 1015:457–463.

Wechsler, T. D., R. A. Brunisholz, G. Frank, F. Suter, and H. Zuber. 1987. The complete amino-acid-sequence of the antenna polypeptide B806-866-beta from the cytoplasmic membrane of the green bacterium *Chloroflexus aurantiacus*. *FEBS Lett.* 210:189–194.

Wen, J. Z., H. Zhang, M. L. Gross, and R. E. Blankenship. 2009. Membrane orientation of the FMO antenna protein from *Chlorobaculum tepidum* as determined by mass spectrometry-based footprinting. *Proc. Natl. Acad. Sci. U. S. A.* 106:6134–6139.

Wendling, M., M. A. Przyjalgowski, D. Gulen, S. I. Vulto, T. J. Aartsma, R. van Grondelle, and H. van Amerongen. 2002. The quantitative relationship between structure and polarized spectroscopy in the FMO complex of *Prosthecochloris aestuarii*: Refining experiments and simulations. *Photosynth. Res.* 71:99–123.

Wendling, M., T. Pullerits, M. A. Przyjalgowski, S. E. Vulto, T. J. Aartsma, R. van Grondelle, and H. van Amerongen. 2000. Electron-vibrational coupling in the Fenna-Matthews-Olson complex of *Prosthecochloris aestuarii* determined by temperature-dependent absorption and fluorescence line-narrowing measurements. *J. Phys. Chem. B* 104:5825–5831.

Wolynes, P. G. 2009. Some quantum weirdness in physiology. *Proc. Natl. Acad. Sci. U. S. A.* 106:17247–17248.

Wraight, C. A. and R. K. Clayton. 1974. Absolute quantum efficiency of bacteriochlorophyll photooxidation in reaction centers of *Rhodopseudomonas spheroides*. *Biochim. Biophys. Acta* 333:246–260.

Xin, Y. Y., S. Lin, G. A. Montano, and R. E. Blankenship. 2005. Purification and characterization of the B808-866 light-harvesting complex from green filamentous bacterium *Chloroflexus aurantiacus*. *Photosynth. Res.* 86:155–163.

Xin, Y. Y., J. Pan, A. M. Collins, S. Lin, and R. E. Blankenship. 2012. Excitation energy transfer and trapping dynamics in the core complex of the filamentous photosynthetic bacterium *Roseiflexus castenholzii*. *Photosynth. Res.* 111:149–156.

Yakovlev, A. G., A. S. Taisova, and Z. G. Fetisova. 2002. Light control over the size of an antenna unit building block as an efficient strategy for light harvesting in photosynthesis. *FEBS Lett.* 512:129–132.

Yamada, M., H. Zhang, S. Hanada, K. V. P. Nagashima, K. Shimada, and K. Matsuura. 2005. Structural and spectroscopic properties of a reaction center complex from the chlorosome-lacking filamentous anoxygenic phototrophic bacterium *Roseiflexus castenholzii*. *J. Bacteriol.* 187:1702–1709.

Zhou, W., R. LoBrutto, S. Lin, and R. E. Blankenship. 1994. Redox effects on the bacteriochlorophyll-*a* containing Fenna-Matthews-Olson protein from *Chlorobium tepidum*. *Photosynth. Res.* 41:89–96.

Light-harvesting complexes in chlorophyll c–containing algae

CLAUDIA BÜCHEL

8.1 INTRODUCTION

Cyanobacteria, green algae, and higher plants are usually well studied in photosynthesis research, but our knowledge about all other eukaryotic photosynthetic algae is far less advanced. Evolutionarily, the eukaryotic phototrophs fall into four main groups: (1) glaucophytes, which, in a simplified view, resemble eukaryotic cells with a cyanobacterium within; (2) the red algae, which also contain phycobilisomes as major antenna complexes comparable to those of Cyanophyta; (3) the "green lineage," comprising green alga, mosses, ferns, and higher plants, but also Euglenophyta and Chlorarachniophyta; and (4) a rather diverse group called chromalveolates. Whereas glaucophytes, green and red algae evolved after a primary endosymbiosis between a cyanobacterium and a host eukaryote, chromalveolates are the result of one or several secondary endosymbioses between two eukaryotes, one of them already photosynthetic and most probably related to extant red algae (Archibald and Keeling 2002). This symbiont became reduced to almost only the chloroplast, which is surrounded by more than two membranes, whereby the inner two are the canonical chloroplast envelope membranes known from all other eukaryotic phototrophs. Additionally, one or two further membranes surround the secondary plastids, and the outer one is connected to the nuclear endoplasmic reticulum. The space between the two inner membranes and the outer ones represents the original cytosol of the endosymbiont and contains a nucleomorph in some cases, for example, in cryptophytes. This nucleomorph is a remnant of the symbionts' nucleus, and if present, the so-called periplastid compartment also contains active eukaryotic ribosomes (Maier et al. 1991).

Members of chromalveolates are cryptophytes, haptophytes, heterokonts (including among others brown algae and diatoms), as well as alveolates, which include the photosynthetic dinoflagellates (Dinophyta), and a newly discovered group, the Chromeridae. The current knowledge about light-harvesting systems in most of these groups is very limited, with by far the most research done on diatoms and dinoflagellates. In this chapter, a general overview about the pigments and the genes for antenna proteins will be given first, followed by the known light-harvesting systems of chromalveolates, with a special focus on cryptophytes, dinophytes, and diatoms. The reader is also referred to the excellent review of MacPherson and Hiller (2003).

8.2 LIGHT-HARVESTING SYSTEMS IN THE DIFFERENT CHLOROPHYLL c–CONTAINING ALGAL GROUPS

8.2.1 Pigments

All photosynthetic eukaryotes use chlorophyll (Chl) a as photosynthetic pigment, accompanied by several accessory chromophores (see also Chapters 1 and 3). In chromalveolates, no Chl b is present and is mostly replaced by different forms of Chl c. This accounts for the synonym "Chl c–containing algae" for chromalveolates, although some do not even contain a second Chl besides Chl a, and prasinophytes—a group related to the green lineage—contain Chl c– or Chl c–like pigments, as well (Six et al. 2005, Wilhelm and Lenartz-Weiler 1987). In all types of Chl c, the phytol ester is lacking and their spectral signature is dominated by a huge Soret band absorption around 465 nm in organic solvent, with about three times the extinction coefficient of Chl a in its Soret maximum (Figure 8.1). In contrast, the Q_Y absorption at around 630 nm is minor, with only about a quarter of the extinction of Chl a in its Q_Y band. Three different kinds of Chl c exist, distinguished by their residues at the porphyrin ring. In Chls c_2 and c_3, this residue contains a double bond which is part of the conjugated system. Thus, these absorb at slightly longer wavelengths as compared to Chl c_1. In

haptophytes Chl c_3 and in dinoflagellates Chl c_2 were described as the major Chl c, whereas diatoms usually contain Chl c_1 and some Chl c_2 as well (Fawley 1989, Jeffrey and Humphrey 1975, Kraay et al. 1992). The two chlorophylls are accompanied by different xanthophylls. In diatoms, fucoxanthin (Fx) is the major carotenoid, whereas in dinoflagellates peridinin (Per) takes its function. Both are peculiar with a carbonyl moiety in conjugation with the polyene backbone (Damjanović et al. 2000, Frank et al. 2000, Katoh et al. 1991, Zigmantas et al. 2004). A special case are cryptophytes, where phycobilins, but not in the form of phycobilisomes, are located in the thylakoid lumen as accessory pigments. In addition, the carotenoid alloxanthin is present (Table 8.1).

In the Per- or Fx-containing complexes, these carotenoids (Car) are the key pigments, usually found in equal or even higher amounts than Chls. The spectroscopic features of a carotenoid are determined by the symmetry of its conjugated chain, resulting in a forbidden transition from the ground state S_0 to the first excited state S_1, but an allowed transition into the S_2 state. This transition is responsible for the strong absorption of carotenoids in the 400–500 nm region, with a typical three-peak structure characteristic for each carotenoid in a given solvent (see diadinoxanthin spectrum in Figure 8.1). Symmetric carotenoids typically have an extremely weak fluorescence from the S_1 state, although this emission is stronger in the case of Car with a carbonyl moiety like Per (Mimuro et al. 1992), Fx (Katoh et al. 1991), and siphonoxanthin, whereby the latter will not be considered here. This is most probably due to the strong deviation from perfect symmetry and is also reflected by the fact that the fluorescence quantum yield and the shape of the absorption spectrum are extremely influenced by the polarity of the solvent. The absorption in polar solvents for Fx and Per resembles the "typical" Car spectrum, whereas in nonpolar solvent, the three peaks are lacking and are replaced by one broad, more featureless absorption band (see Fx spectrum in Figure 8.1). For Per, the fluorescence quantum yield is 10^{-5} in polar solvents and increases to 10^{-3} in nonpolar environments. The latter value is about the same for Fx in ethanol (Shreve et al. 1991). The change in fluorescence quantum yield is accompanied by a change in fluorescence lifetime. This was first studied for Per, where the lifetime of the lowest excited state was demonstrated to be around 160 ps in a nonpolar solvent, but shorter than 10 ps in polar solvents (Bautista et al. 1999b, Frank et al. 2000, Zigmantas et al. 2001, 2004),

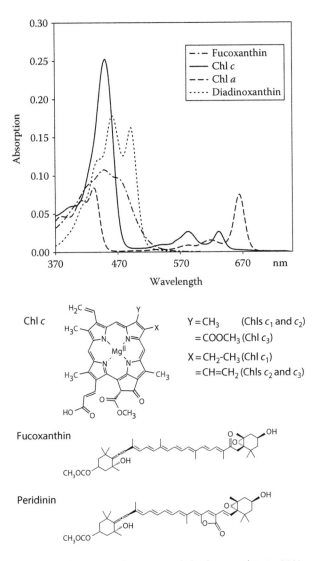

Figure 8.1 Spectra of 1 mM each of Chl a, Chl c_2, Fx, and diadinoxanthin in 80% acetone are shown in the upper panel. In the lower panel, the formulas for Chl c, Fx, and Per are given.

and is around 40 ps for Fx in ethanol (Shreve et al. 1991). This extreme shortening of lifetimes depending on the polarity of the solvent was explained by another excited state with charge-transfer character, an intramolecular charge-transfer state called S_{ICT} state. The S_{ICT} is stabilized in polar solvents and is supposed to be close to the S_1 state or even coupled to it, both for Per and Fx (Papagiannakis et al. 2004, 2006, Premvardhan et al. 2008, Vaswani et al. 2003, Zigmantas et al. 2003). For both Car, however, it was shown that also the S_2 state has considerable charge-transfer character (Premvardhan et al. 2005, 2008).

Chl c–containing algae also contain pigments involved in photoprotection. Whereas the xanthophyll

cycle of, for example, brown algae is the one known from higher plants, that is, the violaxanthin–antheraxanthin–zeaxanthin cycle, many groups like diatoms use an alternative cycle, consisting of diadinoxanthin and diatoxanthin (Table 8.1). However, under extreme conditions the classical cycle can be found as well (Lohr and Wilhelm 1999). Diadinoxanthin is the monoepoxidized form of diatoxanthin, and like in the violaxanthin cycle, the amounts depend on the light intensity the organism experiences. The ratio between the two xanthophylls is regulated via the acidification of the thylakoid lumen, which influences the activities of the de-epoxidase and the epoxidase (Goss et al. 2006).

Table 8.1 Pigment composition of the different algal groups

Organism	Chl c	Main carotenoid bound to intrinsic LHCs	Phycobilins	Main xanthophyll cycle
Cryptophyta	c_2	Alloxanthin	×	–
Dinophyta	c_2	Peridinin (fucoxanthin)	–	?
Haptophyta	c_3	Fucoxanthin	–	Diadino-/diatoxanthin
Heterokonta				
Chrysophyceae	+	Fucoxanthin	–	Viola-/zeaxanthin
Xanthophyceae	+	Heteroxanthin	–	?
Eustigmatophyceae	–	Violaxanthin	–	?
Phaeophyceae	c_1, c_2	Fucoxanthin	–	Viola-/zeaxanthin
Bacillariophyceae	c_1, c_2	Fucoxanthin	–	Diadino-/diatoxanthin

Note: x: present in the organism; –: not present in the organism; + indicates that no reference in the literature is available as to which Chl c is present.

8.2.2 Overview of the light-harvesting proteins and their genes

All members of chromalveolates contain membrane intrinsic light-harvesting systems, which belong to the light-harvesting complex (LHC) superfamily (Durnford et al. 1999), characterized by three membrane-spanning α-helices. These are supposed to be arranged in a way similar to LHCII of higher plants (Liu et al. 2004, Standfuss et al. 2005), although no molecular structure is currently available. Like in higher plants, LHC proteins are encoded in the nuclear genome. This implies that a second gene transfer has taken place after the secondary endosymbiosis, from the nucleomorph to the nucleus (Deane et al. 2000). Many reviews deal with the high numbers of lhc genes of Chl c–containing algae and their phylogenetic relationships (for recent ones, see Boldt et al. 2012, Dittami et al. 2010, Green 2011, Hoffman et al. 2011, Neilson and Durnford 2010). LHCs can be roughly divided into several multigene families, (1) the LHCs most closely related to those of red algae called lhcr, (2) the so-called LI818 or lhcSR or lhcx genes related to green algal lhcSR (Richard et al. 2000), (3) a small group called lhcz, and (4) the major group of LHCs, named lhcf. In red algae, the likely ancestors of the chloroplasts of this group, membrane-intrinsic LHCs are restricted to photosystem (PS) I. Accordingly, proteins encoded by lhcr genes in chromalveolates were expected to serve as PS I antennae, which has been proven for diatoms (Veith and Büchel 2007, Veith et al. 2009). However, in cryptophytes there is evidence that lhcr gene products may also work as PS II antennae (see Section 8.3.1). LhcSR (Lhcx) proteins were early on associated with light stress, since they are up-regulated under high light in diatoms (Eppard and Rhiel 1998, see below). No homologue to PsbS found in any of the chromalveolates.

Only cryptophytes and dinophytes contain additional, water-soluble antenna systems in addition to the membrane-intrinsic antennae. Whereas the water-soluble antenna Per chlorophyll protein (PCP) of dinophytes is unique, the lumenal phycobilins of cryptophytes are related to the proteins of the phycobilisome of cyanobacteria and red algae.

8.3 LIGHT-HARVESTING SYSTEMS OF SPECIFIC GROUPS OF Chl c–CONTAINING ALGAE

8.3.1 Cryptophytes

As mentioned earlier, cryptophytes are special in using phycobiliproteins for light harvesting, which are located in the thylakoid lumen. In addition, cryptophytes contain membrane-intrinsic LHCs, but only Lhcr and Lhcz, according to Neilson and Durnford (2010) and Doust et al. (2004).

The intrinsic LHCs contain Chl a, Chl c_2, and alloxanthin. Early reports associated these LHCs mainly with PS II (Ingram and Hiller 1983, Lichtlé et al. 1980, 1987). In a later study, 10 different light-harvesting proteins could be identified, some with a specific attribution to either PS II or PS I (Bathke et al. 1999). The PS II–specific polypeptides were more abundant and showed a higher Chl a/Chl c ratio of 4.4 as compared to those specific for PS I with 3.4 Chl a/Chl c. This was later confirmed by Kaňa et al. (2012)

for *Rhodomonas salina*. However, the sequences of the biochemically isolated light-harvesting proteins were not determined, and thus the link to the genetic information is missing. In addition, no molecular structure of any of the LHCs is available so far. Early on, PS I complexes containing LHC polypeptides had been isolated (Hiller et al. 1992, Rhiel et al. 1987). Such a complex composed of a monomeric core and six to eight LHC monomers was analyzed using single-particle analysis. The LHC were located on both sides of the monomeric PS I core in contrast to the arrangement in higher plants (Kereïche et al. 2008). In addition, these complexes lacked the red-emitting Chl *a* species typical for LHCI of higher plants.

Phycobiliproteins are water-soluble proteins with a chromophore covalently attached to the protein backbone. In cryptophytes, either phycoerythrin (PE) or phycocyanin (PC) is found. Since the amino acids binding the chromophores are not conserved, these phycobiliproteins show different absorption maxima in different cryptophyte species (Wedemayer et al. 1996). Best studied are phycoerythrin 545 (PE545) isolated from *Rhodomonas* CS24 (Figure 8.2) and phycocyanin 645 (PC645) from *Chroomonas* CCMP270. The structure of PE545 was solved in 1999 by Wilk et al. (1999) and subsequently with an ultrahigh

resolution of around 1 Å (Doust et al. 2004). PE545 is a heterodimer built of (α1β) (α2β) subunits. Three PEs are bound per β-subunit, and one 15,16-dihydrobiliverdin (DBV) per α-monomer. Both absorb in the green spectral region, with maxima at 545 and 569 nm, respectively. The PE545 complex emits fluorescence mainly at 585 nm with lesser emissions at 648 and 675 nm. Although PE545 is a rather symmetric molecule, the environment of two of the PE molecules is not identical due to the different α-subunits in their vicinity, and the same holds true for the two DBVs directly bound by α1 or α2, respectively (Figure 8.2). Based on steady-state fluorescence and circular dichroism (CD) spectroscopy, as well as time-resolved anisotropy measurements, a model for the intramolecular excitation energy transfer was proposed where the more blue-absorbing PEs funnel their energy to the central pair of PE bound at the interface of the two monomers. From there, energy is transferred to the final emitters, DBV. For the latter, two chromophores with extreme differences in fluorescence yield were found, leading to the hypothesis of only one being the final emitter (Doust et al. 2004). Later experiments on whole cells, however, revealed that both DBVs participate in the energy transfer to the photosystems (Doust et al. 2006, van der Weij-de Wit et al. 2006).

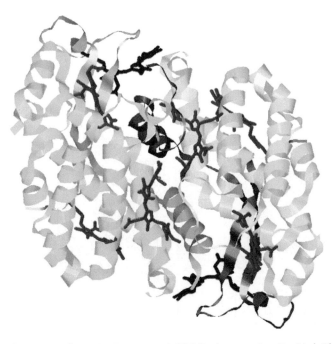

Figure 8.2 Structure of PE545 (pdb 1XF6, Doust et al. 2004), drawn using RasMol. Phycoerythrin is shown in red, 15,16-dihydrobiliverdin in black. The β-subunits are shown in dark blue, whereas the α-subunits are given in green and light-blue, respectively.

The structure of PC645 from *Chroomonas* CCMP270 became available in 2004 (Doust et al. 2006), and it shows very similar features to PE545 concerning the protein and chromophore locations. In contrast to PE545, three different chromophores are bound: two DBVs, two mesobiliverdins (MBVs), and four PCs. The MBVs are located in the place occupied by the DBVs in PE545, that is, they constitute the central pair of bilins. PC645 has a longer wavelength absorption compared to PE545, with absorption bands at 585 and 645 nm, and a shoulder at 620 nm. The shortest wavelength peak is due to the two DBVs, which are thus the bluest pigments in PC645, albeit being the reddest ones in PE545. The red peak at 645 nm is assigned to the two MBVs, and the shoulder at 620 nm to the four PCs. The fluorescence emission spectrum has a peak at 662 nm at room temperature. In a recent analysis, Marin et al. (2011) proposed a model for the excitation energy transfer in the complex, whereby the main conclusions of Doust et al. (2006) were confirmed: an asymmetrical pathway favoring excitation energy transfer to MBV on one side of the complex as shown for DBV in PE545.

In 2010, Collini et al. (2010) published evidence for long-lasting excitonic oscillations with distinct correlations even at room temperature in PE545 and PC645. This study has since fuelled research on the quantum-coherent sharing of electronic excitations under *in vivo* conditions, assumed to increase the efficiency of light harvesting and channelling excitation to preferential sites (Kolli et al. 2012).

Cryptophytes show a lateral differentiation of thylakoid membranes into grana and stroma regions, thus PS II (grana) and PS I (stroma) are spatially separated (Figure 8.3 Lichtlé et al. 1992). Phycobilins are densely packed throughout the lumen, thus principally able to serve PS II as well as PS I. A sequential energy transfer from PE via LHC to PS II, but no energy transfer from PE to PS I, was proposed by Lichtlé et al. (1980) already in 1980 and later by Mimuro et al. (1998). An almost equal distribution of excitation energy from phycobilins to PS II and PS I was later reported for *Rhodomonas* CS24 by van der Weij-de Wit et al. (2006), using intact cells for time-resolved fluorescence and fluorescence anisotropy measurements. Due to the lumenal location of phycobilins, this would require excitation energy transfer through the membrane-intrinsic LHCs at least in case of PS II, because of the large protein mass of the water-splitting complex on the lumenal side. The equal energy distribution between the photosystems was later validated for *Chroomonas* (van der Weij-de Wit et al. 2008).

Like most photosynthetic organisms, cryptophytes protect the photosystems against too much excitation energy by a mechanism called nonphotochemical quenching (NPQ), a reduction in fluorescence by increase in heat dissipation. A quenching dependent on the transthylakoid pH gradient (qE, energy quenching) was recently demonstrated (Kaňa et al. 2012). However, cryptophytes do not display a xanthophyll cycle, that is, the light-dependent de-epoxidation of a carotenoid and *vice versa*, which is usually important for NPQ processes. The membrane-intrinsic LHC was shown to

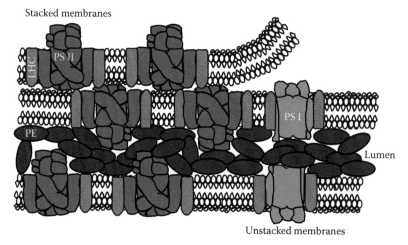

Figure 8.3 Schematic representation of the light-harvesting organization in cryptophytes. (Adapted from van der Weij-de Wit, C.D. et al., *J. Phys. Chem. B*, 110, 25066, 2006.)

change the fluorescence yield, depending on the pH, and pH-sensing residues were demonstrated by the use of DCCD (Kaňa et al. 2012). As pointed out earlier, cryptophytes do not contain genes of the *lhcSR* (*lhcx*) family, thus a member of either *lhcr* or *lhcz* has to be involved in this protective mechanism.

Classical state transition, that is, the redistribution of excitation energy between PS I and PS II under illumination conditions that preferentially excite one of the photosystems, could not be proven for cryptophytes. However, preferential excitation of PS II seemed to decrease the probability for excitation energy transfer from PE into PS II and was thus interpreted as a protection mechanism (Snyder and Biggins 1987).

8.3.2 Dinophytes

Dinophytes contain two antenna systems, whereby one is intrinsic and a member of the LHC protein family (Durnford et al. 1999), that is, assumed to fold with three membrane-spanning α-helices in a similar arrangement to LHCII, whereas the other is extrinsically located. The latter is PCP, which is a rare example of a carotenoid-containing soluble antenna protein in eukaryotes.

The water-soluble PCP is probably located on the lumenal side of the thylakoid membrane according to the gene sequence, which predicts a lumenal-targeting sequence (Norris and Miller 1994). Dinoflagellates contain many genes for PCPs per cell, coding for slightly heterogeneous isoforms. In some cases, the final protein is transcribed as one protein of around 32 kDa, or it is a dimer of 15 kDa monomers (MacPherson and Hiller 2003).

The structure of PCP from the dinoflagellate *Amphidinium carterae* was solved to 2 Å resolution in 1996 (Hofmann et al. 1996). In the crystal, the monomers are assembled into trimers, which is probably also the *in vivo* configuration (Hofmann et al. 1996). The monomers show a twofold symmetry of subdomains resembling a boat binding eight Per and two Chl *a* molecules each (Figure 8.4). Each Chl *a* is surrounded by four Per in a dense packing of pigments, where the conjugated parts of the Pers are in van der Waals contact to the Chl porphyrins. The Per absorb up to 550 nm *in vivo* with a peak at 490 nm, and efficiency in excitation energy transfer from Per to Chl *a* is about 90% (Bautista et al. 1999b, Krueger et al. 2001, Linden et al. 2004, Zigmantas et al. 2002). Thus, mainly Car are used for light harvesting, in contrast to, for example, higher plant LHCII. The study of pigment binding and pigment interaction was greatly facilitated by the ability to reconstitute heterologously expressed protein with purified pigments and, thus, by the chance to study the influence of mutated pigment-binding sites (Miller et al. 2005). Besides this main form of PCP, other variants exist, which will not be considered here.

As mentioned earlier, Per is special in having a carbonyl moiety in conjugation with the polyene backbone, which alters its spectral properties. Usually, carotenoids are able to transfer energy from both their lowest singlet excited states, that is, S_1 and S_2, whereby direct accession of S_1 via absorption

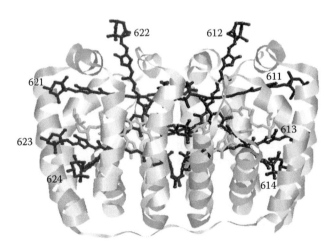

Figure 8.4 Structure of PCP (pdb 1PPR, Hofmann et al. 1996), drawn using RasMol. Pers are shown in red and blue and numbered according to pdb 1PPR. Chl a molecules are drawn in green.

is forbidden due to the symmetry of Car. In Per and other Car containing a conjugated carbonyl, an additional excited state with an intramolecular charge-transfer character exists. This ICT state can be coupled to the S_1 state and plays a major role in Car–Chl energy transfer (Bautista et al. 1999a, Papagiannakis et al. 2004, Premvardhan et al. 2005, Vaswani et al. 2003, Zigmantas et al. 2001, 2003, 2004). The protein in PCP provides a moderately polar environment for the Per molecules, stabilizing the S_{ICT} state, which is thought to be a key factor for the high efficiency of excitation energy transfer. Nonlinear experiments demonstrated that the S_{ICT} is isoenergetic or slightly above the Q_Y band of Chl a (Krikunova et al. 2006).

When Per is excited into the allowed level of S_2, direct transfer to the Q_X band of Chl a is occurring on an ultrafast timescale (~150 fs). This transfer accounts for about 25%–40% of the energy and is in competition with internal conversion to S_1. In 2010, Bonetti et al. (2010), using midinfrared spectroscopy, could distinguish the S_{ICT} and the S_1 states on two different Pers by their lactone bleaches. The transfer from S_{ICT} to Chl a Q_Y, with a time constant of 2 ps, was the predominant transfer route from these two low-lying excited states. The S_1 state relaxed mainly via internal conversion to the ground state, and by slow excitation energy transfer to Chl a. Per is also an excellent quencher of ^{T}Chl a (Bautista et al. 1999b, Kleima et al. 2000). In conclusion, the S_{ICT} was interpreted to be localized mainly on Per621/611 and Per623/613, the S_1 state on Per622/612, and the triplet state on Per624/614 (Bonetti et al. 2010). Further details on excitation energy transfer in PCPs can be found in the reviews by Polívka et al. (2007) and Polívka and Frank (2010).

The thylakoids of dinoflagellates do not show a differentiation into grana and stroma lamellae. The presumably trimeric PCP has been assumed to be localized close to the intrinsic LHC (see below), in a distance allowing Förster energy transfer (Hofmann et al. 1996). Direct transfer into PS II has also been suggested from transient absorption measurements on whole cells (Mimuro et al. 1990b). Both assumptions are fuelled by the differences between dinoflagellate LHC/PS II gene sequences and the related sequences of higher plant proteins, especially concerning membrane-extrinsic parts that could provide possible docking sites. This is hypothesized for PS I as well (MacPherson and Hiller 2003).

Two major groups of dinoflagellates exist; one group obtained their plastids directly from a secondary endosymbiosis with a rhodophyte-like ancestor, whereas the others underwent a tertiary endosymbiosis with a relative of haptophytes as symbionts. Due to this different history, the former contains Per-binding intrinsic LHC, whereas in the latter Fx is found. At a genetic level, this different phylogeny is reflected as well. The membrane-intrinsic LHCs of dinoflagellates can be classified as either Lhcr, or belonging to the main LHC family (Lhcf), or as Lhcx, whereby the latter sequences are only found in haptophyte-derived dinoflagellates (Boldt et al. 2012, Hoffman et al. 2011).

Biochemically, only the Per-containing complexes have been studied so far. A membrane-intrinsic LHC with subunits of about 19 kDa isolated in 1993 from *Amphidinium carterae* was found to bind 7 Chl a, 4 Chl c_2, 10–12 Per, and 2 diadinoxanthin molecules (Hiller et al. 1993). Recently, this pigment number was disputed, based on the fact that, unlike in LHCII, the Q_Y band of Chl a absorption is not redshifted, indicating less interaction of Chl a and thus a less tight packing (Polívka et al. 2006). However, these numbers indicate that carotenoids play a major role in light harvesting, proven by excitation energy transfer from Per to Chl (Hiller et al. 1993, Iglesias-Prieto et al. 1993, Polívka et al. 2006). The Pers transfer their energy to Chl a with about 90% efficiency (Polívka et al. 2006). Like in PCP, at least the Pers absorbing more to the red are characterized by a S_1/S_{ICT} state with significant charge-transfer character (Šlouf et al. 2013). Chl c_2 is not only even more efficient at nearly 100% but transfer is also very fast with a 1.4 ps time component, and probably an additional component faster than 100 fs (Polívka et al. 2006). There is no evidence of Per-to-Chl-c_2 transfer, although a recent study showed that they have to be arranged quite closely (Di Valentin et al. 2010), since quenching not only of ^{T}Chl a but also of ^{T}Chl c by Per is very effective. Diadinoxanthin, one of the xanthophyll cycle pigments, contributes only marginally to excitation energy transfer, so a probable role in photoprotection is assumed (Polívka et al. 2006). However, no data demonstrating a direct role of diadinoxanthin are available so far. Two of the Pers, assumed to be arranged like the central luteins in LHCII, are the efficient quenchers of ^{T}Chl, albeit to a lesser extent than the luteins in LHCII (Di Valentin et al. 2010). For the intrinsic complex from another dinoflagellate species, *Symbiodinium*, other pigment stoichiometries were reported with only four Chl *a*, six Chl *c*, six Per, and two diadinoxanthin molecules (Niedzwiedzki et al. 2014), although absorption spectra do not differ much from those reported from *A. carterae*. *Symbiodinium* is a large group of dinoflagellates living as endosymbionts in the endoderm of corals, and attention has focused on these algae since elevated sea temperature as well as high illumination leads to coral bleaching. Although quenching of ^TChl by Per was demonstrated to be very effective for the LHC from *A. carterae* (Di Valentin et al. 2010), Niedzwiedzki et al. (2014) could not demonstrate an equal role of Per in the complexes of *Symbiodinium*. In these coral symbionts, photoprotection *via* a so-called "super-quenching" state due to high spillover rates was identified (Slavov et al. 2016). Like for all Chl *c*–containing intrinsic antenna proteins, a molecular structure is still missing; thus, the question of the precise pigment arrangement is still open. In addition, the isolated complexes were not analyzed for their sequence, and nothing is known about the attribution to one of the photosystems.

8.3.3 Haptophytes

The haptophytes studied most are *Emiliania huxleyi* and *Isochrysis galbana*. Haptophytes contain only intrinsic LHCs. At gene level, *lhcf-*, *lhcr-*, *lhcz-*, and *lhcSR* (*lhcx*)-like sequences were found (Hoffman et al. 2011, Neilson and Durnford 2010). The LHC are supposed to bind Chl *a*, Chl c_3, and probably Fx, whereby also Chl *c* esterified with monogalactosyldiacylglyceride, 19′-acyloxy derivatives of Fx, and the xanthophyll cycle pigments diadinoxanthin and diatoxanthin were found in whole cells (Garrido and Zapata 1998). High light–acclimated cultures accumulate more of the xanthophyll cycle pigments (Ragni et al. 2008) and protect PS II more efficiently by NPQ than low light–grown cells. To our knowledge, no biochemical or spectroscopical studies on LHCs from haptophytes are published so far.

8.3.4 Heterokonts

Many groups of algae belong to the group of heterokonts, but knowledge about their light-harvesting systems or even photosynthesis is not equally distributed between the groups. Most studied are the Bacillariophyceae (diatoms), and some biochemical and spectroscopic data exist from Phaeophyceae (brown algae) and Xanthophyceae. The latter, in addition to Eustigmatophyceae and Chrysophyceae, will be shortly reviewed first, but the main focus of this chapter will be on diatoms.

All heterokonts contain only membrane-intrinsic LHCs, which belong to the LHC protein family, that are predicted to have three membrane-spanning α-helices arranged in a similar manner as in LHCII. However, polypeptides are usually smaller, and thus C and N termini are shorter; the same holds for the helix-connecting loops. All these complexes contain Chl *a* and Chl *c*, but stoichiometry as well as accessory carotenoids vary between the groups (Table 8.1). All heterokonts show a typical thylakoid membrane structure, whereby bands of three thylakoids (i.e., six thylakoid membranes surrounding three lumens) run along the whole plastid. No data exist about the precise distance of membranes inside these bands, and nothing much is known about the distribution of photosynthetic complexes.

8.3.4.1 CHRYSOPHYCEAE

Information about chrysophycean LHCs is scarce. Lichtlé et al. (1995) characterized a LHC from *Giraudyopsis stellifer*. The LHC contained Chl *a*, Chl *c*, Fx, and violaxanthin, like the complexes of brown algae (see Section 8.3.4.4). Immunocytochemical labelling confirmed that there is no segregation of LHC between the three thylakoids. This alga displayed a violaxanthin to zeaxanthin cycle like higher plants, which also efficiently protected the PS II reaction centers against photoinhibition (Lichtlé et al. 1995). Interestingly, the chrysophyte *Ochromonas danica* is the only heterokont alga where wavelength-dependent state transitions were detected (Gibbs and Biggins 1989) and correlated with the phosphorylation status of LHCs (Gibbs and Biggins 1991).

8.3.4.2 EUSTIGMATOPHYCEAE

Eustigmatophyceae belong to heterokonts but do not possess Chl *c*. Sukenik et al. (1992) isolated an LHC that binds Chl *a*, violaxanthin, and vaucheriaxanthin ester. From another species, Arsalane et al. (1992) isolated a similar complex and the Car/Chl *a* numbers were somehow similar with around three times more Chl *a* than violaxanthin, and about six times more Chl *a* than vaucheriaxanthin ester.

Recently, the LHC complexes of *Nannochloropsis* named violaxanthin-chlorophyll a binding protein (VCP) were analyzed in detail. The complexes were all trimeric and composed of different LHC polypeptides of the Lhcf type (Litvín et al. 2016). The overall excitation energy transfer from carotenoids to Chl *c* is very efficient with over 90%, due to two different pools of carotenoids that transfer efficiently from S$_2$ (Keşan et al. 2016). Thus, VCPs show high homologies to Fx chlorophyll proteins (FCPs) from diatoms (see Section 8.3.4.5). Although these antenna proteins bind different pigments, recent results suggested that two central carotenoids are comparable in localization and function to the luteins of plant LHCII and play a photoprotective role in ^{T}Chl quenching (Carbonera et al. 2014). In PS I complexes, five LHC proteins (Lhcr4–Lhcr8) were reported, with four of them in the same position as in higher plant photosystems and one on the other side of the core (Alboresi et al. 2017), resembling the structure reported from Cryptophytes. In another study, about double the amount of LHC monomers were found, but with similar localizations (Bína et al. 2017).Thus, eustigmatophyctes like *Nannochloropsis* are unusual chromalveolates in (1) not having Chl *c* and (2) having less carotenoids than Chl bound to their LHC.

8.3.4.3 XANTHOPHYCEAE

Mainly two xanthophyceaen algae were studied in photosynthesis research, *Pleurochloris meiringensis* and *Xanthonema debile*. No genome or even partial sequences are available until now, and thus only biochemical data on complexes isolated from the native organism are available. *P. meiringensis* was shown to contain two antenna systems, whereby one is associated with PS I and contains two polypeptides of about 17 and 21 kDa, with the main antenna complex containing subunits of about 22 kDa (Büchel and Wilhelm 1993). Later, using *X. debile*, Gardian et al. (2011) were able to isolate two pools of photosynthetic proteins, one containing photosystems and the other LHCs. Using electron microscopy, it was shown that the PS fraction contained PS II cores and other complexes resembling the monomeric PS I complexes of higher plants. The LHC band contained complexes of different oligomeric states, trimeric LHC, and oligomers out of seven LHC trimers. No further biochemical separation was carried out. Xanthophyte LHC binds Chl *a*, Chl *c*, diadinoxanthin, diatoxanthin, and heteroxanthin

as main carotenoids, accompanied by vaucheriaxanthin ester (Büchel and Wilhelm 1993, Gardian et al. 2011). In contrast to the groups containing Fx or Per, and like the Chl *c*–less eustigmatophyceae, xanthophytes have a Car/Chl ratio of about 0.5, that is, less carotenoid than Chl. Like for Chl *c*–containing LHC, the CD spectra are dominated by one negative band in the Q$_Y$ region of Chl *a*, that is, no excitonic interactions could be identified. In contrast to Fx-containing LHCs, there are only weak split signals in the Chl *a* Soret and carotenoid region (Büchel and Garab 1997). The Car-to-Chl-*a* transfer was calculated to have an efficiency of about 85% around 490 nm for *P. meiringensis* (Büchel et al. 1998). For the LHC of *X. debile*, only 60% overall efficiency was measured (Durchan et al. 2012), whereby the pathway from the Car S$_2$ state was more efficient when exciting at 510 nm. Also the Car S$_1$-to-Chl transfer depended on excitation wavelength, with transfer being much faster after 510 nm excitation (1.5 ps as compared to 3.4 ps after excitation at 490 nm). These features were attributed to two pools of differently bound diadinoxanthin molecules (Durchan et al. 2012). No long-living ^{T}Chl *a* could be detected in tripletminus-singlet spectra of thylakoids, demonstrating efficient quenching of Chl triplets (Büchel et al. 1998) via triplet–triplet transfer to Car (Durchan et al. 2012). Wavelength-dependent state transitions could not be proven (Büchel and Wilhelm 1990).

8.3.4.4 PHAEOPHYCEAE (BROWN ALGAE)

Phaeophyceae (brown algae) possess membraneintrinsic LHCs, which are also called Fx chlorophyll proteins (FCP) due to their pigmentation. Chl *a*, Chl *c*, and Fx are the major light-harvesting chromophores. Thus, the brown alga antennae are very similar to those of diatoms, but one major difference has to be pointed out. Whereas diatom FCPs contain diadinoxanthin and diatoxanthin, brown algal FCPs bind violaxanthin, which can be converted to zeaxanthin under high light (Ocampo-Alvarez et al. 2013). Thus, features related to the light harvesting are similar between brown algae and diatoms, but features related to photoprotection differ, which will be discussed in the context of diatoms (see below).

With the introduction of mild detergents in the mid-1980s, FCPs were isolated from different species of brown algae, including *Laminaria* (De Martino et al. 2000, Passaquet et al. 1991), *Fucus* (Caron et al. 1985, Passaquet et al. 1991), *Pelvetia*

(De Martino et al. 1997), *Petalonia* (Katoh and Ehara 1990), and *Dictyota* (Katoh et al. 1989). The Chl *a*–to–Chl *c*–to–Fx ratios were about 6:2:8 (De Martino et al. 2000, Pascal et al. 1998) or 13:3:10 plus one violaxanthin (Katoh et al. 1989), that is, the Car-to-Chl ratio is much higher than in higher plants. The absorption spectrum is thus dominated by Fx. Two different populations of Fx in FCPs could be identified by absorption spectra and their second derivatives, differing in their bathochromic shift upon binding to the protein (Katoh et al. 1989). One of them absorbs up to 570 nm in the FCP complex. In addition, two different populations of Chl *c* were also identified (Mimuro et al. 1990a). Using resonance Raman spectroscopy, Pascal et al. (1998) analyzed these clearly different Chl *c* populations and found one of them is in a polar or only weakly interacting environment. Most of the Fx are in all-trans configuration and one of them turned out to be highly twisted (Pascal et al. 1998). The CD spectrum of FCPs is characterized by a huge split in the Soret region ((+)440 nm and (−)475 nm), attributed by Mimuro et al. (1990a) to Chl *a* and Fx. No ultrafast measurements on the excitation energy transfer between pigments are available so far.

In 2010, the first complete genome of a brown alga (*Ectocarpus siliculosus*) became available (Cock et al. 2010). The genome codes for many LHC proteins, and brown algae contain genes for the *lhcf*, *lhcr*, *lhcSR* (*lhcx*), and *lhcz* groups. So far, no biochemical data on certain gene products are available.

As pointed out earlier, brown algae possess a xanthophyll cycle, where violaxanthin is converted to zeaxanthin under high light conditions. Testing blades of the giant kelp *Macrocystis pyrifera* collected either from 6 m depth or from the region close to the surface of the water, Ocampo-Alvarez et al. (2013) could show that the kinetics of the induction of NPQ depended on the amounts of xanthophyll cycle pigments, whereas the amount of NPQ was correlated to the de-epoxidation ratio.

No PS II complexes with FCPs bound have been isolated so far, but PS I preparations containing functional FCP antenna were analyzed (Berkaloff et al. 1990). Biochemical and spectroscopic differences between the main pool of FCP and the PS I antennae were small, and at that time different LHC polypeptides could not be identified due to the lack of sequence information. In addition, the question about possible PS II supercomplexes and the regulation of excitation energy distribution

between photosystems remains open so far. Fork et al. (1991) could not detect wavelength-dependent redistributions of antenna complexes in the giant kelp *Macrocystis*, that is, no state transitions could be proven.

8.3.4.5 BACILLARIOPHYCEAE (DIATOMS)

Diatom LHCs, like those of brown algae, are often referred to as FCPs. The members of the Lhcf group are very abundant in the thylakoid membranes of diatoms and were the first to be identified at gene level (Bhaya and Grossman 1993, Eppard and Rhiel 1998, 2000, Eppard et al. 2000). Diatoms also possess Lhcr proteins (Durnford et al. 1996) and Lhcx, the proteins related to LhcSR (former LI818) proteins of *Chlamydomonas reinhardtii* that are involved in light protection (Eppard and Rhiel 1998, Peers et al. 2009, Zhu and Green 2010). A fourth group called Lhcz was also genetically identified, but no biochemical data or functional data are available so far (Neilson and Durnford 2010).

Like in brown algae, the genomes of diatoms contain large numbers of *lhc* genes. For *Phaeodactylum tricornutum*, 17 *lhcf* genes, 14 *lhcr*, and 4 *lhcx* are annotated (Bowler et al. 2008), and for *Thalassiosira pseudonana*, 11 *lhcf*, 14 *lhcr*, and 7 *lhcx* genes are described (Armbrust et al. 2004). All are expressed according to EST data, or their expression was proven otherwise (Nymark et al. 2009). The homology to higher plant LHCs is only significant in the membrane-spanning helices, especially for helices 1 and 3, whereas helix 2 shows stronger deviations. In addition, FCPs are generally smaller than LHC proteins due to smaller loops and termini. Thus, molecular weights are in the range of 18–21 kDa (Eppard and Rhiel 1998). As a consequence, the proteins are even more hydrophobic than higher plant LHCs.

Diatom FCPs are the best-studied group of light-harvesting proteins in heterokonts, a work which started by isolating mixed fractions or "pools" of FCPs and later subfractionating these pools into different FCP complexes (Alberte et al. 1981, Beer et al. 2006, 2011, Berkaloff et al. 1990, Brakemann et al. 2006, Brown 1988, Büchel 2003, Caron and Brown 1987, Fawley and Grossman 1986, Friedman and Alberte 1984, 1986, Grouneva et al. 2011, Gugliemelli 1984, Guglielmi et al. 2005, Gundermann and Büchel 2008, Lavaud et al. 2003, Lepetit et al. 2007, Owens 1986, 1988, Owens and Wold 1986). Diatoms are divided into several subgroups and the species most used in diatom research either belong to the so-called centric diatoms like *T. pseudonana* and *Cyclotella meneghiniana* or are pennates like the widely used *P. tricornutum*. Lately, differences in the macromolecular organization of the antenna systems between pennates and centrics became obvious (Figure 8.5). In *P. tricornutum* (pennate), only trimeric complexes were found when separating FCPs from photosystems (Grouneva et al. 2011). This trimeric fraction contained three major trimers, composed of Lhcf5, Lhcf10/Lhcf2, and Lhcf4 with different interaction partners, respectively (Gundermann et al. 2013). No members of the other LHC families (Lhcr or Lhcx) could be found in the trimers (Grouneva et al. 2011, Gundermann et al. 2013). This is in contrast to *Cyclotella meneghiniana*, a centric diatom closely related to *T. pseudonana*. Here, Lhcx (Fcp6) proteins accompanied by Lhcf (Fcp2) proteins constitute the major trimeric complex, named FCPa. In addition, a nonameric complex named FCPb was isolated, which differed significantly in polypeptide composition from FCPa, since it was composed solely of Lhcf polypeptides, most probably Fcp5 (Beer et al. 2006, 2011, Büchel 2003, Gundermann and Büchel 2008, Röding et al. 2016). This supramolecular structure seems to be specific for centrics, since analogous proteins were later found in trimeric and oligomeric FCPs of *T. pseudonana* as well (Grouneva et al. 2011). In addition, Nagao et al. (2012, 2013) isolated a probably trimeric FCP complex from the centric diatom *Chaetoceros gracilis* and a higher oligomeric form. The polypeptide composition was not investigated, but like in the other two centric diatoms, the oligomeric complex was built from polypeptides of different molecular weights compared to the trimeric complexes.

Diatom FCPs bind Chl a, Chl c_1, and Chl c_2 (Fawley 1989, Jeffrey and Humphrey 1975, Kraay et al. 1992). The major carotenoid Fx is accompanied by substoichiometric amounts of diadinoxanthin and diatoxanthin, whereby the Chl:Car ratio is about one (Beer et al. 2006, Papagiannakis et al. 2005). Fx resembles Per in having a carbonyl moiety in conjugation with the polyene backbone (Damjanović et al. 2000, Frank et al. 2000, Katoh et al. 1991, Zigmantas et al. 2004). Like already described for brown algae, Fx displays an extreme bathochromic shift upon protein binding, extending the absorption from 390 nm up to 580 nm (Premvardhan et al. 2009). This huge

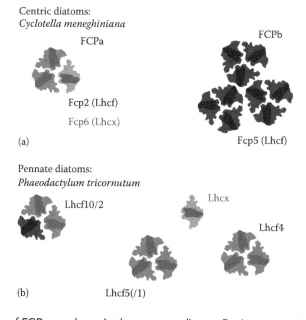

Centric diatoms:
Cyclotella meneghiniana

FCPa

FCPb

Fcp2 (Lhcf)

Fcp6 (Lhcx)

(a)

Fcp5 (Lhcf)

Pennate diatoms:
Phaeodactylum tricornutum

Lhcf10/2

Lhcx

Lhcf4

(b)   Lhcf5(/1)

Figure 8.5 Organization of FCP complexes in the pennate diatom *P. tricornutum* **(a)** and the centric diatom *C. meneghiniana* **(b)**. Labels refer to the main polypeptides only. In the FCPa trimers of *C. meneghiniana*, besides Lhcf (brown) also Lhcx polypeptides (red) are present, which have so far not been reported for FCP trimers of *P. tricornutum*.

range is accomplished by different shifts of the various Fx populations found in a FCP monomer. Using Stark and resonance Raman spectroscopies, more "blue-," "green-," and "red-" absorbing Fx molecules could be detected in FCPa as well as in FCPb from *C. meneghiniana* (Premvardhan et al. 2008, 2009, 2010). "Blue" and "red" Fx were also demonstrated for whole cells using electrochromic shift measurements (Szabó et al. 2010). Like for Per, excitation energy transfer from Fx to Chl *a* was proven to proceed mainly via the S_1/S_{ICT} state in FCPs, which is responsible for about 60% of the transfer to Chl *a* (Gildenhoff et al. 2010a, Papagiannakis et al. 2005). Upon photon absorption by Fx in solvent, a huge change in the static dipole moment of 17 D takes place, indicating photoinduced charge transfer of the Fx molecules. When Fx in FCPs is examined, two populations with changes of 17 D and up to 40 D can be distinguished, again underlining the different properties of the bound Fx molecules (Premvardhan et al. 2008). In intact systems, the transfer to Chl *a* from Fx is extremely fast. Gildenhoff et al. (2010a) determined lifetimes of <150 fs for the Fx S_2 state in FCPa, transferring directly into the Q_X state of Chl *a*. For the relaxed S_1/S_{ICT} state (transferring into Q_Y) 2.6/4.2 ps and for the unrelaxed state 0.6/0.9 ps were determined, whereby the longer values represent the lifetimes of the "red" Fx molecules and the shorter ones those of the "blue" or "green" Fx. Thus, when exciting the "blue/green" and the "red" Fx molecules to different extents, the observed excitation transfer dynamics change. In addition, when exciting the bluer Fx molecules of FCPa, an additional time constant of around 25 ps was found. This additional lifetime was assigned to the intrinsic lifetime of the blue-absorbing Fx engaged in Fx–Fx excitation energy transfer (Gildenhoff et al. 2010a). Using anisotropy measurements, it was concluded that one "red" Fx and two of the "blue/green" Fx transfer their energy directly to Chl *a*, whereas a further "blue/green" Fx is depending on another Fx molecule for excitation energy transfer to Chl *a* (Gildenhoff et al. 2010b). In addition, Fx is not only able to provide energy to Chl *a* but also acts as an efficient quencher of Chl *a* triplets (Di Valentin et al. 2012). No transfer from Fx to Chl *c* could be observed within the limit of the instrumentation (<100 fs) for *C. meneghiniana* FCPs (Gildenhoff et al. 2010a, Papagiannakis et al. 2005), giving rise to the assumption that Fx transfers its absorbed energy directly (or via

other Fx) to Chl *a* molecules. On the other hand, this implies that Chl *c* absorption will lead to direct and efficient transfer into Chl *a* (Gildenhoff et al. 2010a, Papagiannakis et al. 2005, Premvardhan et al. 2009), which was recently demonstrated using 2D spectroscopy (Butkus et al. 2015, Gelzinis et al. 2015, Songaila et al. 2013).

When comparing FCPa and FCPb of *C. meneghiniana*, the trimeric FCPa has the more efficient energy transfer. On the other hand, the oligomeric FCPb is intrinsically less fluorescent (Gundermann and Büchel 2008), and thus some of the lifetimes for FCPa mentioned previously are even shorter in FCPb. In addition, for both "blue" and "red" Fx of FCPb, Fx–Fx transfer was detected (Gildenhoff et al. 2010a).

Generally, the pigment ratio of FCPs is around 3–4 Chl *a* to 1 Chl *c* to 3–4 Fx molecules, depending on isolation procedure and species. The FCPs of the pennate diatom *P. tricornutum* are characterized by a higher amount of Fx as compared to the centric *C. meneghiniana* (Beer et al. 2006, Gundermann and Büchel 2012, Joshi-Deo et al. 2010, Lepetit et al. 2007, Papagiannakis et al. 2005). In all cases, the Q_Y absorption of FCPs is at a relatively short wavelength (~671 nm). Long wavelength–absorbing Chl *a* molecules result mainly from excitonic interactions, and thus it was argued that the Chl *a* of FCPs have to be further apart than in LHCII. This argument was strengthened by CD spectra, where no excitonic interactions are visible in the Q_Y region (Büchel 2003, Joshi-Deo et al. 2010, Szabó et al. 2008). However, using resonance Raman spectroscopy, two differently bound Chl *c* molecules were identified in FCPa as well as in FCPb by their signature ring-breathing modes at ~1360 cm^{-1} (Premvardhan et al. 2010). Thus, most likely around eight Chl *a*, two Chl *c*, and up to eight Fx molecules are bound per FCP monomer.

Based on the LHCII structure as a template, a model for pigment binding was developed (Premvardhan et al. 2010). According to this model, the central luteins of LHCII should be replaced by ("blue") Fx molecules in FCPa and FCPb. According to the protein sequences, seven Chl-binding sites are conserved (a602, a603, b609, a610, a612, a613, and a614; nomenclature according to Liu et al. 2004) and are most probably occupied by Chls in FCPs as well. Three other Chl-binding sites (a605, a611, and a604) cannot be attributed with certainty, but for symmetry reasons, most probably contain Chls in FCPs as well (for further reasons, see Premvardhan et al. 2008). From the total ten Chl sites, six (a602, a 603, a610, a611, a612, and b605) are likely to be occupied by Chl *a*. From the other four sites, most probably b609 and either a613 or a614 are Chl *c*–binding sites, leaving the remaining two for Chl *a*. Chls a614 and a613, which are rather close together, but further away from other Chls, have to be occupied by Chl *c* and Chl *a* in order to break the excitonic couple seen in LHCII for these two Chl *a* molecules (Georgakopoulou et al. 2007). The Fx molecules would be located in the lutein sites as said previously, but also in the neoxanthin site and the remaining space, mostly around helix 2, which would suffice to bind the additional Car molecules. The violaxanthin site might be the one binding diadinoxanthin in Lhcx polypeptides, but no experimental evidence is available so far.

Under conditions of increased light intensities, diadinoxanthin is de-epoxidized to diatoxanthin in the xanthophyll cycle (Lavaud et al. 2002, Lohr and Wilhelm 1999) as an important part of NPQ induction. The amount of total diadinoxanthin and diatoxanthin found in the various FCP preparations depends not only on the preillumination of the cells but also on the isolation method. Generally, the more lipids that are retained in the samples, the more xanthophyll cycle pigments are found as well (Beer et al. 2006, Büchel 2003, Lepetit et al. 2010). However, this does not necessarily argue for a localization of all xanthophyll pigments in the lipid phase that is not bound to the protein, since additional pigments synthesized under high light conditions were demonstrated to adopt a more twisted configuration *in vivo*, which is only possible upon protein binding (Alexandre et al. 2014). Using more rigid methods that remove almost all lipids, diadinoxanthin or diatoxanthin is still found in FCPa preparations, and the amount depends on the presence of Fcp6, a Lhcx protein (Beer et al. 2006, 2011). The amount of Lhcx in turn is dependent on the light intensity during growth.

Diatoms display a huge, pH-dependent NPQ, and the de-epoxidation ratio of diadinoxanthin to diatoxanthin is correlated with it (Lavaud et al. 2002, 2003, Ruban et al. 2004). As mentioned previously, diatoms do not possess PsbS, but Lhcx proteins were demonstrated to be involved in NPQ in centrics (Zhu and Green 2010) as well as in pennates (Bailleul et al. 2010). These proteins are up-regulated during prolonged high light treatment (Bailleul

et al. 2010, Becker and Rhiel 2006, Beer et al. 2006, Janssen et al. 2001, Lepetit et al. 2010, Nymark et al. 2009, Oeltjen et al. 2002, 2004, Zhu and Green 2010). However, Lhcx proteins from pennates and centrics differ in one crucial point, whereas Lhcx polypeptides are constituents of trimeric FCP complexes in centrics (Beer et al. 2006); no complexes containing Lhcx proteins (except PS I) are reported for pennates (Grouneva et al. 2011, Gundermann et al. 2013). Thus, theories about the action of Lhcx proteins differ depending on the species analyzed. Lepetit et al. (2012), based on the work of Goss et al. (2006) and Bailleul et al. (2010), proposed a model for Lhcx1 in *P. tricornutum* in analogy to PsbS in higher plants (Ruban et al. 2012, Szabò et al. 2005), where Lhcx1 does not bind diadinoxanthin or diatoxanthin but acts on FCPs by inducing a conformational change, which in turn leads to reduced fluorescence yield of FCPs. However, no experimental evidences for this induction are available. A role in the interaction of proteins in the thylakoid membrane was also attributed to Lhcx1 from a centric diatom, *T. pseudonana*, because its amount did not change dramatically upon short high light stress, in contrast to Lhcx6 (Zhu and Green 2010). No protein comparable to Lhcx6 is described so far in *C. meneghiniana*. The protein found in FCPa trimers closely resembles Lhcx1, and its amount does indeed change under prolonged high light conditions (Beer et al. 2006). Gundermann and Büchel (2008) were able to show that the Fcp6(Lhcx1) containing FCPa of *C. meneghiniana* changes its fluorescence yield depending on diatoxanthin and Fcp6 content, implying a function of Lhcx1 in the regulation of fluorescence emission in centrics. The influence of protein distance, pH, and content of xanthophyll cycle pigments on the Lhcx containing FCPa was also analyzed in proteoliposomes (Gundermann and Büchel 2012). Indeed, FCPa protein aggregation led to reduced fluorescence yields, which was in addition strongly influenced by the pH. Diatoxanthin independently enhanced the reduction of fluorescence emission in accordance with *in vivo* data (Büchel 2014, Chukhutsina et al. 2014). It had already been demonstrated that FCPb is changing fluorescence yield depending on aggregation, but not according to diatoxanthin content or pH (Gundermann and Büchel 2008). Recently, using transient absorption spectroscopy, two quenching channels active in quenched FCPa complexes were detected that are characterized by differing rate constants and distinct spectroscopic signatures. One channel was associated with a faster quenching rate (16 ns⁻¹), with virtually no difference in spectral shape compared to the bulk unquenched chlorophylls, while a second channel was associated with a slower quenching rate (2.7 ns⁻¹) and exhibited an increased population of red-emitting states (Ramanan et al. 2014). Also, using Stark fluorescence spectroscopy, an additional redshifted fluorescence emission band in FCPa could be detected upon quenching, which is missing in FCPb (Wahadoszamen et al. 2013). Using antisense mutants of Fcp6 in *C. meneghiniana*, it could be demonstrated that Fcp6, however, is mainly influencing the interaction between different FCPa complexes, that is, their aggregation (Ghazaryan et al. 2016). Thus, it seems that in centric diatoms, a constituent of the trimeric FCPa complexes is an active player in NPQ, whereas in pennates at least Lhcx1 is working in a different manner. This is rather surprising, since Lhcx1 of *P. tricornutum* is most closely related to Fcp6 of *C. meneghiniana*. Differences mainly concern helix 2 of those proteins, which might directly relate to the pigment-binding capacities of Lhcx1 of *P. tricornutum*. In addition, the Lhcf proteins are quite different between the two groups of diatoms (Gundermann et al. 2013), which might lead to different protein–protein interactions especially concerning Lhcx–Lhcf assemblies.

If grown under red light only, the pennate diatom *P. tricornutum* displays an additional long wavelength fluorescence at 710 nm at room temperature. This is accompanied by an increase in the Lhcf15 content. Herbstová et al. (2015) were able to isolate an additional antenna complex, which is only present under those special red light conditions and is emitting fluorescence at longer wavelength compared to all other FCP complexes. Lhcf15 was identified as major constituent. This red light adaptation induced further changes in the chloroplast, including PS I-only areas and a massive increase in thylakoid bands (Bína et al. 2016).

Nothing much is known about the attribution of the different FCPs to the two photosystems and their supramolecular structure. PS I supercomplexes were isolated early (Berkaloff et al. 1990) and from several organisms (Brakemann et al. 2006, Grouneva et al. 2011, Ikeda et al. 2008, Veith and Büchel 2007, Veith et al. 2009). Since gene sequences became available, Lhcr proteins were always supposed to serve as PS I antennas, which could indeed

be shown at protein level as well (Grouneva et al. 2011, Veith et al. 2009). There is still some controversy as to which and how many Lhcf proteins besides several Lhcr proteins are present in the PS I antenna (Brakemann et al. 2006, Grouneva et al. 2011, Ikeda et al. 2013, Juhas and Büchel 2012, Lepetit et al. 2010, Veith and Büchel 2007, Veith et al. 2009). Only in pennates, Lhcx polypeptides were found in the PS I antennae as well (Grouneva et al. 2011). Like in higher plants, PS I of diatoms is a monomer (Ikeda et al. 2013, Veith and Büchel 2007), but no data on the arrangement of the antenna proteins are available until now. The same holds for PS II antennae composition and arrangement, since supercomplexes as can be isolated from higher plants (Boekema et al. 1995) have not been obtained yet. Nagao et al. (2007, 2010) were able to isolate PS II complexes from *C. gracilis*, which still contained FCP proteins, but unfortunately these FCPs were too loosely bound to allow for further analysis. Thus, no proof for minor LHCs like CP24, CP26, or CP29 is available so far. In accordance with what was published for brown algae, no state transitions could be detected in diatoms (Owens 1986).

8.3.4.6 OTHER HETEROKONTS

As mentioned, heterokonts include some other groups than those described earlier. The raphidophyte *Heterosigma carterae* was studied by Durnford and Green (1994). The organism contains Chl *a*, Chl *c*, and Fx. One major LHC fraction consisting of 19.5 kDa polypeptides and a PS I fraction containing 12 other LHCs were found.

Lately, the pelagophyte *Aureococcus anophageferens* has been studied in more detail (Alami et al. 2012). This species is responsible for large algal blooms in the northern Atlantic, impeding the shellfish industry. The organism contains Chl *a*, Chl *c₂*, Chl *c₃*, Fx, diadinoxanthin, diatoxanthin, and ß-carotene. Under high light, some of the Fx are converted to 19′-butanoyloxyfucoxanthin, a much more bulky molecule compared to Fx. Unfortunately, no preparation of LHC was reported. Thus, the localization of the Fx derivate remains unknown.

8.3.5 Chromerida

Lately, studies on the light-harvesting system of *Chromera velia*, the only member of this group, were published. Although only limited data has been reported so far, we will briefly review it here since *C. velia* is a very peculiar photosynthetic organism from an evolutionary viewpoint. Pan et al. (2012) analyzed the *lhc* gene sequences and found *lhcf*, *lhcr*, and *lhcx* genes, in accordance with the fact that the organisms are related to dinoflagellates. However, pigmentation is different since the algae contain an isofucoxanthin-like pigment as the major carotenoid in addition to violaxanthin, but no Chl *c*. These pigments were identified in LHC complexes by Tichy et al. (2013). The isofucoxanthin-like pigment absorbs in the green spectral region, and thus the absorption spectra of LHC complexes show similarities to those of FCPs of diatoms. The Car-to-Chl ratio, however, is not as high as in FCPs but resembles the ratio found in the LHCs of xanthophytes (0.45:1). Despite an isofucoxanthin-like pigment, showing similar absorption features to Fx, the huge split in the Soret region typically found in Fx-containing complexes is missing, and the CD spectra resemble those of xanthophytes. Trimeric and complexes of higher oligomeric states were identified by electron microscopy. In addition, a specific PS I antenna containing relatively more violaxanthin was isolated as well (Tichy et al. 2013). Like seen for the diatom *P. tricornutum*, also *C. velia* has a special, red light–induced LHC complex (Bína et al. 2014) that serves as PS II antenna (Kotabová et al. 2014).

8.4 CONCLUSIONS

Whereas the knowledge about gene sequences coding for LHCs in Chl *c*–containing antennae has tremendously improved during the last years, biochemical and spectroscopic knowledge is lacking far behind. At genetic level, three major groups of intrinsic LHC were analyzed, whereby members of one group, Lhcr, are assumed to serve PS I in excitation energy transfer due to their similarity with red algae LHCI. This was proven for diatoms, although the PS I antenna consists of other polypeptides as well. In cryptophytes, Lhcr also constitutes the PS II antenna. The Lhcx proteins, assumed to work in photoprotection, were shown to be involved in NPQ in diatoms, but localization, pigmentation, and oligomerization seem to differ even between the two major diatom groups, and no data are available for the other algal groups containing Lhcx proteins. Within the Lhcf group, some polypeptides seem to be organized in trimers,

whereas others facilitate the build of higher oligomers at least in centric diatoms. Thus, the direct correlation between protein sequence and protein function is impossible. This also becomes obvious when considering the different carotenoids that are bound to very similar proteins within the various algae groups.

In addition, some Chl *c*–containing algae contain membrane-extrinsic light-harvesting proteins, which differ completely from other known antenna systems. Whereas the cryptophytes contain lumenal associated phycobilins, the PCP of dinoflagellates is one of the rare examples of carotenoid-containing soluble antenna proteins. The PCPs are thus ideally suited to study Per, one of the three important Car containing a keto group coupled into the conjugated π–electron system rendering Per, as well as Fx, extremely efficient in excitation energy transfer to Chl *a*.

Chl *c*–containing algae contribute significantly to the worldwide marine photosynthetic productivity and are thus as such very interesting objects of research. However, this group is vast and very heterogeneous, and our knowledge about their function is scarce. Much remains to be discovered, and it may be anticipated that, as yet uncharacterized, unusual antenna systems might harbor some surprises, not to mention unique insights, on the principles of efficient excitation energy transfer, and mechanisms of photoprotection.

REFERENCES

Alami, M., D. Lazar, and B. R. Green. 2012. The harmful alga *Aureococcus anophagefferens* utilizes 19′-butanoyloxyfucoxanthin as well as xanthophyll cycle carotenoids in acclimating to higher light intensities. *Biochim. Biophys. Acta* 1817: 1557–1564.

Alberte, R. S., A. L. Friedman, D. L. Gustafson, M. S. Rudnick, and H. Lyman. 1981. Light-harvesting systems of brown algae and diatoms. Isolation and characterization of chlorophyll a/c and chlorophyll a/fucoxanthin pigment-protein complexes. *Biochim. Biophys. Acta* 635: 304–316.

Alboresi, A., C. Le Quiniou, S. K. N. Yadav et al. 2017. Conservation of core complex subunits shaped the structure and function of photosystem I in the secondary endosymbiont alga *Nannochloropsis gaditana*. *New Phytol.* 213: 714–726.

Alexandre, M. T. A., K. Gundermann, A. A. Pascal, R. van Grondelle, C. Büchel, and B. Robert. 2014. Probing the carotenoid content of intact *Cyclotella* cells by resonance Raman spectroscopy. *Photosynth. Res.* 119: 273–281.

Archibald, J. M. and P. J. Keeling. 2002. Recycled plastids: A 'green movement' in eukaryotic evolution. *Trends Genet.* 18: 577–584.

Armbrust, E. V., J. A. Berges, C. Bowler et al. 2004. The genome of the diatom *Thalassiosira pseudonana*: Ecology, evolution, and metabolism. *Science* 306: 79–86.

Arsalane, W., B. Rousseau, and J. C. Thomas. 1992. Isolation and characterization of native pigment-protein complexes from two Eustigmatophyceae. *J. Phycol.* 28: 32–36.

Bailleul, B., A. Rogato, A. De Martino et al. 2010. An atypical member of the light-harvesting complex stress-related protein family modulates diatom responses to light. *Proc. Natl. Acad. Sci. USA* 107: 18214–18219.

Bathke, L., E. Rhiel, W. E. Krumbein, and J. Marquardt. 1999. Biochemical and immunochemical investigations on the light-harvesting system of the Cryptophyte *Rhodomonas* sp.: Evidence for a photosystem I specific antenna. *Plant Biol.* 1: 516–523.

Bautista, J. A., R. E. Connors, B. B. Raju et al. 1999a. Excited-state properties of peridinin: Observation of a solvent dependence of the lowest excited singlet state lifetime and spectral behavior unique among carotenoids. *J. Phys. Chem. B* 103: 8751–8758.

Bautista, J. A., R. G. Hiller, F. P. Sharples, D. Gosztola, M. Wasielewski, and H. A. Frank. 1999b. Singlet and triplet energy transfer in the peridinin-chlorophyll a protein from *Amphidinium carterae*. *J. Phys. Chem. A* 103: 2267–2273.

Becker, F. and E. Rhiel. 2006. Immuno-electron microscopic quantification of the fucoxanthin chlorophyll a/c binding polypeptides Fcp2, Fcp4, and Fcp6 of *Cyclotella cryptica* grown under low- and high-light intensities. *Int. Microbiol.* 9: 29–36.

Beer, A., K. Gundermann, J. Beckmann, and C. Büchel. 2006. Subunit composition and pigmentation of fucoxanthin-chlorophyll proteins in diatoms: Evidence for a subunit involved in diadinoxanthin and diatoxanthin binding. *Biochemistry* 45: 13046–13053.

Beer, A., M. Juhas, and C. Büchel. 2011. Influence of different light intensities and different iron nutrition on the photosynthetic apparatus in the diatom *Cyclotella meneghiniana* (Bacillariophyceae). *J. Phycol.* 47: 1266–1273.

Berkaloff, C., L. Caron, and B. Rousseau. 1990. Subunit organization of PS I particles from brown algae and diatoms: Polypeptide and pigment analysis. *Photosynth. Res.* 23: 181–193.

Bhaya, D. and A. R. Grossman. 1993. Characterization of gene clusters encoding the fucoxanthin chlorophyll proteins of the diatom *Phaeodactylum tricornutum*. *Nucl. Acids Res.* 21: 4458–4466.

Bína, D., Z. Gardian, M. Herbstová, and R. Litvín. 2017. Modular antenna of photosystem I in secondary plastids of red algal origin: A *Nannochloropsis oceanica* case study. *Photosynth. Res.* 131: 255–266.

Bína, D., Z. Gardian, M. Herbstová et al. 2014. Novel type of red-shifted chlorophyll a antenna complex from *Chromera velia*: II. Biochemistry and spectroscopy. *Biochim. Biophys. Acta* 1837: 802–810.

Bína, D., M. Herbstová, Z. Gardian, F. Vácha, and R. Litvin. 2016. Novel structural aspect of the diatom thylakoid membrane: Lateral segregation of photosystem I under red-enhanced illumination. *Sci. Rep.* 6: 25583.

Boekema, E. J., B. Hankamer, D. Bald et al. 1995. Supramolecular structure of photosystem II complex from green plants and cyanobacteria. *Proc. Natl. Acad. Sci. USA* 92: 175–179.

Boldt, L., D. Yellowlees, W. Leggat, and A. Webber. 2012. Hyperdiversity of genes encoding integral light-harvesting proteins in the dinoflagellate *Symbiodinium* sp. *PLoS ONE* 7: e47456.

Bonetti, C., M. T. A. Alexandre, I. H. M. van Stokkum et al. 2010. Identification of excited-state energy transfer and relaxation pathways in the peridinin–chlorophyll complex: An ultrafast mid-infrared study. *Phys. Chem. Chem. Phys.* 12: 9256.

Bowler, C., A. E. Allen, J. H. Badger et al. 2008. The *Phaeodactylum* genome reveals the evolutionary history of diatom genomes. *Nature* 456: 239–244.

Brakemann, T., W. Schlörmann, J. Marquardt, M. Nolte, and E. Rhiel. 2006. Association of fucoxanthin chlorophyll a/c-binding polypeptides with photosystems and phosphorylation in the centric diatom *Cyclotella cryptica*. *Protist* 157: 463–475.

Brown, J. S. 1988. Photosynthetic pigment organization in diatoms (Bacillariophyceae). *J. Phycol.* 24: 96–102.

Büchel, C. 2003. Fucoxanthin-chlorophyll proteins in diatoms: 18 and 19 kDa subunits assemble into different oligomeric states. *Biochemistry* 42: 13027–13034.

Büchel, C. 2014. Fucoxanthin-chlorophyll-proteins and non-photochemical fluorescence quenching of diatoms. In *Non-Photochemical Quenching and Energy Dissipation in Plants, Algae and Cyanobacteria* eds. B. Demmig-Adams, G. Garab, W. Adams III, and Govindjee, pp. 259–275. Dordrecht, the Netherlands: Springer Verlag.

Büchel, C. and G. Garab. 1997. Organization of the pigment molecules in the chlorophyll a/c light-harvesting complex of *Pleurochloris meiringensis* (Xanthophyceae). Characterization with circular dichroism and absorbance spectroscopy. *J. Photoch. Photobio. B* 37: 118–124.

Büchel, C., K. R. Naqvi, and T. B. Melø. 1998. Pigment-pigment interactions in thylakoids and LHCII of chlorophyll a/c containing alga *Pleurochloris meiringensis*: Analysis of fluorescence excitation and triplet-minus-singlet spectra. *Spectrochim. Acta A* 54: 719–726.

Büchel, C. and C. Wilhelm. 1990. Wavelength independent state transitions and light regulated chlororespiration as mechanisms to control the energy status in the chloroplast of *Pleurochloris meiringensis*. *Plant Physiol. Bioch.* 28: 307–314.

Büchel, C. and C. Wilhelm. 1993. Isolation and characterization of a photosystem I-associated antenna (LHC I) and a photosystem I-core complex from the chlorophyll c-containing alga *Pleurochloris meiringensis* (Xanthophyceae). *J. Photoch. Photobio. B* 20: 87–93.

Butkus, V., A. Gelzinis, R. Augulis et al. 2015. Coherence and population dynamics of chlorophyll excitations in FCP complex: Two-dimensional spectroscopy study. *J. Chem. Phys.* 142: 212414.

Carbonera, D., A. Agostini, M. Di Valentin et al. 2014. Photoprotective sites in the violaxanthin-chlorophyll a binding protein (VCP) from *Nannochloropsis gaditana*. *Biochim. Biophys. Acta* 1837: 1235–1246.

Caron, L. and J. S. Brown. 1987. Chlorophyll-carotenoid protein complexes from the diatom, *Phaeodactylum tricornutum*: Spectrophotometric, pigment and polypeptide analyses. *Plant Cell Physiol.* 28: 775–785.

Caron, L., J. P. Dubacq, C. Berkaloff, and H. Jupin. 1985. Subchloroplast fractions from the brown alga *Fucus serratus*. Phosphatidylglycerol contents. *Plant Cell Physiol.* 26: 131–139.

Chukhutsina, V. U., C. Büchel, and H. van Amerongen. 2014. Disentangling two non-photochemical quenching processes in *Cyclotella meneghiniana* by spectrally-resolved picosecond fluorescence at 77 K. *Biochim. Biophys. Acta* 1837: 899–907.

Cock, J. M., L. Sterck, P. Rouzé et al. 2010. The *Ectocarpus* genome and the independent evolution of multicellularity in brown algae. *Nature* 465: 617–621.

Collini, E., C. Y. Wong, K. E. Wilk, P. M. G. Curmi, P. Brumer, and G. D. Scholes. 2010. Coherently wired light-harvesting in photosynthetic marine algae at ambient temperature. *Nature* 463: 644–647.

Damjanović, A., T. Ritz, and K. Schulten. 2000. Excitation transfer in the peridinin-chlorophyll-protein of *Amphidinium carterae*. *Biophys. J.* 79: 1695–1705.

De Martino, A., D. Douady, M. Quinet-Szely et al. 2000. The light-harvesting antenna of brown algae. Highly homologous proteins encoded by a multigene family. *Eur. J. Biochem.* 267: 5540–5549.

De Martino, A., D. Douady, B. Rousseau, J.-C. Duval, and L. Caron. 1997. Characterization of two light-harvesting subunits isolated from the brown alga *Pelvetia canaliculata*: Heterogeneity of xanthophyll distribution. *Photochem. Photobiol.* 66: 190–197.

Deane, J. A., M. Fraunholz, V. Su et al. 2000. Evidence for nucleomorph to host nucleus gene transfer: Light-harvesting complex proteins from cryptomonads and chlorarachniophytes. *Protist* 151: 239–252.

Di Valentin, M., C. Büchel, G. M. Giacometti, and D. Carbonera. 2012. Chlorophyll triplet quenching by fucoxanthin in the fucoxanthin-chlorophyll protein from the diatom *Cyclotella meneghiniana*. *Biochem. Biophys. Res. Commun.* 427: 637–641.

Di Valentin, M., E. Salvadori, G. Agostini et al. 2010. Triplet-triplet energy transfer in the major intrinsic light-harvesting complex of *Amphidinium carterae* as revealed by ODMR and EPR spectroscopies. *Biochim. Biophys. Acta* 1997: 1759–1767.

Dittami, S. M., G. Michel, J. Collén, C. Boyen, and T. Tonon. 2010. Chlorophyll-binding proteins revisited—A multigenic family of light-harvesting and stress proteins from a brown algal perspective. *BMC Evol. Biol.* 10: 365.

Doust, A. B., Marai, C. N. J., S. J. Harrop, K. E. Wilk, P. M. G. Curmi, and G. D. Scholes. 2004. Developing a structure-function model for the cryptophyte phycoerythrin 545 using ultrahigh resolution crystallography and ultrafast laser spectroscopy. *J. Mol. Biol.* 344: 135–153.

Doust, A. B., K. E. Wilk, P. M. G. Curmi, and G. D. Scholes. 2006. The photophysics of cryptophyte light-harvesting. *J. Photochem. Photobiol. A* 184: 1–17.

Durchan, M., J. Tichý, R. Litvín et al. 2012. Role of carotenoids in light-harvesting processes in an antenna protein from the chromophyte *Xanthonema debile. J. Phys. Chem. B* 116: 8880–8889.

Durnford, D. G., R. Aebersold, and B. R. Green. 1996. The fucoxanthin-chlorophyll proteins from a chromophyte alga are part of a large multigene family: Structural and evolutionary relationships to other light-harvesting antennae. *Mol. Gen. Genet.* 253: 377–386.

Durnford, D. G., J. A. Deane, S. Tan, G. I. McFadden, E. Gantt, and B. R. Green. 1999. A phylogenetic assessment of the eukaryotic light-harvesting antenna proteins, with implications for plastid evolution. *J. Mol. Evol.* 48: 59–68.

Durnford, D. G. and B. R. Green. 1994. Characterization of the light-harvesting proteins of the chromophytic alga, *Olisthodiscus luteus* (*Heterosigma carterae*). *Biochim. Biophys. Acta* 1184: 118–123.

Eppard, M., W. E. Krumbein, A. von Haesler, and E. Rhiel. 2000. Characterization of fcp4 and fcp12, two additional genes encoding light harvesting proteins of *Cyclotella cryptica* (Bacillariophyceae) and phylogenetic analysis of this complex gene family. *Plant Biol.* 2: 283–289.

Eppard, M. and E. Rhiel. 1998. The genes encoding light-harvesting subunits of *Cyclotella cryptica* (Bacillariophyceae) constitute a complex and heterogeneous family. *Mol. Gen. Genet.* 260: 335–345.

Eppard, M. and E. Rhiel. 2000. Investigation on gene copy number, introns and chromosomal arrangements of genes encoding the fucoxanthin chlorophyll a/c-binding proteins of the centric diatom *Cyclotella cryptica. Protist* 151: 27–39.

Fawley, M. W. 1989. A new form of chlorophyll c involved in light-harvesting. *Plant Physiol.* 91: 727–732.

Fawley, M. W. and A. R. Grossman. 1986. Polypeptides of a light-harvesting complex of the diatom *Phaeodactylum tricornutum* are synthesized in the cytoplasm of the cell as precursors. *Plant Physiol.* 81: 149–155.

Fork, D. C., S. K. Herbert, and S. Malkin. 1991. Light energy distribution in the brown alga *Macrocystis pyrifera* (giant kelp). *Plant Physiol.* 95: 731–739.

Frank, H. A., J. A. Bautista, J. Josue et al. 2000. Effect of the solvent environment on the spectroscopic properties and dynamics of the lowest excited states of carotenoids. *J. Phys. Chem. B* 104: 4569–4577.

Friedman, A. L. and R. S. Alberte. 1984. A diatom light-harvesting pigment-protein complex. *Plant Physiol.* 76: 483–489.

Friedman, A. L. and R. S. Alberte. 1986. Biogenesis and light regulation of the major light harvesting chlorophyll-protein of diatoms. *Plant Physiol.* 80: 43–51.

Gardian, Z., J. Tichý, and F. Vácha. 2011. Structure of PSI, PSII and antennae complexes from yellow-green alga *Xanthonema debile. Photosynth. Res.* 108: 25–32.

Garrido, J. L. and M. Zapata. 1998. Detection of new pigments from *Emiliania huxleyi* (Prymnesiophyceae) by high-performance liquid chromatography, liquid chromatography-mass spectrometry, visible spectroscopy, and fast atom bombardment mass spectrometry. *J. Phycol.* 34: 70–78.

Gelzinis, A., V. Butkus, E. Songaila et al. 2015. Mapping energy transfer channels in fucoxanthin-chlorophyll protein complex. *Biochim. Biophys. Acta* 1847: 241–247.

Georgakopoulou, S., G. van der Zwan, R. Bassi, R. van Grondelle, H. van Amerongen, and R. Croce. 2007. Understanding the changes in the circular dichroism of light harvesting complex II upon varying its pigment composition and organization. *Biochemistry* 46: 4745–4754.

Ghazaryan, A., P. Akhtar, G. Garab, P. H. Lambrev, and C. Büchel. 2016. Involvement of the Lhcx protein Fcp6 of the diatom *Cyclotella meneghiniana* in the macro-organization and structural flexibility of thylakoid membranes. *Biochim. Biophys. Acta* 1857: 1373–1379.

Gibbs, P. B. and J. Biggins. 1989. Regulation of the distribution of excitation energy in *Ochromonas danica*, an organism containing a chlorophyll a/c/carotenoid light harvesting antenna. *Photosynth. Res.* 21: 81–91.

Gibbs, P. B. and J. Biggins. 1991. *In vivo* and *in vitro* protein phosphorylation studies on *Ochromonas danica*, an alga with chlorophyll a/c/fucoxanthin binding protein. *Plant Physiol.* 97: 388–395.

Gildenhoff, N., S. Amarie, K. Gundermann, A. Beer, C. Büchel, and J. Wachtveitl. 2010a. Oligomerization and pigmentation dependent excitation energy transfer in fucoxanthin-chlorophyll proteins. *Biochim. Biophys. Acta* 1797: 543–549.

Gildenhoff, N., J. Herz, K. Gundermann, C. Büchel, and J. Wachtveitl. 2010b. The excitation energy transfer in the trimeric fucoxanthin-chlorophyll protein from *Cyclotella meneghiniana* analyzed by polarized transient absorption spectroscopy. *Chem. Phys.* 373: 104–109.

Goss, R., E. A. Pinto, C. Wilhelm, and M. Richter. 2006. The importance of a highly active and Δ-pH-regulated diatoxanthin epoxidase for the regulation of the PS II antenna function in diadinoxanthin cycle containing algae. *J. Plant Physiol.* 163: 1008–1021.

Green, B. R. 2011. After the primary endosymbiosis: An update on the chromalveolate hypothesis and the origins of algae with Chl c. *Photosynth. Res.* 107: 103–115.

Grouneva, I., A. Rokka, and E.-M. Aro. 2011. The thylakoid membrane proteome of two marine diatoms outlines both diatom-specific and species-specific features of the photosynthetic machinery. *J. Proteome Res.* 10: 5338–5353.

Guglielmi, G., J. Lavaud, B. Rousseau, A.-L. Etienne, J. Houmard, and A. V. Ruban. 2005. The light-harvesting antenna of the diatom *Phaeodactylum tricornutum*. Evidence for a diadinoxanthin-binding subcomplex. *FEBS J.* 272: 4339–4348.

Gugliemelli, A. 1984. Isolation and characterization of pigment-protein particles from the light-harvesting complex of *Phaeodactylum tricornutum*. *Biochim. Biophys. Acta* 766: 45–50.

Gundermann, K. and C. Büchel. 2008. The fluorescence yield of the trimeric fucoxanthin-chlorophyll-protein FCPa in the diatom *Cyclotella meneghiniana* is dependent on the amount of bound diatoxanthin. *Photosynth. Res.* 95: 229–235.

Gundermann, K. and C. Büchel. 2012. Factors determining the fluorescence yield of fucoxanthin-chlorophyll complexes (FCP) involved in non-photochemical quenching in diatoms. *Biochim. Biophys. Acta* 1817: 1044–1052.

Gundermann, K., M. Schmidt, W. Weisheit, M. Mittag, and C. Büchel. 2013. Identification of several sub-populations in the pool of light harvesting proteins in the pennate diatom *Phaeodactylum tricornutum*. *Biochim. Biophys. Acta* 1827: 303–310.

Herbstová, M., D. Bína, P. Koník, Z. Gardian, F. Vácha, and R. Litvín. 2015. Molecular basis of chromatic adaptation in pennate diatom *Phaeodactylum tricornutum*. *Biochim. Biophys. Acta* 1847: 534–543.

Hiller, R. G., C. D. Scaramuzzi, and J. Breton. 1992. The organisation of photosynthetic pigments in a cryptophyte alga: A linear dichroism study. *Biochim. Biophys. Acta* 1102: 360–364.

Hiller, R. G., P. M. Wrench, A. P. Gooley, G. Shoebridge, and J. Breton. 1993. The major intrinsic light-harvesting protein of *Amphidinium*: Characterization and relation to other light-harvesting proteins. *Photochem. Photobiol.* 57: 125–131.

Hoffman, G. E., Sanchez Puerta, M. V., and C. F. Delwiche. 2011. Evolution of light-harvesting complex proteins from Chl c-containing algae. *BMC Evol. Biol.* 11: 101.

Hofmann, E., P. M. Wrench, F. P. Sharples, R. G. Hiller, W. Welte, and K. Diederichs. 1996. Structural basis of light harvesting by carotenoids: Peridinin–chlorophyll-protein from *Amphidinium carterae*. *Science* 272: 1788–1791.

Iglesias-Prieto, R., N. S. Govind, and R. K. Trench. 1993. Isolation and characterization of three membrane-bound chlorophyll-protein complexes from four dinoflagellate species. *Philos. Trans. R. Soc. B* 340: 381–392.

Ikeda, Y., M. Komura, M. Wanatabe et al. 2008. Photosystem I complexes associated with fucoxanthin-chlorophyll-binding proteins from a marine centric diatom, *Chaetoceros gracilis*. *Biochim. Biophys. Acta* 1777: 351–361.

Ikeda, Y., A. Yamagishi, M. Komura et al. 2013. Two types of fucoxanthin-chlorophyll-binding proteins I tightly bound to the photosystem I core complex in marine centric diatoms. *Biochim. Biophys. Acta* 1827: 529–539.

Ingram, K. and R. G. Hiller. 1983. Isolation and characterization of a major chlorophyll a/c_2 light-harvesting protein from a *Chroomonas* species (Cryptophyceae). *Biochim. Biophys. Acta* 722: 310–319.

Janssen, M., L. Bathke, J. Marquardt, W. E. Krumbein, and E. Rhiel. 2001. Changes in the photosynthetic apparatus of diatoms in response to low and high light intensities. *Int. Microbiol.* 4: 27–33.

Jeffrey, S. W. and G. F. Humphrey. 1975. New spectrometric equations for determining chlorophyll a, b, c1 and c2 in higher plants, algae and natural phytoplankton. *Biochem. Physiol. Pflz.* 167: 191–194.

Joshi-Deo, J., M. Schmidt, A. Gruber et al. 2010. Characterization of a trimeric light-harvesting complex in the diatom *Phaeodactylum tricornutum* built of FcpA and FcpE proteins. *J. Exp. Bot.* 61: 3079–3087.

Juhas, M. and C. Büchel. 2012. Properties of photosystem I antenna protein complexes of the diatom *Cyclotella meneghiniana*. *J. Exp. Bot.* 63: 3673–3681.

Kaňa, R., E. Kotabová, R. Sobotka, and O. Prášil. 2012. Non-photochemical quenching in cryptophyte alga *Rhodomonas salina* is located in chlorophyll a/c antennae. *PLoS ONE* 7: e29700.

Katoh, H., M. Mimuro, and S. Takaichi. 1989. Light-harvesting particles isolated from a brown alga, *Dictyota dichotoma*. A supramolecular assembly of fucoxanthin-chlorophyll-protein complexes. *Biochim. Biophys. Acta* 976: 233–240.

Katoh, T. and T. Ehara. 1990. Supramolecular assembly of fucoxanthin-chlorophyll-protein complexes isolated from a brown alga, *Petalonia fascia*. Electron microscopic studies. *Plant Cell Physiol.* 31: 439–447.

Katoh, T., U. Nagashima, and M. Mimuro. 1991. Fluorescence properties of the allenic carotenoid fucoxanthin: Implication for energy transfer in photosynthetic pigment systems. *Photosynth. Res.* 27: 221–226.

Kereïche, S., R. Kouřil, G. T. Oostergetel et al. 2008. Association of chlorophyll a/c2 complexes to photosystem I and photosystem II in the cryptophyte Rhodomonas CS24. *Biochim. Biophys. Acta* 1777: 1122–1128.

Keşan, G., R. Litvín, D. Bína, M. Durchan, V. Šlouf, and T. Polívka. 2016. Efficient light-harvesting using non-carbonyl carotenoids: Energy transfer dynamics in the VCP complex from *Nannochloropsis oceanica*. *Biochim. Biophys. Acta* 1857: 370–379.

Kleima, F. J., E. Hofmann, B. Gobets et al. 2000. Förster excitation energy transfer in peridinin-chlorophyll-a-protein. *Biophys. J.* 78: 344–353.

Kolli, A., E. J. O'Reilly, G. D. Scholes, and A. Olaya-Castro. 2012. The fundamental role of quantized vibrations in coherent light harvesting by cryptophyte algae. *J. Chem. Phys.* 137: 174109.

Kotabová, E., J. Jarešová, R. Kaňa, R. Sobotka, D. Bína, and O. Prášil. 2014. Novel type of red-shifted chlorophyll a antenna complex from *Chromera velia*. I. Physiological relevance and functional connection to photosystems. *Biochim. Biophys. Acta* 1837: 734–743.

Kraay, G. W., M. Zapata, and M. J. W. Veldhuis. 1992. Separation of chlorophylls c_1, c_2, and c_3 of marine phytoplankton by reversed-phase-C18-high-performance liquid chromatography. *J. Phycol.* 28: 708–712.

Krikunova, M., H. Lokstein, D. Leupold, R. G. Hiller, and B. Voigt. 2006. Pigment-pigment interactions in PCP of *Amphidinium carterae* investigated by nonlinear polarization spectroscopy in the frequency domain. *Biophys. J.* 90: 261–271.

Krueger, B., S. S. Lampoura, I. H. M. van Stokkum et al. 2001. Energy transfer in the peridinin chlorophyll a protein of *Amphidinium carterae* studied by polarized transient absorption and target analysis. *Biophys. J.* 80: 2843–2855.

Lavaud, J., B. Rousseau, and A.-L. Etienne. 2003. Enrichment of the light-harvesting complex in diadinoxanthin and implications for the non-photochemical fluorescence quenching in diatoms. *Biochemistry* 42: 5802–5808.

Lavaud, J., B. Rousseau, H. J. van Gorkom, and A.-L. Etienne. 2002. Influence of the diadinoxanthin pool size on photoprotection in the marine planktonic diatom *Phaeodactylum tricornutum*. *Plant Physiol.* 129: 1398–1406.

Lepetit, B., R. Goss, T. Jakob, and C. Wilhelm. 2012. Molecular dynamics of the diatom thylakoid membrane under different light conditions. *Photosynth. Res.* 111: 245–257.

Lepetit, B., D. Volke, M. Gilbert, C. Wilhelm, and R. Goss. 2010. Evidence for the existence of one antenna-associated lipid-dissolved and two protein-bound pools of diadinoxanthin cycle pigments in diatoms. *Plant Physiol.* 154: 1905–1920.

Lepetit, B., D. Volke, M. Szabó et al. 2007. Spectroscopic and molecular characterization of the oligomeric antenna of the diatom *Phaeodactylum tricornutum*. *Biochemistry* 46: 9813–9822.

Lichtlé, C., W. Arsalane, J.-C. Duval, and C. Passaquet. 1995. Characterization of the light-harvesting complex of *Giraudyopsis stellifer* (Chrysophyceae) and effects of light stress. *J. Phycol.* 31: 380–387.

Lichtlé, C., J.-C. Duval, and Y. Lemoine. 1987. Comparative biochemical, functional and ultrastructural studies of photosystem particles from a Cryptophyceae: *Cryptomonas rufescens*; isolation of an active phycoerythrin particle. *Biochim. Biophys. Acta* 894: 76–90.

Lichtlé, C., H. Jupin, and J.-C. Duval. 1980. Energy transfers from photosystem II to photosystem I in *Cryptomonas rufescens* (Cryptophyceae). *Biochim. Biophys. Acta* 591: 104–112.

Lichtlé, C., A. Spilar, and J.-C. Duval. 1992. Immunogold localization of light-harvesting and photosystem I complexes in the thylakoids of *Fucus serratus* (Phaeophyceae). *Protoplasma* 166: 99–106.

Linden, P. A., J. Zimmermann, T. Brixner et al. 2004. Transient absorption study of peridinin and peridinin-chlorophyll a-protein after two-photon excitation. *J. Phys. Chem. B* 108: 10340–10345.

Litvín, R., D. Bína, M. Herbstová, and Z. Gardian. 2016. Architecture of the light-harvesting apparatus of the eustigmatophyte alga *Nannochloropsis oceanica*. *Photosynth. Res.* 130: 137–150.

Liu, Z., H. Yan, K. Wang et al. 2004. Crystal structure of spinach major light-harvesting complex at 2.72 Å resolution. *Nature* 428: 287–292.

Lohr, M. and C. Wilhelm. 1999. Algae displaying the diadinoxanthin cycle also possess the violaxanthin cycle. *Proc. Natl. Acad. Sci. USA* 96: 8784–8789.

MacPherson, A. N. and R. G. Hiller. 2003. Light-harvesting systems in chlorophyll c-containing algae. In *Light-Harvesting Antennas in Photosynthesis* eds. B. R. Green and W. W. Parson, pp. 323–352. Dordrecht, the Netherlands: Kluwer Academic Publishers.

Maier, U.-G., C. J. B. Hofmann, S. Eschbach, J. Wolters, and G. L. Igloi. 1991. Demonstration of nucleomorph-encoded eukaryotic small subunit ribosomal RNA in cryptomonads. *Mol. Gen. Genet.* 230: 155–160.

Marin, A., A. B. Doust, G. D. Scholes et al. 2011. Flow of excitation energy in the cryptophyte light-harvesting antenna phycocyanin 645. *Biophys. J.* 101: 1004–1013.

Miller, D. J., J. Catmull, R. Puskeiler, H. Tweedale, F. P. Sharples, and R. G. Hiller. 2005. Reconstitution of the peridinin-chlorophyll a protein (PCP): Evidence for functional flexibility in chlorophyll binding. *Photosynth. Res.* 86: 229–240.

Mimuro, M., T. Katoh, and H. Kawai. 1990a. Spatial arrangement of pigments and their interaction in the fucoxanthin/chlorophyll a/c protein assembly (FCPA) isolated from the brown alga *Dictyota dichotoma*. Analysis by means of polarized spectroscopy. *Biochim. Biophys. Acta* 1015: 450–456.

Mimuro, M., U. Nagashima, S. Takaichi, Y. Nishimura, I. Yamazaki, and T. Katoh. 1992. Molecular structure and optical properties of carotenoids for the in vivo energy transfer function in the algal photosynthetic pigment system. *Biochim. Biophys. Acta* 1098: 271–274.

Mimuro, M., N. Tamai, T. Ishimaru, and I. Yamazaki. 1990b. Characteristic fluorescence components in photosynthetic pigment system of a marine dinoflagellate, *Protogonyaulax tamarensis*, and excitation energy flow among them. Studies by means of steady-state and time-resolved fluorescence spectroscopy. *Biochim. Biophys. Acta* 1016: 280–287.

Mimuro, M., N. Tamai, A. Murakami et al. 1998. Multiple pathways of excitation energy flow in the photosynthetic pigment system of a cryptophyte, *Cryptomonas* sp. (CR-1). *Phycol. Res.* 46: 155–164.

Nagao, R., A. Ishii, O. Tada et al. 2007. Isolation and characterization of oxygen-evolving thylakoid membranes and photosystem II particles from a marine diatom *Chaetoceros gracilis*. *Biochim. Biophys. Acta* 1767: 1353–1362.

Nagao, R., T. Tomo, E. Noguchi et al. 2010. Purification and characterization of a stable oxygen-evolving photosystem II complex from a marine centric diatom, *Chaetoceros gracilis*. *Biochim. Biophys. Acta* 1797: 160–166.

Nagao, R., T. Tomo, E. Noguchi et al. 2012. Proteases are associated with a minor fucoxanthin chlorophyll *a/c*-binding protein from the diatom, *Chaetoceros gracilis*. *Biochim. Biophys. Acta* 1817: 2110–2117.

Nagao, R., M. Yokono, S. Akimoto, and T. Tomo. 2013. High excitation energy quenching in fucoxanthin chlorophyll *a/c*-binding protein complexes from the diatom *Chaetoceros gracilis*. *J. Phys. Chem. B* 117: 6888–6895.

Neilson, J. A. D. and D. G. Durnford. 2010. Structural and functional diversification of the light-harvesting complexes in photosynthetic eukaryotes. *Photosynth. Res.* 106: 57–71.

Niedzwiedzki, D. M., J. Jiang, C. S. Lo, and R. E. Blankenship. 2014. Spectroscopic properties of the chlorophyll *a*-chlorophyll c_2-peridinin-protein-complex (acpPC) from the coral symbiotic dinoflagellate *Symbiodinium*. *Photosynth. Res.* 120: 125–139.

Norris, B. J. and D. J. Miller. 1994. Nucleotide sequence of a cDNA clone encoding the precursor of the peridinin-chlorophyll *a*-binding protein from the dinoflagellate *Symbiodinium* sp. *Plant Mol. Biol.* 24: 673–677.

Nymark, M., K. C. Valle, T. Brembu et al. 2009. An integrated analysis of molecular acclimation to high light in the marine diatom *Phaeodactylum tricornutum*. *PLoS ONE* 4: e7743.

Ocampo-Alvarez, H., E. García-Mendoza, and Govindjee. 2013. Antagonist effect between violaxanthin and de-epoxidated pigments in nonphotochemical quenching induction in the qE deficient brown alga *Macrocystis pyrifera*. *Biochim. Biophys. Acta* 1827: 427–437.

Oeltjen, A., W. E. Krumbein, and E. Rhiel. 2002. Investigations on transcript sizes, steady state mRNA concentrations and diurnal expression of genes encoding fucoxanthin chlorophyll a/c light harvesting polypeptides in the centric diatom *Cyclotella cryptica*. *Plant Biol.* 4: 250–257.

Oeltjen, A., J. Marquardt, and E. Rhiel. 2004. Differential circadian expression of genes fcp2 and fcp6 in *Cyclotella cryptica*. *Int. Microbiol.* 7: 127–131.

Owens, T. G. 1986. Light-harvesting function in the diatom *Phaeodactylum tricornutum*—II. Distribution of excitation energy between photosystems. *Plant Physiol.* 80: 739–746.

Owens, T. G. 1988. Light-harvesting antenna systems in the chlorophyll a/c-containing algae. In *Light-Energy Transduction in Photosynthesis: Higher Plants and Bacterial Models* eds. S. E. Stevens and D. A. Bryant, pp. 122–136. Rockville, MD: American Society of Plant Physiologists.

Owens, T. G. and E. R. Wold. 1986. Light-harvesting function in the diatom *Phaeodactylum tricornutum*—I. Isolation and characterization of pigment-protein complexes. *Plant Physiol.* 80: 732–738.

Pan, H., J. Šlapeta, D. Carter, and M. Chen. 2012. Phylogenetic analysis of the light-harvesting system in *Chromera velia*. *Photosynth. Res.* 111: 19–28.

Papagiannakis, E., D. S. Larsen, I. H. M. van Stokkum, M. Vengris, R. G. Hiller, and R. van Grondelle. 2004. Resolving the excited-state equilibrium of peridinin in solution. *Biochemistry* 43: 15303–15309.

Papagiannakis, E., I. H. M. van Stokkum, H. Fey, C. Büchel, and R. van Grondelle. 2005. Spectroscopic characterization of the excitation energy transfer in the fucoxanthin-chlorophyll protein of diatoms. *Photosynth. Res.* 86: 241–250.

Papagiannakis, E., M. Vengris, D. S. Larsen, I. H. M. van Stokkum, R. G. Hiller, and R. van Grondelle. 2006. Use of ultrafast dispersed pump-dump-probe and pump-repump-probe spectroscopies to explore the light-induced dynamics of peridinin in solution. *J. Phys. Chem. B* 110: 512–521.

Pascal, A. A., L. Caron, B. Rousseau, K. Lapouge, J.-C. Duval, and B. Robert. 1998. Resonance Raman spectroscopy of a light-harvesting protein from the brown alga *Laminaria saccharina*. *Biochemistry* 37: 2450–2457.

Passaquet, C., J. C. Thomas, L. Caron, N. Hauswirth, F. Puel, and C. Berkaloff. 1991. Light-harvesting complexes in brown algae. Biochemical characterization and immunological relationships. *FEBS Lett.* 280: 21–26.

Peers, G., T. B. Truong, E. Ostendorf et al. 2009. An ancient light-harvesting protein is critical for the regulation of algal photosynthesis. *Nature* 462: 518–521.

Polívka, T. and H. A. Frank. 2010. Molecular factors controlling photosynthetic light harvesting by carotenoids. *Acc. Chem. Res.* 43: 1125–1134.

Polívka, T., R. G. Hiller, and H. A. Frank. 2007. Spectroscopy of the peridinin-chlorophyll-a protein: Insight into light-harvesting strategy of marine algae. *Arch. Biochem. Biophys.* 458: 111–120.

Polívka, T., I. H. M. van Stokkum, D. Zigmantas, R. van Grondelle, V. Sundström, and R. G. Hiller. 2006. Energy transfer in the major intrinsic light-harvesting complex from *Amphidinium carterae*. *Biochemistry* 45: 8516–8526.

Premvardhan, L., L. Bordes, A. Beer, C. Büchel, and B. Robert. 2009. Carotenoid structures and environments in trimeric and oligomeric fucoxanthin chlorophyll a/c$_2$ proteins from resonance Raman spectroscopy. *J. Phys. Chem. B* 113: 12565–12574.

Premvardhan, L., E. Papagiannakis, R. G. Hiller, and R. van Grondelle. 2005. The charge-transfer character of the S0-S2 transition in the carotenoid peridinin is revealed by Stark spectroscopy. *J. Phys. Chem. B* 109: 15589–15597.

Premvardhan, L., B. Robert, A. Beer, and C. Büchel. 2010. Pigment organization in fucoxanthin chlorophyll a/c$_2$ proteins (FCP) based on resonance Raman spectroscopy and sequence analysis. *Biochim. Biophys. Acta* 1797: 1647–1656.

Premvardhan, L., D. J. Sandberg, H. Fey, R. R. Birge, C. Büchel, and R. van Grondelle. 2008. The charge-transfer properties of the S$_2$ state of fucoxanthin in solution and in fucoxanthin chlorophyll-a/c$_2$ protein (FCP) based on Stark spectroscopy and molecular-orbital theory. *J. Phys. Chem. B* 112: 11838–11853.

Ragni, M., R. L. Airs, N. Leonardos, and R. J. Geider. 2008. Photoinhibition of PSII in *Emiliania huxleyi* (Haptophyta) under high light stress: The roles of photoacclimation, photoprotection, and photorepair. *J. Phycol.* 44: 670–683.

Ramanan, C., R. Berera, K. Gundermann, I. H. M. van Stokkum, C. Büchel, and R. van Grondelle. 2014. Exploring the mechanism(s) of energy dissipation in the light harvesting complex of the photosynthetic algae *Cyclotella meneghiniana*. *Biochim. Biophys. Acta* 1837: 1507–1513.

Rhiel, E., E. Mörschel, and W. Wehrmeyer. 1987. Characterization and structural analysis of a chlorophyll a/c light-harvesting complex and of photosystem I particles isolated from thylakoid membranes of *Cryptomonas maculata* (Cryptophyceae). *Eur. J. Cell Biol.* 43: 82–92.

Richard, C., H. Ouellet, and M. Guertin. 2000. Characterization of the LI818 polypeptide from the green unicellular alga *Chlamydomonas reinhardtii*. *Plant Mol. Biol.* 42: 303–316.

Röding, A., E. Boekema, and C. Büchel. 2016. The structure of FCPb, a light harvesting complex in the diatom *Cyclotella meneghiniana*. *Photosynth. Res.* DOI:10.1007/s11120-016-0328-9.

Ruban, A. V., M. P. Johnson, and C. D. P. Duffy. 2012. The photoprotective molecular switch in the photosystem II antenna. *Biochim. Biophys. Acta* 1817: 167–181.

Ruban, A. V., J. Lavaud, B. Rousseau, G. Guglielmi, P. Horton, and A.-L. Etienne. 2004. The super-excess energy dissipation in diatom algae: Comparative analysis with higher plants. *Photosynth. Res.* 82: 165–175.

Shreve, A. P., J. K. Trautman, T. G. Owens, and A. C. Albrecht. 1991. A femtosecond study of electronic state dynamics of fucoxanthin and implications for photosynthetic carotenoid-to-chlorophyll energy transfer mechanisms. *Chem. Phys.* 154: 171–178.

Six, C., A. Z. Worden, F. Rodriguez, H. Moreau, and F. Partensky. 2005. New insights into the nature and phylogeny of prasinophyte antenna proteins: *Ostreococcus tauri*, a case study. *Mol. Biol. Evol.* 22: 2217–2230.

Slavov, C., V. Schrameyer, M. Reus et al. 2016. "Super-quenching" state protects *Symbiodinium* from thermal stress: Implications for coral bleaching. *Biochim. Biophys. Acta* 1857: 840–847.

Šlouf, V., M. Fuciman, S. Johanning, E. Hofmann, H. A. Frank, and T. Polívka. 2013. Low-temperature time-resolved spectroscopic study of the major light-harvesting complex of *Amphidinium carterae*. *Photosynth. Res.* 117: 257–265.

Snyder, U. K. and J. Biggins. 1987. Excitation-energy redistribution in the cryptomonad alga *Cryptomonas ovata*. *Biochim. Biophys. Acta* 892: 48–55.

Songaila, E., R. Augulis, A. Gelzinis et al. 2013. Ultrafast energy transfer from chlorophyll c_2 to chlorophyll *a* in fucoxanthin-chlorophyll protein complex. *J. Phys. Chem. Lett.* 4: 3590–3595.

Standfuss, J., A. C. T. van Scheltinga, M. Lamborghini, and W. Kühlbrandt. 2005. Mechanisms of photoprotection and nonphotochemical quenching in pea light-harvesting complex at 2.5 Å resolution. *EMBO J.* 24: 919–928.

Sukenik, A., A. Livne, A. Neori, Y. Z. Yacobi, and D. Katcoff. 1992. Purification and characterization of a light-harvesting chlorophyll-protein complex from the marine eustigmatophyte *Nannochloropsis* sp. *Plant Cell Physiol.* 33: 1041–1048.

Szabò, I., E. Bergantino, and G. M. Giacometti. 2005. Light and oxygenic photosynthesis: Energy dissipation as a protection mechanism against photo-oxidation. *EMBO Rep.* 6: 629–634.

Szabó, M., B. Lepetit, R. Goss, C. Wilhelm, L. Mustárdy, and G. Garab. 2008. Structurally flexible macro-organization of the pigment-protein complexes of the diatom *Phaeodactylum tricornutum*. *Photosynth. Res.* 95: 237–245.

Szabó, M., L. Premvardhan, B. Lepetit, R. Goss, C. Wilhelm, and G. Garab. 2010. Functional heterogeneity of the fucoxanthins and fucoxanthin-chlorophyll proteins in diatom cells revealed by their electrochromic response and fluorescence and linear dichroism spectra. *Chem. Phys.* 373: 110–114.

Tichy, J., Z. Gardian, D. Bina et al. 2013. Light harvesting complexes of *Chromera velia*, photosynthetic relative of apicomplexan parasites. *Biochim. Biophys. Acta* 1827: 723–729.

Vaswani, H. M., C.-P. Hsu, M. P. Head-Gordon, and G. R. Fleming. 2003. Quantum chemical evidence for an intramolecular charge-transfer state in the carotenoid peridinin of peridinin-chlorophyll-protein. *J. Phys. Chem. B* 107: 7940–7946.

Veith, T., J. Brauns, W. Weisheit, M. Mittag, and C. Büchel. 2009. Identification of a specific fucoxanthin-chlorophyll protein in the light harvesting complex of photosystem I in the diatom *Cyclotella meneghiniana*. *Biochim. Biophys. Acta* 1787: 905–912.

Veith, T. and C. Büchel. 2007. The monomeric photosystem I-complex of the diatom *Phaeodactylum tricornutum* binds specific fucoxanthin chlorophyll proteins (FCPs) as light-harvesting complexes. *Biochim. Biophys. Acta* 1767: 1428–1435.

Wahadoszamen, M., A. Ghazaryan, H. E. Cingil et al. 2013. Stark fluorescence spectroscopy reveals two emitting sites in the dissipative state of FCP antennas. *Biochim. Biophys. Acta* 1837: 193–200.

Wedemayer, G. J., D. G. Kidd, and A. N. Glazer. 1996. Cryptomonad biliproteins: Bilin types and locations. *Photosynth. Res.* 48: 163–170.

van der Weij-de Wit, C. D., A. B. Doust, I. H. M. van Stokkum et al. 2006. How energy funnels from the phycoerythrin antenna complex to photosystem I and photosystem II in cryptophyte *Rhodomonas* CS24 cells. *J. Phys. Chem. B* 110: 25066–25073.

van der Weij-de Wit, C. D., A. B. Doust, I. H. M. van Stokkum et al. 2008. Phycocyanin sensitizes both photosystem I and photosystem II in cryptophyte *Chroomonas* CCMP270 cells. *Biophys. J.* 94: 2423–2433.

Wilhelm, C. and I. Lenartz-Weiler. 1987. Energy transfer and pigment composition in three chlorophyll b-containing light harvesting complexes isolated from *Mantoniella squamata* (Prasinophyceae), *Chlorella fusca* (Chlorophyceae) and *Sinapis alba*. *Photosynth. Res.* 13: 125–141.

Wilk, K. E., S. J. Harrop, L. Jankova et al. 1999. Evolution of a light-harvesting protein by addition of new subunits and rearrangement of conserved elements: Crystal structure of a cryptophyte phycoerythrin at 1.63-Å resolution. *Proc. Natl. Acad. Sci. USA* 96: 8901–8906.

Zhu, S.-H. and B. R. Green. 2010. Photoprotection in the diatom *Thalassiosira pseudonana*: Role of LI818-like proteins in response to high light stress. *Biochim. Biophys. Acta* 1797: 1449–1457.

Zigmantas, D., R. G. Hiller, F. P. Sharples, H. A. Frank, V. Sundström, and T. Polívka. 2004. Effect of a conjugated carbonyl group on the photophysical properties of carotenoids. *Phys. Chem. Chem. Phys.* 6: 3009–3016.

Zigmantas, D., R. G. Hiller, V. Sundström, and T. Polívka. 2002. Carotenoid to chlorophyll energy transfer in the peridinin-chlorophyll-a-protein complex involves an intramolecular charge transfer state. *Proc. Natl. Acad. Sci. USA* 99: 16760–16765.

Zigmantas, D., R. G. Hiller, A. Yartsev, V. Sundström, and T. Polívka. 2003. Dynamics of excited states of the carotenoid peridinin in polar solvents: Dependence on excitation wavelength, viscosity, and temperature. *J. Phys. Chem. B* 107: 5339–5348.

Zigmantas, D., T. Polívka, R. G. Hiller, A. Yartsev, and V. Sundström. 2001. Spectroscopic and dynamic properties of the peridinin lowest singlet excited states. *J. Phys. Chem. A* 105: 10296–10306.

Reaction centers: Structure and mechanism

MICHAEL R. JONES

9.1 INTRODUCTION: REACTION CENTERS

In plants, algae, and photosynthetic bacteria, reaction center (RC) pigment proteins constitute the end point of excitation energy flow through the light-harvesting regions of the photosystem. Conversion of a "primary electron donor" (bacterio)chlorophyll ((B)Chl) species in the RC to its first singlet excited state triggers electron transfer in a few picoseconds to an adjacent acceptor (bacterio)chlorin. This primary electron donor is found embedded within the RC protein close to one side of the energy-transducing membrane, and its photooxidation creates a radical pair. Further steps of electron or hole transfer sequentially increase the distance between the two radicals, stabilizing charge separation across the membrane and facilitating electron transfer interactions with other mobile components of the photosystem.

The first few steps of this charge separation process are extremely rapid in order to minimize slower loss processes, including transfer of the singlet excited state from the primary electron donor back into the surrounding antenna (detrapping). In normally functioning RCs, the extent of loss processes is very low such that when combined with very rapid and efficient energy transfer through the antenna, light-powered charge separation is achieved with a very high "quantum yield" (defined as event per photon absorbed). However, a consequence of this high quantum efficiency is that the energy efficiency of the overall process is not particularly high (see Barber 2009 for a discussion of the true efficiency of oxygenic photosynthesis). A great deal of the energy of the photons absorbed by antenna pigments can be lost as the excited electronic state is "funnelled" to the RC primary electron donor, which usually occupies the low energy edge of the absorbance spectrum of the photosystem. In addition, as each step of charge separation is potentially reversible, drops in free energy as the radical pair evolves are required to ensure that each step is in practice irreversible. As a result, a large part of the energy arriving in the RC is released as heat rather than being conserved in the form of a proton electrochemical gradient, or a high ATP/ADP or NADPH/NADP$^+$ ratio.

Another marked feature of RCs that contributes to the high quantum yield of charge separation is their ability to form a radical pair on immediately adjacent pigments without significant charge recombination to the ground state. To achieve this, the initial picosecond charge separation is a two-step process involving at least three (bacterio) chlorin cofactors, the second step moving the initially formed electron and hole further apart before they can recombine. An intriguing aspect of this is that the free energies of the first and second radical pairs are relatively high, and so there is a large driving force for wasteful charge recombination to the ground state. As the RC protein provides a relatively low–reorganization energy environment within which to carry out electron transfer (Parson et al. 1998), recombination could be relatively slow due to the activation energy that results from an imbalance between a large driving force and low reorganization energy—a manifestation of the "Marcus inverted region" for biological electron transfer (Marcus 1993). Whether this is indeed the case is the subject of ongoing investigation, as is the general question of how protein dynamics influence charge separation and charge recombination reactions (see Section 9.5). Once electron transfer begins to involve cofactors other than the initial (bacterio)chlorins, each step of forward electron transfer tends to be much faster than the competing recombination reaction, a feature that also contributes to the irreversibility of the overall process and the high quantum yield.

Although RCs are found in very diverse photosynthetic organisms, the basic structure within which the first few steps of charge separation take place is generally similar in all characterized systems (Fromme et al. 1996; Schubert et al. 1998;

Heathcote et al. 2002; Nelson and Yocum 2006; Sadekar et al. 2006; Heathcote and Jones 2012; Caffarri et al. 2014; Nelson and Junge 2015). This fits with the common mechanistic strategy of all RC pigment proteins, to convert an excited electronic state into a metastable radical pair state with almost no loss of quantum yield. However, there are differences in the precise mechanism through which this is achieved, some of which are imposed by the additional requirement of primary electron donor (B)Chls to act as acceptors of energy from a range of antenna pigments. Other differences are imposed by the diverse chemical natures of the longer timescale donors and acceptors of electrons, such as quinones, iron–sulfur centers, and cytochromes. Variety within the structures that carry out the initial charge separation comes from (1) differences in the chemical properties of the various (B) Chls found in plants, algae, cyanobacteria, and the range of anoxygenic photosynthetic bacteria; (2) differences in the positioning of these (B)Chls relative to one another within the protein scaffold; and (3) differences in the physicochemical interactions between these (B)Chls and the surrounding protein scaffold.

The following looks at the process of charge separation within the three best characterized RCs, Photosystems I and II of oxygenic photosynthesis and the complex from the purple photosynthetic bacteria. After comparing their structures, the chapter goes on to consider what is known about the mechanism of initial charge separation within each RC, and how variation in the detailed mechanism correlates with the properties of the surrounding light harvesting and electron transfer systems.

9.2 STRUCTURES OF THE CHARGE-SEPARATING CHAINS

9.2.1 The purple bacterial reaction center

For many years, our understanding of the molecular arrangement of the cofactors that carry out the first, rapid steps of photochemical charge separation was strongly influenced by the groundbreaking x-ray crystal structures of the RCs from the purple bacteria *Blastochloris* (*Blc.*—formerly *Rhodopseudomonas*) *viridis* (Deisenhofer et al. 1984, 1985, 1995) and *Rhodobacter* (*Rba.*) *sphaeroides* (Allen et al. 1986, 1987a,b, 1988; Yeates et al.

1988; Chang et al. 1991; Ermler et al. 1994a,b). In the years that followed publication of these structures, the latter organism became the more favored vehicle for structure/function studies. This occurred for a variety of reasons, not the least of which was the ease with which variants with mutations in the RC genes could be grown under nonphotosynthetic conditions. The structures of these two RCs have been described in detail and reviewed extensively (Woodbury and Allen 1995; Hoff and Deisenhofer 1997; Allen and Williams 1998; Zinth and Wachtveitl 2005; Jones 2009; Heathcote and Jones 2012), and only an outline description of particular features is given here. In the Protein Data Bank, there are currently 121 structures for wild-type or engineered RCs from *Rba. sphaeroides*, *Blc. viridis*, and *Thermochromatium* (*Tch.*) *tepidum*. In addition, there are four structures for the RC–LH1 complex from *Tch. tepidum* at resolutions between 3.0 and 3.3 Å (Niwa et al. 2014), a structure for the *Rhodopseudomonas palustris* RC–LH1 at a resolution of 4.8 Å (Roszak et al. 2003) and a

structure for the dimeric *Rba. sphaeroides* RC-LH1 at a resolution of 7.8 Å (Qian et al. 2013).

The charge-separating unit within the *Rba. sphaeroides* RC comprises four bacteriochlorophyll *a* (BChl *a*), two bacteriopheophytin *a* (BPhe *a*) and two ubiquinones positioned by the protein scaffold around an axis of twofold symmetry in two membrane-spanning branches. The arrangement of the redox-active macrocycles of these cofactors is shown in Figure 9.1a; for clarity, this diagram does not show the 50- and 20-carbon side chains that render ubiquinone and BChl/BPhe, respectively, lipid soluble.

Two of the BChls, labeled P_A and P_B, straddle the symmetry axis, which runs perpendicular to the plane of the membrane, and are located close to the periplasmic side of the bacterial energy-transducing membrane. The tetrapyrrole macrocycles of these BChls are approximately parallel to one another and partially overlap, with direct atomic contact in the overlap region. As a result, the π electron systems of the two BChls mix and they behave photochemically as a dimer—the so-called "special pair." The two

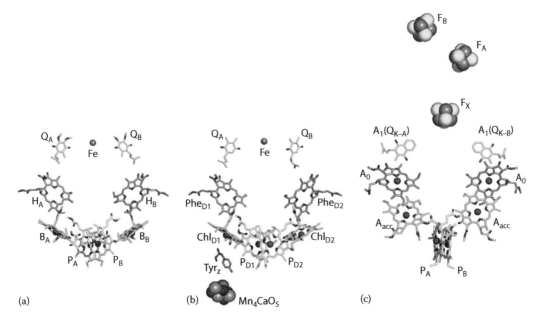

(a)   (b)   (c)

Figure 9.1 Arrangement of electron transfer cofactors in PSI and PSII from cyanobacteria and the *Rba. sphaeroides* RC. **(a)** The *Rba. sphaeroides* RC (PDB file 3ZUW—Gibasiewicz et al. 2011), with the BChls, BPhes, and ubiquinones shown as sticks (carbon—multiple colors, nitrogen—blue, oxygen—red). Hydrocarbon side chains have been removed for clarity. Mg atoms of the BChls are shown as magenta spheres, and the iron atom as an orange sphere. **(b)** The PSII RC from *T. elongatus* (PDB entry 3ARC—Umena et al. 2011). Representation is as for (a) with the Mn_4CaO_5 cluster shown as olive (manganese), light blue (calcium), and red (oxygen) spheres. **(c)** The PSI RC from *T. elongatus* (PDB file 1JB0—Jordan et al. 2001). Representation is as for (a) with the Fe_4S_4 centers shown as orange (iron) and yellow (sulfur) spheres.

BChls display exciton coupling, a principal manifestation of which is a splitting of the lowest energy absorbance band they would have as monomers (with a maximum at around 800–815 nm) into a minor component at ≈815 nm and a dominant component at between 865 and 870 nm, depending on conditions. An early observation was that this latter band bleaches when an excited or cation state is formed, leading to the pigment(s) responsible being dubbed "P870." Similar assignments were subsequently made for other RCs (see Parson 2003 for an account of the development of ideas on these RC components in the 1950s and 1960s). In *Blc. viridis*, which contains BChl *b* rather than BChl *a*, the equivalent pair of BChls is referred to as P960, for hopefully obvious reasons.

Although the two halves of the special pair are chemically identical, differences in their detailed geometry and their interactions with the surrounding protein produce an internal asymmetry, such that unpaired electrons are not shared equally between the two halves. When in the $P870^+$ state some 68% of the unpaired valence electron is localized on the P_A half of the dimer (Lendzian et al. 1993), or in other words some 68% of the positive charge of the cation is located on the P_A half of the dimer. This asymmetric distribution in seen in other species of purple bacteria and can be varied through site-directed mutagenesis in the vicinity of the P870 BChls (Rautter et al. 1995; Artz et al. 1997; Müh et al. 2002). Excitonic coupling between the two halves of P870, calculated at 550 cm^{-1} (Scherer and Fischer 1986), is much stronger than coupling between the P870 BChls and the adjacent accessory BChls or between each of the latter and the adjacent BPhe (calculated to be in the range 136–166 cm^{-1} (Scherer and Fischer 1986)). Coupling between the two BChl *b* of the P960 special pair in the *Blc. viridis* RC was calculated to be even stronger at 900 cm^{-1}, with couplings with and between the monomeric cofactors being much weaker below 200 cm^{-1} (Knapp et al. 1985; Zinth et al. 1985).

Immediately adjacent and edge-on to the special pair are two monomeric BChls, B_A and B_B, which in turn are adjacent and edge-on to two BPhes, H_A and H_B (Figure 9.1a). The monomeric BChls are sometimes called "accessory" BChls, and that term will be used in the following as it is useful to identify structurally equivalent Chls in other RCs where there are additional monomeric Chls. BPhe *a* is chemically identical to BChl *a* with the exception

that the central Mg^{2+} is replaced by two hydrogen ions, a difference that shifts its absorbance bands to higher energy and, more importantly, raises its potential for one electron reduction. This enables each BPhe to act, in principle, as an effective electron acceptor from the adjacent monomeric BChl. Each membrane-spanning chain of cofactors is terminated by a ubiquinone-10, termed Q_A and Q_B, located close to the cytoplasmic side of the bacterial membrane. Between the two quinones is an iron atom that is thought to play a mainly structural role, but there is ongoing interest in whether it participates in electron transfer as a discrete intermediate, with evidence both for and against (Remy and Gerwert 2003; Hermes et al. 2006; Breton 2007). There are also questions over the role it plays in electron tunnelling between the two quinones (Burggraf and Koslowski 2011), and over the extent to which it is involved in motions in this region of the protein that influence electron transfer (Orzechowska et al. 2010).

The cofactors depicted in Figure 9.1a are held in place by a protein scaffold formed principally from two related polypeptides termed L and M. Each has five membrane-spanning α-helices that are connected by short helices and loops at either side of the membrane, and L and M are arranged in a pseudosymmetrical fashion around the same symmetry axis that relates the two branches of cofactors. The BChl and BPhe cofactors are located at the interface between the L and M polypeptides, whereas the Q_A and Q_B quinones are enveloped in a binding pocket formed by helix and loop regions of the M and L polypeptides, respectively. In all purple bacteria, a third polypeptide, termed H, is also present. This has an extramembrane domain that caps the cytoplasmic faces of the L/M heterodimer and a single α-helix that extends across the membrane. The principal function of the H-polypeptide is to assist in dictating the correct pattern of protonation of the Q_A and Q_B quinones during electron transfer between the two (Okamura et al. 2000; Wraight 2004; Wraight and Gunner 2009) and it may also serve to insulate the quinones from the cytoplasm to prevented unwanted, unproductive electron transfer reactions. In some species, including *Blc. viridis*, a fourth polypeptide on the periplasmic side of the membrane encases four heme cofactors involved in the delivery of electrons to the photooxidized special pair (Deisenhofer et al. 1984, 1985, 1995; Ortega and Mathis 1993) (and see

Vermeglio et al. 2012 for a discussion of this additional component).

On a technical note, the nomenclature for the RC cofactors is somewhat varied, with some in the field preferring use of the subscripts L and M rather than A and B to describe the bacteriochlorin cofactors on different branches. There are other complexities too—see Jones (2009) for a more long-winded account.

9.2.2 Photosystem II RC

Superficially, the arrangement of charge-separating cofactors in Photosystem II (PSII) is similar to that in the purple bacterial reaction center (pbRC) (see Nelson and Yocum 2006; Kern and Renger 2007; Renger 2010; Renger and Schlodder 2010; Renger and Schlodder 2011; Cardona et al. 2012; Müh et al. 2012; Mamedov et al. 2015; Nelson and Junge 2015; Barber 2016 for reviews). High-resolution x-ray crystal structures have been reported for PSII complexes from two thermophilic cyanobacteria, *Thermosynechococcus* (*T.*) *elongatus* (Zouni et al. 2001; Biesiadka et al. 2004; Ferreira et al. 2004; Loll et al. 2005; Guskov et al. 2009; Broser et al. 2010, 2011; Hellmich et al. 2014; Ayyer et al. 2016; Young et al. 2016) and *T. vulcanus* (Kamiya and Shen 2003; Kawakami et al. 2009; Umena et al. 2011; Koua et al. 2013; Suga et al. 2015, 2017; Tanaka et al. 2017). At the heart of PSII, a pseudosymmetrical heterodimer formed by two related polypeptides, D1 and D2, holds in place two branches of cofactors arranged around an axis of twofold symmetry (Figure 9.1b). These comprise four chlorophyll *a* (Chl *a*), two pheophytin *a* (Phe *a*), and two plastoquinones and so are chemically somewhat different from the BChl, BPhe, and ubiquinone found in purple bacteria. As can be seen by comparing Figure 9.1b with Figure 9.1a, the arrangement of cofactors in PSII follows the same general blueprint as for the pbRC, with two Chls arranged parallel to one another close to one side of the membrane (P_{D1}, P_{D2}), adjacent to two monomeric Chls (Chl_{D1}, Chl_{D2}), two Phes (Phe_{D1}, Phe_{D2}), and two quinones (Q_A, Q_B) that are separated by an iron atom that sits on the symmetry axis.

Although the general arrangement of cofactors is similar in PSII and the pbRC, there are some key differences, which relate to the mechanism of charge separation. Perhaps the most noteworthy are small shifts in the positions of the P_{D1} and P_{D2} Chls relative to those of their counterparts in the pbRC, such

that the degree of coupling between the two is much weaker (for more detail on this, see Section 9.3.4). As a result, the PSII RC does not have a "special pair" of Chls, something that has important ramifications for the mechanism of charge separation and the interaction of the resulting Chl cation with the adjacent donor-side electron transfer chain.

Although at their cores they display obvious similarities, the PSII RC is a much larger entity than its purple bacterial counterpart. In addition to the D1/D2 polypeptide heterodimer and associated two branches of electron transfer cofactors, the structurally characterized cyanobacterial version of PSII contains 18 additional polypeptides, 31 additional Chls, 11 carotenoids (β-carotene), 2 hemes, a bound bicarbonate ion, and several manganese, calcium, and chloride ions (Umena et al. 2011). Of particular note, two intramembrane polypeptides, termed CP43 and CP47, are arranged on either side of the D1/D2 heterodimer and bind 16 and 13 light-harvesting Chls, respectively, along with a total of 11 β-carotenes (Umena et al. 2011). On the luminal side of the membrane, extrinsic protein subunits scaffold a Mn_4CaO_5 cluster, which is the site of water oxidation (Zouni et al. 2001; Kamiya and Shen 2003; Biesiadka et al. 2004; Ferreira et al. 2004; Loll et al. 2005; Guskov et al. 2009; Kawakami et al. 2009; Umena et al. 2011; Koua et al. 2013; Suga et al. 2015, 2017; Young et al. 2016; Tanaka et al. 2017), the liberated electrons being used to reduce the Chl cation that is photogenerated by charge separation in the central electron transfer chain (see Section 9.3.4).

9.2.3 Photosystem I RC

Photosystem I (PSI) is an even larger and more elaborate complex than PSII (see Chitnis 2001; Grotjohann and Fromme 2005; Nelson and Yocum 2006; Jensen et al. 2007; Amunts and Nelson 2008, 2009; Renger 2010; Busch and Hippler 2011; Kargul et al. 2012; Caffarri et al. 2014; Nelson and Junge 2015 for reviews). The first high-resolution x-ray crystal structure for PSI was that of the trimeric complex from *T. elongatus* (Fromme et al. 2001; Jordan et al. 2001). Each monomer comprises 12 polypeptide chains that scaffold 96 molecules of Chl *a*, 22 carotenoids, 2 phylloquinones, and 3 Fe_4S_4 iron–sulfur centers. Despite this greater complexity, at the heart of PSI, the same generic RC architecture is found (Figure 9.1c), with two symmetrically arranged membrane-spanning branches of chlorin and quinone cofactors encased by a heterodimeric protein scaffold (Fromme et al. 2001; Jordan et al. 2001). On the luminal side of the membrane, one Chl a (P_A) and one Chl a' (P_B) are arranged in a similar way to their P_{D1}/P_{D2} and P_A/P_B counterparts in PSII and the pbRC (for more details, see Section 9.3.4). Chl a' is an epimer of Chl a in which $-COOCH_3$ and $-H$ groups at carbon $C13^2$ are swapped relative to their positions in Chl a (Webber and Lubitz 2001).

Adjacent to this Chl a/Chl a' pair are two monomeric Chls (termed A or A_{acc}), then two further monomeric Chls approximately in the positions occupied by the Phes in PSII (termed A_0), then two phylloquinones (termed A_1 or Q_k). The iron atom that is a feature of the quinone region of PSII and the pbRC is not present in PSI, but instead a Fe_4S_4 cluster is located on the symmetry axis in a position suitable for accepting electrons from one or both phylloquinones. This is the first of three such clusters that form an electron transfer chain that extends away from the membrane on the stromal side (Figure 9.1c). The central charge-separating electron transfer cofactors are held in place by a pseudosymmetrical heterodimer formed by the C-terminal regions of two polypeptides, PsaA and PsaB. Each of these regions has the five membrane-spanning α-helices that are characteristic of the L/M and D1/D2 heterodimers found in the pbRC and PSII.

One point that is apparent from the comparison in Figure 9.1 is that the very faithful reproduction of the positions and orientations of the eight symmetrically arranged cofactors between the pbRC and PSII does not extend to PSI. The view of each complex in Figure 9.1 is perpendicular to the symmetry axis and maximizes the separation of the two quinones in each structure to emphasize the two membrane-spanning cofactor branches. In PSII and the pbRC, this results in an edge-on orientation for the accessory (B)Chls and a largely face-on orientation for the P_{D1}/P_{D2} and P_A/P_B (B)Chls (Figure 9.1a and b). In the case of PSI, the same view preserves the orientation of the quinones and A_0 Chls, but the accessory Chls (A_{acc}) are changed from edge-on to face-on and are now roughly parallel to the A_0 Chls (Figure 9.1c). The pair of P_A/P_B Chls in PSI are rotated around the vertical axis from right to left such that they are largely edge-on. This indicates that the highly efficient photochemical charge separation achieved by RCs does not require a very precise and immutable arrangement of (bacterio)chlorin cofactors but rather can be carried out within a variety of structures.

As with PSII, most of the Chl cofactors in PSI play a light-harvesting role. The majority are held in place by N-terminal six-helix domains of PsaA and PsaB, and these are structurally equivalent to the CP43 and CP47 proteins found in PSII. The remaining Chls are held in place by additional polypeptides, most of which have either one or two membrane-spanning a-helices. The PsaA/PsaB heterodimer, together with multiple minor polypeptides, are associated with large numbers of additional light-harvesting complexes, and structural information at increasingly high resolutions has been obtained on large protein assemblies from pea chloroplasts in which a monomeric PSI is associated with four light-harvesting complexes (Ben-Shem et al. 2003; Amunts et al. 2007, 2010; Qin et al. 2015; Mazor et al. 2015, 2017).

9.3 MECHANISMS OF PHOTOCHEMICAL CHARGE SEPARATION

9.3.1 General principles

As with appreciation of the molecular structures, understanding of the general mechanism of charge separation was first developed in the pbRC, with ultrafast absorbance difference spectroscopy playing a crucial role. In fact, a plethora of spectroscopic and computational techniques have been applied to RCs in order to try to understand the electronic structures of the different arrangements of cofactors, the roles individual components play during charge separation, and both static and dynamic aspects of cofactor–protein–solvent interactions (see Hoff and Deisenhofer 1997 for a comprehensive account of the application of spectroscopic techniques to the structure and mechanism of the pbRC).

In the simplest terms, formation of the first singlet excited state of a particular (B)Chl species within the central electron transfer chain of an RC induces the transfer of an electron from this primary electron donor to a neighboring acceptor, which is also a (B)Chl molecule. This reaction occurs within a few picoseconds of formation of the primary donor excited state and is followed by an equally rapid second step of electron transfer that moves the cation and anion of the radical pair further apart before they can recombine. Subsequent steps of electron transfer on the donor (cation) and acceptor (anion) sides of the RC are successively slower as the increasingly stabilized radical pair evolves. As outlined earlier, the first two steps of charge separation have to be very rapid in order to compete with detrapping of excitation energy back into the surrounding antenna, and to achieve this, nature utilizes various arrangements of three (bacterio)chlorin species that form a very closely spaced chain. The close spacing ensures the extremely rapid electron transfer that is required, and unpicking the sequence of events among these (bacterio)chlorins has proven to be a major technical challenge, particularly in RCs such as PSI where the excess of antenna Chls cannot be physically separated from the small number of Chls that carry out charge separation at the heart of the macromolecule.

9.3.2 Charge separation in the pbRC

Charge separation in the pbRC is now well understood in terms of "what happens" (see Woodbury and Allen 1995; Hoff and Deisenhofer 1997; Allen and Williams 1998; Zinth and Wachtveitl 2005; Jones 2009; Heathcote and Jones 2012 for reviews), but many aspects of "how it happens" are the subject of ongoing investigation.

The primary electron donor in the *Rba. sphaeroides* RC is the P_A/P_B BChl special pair. As outlined earlier, a consequence of the close interaction between the two halves of the special pair is a shifting of their lowest energy Q_y absorbance band to a significantly lower energy than for the monomeric BChls of the RC, such that it appears as a discrete feature in the absorbance spectrum.

Figure 9.2 shows the absorbance spectrum of the *Rba. sphaeroides* RC; the Q_y absorbance band attributable to the two monomeric BChls is at a higher energy (≈ 800 nm) than that of the P870 dimer, with the band attributable to the two BPhes at a higher energy still (≈ 760 nm). Such an arrangement creates a relatively steep gradient for excitation energy transfer among the RC bacteriochlorins, where energy acquired directly by the BPhes or accessory BChls would be expected to be passed in an energetically favorable fashion to the P870 dimer. Spacing of the absorbance bands of the special pair, accessory BChls and BPhes in this way greatly helps interpretation of spectral changes that accompany charge separation.

Many years of analysis of the *Rba. sphaeroides* RC, employing a large number of time-resolved techniques and informed by insights from x-ray crystallography, steady-state spectroscopy, and computation, have revealed the sequence of electron transfer steps during charge separation (Woodbury and Allen 1995; Zinth and Wachtveitl 2005; Jones 2009; Heathcote and Jones 2012). P870*, the singlet excited state of P870, has a lifetime of 3–5 ps and decays principally through donation of an electron to the neighboring monomeric BChl B_A, forming a $P870^+B_A^-$ radical pair. This pair has an even shorter lifetime of around 1 ps, the electron being passed to the adjacent BPhe to form the pair $P870^+H_A^-$. The rapidity of this second reaction is such that the $P870^+B_A^-$ state does not accumulate at high levels, making it difficult to detect, and for

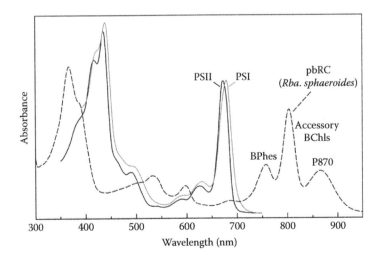

Figure 9.2 Absorbance spectra of the *Rba. sphaeroides* RC and monomeric PSI and PSII complexes from *Synechocystis* PCC6803.

many years the formation of a discrete $P870^+B_A^-$ intermediate was an open question. However, it is now generally accepted that B_A^- does form as a genuine intermediate (Zinth and Wachtveitl 2005). Membrane-spanning charge separation is completed by transfer of the electron from H_A to the Q_A ubiquinone with a lifetime around 200 ps, forming a $P870^+Q_A^-$ radical pair.

One of the notable features of this charge separation is the effectively exclusive use of the so-called A-branch of cofactors (B_A, H_A, and Q_A) to transfer the electron across the membrane (Kellogg et al. 1989; Heller et al. 1995), despite the presence of a second potential route involving the symmetrically located cofactors. The cause of this functional asymmetry appears to be multifaceted, with no single structural feature of the protein/cofactor matrix dictating exclusive use of the A-branch, and attempts are ongoing to activate electron transfer along the inactive B-branch to get a better understanding of this (Kressel et al. 2014; Faries et al. 2016). The reason for this asymmetry becomes apparent when the properties of the RC quinones are considered, as although the locations of the two quinone redox-active headgroups relative to the symmetry axis are rather similar, the roles they play are very different (Okamura et al. 2000; Wraight 2004; Wraight and Gunner 2009). The function of Q_A is to undergo a one-electron cycle in which the electron delivered from H_A is passed on to Q_B, and to achieve this, the Q_A headgroup is permanently associated with its intraprotein binding site. In contrast, the role of Q_B is to collect two electrons from Q_A, to acquire two protons from the cytoplasm, and to deliver these electrons and protons in the form of dihydroubiquinone (ubiquinol) to a site of quinol oxidation in the cytochrome bc_1 complex. To achieve this, the Q_B site is more open to allow binding and release of oxidized and reduced quinones, with protein structural features that facilitate delivery of protons to the quinone headgroup as it undergoes reduction. There is, therefore, specialization in the two cofactor branches, with the A-branch optimized for picosecond membrane-spanning charge separation to form $P870^+Q_A^-$, and the quinone of the B-branch optimized for the slower charge accumulation and protonation required for the formation of ubiquinol.

One point for consideration is why the B-branch BChl and BPhe have been retained if only the A-branch is used for charge separation

(see the discussion in Watson et al. 2005). The B_B BChl is known to play a role in photoprotection in response to the formation of triplet states that can sensitize singlet oxygen (Cogdell et al. 2000), passing triplet energy to a carotenoid in the RC (not shown in Figure 9.1a). It also mediates singlet state transfer from the same carotenoid to P870 (Lin et al. 2003). The reason the H_B cofactor is present and is a specialist BPhe rather than BChl is less clear, and it can be excluded from the structure through mutation of its binding pocket without any obvious consequences (Watson et al. 2005). Possibly, it has been retained as BPhe so that any electrons from $P870^*$ that stray down the B-branch reach the Q_B quinone with the same efficiency as the majority A-branch population.

9.3.3 Other fast energy and electron transfer events in the pbRC

The description given outlines what happens when excited state energy enters the RC via the P870 BChls, but in principle, the two accessory BChls and two BPhes can also directly absorb light energy and so act as internal light-harvesting pigments. Initially, it was assumed that formation of the singlet excited state of B_A, H_A, B_B, or H_B would simply trigger ultrafast energy transfer to the lower energy $P870^*$ state, and efforts were made to measure the rates of these reactions and distinguish processes on the two cofactor branches (Stanley et al. 1996; Arnett et al. 1999; King et al. 2000). However, it subsequently became apparent that direct formation of B_A^* also initiates charge separation reactions not involving formation of $P870^*$ but rather proceeding directly through initial $P870^+B_A^-$ or $B_A^+H_A^-$ radical pairs, both of which rapidly decay through electron or hole transfer, respectively, to form $P870^+H_A^-$ (van Brederode et al. 1997a,b, 1999a,b; van Brederode and van Grondelle 1999). Such "dimer-independent" mechanisms for charge separation are unlikely to be of physiological relevance for the pbRC, as energy transfer from the bacterial antenna to the RC overwhelmingly involves direct formation of $P870^*$ (see Section 9.3.4), but at the time these events were identified, it was pointed out that an equivalent mechanism could be relevant to RCs that do not contain a discrete "special pair," such as PSII (van Brederode et al. 1999a,b).

9.3.4 Charge separation in PSII

Delivery of excitation energy to the PSII RC oxidizes a Chl species absorbing at 680 nm, and so the primary electron donor in PSII is generally referred to as P680 (Barber and Archer 2001). In overall terms, the mechanism of subsequent membrane-spanning charge separation is equivalent to that in the pbRC, an electron migrating to the Q_B quinone via a single route (A-branch) formed by Phe_{D1} and Q_A (see Nelson and Yocum 2006; Kern and Renger 2007; Rappaport and Diner, 2008; Renger and Renger 2008; Renger 2010; Renger and Schlodder 2010,2011; Cardona et al. 2012; Müh et al. 2012; Mamedov et al. 2015; Nelson and Junge 2015; Barber 2016 for reviews). The two RCs differ strongly with regard to how the resulting primary donor cation is reduced, however. In the pbRC, the P870⁺ pair is reduced by a *c*-type cytochrome, whereas in PSII, the P680⁺ Chl species is reduced by a tyrosine side chain (Tyr_Z) that in turn receives an electron from the Mn_4CaO_5 water-oxidizing center (Figure 9.1b).

Despite similarities in the general arrangement of (B)Chl cofactors, experimental investigation has established that the PSII RC does not have a strongly coupled Chl pair that is directly equivalent to the P870 species in the pbRC (see Durrant et al. 1995; Dekker and van Grondelle 2000 for discussions of the relevant literature). Prior to determination of the x-ray crystal structure of the PSII RC, a general view developed that the precise arrangement of the two structurally equivalent Chls must be subtly different such that the excitonic coupling between the two is weaker than is the case in the pbRC. The subtle difference subsequently uncovered through x-ray crystallography can be seen by comparing Figure 9.3b with Figure 9.3a; both panels show approximately the same view of the (B)Chl on the left. In the case of P870, there is almost perfect overlap of one of the four pyrrole rings that make up each BChl tetrapyrrole, the two macrocycles being in atomic contact in the overlap region. In contrast, in PSII, there is a rotation of the two structurally equivalent Chls in opposing directions such that, although they are still in atomic contact, the nature of their overlap region is significantly altered. This rotation and altered overlap significantly reduces the strength of the electronic coupling between the two Chls (see Renger and Schlodder 2010, 2011; Cardona et al. 2012 for detailed discussions of this).

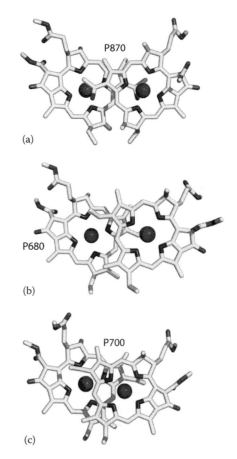

(a)

(b)

(c)

Figure 9.3 Special pairs and not-so-special pairs, aligned to the (bacterio)chlorin on the left. Representation is as for Figure 9.1. **(a)** The P870 special pair BChls in the *Rba. sphaeroides* RC (PDB file 3ZUW—Gibasiewicz et al. 2011). **(b)** The structurally—equivalent Chls in the cyanobacterial PSII (PDB entry 3ARC—Umena et al. 2011); P680 is localized on the Chl on the left. **(c)** The P700 pair of Chls in the cyanobacterial PSI (PDB file 1JB0—Jordan et al. 2001).

Calculations predicting that the coupling between adjacent cofactors is of a similar strength among the four Chls and two Phes of PSII (relatively weak at <120 cm⁻¹) led to the proposal of a multimer model for the excited state of P680, which envisaged exciton transitions that are delocalized over several weakly coupled cofactors (Durrant et al. 1995). A number of variations and elaborations of this multimer model have been proposed (see Raszewski et al. 2008 and the discussion therein).

This lack of a discrete, strongly excitonically coupled Chl dimer in PSII is evident from its absorbance spectrum, which is dominated by a single band with a maximum at ≈ 675 nm attributable to the Q_y transitions of all of the Chl and Phe cofactors (Figure 9.2). This spectrum is very different in character from that of the pbRC, where the absorbances of the BPhes, monomeric BChls, and dimeric BChls are well separated. The inference from the spectrum of PSII is that the steep energy gradient that focuses excitation energy on the P870 dimer in the pbRC is absent in PSII, complicating identification of the Chl that carries the excited state that triggers charge separation. If the energies of the Chls and Phes in the PSII RC are all approximately equal, with similar strength couplings between neighbors, then within a population of RCs, the excitation energy arriving from the antenna would be expected to be distributed among the six cofactors rather than being uniformly focused on a single species. This raises the possibility that charge separation could be initiated from a number of different near-isoenergetic excited states, with parallel pathways of electron and hole transfer being followed during the first few picoseconds after excitation before a common $P680^+Phe_{D1}^-$ product is formed at longer times (Dekker and van Grondelle 2000). A consequence would be that, unlike in the pbRC where the $P870^*$ excited pair gives rise directly to the $P870^+$ cation pair, in PSII it could be the case that the excited Chl that triggers charge separation is not the same molecule as the Chl cation that initiates electron donation from the water-oxidizing center. As outlined previously, a precedent for this has been seen in the pbRC, where B_A^* initiates charge separation reactions that produce the metastable pair $P870^+H_A^-$, the hole rapidly migrating from B_A^+ to P870 (van Brederode et al. 1997a,b, 1999a,b; van Brederode and van Grondelle 1999).

In tandem with emerging information on the structure of PSII, a variety of approaches have been used to characterize the first steps in charge separation. Two research groups employing ultrafast absorbance difference spectroscopy have produced experimental data on PSII RCs at room temperature that support Phe_{D1} as the initial electron acceptor, with the implication that initial charge separation forms $Chl_{D1}^+Phe_{D1}^-$ with subsequent hole migration to the adjacent P_{D1} Chl (Groot et al. 2005; Holzwarth et al. 2006a). Both studies postulated a faster first charge separation step followed by a slower (<20 ps) second step, though there was no consensus on precise lifetimes or, in particular,

whether the formation of $Chl_{D1}^+Phe_{D1}^-$ is a subpicosecond process. One of these studies employed ultrafast visible pump/midinfrared probe absorbance difference spectroscopy (Groot et al. 2005), and its findings were supported by subsequent studies using complexes with site-specific mutations (Di Donato et al. 2008).

Following this, evidence for multiple parallel pathways for charge separation operating at low temperature was obtained, and the possibility of these also being present at room temperature has been discussed (Novoderezhkin et al. 2007; Romero et al. 2010). In this case, electron transfer was proposed to be initiated from either $(Chl_{D1}Phe_{D1})^*$ or $(P_{D1}P_{D2})^*$ exciton states, where the excited state is not localized on a single cofactor, and the initial radical pair was concluded to be $Chl_{D1}^+Phe_{D1}^-$ or $P_{D1}^+P_{D2}^-$, respectively (Novoderezhkin et al. 2007; Romero et al. 2010). Stark spectroscopy has indicated that both of these exciton states have considerable charge transfer character, that is, $(Chl_{D1}^{\delta+}Phe_{D1}^{\delta-})^*$ and $(P_{D1}^{\delta+}P_{D2}^{\delta-})^*$ (Romero et al. 2012, 2017).

This picture of charge separation in PSII is not universally accepted, an alternative model having been put forward that has initial charge separation to form $P680^+Chl_{D1}^-$, with reduction of Phe_{D1} as the second step (Shelaev et al. 2008, 2011) (reviewed in Semenov et al. 2011; Nadtochenko et al. 2014; Mamedov et al. 2015).

It is probably fair to say that a universally agreed view of the mechanism of initial charge separation in the PSII RC is yet to emerge, particularly with regard to the precise identity of the excited state that triggers the process, although the majority of studies seem to favor $Chl_{D1}^+Phe_{D1}^-$ being the initial radical pair. Where there is agreement on a sequence of states, as yet there is also a lack of consensus on the lifetimes of the first and second steps of charge separation. Given that the PSII RC is characterized by multiple excited and radical pair states that are very close to one another in energy, it may well prove that there are indeed multiple parallel mechanisms that can occur within this particular arrangement of Chls and Phes, with different relative contributions in RCs from different sources or under different conditions.

One point that does seem to be agreed upon is that the P680 cation is localized on the P_{D1} Chl, which is located closest to Tyr_Z and the water-oxidizing center (Diner et al. 2001).

9.3.5 Charge separation in PSI

The primary electron donor in the PSI RC is a Chl a/Chl a′ pair denoted P700 that occupies a position that is structurally equivalent to the P870 BChls in the pbRC (Käss et al. 2001; Webber and Lubitz 2001; Poluektov et al. 2002). The strength of the excitonic coupling between the two halves of P700 has been a matter of debate, and it is likely to be intermediate between the strong coupling seen for P870 and the weak coupling seen for the P_{D1} and P_{D2} BChls in PSII (Webber and Lubitz 2001). The x-ray crystal structure of this pair is compared with the equivalent (B)Chls in PSII and pbRC in Figure 9.3. The structure of the overlap region that gives rise to P700 is different again from those that give rise to P870 and P680, the two Chls having undergone a translation in opposing directions relative to the positions of the equivalent Chls in PSII. Consistent with an intermediate coupling, the overlap between the P700 Chls is more extensive than is the case for the two equivalent Chls in PSII, but the pyrrole rings are less perfectly aligned than is the case for the P_A and P_B BChls in the pbRC.

As with PSII, in PSI the absorbance properties of the P700 primary electron donor are not substantially redshifted from those of the remaining Chls in the RC and surrounding antenna. In the absorbance spectrum of monomeric PSI in Figure 9.2, the P700 Chls represent only two of the 96 Chls in each PSI monomer, and the spectrum illustrates the strong overlap between the absorbance of the bulk of antenna Chls and that of the P700 Chls at the core of the complex.

On an ultrafast timescale, photoexcitation of PSI produces the radical pair $P700^+A_0^-$, where A_0 is a monomeric Chl located approximately halfway across the membrane, in a position structurally equivalent to the acceptor Phe in PSII and the acceptor BPhe in the pbRC (see Brettel and Leibl 2001; Chitnis 2001; Renger 2010; Semenov et al. 2011; Kargul et al. 2012; Caffarri et al. 2014; Mamedov et al. 2015 for reviews of charge separation in PSI). Intermediate between the P700 pair and each Chl that is a candidate for A_0 is an additional monomeric Chl that is structurally equivalent to Chl_{D1}/Chl_{D2} found in PSII and the accessory B_A/B_B BChls in the pbRC. The positions of these monomeric Chls suggest strongly a role in charge separation between P700 and the A_0 acceptor. The initially formed $P700^+A_0^-$ radical pair decays in around 30 ps by electron transfer to the A_1 phylloquinone, and this in turn passes the electron on to the F_X iron–sulfur center on a nanosecond timescale. Charge separation is further stabilized by transfer of the electron through the F_A and F_B iron–sulfur centers to a water-soluble ferredoxin, while $P700^+$ is reduced by a water-soluble cytochrome or plastocyanin redox protein on the opposite side of the membrane.

It is probably fair to say that, as with PSII, thinking about the mechanism of charge separation in PSI was strongly influenced by the relatively detailed structural and mechanistic information on the pbRC that was available for many years. The strength of the paradigm of strongly asymmetric charge separation within a pseudosymmetric arrangement of RC cofactors was such that there was a widespread assumption that the same basic feature found in the pbRC and PSII would also be found in PSI. It was something of a surprise therefore when spectroscopic data began to emerge that suggested that charge separation proceeds along both cofactor branches in PSI.

The first hints that charge separation in PSI may be "bidirectional" came from the finding that the kinetics of reoxidation of the A_1 anion are biexponential, with lifetimes of 18–25 ns and 150–160 ns (Mathis and Setif 1988; Setif and Brettel 1993; Joliot and Joliot 1999). As outlined earlier, the A_1 acceptor is identified with the phylloquinones located between the Chls that conduct the initial charge separation and the F_X iron–sulfur center that sits on the symmetry axis of the complex. Initially, this biexponential decay was attributed to sample heterogeneity, or a low equilibrium constant between the A_1 donor and F_X acceptor (Mathis and Setif 1988; Setif and Brettel 1993), but subsequently it was attributed to parallel reoxidation of both phylloquinones by F_X at different rates within the PSI population (Joliot and Joliot 1999). This idea that both branches of cofactors in PSI could be facilitating charge separation at somewhat different rates was tested in Chlamydomonas reinhardtii by either single or double mutations of two tryptophan residues, each of which engages in a π-stacking interaction with one of the phylloquinones (Guergova-Kuras et al. 2001). The outcome was attribution of the fast and slow phases of reoxidation of the A_1 acceptor to the Q_k-B and Q_k-A phylloquinone, respectively, and further support for this concept of both branches of cofactors in PSI being capable of conducting charge separation was provided through EPR spectroscopy of A_1^- oxidation in mutant PSI RCs (Muhiuddin et al. 2001; Purton et al. 2001; Fairclough et al. 2003). There have since been multiple reports confirming electron transfer along both branches (e.g., see Ramesh et al. 2004; Bautista et al. 2005; Poluektov et al. 2005; Santabarbara et al. 2005, 2006; Ali et al. 2006; Li et al. 2006; Giera et al. 2009, 2010).

With the benefit of hindsight, it seems entirely rational that either chain of cofactors should be capable of membrane-spanning charge separation in PSI. As outlined earlier, in PSII and the pbRC, there is specialization in the roles of the two quinones, and it would appear that the Q_A quinone on the A-branch of cofactors has evolved for the facilitation of highly efficient charge separation, while the Q_B binding site on the B-branch has evolved for a controlled double reduction and double protonation of a mobile quinone. However, in PSI the function of the A_1 acceptor is to pass a single electron on to the F_X iron–sulfur center (Moenne-Loccoz et al., 1994), a role that could potentially be played by either phylloquinone. As a result, there is no need for exclusive use of only one cofactor branch for charge separation in PSI. An open question, however, is why electron transfer seems to proceed from P700 to F_X at somewhat different rates along the two available branches, and why one branch (A) is favored over the other (see Kargul et al. 2012 for a review).

Turning to the mechanism of the initial charge separation, examination of this by transient absorption spectroscopy is complicated in PSI by the presence of a large pool of antenna Chls that cannot be removed by protein purification. This presents the challenge of distinguishing charge separation among the Chl cofactors of the electron transfer chains from energy transfer events within the antenna regions of PSI that include transfer of excitation energy into the electron transfer chain (Müller et al. 2003; Di Donato et al. 2011). It was established that the $P700^+A_1^-$ state appears in around 100 ps after excitation of PSI (Holzwarth et al. 2006b), but consideration of the structure of the electron transfer chain and analogies with other RCs indicates that one, or more probably two, electron transfer events precede formation of $P700^+A_1^-$.

Two research groups have presented evidence that the initial charge separation in PSI takes place on both branches, between the A_{acc} accessory Chl and the adjacent A_0 Chl, the second step being hole transfer to the P700 Chl pair. This evidence was based on ultrafast studies of PSI preparations bearing site-specific mutations (Holzwarth et al. 2006b; Müller et al. 2010) or on ultrafast visible pump/midinfrared probe spectroscopy (Di Donato et al. 2011). This represents a direct parallel with one of the proposed mechanisms for charge separation in PSII, and the B_A^*-driven mechanism observable in the *Rba. sphaeroides* RC (see Section 9.3.3). Both studies indicated a very fast (subpicosecond) first step of charge separation with a slower (6–20 ps) hole transfer to P700.

This view of charge separation in PSI is not universally accepted, with data from ultrafast pump-probe spectroscopy being interpreted as showing initial charge separation between P700 as donor and the accessory Chl as acceptor, reduction of the A_0 Chl occurring only as a second step (Semenov et al. 2011). In addition, it has been proposed that a delocalized excited state covering both P700 and A_0 could lead directly, within 100 fs, to a $P700^+A_0^-$ state (Shelaev et al. 2010, Semenov et al. 2012; Mamedov et al. 2015).

9.4 VARIATIONS ON A THEME

9.4.1 The atypical paradigm

As outlined earlier, the picture of charge separation in the pbRC that emerged in the 1980s and 1990s had a strong influence on unraveling of the mechanisms of this process in PSI and PSII. Much of the research on these latter systems addressed similar themes, including the identity and electronic structure of the excited and cation forms of the primary donor species and the roles of monomeric Chls and Phes during the first electron transfer steps. Although there are still points of contention, the most widely accepted view of charge separation in PSI and PSII is that it is initiated by photooxidation of one (or both) accessory Chl(s) to form the radical pair $A_{acc}^+A_0^-$ or $Chl_{D1}^+Phe_{D1}^-$, respectively, with transfer of the cation to P700 or P680, respectively, as the second step. The pbRC is also capable of separating charge by the same mechanism if the accessory B_A BChl is excited directly, with $B_A^+H_A^-$ forming before oxidation of P870, but this is not the dominant process in native RCs because this B_A^*-driven charge separation takes place in parallel with favorable subpicosecond energy transfer to the longer-wavelength P870 special pair.

This general picture of charge separation being initiated among monomeric (B)Chl species in all RCs, with the special pair in the pbRC providing an exception to this theme, is somewhat at odds with the textbook descriptions of photochemical charge separation. The accounts of the pbRC that appear in most general Biochemistry textbooks (or at least the half dozen the author checked) focus exclusively on charge separation that is initiated by P870* (or more commonly P960* in the *Blc. viridis* RC), and descriptions of charge separation in PSII and PSI are couched in similar terms. The general line taken is that P680* initiates charge separation to form P680⁺Phe⁻ in PSII, and P700* initiates charge separation to form P700⁺A₀⁻ in PSI, with some text books referring in words and/or pictures to both P680 and P700 as Chl pairs, or even "special pairs." However, much of the experimental data accumulated on these RCs points at this description being somewhat misleading, and it is becoming apparent that the charge separation that has been characterized in such detail in the *Rba. sphaeroides* RC does not represent a mechanistic blueprint for the whole of photosynthesis, but rather represents a particular development of a more general process that does not require the involvement of a pair of (B)Chls, whether special or not so special.

9.4.2 Source of the atypical paradigm: The needs of the primary energy acceptor

If a special pair of BChls is not necessarily needed to initiate charge separation, then why is it present in the pbRC? The answer can be found in considering the second role that P870 plays, as the acceptor of excited state energy flowing in from the surrounding antenna. In all characterized purple bacteria, the RC is surrounded in the membrane by an approximately cylindrical LH1 antenna protein that contains a ring of excitonically coupled BChls. The dominant, lowest-energy absorbance transition of these LH1 BChls is around 875 nm in *Rba. sphaeroides* (Figure 9.4), and their first singlet excited state is the immediate donor of energy to the BChls of the RC. As individual monomeric BChl *a* molecules in a protein environment absorb at around 800 nm to avoid the final step in energy transfer being energetically unfavorable, the absorbance of the receiving BChl species in the RC is strongly redshifted to 870 nm by arranging the two BChls closest to the symmetry axis as a partially overlapping, optimally aligned, excitonically coupled pair (Figure 9.4). This greatly increases the Förster overlap between the emission of the LH1 BChl energy donors and the absorbance of the RC BChl energy acceptors, facilitating effective energy transfer between the two. As the special pair of BChls has the lowest-energy excited state in the RC system, it is logical that charge separation should be initiated when P870* is formed. Figure 9.4 compares the absorbance spectrum of the *Rba. sphaeroides* RC with that of the larger complex it forms with the LH1 antenna protein; the absorbance band of the RC special pair is redshifted to underlie the strong absorbance band of the light-harvesting BChls.

An additional point to note is that there is no *a priori* reason why charge separation has to be

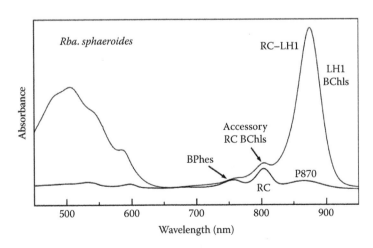

Figure 9.4 Absorbance spectra of RCs and RC–LH1 complexes from *Rba. sphaeroides*. In the latter, the absorbance band of the multiple LH1 BChls obscures that of the pair of P870 BChls in the RC.

initiated at one side of the membrane in a RC, and indeed in PSI and PSII, the initial charge separation is followed by the electron and hole migrating in opposite directions. However, selecting the two BChls in the pbRC closest to the periplasmic side of the membrane to bring together to form the energy-receiving special pair rather than symmetrical cofactors deeper into the membrane brings about additional benefits. First, because BPhe absorbs at higher energy than BChl, having the special pair at one end of the electron transfer chain sets up a simple energy gradient where any energy acquired directly by the BPhe cofactors can pass to the adjacent accessory BChl and on to the special pair. In addition, once electron transfer is initiated at the special pair, the cofactors carrying the electron across the membrane are well separated from their symmetrical counterparts in the "inactive branch," preventing unwanted cross-branch electron transfer reactions that could adversely affect the quantum yield of the system.

As is evident in Figure 9.2, in both PSII and PSI, the antenna and RC Chls absorb in the same wavelength region, the lowest-energy transition giving rise to a single absorbance band between 650 and 700 nm. Given this, the need to have a strongly red-shifted Chl species to act as an acceptor for energy flowing in from the antenna is not present in PSII or PSI. As there is no need for a Chl special pair to satisfy requirements for energy transfer, the degree of structural overlap and spacing between the two Chls structurally equivalent to the special pair BChls in the pbRC is free to be dictated by other factors, such as the required oxidizing potential of the P680 or P700 cation, the strength of excitonic coupling with other cofactors that will affect energy migration and localization among the RC chlorins and electronic couplings that will affect electron transfer.

9.4.3 Extreme high-potential photochemistry: PSII

The major challenge for charge separation in PSII is to generate a cation with sufficient oxidizing power to induce electron transfer from a Tyr_Z, forming a deprotonated tyrosyl radical, which is the causative agent of water oxidation in the Mn_4CaO_5 cluster. This P680$^+$ species is believed to be localized on the P_{D1} Chl and has a potential of over 1.2 V compared to $\approx$0.45–0.5 V for P870$^+$ in the structurally similar *Rba sphaeroides* RC. A factor relevant to this $\approx$0.7 V

difference is the use of Chl rather than BChl as the photoactive pigment, but this actually plays a relatively minor role as there is only a $\approx$160 mV difference in the potentials of BChl a/BChl a^+ and Chl a/Chl a^+ in organic solvent (Ishikita et al. 2006). In PSI, which also employs Chl, the oxidation potential of P700 is similar to that of P870 at around 0.45–0.5 V, and very different to that of P680 in PSII.

Although the use of Chl makes some contribution, it is thought that the very high oxidizing potential of P680 arises through a combination of localizing the cation on a single Chl rather than a Chl pair and electrostatic interactions between the single P680$^+$ Chl and the Mn_4CaO_5 cluster, dipoles in the protein backbone, and other surrounding structural features (Ishikita et al. 2006). These electrostatic interactions, most of which are not replicated in the pbRC, make the main contribution to pushing up the oxidation potential of P680 by around 500 mV. An additional point to note is that it is not only the P_{D1} Chl in PSII that has a very high oxidation potential because in order to localize the cation on this Chl, it is necessary for the remaining five Chl and Phe cofactors to be even more oxidizing than P680. As a result, it may be that some local interactions actually selectively reduce the redox potential of P_{D1}^+ relative to the adjacent chlorins to ensure the cation localizes on the Chl closest to Tyr_Z (Ishikita et al. 2006).

9.4.4 Extreme low-potential photochemistry: PSI

PSI operates at the opposite end of the biological redox scale to PSII, using the same molecule—chlorophyll—to generate electrons with a low enough potential (below about −0.45 V) to reduce first ferredoxin and then NADP$^+$. In PSI, the P700$^+$ species is associated with the P_A and P_B Chls, though these do not form a strongly excitonically coupled pair as in the pbRC, and its oxidation potential of $\approx$0.5 V is much more modest than the very high potential of P680$^+$. To achieve this, it is thought that electrostatic interactions with the protein environment have evolved to lower the oxidation potential of P700$^+$ (Ishikita et al. 2006), placing it in a region suitable for electron acceptance from plastocyanin or a c-type cytochrome.

Special adaptation in PSI concerns mainly the carriers of the translocating electron, the potentials of the A_0 Chl(s) and A_1 phylloquinone(s) being estimated at −1.05 and −0.82 V, respectively (see Brettel and Leibl 2001; Fromme et al. 2001; Heathcote and Jones 2012 for reviews). In the case of the phylloquinones, factors contributing to a very low reduction potential may include asymmetric hydrogen bonding to the protein, π–π stacking with a nearby tryptophan, and a relatively hydrophobic binding pocket buried deep within the membrane interior (Jordan et al. 2001). In the case of A_0, use of Chl rather than Phe or BPhe at this position in the electron transfer chain results in a much lower operating potential at this stage in charge separation, and other structural features are likely to lower this potential further. These include the use of a methionine side chain to provide the axial ligand to the pentacoordinated Mg^{2+} of the A_0 macrocycle (Fromme et al. 2001; Jordan et al. 2001). In the case of the preceding A Chl cofactor, which presumably has an even lower redox potential, this fifth ligand is provided by a water molecule that is also hydrogen-bonded to an asparagine side chain (Fromme et al. 2001; Jordan et al. 2001).

9.4.5 What about P865, P840, P798, and the other P840?

Thus far, this chapter has focused on the three most heavily studied RCs and has overlooked other members of this protein family from the green sulfur bacteria, heliobacteria, green acidobacteria, and green filamentous bacteria. The last of these has a single RC that is similar to the BChl-containing L/M heterodimer found in purple bacteria but lacks an equivalent of the H-polypeptide (Pierson and Thornber 1983; Pierson et al. 1983; Shiozawa et al. 1987). The other structural deviation is the presence of BPhe a rather than BChl a at the accessory position (B_B) on the inactive cofactor branch (Bruce et al. 1982; Pierson and Thornber 1983), but the significance of this is not clear. The mechanism of charge separation from the P865 primary electron donor BChl a special pair found in this RC is assumed to be similar to that in purple bacteria.

The green sulfur bacteria (*Chlorobi*) and heliobacteria (*Firmicutes*) both have an RC similar to PSI, but which intriguingly is homodimeric rather than heterodimeric (Büttner et al. 1992a,b; Liebl et al. 1993; Permentier et al. 2000; Hauska et al. 2001; Neerken and Amesz 2001). This feature has added much fuel to ongoing, entertaining debates about the evolutionary relationships between the extant RCs and their origins (see Hohmann-Marriott and Blankenship 2011 for a review). Not a great deal

is known about the details of charge separation in these RCs, which are particularly sensitive to oxygen damage, but it is likely to follow a similar mechanism to PSI. In green sulfur bacteria, the primary electron donor is usually called P840, is formed from a pair of BChl a' (the C-13^2 epimer of BChl a), and has an oxidation potential of 0.23 V, whereas in heliobacteria, it is termed P798, is formed from two BChl g epimers (BChl g'), and has an oxidation potential of about 0.24 V (see Ohashi et al. 2010 for a detailed review). Consistent with being embedded in a homodimeric protein scaffold, both P798 and P840 show an equal distribution of the cation between the two constituent bacteriochlorins. Both types of homodimeric RC bind far fewer pigment molecules than is the case for the related PSI RC from oxygenic phototrophs, specifically 14 BChl a, two BChl a', four Chl a derivatives (Chl a esterified with Δ2,6-phytadienol) and two carotenoids per RC in green sulfur bacteria, and 35–40 BChl g, two BChl g', and two Chl a derivatives (8^1-OH-Chl a esterified with farnesol) in heliobacteria (Ohashi et al. 2010). In both RCs, the Chl a derivatives are candidates for the electron acceptor A_0 (Ohashi et al. 2010).

This type of homodimeric, PSI-like RC is also present in a phylogenetically distinct bacterium *Chloroacidobacterium thermophilum* (Bryant et al. 2007; Tsukatani et al. 2012). The primary electron donor in this RC is denoted P840. This new RC contains two molecules of BChl a' (the C-13^2 epimer of BChl a) that have zinc as the central metal rather than magnesium (Zn-BChl a'), in addition to an estimated 13 BChl a and 8 Chl a per RC. It has been speculated that these Zn-BChl a' may form the P840 special pair or be involved in charge separation as electron acceptors (Tsukatani et al. 2012). Photosynthetic bacteria that synthesize Zn-BChl either naturally (Wakao et al. 1996) or in response to mutation (Jaschke et al. 2011) have been described previously, but *Chloroacidobacterium thermophilum* is the first recorded instance of a bacterium that incorporates both the Zn- and Mg-containing variants into the RC protein.

9.5 SUMMARY AND OUTLOOK

As, hopefully, explained in this chapter, the general mechanism of charge separation is similar in the three types of photoreaction center that have been characterized in detail to date, but there are variations on the theme that can be explained by considering how the components of the RC interact with the surrounding excess of pigments that harvest sunlight and with chemically varied electron donors and acceptors. The detail in which many aspects of charge separation in PSI and PSII are understood is beginning to approach that in the pbRC, although there are still points of contention.

As to the future, one important aspect of RC function that remains a challenge is to understand the role that protein dynamics play during and immediately after each step in the electron transfer process. Ultrafast laser spectroscopy provides a means of examining the types of nuclear motions that take place during picosecond timescale reactions and are gradually uncovering the nature of these motions and the extent to which they influence, or merely report on, the electron transfer process (Vos et al. 1993, 1994; Wang et al. 2007; Guo et al. 2012; Gibasiewicz et al. 2013; Pan et al. 2013; Zhu et al. 2013). The application of two-dimensional electronic spectroscopy is throwing new light on the roles played by electronic coherence, vibrational coherence, and mixed electronic/vibrational coherence during charge separation in PSII RCs (Myers et al. 2010; Romero et al. 2014) (and see Romero et al. 2017 for a review). There is also evidence of protein conformational changes that take place on slower timescales that seem to influence the overall charge separation process, and these remain poorly understood (Kleinfeld et al. 1984; Graige et al. 1998; Fritzsch et al. 2002; Nagy et al. 2008; Francia et al. 2009; Deshmukh et al. 2011). Despite many years of study at the interfaces between biology, chemistry, and physics, there is still much to learn about the structures, mechanisms, and dynamics of photoreaction centers.

REFERENCES

Ali, K., Santabarbara, S., Heathcote, P., Evans, M. C. W., Purton, S. 2006. Bidirectional electron transfer in photosystem I: Replacement of the symmetry-breaking tryptophan close to the PsaB-bound phylloquinone (A_{1B}) with a glycine residue alters the redox properties of A_{1B} and blocks forward electron transfer at cryogenic temperatures. *Biochim. Biophys. Acta* 1757: 1623–1633.

Allen, J. P., Feher, G., Yeates, T. O., Komiya, H., Rees, D. C. 1987a. Structure of the reaction center from *Rhodobacter sphaeroides* R-26: The cofactors. *Proc. Natl. Acad. Sci. USA* 84: 5730–5734.

Allen, J. P., Feher, G., Yeates, T. O., Komiya, H., Rees, D. C. 1987b. Structure of the reaction center from *Rhodobacter-sphaeroides* R-26: The protein subunits. *Proc. Natl. Acad. Sci. USA* 84: 6162–6166.

Allen, J. P., Feher, G., Yeates, T. O., Komiya, H., Rees, D. C. 1988. Structure of the reaction center from *Rhodobacter sphaeroides* R-26: Protein-cofactor (quinones and iron(II)) interactions. *Proc. Natl. Acad. Sci. USA* 85: 8487–8491.

Allen, J. P., Feher, G., Yeates, T. O., Rees, D. C., Deisenhofer, J., Michel, H., Huber, R. 1986. Structural homology of reaction centers from *Rhodopseudomonas sphaeroides* and *Rhodopseudomonas viridis* as determined by x-ray diffraction. *Proc. Natl. Acad. Sci. USA* 83: 8589–8593.

Allen, J. P., Williams, J. C. 1998. Photosynthetic reaction centers. *FEBS Lett.* 438: 5–9.

Amunts, A., Drory, O., Nelson, N. 2007. The structure of a plant photosystem I supercomplex at 3.4 Å resolution. *Nature* 447: 58–63.

Amunts, A., Nelson, N. 2008. Functional organization of a plant photosystem I: Evolution of a highly efficient photochemical machine. *Plant Physiol. Biochem.* 46: 228–237.

Amunts, A., Nelson, N. 2009. Plant photosystem I design in the light of evolution. *Structure* 17: 637–650.

Amunts, A., Toporik, H., Borovikova, A., Nelson, N. 2010. Structure determination and improved model of plant photosystem I. *J. Biol. Chem.* 285: 3478–3486.

Arnett, D. C., Moser, C. C., Dutton, P. L., Scherer, N. F. 1999. The first events in photosynthesis: Electronic coupling and energy transfer dynamics in the photosynthetic reaction center from *Rhodobacter sphaeroides*. *J. Phys. Chem. B* 103: 2014–2032.

Artz, K., Williams, J. C., Allen, J. P., Lendzian, F., Rautter, J., Lubitz, W. 1997. Relationship between the oxidation potential and electron spin density of the primary electron donor in reaction centers from *Rhodobacter sphaeroides*. *Proc. Natl. Acad. Sci. USA* 94: 13582–13587.

Ayyer, K., Yefanov, O. M., Oberthur, D. et al. 2016. Macromolecular diffractive imaging using imperfect crystals. *Nature* 530: 202–206.

Barber, J. 2009. Photosynthetic energy conversion: Natural and artificial. *Chem. Soc. Rev.* 38: 185–196.

Barber, J. 2016. Photosystem II: The water splitting enzyme of photosynthesis and the origin of oxygen in our atmosphere. *Q. Rev. Biophys.* 49: e14.

Barber, J., Archer, M. D. 2001. P680, the primary electron donor of photosystem II. *J. Photochem. Photobiol. A: Chem.* 142: 97–106.

Bautista, J. A., Rappaport, F., Guergova-Kuras, M. et al. 2005. Biochemical and biophysical characterization of photosystem I from phytoene desaturase and ξ-carotene desaturase deletion mutants of *Synechocystis* sp. PCC 6803. *J. Biol. Chem.* 280: 20030–20041.

Ben-Shem, A., Frolow, F., Nelson, N. 2003. Crystal structure of plant photosystem I. *Nature* 426: 630–635.

Biesiadka, J., Loll, B., Kern, J., Irrgang, K. D., Zouni, A. 2004. Crystal structure of cyanobacterial photosystem II at 3.2 angstrom resolution: A closer look at the Mn-cluster. *Phys. Chem. Chem. Phys.* 6: 4733–4736.

Breton, J. 2007. Steady-state FTIR spectra of the photoreduction of Q_A and Q_B in *Rhodobacter sphaeroides* reaction centers provide evidence against the presence of a proposed transient electron acceptor X between the two quinones. *Biochemistry* 46: 4459–4465.

Brettel, K., Leibl, W. 2001. Electron transfer in photosystem I. *Biochim. Biophys. Acta Bioenerg.* 1507: 100–114.

Broser, M., Gabdulkhakov, A., Kern, J. et al. 2010. Crystal structure of monomeric photosystem II from *Thermosynechococcus elongatus* at 3.6-angstrom resolution. *J. Biol. Chem.* 285: 26255–26262.

Broser, M., Glockner, C., Gabdulkhakov, A. et al. 2011. Structural basis of cyanobacterial photosystem II Inhibition by the herbicide terbutryn. *J. Biol. Chem.* 286: 15964–15972.

Bruce, B. D., Fuller, R. C., Blankenship, R. E. 1982. Primary photochemistry in the facultatively aerobic green photosynthetic bacterium *Chloroflexus aurantiacus*. *Proc. Natl. Acad. Sci. USA* 79: 6532–6536.

Bryant, D. A., Costas, A. M. G., Maresca, J. A. et al. 2007. *Candidatus Chloracidobacterium thermophilum*: An aerobic phototrophic acidobacterium. *Science* 317: 523–526.

Burggraf, F., Koslowski, T. 2011. The simulation of interquinone charge transfer in a bacterial photoreaction center highlights the central role of a hydrogen-bonded non-heme iron complex. *Biochim. Biophys. Acta Bioenerg.* 1807: 53–58.

Busch, A., Hippler, M. 2011. The structure and function of eukaryotic photosystem I. *Biochim. Biophys. Acta Bioenerg.* 1807: 864–877.

Büttner, M., Xie, D. L., Nelson, H., Pinther, W., Hauska, G., Nelson, N. 1992a. Photosynthetic reaction center genes in green sulfur bacteria and in photosystem-1 are related. *Proc. Natl. Acad. Sci. USA* 89: 8135–8139.

Büttner, M., Xie, D. L., Nelson, H., Pinther, W., Hauska, G., Nelson, N. 1992b. The photosystem-I-like P840-reaction center of green s-bacteria is a homodimer. *Biochim. Biophys. Acta* 1101: 154–156.

Caffarri, S., Tibiletti, T., Jennings, R. C., Santabarbara, S. 2014. A comparison between plant photosystem I and photosystem II architecture and functioning. *Curr. Protein Pept. Sci.* 15: 296–331.

Cardona, T., Sedoud, A., Cox, N., Rutherford, A. W. 2012. Charge separation in photosystem II: A comparative and evolutionary overview. *Biochim. Biophys. Acta Bioenerg.* 1817: 26–43.

Chang, C.-H., El-Kabbani, O., Tiede, D., Norris, J., Schiffer, M. 1991. Structure of *Rhodopseudomonas sphaeroides* R-26 reaction center. *Biochemistry* 30: 5352–5360.

Chitnis, P. R. 2001. Photosystem I: Function and physiology. *Annu. Rev. Plant. Physiol. Plant. Mol. Biol.* 52: 593–626.

Cogdell, R. J., Howard, T. D., Bittl, R., Schlodder, E., Geisenheimer, I., Lubitz, W. 2000. How carotenoids protect bacterial photosynthesis. *Philos. Trans. R. Soc. Lond. B* 355: 1345–1349.

Deisenhofer, J., Epp, O., Miki, K., Huber, R., Michel, H. 1984. X-ray structure analysis of a membrane protein complex: Electron-density map at 3 Å resolution and a model of the chromophores of the photosynthetic reaction center from *Rhodopseudomonas viridis*. *J. Mol. Biol.* 180: 385–398.

Deisenhofer, J., Epp, O., Miki, K., Huber, R., Michel, H. 1985. Structure of the protein subunits in the photosynthetic reaction center of *Rhodopseudomonas viridis* at 3 Angstrom resolution. *Nature* 318: 618–624.

Deisenhofer, J., Epp, O., Sinning, I., Michel, H. 1995. Crystallographic refinement at 2.3-angstrom resolution and refined model of the photosynthetic reaction-center from *Rhodopseudomonas viridis*. *J. Mol. Biol.* 246: 429–457.

Dekker, J. P., van Grondelle, R. 2000. Primary charge separation in photosystem II. *Photosynth. Res.* 63: 195–208.

Deshmukh, S., Williams, J. C., Allen, J. P., Kalman, L. 2011. Light-induced conformational changes in photosynthetic reaction centers: Dielectric relaxation in the vicinity of the dimer. *Biochemistry* 50: 340–348.

Di Donato, M., Cohen, R. O., Diner, B. A., Breton, J., van Grondelle, R., Groot, M. L. 2008. Primary charge separation in the photosystem II core from *Synechocystis*: A comparison of femtosecond visible/midinfrared pump-probe spectra of wild-type and two P680 mutants. *Biophys. J.* 94: 4783–4795.

Di Donato, M., Stahl, A. D., van Stokkum, I. H. M., van Grondelle, R., Groot, M. L. 2011. Cofactors involved in light-driven charge separation in photosystem I identified by subpicosecond infrared spectroscopy. *Biochemistry* 50: 480–490.

Diner, B. A., Schlodder, E., Nixon, P. J. et al. 2001. Site-directed mutations at D1-His198 and D2-His197 of photosystem II in *Synechocystis* PCC 6803: Sites of primary charge separation and cation and triplet stabilization. *Biochemistry* 40: 9265–9281.

Durrant, J. R., Klug, D. R., Kwa, S. L. S., van Grondelle, R., Porter, G., Dekker, J. P. 1995. A multimer model for P680, the primary electron donor of photosystem II. *Proc. Natl. Acad. Sci. USA* 92: 4798–4802.

Ermler, U., Fritzsch, G., Buchanan, S. K., Michel, H. 1994a. Structure of the photosynthetic reaction-center from *Rhodobacter sphaeroides* at 2.65-angstrom resolution: Cofactors and protein-cofactor interactions. *Structure* 2: 925–936.

Ermler, U., Michel, H., Schiffer, M. 1994b. Structure and function of the photosynthetic reaction center from *Rhodobacter sphaeroides*. *J. Bioenerg. Biomembr.* 26: 5–15.

Fairclough, W. V., Forsyth, A., Evans, M. C. W., Rigby, S. E. J., Purton, S., Heathcote, P. 2003. Bidirectional electron transfer in photosystem I: Electron transfer on the PsaB side is essential for photoautotrophic growth in *Chlamydomonas reinhardtii*. *Biochim. Biophys. Acta Bioenerg.* 1606: 43–55.

Faries, K. M., Kressel, L. L., Dylla, N. P. et al. 2016. Optimizing multi-step B-side charge separation in photosynthetic reaction centers from *Rhodobacter capsulatus*. *Biochim. Biophys. Acta* 1857: 150–159.

Ferreira, K. N., Iverson, T. M., Maghlaoui, K., Barber, J., Iwata, S. 2004. Architecture of the photosynthetic oxygen-evolving centre. *Science* 303: 1831–1838.

Francia, F., Malferrari, M., Sacquin-Mora, S., Venturoli, G. 2009. Charge recombination kinetics and protein dynamics in wild type and carotenoid-less bacterial reaction centers: Studies in trehalose glasses. *J. Phys. Chem. B* 113: 10389–10398.

Fritzsch, G., Koepke, J., Diem, R., Kuglstatter, A., Baciou, L. 2002. Charge separation induces conformational changes in the photosynthetic reaction centre of purple bacteria. *Acta Crystallogr. D: Biol. Crystallogr.* 58: 1660–1663.

Fromme, P., Jordan, P., Krauss, N. 1996. Structure of photosystem-I at 4.5 Å resolution: A short review including evolutionary aspects. *Biochim. Biophys. Acta* 1275: 76–83.

Fromme, P., Jordan, P., Krauss, N. 2001. Structure of photosystem I. *Biochim. Biophys. Acta Bioenerg.* 1507: 5–31.

Gibasiewicz, K., Pajzderska, M., Dobek, A. et al. 2013. Analysis of the temperature-dependence of $P^+H_A^-$ charge recombination in the *Rhodobacter sphaeroides* reaction center suggests nanosecond temperature-independent protein relaxation. *Phys. Chem. Chem. Phys.* 15: 16321–16333.

Gibasiewicz, K., Pajzderska, M., Potter, J. A. et al. 2011. Mechanism of recombination of the $P^+H_A^-$ radical pair in mutant *Rhodobacter sphaeroides* reaction centers with modified free energy gaps between $P^+B_A^-$ and $P^+H_A^-$. *J. Phys. Chem. B* 115: 13037–13050.

Giera, W., Gibasiewicz, K., Ramesh, V. M., Lin, S., Webber, A. 2009. Electron transfer from A_0 to A_1 in photosystem I from *Chlamydomonas reinhardtii* occurs in both the A and B branch with 25–30-ps lifetime. *Phys. Chem. Chem. Phys.* 11: 186–191.

Giera, W., Ramesh, V. M., Webber, A. N., van Stokkum, I., van Grondelle, R., Gibasiewicz, K. 2010. Effect of the P700 pre-oxidation and point mutations near A_0 on the reversibility of the primary charge separation in photosystem I from *Chlamydomonas reinhardtii. Biochim. Biophys. Acta* 1797: 106–112.

Graige, M. S., Feher, G., Okamura, M. Y. 1998. Conformational gating of the electron-transfer reaction $Q_A^- Q_B \rightarrow Q_A Q_B^-$ in bacterial reaction centers of *Rhodobacter sphaeroides* determined by a driving force assay. *Proc. Natl. Acad. Sci. USA* 95: 11679–11684.

Groot, M. L., Pawlowicz, N. P., van Wilderen, L. J. G. W., Breton, J., van Stokkum, I. H. M., van Grondelle, R. 2005. Initial electron donor and acceptor in isolated photosystem II reaction centers identified with femtosecond mid-IR spectroscopy. *Proc. Natl. Acad. Sci. USA* 102: 13087–13092.

Grotjohann, I., Fromme, P. 2005. Structure of cyanobacterial photosystem I. *Photosynth. Res.* 85: 51–72.

Guergova-Kuras, M., Boudreaux, B., Joliot, A., Joliot, P., Redding, K. 2001. Evidence for two active branches for electron transfer in photosystem I. *Proc. Natl. Acad. Sci. USA* 98: 4437–4442.

Guo, Z., Woodbury, N. W., Pan, J., Lin, S. 2012. Protein dielectric environment modulates the electron-transfer pathway in photosynthetic reaction centers. *Biophys. J.* 103: 1979–1988.

Guskov, A., Kern, J., Gabdulkhakov, A., Broser, M., Zouni, A., Saenger, W. 2009. Cyanobacterial photosystem II at 2.9 Å resolution and role of quinones, lipids, channels and chloride. *Nat. Struct. Mol. Biol.* 16: 334–342.

Hauska, G., Schoedl, T., Remigy, H., Tsiotis, G. 2001. The reaction center of green sulfur bacteria. *Biochim. Biophys. Acta* 1507: 260–277.

Heathcote, P., Fyfe, P. K., Jones, M. R. 2002. Reaction centres: Structure and mechanism in biological solar power. *Trends Biochem. Sci.* 27: 79–87.

Heathcote, P. Jones, M. R. 2012. The structure-function relationships of photosynthetic reaction centres. In *Comprehensive Biophysics*, eds. E. H. Egelman, S. Ferguson, vol. 8, pp. 115–144. Academic Press, Oxford, U.K.

Heller, B. A., Holten, D., Kirmaier, C. 1995. Control of electron-transfer between the L-side and M-side of photosynthetic reaction centers. *Science* 269: 940–945.

Hellmich, J., Bommer, M., Burkhardt, A. et al. 2014. Native-like photosystem II superstructure at 2.44 angstrom resolution through detergent extraction from the protein crystal. *Structure* 22: 1607–1615.

Hermes, S., Bremm, O., Garczarek, F. et al. 2006. A time-resolved iron-specific x-ray absorption experiment yields no evidence for an $Fe^{2+} \rightarrow Fe^{3+}$ transition during $Q_A^- \rightarrow Q_B$ electron transfer in the photosynthetic reaction center. *Biochemistry* 45: 353–359.

Hoff, A. J., Deisenhofer, J. 1997. Photophysics of photosynthesis: Structure and spectroscopy of reaction centres of purple bacteria. *Phys. Rep.* 287: 2–247.

Hohmann-Marriott, M. F., Blankenship, R. E. 2011. Evolution of photosynthesis. *Annu. Rev. Plant Biol.* 62: 515–548.

Holzwarth, A. R., Müller, M. G., Niklas, J., Lubitz, W. 2006a. Ultrafast transient absorption studies on photosystem I reaction centers from *Chlamydomonas reinhardtii*. 2: Mutations near the P700 reaction center chlorophylls provide new insight into the nature of the primary electron donor. *Biophys. J.* 90: 552–565.

Holzwarth, A. R., Muller, M. G., Reus, M., Nowaczyk, M., Sander, J., Rogner, M. 2006b. Kinetics and mechanism of electron transfer ion intact photosystem II and in isolated reaction center: Pheophytin is the primary electron acceptor. *Proc. Natl. Acad. Sci. USA* 103: 6895–6900.

Ishikita, H., Saenger, W., Biesiadka, J., Loll, B., Knapp, E. W. 2006. How photosynthetic reaction centers control oxidation power in chlorophyll pairs P680, P700, and P870. *Proc. Natl. Acad. Sci. USA* 103: 9855–9860.

Jaschke, P. R., Hardjasa, A., Digby, E. L., Hunter, C. N., Beatty, J. T. 2011. A BchD (magnesium chelatase) mutant of *Rhodobacter sphaeroides* synthesizes zinc bacteriochlorophyll through novel zinc-containing intermediates. *J. Biol. Chem.* 286: 20313–20322.

Jensen, P. E., Bassi, R., Boekema, E. J. et al. 2007. Structure, function and regulation of plant photosystem I. *Biochim. Biophys. Acta Bioenerg.* 1767: 335–352.

Joliot, P., Joliot, A. 1999. *In vivo* analysis of the electron transfer within photosystem-I: Are the two phylloquinones involved? *Biochemistry* 38: 11130–11136.

Jones, M. R. 2009. The petite purple photosynthetic powerpack. *Biochem. Soc. Trans.* 37: 400–407.

Jordan, P., Fromme, P., Witt, H. T., Klukas, O., Saenger, W., Krauss, N. 2001. Three-dimensional structure of cyanobacterial photosystem I at 2.5Å resolution. *Nature* 411: 909–917.

Kamiya, N., Shen, J.-R. 2003. Crystal structure of oxygen-evolving photosystem II from *Thermosynechococcus vulcanus* at 3.7-angstrom resolution. *Proc. Natl. Acad. Sci. USA* 100: 98–103.

Kargul, J., Olmos, J. D. J., Krupnik, T. 2012. Structure and function of photosystem I and its application in biomimetic solar-to-fuel systems. *J. Plant Physiol.* 169: 1639–1653.

Käss, H., Fromme, P., Witt, H. T., Lubitz, W. 2001. Orientation and electronic structure of the primary donor radical cation P700$^+$ in photosystem I: A single crystals EPR and ENDOR study. *J. Phys. Chem. B* 105: 1225–1239.

Kawakami, K., Umena, Y., Kamiya, N., Shen, J.-R. 2009. Location of chloride and its possible functions in oxygen-evolving photosystem II revealed by x-ray crystallography. *Proc. Natl. Acad. Sci. USA* 106: 8567–8572.

Kellogg, E. C., Kolaczkowski, S., Wasielewski, M. R., Tiede, D. M. 1989. Measurement of the extent of electron-transfer to the bacteriopheophytin in the M-subunit in reaction centers of *Rhodopseudomonas viridis*. *Photosynth. Res.* 22: 47–59.

Kern, J., Renger, G. 2007. Photosystem II: Structure and mechanism of the water:plastoquinone oxidoreductase. *Photosynth. Res.* 94: 183–202.

King, B. A., McAnaney, T. B., deWinter, A., Boxer, S. G. 2000. Excited state energy transfer pathways in photosynthetic reaction centers. 3. Ultrafast emission from the monomeric bacteriochlorophylls. *J. Phys. Chem. B* 104: 8895–8902.

Kleinfeld, D., Okamura, M. Y. Feher, G. 1984. Electron transfer kinetics in photosynthetic reaction centers cooled to cryogenic temperatures in the charge-separated state: Evidence for light-induced structural changes. *Biochemistry* 23: 5780–5786.

Knapp, E. W., Fischer, S. F., Zinth, W. et al. 1985. Analysis of optical spectra from single crystals of *Rhodopseudomonas viridis* reaction centers. *Proc. Natl. Acad. Sci. USA* 82: 8463–8467.

Koua, F. H. M., Umena, Y., Kawakami, K., Kamiya, N., Shen, J. R. 2013. Structure of Sr-substituted photosystem II at 2.1 Å resolution and its implications in the mechanism of water oxidation. *Proc. Natl. Acad. Sci. USA* 110: 3889–3894.

Kressel, L., Faries, K. M., Wander, M. J., Zogzas, C. E., Mejdrich, R. J., Hanson, D. K., Holten, D., Laible, P. D., Kirmaier, C. 2014. High yield of secondary B-side electron transfer in mutant *Rhodobacter capsulatus* reaction centers. *Biochim. Biophys. Acta* 1837: 1892–1903.

Lendzian, F., Huber, M., Isaacson, R. A. et al. 1993. The electronic-structure of the primary donor cation-radical in *Rhodobacter-sphaeroides* R-26: ENDOR and triple-resonance studies in single-crystals of reaction centers. *Biochim. Biophys. Acta* 1183: 139–160.

Li, Y., van der Est, A., Lucas, M. G. et al. 2006. Directing electron transfer within photosystem I by breaking H-bonds in the cofactor branches. *Proc. Natl. Acad. Sci. USA* 103: 2144–2149.

Liebl, U., Mockensturm-Wilson, M., Trost, J. T., Brune, D. C., Blankenship, R. E., Vermaas, W. 1993. Single core polypeptide in the reaction-center of the photosynthetic bacterium *Heliobacillus-mobilis*: Structural implications and relations to other photosystems. *Proc. Natl. Acad. Sci. USA* 90: 7124–7128.

Lin, S., Katilius, E., Taguchi, A. K. W., Woodbury, N. W. 2003. Excitation energy transfer from carotenoids to bacteriochlorophyll in the photosynthetic purple bacterial reaction centre of *Rhodobacter sphaeroides*. *J. Phys. Chem. B* 107: 14103–14108.

Loll, B., Kern, J., Saenger, W., Zouni, A., Biesiadka, J. 2005. Towards complete cofactor arrangement in the 3.0 angstrom resolution structure of photosystem II. *Nature* 438: 1040–1044.

Mamedov, M., Nadtochenko, G. V., Semenov, A. 2015. Primary electron transfer processes in photosynthetic reaction centers from oxygenic organisms. *Photosynth. Res.* 125: 51–63.

Marcus, R. A. 1993. Electron-transfer reactions in chemistry: Theory and experiment. *Rev. Mod. Phys.* 65: 599–610

Mathis, P., Setif, P. 1988. Kinetic-studies on the function of A_1 in the photosystem I reaction center. *FEBS Lett.* 237: 65–68.

Mazor, Y., Borovikova, A., Caspy, I., Nelson, N. 2017. Structure of the plant photosystem I supercomplex at 2.6 Å resolution. *Nat. Plants* 3: 17014.

Mazor, Y., Borovikova, A., Nelson, N. 2015. The structure of plant photosystem I super-complex at 2.8 Å resolution. *Elife* 4: e07433.

Moenne-Loccoz, P., Heathcote, P., MacLachlan, D. J., Berry, M. C., Davis, I. H., Evans, M. C. W. 1994. Path of electron transfer in photosystem I: Direct evidence of forward electron transfer from A_1 to Fe-S_X. *Biochemistry* 33: 10037–10042.

Müh, F., Glöckner, C., Hellmich, J., Zouni, A. 2012. Light-induced quinone reduction in photosystem II. *Biochim. Biophys. Acta* 1817: 44–65.

Müh, F., Lendzian, F., Roy, M., Williams, J. C., Allen, J. P., Lubitz, W. 2002. Pigment-protein interactions in bacterial reaction centers and their influence on oxidation potential and spin density distribution of the primary donor. *J. Phys. Chem. B* 106: 3226–3236.

Muhiuddin, I. P., Heathcote, P., Carter, S., Purton, S., Rigby, S. E., Evans, M. C. 2001. Evidence from time resolved studies of the P700^{+}$A_1$$^{-}$ radical pair for photosynthetic electron transfer on both the PsaA and PsaB branches of the photosystem I reaction centre. *FEBS Lett.* 503: 56–60.

Müller, M. G., Niklas, J., Lubitz, W., Holzwarth, A. R. 2003. Ultrafast transient absorption studies on photosystem I reaction centers from *Chlamydomonas reinhardtii*. 1. A new interpretation of the energy trapping and early electron transfer steps in photosystem I. *Biophys. J.* 85: 3899–3922.

Müller, M. G., Slavov, C., Luthra, R., Redding, K. E., Holzwarth, A. R. 2010. Independent initiation of primary electron transfer in the two branches of the photosystem I reaction center. *Proc. Natl. Acad. Sci. USA* 107: 4123–4128.

Myers, J. A., Lewis, K. L. M., Fuller, F. D., Tekavec, P. F., Yocum, C. F., Ogilvie, J. P. 2010. Two-dimensional electronic spectroscopy of the D1–D2–cyt b559 photosystem II reaction center complex. *J. Phys. Chem. Lett.* 1: 2774–2780.

Nadtochenko, V. A., Shelaev, I. V., Mamedov, M. D., Ya Shkuropatov, A., Yu Semenov, A., Shuvalov, V. A. 2014. Primary radical ion pairs in photosystem II core complexes. *Biochemistry (Mosc.)* 79: 197–204.

Nagy, L., Maroti, P., Terazima, M. 2008. Spectrally silent light induced conformation change in photosynthetic reaction centers. *FEBS Lett.* 582: 3657–3662.

Neerken, S., Amesz, J. 2001. The antenna reaction center complex of heliobacteria: Composition, energy conversion and electron transfer. *Biochim. Biophys. Acta* 1507: 278–290.

Nelson, N., Junge, W. 2015. Structure and energy transfer in photosystems of oxygenic photosynthesis. *Annu. Rev. Biochem.* 84: 659–683.

Nelson, N., Yocum, C. F. 2006. Structure and function of photosystems I and II. *Annu. Rev. Plant Biol.* 57: 521–565.

Niwa, S., Yu, L. J., Takeda, K., Hirano, Y., Kawakami, T., Wang-Otomo, Z. Y., Miki, K. 2014. Structure of the LH1-RC complex from *Thermochromatium tepidum* at 3.0 angstrom. *Nature* 508: 228–232.

Novoderezhkin, V. I., Dekker, J. P., van Grondelle, R. 2007. Mixing of exciton and charge-transfer states in photosystem II reaction centers: Modeling of stark spectra with modified Redfield theory. *Biophys. J.* 93: 1293–1311.

Ohashi, S., Iemura, T., Okada, N. et al. 2010. An overview on chlorophylls and quinones in the photosystem I-type reaction centers. *Photosynth. Res.* 104: 305–319.

Okamura, M. Y., Paddock, M. L., Graige, M. S., Feher, G. 2000. Proton and electron transfer in bacterial reaction centers. *Biochim. Biophys. Acta* 1458: 148–163.

Ortega, J. M., Mathis, P. 1993. Electron-transfer from the tetraheme cytochrome to the special pair in isolated reaction centers of *Rhodopseudomonas-viridis*. *Biochemistry* 32: 1141–1151.

Orzechowska, A., Lipinska, M., Fiedor, J. et al. 2010. Coupling of collective motions of the protein matrix to vibrations of the non-heme iron in bacterial photosynthetic reaction centers. *Biochim. Biophys. Acta Bioenerg.* 1797: 1696–1704.

Pan, J. Saer, R. G. Lin, S., Guo, Z. Beatty, J. T. Woodbury, N. W. 2013. The protein environment of the bacteriopheophytin anion modulates charge separation and charge recombination in bacterial reaction centers. *J. Phys. Chem. B* 24: 7179–7189.

Parson, W. W. 2003. Electron donors and acceptors in the initial steps of photosynthesis in purple bacteria: A personal account. *Photosynth. Res.* 76: 81–92.

Parson, W. W., Chu, Z. T., Warshel, A. 1998. Reorganization energy of the initial electron-transfer step in photosynthetic bacterial reaction centers. *Biophys. J.* 74: 182–191.

Permentier, H. P., Schmidt, K. A., Kobayashi, M. et al. 2000. Composition and optical properties of reaction centre core complexes from the green sulfur bacteria *Prosthecochloris aestuarii* and *Chlorobium tepidum*. *Photosynth. Res.* 64: 27–39.

Pierson, B. K., Thornber, J. P. 1983. Isolation and spectral characterization of photochemical reaction centers from the thermophilic green bacterium *Chloroflexus aurantiacus* strain J-10-f1. *Proc. Natl. Acad. Sci. USA* 80: 80–84.

Pierson, B. K., Thornber, J. P., Seftor, R. E. B. 1983. Partial purification, subunit structure and thermal stability of the photochemical reaction center of the thermophilic green bacterium *Chloroflexus aurantiacus*. *Biochim. Biophys. Acta Bioenerg.* 723: 322–326.

Poluektov, O. G., Paschenko, S. V., Utschig, L. M., Lakshmi, K. V., Thurnauer, M. C. 2005. Bidirectional electron transfer in photosystem I: Direct evidence from high-frequency time-resolved EPR spectroscopy. *J. Am. Chem. Soc.* 127: 11910–11911.

Poluektov, O. G., Utschig, L. M., Schlesselman, S. L. et al. 2002. Electronic structure of the P 700 special pair from high-frequency electron paramagnetic resonance spectroscopy. *J. Phys. Chem. B* 106: 8911–8916.

Purton, S., Stevens, D. R., Muhiuddin, I. P. et al. 2001. Site-directed mutagenesis of PsaA residue W693 affects phylloquinone binding and function in the photosystem I reaction center of *Chlamydomonas reinhardtii*. *Biochemistry* 40: 2167–2175.

Qian, P., Papiz, M. Z., Jackson, P. J., Brindley, A. A., Ng, I. W., Olsen, J. D., Dickman, M. J., Bullough, P. A., Hunter, C. N. 2013. Three-dimensional structure of the *Rhodobacter sphaeroides* RC-LH1-PufX complex: Dimerization and quinone channels promoted by PufX. *Biochemistry* 52: 7575–7585.

Qin, X., Suga, M., Kuang, T., Shen, J.-R. 2015. Structural basis for energy transfer pathways in the plant PSI-LHCI supercomplex. *Science* 348: 989–995.

Ramesh, V. M., Gibasiewicz, K., Lin, S., Bingham, S. E., Webber, A. N. 2004. Bidirectional electron transfer in photosystem I: Accumulation of A_0^- in A-side or B-side mutants of the axial ligand to chlorophyll A_0. *Biochemistry* 43: 1369–1375.

Rappaport, F., Diner, B. A. 2008. Primary photochemistry and energetics leading to the oxidation of the $(Mn)_4Ca$ cluster and to the evolution of molecular oxygen in photosystem II. *Coord. Chem. Rev.* 252: 259–272.

Raszewski, G., Diner, B. A., Schlodder, E., Renger, T. 2008. Spectroscopic properties of reaction center pigments in photosystem II core complexes: Revision of the multimer model. *Biophys. J.* 95: 105–119.

Rautter, J., Lendzian, F., Schulz, C. et al. 1995. ENDOR studies of the primary donor cation-radical in mutant reaction centers of *Rhodobacter-sphaeroides* with altered hydrogen-bond interactions. *Biochemistry* 34: 8130–8143.

Remy, A., Gerwert, K. 2003. Coupling of light-induced electron transfer to proton uptake in photosynthesis. *Nat. Struct. Biol.* 10: 637–644.

Renger, G. 2010. The light reactions of photosynthesis. *Curr. Sci.* 98: 1305–1319.

Renger, G., Renger, T. 2008. Photosystem II: The machinery of photosynthetic water splitting. *Photosynth. Res.* 98: 53–80.

Renger, T., Schlodder, E. 2010. Primary photophysical processes in photosystem II: Bridging the gap between crystal structure and optical spectra. *ChemPhysChem* 11: 1141–1153.

Renger, T., Schlodder, E. 2011. Optical properties, excitation energy and primary charge transfer in photosystem II: Theory meets experiment. *J. Photochem. Photobiol. B: Biol.* 104: 126–141.

Romero, E., Diner, B. A., Nixon, P. J., Coleman, W. J., Dekker, J. P., van Grondelle, R. 2012. Mixed exciton–charge-transfer states in photosystem II: Stark spectroscopy on site-directed mutants. *Biophys. J.* 103: 185–194.

Romero, E., Augulis, R., Novoderezhkin, V. I., Ferretti, M., Thieme, J., Zigmantas, D., van Grondelle, R. 2014.

Quantum coherence in photosynthesis for efficient solar-energy conversion. *Nat. Phy.* 10: 677–683.

Romero, E., Novoderezhkin, V. I., van Grondelle, R. 2017. Quantum design of photosynthesis for bio-inspired solar-energy conversion. *Nature* 543: 355–365.

Romero, E., van Stokkum, I. H. M., Novoderezhkin, V. I., Dekker, J. P., van Grondelle, R. 2010. Two different charge separation pathways in photosystem II. *Biochemistry* 49: 4300–4307.

Roszak, A. W., Howard, T. D., Southall, J., Gardiner, A. T., Law, C. J., Isaacs, N. W., Cogdell, R. J. 2003. Crystal structure of the RC-LH1 core complex from *Rhodopseudomonas palustris*. *Science* 302: 1969–1972.

Sadekar, S., Raymond, J., Blankenship, R. E. 2006. Conservation of distantly related membrane proteins: Photosynthetic reaction centers share a common structural core. *Mol. Biol. Evol.* 23: 2001–2007.

Santabarbara, S., Kuprov, I., Fairclough, W. V. et al. 2005. Bidirectional electron transfer in photosystem I: Determination of two distances between P 700$^+$ and A$_1^-$ in spin-correlated radical pairs. *Biochemistry* 44: 2119–2128.

Santabarbara, S., Kuprov, I., Hore, P. J., Casal, A., Heathcote, P., Evans, M. C. 2006. Analysis of the spin-polarized electron spin echo of the P 700$^+$ A$_1^-$ radical pair of photosystem I indicates that both reaction center subunits are competent in electron transfer in cyanobacteria, green algae, and higher plants. *Biochemistry* 45: 7389–7403.

Scherer, P. O. J., Fischer, S. F. 1986. On the stark-effect for bacterial photosynthetic reaction centers. *Chem. Phys. Lett.* 131: 153–159.

Schubert, W. -D., Klukas, O., Saenger, W., Witt, H. T., Fromme, P., Krauss, N. 1998. A common ancestor for oxygenic and anoxygenic photosynthetic systems: A comparison based on the structural model of photosystem-I. *J. Mol. Biol.* 280: 297–314.

Semenov, A. Y., Kurashov, V. N., Mamedov, M. D. 2011. Transmembrane charge transfer in photosynthetic reaction centers: Some similarities and distinctions. *J. Photochem. Photobiol. B: Biol.* 104: 326–332.

Semenov, A. Y., Shelaev, I. V., Gostev, F. E. et al. 2012. Primary steps of electron and energy transfer in photosystem I: Effect of excitation pulse wavelength. *Biochemistry (Mosc.)* 77: 1011–1020.

Setif, P., Brettel, K. 1993. Forward electron-transfer from phylloquinone-A$_1$ to iron-sulfur centers in spinach photosystem I. *Biochemistry* 32: 7846–7854.

Shelaev, I. V., Gostev, F. E., Mamedov, M. D. et al. 2010. Femtosecond primary charge separation in *Synechocystis* sp. PCC 6803 photosystem I. *Biochim. Biophys. Acta* 1797: 1410–1420.

Shelaev, I. V., Gostev, F. E., Nadtochenko, V. A. et al. 2008. Primary light-energy conversion in tetrameric chlorophyll structure of photosystem II and bacterial reaction centers: II. Femto- and picosecond charge separation in PSII D1/D2/Cyt b559 complex. *Photosynth. Res.* 98: 95–103.

Shelaev, I. V., Gostev, F. E., Vishnev, M. I. et al. 2011. Alternative electron donors P680 and Chl D1 in photosystem II reaction centers. *Photochem. Photobiol.* 104: 44–50.

Shiozawa, J. A., Lottspeich, F., Feick, R. 1987. The photochemical reaction center of *Chloroflexus aurantiacus* is composed of two structurally similar polypeptides. *Eur. J. Biochem.* 167: 595–600.

Stanley, R. J., King, B., Boxer, S. G. 1996. Excited state energy transfer pathways in photosynthetic reaction centers. 1. Structural symmetry effects. *J. Phys. Chem.* 100: 12052–12059.

Suga, M., Akita, F., Hirata, K. et al. 2015. Native structure of photosystem II at 1.95 angstrom resolution viewed by femtosecond x-ray pulses. *Nature* 517: 99–103.

Suga, M., Akita, F., Sugahara, M. et al. 2017. Light-induced structural changes and the site of O=O bond formation in PSII caught by XFEL. *Nature* 543: 131–135.

Tanaka, A., Fukushima, Y., Kamiya, N. 2017. Two different structures of the oxygen-evolving complex in the same polypeptide frameworks of photosystem II. *J. Am. Chem. Soc.* 139: 1718–1721.

Tsukatani, Y., Romberger, S. P., Golbeck, J. H., Bryant, D. A. 2012. Isolation and characterization of homodimeric type-I reaction center complex from *Candidatus Chloracidobacterium thermophilum*, an aerobic chlorophototroph. *J. Biol. Chem.* 287: 5720–5732.

Umena, Y., Kawakami, K., Shen, J. -R., Kamiya, N. 2011. Crystal structure of oxygen-evolving photosystem II at a resolution of 1.9 Å. *Nature* 473: 55–60.

van Brederode, M. E., Jones, M. R. van Grondelle, R. 1997b. Fluorescence excitation spectra of membrane-bound photosynthetic reaction centers of *Rhodobacter sphaeroides* in which tyrosine M210 residue is replaced by tryptophan: Evidence for a new pathway of charge separation. *Chem. Phys. Lett.* 268: 143–149.

van Brederode, M. E., Jones, M. R., van Mourik, F., van Stokkum, I. H. M., van Grondelle, R. 1997a. A new pathway for transmembrane electron transfer in photosynthetic reaction centers of *Rhodobacter sphaeroides* not involving the excited special pair. *Biochemistry* 36: 6855–6861.

van Brederode, M. E., van Grondelle, R. 1999. New and unexpected routes for ultrafast electron transfer in photosynthetic reaction centers. *FEBS Lett.* 455: 1–7.

van Brederode, M. E., van Mourik, F., van Stokkum, I. H. M., Jones, M. R., van Grondelle, R. 1999a. Multiple pathways for ultrafast transduction of light energy in the photosynthetic reaction center of *Rhodobacter sphaeroides*. *Proc. Natl. Acad. Sci. USA* 96: 2054–2059.

van Brederode, M. E., van Stokkum, I. H. M., Katilius, E., van Mourik, F., Jones, M. R., van Grondelle, R. 1999b. Primary charge separation routes in the Bchl:Bphe heterodimer reaction centres of *Rhodobacter sphaeroides*. *Biochemistry* 38: 7545–7555.

Vermeglio, A., Nagashima, S., Alric, J., Arnoux, P., Nagashima, K. V. P. 2012. Photo-induced electron transfer in intact cells of *Rubrivivax gelatinosus* mutants deleted in the RC-bound tetraheme cytochrome: Insight into evolution of photosynthetic electron transport. *Biochim. Biophys. Acta Bioenerg.* 1817: 689–696.

Vos, M. H., Jones, M. R., Hunter, C. N., Breton, J., Martin, J. L. 1994. Coherent nuclear-dynamics at room temperature in bacterial reaction centers. *Proc. Natl. Acad. Sci. USA* 91: 12701–12705.

Vos, M. H., Rappaport, F., Lambry, J. C., Breton, J., and Martin, J. L. 1993. Visualization of coherent nuclear motion in a membrane-protein by femtosecond spectroscopy. *Nature* 363: 320–325.

Wakao, N., Yokoi, N., Isoyama, N. et al. 1996. Discovery of natural photosynthesis using zinc-containing bacteriochlorophyll in an aerobic bacterium *Acidiphilium rubrum*. *Plant Cell Physiol.* 37: 889–893.

Wang, H. Y., Lin, S., Allen, J. P. et al. 2007. Protein dynamics control the kinetics of initial electron transfer in photosynthesis. *Science* 316: 747–750.

Watson, A. J., Fyfe, P. K., Frolov, D. et al. 2005. Replacement or exclusion of the B-branch bacteriopheophytin in the purple bacterial reaction centre: The H_B cofactor is not required for assembly or core function of the *Rhodobacter sphaeroides* complex. *Biochim. Biophys. Acta: Bioenerg.* 1710: 34–46.

Webber, A. N., Lubitz, W. 2001. P700: The primary electron donor of photosystem I. *Biochim. Biophys. Acta Bioenerg.* 1507: 61–79.

Woodbury, N. W., Allen, J. P. 1995. The pathway, kinetics and thermodynamics of electron transfer in wild type and mutant reaction centers of purple nonsulfur bacteria. In *Anoxygenic Photosynthetic Bacteria*, eds. R. E. Blankenship, M. T. Madigan, C. E. Bauer, Advances in Photosynthesis 2, pp. 527–557. Kluwer Academic Publishers, Dordrecht, the Netherlands.

Wraight, C. A. 2004. Proton and electron transfer in the acceptor quinone complex of photosynthetic reaction centers from *Rhodobacter sphaeroides*. *Front. Biosci.* 9: 309–337.

Wraight, C. A., Gunner, M. R. 2009. The acceptor quinones of purple photosynthetic bacteria-structure and spectroscopy. In *The Purple Phototrophic Bacteria*, eds. C. N. Hunter, F. Daldal, M. C. Thurnauer, J. T. Beatty, Advances in Photosynthesis and Respiration, vol. 28, pp. 379–405. Springer, Dordrecht, the Netherlands.

Yeates, T. O., Komiya, H., Chirino, A., Rees, D. C., Allen, J. P., Feher, G. 1988. Structure of the reaction center from *Rhodobacter-sphaeroides* R-26 and 2.4.1-protein-cofactor (bacteriochlorophyll, bacteriopheophytin, and carotenoid) interactions. *Proc. Natl. Acad. Sci. USA* 85: 7993–7997.

Young, I. D., Ibrahim, M., Chatterjee, R. et al. 2016. Structure of photosystem II and substrate binding at room temperature. *Nature* 540: 453–457.

Zhu, J., van Stokkum, I. H. M., Paparelli, L., Jones, M. R., Groot, M. L. 2013. Early bacteriopheophytin reduction in charge separation in reaction centers of *Rhodobacter sphaeroides*. *Biophys. J.* 104: 2493–2502.

Zinth, W., Knapp, E. W., Fischer, S. F., Kaiser, W., Deisenhofer, J., Michel, H. 1985. Correlation of structural and spectroscopic properties of a photosynthetic reaction center. *Chem. Phys. Lett.* 119: 1–4.

Zinth, W., Wachtveitl, J. 2005. The first picoseconds in bacterial photosynthesis—Ultrafast electron transfer for the efficient conversion of light energy. *ChemPhysChem* 6: 871–880.

Zouni, A., Witt, H. T., Kern, J. et al. 2001. Crystal structure of photosystem-II from *Synechococcus elongatus* at 3.8 Å resolution. *Nature* 409: 739–743.

Organization of photosynthetic membrane proteins into supercomplexes

EGBERT J. BOEKEMA AND DMITRY A. SEMCHONOK

10.1 INTRODUCTION

Photosystem I (PSI) and photosystem II (PSII) are the key proteins of the light reactions of oxygenic photosynthesis. They are universally distributed throughout prokaryotes and eukaryotes capable of oxygenic photosynthesis. They are present in green (higher) plants, macroalgae, diatoms, dinoflagellates, and the oxyphotobacteria (cyanobacteria and prochlorophytes), but not in archaea. Among these organisms, the core complexes of PSI and PSII are relatively well preserved compared to the peripheral antenna. In this chapter, we want to address the topic of how PSI and PSII are structurally organized in oxygenic photosynthesis. This is relevant, because the function of the photosystems and their particular role in the light reactions of photosynthesis cannot be understood without a detailed knowledge of the structure, including the peripheral antenna. This also requires knowledge of how PSI and PSII work together in the photosynthetic membrane with other protein complexes, such as the cytochrome b6f complex (Cyt b6f) and NDH/NDH-1 to perform the primary light reactions in an optimal way.

There is an increasing emphasis on the interaction of PSI and PSII complexes into higher-order assemblies, which have been named supercomplexes (Dekker and Boekema 2005). We will discuss several examples of such large assemblies and focus on the variation in the membrane-bound peripheral antenna. Some other aspects, such as the spatial separation of PSI and PSII in the photosynthetic (thylakoid) membrane and the way the thylakoid membranes are spatially organized, are not discussed in detail.

Further, during evolution, green plants and lower organisms adapted different strategies to deal with diverse environmental conditions, such as strongly fluctuating light conditions or certain types of stress conditions. To deal with such conditions, organisms are able to structurally change their photosynthetic proteins. For instance, by changing the interaction between PSI and Cyt *b6f*, resulting in a shift of the ratio of cyclic to noncyclic electron flow. This shift influences the ratio of ATP/NADPH production. The structural aspects of the two supercomplexes, involved in electron flow regulation, are also discussed.

10.1.1 Organization of photosystems I and II in Supercomplexes

In oxygenic photosynthesis, supercomplexes play a dual role. Their main function is to optimize light harvesting. The capture of sunlight basically implies that excitation energy, trapped by multiple chlorophyll molecules and bound to antenna protein complexes, is efficiently transferred to the reaction centers in the core parts of PSI and PSII. To enlarge the antenna, multiple copies of light-harvesting proteins are associated with PSI and PSII forming supercomplexes. A complicating factor is the ever-changing environment. This means that extension of the antenna to the limit is not the only thing that counts for an organism. Long-term survival is more relevant. Extension of the antenna to the limit is only relevant in situations where severe excess damaging of light does not occur. In the deep sea, at the bottom of the ocean, excess light never occurs, and in the darkness, specialized prokaryotes are able, by the help of a giant antenna system called the chlorosome, to harvest the little light quanta emitted as thermal heat from black smokers that exist due to volcanic events occurring in the ocean floor (see Chapter 7). But this is an exception to the rule. Organisms exposed to the atmosphere or living in shallow water have to quickly react to changing light conditions, to prevent overreduction of their reaction centers. This sets constraints to the size and organization of the PSI and PSII antennae. One way is to disconnect their peripheral antenna partly or completely at short timescales. This makes a discussion how PSI and PSII are organized into supercomplexes relevant.

The mechanistic details of excess of light and other stress factors are discussed in Chapter 11. Here, we just give an overview of the structures of PSI and PSII with their peripheral antenna proteins supercomplexes, relevant for light harvesting.

Another aspect of regulation, besides the regulation of light, deals with the electron flow through two major components of the light reactions, PSI and Cyt *b6f*. If we overlook the bioenergetics reactions in the chloroplast, there are some interesting aspects concerning the flow of electrons. Several large membrane proteins are involved in the catalysis, and electrons are carried either by small, membrane-soluble carriers (plastoquinone, or PQ) or by small proteins (cytochrome c, ferredoxin, plastocyanin). This is common textbook knowledge. However, it became only recently clear that PSI associates into supercomplexes with the Cyt *b6f* complex and NDH to optimize and regulate electron flow, for instance, to adapt the NADPH and ATP production. Both compounds are essential for the metabolic processes but are necessary in variable amounts. The way PSI is organized in a particular supercomplex with Cyt *b6f* is suggested to mediate the balance between NADPH and ATP.

The supercomplexes in which either PSI or PSII is involved will be discussed in the following sections.

10.2 PHOTOSYSTEM I SUPERCOMPLEXES

If we consider all PSI supercomplexes from oxygenic photosynthesis and make a comparison with those of PSII, it turns out that there is much more variation in PSI. Most of it has to do with the peripheral antenna. The peripheral antenna is composed of multiple copies of light-harvesting antenna (Lhc) proteins, and the number of associated proteins is variable between distant groups of species (algae, plants). There is no high-resolution structural information of any of these supercomplexes, except for the plant PSI supercomplex, but from low-resolution electron microscopy, applied to negatively stained supercomplexes, we have a rather good impression how PSI supercomplexes are composed in cyanobacteria, various algae, and green plants. An overview of the different types of PSI complexes is presented in Figure 10.1, and the various types of supercomplexes are discussed in the following.

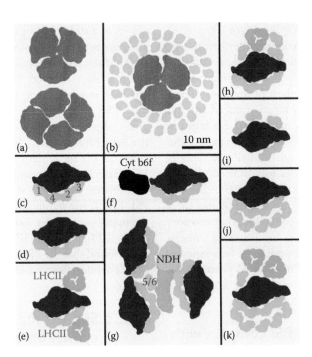

Figure 10.1 An overview of photosystem I (PSI) complexes and supercomplexes from oxygenic photosynthetic organisms. Complexes are drawn schematically, with core parts in pink or red, peripheral antenna proteins belonging to the LHC superfamily in bright and dark green, the IsiA antenna protein in gray-blue, and the NDH complex in blue. **(a)** Trimeric and tetrameric photosystem I, as can be found in cyanobacteria. **(b)** PSI–IsiA supercomplex, typically found in cyanobacteria under stress conditions, such as long-term iron stress. Trimeric PSI can be surrounded by 1–2 rings of IsiA, consisting of 18 and 25 IsiA copies, respectively. **(c)** Green plant PSI, in which the positions of the four Lhca proteins (Lhca1–4, in blue) were established by solving its high-resolution structure. **(d)** Small PSI complex from the moss *Physcomitrella*, in which four Lhca proteins are attached on one side of the core part. **(e)** Green plant PSI–LHCII supercomplex under state transition, in which the PSI complex can bind a LHCII trimer in state 2 (upper LHCII trimer) and sometimes even a second one (lower LHCII trimer). **(f)** Plant PSI–Cyt b6f supercomplex, with the dimeric Cyt b6f complex attached to Lhca1. **(g)** PSI–NDH supercomplex, in which a central NDH complex can bind several PSI complexes. The substoichiometric antenna proteins Lhca5 and Lhca6 are thought to help in binding PSI to NDH. **(h)** PSI from the cryptophyte *Rhodomonas*, which binds an additional antenna moiety on the opposite side of the common antenna. **(i)** PSI from the heterokont alga *Nannochloropsis gaditana*, a complex with five different Lhca copies, of which only two bind at the lower side. **(j)** PSI from the green alga *Chlamydomonas*, a large complex with nine different Lhca copies. **(k)** *Chlamydomonas* PSI supercomplex under state transition, in which the PSI complex can bind two LHCII trimers and a single Lhca copy, possibly CP29, in state 2.

10.2.1 Supercomplexes in cyanobacteria

Oxygenic photosynthesis started in cyanobacteria. In most cyanobacteria, PSI comprises about 12 different subunits, and together with its pigment cofactors, a PSI monomer has a mass of about 330 kDa. Due to the initial characterization of a trimeric PSI complex in cyanobacteria, including the determination of a high-resolution structure (Jordan et al. 2001),

it has been widely accepted that the trimer (Figure 10.1a, top) is the only multimeric form of PSI found in cyanobacteria. However, tetrameric forms of PSI association (Figure 10.1a, bottom) were also found, as reported in *Anabaena* sp. PCC 7120 (Watanabe et al. 2014) and *Chroococcidiopsis* sp. TS-821 (Li et al. 2014). It is likely that there are quite a few other cyanobacteria with a tetrameric structure.

Cyanobacterial PSI and PSII can associate with phycobilisomes, which are large multisubunit, water-soluble antenna complexes (see Chapter 5). For PSI tetramers, the way of interaction with a small type of phycobilisome was established (Watanabe et al. 2014). Further, some groups of cyanobacteria can extend their peripheral antenna within the membrane. The stress-induced antenna protein IsiA of 35 kDa can surround a trimeric PSI complex with 18 copies, enlarging the light-harvesting capacity in a substantial way. This is, for instance, the case under low-iron conditions. This is a way of adaptation. As iron, is indispensable for functioning of PSI, because it has three iron–sulfur centers involved in electron transport. Under continuous low-iron conditions, a trimeric PSI particle can be surrounded by two rings of IsiA (Figure 10.1b), with in total 43 protein copies (Chauhan et al. 2011). In this way, iron is optimally used.

10.2.2 Green plant PSI complex with four Lhca antenna proteins

Algae and plants have no IsiA but other antenna complexes, which are all members of the LHC superfamily (see Chapters 4). Plant light-harvesting II (LHCII) was the very first antenna protein of which a high resolution structure became available (Liu et al. 2004; Standfuss et al. 2005). Typically, they are composed of three membrane-spanning helices and several pigments with a total mass around 40 kDa. In negative stain maps of PSI or PSII-associated LHCII trimers can be discerned, but the features of single LHC copies remain often obscured. There is, however, a high-resolution green plant PSI structure determined by X-ray diffraction (Mazor et al. 2015; Qin et al. 2015). It precisely shows how the PSI core associates with four Lhca proteins (Lhca1–4) on one side. This high-resolution map is useful for comparison with PSI complexes from other organisms on the lower-resolution scale. Therefore, a simple scheme is presented (Figure 10.1c). It is perhaps not a coincidence that the small PSI complex from the moss *Physcomitrella patens* (Busch et al. 2013) has the same outline (Figure 10.4d), because *Physcomitrella* is considered to be one of the first plants that lived outside water habitats and thus in some way an ancestor of the current large group of green plants. However, it was recently shown that substantially larger PSI particles can be purified as well, in which, besides a LHCII trimer, several

copies of a single Lhca protein are bound (R. Bassi and colleagues, unpublished data).

Another type of plant supercomplex variation is the state transition complex. State transitions are rearrangements of the photosynthetic apparatus that occur on short timescales in plants, in green algae (see below) and cyanobacteria. On the structural side, in plants, the rearrangement concerns the movement of a LHCII trimer toward PSI (state 2), away from the pool of LHCII that is transferring its excitation energy to PSII. The plant PSI–LHCII supercomplex has one trimer bound in the upper part, where PsaH is located, as shown in Figure 10.1e (upper LHCII trimer). More recently, it was found that the PSI complex can bind a second LHCII trimer (lower LHCII trimer) (Yadav et al. 2017a).

10.2.3 Supercomplexes with cytochrome b6f and NDH

The main function of the PSI and PSII complexes is the conversion of light energy into chemically fixed energy, and the PSI supercomplexes clearly contribute to this. But besides excitation energy flow, there is also electron transport in primary photosynthesis. It has some implications on PSI supercomplex formation, and two specific supercomplexes that optimize and regulate electron transport are briefly discussed here. In the primary reactions of photosynthesis, there are two ways of electron flow: linear electron flow (LEF) and cyclic electron flow (CEF) (see Johnson 2011 for a review). Currently, two cyclic pathways are known. They depend on either the PROTON GRADIENT REGULATION5 (PGR5) and PGR5-LIKE1 (PGRL1) complex (DalCorso et al. 2008) or the NAD(P)H dehydrogenase (NDH) complex (Shikanai et al. 1998). The PGR5-dependent pathway is efficient in the control of the ATP/NADPH ratio and essential to induce the dissipation of energy by nonphotochemical quenching. The NDH-dependent pathway is important under environmental stress conditions and has no role in nonphotochemical quenching. For both cyclic pathways, there are PSI supercomplexes.

The first evidence for a very remarkable supercomplex, composed of PSI and NDH, was found some years ago by blue-native gel electrophoresis (Peng et al. 2008). Later on, the existence was confirmed by electron microscopy (Kouril et al. 2014). It appeared to be composed of a central NDH complex, flanked by two PSI complexes. NDH is a proton-pumping complex, which can be found in many prokaryotes, such as *E. coli*, where it is called NDH-1. The core of the NDH protein is structurally related to mitochondrial complex I. Later on, single particle electron microscopy showed the existence of a supercomplex family of PSI–NDH particles. The central NDH complex can bind up to 5 PSI complexes in an ordered way. A supercomplex with three copies is shown in Figure 10.1g. In some unknown way, the low-abundant light-harvesting complexes Lhca5 and Lhca6 are involved in the binding of PSI complexes (Yadav et al. 2017a). A possible position is indicated. Probably, the supercomplex has to do with fine-tuning electron transport by the small carrier ferredoxin, which receives electrons directly from PSI and docks on the upper part of NDH.

In another supercomplex, the cytochrome *b6f* complex is associated with PSI. The first evidence for such a complex was obtained in the model alga *Chlamydomonas reinhardtii* (*C. reinhardtii*) (Iwai et al. 2010). The complex contained, besides PSI and Cyt *b6f*, several smaller components: LHCII, ferredoxin-NADP$^+$ reductase, and PGRL1. It is, however, not yet structurally characterized. How and where the cytochrome *b6f* is attached to PSI remains an open question. Recently, however, a PSI–Cyt *b6f* complex was structurally characterized from *Arabidopsis thaliana* (Yadav et al. 2017a). In this complex, the Cyt *b6f* dimer is attached with its short end to PSI at the site of Lhca1.

The PSI–Cyt *b6f* complex functions in CEF, in which only PSI and Cyt *b6f* are involved as the major players, without PSII, which is only part of LEF. This pathway generates extra ATP. Control of the LEF and CEF can adjust the ratio of the produced ATP/NADPH, required under changing metabolic and environmental conditions. Under certain conditions, CEF appears to contribute substantially to photosynthetic electron flow, for instance, during induction of photosynthesis and under stress conditions like drought, high light, and extreme temperatures (reviewed in Johnson 2011). The organization of PSI and Cyt *b6f* in a supercomplex may improve the regulation of electron transport, which is mediated by the small soluble protein plastocyanin. This small protein transports the electrons by diffusion, which is a random movement and thus much limited by distance. In the supercomplex, the distance between the plastocyanin-binding regions of PSI and Cyt *b6f* is optimized (Yadav et al. 2017a).

10.2.4 PSI supercomplexes of eukaryotic algae

Several groups of organisms have, in addition to the standard four Lhca proteins, an extended peripheral antenna. PSI from the cryptophyte *Rhodomonas* CS24 has an additional antenna moiety on the opposite side of the common antenna (Kereïche et al. 2008), probably consisting of monomeric antenna proteins (Figure 10.1h). PSI from the heterokont alga *Nannochloropsis gaditana* has also antenna complexes bound in the upper part of PSI, where PsaH is located (Figure 10.1l). But as a surprise, it only has two Lhca copies bound at the lower side, at the position of Lhca2 and Lhca3 (Alboresi et al. 2017). This is unique, because in green plant PSI, the presence of Lhca1 and Lhca4 is always 100%. Only in mutants where Lhca1 and Lhca4 are not expressed, these subunits are absent and not replaced by other Lhca proteins (Wientjes et al. 2009).

The standard PSI complex from the model green alga *C. reinhardtii* is a large complex with nine different Lhca copies (Drop et al. 2011). All are bound on one side of the complex, as schematically presented in Figure 10.1j. It is not precisely known where the nine copies of the nine different proteins are located. This question can only be solved in the future by a high-resolution structure, for instance, from cryo-electron microscopy (cryo-EM). PSI from the red alga *Galdieria sulphuraria* appears to be structurally homologous to *C. reinhardtii* PSI (Thangaraj et al. 2011). Both PSI complexes have the same surface, although the shapes are not fully identical. This means that red algae could have the same set of nine Lhca proteins, arranged in a same way.

A final supercomplex is the *C. reinhardtii* PSI supercomplex in state 2. State transition is an adaptation of the photosystems to a change in light. Under unfavorable PSI light, in state 2, LHCII is reallocated and bound to PSI, enlarging its photosynthetic capacity. It was shown by electron microscopy that the standard PSI particle can bind two LHCII trimers and a single Lhcb copy, possibly

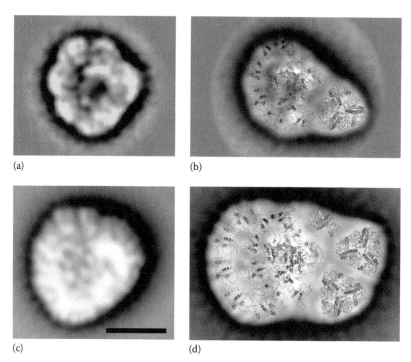

Figure 10.2 Comparison of maps of photosystem I complexes from plants and green algae, obtained by single particle electron microscopy. Under state transition, both plant and algal PSI can extend their antenna by binding additional antenna proteins. **(a)** Plant PSI complex under normal physiological conditions (state 1). **(b)** Pseudo-atomic model of the PSI complex in state 2 with wire model representations (left), with an additional LHCII trimer (in green) attached at the right. **(c)** Green algal PSI in state 1. **(d)** Pseudo-atomic model of the supercomplex of the PSI structure including 9 Lhc proteins (left), two LHCII trimers (dark green), and CP29 (pale green) at the right. Scale bar equals 10 nm. (Adapted from Drop, B. et al., *Biochim. Biophys. Acta*, 1837, 63, 2014a.)

CP29, in state 2 (Drop et al. 2014b). One of the trimers binds in a similar position as in plants (Figure 10.1k).

Some modeling helps to get further insight, even at low resolution. High-resolution structures can be modeled into the negative stain maps, to get an impression how the *C. reinhardtii* complexes are arranged. The idea is that by modeling higher-resolution data, for instance, from X-ray diffraction, into lower-resolution EM maps, a substantial gain in resolution can be achieved (Rossmann et al. 2005). This means that at the level of small subunits, or helices within a LHCII trimer, we have an impression of how a supercomplex is composed. Of course, this is still one step lower than to have an impression where chlorophylls are located and how far chlorophylls are separated. The best maps and their modeling of green plants and *C. reinhardtii* PSI, obtained by negatively stained single particle electron microscopy, are presented in Figure 10.2.

10.3 PSII SUPERCOMPLEXES

PSII is present mainly in dimeric form in the membrane, each monomer consisting of at least 30 subunits, depending on the organism (Shi et al. 2012). It is organized in two moieties: the core complex and the antenna system (Dekker and Boekema 2005). Most green plants have six different light-harvesting proteins. Lhcb1, Lhcb2, and Lhcb3 are components of LHCII and form heterotrimers. The minor antenna components Lhcb4 (CP29), Lhcb5 (CP26), and Lhcb6 (CP24) connect the LHCII trimers to the dimeric core (see Chapter 4).

10.3.1 Plant PSII supercomplexes

PSII can bind multiple copies of these peripheral antenna complexes in a number of smaller and larger supercomplexes. They consist of a dimeric core ("C_2") complex to which a variable number of light-harvesting antenna subunits are attached. In the model plant *A. thaliana*, 2–4 copies of LHCII trimers are attached under normal to low-light conditions. An inner trimer ("S" trimer) is connected to the core via two minor antenna complexes (CP26 and CP29; Figure 10.2a). Attachment of another, more peripheral trimer ("M" trimer) is mediated by CP24, a minor antenna protein, which is present in all green plants but absent in green algae. A supercomplex with four trimers is called a $C_2S_2M_2$ supercomplex. Many supercomplexes, however, do not have two full sets of peripheral antenna subunits. This is especially the case under high-light conditions, where the smaller C_2S_2 and C_2S_2M complexes often dominate.

10.3.2 PSII supercomplexes lacking CP24

Until recently, it was thought that in the model green alga *C. reinhardtii*, the C_2S_2 supercomplex was the largest possible PSII particle, because of the absence of CP24. It was considered that CP24 would be necessary to bind the M trimer. However, nature proved otherwise, and under mild membrane solubilization conditions, a particle was purified from *Chlamydomonas* grown under low-light conditions, which is even larger than the $C_2S_2M_2$ supercomplex from *Arabidopsis*, because it can contain up to six trimers (Tokutsu et al. 2012; Drop et al. 2014a).

The innermost S trimer is attached in a same position as in plants, but the M trimer and a novel trimer, called N, are in a unique position. This is depicted in Figure 10.2b, which shows a C_2S_2MN supercomplex, lacking a second copy of the M and N trimers on the upper right side. Green algae are not unique in their absence of CP24, because also some of the more primitive plants, such as spruce, do not contain CP24. The C_2S_2 particle of Norway spruce (*Picea abies*) was recently investigated. This complex has a very similar peripheral antenna as *C. reinhardtii*, because the S and M trimers appear to be oriented in the same way (Figure 10.3c).

10.3.3 Toward higher resolution

The maps presented in Figure 10.3 have been obtained by processing negatively stained EM specimens. The negative stain technique applied for producing these EM maps is especially useful in cases where mixtures of supercomplexes are present, because it provides high contrast. They have a resolution of about 1.5 nm. This is not enough to see the small subunits with a single membrane-spanning α-helix. On the other hand, the center of the LHCII trimers is directly visible in these maps. Thus, such maps can be used to fit the higher-resolution structure of LHCII and other light-harvesting components in a rather accurate way (Caffarri et al. 2009).

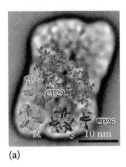 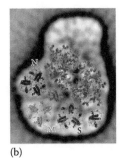 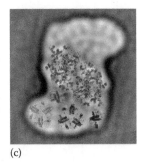

(a) (b) (c)

Figure 10.3 A comparison of photosystem II supercomplexes from plants and green algae. Density maps obtained by electron microscopy were used to model the various components at the subunit level. **(a)** Model of the largest PSII supercomplex from the plant *Arabidopsis thaliana* obtained by single particle electron microscopy. The $C_2S_2M_2$ supercomplex exists of a dimeric core complex (ochre) flanked on both sides by a peripheral antenna consisting of an inner LHCII trimer (S trimer) and a more peripheral LHCII trimer (M trimer) plus single copies of the three minor antenna proteins CP24, Cp26 and CP29. **(b)** Model of the largest $C_2S_2M_2N_2$ PSII supercomplex from the green alga *C. reinhardtii*. The positions of the three different types of LHCII trimers are indicated (S,M,N), as well as the two minor antenna complexes CP26 and CP29 (purple and pink). The main difference with *Arabidopsis* PSII, besides the absence of CP24, is the extra N trimer and the different orientation of the M trimer. The N trimer is called L trimer in the papers of Minagawa's group **(c)** EM map of Norway spruce C_2S_2 PSII.

This yields pseudo-atomic models, which give an impression how subunits are arranged.

Near-atomic resolution data of supercomplexes can be provided by cryo-EM. For many years, structure determination of biological macromolecules by cryo-EM limited to large complexes or low-resolution models. With recent advances in electron detection and image processing, the resolution by cryo-EM is now beginning to rival X-ray crystallography (Bai et al. 2014). Recently, the first 3D models of plant PSII appeared. The models of spinach C_2S_2 and *Arabidopsis* $C_2S_2M_2$ supercomplexes at 3.2 and 5.3 Å resolution, respectively, give the precise position of all membrane-embedded subunits and of the chlorophyll pigments (Wei et al. 2016; van Bezouwen et al. 2017). This is a significant step forward, although the resolution is not yet high enough to assign all Chl *a* and Chl *b* molecules. Further improvement of the resolution is possible by extending the data sets, and it will be a matter of time before the models will be refined to a level that also carotenoids and lipids become fully visible.

If we compare the new models derived from high-resolution cryo-EM data with the old models based on negative stain maps, it is obvious that the positions of some components like CP26 and CP29 differ. Most of the mismatch of the old model is in a different rotational position. For CP26 and CP29, the mismatch is about 40° and 15°, respectively. This is understandable because the negative stain, which does not penetrate the hydrophobic interior of PSII, does not outline the very flat light-harvesting subunits. It is expected that over time negative staining will be fully abandoned, at least for those supercomplexes that can be purified in enough quantity and homogeneity.

All further variation of PSII supercomplexes is within the set of the six Lhcb proteins. There may be particles that have an M trimer attached, without an S trimer, and so on. There are also other proteins loosely attached that function in energy quenching. One is PsbS, of which the binding site remains to be elucidated. A combination of ΔpH and zeaxanthin increases the proportion of PsbS bound to the minor LHCII antenna complex proteins Lhcb4, Lhcb5, and Lhcb6 (Sacharz et al. 2017), but structural details are not yet available. On the other side, it was found by electron microscopy that the quenching protein LHCSR3 from *C. reinhardtii* binds at several sites on C_2S_2 supercomplexes, mostly around CP26 (Yadav et al. 2017b).

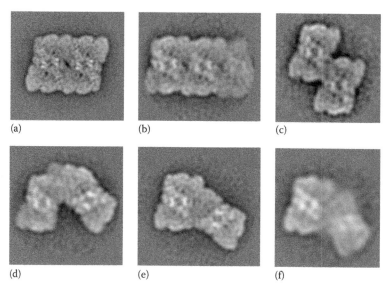

(a) (b) (c)

(d) (e) (f)

Figure 10.4 A gallery of photosystem II megacomplexes from *Arabidopsis thaliana* obtained by single particle electron microscopy (Nosek et al. 2017). **(a–c)** Megacomplexes in which the supercomplexes are arranged in a parallel way, without making an angle. **(d–f)** Megacomplexes in which the supercomplexes are arranged in a nonparallel way, making a specific angle.

10.3.4 Arrangement of supercomplexes in megacomplexes

Concerning PSII, the supercomplex formation is not the end of the higher order of association. One or more supercomplexes can associate in specific ways into megacomplexes and further into crystalline or semicrystalline arrays in the photosynthetic membrane. The stacked photosynthetic membrane is only partially filled with PSII crystals. It has been known for a while that the spinach C_2S_2M and *Arabidopsis* $C_2S_2M_2$ supercomplexes are present as semicrystalline arrays, but only for a minor fraction (Boekema et al. 2000; Yakushevska et al. 2001). It was always considered that the majority of the PSII complexes are randomly organized. But a recent analysis indicated that a major part of the seemingly randomly distributed PSII supercomplexes is also arranged in PSII megacomplexes (Nosek et al. 2017), but in such a way that the two supercomplexes make an angle in the membrane (Figure 10.4c–e). The organization of the membrane is a topic on its own, but the particular megacomplex organization indicates that in fact much can be described in terms of the now commonly accepted supercomplexes.

REFERENCES

Alboresi, A., Le Quiniou, C., Yadav, K. N. S., Scholz, M., Meneghesso, A., Gerotto, C., Simionato, D. et al. 2017. Conservation of core complex subunits shaped the structure and function of photosystem I in the secondary endosymbiont alga *Nannochloropsis gaditana*. *New Phytol.* 213: 714–726.

Bai, X. C., McMullan, G., and Scheres, S. H. W. 2014. How cryo-EM is revolutionizing structural biology. *Trends Biochem. Sci.* 40: 49–57.

Boekema, E. J., van Breemen, J. F. L., van Roon, H., and Dekker, J. P. 2000. Arrangement of PSII supercomplexes in crystalline macrodomains within the thylakoid membrane of green plants. *J. Mol. Biol.* 301: 1123–1133.

Busch, A., Petersen, J., Webber-Birungi, M. T., Powikrowska, M., Münter Lassen, L. M., Naumann-Busch, B., Zygadlo Nielsen, A.

et al. 2013. The composition and structure of photosystem I in the moss *Physcomitrella patens*. *J. Exp. Bot.* 64: 2689–2699.

Caffarri, S., Kouril, R., Kereïche, S., Boekema, E. J., and Croce, R. 2009. Functional architecture of higher plant photosystem II supercomplexes. *EMBO J.* 28: 3052–3063.

Chauhan, D., Folea, I. M., Kouřil, R., Lubner, C., Wolfe-Simon, F., Golbeck, J., Boekema, E. J., and Fromme, P. 2011. Unraveling of new photosynthetic strategies for adaptation to low iron environments: The optimization of both antenna and electron transfer in an IsiA-photosystem I-Supercomplex with complete double rings. *Biochemistry* 50: 686–692.

DalCorso, G., Pesaresi, P., Masiero, S., Aseeva, E., Schünemann, D., Finazzi, G., Joliot, P., Barbato, R., and Leister, D. 2008. A complex containing PGRL1 and PGR5 is involved in the switch between linear and cyclic electron flow in Arabidopsis. *Cell* 132: 273–285.

Dekker J. P. and Boekema E. J. 2005. Supermolecular organization of the thylakoid membrane proteins in green plants. *Biochim. Biophys. Acta* 1706: 12–39.

Drop, B., Webber-Birungi, M., Fusetti, F., Kouřil, R., Redding, K. E., Boekema, E. J., and Croce, R. 2014a. Functional architecture of photosystem I of *Chlamydomonas reinhardtii*. *J. Biol. Chem.* 286: 44878–44887.

Drop, B., Webber-Birungi, M., Yadav, N. K. S., Filipowicz-Szymanska, A., Fusetti, F., Boekema, E. J., and Croce, C. 2014a. Light-harvesting complex II (LHCII) and its super-molecular organization in *Chlamydomonas reinhardtii*. *Biochim. Biophys. Acta* 1837: 63–72.

Drop, B., Yadav K. N. S., Boekema, E. J., and Croce, R. 2014b. Consequences of state transitions on the structural and functional organization of photosystem I in the green alga *Chlamydomonas reinhardtii*. *Plant J.* 78: 181–191.

Iwai, M., Takizawa, K., Tokutsu, R., Okamura, A., Takahashi, Y., and Minagawa, J. 2010. Isolation of the elusive supercomplex that drives cyclic electron flow in photosynthesis. *Nature* 464: 1210–1213.

Johnson, G. N. 2011. Physiology of PSI cyclic electron transport in higher plants. *Biochim. Biophys. Acta* 1807: 384–389.

Jordan, P., Fromme, P., Witt, H. T., Klukas, O., Saenger, W., and Krauss, N. 2001. Three-dimensional structure of cyanobacterial photosystem I at 2.5 Å resolution. *Nature* 411: 909–917.

Kereïche, S., Kouřil, R., Oostergetel, G. T., Fusetti, F., Boekema, E. J., Doust, A. B., van der Weij-de Wit, C. D., and Dekker, J. P. 2008. Association of chlorophyll *a*/*c2* complexes to photosystem I and photosystem II in the cryptophyte *Rhodomonas* CS24. *Biochim. Biophys. Acta* 1777: 1122–1128.

Kouril, R., Strouhal, O., Nosek, L., Lenobel, R., Chamrad, I., Boekema, E., Sebela, M., and Ilik, P. 2014. Structural characterization of a plant photosystem I and NAD(P)H dehydrogenase supercomplex. *Plant J.* 77: 568–576.

Li, M., Semchonok, D. A., Boekema, E. J., and Bruce, B. D. 2014. Characterization and evolution of tetrameric photosystem I from the cyanobacteria *Chroococcidiopsis* sp. TS-821. *Plant Cell* 26: 1230–1245.

Liu, Z. F., Yan, H. C., Wang, K. B., Kuang, T. Y., Zhang, J. P., Gui, L. L., An, X. M., and Chang, W. R. 2004. Crystal structure of spinach major light-harvesting complex at 2.72 Å resolution. *Nature* 428: 287–292.

Mazor, Y., Borovikova, A., and Nelson, N. 2015. The structure of plant photosystem I super-complex at 2.8 Å resolution. *elife* 4: e07433.

Nosek, L., Semchonok, D., Boekema, E. J., Ilik, P., and Kouril, R. 2017. Structural variability of plant photosystem II megacomplexes in thylakoid membranes. *Plant J.* 89: 104–111.

Peng, L., Shimizu, H., and Shikanai, T. 2008. The chloroplast NAD(P)H dehydrogenase complex interacts with photosystem I in Arabidopsis. *J. Biol. Chem.* 283: 34873–34879.

Qin, X., Suga, M., Kuang, T., and Shen, J. R. 2015. Structural basis for energy transfer pathways in the plant PSI-LHCI supercomplex. *Science* 348: 989–995.

Rossmann, M. G., Morais, M. C., Leimann, P. G., and Zhang, W. 2005. Combining X-ray crystallography and electron microscopy. *Structure* 13: 355–362.

Sacharz, J., Giovagnetti, V., Ungerer, P., Mastroianni, G., and Ruban, A. V. 2017. The xanthophyll cycle affects reversible interactions between PsbS and light-harvesting complex II to control non-photochemical quenching. *Nat. Plants* 3: 16225.

Shi, L. X., Hall, M., Funk, C., and Schröder, W. P. 2012. Photosystem II, a growing complex: Updates on newly discovered components and low molecular mass proteins. *Biochim. Biophys. Acta* 1817: 13–25.

Shikanai, T., Endo, T., Hashimoto, T., Yamada, Y., Asada, K., and Yokota, A. 1998. Directed disruption of the tobacco ndhB gene impairs cyclic electron flow around photosystem I. *Proc. Natl. Acad. Sci. U. S. A.* 95: 9705–9709.

Standfuss, J., van Scheltinga, A. C. T., Lamborghini, M., and Kühlbrandt, W. 2005. Mechanisms of photoprotection and nonphotochemical quenching in pea light-harvesting complex at 2.5 Å resolution. *EMBO J.* 24: 919–928.

Thangaraj, B., Jolley, C. C., Sarrou, I., Bultema, J. B., Greylak, J., Whitelegge, J. P., Lin, S. et al. 2011. Efficient light harvesting in a dark, hot, acidic environment: The structure and function of PSI-LHCI from *Galdieria sulphuraria*. *Biophys. J.* 100: 135–143.

Tokutsu, R., Kato, N., Bui, K. H., Ishikawa, T., and Minagawa, J. 2012. Revisiting the supramolecular organization of photosystem II in *Chlamydomonas reinhardtii*. *J. Biol. Chem.* 287: 31574–31581.

van Bezouwen, L. S., Caffarri, S., Kale, R. S., Kouřil, R., Thunnissen, A. M. W. H., Oostergetel, G. T., and Boekema, E. J. 2017. Subunit and chlorophyll organization of the plant photosystem II supercomplex. *Nat. Plants* 3: 17080.

Watanabe, M., Semchonok, D. A., Webber-Birungi, M., Ehir, S., Kondo, K., Narikawa, R., Ohmori, M., Boekema E. J., and Ikeuchi, M. 2014. Attachment of phycobilisomes in an antenna–photosystem I supercomplex in cyanobacteria. *Proc. Natl. Acad. Sci. U. S. A.* 111: 2512–2517.

Wei, X., Su, X., Cao, P., Liu, X., Chang, W., Li, M., Zhang, X., and Liu, Z. 2016. Structure of spinach photosystem II-LHCII at a of 3.2 Å resolution. *Nature* 534: 69–76.

Wientjes, E., Oostergetel, G. T., Jansson, S., Boekema, E. J., and Croce, R. 2009. The role of LHCA complexes in the supramolecular organization of photosystem I. *J. Biol. Chem.* 284: 7800–7807.

Yadav, K. N. S., Semchonok, D. A., Nosek, L., Kouřil, R., Fucile, G., Boekema, E. J., and Eichacker, L. A. 2017a. Supercomplexes of plant photosystem I with cytochrome b6f, light-harvesting complex II and NDH. *Biochim. Biophys. Acta* 1858: 12–20.

Yadav, S. K. N., Semchonok, D. A., Xu, P., Drop, B., Croce, R., and Boekema, E. J. 2017b. Interaction between the photoprotective protein LHCSR3 and C2S2 photosystem II supercomplex in *Chlamydomonas reinhardtii*. *Biochim. Biophys. Acta* 1858: 379–385.

Yakushevska, A. E., Jensen, P. E., Keegstra, W., van Roon, H., Scheller, H. V., Boekema, E. J., and Dekker, J. P. 2001. Supermolecular organization of photosystem II and its associated light-harvesting antenna in *Arabidopsis thaliana*. *Eur. J. Biochem.* 268: 6020–6021.

Photoprotective excess energy dissipation

ALBERTA PINNOLA, DIANA KIRILOVSKY, AND ROBERTO BASSI

ABBREVIATIONS

APC	Allophycocyanin
ATP	Adenosine triphosphate
C_2S_2	Photosystem II supercomplex including two core complexes and two LHCII trimers
Car(s)	Carotenoid(s)
C–B–B cycle	Calvin–Benson-Bassham cycle
Chl(s)	Chlorophyll(s)
Ddx	Diadinoxanthin
Dtx	Diatoxanthin
EED	Excess Energy Dissipation
ET	Electron transport
FLM	Fluorescence lifetime measurements
F_m	Maximal fluorescence of dark-adapted sample
F_m'	Maximal fluorescence of light-exposed sample
FRP	Fluorescence recovery protein
hECN	3′-hydroxyechinenone
KO	Knockout
LHC	Light-harvesting complex
Lhca (Lhcb)	Light-harvesting complex polypeptide of photosystem I (or II, respectively)
LHCSR	Light-harvesting complex stress-related

Lut	Lutein
NADPH	Nicotinamide adenine dinucleotide phosphate
NPQ	Nonphotochemical quenching
OCP (OCP^o, OCP^r)	Orange carotenoid protein (its orange [o] and red [r] form, respectively)
PB	Phycobilisome
PC	Phycocyanin
PE	Phycoerythrin
PEC	Phycoerythrocyanin
P$_i$	Inorganic phosphate
PSBS	Photosystem II subunit S
PSI (PSII)	Photosystem I (or II, respectively)
qE	Energy-dependent quenching
qI	Photoinhibitory quenching
qM	Decrease in fluorescence yield due to chloroplast relocation
qZ	Zeaxanthin-dependent quenching
RC	Reaction center
ROS	Reactive oxygen species
Viola	Violaxanthin
WT	Wild type
Zea	Zeaxanthin
ΔpH	Proton gradient

11.1 SUMMARY

All oxygenic photosynthetic organisms possess a variety of mechanisms for the regulation of light-harvesting efficiency in response to variable light intensity. Among these, excess energy dissipation (EED) transforms excitation from light absorbed in excess into heat. This process is usually named from an operational definition: Nonphotochemical quenching (NPQ) of chlorophyll fluorescence. Heat dissipation has a photoprotective effect since it prevents the formation of Reactive Oxygen Species (ROS), and yet it can be reversed for photosynthesis to resume when light intensity is brought back to normal intensity. Although present among all oxygenic photosynthetic organisms, NPQ is activated through distinct molecular effectors depending on the taxa. In cyanobacteria, it is operated by the orange carotenoid protein (OCP), which is directly activated by light. In unicellular eukaryotes, such as green algae and other algal groups, NPQ activity depends on a light-harvesting complex (LHC)-like protein called LHC stress-related (LHCSR). In land plants, such as

Arabidopsis thaliana, NPQ depends on a related protein: Photosystem II subunit S (PSBS). LHCSR and PSBS respond to lumenal pH and are activated via protonation of acidic residues essential for activity. The mode through which protonation of the trigger proteins is translated into NPQ is the matter of lively debate. In the case of LHCSR, the protein binds both chlorophylls (Chl) and carotenoids (Cars) and its lifetime is reduced by acidification, suggesting that its interaction with the Photosystem II (PSII) antenna system is likely to cause excitation energy trapping and dissipation within the LHCSR protein itself. The case of PSBS is more obscure since this polypeptide does not have pigment binding sites that would allow for direct energy dissipation, although it does have protonatable sites essential for NPQ activity. This suggests the quenching site might be induced within PSBS-interacting proteins belonging to the PSII antenna system. Transduction of the pH signal into quenching has been proposed to occur through PSBS-induced reorganization of thylakoid membrane complexes. NPQ is of crucial importance for the control of productivity of crops and algae, thus the understanding of molecular mechanisms underlying this function is critical for domesticating unicellular algae for food and fuel production and further increasing crop productivity.

11.2 INTRODUCTION: ALL OXYGENIC PHOTOSYNTHETIC ORGANISMS EXHIBIT NPQ ACTIVITY

Oxygenic photosynthetic reaction centers harness solar energy for driving electrons from water to NADP$^+$. Electron flow is coupled to H$^+$ transfer from the chloroplast stroma to the thylakoid *lumen*, building a proton gradient for ATP synthesis. ATP and NADPH are then exploited for carbon fixation. Photosynthetic organisms are exposed to strong and rapid changes in light intensity, temperature, and water availability in their natural environment. In response to some of these conditions, limitation in photochemical quenching leads to increased Chl excited state (^{1}Chl*) lifetimes and an increased probability of Chl *a* triplet formation (^{3}Chl*) by intersystem crossing. Chl triplets react with molecular oxygen (^{3}O$_2$) to yield potentially harmful ROS, which are responsible for photoinhibition and oxidative stress (Barber and Andersson, 1992) (Figure 11.1).

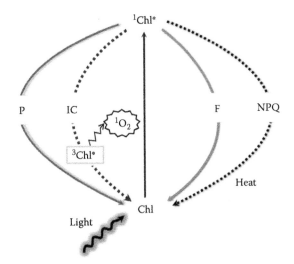

Figure 11.1 The fate of excited chlorophyll (Chl) singlet states. Upon absorption of a photon, Chl is promoted to its first excited state (^{1}Chl*). The return to the ground state can occur through multiple competing pathways: the rate of photosynthetic reactions (P) can be measured by a decrease of fluorescence emission (F). In high-light conditions, P saturates and the lifetime of ^{1}Chl* increases: ^{1}Chl* produces triplet chlorophyll (^{3}Chl*) through intersystem crossing (IC). ^{3}Chl* readily reacts with O_2, forming singlet oxygen ($^1O_2$*), a highly toxic/dangerous/unstable ROS. Heat dissipation (NPQ) decreases ^{1}Chl*, thus preventing ^{3}Chl* and ROS formation.

These events are counteracted by various photoprotection mechanisms consisting either of scavenging of ROS (Asada, 1999) or prevention of ROS formation through downregulation of triplet yield (Dall'Osto et al., 2012), quenching of ^{3}Chl* by xanthophylls (Havaux and Niyogi, 1999), or dissipation of excess ^{1}Chl* (Niyogi, 2000; Külheim et al., 2002). Among regulatory mechanisms, EED of ^{1}Chl* into heat is of major importance. This process has received the early operational definition of NPQ from "Nonphotochemical Quenching of Chl fluorescence" since it is usually estimated from the decrease of fluorescence yield of photosynthetic systems exposed to excess light. In these conditions, photosynthesis is saturated and therefore further quenching (additional to the one induced by photochemical reactions) can only be produced nonphotochemically. The ability to activate an NPQ response has been observed in all the phyla performing oxygenic photosynthesis including cyanobacteria, algae, and plants. Despite its widespread occurrence, NPQ has distinct features in different taxa, possibly shaped by adaptation to different environments (Figure 11.2). In cyanobacteria, NPQ depends on the presence of a soluble Car-binding protein, OCP (Wilson et al., 2006). In eukaryotic algae and plants, NPQ requires specific members of the extended LHC protein family (see Chapters 4 and 8).

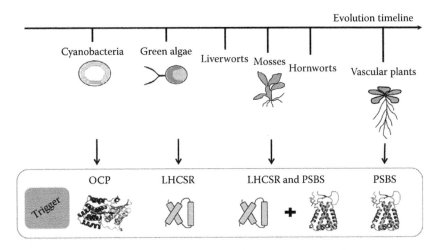

Figure 11.2 Protein subunits indispensable for triggering NPQ response through evolution. NPQ response is present in all organisms performing oxygenic photosynthesis including cyanobacteria, algae, and plants and yet evolved. In cyanobacteria, NPQ is triggered by OCP, in green algae by LHCSR; mosses have both LHCSR and Photosystem II Subunit S (PSBS) proteins, while in higher plants, PSBS only is found. For OCP and PSBS, crystal structures have been resolved.

In particular, the LHCSR protein is essential in green algae and diatoms; mutants depleted in this protein have a reduced level of NPQ (Peers et al., 2009; Bailleul et al., 2010). In vascular plants, NPQ instead depends on the presence of the PSBS (Li et al., 2000; Kasajima et al., 2011). PSBS is an LHC-like protein that possesses special features, such as four transmembrane helices rather than three as in most other eukaryotic antenna proteins, and the lack of well-defined Chl binding motifs in its sequence (Li et al., 2000; Fan et al., 2015).

11.2.1 How to detect and measure NPQ

NPQ is defined as the quenching of Chl singlet excited states, the same chemical species that yield Chl fluorescence. Thus, NPQ can be detected as a decrease of Chl a (and/or bilin) fluorescence intensity. Since the fluorescence yield of the photosynthetic apparatus is variable depending on the redox state of PSII electron acceptors, NPQ is usually assessed in light conditions that saturate electron transport (ET). Temperature and CO_2 availability can affect the ET rate and therefore the light intensity at which photosynthesis is saturated.

11.2.1.1 PULSE AMPLITUDE MODULATED FLUOROMETRY MEASUREMENTS

Pulse amplitude–modulated fluorometric measurements (PAM) are the most popular type of NPQ measurements (Figure 11.3 and see Chapter 23). PAM measurements are based on the determination of F_m (the maximal amplitude of fluorescence obtained from a photosynthetic tissue or cell by shining a light flash with saturating intensity) in dark-adapted samples and following the decrease in F_m' (F_m measured in light-exposed samples) during exposure to actinic light (Figure 11.3a). The normalized difference $(F_m - F_m')/F_m'$ is called "quenching" (Figure 11.3b), and its time course can be measured with different actinic lights as well as during relaxation in low-light or dark conditions. The measurement of NPQ by pulse amplitude–modulated fluorometry relies on the assumption that the rate of photon absorption remains the same during the transition from light to dark and the following period of measurement. Although this might appear obvious, in fact it is not, due to reorganization of chloroplast distribution within the cell caused by activation of blue light receptor *phototropin* (Kasahara et al., 2002), thus leading to differences in optical density and fluorescence emission unrelated to quenching events (Cazzaniga et al., 2013). In cyanobacteria, due to their particular

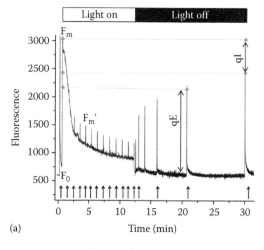

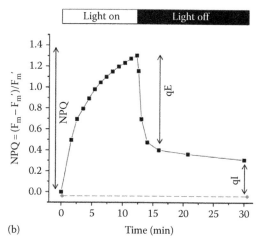

(a)

(b)

Figure 11.3 **(a)** Pulse amplitude fluorometry measurement from a *N. tabacum* leaf disk. NPQ can be measured as the difference between F_m (the fluorescence measured in dark-adapted state upon a saturating light pulse) (arrows on the *x* axis) and F_m measured at different times during illumination (F_m'). Upon switching off the actinic light, F_m' recovers within a few minutes, reflecting relaxation of the qE component of NPQ. **(b)** NPQ kinetic as obtained by analyzing the fluorescence data in **(a)** according to NPQ = $(F_m - F_m')/F_m'$.

antenna, the phycobilisomes (PBs), the measurements and calculations of NPQ are slightly different and are further described in Kirilovsky (2015).

11.2.1.2 FLUORESCENCE LIFETIME MEASUREMENTS

A more accurate measurement of quenching can be performed by Fluorescence Lifetime Measurements (FLM) since FLM is independent of the concentration of the fluorophore (Lakowicz, 1999). NPQ measurement by FLM have been introduced by Gilmore and coworkers (Gilmore et al., 1995) and further improved by Holzwarth et al. (Lambrev et al., 2010) with the introduction of time-resolved spectral analysis allowing for the resolution and identification of multiple quenching species (Nilkens et al., 2010). The limitation of this method is due to the limited time resolution for each measurement, thus restricting the analysis to steady-state unquenched or fully quenched states. Higher time resolution has been obtained by single wavelength analysis (Sylak-Glassman et al., 2014). Despite the higher performance of the different lifetime methods, the pulse fluorometry is still the most widely used method because of its low cost for hardware and easy handling.

11.2.1.3 LOW-TEMPERATURE FLUORESCENCE SPECTROSCOPY

At room temperature, the fluorescence yield of PSII is far higher than that of PSI, thus the measurements of the fluorescence yield of leaves and algal cells is dominated by the PSII emission. At low temperature (77 K), the fluorescence yields of PSI and PSII are similar (Cho et al., 1966), and the two emissions can be distinguished by the different wavelength maxima, namely 685 and 695 nm for PSII and 725–740 nm for PSI, depending on the species. Rapid freezing of plant tissues in liquid nitrogen preserves the fluorescence yield characteristics of the physiological state determined by light and metabolic conditions (Krause et al., 1983; Pinnola et al., 2015a), and allows for fluorometric analysis of emissions from both PSs (Figure 11.4a and b). A problem with this type of measurement is the irreproducible optical properties that the samples assume during freezing, which makes it difficult to compare with other samples. To overcome this problem, fluorescence lifetime analysis has been applied (Chukhutsina et al., 2014) or the Green Fluorescent Protein (GFP) fluorophore was added as an internal standard

(Pinnola et al., 2015a). Although low-temperature analysis is a static method, quenching kinetics can be reconstructed by freezing samples at different times during light treatment and dark recovery (Figure 11.4c and d) (Pinnola et al., 2015a).

11.3 NPQ IN PROKARYOTES

In cyanobacteria, the membrane intrinsic light-harvesting antenna system (LHC) is absent. Instead, light is absorbed by a large extramembrane complex known as the phycobilisome (PB) (see Chapter 5). PBs are composed of several types of colored proteins (red, violet, and blue), named phycobiliproteins, that covalently attach bilins (open tetrapyrrole chains) and require linker peptides for organization and function (for reviews about PBs, see Glazer, 1984; Grossman et al., 1993; MacColl, 1998; Tandeau de Marsac, 2003; Adir, 2005 and chapter Adir). PBs have a trimeric core (Figure 11.5) constituted by allophycocyanin (APC, blue) from which rods radiate. In most freshwater species, rods contain only phycocyanin (PC, blue), while in marine species, phycoerythrin (PE, red) and phycoerythrocyanin (PEC, violet) are present in the distal ends of the rods. These complexes are attached to the outer surface of the thylakoid membranes (Gantt and Conti, 1966a,b) via the large, chromophorylated core membrane linker protein ApcE (Redlinger and Gantt, 1982), which also serves as the terminal energy emitter. Two other chromophorylated proteins, ApcD and ApcF, also function as terminal emitters. Harvested light energy is transferred from the terminal emitters to the Chls of PSII and PSI (Mullineaux, 1992; Rakhimberdieva et al., 2001).

In cyanobacteria, as in plants and algae, the NPQ mechanism is accompanied by a diminution of PSII-related fluorescence emission; however, the molecular processes behind this NPQ mechanism are simpler and completely different from those existing in plants and algae (see reviews: [Kirilovsky and Kerfeld, 2012, 2013, 2016; Kerfeld and Kirilovsky, 2013; Kirilovsky, 2015]).

Three elements are essential and sufficient for the cyanobacterial NPQ photoprotective mechanism: the OCP, the FRP, and the PBs (Gwizdala et al., 2011). The OCP is a soluble protein of 35 kDa, which binds a Car molecule, the 3′-hydroxyechinenone (hECN) (Wu and Krogmann, 1997; Kerfeld et al., 2003).

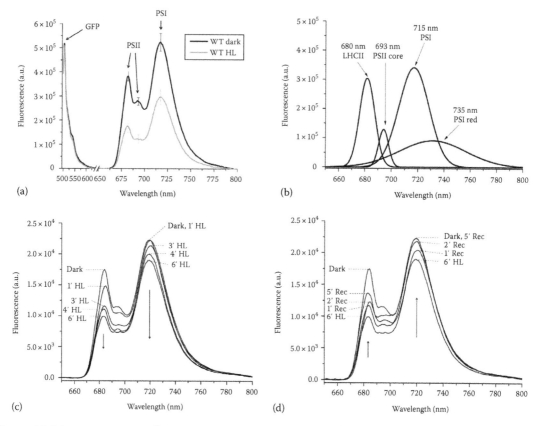

Figure 11.4 Low-temperature fluorescence spectra using Green Fluorescent Protein (GFP) as method to measure quenching. **(a)** 77 K fluorescence emission spectra of chloroplasts maintained in the dark (black) or exposed to high light (HL) (850 μmol photons m^{-2} s^{-1} for 10 min [gray]) before rapidly freezing in liquid nitrogen. GFP was added in the samples as internal standard. **(b)** Spectra deconvolution using four Gaussians peaking at 682, 693, 715, and 735 nm. Each peak corresponds to LHCII (682 nm), PSII core (693 nm), and PSI-LHCI (715 + 735 nm). **(c)** 77 K spectra of chloroplasts exposed in HL for different times. **(d)** 77 K spectra of chloroplast exposed in HL for 6 minutes (6′ HL) and then recovered for different times in the dark (Rec). Arrows indicate the direction on changes in the amplitude of the peaks. Spectra were obtained upon excitation at 475 nm and were normalized to the amplitude of the 513 nm emission peak of GFP. a.u., arbitrary units. (From Pinnola, A. et al., *Plant Cell*, 27(11), 3213, 2015a. Copyright American Society of Plant Biologists.)

The OCP is composed of two domains: an α-helical N-terminal domain (residues 15–165) unique for cyanobacteria and an α/β C-terminal domain (residues 190–317) that is member of the nuclear transport factor 2-fold superfamily (Figure 11.6a) (Kerfeld et al., 2003). The Car hECN spans both domains of the protein (Kerfeld et al., 2003; Wilson et al., 2010), and it is in all-trans configuration (Kerfeld et al., 2003; Polívka et al., 2005; Wilson et al., 2008). The two domains are connected by a long flexible linker (about 25 residues). In addition, they interact through two regions: the first 19 amino acids of the OCP and an interface, across which the

Car spans the protein. The closed conformation of the OCP is stabilized by hydrogen bonds and van der Waals interactions existing at the interface between the N-terminal and C-terminal domains (including Arg155-Glu244) and between the N-terminal arm (the first 19 amino acids) and the C-terminal domain (Kerfeld et al., 2003; Wilson et al., 2010). The close structure of the protein avoids a large exposure of the Car to the solvent. The FRP is a soluble protein of 13 kDa without any chromophore (Boulay et al., 2010). Its secondary structure consists only of α-helices, and its tertiary structure seems to be unique (Sutter et al., 2013). The FRP is active as a dimer. On one

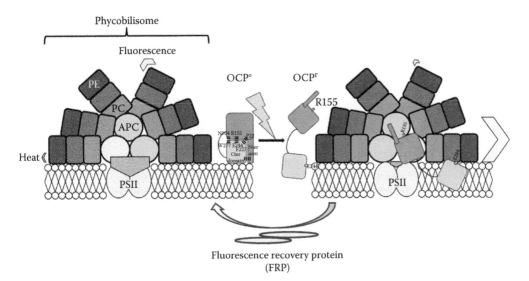

Figure 11.5 Working model of the Orange Carotenoid Protein (OCP)-related photoprotective mechanism. In darkness and under low irradiance, the OCP is predominantly in the orange form (OCP°) and it is not (or weakly) attached to the phycobilisomes (PBs). Absorption of blue-green light by the carotenoid induces changes in the conformation of the carotenoid, provoking the breakage of the hydrogen bonds of the carotenoid carbonyl with Tyr201 and Trp288, leading to the translocation of the carotenoid into the N-terminal domain and structural changes in the C-terminal domain provoking the opening of the OCP. This converts the orange OCP into the red active OCP (OCPʳ). The N-terminal domain of the OCPʳ binds to one of the basal cylinders of the PBs inducing the total quenching of fluorescence PBs. The first site of quenching seems to be one bilin emitting at 660 nm. The Fluorescence Recovery Protein (FRP) as a dimer binds to the C-terminal of the OCPʳ, helps to detach the OCP from the PBs, and accelerates the OCPʳ to OCP° conversion. APC, allophycocyanin; PC, phycocyanin; PE, phycoerythrin.

side of the FRP dimer, a stretch of 100% conserved amino acid residues forms a network of hydrogen bonds between the two monomers, stabilizing the dimer (Figure 11.6b) (Sutter et al., 2013).

A major difference between the eukaryotic NPQ and the cyanobacterial NPQ is the mechanism that triggers the energy dissipation function. In plants and algae, the activating signal is the low lumenal pH induced by excess ET *vs* ATP/NADPH-consuming reactions. In cyanobacteria, NPQ activation is directly induced by strong blue-green or white light (El Bissati et al., 2000; Wilson et al., 2006). Thus, while in plants and algae, the NPQ mechanism is induced under all conditions generating an overreduction of the photosynthetic ET chain (low CO_2, nutrient starvation, low temperature, high light), in cyanobacteria, the NPQ mechanism is induced only under high-light conditions. Nevertheless, most of the stress conditions induce the expression of the *ocp* gene. When the cellular OCP concentration is high, the induction of the NPQ mechanism occurs at lower light intensities.

The OCP is a photoactive protein (Wilson et al., 2008); absorption of blue-green light induces conformational changes in the Car and in the protein converting the inactive orange form (OCP°) into an active red form (OCPʳ) (Figure 11.5). Since this conversion has a very low yield, accumulation of the active red form occurs only under high-irradiance conditions when photoprotection is needed. Only the red form is able to associate with the PBs and induce quenching (Wilson et al., 2008; Gwizdala et al., 2011). The OCP binding is light-independent: once OCP is photoactivated and converted to the red form, it can bind to PBs even in darkness (Gwizdala et al., 2011).

Light absorption by the Car and the concomitant change of its conformation provokes a break of the hydrogen bonds between its keto group and Tyr201 and Trp288 of the protein (Wilson et al., 2011; Liu et al., 2004). This induces a subsequent reorganization of the β-sheet core, leading to the breakage of the amino acid interactions at the mean interface of the two domains between the two domains

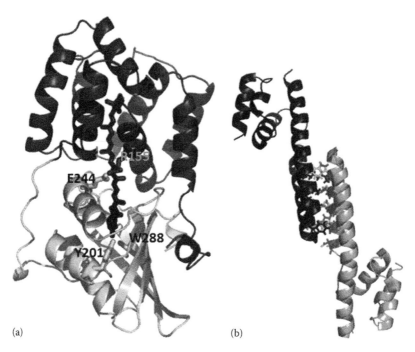

(a)   (b)

Figure 11.6 The structure of the dark orange OCP **(a)** and of the FRP dimer **(b)**. **(a)** The OCP is composed by an a-helix N-terminal domain (dark gray) and an α helix/β sheet C-terminal domain (light gray) relied upon by a flexible linker. The Car shown as sticks is hydrogen bonded to the protein via its carbonyl that interacts with Tyr201 and Trp288. The salt bond between Arg155 and E244 is important for the stabilization of the closed form of the OCP, and Arg 155 is essential for OCP binding to PBS. **(b)** FRP dimer. The Arg60, Asp54, and Trp50 that are essential for the FRP activity are shown as sticks.

in their interface and at the interface between the N-terminal arm and the C-terminal domain (Wilson et al., 2012; Liu et al., 2014; Gupta et al., 2015; Maksimov et al., 2015). Large changes in the tertiary structure of the protein are observed upon photoactivation since the opening of the protein causes a complete separation of the two domains (Gupta et al., 2015). In contrast, the secondary structure of each domain is only slightly altered during photoactivation, with the exception of the unfolding of the N-terminal α-helix (Gupta et al., 2015; Leverenz et al., 2015). The opening of the protein exposes the Arg155, which is essential for the OCP binding to PBs (Wilson et al., 2012).

The breakage of the hydrogen bonds between the Car carbonyl and the protein also allows the free movement of the Car. A substantial translocation of the Car (12 Å) deeper into the N-terminus occurs during photoactivation (Leverenz et al., 2015). When OCPr is partially proteolyzed, the N-terminal and C-terminal domains can be isolated separately. The Car remains associated with the N-terminal domain, which is able to interact with isolated PBs

and quench their fluorescence (Leverenz et al., 2014). Thus, the N-terminal domain is the active module of OCP and the C-terminal domain is the regulator of its activity (Leverenz et al., 2014) (Figure 11.5).

OCP binds the core of the PBs to one of its basal cylinders (Jallet et al., 2012; Zhang et al., 2014). All PBs contain two basal APC cylinders formed by two trimers emitting at 660 nm and two trimers emitting at 680 nm. The latter contains the ApcD subunit or the ApcF-ApcE subunit. The open structure of the OCPr protein and the possible interaction between Arg155 and the negative charges around one of the bilin chromophores of the core of the PB permits a closer interaction between the Car and the bilin, allowing an efficient energy (or charge) transfer from the excited bilin to the Car (Tian et al., 2011; Berera et al., 2013). The primary site of quenching in the PBs is at one of APC bilins emitting at 660 nm (Tian et al., 2011). This quenching is very fast and efficient and significantly decreases the rate of energy transfer to the reaction centers (Tian et al., 2011, 2012;

Jallet et al., 2012; Zhang et al., 2014). The monomolecular interaction of OCP with the PBs results in an almost complete quenching of its fluorescence (Gwizdala et al., 2011; Tian et al., 2011).

Several models of the OCP–PBs interaction were proposed (Stadnichuk et al., 2012, 2015; Zhang et al., 2014). Two of them were proposed before the opening of the protein, and the specific interaction of the N-terminal domain of the OCP with the PBs was demonstrated. Cross-linking experiments suggested that OCP binds between one APC660 trimer and the APC680 trimer containing the ApcF and ApcE subunits in one of the basal core cylinders (Zhang et al., 2014). Another model proposes that OCP interacts with one of the external bilins of the trimer containing ApcE (Stadnichuk et al., 2012, 2015). The third model proposes that the N-terminal domain is buried in the interface between the external trimer containing ApcD and the trimer containing ApcF-ApcE, while the C-terminal domain floats in solution on the outer surface of the cylinder (Harris et al., 2016). Currently, the site of OCP binding in the PBs still remains an open question.

In conditions where quenching is no longer imperative, the active OCP^r is converted back to the inactive orange form (OCP^o). *In vivo*, the detachment of OCP^r from PBs and its conversion into the orange form need the action of the FRP (Boulay et al., 2010; Sutter et al., 2013). This conversion occurs in darkness and is temperature dependent, being faster at higher temperatures (Wilson et al., 2008). FRP interacts with the C-terminal domain of the OCP, and Arg60, Asp50, and Trp54 amino acid residues are essential for FRP activity (Sutter et al., 2013). Due to the action of FRP, the concentration of OCP^r is null or very low in the dark, and recovery of the lost fluorescence and of the full size of antenna are observed.

In conclusion, in cyanobacteria the amplitude of energy quenching depends on the concentration of the active OCP^r and its affinity for PB. The concentration of OCP^r primarily depends on the concentration of OCP in the cell and the light intensity. However, it also depends on the stability of the unbound OCP^r and the concentration of FRP. The metastable OCP^r can reverse alone to the orange form or it can interact with an FRP molecule, which facilitates the rapid conversion to the inactive form (Gwizdala et al., 2011; Kuzminov et al., 2012; Wilson et al., 2012).

11.4 NPQ IN EUKARYOTES

Three components have been resolved that contribute to fluorescence decrease in eukaryotic organisms: energy-dependent quenching (qE), zeaxanthin-dependent quenching (qZ), and photoinhibitory quenching (qI). A further component was recently recognized to affect fluorescence yield, although its mechanism is not of quenching type (qM). The first type, qE, depends on lumen acidification and develops within seconds upon an increase in light intensity (Niyogi, 2000). qE relaxes on a 1–2 minute timescale in the dark (Demmig-Adams et al., 1996; Horton, 1996). Zeaxanthin (Zea) is accumulated within several minutes of exposure to excess light and causes enhancement of NPQ (Horton, 1996; Nilkens et al., 2010; Förster et al., 2011; Jahns and Holzwarth, 2012; Nichol et al., 2012). If Zea is already present before illumination (e.g., mutants constitutively accumulating Zea or leaves recently exposed to excess light), a faster qE rise is induced upon exposure to light (see below). Zea is also responsible for a slower fluorescence recovery component rising in parallel with back conversion of Zea into violaxanthin (Viola), which takes up to 1 hour in lab experiments (Dall'Osto et al., 2005; Reinhold et al., 2008). qM is the most recently reported component that decreases the fluorescence yield of leaves upon exposure to excess light. Rather than a genuine quenching, this component depends on the lower photon absorption caused by chloroplast movement away from excess light and toward the cell walls, where they align parallel to the incident light direction. The resulting decrease of light absorption contributes to photoprotection (Kasahara et al., 2002; Cazzaniga et al., 2013; Dall'Osto et al., 2014). The slowest quenching component, qI, is associated with PSII photoinhibition and occurs under a variety of conditions of sink limitation. Its relaxation requires hours to days and is sensitive to chloroplast translation inhibitors (Walters and Horton, 1991).

11.4.1 Eukaryotic NPQ is a feedback response

The largest and fastest component of NPQ is qE. Its activation depends on the establishment of a pH gradient across the thylakoid membrane, as shown by its inhibition using nigericin, which

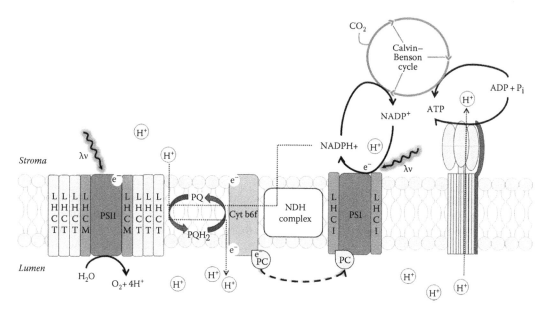

Figure 11.7 The triggering of nonphotochemical quenching (NPQ). Photosynthetic electron transport is coupled to proton accumulation into the lumen. At the same time, protons are brought back to the stroma by ATPase and the synthesis of ATP during steady-state photosynthesis. ATP and NADPH produced are used by Calvin–Benson–Bassham cycle to fix CO_2 and restore inorganic phosphate (P_i) and ADP, the substrates of ATPase. When downstream reactions are saturated, ATP accumulates while ADP becomes low or absent, thus impairing ATPase activity and causing lumen acidification. NADPH in excess is used to activate cyclic electron transport, thus further decreasing lumenal pH. NPQ activation in the antenna system of Photosystem II (PSII), triggered by protonatable residues, is thus the result of ATP and NADPH feedback. LHCT, trimeric light-harvesting complex II; LHCM, monomeric light-harvesting complex II; PSI (PSII), photosystem I (or II, respectively); LHCI, light-harvesting complex I; NDH complex, NADH dehydrogenase–like complex; Cyt b6f, cytochrome b6f; PC, plastocyanin; PQ, plastoquinone; PQH_2, plastoquinol; λv, photon.

disrupts membrane integrity (Walters and Horton, 1991; Gilmore et al., 1998). The NPQ dependence on lumenal pH generates a feedback loop between the production and consumption of the products of light reactions. As depicted schematically in Figure 11.7, photosynthetic complexes in thylakoid membranes convert light energy into NADPH and concomitantly build up a proton gradient (ΔpH) for ATP synthesis. ATP is then used by the Calvin–Benson–Bassham (C–B–B) cycle for CO_2 fixation, yielding inorganic phosphate (P_i) and ADP, which supports ATPase activity. Additional supply of P_i to the chloroplast is coupled to the export of triose phosphates as products of the C–B–B cycle. Any excess in light absorption rate with respect to the cell's ability to hydrolyze ATP causes a depletion of ADP + P_i in the chloroplast, limiting ATPase activity and consequently proton export from the thylakoid lumen. The imbalance between light absorption rate and use of ATP translates into an increase in proton concentration in the lumen, which is the major signal activating photoprotection responses. NADPH also accumulates and is used to activate cyclic ET (Munekage et al., 2004), thus further decreasing lumenal pH. Increased lumenal proton concentration protonates exposed residues in PSBS and LHCSR proteins, which triggers NPQ activation in the antenna system of PSII. This feedback mechanism aims to adjust light harvesting efficiency to the cell's capacity to use the products of photochemical reactions. On a longer timescale, the consumption of reduced carbon by the plant's sinks also influences NPQ activity. The NPQ mechanism of eukaryotes is thus distinct from that implemented in cyanobacteria, which depends only on the photoconversion of OCP and is independent of ΔpH or the rate of downstream metabolic reactions.

11.4.1.1 LHCSR IS RESPONSIBLE FOR NPQ ACTIVITY IN LOWER PLANTS AND ALGAE

NPQ in the green alga *Chlamydomonas reinhardtii* (Chlorophyta) depends on the LHC-like protein LHCSR (also known as Li818 in an older nomenclature). A similar protein in diatoms is called LHCX, and its involvement in NPQ was shown by genetic analysis in two species of diatoms (Gagné and Guertin, 1992; Peers et al., 2009; Bailleul et al., 2010; Zhu and Green, 2010) (see chapter Büchel). LHCSR-like sequences have been identified in all eukaryotic algal groups, suggesting that algae rely on LHCSR-like proteins for NPQ. While no LHCSR genes are present in any known vascular plant, LHCSR has been identified in lower plants, for example, in the moss *Physcomitrella patens* (Alboresi et al., 2008). The generation of knockout (KO) lines specifically depleted in LHCSR demonstrated that this protein is the major component responsible for NPQ activity in mosses (Alboresi et al., 2010; Gerotto et al., 2012).

The level of LHCSR accumulation has been reported to strongly depend on light intensity. While LHCSR is constitutively present even in low-light acclimated cells in some algal species (Zhu and Green, 2010; Gerotto et al., 2011), in other species, such as *C. reinhardtii*, acclimation to light environment has a large influence on NPQ activity, through accumulation of LHCSR in high light (Peers et al., 2009) or salt stress (Azzabi et al., 2012). In all species, including the moss *P. patens*, modulation of LHCSR accumulation by growth conditions seems to be a common strategy to optimize NPQ amplitude, suggesting there might be a price to pay for LHCSR accumulation in low light.

11.4.1.1.1 LHCSR proteins

The LHCSR protein has not been purified from algae so far, and current knowledge on its properties derive from the recombinant protein refolded with pigments *in vitro* (Bonente et al., 2011; Liguori et al., 2013). The moss LHCSR1 was expressed in heterologous plant systems and partially purified from moss (Pinnola et al., 2013, 2015b). Like most eukaryotic LHC antennae (see Chapter 4), LHCSR has three predicted transmembrane helices and binds pigments, both Chl and Car (Figure 11.8a) (Bonente et al., 2011). LHCSR/LHCX activity has been shown in organisms with different pigment composition: diatoms, where the major

Car is fucoxanthin, and green algae, where lutein (Lut) is the most abundant xanthophyll (see also Chapter 8). LHCSR is thus likely to be flexible in binding pigments, although it does not bind neoxanthin (Bonente et al., 2011; Pinnola et al., 2015b). Its presence in organisms with different accessory Chl (Chl *c* and *b*) suggests it can also bind different Chl types. Alternatively, it has been suggested that Chl *a* is the only required porphyrin, based on the substoichiometric Chl *b* content present in LHCSR (Pinnola et al., 2015b) and the ability to refold *in vitro* with Chl *a* only (Bonente et al., 2011). One major difference between antenna proteins and LHCSR is the enhanced ability of LHCSR to dissipate energy as heat (Figure 11.8b). In fact, even in detergent solution, LHCSR shows a shorter fluorescence lifetime in respect to other LHCs, suggesting it efficiently dissipates energy as heat (Figure 11.8c) (Bonente et al., 2011). This observation suggests that LHCSR operates in thermal energy dissipation by receiving energy from nearby antenna complexes and catalyzing the dissipation of excited states into heat. Since this ability to dissipate energy is present in the isolated protein *in vitro* at neutral pH, LHCSR appears to thermally dissipate a fraction of its excitation energy constitutively. Upon illumination with high light, the decrease in lumenal pH further increases this quenching activity (Bonente et al., 2011).

11.4.1.1.2 Additional factors involved in LHCSR-dependent NPQ: Carotenoids and Lhcb subunits

While LHCSR is indispensable for NPQ in eukaryotic algae, other factors do modulate its activity. Cars play a major role in NPQ in both plants and algae (Niyogi et al., 1997, 2001). Car composition of photosynthetic organisms is not constant but depends on environmental conditions. In particular, xanthophyll cycles reversibly convert Viola to Zea in green algae and plants and diadinoxanthin (Ddx) to diatoxanthin (Dtx) in diatoms in high light. These reactions are typically slowly reversed when the organisms return to low light. The xanthophyll cycle in plants and algae is operated by violaxanthin de-epoxidase and activated by low lumenal pH, like NPQ induction (Arnoux et al., 2009). The de-epoxidated xanthophylls apparently affect NPQ in several organisms, with their level being correlated with NPQ amplitude. In diatoms,

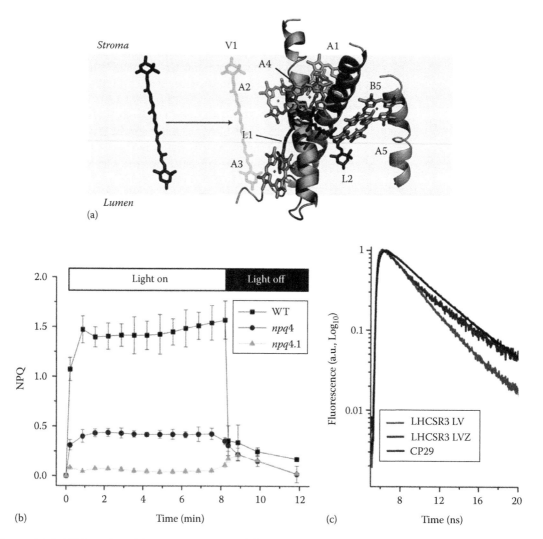

Figure 11.8 NPQ and LHCSR protein. **(a)** Model showing chlorophyll and xanthophyll chromophores bound to different sites in LHCSR3 from *Chlamydomonas reinhardtii*. The model was built by a homology based on the crystal structure of LHCII by Liu et al. (2004). **(b)** NPQ measured in *C. reinhardtii* cells of WT (CW15), *npq4* mutant depleted in LHCSR3.1/3.2 isoforms, but still retaining the LHCSR1 isoform and *npq4.1* mutant depleted in LHCSR3.1, LHCSR3.2, and LHCSR1 isoforms. **(c)** Comparison of fluorescence lifetime decays of LHCSR and an antenna complex (CP29). a.u., arbitrary units. (a and c: From Bonente, G. et al., *PLoS Biol.*, 9, e1000577, 2011.)

NPQ activity is closely associated with high Dtx level, suggesting that this pigment affects LHCSR quenching activity (Ruban et al., 2004; Bailleul et al., 2010). The same holds for the moss *P. patens*, where LHCSR-dependent NPQ is strongly upregulated by Zea (Pinnola et al., 2013). In contrast, in other species, the influence of the xanthophyll cycle may be small, as in the case of *C. reinhardtii*, where the *npq1* mutant, which is unable to synthesize Zea, exhibits NPQ activity similar to wild type (WT). This suggests that the LHCSR quenching ability in this species does not respond to Zea (Bonente et al., 2011). In agreement with these data, it was found that the properties of LHCSR proteins differ depending on species and may well modulate the features of the NPQ response *in vivo*. Free-floating flagellate unicellular algae show Zea-independent NPQ, while species forming biofilms on substrate show a strong dependence on Zea (Quaas et al., 2015).

Besides xanthophylls, interactions with neighboring LHC proteins modulate NPQ. In *C. reinhardtii*, depletion of LhcbM1 and LhcbM6, which are among the major components of the PSII antenna complex, caused a major decrease in NPQ activity (Elrad et al., 2002). However, depletion of other, closely related PSII antenna, LhcbM2 and LhcbM7, has no effect on NPQ activity (Ferrante et al., 2012). Although direct experimental evidence is still missing, this result suggests that LHCSR may require LhcbM1 as a docking site for excitation energy spillover upon lumen acidification. Also supporting this view is the report that in mosses LHCSR1 is localized in the stroma membranes, together with an LHCII subpopulation that serves as supplemental antenna for PSI (Pinnola et al., 2015a). When LHCSR1 from the moss *P. patens* is expressed in tobacco, its activity is five times lower, consistent with a similarly decreased LHCII abundance in stromal membranes (Pinnola et al., 2015b).

11.4.1.2 NPQ ACTIVITY IN VASCULAR PLANTS IS CATALYZED BY PSBS

Genomes of vascular plants do not contain *lhcsr* genes (Alboresi et al., 2008). Instead, they include a different LHC-like protein, PSBS, which is responsible for NPQ activation in plants (Li et al., 2000). Mutants lacking PSBS are, in fact, unable to activate the NPQ response. Clear evidence for PSBS-dependent qE activity has been found in several plant species, including *A. thaliana*, rice, and the moss *P. patens*, based on the phenotype of PSBS-less mutants. Genes for PSBS are absent in diatoms and red algae but have been identified in all Chlorophyta genomes sequenced so far, suggesting a monophyletic origin in green algae and plants of *Viridiplantae* (Gerotto and Morosinotto, 2013). However, even in unicellular algae carrying *psbs* genes, the encoded polypeptides did not accumulate under continuous illumination (Bonente et al., 2008b; Engelken et al., 2010; Gerotto and Morosinotto, 2013) and were detected only transiently upon dark to light transition (Correa-Galvis et al., 2016; Tibiletti et al., 2016). The moss *P. patens*, considered an intermediate in the evolution from algae to higher plants, contains both PSBS and LHCSR, which are active in inducing NPQ (Alboresi et al., 2010). This suggests that PSBS evolved during the land colonization of photosynthetic organisms in order to meet the demands of NPQ in a terrestrial environment.

Recently, it was shown that accumulation of PSBS requires the photoreceptor UVR8, which is sensitive to the UVB light available on land, while LHCSR requires phototropins, activated by the blue light available in water (Petroutsos et al., 2016). PSBS became the only responsible NPQ mechanism for qE in vascular plants. Consistently, PSBS-dependent NPQ activity might also be present in *Charales* and *Coleochaetales*, which are algal taxa derived from plant ancestors (Gerotto and Morosinotto, 2013).

11.4.1.2.1 Properties of the PSBS protein

PSBS is evolutionarily related to the LHC proteins and yet has unique features, such as the presence of four transmembrane helices rather than three as in most members of the LHC family (Engelken et al., 2010) (Figure 11.9a). Similar to that of LHC proteins, PSBS structure is characterized by a central supercoil composed of two transmembrane α-helices. However, the third helix, C, which in LHCs is shorter and perpendicular to the membrane plane, has two symmetrical helices tilted by approximately 30° from the membrane normal and located closer to the central supercoil, closing the gap otherwise filled by pigment molecules (Fan et al., 2015).

PSBS lacks most of the conserved Chl ligands present in LHC proteins (Dominici et al., 2002). Only two arginine residues, also known to stabilize protein structure by participating in salt bridges between adjacent transmembrane helices (Remelli et al., 1999), are conserved, suggesting structural stabilization rather than pigment binding might be the reason for their conservation in PSBS. The pigment-binding capacity of PSBS has been the subject of debate since the identification of its key role in NPQ. Binding of both Chl and xanthophyll was first proposed (Funk et al., 1995). Alternatively, PSBS was suggested to bind Zea only (Aspinall-O'Dea et al., 2002) or no pigments at all (Dominici et al., 2002; Bonente et al., 2008a). Recently, X-ray crystallography has shown no pigment binding to the individual PSBS monomer, while a single Chl was bound at the outer surface, possibly as an aspecific absorption (Figure 11.9c) (Fan et al., 2015). PSBS is the sensor for NPQ activation upon lumen acidification, detected through the protonation of two glutamate residues (Li et al., 2000). Mutants of *A. thaliana* carrying substitutions with

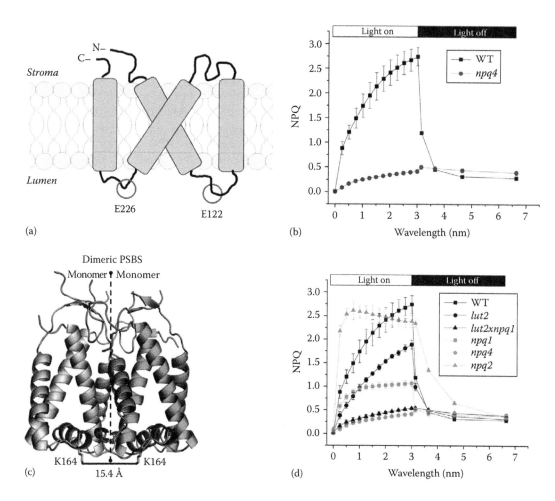

Figure 11.9 NPQ and PSBS protein. **(a)** Scheme of PSBS structure with protonatable residues (gray circles). **(b)** NPQ measured in *A. thaliana* WT (black squares) and *npq4* mutant depleted in PSBS (gray circles). **(c)** Crystal structures of the PSBS protein in its dimeric form. (From Fan, M. et al., *Nat. Struct. Mol. Biol.*, 22, 729, 2015.) **(d)** NPQ dependence on additional factors. NPQ measured in *A. thaliana* WT (black squares), *npq4* mutant depleted in PSBS (gray circles), *lut2* KO (unable to synthesize lutein) (back circles), *npq1* KO (unable to synthesize zeaxanthin) (gray squares), *npq2* KO (accumulate zeaxanthin constitutively) (gray triangles), and *lut2xnpq1* KO double mutant (black triangles).

nonprotonatable glutamine residues showed no qE despite accumulation of the PSBS protein to WT levels or higher (Li et al., 2000, 2004).

11.4.1.2.2 Additional factors involved in PSBS-dependent quenching: Carotenoids and Lhcb subunits

While fundamental for the activation of qE, PSBS is not sufficient. For full NPQ activation, the xanthophylls Zea and Lut are also required (Niyogi et al., 2001). Lack of Zea, as in the *npq1* mutant,

decreases qE to about 30% of the WT level in *A. thaliana*, while constitutive Zea accumulation, as in the *npq2* mutant, makes the onset of qE faster (Figure 11.9d). Since Zea formation from Viola is typically faster than the reverse reaction to Viola catalyzed by Zea-epoxidase, Zea remains present for about 1 hour after returning to low light in the model plant *A. thaliana*. The presence of this slowly epoxidized Zea pool is responsible for the component of NPQ named qZ (Dall'Osto et al., 2005; Nilkens et al., 2010). qZ is likely due to Zea binding to antenna complexes, with Lhcb5 being particularly important for this slow component (Dall'Osto et al., 2005), although recent results have challenged this view (Xu et al., 2015). Recently, it was shown that overexpression of xanthophyll cycle enzymes in tobacco increases the re-epoxidation reaction upon high light and thus recovery from quenching (Kromdijk et al., 2016) with an increase in biomass accumulation. Besides Zea, Lut is also a player in NPQ: Lut-deficient mutants (*lut2* KO) exhibit slower initial NPQ kinetics and a reduced NPQ amplitude over longer periods of time (Figure 11.9d). Overaccumulation of Lut, as in the *szl1xnpq1* KO, compensates in part for the lack of Zea in this double mutant (Li et al., 2009). Consistently, lack of both Lut and Zea, as in *npq1xlut2* KO, yields a null NPQ phenotype similar to the PSBS KO mutant (*npq4*) (Figure 11.9d). Since PSBS binds no pigments, the strong effect of Zea and Lut is hardly explained without the involvement of other xanthophyll-binding proteins. Alternatively, the xanthophyll-dependent effect on NPQ likely occurs through their binding to LHC proteins that are well known to constitutively bind Lut as a major component (Bassi et al., 1993) and undergo exchange of Viola with Zea *in vitro* (Morosinotto et al., 2002). This view has been recently challenged by (Xu et al., 2015) who only found Zea bound to external sites, supporting the hypothesis that conformational changes in LHC proteins may be induced by Zea free in the lipid phase (Ilioaia et al., 2013; Ware et al., 2015). Consistent with a major role of antenna proteins in NPQ activation, extensive depletion of Lhcb proteins strongly reduces NPQ activity (Havaux et al., 2007), while selective deletion of individual Lhcb gene products yields a more mild decrease in NPQ, suggesting a functional redundancy within this protein subfamily. In particular, deletion of monomeric complexes close to the PSII core complex produces a delayed NPQ, while depletion of LHCII component Lhcb1 produces a modest decrease in NPQ amplitude (Andersson et al., 2003) and no decrease at all when Lhcb2 or Lhcb3 were targeted (Andersson et al., 2003; Damkjaer et al., 2009; Leoni et al., 2013; Pietrzykowska et al., 2014). Deletion of Lhcb4 and Lhcb6, the monomeric Lhcb proteins connecting PSII core to the outer trimeric LHCII antenna, produced the strongest effect on qE, while targeting of Lhcb5 affected the slowly relaxing qZ component (de Bianchi et al., 2008, 2011). Together, these results

suggest that the most rapid phase of qE is catalyzed by monomeric LHCs while Lhcb1 is involved in less rapid (>4 minutes) qE establishment.

11.4.1.2.3 Mechanism of PSBS activity: Modulation of thylakoid structural organization

Despite intense effort by many labs, the molecular mechanism for PSBS function has proved to be elusive. PSBS is unlikely to be the site of the quenching reaction(s) dissipating Chl excited states due to lack of pigment cofactors. Yet in the absence of PSBS, other *npq* mutants, including *npq2* and *szl1*, overaccumulating Zea and Lut, respectively, yield an NPQ null phenotype, implying an essential role for PSBS (Li et al., 2009). Recent proposals suggest that PSBS is active as a "facilitator" of the quenching reaction that occurs in Lhcb proteins rather than in PSBS itself (Betterle et al., 2009; Johnson et al., 2011). Antenna complexes associated with PSII have been shown to undergo conformational transitions between states with different lifetimes, suggesting they switch between a light-harvesting and a dissipative state (Moya et al., 2001; Krüger et al., 2011, 2012; Liguori et al., 2015). In their conformation with long lifetimes (highly efficient in light harvesting), Lhcb complexes transfer energy to the PSII reaction center, while in their short-lifetime conformation, a significant fraction of absorbed energy is dissipated as heat before being trapped by the PSII reaction center, thus reducing the probability of light-induced damage under excess-light conditions.

Although changes in fluorescence lifetime and yield in isolated LHC proteins have been experimentally assessed (Moya et al., 2001; Krüger et al., 2011, 2012), the state of individual LHC molecules within the thylakoid membranes and the factors affecting it are not easily accessible (van Oort et al. 2010). Early work with Lhcb proteins reconstituted in liposomes showed that the probability of protein–protein interactions is a powerful modulator of Chl fluorescence lifetime (Moya et al., 2001). More recent analysis has shown that LHCII in PSII-depleted thylakoids exhibits shorter fluorescence lifetimes than in WT, which is further decreased by high-light treatment (Tian et al., 2015).

The finding that PSBS is essential for the reorganization of PSII–LHCII supercomplexes within grana membranes (Betterle et al., 2009; Johnson et al., 2011) provides a model for PSBS-mediated

NPQ activation. In low light, Viola-binding Lhcb proteins are organized in supercomplexes locked in their long-lifetime (light-harvesting) conformation. Lumen acidification, triggering the activation of PSBS, causes dissociation of the external layer of the PSII–LHCII antenna system, which clusters into distinct membrane domains containing either C_2S_2 particles (PSII dimer + Lhcb4 + Lhcb5 + LHCII, see chapter Croce) or the LHCII antenna plus Lhcb6 (Betterle et al., 2009; Johnson et al., 2011). Chl fluorescence lifetime analysis *in vivo* has identified two different quenching sites, each associated to one of the two compartments, suggesting that altered pigment–protein interactions resulting from reorganization of the thylakoid membrane lead the transition into the quenching state (Miloslavina et al., 2011). This model can also explain the enhancing effect of Zea on NPQ. Zea synthesis, also activated by low lumenal pH, enhances the rate of supercomplex dissociation (Betterle et al., 2009) and binds to Lhcb proteins, thereby enhancing their quenching activity (Moya et al., 2001; Lambrev et al., 2010).

11.4.1.3 PSBS- *VERSUS* LHCSR-DEPENDENT NPQ: DIFFERENCES AND SIMILARITIES

PSBS and LHCSR have common features (Figures 11.8 and 11.9), that is, they are both LHC-like transmembrane thylakoid proteins essential for qE and indispensable for NPQ activity. In addition, both proteins have protonatable residues (Li et al., 2004; Bonente et al., 2008a; Ballottari et al., 2015) that are likely to detect the low lumenal pH (Li et al., 2002). Both PSBS- and LHCSR-dependent mechanisms are activated by low lumenal pH and their activities are both modulated by specific xanthophylls. Finally, both LHCSR and PSBS have a dimeric organization (Bergantino et al., 2003; Pinnola et al., 2013, 2015b; Fan et al., 2015). For PSBS, low pH was proposed to induce monomerization (Bergantino et al., 2003), a hypothesis that is so far not supported by structural data (Fan et al., 2015). Differences include presence of Chl and xanthophyll ligands in LHCSR but not in PSBS, implying that the former, but not the latter, might be the actual site of the quenching activity. PSBS is permanently located in grana thylakoids, in close contact with PSII supercomplexes (Harrer et al., 1998), while LHCSR localizes in stroma membranes and grana margins in *P. patens* (Pinnola et al., 2015a) and, according to its ready

extractability with low detergent concentrations, in *C. reinhardtii* as well (Richard et al., 2000). The isolation of a PSII–LHCII supercomplex including LHCSR from *C. reinhardtii* (Tokutsu et al., 2012) supports its localization in grana margin domains. A remarkable property of both PSBS and LHCSR is their weak interactions with photosynthetic complexes. Most reports have found a large majority of these proteins migrate as monomers or dimers in green gels and sucrose gradients, unassociated with other pigment-binding proteins (however, see Harrer et al., 1998; Tokutsu et al., 2012; Teardo et al., 2007), and affinity studies with tagged PSBS (Gerotto et al., 2015) have yielded a large number of possible PSBS interactors including the major LHCII antenna. Functional studies are consistent with LHCSR interaction with a PSI-bound LHCII population (Pinnola et al., 2015a), while PSBR was needed for interaction of LHCSR with PSII (Xue et al., 2015).

11.4.1.3.1 Toward a molecular model for excess energy dissipation in eukaryotes

While the mechanism for NPQ in cyanobacteria is well defined based on protein–protein interactions in the soluble phase, which are highly reproducible *in vitro*, studies on eukaryotic systems have been less conclusive. Nevertheless, milestones have been reached in defining molecular architectures where quenching reactions are activated:

1. PSBS and LHCSR need antenna protein interactors for quenching activation. This is clear from the strong decrease in NPQ activity in the Ch1 mutants lacking LHC proteins and retaining PSBS (Havaux et al., 2007) and from the *C. reinhardtii* cbs3 mutant lacking LHC proteins but retaining LHCSR (Bonente et al., 2011), as well as from the findings of nearest-neighbor analysis. This implies triggers cannot directly interact with PSII core complexes and need LHC antennae partners as binding site(s).
2. The requirement for LHC proteins is specific: only selected LHC proteins are competent as LHC functional interactors, namely Lhcb1, Lhcb4, and Lhcb6 (but not Lhcb2, Lhcb3, or Lhcb5) in plants (Leoni et al., 2013; Pietrzykowska et al., 2014) and LhcbM1 and LhcbM6 (but not LhcbM2 or LhcbM7) in *C. reinhardtii* (Elrad et al., 2002; Ferrante et al., 2012).

3. In the absence of PSBS or LHCSR, all model systems so far analyzed are unable to perform qE (Li et al., 2000; Peers et al., 2009; Alboresi et al., 2010; Dinc et al., 2016), implying that the LHC proteins alone are unable to undergo quenching in the thylakoid membrane to a significant extent despite a low lumenal pH induced in excess light.

4. The reorganization of thylakoid pigment-binding proteins cannot be the only factor responsible for NPQ. In fact, if this was the case, mutations leading to partial or complete disruption of PSII supercomplexes similar to the effect of PSBS on WT upon high-light treatment (Betterle et al., 2009; Johnson et al., 2011) would lead to constitutive quenching, which is not the case (de Bianchi et al., 2008, 2011).

Based on the given results and considering that both LHCSR and PSBS can dimerize (Bergantino et al., 2003; Fan et al., 2015; Pinnola et al., 2015b), a tentative model for their activity can be proposed. Activation of NPQ depends on low pH–induced monomerization followed by interaction of the resulting monomers with specific neighbor LHC proteins, which induces either a conformational change to a quenched state in these LHC interactors, as known in the case of PSBS, and/or a spillover of excitation energy quenched by the short lifetime conformation in the case of LHCSR (Bonente et al., 2011; Liguori et al., 2013). It is not clear how much the LHCSR–LHC interaction allows for excitation energy transfer to LHCSR only or if a conformational change to a quenching state is also induced in the LHC. However, in the case of the pigmentless PSBS, this must be the latter case. Certainly, the capacity for LHC proteins to undergo quenching upon aggregation in low detergent/low pH has been largely documented in both plant and algal LHC proteins (Ruban et al., 1996; Ballottari et al., 2010; Tian et al., 2015).

11.4.1.3.2 On the physical mechanism(s) of quenching reactions

Several models have been proposed for the physical mechanism of quenching:

Model 1: Aggregation-dependent LHCII quenching. This model was proposed based on the early evidence that aggregation of isolated LHC proteins, induced by low detergent concentration and/or low pH, causes a decrease in Chl fluorescence lifetime (Walters and Horton, 1991; Horton, 2014). Aggregation was later shown to be instrumental in catalyzing conformational change(s) within the LHCII protein, and the spectral signatures associated with this event were interpreted to indicate formation of a tight interaction between Lut and a Chl *a* molecule (Ruban et al., 2007). Extrapolation to NPQ *in vivo* relies on similarities between spectral changes in leaves upon NPQ induction and those upon aggregation of the isolated protein (Pascal et al., 2005; Ruban et al., 2007). Yet, LHCII aggregation quenching is insensitive to the presence or absence of Zea in the preparation (Ballottari et al., 2010), while most of NPQ *in vivo* is dependent on Zea synthesis as shown by *npq1* mutants (Niyogi et al., 1998).

Model 2: Charge-transfer (CT) quenching mechanism. In this model, qE activation involves a charge separation between a Chl–Zea heterodimer that produces a transient Zea radical cation (Zea$^+$) that rapidly (50–200 ps) relaxes to the uncharged state. The spectroscopic signal of Zea$\cdot^+$ was detected in isolated monomeric Lhcbs (Avenson et al., 2008), and mutation analysis of Chl-binding sites in Lhcb4 enabled the proposal that a specific Chl pair (Chl A5/603 and Chl B5/609) is involved in this charge transfer event (Ahn et al., 2008; Walla et al., 2014). The charge transfer model is also based on the observation that Viola in the monomeric Lhcb antenna complexes can be replaced by Zea (Morosinotto et al., 2002; Dall'Osto et al., 2005). Both Lhcb4 and Lhcb6 subunits that are involved in the PSBS-dependent NPQ and LHCSR for the algal-type mechanism were observed to undergo transient formation of radical cations (Holt et al., 2005; Ahn et al., 2008; Bonente et al., 2011), while the major trimeric LHCII complex did not.

Besides these two models that have been detailed at the molecular level, additional observation needs to be integrated for the construction of a general view of photosynthetic regulation by quenching reactions.

Ultrafast Chl fluorescence kinetic analysis of intact leaves of *A. thaliana* (Miloslavina et al., 2008, 2011) provides evidence that at least two different quenching sites contribute to qE *in vivo*: a PSBS-dependent site located in LHCII that becomes detached from PSII and aggregates upon illumination and a PSBS-independent site located within the minor antennae associated with PSII. Furthermore, Car S_1–Chl excited state coupling was measured in isolated LHCII and correlated with qE amplitude *in vivo* in mutants such as *npq1*, *npq2*, and *lut2* (Bode et al., 2009). On this basis, it was proposed that a short-lived, low-energy excitonic Car–Chl state, formed upon lumen acidification, may also function *in vivo* as an EED valve.

11.5 CONCLUDING REMARKS: WHY DID NPQ TRIGGERING EVOLVE FROM LHCSR TO PSBS?

Plants started colonizing land about 450 million years ago and evolved new morphophysiological traits to adapt to the conditions of the terrestrial environment: flooding/desiccation cycles, low and high temperatures, and increased exposure to UV radiation and excess visible light (Waters, 2003; Becker and Marin, 2009). Oxygen is more abundant and diffuses faster in the atmosphere than in water (Scott and Glasspool, 2006) and acts as a competitor for the fixation of CO_2 in photosynthesis because of the oxygenase activity of ribulose-1,5-bisphosphate carboxylase/oxygenase (Maurino and Peterhansel, 2010) and as a generator of reactive species (Niyogi, 1999). The availability of carbon dioxide, the final acceptor of electrons extracted from water by the oxygen-evolving complex, depends on the stomatal opening in plants and may become restricted when water availability is reduced under climatically stressful conditions (Waters, 2003). These conditions increase the probability of harvesting light in excess of what can be utilized by the photochemical reactions, thus requiring optimization of photoprotective mechanisms in plants.

It may be asked why plants retained only the PSBS-dependent mechanism of NPQ. The answer probably does not lie in the amplitude of the NPQ response since algae and plants have been shown to be active in inducing NPQ of different amplitudes depending on the species and growth conditions. An alternative explanation can be inferred from the observation that LHCSR shows a level of constitutive quenching, while PSBS does not (Bonente et al., 2011). PSBS, when inactive, has essentially no influence on the photon harvesting efficiency of the antenna proteins, thus allowing plants to use light as efficiently as possible under light-limited conditions (Dall'Osto et al., 2005;

Bonente et al., 2008a). In contrast, LHCSR efficiently dissipates energy even when isolated in detergent solution (Bonente et al., 2011), a condition that induces longer fluorescence lifetimes than those measured from LHC proteins *in vivo* or in lipid membranes (Moya et al., 2001; Miloslavina et al., 2008; van Oort et al., 2010), suggesting that LHCSR may dissipate a fraction of the harvested energy even in low-light conditions. One way to overcome this problem is to accumulate LHCSR only when needed, upon exposure to strong light, as observed in *C. reinhardtii* (Peers et al., 2009). However, this strategy leaves cells vulnerable to radiation damage by any abrupt change in light intensity, conditions where the physiological role of PSBS was shown to be particularly relevant (Külheim et al., 2002; Allorent et al., 2013, 2016). Therefore, the ability of PSBS to finely regulate thermal energy dissipation may provide a selective advantage in the terrestrial environment where fast and pronounced fluctuation of light irradiance is common (Külheim et al., 2002). In this context, it should also be considered that other mechanisms provide protection from constitutive strong illumination in plants, such as excess light-evading chloroplast movements (Brugnoli and Björkman, 1992; Cazzaniga et al., 2013), with a more specific role for NPQ in short-term regulation. Figure 11.10 shows the dependence of NPQ activity on light intensity in *A. thaliana*, with its PSBS-dependent NPQ system, *vs* *C. reinhardtii*, performing NPQ through LHCSR. Both organisms were grown under high-light conditions to compensate for the different kinetics of accumulation of the key proteins during acclimation (Ballottari et al., 2007; Bonente et al., 2012). Here, it can be observed that the NPQ response in *C. reinhardtii* is already activated in low light and saturates at relatively low light intensity. In contrast, *A. thaliana* plants respond proportionally to light intensity, extending their photoprotection effect over a much wider light intensity range. Thermal dissipation levels are very low in limiting light intensity where photo-damage is unlikely. The moss *P. patens* represents an evolutionary intermediate between green algae and higher plants, and it has both LHCSR and PSBS-dependent NPQ. In this organism, the dependence of NPQ activity on light intensity is halfway between that of algae and plants.

Although this relation needs to be extended to a wider range of species before being considered a general property of PSBS- *vs* LHCSR-dependent systems, these differences nonetheless suggest that the different response to light intensity might be

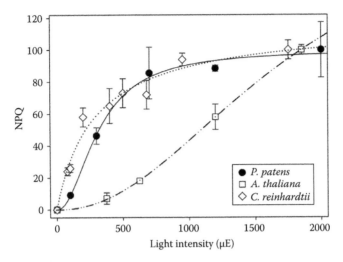

Figure 11.10 Dependence of NPQ activity on light intensity for *A. thaliana*, *C. reinhardtii*, and *P. patens*.

a reason for the transition from algal to plant-like light regulation systems.

ACKNOWLEDGMENTS

AP and RB thank the EEC projects "Sunbiopaths," "Harvest," and "Accliphot" for supporting research on regulation of photosynthesis in plants and algae. Luca Dall'Osto and Stefano Cazzaniga, UNIVR, are acknowledged for continued discussion.

REFERENCES

Adir, N. (2005). Elucidation of the molecular structures of components of the phycobilisome: Reconstructing a giant. *Photosynth. Res.* **85**: 15–32.

Ahn, T.K., Avenson, T.J., Ballottari, M., Cheng, Y.-C., Niyogi, K.K., Bassi, R., and Fleming, G.R. (2008). Architecture of a charge-transfer state regulating light harvesting in a plant antenna protein. *Science* **320**: 794–797.

Alboresi, A., Caffarri, S., Nogue, F., Bassi, R., and Morosinotto, T. (2008). In silico and biochemical analysis of *Physcomitrella patens* photosynthetic antenna: Identification of subunits which evolved upon land adaptation. *PLoS One* **3**: e2033.

Alboresi, A., Gerotto, C., Giacometti, G.M., Bassi, R., and Morosinotto, T. (2010). *Physcomitrella patens* mutants affected on heat dissipation clarify the evolution of photoprotection mechanisms upon land colonization. *Proc. Natl. Acad. Sci. U. S. A.* **107**: 11128–11133.

Allorent, G., Lefebvre-Legendre, L., Chappuis, R., Kuntz, M., Truong, T.B., Niyogi, K.K., Ulm, R., and Goldschmidt-Clermont, M. (2016). UV-B photoreceptor-mediated protection of the photosynthetic machinery in *Chlamydomonas reinhardtii*. *Proc. Natl. Acad. Sci. U. S. A.* **113**: 14864–14869.

Allorent, G. et al. (2013). A dual strategy to cope with high light in *Chlamydomonas reinhardtii*. *Plant Cell* **25**: 545–557.

Andersson, J., Wentworth, M., Walters, R.G., Howard, C.A., Ruban, A.V., Horton, P., and Jansson, S. (2003). Absence of the Lhcb1 and Lhcb2 proteins of the light-harvesting complex of photosystem II: Effects on photosynthesis, grana stacking and fitness. *Plant J.* **35**: 350–361.

Arnoux, P., Morosinotto, T., Saga, G., Bassi, R., and Pignol, D. (2009). A structural basis for the pH-dependent xanthophyll cycle in *Arabidopsis thaliana*. *Plant Cell* **21**: 2036–2044.

Asada, K. (1999). The water-water cycle in chloroplasts: Scavenging of active oxygens and dissipation of excess photons. *Annu. Rev. Plant Physiol. Plant Mol. Biol.* **50**: 601–639.

Aspinall-O'Dea, M., Wentworth, M., Pascal, A., Robert, B., Ruban, A., and Horton, P. (2002). *In vitro* reconstitution of the activated zeaxanthin state associated with energy dissipation in plants. *Proc. Natl. Acad. Sci. U. S. A.* **99**: 16331–16335.

Avenson, T.J., Ahn, T.K., Zigmantas, D., Niyogi, K.K., Li, Z., Ballottari, M., Bassi, R., and Fleming, G.R. (2008). Zeaxanthin radical cation formation in minor light-harvesting complexes of higher plant antenna. *J. Biol. Chem.* **283**: 3550–3558.

Azzabi, G., Pinnola, A., Betterle, N., Bassi, R., and Alboresi, A. (2012). Enhancement of non photochemical quenching in the Bryophyte *Physcomitrella patens* during acclimation to salt and osmotic stress. *Plant Cell Physiol.* **53(10)**: 1815–1825.

Bailleul, B., Rogato, A., de Martino, A., Coesel, S., Cardol, P., Bowler, C., Falciatore, A., and Finazzi, G. (2010). An atypical member of the light-harvesting complex stress-related protein family modulates diatom responses to light. *Proc. Natl. Acad. Sci. U. S. A.* **107**: 18214–18219.

Ballottari, M., Dall'Osto, L., Morosinotto, T., and Bassi, R. (2007). Contrasting behavior of higher plant photosystem I and II antenna systems during acclimation. *J. Biol. Chem.* **282**: 8947–8958.

Ballottari, M., Girardon, J., Betterle, N., Morosinotto, T., and Bassi, R. (2010). Identification of the chromophores involved in aggregation-dependent energy quenching of the monomeric photosystem II antenna protein Lhcb5. *J. Biol. Chem.* **285**: 28309–28321.

Ballottari, M., Truong, T.B., De Re, E., Erickson, E., Stella, G.R., Fleming, G.R., Bassi, R., and Niyogi, K.K. (2016). Identification of pH-sensing sites in the Light Harvesting Complex Stress-Related 3 protein essential for triggering non-photochemical quenching in *Chlamydomonas reinhardtii*. *J. Biochem.* **291**(14): 7334–7346.

Barber, J. and Andersson, B. (1992). Too much of a good thing: Light can be bad for photosynthesis. *Trends Biochem. Sci.* **17**: 61–66.

Bassi, R., Pineau, B., Dainese, P., and Marquardt, J. (1993). Carotenoid-binding proteins of photosystem II. *Eur. J. Biochem.* **212**: 297–303.

Becker, B. and Marin, B. (2009). Streptophyte algae and the origin of embryophytes. *Ann. Bot.* **103**: 999–1004.

Bergantino, E., Segalla, A., Brunetta, A., Teardo, E., Rigoni, F., Giacometti, G.M., and Szabò, I. (2003). Light- and pH-dependent structural changes in the PsbS subunit of photosystem II. *Proc. Natl. Acad. Sci. U. S. A.* **100**: 15265–15270.

Betterle, N., Ballottari, M., Zorzan, S., de Bianchi, S., Cazzaniga, S., Dall'Osto, L., Morosinotto, T., and Bassi, R. (2009). Light-induced dissociation of an antenna hetero-oligomer is needed for non-photochemical quenching induction. *J. Biol. Chem.* **284**: 15255–15266.

Berera, R., Gwizdala, M., van Stokkum, I.H., Kirilovsky, D., and van Grondelle, R. (August 8, 2013). Excited states of the inactive and active forms of the orange carotenoid protein. *J. Phys. Chem. B.* **117**(31): 9121–9128. doi: 10.1021/jp307420p. Epub 2013 Jul 29.

Bode, S., Quentmeier, C.C., Liao, P.-N., Hafi, N., Barros, T., Wilk, L., Bittner, F., and Walla, P.J. (2009). On the regulation of photosynthesis by excitonic interactions between carotenoids and chlorophylls. *Proc. Natl. Acad. Sci. U. S. A.* **106**: 12311–12316.

Bonente, G., Ballottari, M., Truong, T.B., Morosinotto, T., Ahn, T.K., Fleming, G.R., Niyogi, K.K., and Bassi, R. (2011). Analysis of LhcSR3, a protein essential for feedback de-excitation in the green alga *Chlamydomonas reinhardtii*. *PLoS Biol.* **9**: e1000577.

Bonente, G., Howes, B.D., Caffarri, S., Smulevich, G., and Bassi, R. (2008a). Interactions between the photosystem II subunit PsbS and xanthophylls studied *in vivo* and *in vitro*. *J. Biol. Chem.* **283**: 8434–8445.

Bonente, G., Passarini, F., Cazzaniga, S., Mancone, C., Buia, M.C., Tripodi, M., Bassi, R., and Caffarri, S. (2008b). The occurrence of the psbS gene product in *Chlamydomonas reinhardtii* and in other photosynthetic organisms and its correlation with energy quenching. *Photochem. Photobiol.* **84**: 1359–1370.

Bonente, G., Pippa, S., Castellano, S., Bassi, R., and Ballottari, M. (2012). Acclimation of *Chlamydomonas reinhardtii* to different growth irradiances. *J. Biol. Chem.* **287**: 5833–5847.

Boulay, C., Wilson, A., D'Haene, S., and Kirilovsky, D. (2010). Identification of a protein required for recovery of full antenna capacity in OCP-related photoprotective mechanism in cyanobacteria. *Proc. Natl. Acad. Sci. U. S. A.* **107**: 11620–11625.

Brugnoli, E. and Björkman, O. (1992). Chloroplast movements in leaves: Influence on chlorophyll fluorescence and measurements of light-induced absorbance changes related to ΔpH and zeaxanthin formation. *Photosynth. Res.* **32**: 23–35.

Cazzaniga, S., Dall'Osto, L., Kong, S.-G., Wada, M., and Bassi, R. (2013). Interaction between avoidance of photon absorption, excess energy dissipation and zeaxanthin synthesis against photooxidative stress in *Arabidopsis*. *Plant J.* **76**: 568–579.

Cho, F., Spencer, J., and Govindjee (1966). Emission spectra of *Chlorella* at very low temperatures (−269°C to −196°C). *Biochim. Biophys. Acta* **126**(1): 174–176.

Chukhutsina, V.U., Büchel, C., and van Amerongen, H. (2014). Disentangling two non-photochemical quenching processes in *Cyclotella meneghiniana* by spectrally-resolved picosecond fluorescence at 77K. *Biochim. Biophys. Acta* **1837**: 899–907.

Correa-Galvis, V., Redekop, P., Guan, K., Griess, A., Truong, T.B., Wakao, S., Niyogi, K.K., and Jahns, P. (2016). Photosystem II subunit PsbS is involved in the induction of LHCSR protein-dependent energy dissipation in *Chlamydomonas reinhardtii*. *J. Biol. Chem.* **291**: 17478–17487.

Dall'Osto, L., Caffarri, S., and Bassi, R. (2005). A mechanism of nonphotochemical energy dissipation, independent from PsbS, revealed by a conformational change in the antenna protein CP26. *Plant Cell* **17**: 1217–1232.

Dall'Osto, L., Cazzaniga, S., Wada, M., and Bassi, R. (2014). On the origin of a slowly reversible fluorescence decay component in the *Arabidopsis* npq4 mutant. *Philos. Trans. R. Soc. Lond. Ser. B Biol. Sci.* **369**: 20130221.

Dall'Osto, L., Holt, N.E., Kaligotla, S., Fuciman, M., Cazzaniga, S., Carbonera, D., Frank, H.A., Alric, J., and Bassi, R. (2012). Zeaxanthin protects plant photosynthesis by modulating chlorophyll triplet yield in specific light-harvesting antenna subunits. *J. Biol. Chem.* **287**: 41820–41834.

Damkjaer, J.T., Kereïche, S., Johnson, M.P., Kovacs, L., Kiss, A.Z., Boekema, E.J., Ruban, A.V., Horton, P., and Jansson, S. (2009). The photosystem II light-harvesting protein Lhcb3 affects the macrostructure of photosystem II and the rate of state transitions in *Arabidopsis*. *Plant Cell* **21**: 3245–3256.

de Bianchi, S., Betterle, N., Kouril, R., Cazzaniga, S., Boekema, E., Bassi, R., and Dall'Osto, L. (2011). *Arabidopsis* mutants deleted in the light-harvesting protein Lhcb4 have a disrupted photosystem II macrostructure and are defective in photoprotection. *Plant Cell* **23**: 2659–2679.

de Bianchi, S., Dall'Osto, L., Tognon, G., Morosinotto, T., and Bassi, R. (2008). Minor antenna proteins CP24 and CP26 affect the interactions between photosystem II subunits and the electron transport rate in grana membranes of *Arabidopsis*. *Plant Cell* **20**: 1012–1028.

Demmig-Adams, B., Gilmore, A.M., and Adams, W.W. (1996). Carotenoids 3: *In vivo* function of carotenoids in higher plants. *FASEB J.* **10**: 403–412.

Dinc, E., Tian, L., Roy, L.M., Roth, R., Goodenough, U., and Croce, R. (2016). LHCSR1 induces a fast and reversible pH-dependent fluorescence quenching in LHCII in *Chlamydomonas reinhardtii* cells. *Proc. Natl. Acad. Sci. U. S. A.* **113**: 7673–7678.

Dominici, P., Caffarri, S., Armenante, F., Ceoldo, S., Crimi, M., and Bassi, R. (2002). Biochemical properties of the PsbS subunit of photosystem II either purified from chloroplast or recombinant. *J. Biol. Chem.* **277**: 22750–22758.

El Bissati, K., Delphin, E., Murata, N., Etienne, A., and Kirilovsky, D. (2000). Photosystem II fluorescence quenching in the cyanobacterium Synechocystis PCC 6803: Involvement of two different mechanisms. *Biochim. Biophys. Acta* **1457**: 229–242.

Elrad, D., Niyogi, K.K., and Grossman, A.R. (2002). A major light-harvesting polypeptide of photosystem II functions in thermal dissipation. *Plant Cell* **14**: 1801–1816.

Engelken, J., Brinkmann, H., and Adamska, I. (2010). Taxonomic distribution and origins of the extended LHC (light-harvesting complex) antenna protein superfamily. *BMC Evol. Biol.* **10**: 233.

Fan, M., Li, M., Liu, Z., Cao, P., Pan, X., Zhang, H., Zhao, X., Zhang, J., and Chang, W. (2015). Crystal structures of the PsbS protein essential for photoprotection in plants. *Nat. Struct. Mol. Biol.* **22**: 729–735.

Ferrante, P., Ballottari, M., Bonente, G., Giuliano, G., and Bassi, R. (2012). LHCBM1 and LHCBM2/7 polypeptides, components of major LHCII complex, have distinct functional roles in photosynthetic antenna system of *Chlamydomonas reinhardtii*. *J. Biol. Chem.* **287**: 16276–16288.

Förster, B., Pogson, B.J., and Osmond, C.B. (2011). Lutein from deepoxidation of lutein epoxide replaces zeaxanthin to sustain an enhanced capacity for nonphotochemical chlorophyll fluorescence quenching in avocado shade leaves in the dark. *Plant Physiol.* **156**: 393–403.

Funk, C., Schröder, W.P., Napiwotzki, A., Tjus, S.E., Renger, G., and Andersson, B. (1995). The PSII-S protein of higher plants: A new type of pigment-binding protein. *Biochemistry* **34**: 11133–11141.

Gagné, G. and Guertin, M. (1992). The early genetic response to light in the green unicellular alga *Chlamydomonas eugametos* grown under light/dark cycles involves genes that represent direct responses to light and photosynthesis. *Plant Mol. Biol.* **18**: 429–445.

Gantt, E. and Conti, S.F. (1966a). Granules associated with the chloroplast lamellae of *Porphyridium cruentum*. *J. Cell Biol.* **29**: 423–434.

Gantt, E. and Conti, S.F. (1966b). Phycobiliprotein localization in algae. *Brookhaven Symp. Biol.* **19**: 393–405.

Gerotto, C., Alboresi, A., Giacometti, G.M., Bassi, R., and Morosinotto, T. (2011). Role of PSBS and LHCSR in *Physcomitrella patens* acclimation to high light and low temperature. *Plant Cell Environ.* **34**: 922–932.

Gerotto, C., Alboresi, A., Giacometti, G.M., Bassi, R., and Morosinotto, T. (2012). Coexistence of plant and algal energy dissipation mechanisms in the moss *Physcomitrella patens*. *New Phytol.* **196**: 763–773.

Gerotto, C., Franchin, C., Arrigoni, G., and Morosinotto, T. (2015). *In vivo* identification of Photosystem II Light Harvesting Complexes interacting with PHOTOSYSTEM II SUBUNIT S. *Plant Physiol.* **168**: 1747–1761.

Gerotto, C. and Morosinotto, T. (2013). Evolution of photoprotection mechanisms upon land colonization: Evidence of PSBS-dependent NPQ in late *Streptophyte algae*. *Physiol. Plant* **149**(4): 583–598.

Gilmore, A.M., Hazlett, T.L., and van de Govindjee, V.M. (1995). Xanthophyll cycle-dependent quenching of photosystem II chlorophyll a

fluorescence: Formation of a quenching complex with a short fluorescence lifetime. *Proc. Natl. Acad. Sci. U. S. A.* **92**: 2273–2277.

Gilmore, A.M., Shinkarev, V.P., Hazlett, T.L., and Govindjee, G. (1998). Quantitative analysis of the effects of intrathylakoid pH and xanthophyll cycle pigments on chlorophyll a fluorescence lifetime distributions and intensity in thylakoids. *Biochemistry* **37**: 13582–13593.

Glazer, A.N. (1984). Phycobilisome a macromolecular complex optimized for light energy transfer. *Biochim. Biophys. Acta* **768**: 29–51.

Grossman, A.R., Schaefer, M.R., Chiang, G.G., and Collier, J.L. (1993). The phycobilisome, a light-harvesting complex responsive to environmental conditions. *Microbiol. Rev.* **57**: 725–749.

Gupta, S., Guttman, M., Leverenz, R.L., Zhumadilova, K., Pawlowski, E.G., Petzold, C.J., Lee, K.K., Ralston, C.Y., and Kerfeld, C.A. (2015). Local and global structural drivers for the photoactivation of the orange carotenoid protein. *Proc. Natl. Acad. Sci. U. S. A.* **112**: E5567–E5574.

Gwizdala, M., Wilson, A., and Kirilovsky, D. (2011). *In vitro* reconstitution of the cyanobacterial photoprotective mechanism mediated by the Orange Carotenoid Protein in Synechocystis PCC 6803. *Plant Cell* **23**: 2631–2643.

Harrer, R., Bassi, R., Testi, M.G., and Schäfer, C. (1998). Nearest-neighbor analysis of a photosystem II complex from *Marchantia polymorpha* L. (liverwort), which contains reaction center and antenna proteins. *Eur. J. Biochem.* **255**: 196–205.

Harris, D., Tal, O., Jallet, D., Wilson, A., Kirilovsky, D., and Adir, N. (2016). Orange carotenoid protein burrows into the phycobilisome to provide photoprotection. *Proc. Natl. Acad. Sci. U. S. A.* **113**: E1655–E1662.

Havaux, M., Dall'Osto, L., and Bassi, R. (2007). Zeaxanthin has enhanced antioxidant capacity with respect to all other xanthophylls in *Arabidopsis leaves* and functions independent of binding to PSII antennae. *Plant Physiol.* **145**: 1506–1520.

Havaux, M. and Niyogi, K.K. (1999). The violaxanthin cycle protects plants from photooxidative damage by more than one mechanism. *Proc. Natl. Acad. Sci. U. S. A.* **96**: 8762–8767.

Holt, N.E., Zigmantas, D., Valkunas, L., Li, X.-P., Niyogi, K.K., and Fleming, G.R. (2005). Carotenoid cation formation and the regulation of photosynthetic light harvesting. *Science* **307**: 433–436.

Horton, P. (1996). Nonphotochemical quenching of chlorophyll fluorescence. Light as an Energy Source Inf. Carr. *Plant Physiol.* NATO ASI Ser. **287**: 99–111.

Horton, P. (2014). Developments in research on non-photochemical fluorescence quenching: Emergence key ideas, theories and experimental approaches. In Advances in Photosynthesis and Respiration. *Non-Photochemical Quenching and Energy Dissipation in Plants, Algae and Cyanobacteria*, B. Demmig-Adams, G. Garab, W. Adams, III, and Govindjee, eds., Springer, Dordrecht, pp. 73–95.

Ilioaia, C., Duffy, C.D.P., Johnson, M.P., and Ruban, A.V. (2013). Changes in the energy transfer pathways within photosystem II antenna induced by xanthophyll cycle activity. *J. Phys. Chem. B* **117**: 5841–5847.

Jahns, P. and Holzwarth, A.R. (2012). The role of the xanthophyll cycle and of lutein in photoprotection of photosystem II. *Biochim. Biophys. Acta Bioenerg.* **1817**: 182–193.

Jallet, D., Gwizdala, M., and Kirilovsky, D. (2012). ApcD, ApcF and ApcE are not required for the Orange Carotenoid Protein related phycobilisome fluorescence quenching in the cyanobacterium *Synechocystis* PCC 6803. *Biochim. Biophys. Acta* **1817**: 1418–1427.

Johnson, M.P., Goral, T.K., Duffy, C.D.P., Brain, A.P.R., Mullineaux, C.W., and Ruban, A.V. (2011). Photoprotective energy dissipation involves the reorganization of photosystem II light-harvesting complexes in the grana membranes of spinach chloroplasts. *Plant Cell* **23**: 1468–1479.

Kasahara, M., Kagawa, T., Oikawa, K., Suetsugu, N., Miyao, M., and Wada, M. (2002). Chloroplast avoidance movement reduces photodamage in plants. *Nature* **420**: 829–832.

Kasajima, I., Ebana, K., Yamamoto, T., Takahara, K., Yano, M., Kawai-Yamada, M., and Uchimiya, H. (2011). Molecular distinction in genetic regulation of nonphotochemical quenching in rice. *Proc. Natl. Acad. Sci. U. S. A.* **108**: 13835–13840.

Kerfeld, C. and Kirilovsky, D. (2013). Structural, mechanistic and genomic insights into OCP-mediated photoprotection. In Advances in Botanical Research. *Genomics Cyanobacteria*, F. Chauvat and C. Cassier-Chauvat, eds., Oxford: Academic Press, England, U.K. 65, pp. 1–26.

Kerfeld, C.A., Sawaya, M.R., Brahmandam, V., Cascio, D., Ho, K.K., Trevithick-Sutton, C.C., Krogmann, D.W., and Yeates, T.O. (2003). The crystal structure of a cyanobacterial water-soluble carotenoid binding protein. *Structure*, Oxford: Academic Press, England, U.K. **11**: 55–65.

Kirilovsky, D. (2015). Modulating energy arriving at photochemical reaction centers: Orange carotenoid protein-related photoprotection and state transitions. *Photosynth. Res.* **126**: 3–17.

Kirilovsky, D. and Kerfeld, C.A. (2012). The orange carotenoid protein in photoprotection of photosystem II in cyanobacteria. *Biochim. Biophys. Acta Bioenerg.* **1817**: 158–166.

Kirilovsky, D. and Kerfeld, C.A. (2013). The Orange Carotenoid Protein: A blue-green light photoactive protein. *Photochem. Photobiol. Sci.* **12**: 1135–1143.

Kirilovsky, D. and Kerfeld, C.A. (2016). Cyanobacterial photoprotection by the orange carotenoid protein. *Nat. Plants* **2**: 16180.

Krause, G.H., Briantais, J.-M., and Vernotte, C. (1983). Characterization of chlorophyll fluorescence quenching in chloroplasts by fluorescence spectroscopy at 77 K I. ΔpH-dependent quenching. *Biochim. Biophys. Acta Bioenerg.* **723**: 169–175.

Kromdijk, J., Głowacka, K., Leonelli, L., Gabilly, S.T., Iwai, M., Niyogi, K.K., and Long, S.P. (2016). Improving photosynthesis and crop productivity by accelerating recovery from photoprotection. *Science* **354**: 857–861.

Krüger, T.P.J., Ilioaia, C., Johnson, M.P., Ruban, A.V., Papagiannakis, E., Horton, P., and van Grondelle, R. (2012). Controlled disorder in plant light-harvesting complex II explains its photoprotective role. *Biophys. J.* **102**: 2669–2676.

Krüger, T.P.J., Wientjes, E., Croce, R., and van Grondelle, R. (2011). Conformational switching explains the intrinsic multifunctionality of plant light-harvesting complexes. *Proc. Natl. Acad. Sci. U. S. A.* **108**: 13516–13521.

Külheim, C., Agren, J., and Jansson, S. (2002). Rapid regulation of light harvesting and plant fitness in the field. *Science* **297**: 91–93.

Kuzminov, F.I., Karapetyan, N.V., Rakhimberdieva, M.G., Elanskaya, I.V., Gorbunov, M.Y., and Fadeev, V.V. (2012). Investigation of OCP-triggered dissipation of excitation energy in PSI/PSII-less *Synechocystis* sp. PCC 6803 mutant using non-linear laser fluorimetry. *Biochim. Biophys. Acta* **1817**: 1012–1021.

Lakowicz, J.R. (1999). *Principles of Fluorescence Spectroscopy*, 2nd edn., Kluwer Academic/Plenum Publishers, New York.

Lambrev, P.H., Nilkens, M., Miloslavina, Y., Jahns, P., and Holzwarth, A.R. (2010). Kinetic and spectral resolution of multiple nonphotochemical quenching components in *Arabidopsis* leaves. *Plant Physiol.* **152**: 1611–1624.

Leoni, C., Pietrzykowska, M., Kiss, A.Z., Suorsa, M., Ceci, L.R., Aro, E.-M., and Jansson, S. (2013). Very rapid phosphorylation kinetics suggest a unique role for Lhcb2 during state transitions in *Arabidopsis*. *Plant J.* **76**: 236–246.

Leverenz, R.L., Jallet, D., Li, M.-D., Mathies, R.A., Kirilovsky, D., and Kerfeld, C.A. (2014). Structural and functional modularity of the orange carotenoid protein: Distinct roles for the N- and C-terminal domains in cyanobacterial photoprotection. *Plant Cell* **26**: 426–437.

Leverenz, R.L. et al. (2015). PHOTOSYNTHESIS. A 12 Å carotenoid translocation in a photoswitch associated with cyanobacterial photoprotection. *Science* **348**: 1463–1466.

Li, X.P., Björkman, O., Shih, C., Grossman, A.R., Rosenquist, M., Jansson, S., and Niyogi, K.K. (2000). A pigment-binding protein essential for regulation of photosynthetic light harvesting. *Nature* **403**: 391–395.

Li, X.-P., Gilmore, A.M., Caffarri, S., Bassi, R., Golan, T., Kramer, D., and Niyogi, K.K. (2004). Regulation of photosynthetic light harvesting involves intrathylakoid lumen pH sensing by the PsbS protein. *J. Biol. Chem.* **279**: 22866–22874.

Li, X.-P., Muller-Moule, P., Gilmore, A.M., and Niyogi, K.K. (2002). PsbS-dependent enhancement of feedback de-excitation protects photosystem II from photoinhibition. *Proc. Natl. Acad. Sci. U. S. A.* **99**: 15222–15227.

Li, Z., Ahn, T.K., Avenson, T.J., Ballottari, M., Cruz, J.A., Kramer, D.M., Bassi, R., Fleming, G.R., Keasling, J.D., and Niyogi, K.K. (2009). Lutein accumulation in the absence of zeaxanthin restores nonphotochemical quenching in the *Arabidopsis thaliana* npq1 mutant. *Plant Cell* **21**: 1798–1812.

Liguori, N., Periole, X., Marrink, S.J., and Croce, R. (2015). From light-harvesting to photoprotection: Structural basis of the dynamic switch of the major antenna complex of plants (LHCII). *Sci Rep.* **5**: 15661.

Liguori, N., Roy, L.M., Opacic, M., Durand, G., and Croce, R. (2013). Regulation of light harvesting in the geen alga *Chlamydomonas reinhardtii*: The C-terminus of LHCSR Is the knob of a dimmer switch. *J. Am. Chem. Soc.* **135**: 18339–18342.

Liu, Z., Yan, H., Wang, K., Kuang, T., Zhang, J., Gui, L., An, X., and Chang, W. (2004). Crystal structure of spinach major light-harvesting complex at 2.72 A resolution. *Nature* **428**(6980): 287–292.

MacColl, R. (1998). Cyanobacterial phycobilisomes. *J. Struct. Biol.* **124**: 311–334.

Maksimov, E.G. et al. (2015). The signaling state of orange carotenoid protein. *Biophys. J.* **109**: 595–607.

Maurino, V.G. and Peterhansel, C. (2010). Photorespiration: Current status and approaches for metabolic engineering. *Curr. Opin. Plant Biol.* **13**: 249–256.

Miloslavina, Y., de Bianchi, S., Dall'Osto, L., Bassi, R., and Holzwarth, A.R. (2011). Quenching in *Arabidopsis thaliana* mutants lacking monomeric antenna proteins of photosystem II. *J. Biol. Chem.* **286**: 36830–36840.

Miloslavina, Y., Wehner, A., Lambrev, P.H., Wientjes, E., Reus, M., Garab, G., Croce, R., and Holzwarth, A.R. (2008). Far-red fluorescence: A direct spectroscopic marker for LHCII oligomer formation in non-photochemical quenching. *FEBS Lett.* **582**: 3625–3631.

Morosinotto, T., Baronio, R., and Bassi, R. (2002). Dynamics of chromophore binding to Lhc proteins *in vivo* and *in vitro* during operation of the xanthophyll cycle. *J. Biol. Chem.* **277**: 36913–36920.

Moya, I., Silvestri, M., Vallon, O., Cinque, G., and Bassi, R. (2001). Time-resolved fluorescence analysis of the photosystem II antenna proteins in detergent micelles and liposomes. *Biochemistry* **40**: 12552–12561.

Mullineaux, C. (1992). Excitation-energy transfer from phycobilisomes to Photosystem-I and Photosystem-II in a Cyanobacterium *Biochimica and Biophysica Acta (BBA)-Bioenergetics* **1100**(3): 285–292.

Munekage, Y., Hashimoto, M., Miyake, C., Tomizawa, K., Endo, T., Tasaka, M., and Shikanai, T. (2004). Cyclic electron flow around photosystem I is essential for photosynthesis. *Nature* **429**: 579–582.

Nichol, C., Pieruschka, R., Takayama, K., Förster, B., Kolber, Z., Rascher, U., Grace, J., Robinson, S., Pogson, B., and Osmond, B. (2012). Canopy conundrums: Building on the Biosphere 2 experience to scale measurements of inner and outer canopy photoprotection from the leaf to the landscape. *Funct. Plant Biol.* **1**: 1–24.

Nilkens, M., Kress, E., Lambrev, P., Miloslavina, Y., Müller, M., Holzwarth, A.R., and Jahns, P. (2010). Identification of a slowly inducible zeaxanthin-dependent component of non-photochemical quenching of chlorophyll fluorescence generated under steady-state conditions in *Arabidopsis*. *Biochim. Biophys. Acta* **1797**: 466–475.

Niyogi, K.K. (1999). PHOTOPROTECTION REVISITED: Genetic and molecular approaches. *Annu. Rev. Plant Physiol. Plant Mol. Biol.* **50**: 333–359.

Niyogi, K.K. (2000). Safety valves for photosynthesis. *Curr. Opin. Plant Biol.* **3**: 455–460.

Niyogi, K.K., Bjorkman, O., and Grossman, A.R. (1997). Chlamydomonas xanthophyll cycle mutants identified by video imaging of chlorophyll fluorescence quenching. *Plant Cell* **9**: 1369–1380.

Niyogi, K.K., Grossman, A.R., and Björkman, O. (1998). *Arabidopsis* mutants define a central role for the xanthophyll cycle in the regulation of photosynthetic energy conversion. *Plant Cell* **10**: 1121–1134.

Niyogi, K.K., Shih, C., Soon Chow, W., Pogson, B.J., Dellapenna, D., and Björkman, O. (2001). Photoprotection in a zeaxanthin- and lutein-deficient double mutant of *Arabidopsis*. *Photosynth. Res.* **67**: 139–145.

Pascal, A.A., Liu, Z., Broess, K., van Oort, B., van Amerongen, H., Wang, C., Horton, P., Robert, B., Chang, W., and Ruban, A. (2005). Molecular basis of photoprotection and control of photosynthetic light-harvesting. *Nature* **436**: 134–137.

Peers, G., Truong, T.B., Ostendorf, E., Busch, A., Elrad, D., Grossman, A.R., Hippler, M., and Niyogi, K.K. (2009). An ancient light-harvesting protein is critical for the regulation of algal photosynthesis. *Nature* **462**: 518–521.

Petroutsos, D. et al. (2016). A blue-light photoreceptor mediates the feedback regulation of photosynthesis. *Nature* **537**: 563–566.

Pietrzykowska, M., Suorsa, M., Semchonok, D.A., Tikkanen, M., Boekema, E.J., Aro, E.-M., and Jansson, S. (2014). The light-harvesting chlorophyll a/b binding proteins Lhcb1 and Lhcb2 play complementary roles during state transitions in *Arabidopsis*. *Plant Cell* **26**: 3646–3660.

Pinnola, A., Cazzaniga, S., Alboresi, A., Nevo, R., Levin-Zaidman, S., Reich, Z., and Bassi, R. (2015a). Light-harvesting complex stress-related proteins catalyze excess energy dissipation in both photosystems of *Physcomitrella patens*. *Plant Cell* **27**(11): 3213–3227.

Pinnola, A., Dall'Osto, L., Gerotto, C., Morosinotto, T., Bassi, R., and Alboresi, A. (2013). Zeaxanthin binds to light-harvesting complex stress-related protein to enhance nonphotochemical quenching in *Physcomitrella patens*. *Plant Cell* **25**: 3519–3534.

Pinnola, A., Ghin, L., Gecchele, E., Merlin, M., Alboresi, A., Avesani, L., Pezzotti, M., Capaldi, S., Cazzaniga, S., and Bassi, R. (2015b). Heterologous expression of moss LHCSR1: The chlorophyll a-xanthophyll pigment-protein complex catalyzing non-photochemical quenching, in *Nicotiana* sp. *J. Biol. Chem.* **290**: 24340–24354.

Polívka, T., Kerfeld, C.A., Pascher, T., and Sundström, V. (2005). Spectroscopic properties of the carotenoid 3′-hydroxyechinenone in the orange carotenoid protein from the cyanobacterium *Arthrospira maxima*. *Biochemistry* **44**: 3994–4003.

Quaas, T., Berteotti, S., Ballottari, M., Flieger, K., Bassi, R., Wilhelm, C., and Goss, R. (2015). Non-photochemical quenching and xanthophyll cycle activities in six green algal species suggest mechanistic differences in the process of excess energy dissipation. *J. Plant Physiol.* **172**: 92–103.

Rakhimberdieva, M.G., Boichenko, V.A., Karapetyan, N.V., and Stadnichuk, I.N. (2001). Interaction of phycobilisomes with photosystem

II dimers and photosystem I monomers and trimers in the cyanobacterium *Spirulina platensis*. *Biochemistry* **40**: 15780–15788.

Redlinger, T. and Gantt, E. (1982). A M(r) 95,000 polypeptide in *Porphyridium cruentum* phycobilisomes and thylakoids: Possible function in linkage of phycobilisomes to thylakoids and in energy transfer. *Proc. Natl. Acad. Sci. U. S. A.* **79**: 5542–5546.

Reinhold, C., Niczyporuk, S., Beran, K.C., and Jahns, P. (2008). Short-term down-regulation of zeaxanthin epoxidation in *Arabidopsis thaliana* in response to photo-oxidative stress conditions. *Biochim. Biophys. Acta* **1777**: 462–469.

Remelli, R., Varotto, C., Sandonà, D., Croce, R., and Bassi, R. (1999). Chlorophyll binding to monomeric light-harvesting complex. A mutation analysis of chromophore-binding residues. *J. Biol. Chem.* **274**: 33510–33521.

Richard, C., Ouellet, H., and Guertin, M. (2000). Characterization of the LI818 polypeptide from the green unicellular alga *Chlamydomonas reinhardtii*. *Plant Mol. Biol.* **42**: 303–316.

Ruban, A., Lavaud, J., Rousseau, B., Guglielmi, G., Horton, P., and Etienne, A.-L. (2004). The super-excess energy dissipation in diatom algae: Comparative analysis with higher plants. *Photosynth. Res.* **82**: 165–175.

Ruban, A.V., Berera, R., Ilioaia, C., van Stokkum, I.H.M., Kennis, J.T.M., Pascal, A.A., van Amerongen, H., Robert, B., Horton, P., and van Grondelle, R. (2007). Identification of a mechanism of photoprotective energy dissipation in higher plants. *Nature* **450**: 575–578.

Ruban, A.V., Young, A.J., and Horton, P. (1996). Dynamic properties of the minor chlorophyll a/b binding proteins of photosystem II, an *in vitro* model for photoprotective energy dissipation in the photosynthetic membrane of green plants. *Biochemistry* **35**: 674–678.

Scott, A.C. and Glasspool, I.J. (2006). The diversification of Paleozoic fire systems and fluctuations in atmospheric oxygen concentration. *Proc. Natl. Acad. Sci. U. S. A.* **103**: 10861–10865.

Stadnichuk, I.N., Krasilnikov, P.M., Zlenko, D.V., Freidzon, A.Y., Yanyushin, M.F., and Rubin, A.B. (2015). Electronic coupling of the phycobilisome with the orange carotenoid protein and fluorescence quenching. *Photosynth. Res.* **124**: 315–335.

Stadnichuk, I.N., Yanyushin, M.F., Maksimov, E.G., Lukashev, E.P., Zharmukhamedov, S.K., Elanskaya, I.V., and Paschenko, V.Z. (2012). Site of non-photochemical quenching of the phycobilisome by orange carotenoid protein in the cyanobacterium *Synechocystis* sp. PCC 6803. *Biochim. Biophys. Acta* **1817**: 1436–1445.

Sutter, M., Wilson, A., Leverenz, R.L., Lopez-Igual, R., Thurotte, A., Salmeen, A.E., Kirilovsky, D., and Kerfeld, C.A. (2013). Crystal structure of the FRP and identification of the active site for modulation of OCP-mediated photoprotection in cyanobacteria. *Proc. Natl. Acad. Sci. U. S. A.* **110**: 10022–10027.

Sylak-Glassman, E.J., Malnoë, A., De Re, E., Brooks, M.D., Fischer, A.L., Niyogi, K.K., and Fleming, G.R. (2014). Distinct roles of the photosystem II protein PsbS and zeaxanthin in the regulation of light harvesting in plants revealed by fluorescence lifetime snapshots. *Proc. Natl. Acad. Sci. U. S. A.* **111**: 17498–17503.

Tandeau de Marsac, N. (2003). Phycobiliproteins and phycobilisomes: The early observations. *Photosynth. Res.* **76**: 193–205.

Teardo, E., de Laureto, P.P., Bergantino, E., Dalla Vecchia, F., Rigoni, F., Szabò, I., and Giacometti, G.M. (2007). Evidences for interaction of PsbS with photosynthetic complexes in maize thylakoids. *Biochim. Biophys. Acta* **1767**: 703–711.

Tian, L., Dinc, E., and Croce, R. (2015). LHCII Populations in different quenching states are present in the thylakoid membranes in a ratio that depends on the light conditions. *J. Phys. Chem. Lett.* **6**: 2339–2344.

Tian, L., Gwizdala, M., van Stokkum, I.H.M., Koehorst, R.B.M., Kirilovsky, D., and van Amerongen, H. (2012). Picosecond kinetics of light harvesting and photoprotective quenching in wild-type and mutant phycobilisomes isolated from the cyanobacterium Synechocystis PCC 6803. *Biophys. J.* **102**: 1692–1700.

Tian, L., van Stokkum, I.H.M., Koehorst, R.B.M., Jongerius, A., Kirilovsky, D., and van Amerongen, H. (2011). Site, rate, and mechanism of photoprotective quenching in cyanobacteria. *J. Am. Chem. Soc.* **133**: 18304–18311.

Tibiletti, T., Auroy, P., Peltier, G., and Caffarri, S. (2016). *Chlamydomonas reinhardtii* PsbS protein is functional and accumulates rapidly and transiently under high light. *Plant Physiol.* **171**: 2717–2730.

Tokutsu, R., Kato, N., Bui, K.H., Ishikawa, T., and Minagawa, J. (2012). Revisiting the supra-molecular organization of photosystem II in *Chlamydomonas reinhardtii. J. Biol. Chem.* **287**: 31574–31581.

van Oort, B., Alberts, M., de Bianchi, S., Dall'Osto, L., Bassi, R., Trinkunas, G., Croce, R., and van Amerongen, H. (2010). Effect of antenna-depletion in Photosystem II on excitation energy transfer in *Arabidopsis thaliana. Biophys. J.* **98**: 922–931.

Walla, P., Holleboom, C., and Fleming, G. (2014). Electronic carotenoid-chlorophyll interactions regulating photosynthetic light harvesting of higher plants and green algae. In Advances in Photosynthesis and Respiration. *Non-Photochemical Quenching and Energy Dissipation in Plants, Algae and Cyanobacteria*, B. Demmig-Adams, G. Garab, W.W. Adams III, and Govindjee, eds., Springer, Dordrecht, pp. 229–243.

Walters, R.G. and Horton, P. (1991). Resolution of components of non-photochemical chloro-phyll fluorescence quenching in barley leaves. *Photosynth. Res.* **27**: 121–133.

Ware, M.A., Belgio, E., and Ruban, A.V. (2015). Comparison of the protective effectiveness of NPQ in *Arabidopsis* plants deficient in PsbS pro-tein and zeaxanthin. *J. Exp. Bot.* **66**: 1259–1270.

Waters, E.R. (2003). Molecular adaptation and the origin of land plants. *Mol. Phylogenet. Evol.* **29**: 456–463.

Wilson, A., Ajlani, G., Verbavatz, J.-M., Vass, I., Kerfeld, C.A., and Kirilovsky, D. (2006). A soluble carotenoid protein involved in phy-cobilisome-related energy dissipation in cyano-bacteria. *Plant Cell* **18**: 992–1007.

Wilson, A., Gwizdala, M., Mezzetti, A., Alexandre, M., Kerfeld, C.A., and Kirilovsky, D. (2012). The essen-tial role of the N-terminal domain of the orange carotenoid protein in cyanobacterial photoprotec-tion: Importance of a positive charge for phycobili-some binding. *Plant Cell* **24**: 1972–1983.

Wilson, A., Kinney, J.N., Zwart, P.H., Punginelli, C., D'Haene, S., Perreau, F., Klein, M.G., Kirilovsky, D., and Kerfeld, C.A. (2010). Structural determinants underlying photopro-tection in the photoactive orange carotenoid protein of cyanobacteria. *J. Biol. Chem.* **285**: 18364–18375.

Wilson, A., Punginelli, C., Couturier, M., Perreau, F., and Kirilovsky, D. (2011). Essential role of two tyrosines and two tryptophans on the photoprotection activity of the Orange Carotenoid Protein. *Biochim. Biophys. Acta* **1807**: 293–301.

Wilson, A. et al. (2008). A photoactive carotenoid protein acting as light intensity sensor. *Proc. Natl. Acad. Sci. U. S. A.* **105**: 12075–12080.

Wu, Y.P. and Krogmann, D.W. (1997). The orange carotenoid protein of Synechocystis PCC 6803. *Biochim. Biophys. Acta* **1322**: 1–7.

Xu, P., Tian, L., Kloz, M., and Croce, R. (2015). Molecular insights into Zeaxanthin-dependent quenching in higher plants. *Sci Rep* **5**: 13679.

Xue, H., Tokutsu, R., Bergner, S.V., Scholz, M., Minagawa, J., and Hippler, M. (2015). PHOTOSYSTEM II SUBUNIT R is required for efficient binding of LIGHT-HARVESTING COMPLEX STRESS-RELATED PROTEIN3 to photosystem II-light-harvesting supercom-plexes in *Chlamydomonas reinhardtii. Plant Physiol.* **167**: 1566–1578.

Zhang, H., Liu, H., Niedzwiedzki, D.M., Prado, M., Jiang, J., Gross, M.L., and Blankenship, R.E. (2014). Molecular mechanism of photoactiva-tion and structural location of the cyanobacte-rial orange carotenoid protein. *Biochemistry* **53**: 13–19.

Zhu, S.-H. and Green, B.R. (2010). Photoprotection in the diatom *Thalassiosira pseudonana*: Role of Ll818-like proteins in response to high light stress. *Biochim. Biophys. Acta* **1797**: 1449–1457.

PART 3

Light-Harvesting Systems in Action: Energy transfer and electron transport

The exciton concept

LEONAS VALKUNAS, JEVGENIJ CHMELIOV,
AND HERBERT VAN AMERONGEN

12.1 INTRODUCTION

Upon light absorption, molecules undergo a transition from the ground state to one of the excited states. The energy levels of these molecules and thereby their absorption spectrum depend not only on the chemical structure but also on the molecular environment, for instance, a solvent or a protein. In such a molecular environment, the energy levels are usually both shifted and broadened as compared to the vacuum situation. When there is another molecule located nearby, exhibiting similar optical properties/energy levels, both molecules can (strongly) interact through electrostatic interactions in such a way that spectroscopically they behave as a single supermolecule, although they might not even be in "physical" contact. New energy levels are created, transitions to these levels will be (partly) allowed and/or (partly) forbidden, and excitations become intimately shared. Such collective excitations are called excitons, and corresponding absorption and fluorescence properties depend on the relative orientation and position of the interacting molecules. Intermolecular interactions are also the main factors determining excitation energy transfer dynamics, while excitation relaxation is caused by the interaction with vibrational degrees of freedom. Therefore, to describe excitation evolution in molecular systems, both types of interactions have to be taken into account simultaneously. The exciton concept is widely used to explain both steady-state and time-resolved spectroscopic properties of strongly coupled pigments.

In this chapter, an introduction to the physical origin of these excitons will be given and some of their properties, especially the steady-state ones, will be emphasized. First, we provide the exciton representation for a dimer of strongly interacting molecules in a semiquantitative way, discussing energy levels and transitions between them, which will explain changes in the absorption spectrum upon dimerization. This will first be demonstrated for a dimer of identical molecules and then for molecules that differ to some extent. Subsequently, the same description will be provided for large molecular aggregates and molecular crystals.

In fact, the concept of excitons (the Frenkel excitons, to be more precise) was originally developed for molecular crystals. All these excitons share the property that the electron in the excited state and the remaining hole in the ground state are located on the same molecule. However, other types of excitons also exist, such as Wannier–Mott excitons, in which electron and hole move nearly independently, and charge-transfer excitons, where electron and hole reside in each other's neighborhood but on different chromophores. In order to describe exciton dynamics, which is crucial to understand light-harvesting processes in photosynthesis, the interactions of excitons with molecular vibrations and phonons of the protein surrounding are introduced.

12.2 INDIVIDUAL CHROMOPHORES

Spectral properties of a separate molecule are obtained by solving the Schrödinger equation with the corresponding Hamiltonian describing the kinetic and potential energies of electrons and nuclei of the considered molecule. To calculate the electronic energy levels, the so-called adiabatic (or Born–Oppenheimer) approximation is commonly used [1], which is based on the fact that the ratio of electron mass (m_e) and nuclear mass (M_i) is a small parameter (for the lightest atom, the hydrogen, this ratio is just $5 \cdot 10^{-4}$ and is considerably smaller for any heavier nuclei). According to the adiabatic approximation, the molecule is first considered with fixed nuclear positions and is therefore characterized by the following Hamiltonian:

$$\hat{H} = \hat{H}_{el}(r, R) + \hat{V}(r, R), \qquad (12.1)$$

where r and R are the sets of the coordinates of electrons and nuclei of the molecule, respectively, the latter being considered as fixed parameters (nonoperators). The first term, $\hat{H}_{el}(r, R)$, is the electronic Hamiltonian including the kinetic energy of the electrons and the electron–electron Coulomb interaction, while the second term, $\hat{V}(r, R)$, is the potential energy of the electron–nuclei as well as nucleus–nucleus interactions. Both terms are taken at some fixed positions R of the nuclei, and therefore the kinetic energy of the nuclei is disregarded. Thus, by solving the Schrödinger equation for this Hamiltonian, we obtain the energy eigenvalues $\varepsilon^{(i)}(R)$ and eigenfunctions $\psi^{(i)}(r, R)$ for the

electronic subsystem, both being parametrically dependent on the (fixed) nuclei positions R:

$$\hat{H}\psi^{(i)}(r, R) = \varepsilon^{(i)}(R)\psi^{(i)}(r, R). \qquad (12.2)$$

The obtained set of solutions can be used to determine the eigenfunctions of the full Hamiltonian defined by Equation 12.1, which should be supplemented with the operator for the kinetic energy of the nuclei $\hat{T}(R)$. For the description of the molecular energy spectrum, we can use the total wavefunction $\Psi(r, R)$ written as a superposition of electronic eigenfunctions $\psi^{(i)}(r, R)$:

$$\Psi(r, R) = \sum_i \chi_i(R)\psi^{(i)}(r, R) \qquad (12.3)$$

which should satisfy the following Schrödinger equation:

$$\left(\hat{T}(R) + \hat{H}\right)\Psi(r, R) = E\Psi(r, R). \qquad (12.4)$$

The solution of Equation 12.4 determines the expansion coefficients $\chi_i(R)$, which are, in general, not mutually independent because of the so-called nonadiabaticity terms [1,2]. However, by disregarding these terms, we arrive at the "real" adiabatic results, where electronic states numbered as i correspond to the energy levels of the molecule characterized by the eigen functions

$$\Psi_i(r, R) = \chi_i(R)\psi^{(i)}(r, R) \qquad (12.5)$$

and the electronic eigenvalue $\varepsilon^{(i)}(R)$ outlines the potential energy for the nuclear motions in the ith electronic state (see Figure 12.1). Near the minimum/minima of these potential energy curves or surfaces, the harmonic approximation can be used to describe the potential energy. The difference in potential energy of two electronic states (say, the ground state g and the excited state e), $\varepsilon^{(e)}(R) - \varepsilon^{(g)}(R)$, depends on the interaction between the electronic excitation and the nuclear degrees of freedom.

When considering a system of two molecules, their intermolecular interaction has to be taken into account. This interaction is described by the Coulomb coupling between all charged particles (electrons and nuclei) of the molecules n and m,

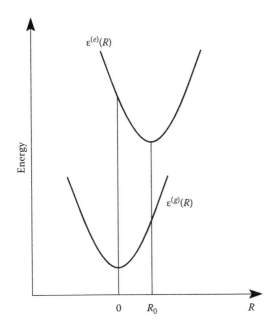

Figure 12.1 Ground- ($\varepsilon^{(g)}$) and excited ($\varepsilon^{(e)}$) state potentials of a single molecule (projection along a single component of nuclear coordinate vector R is shown). Close to the minima of these potentials, harmonic approximation can be used, $\varepsilon^{(g)}(R) \propto R^2$ and $\varepsilon^{(e)}(R) \propto (R - R_0)^2$; here, R_0 is the (possibly nonzero) shift of the excited state potential along the coordinate R.

and in a medium characterized by a dielectric constant ε, it is equal to

$$\hat{V}_{nm} = \eta \iint \frac{\rho_n(r)\rho_m(r')}{|r - r'|} dr dr', \qquad (12.6)$$

with $\eta = (4\pi\varepsilon\varepsilon_0)^{-1}$ and ε_0 being the vacuum permittivity. The charges have been written in the form of charge densities $\rho(r)$. In many practical cases, this interaction is more conveniently expressed in terms of the multipole moments of the respective charge distributions. Taking the center of mass r_n as a reference point for the nth molecule, multiple moments of the charge distribution $\rho_n(r)$ relative to this center can be defined. Thus, the total charge

$$q_n = \int \rho_n(r) dr \qquad (12.7)$$

and the dipole moment

$$\mu_n = \int \rho_n(r)(r - r_n) dr \qquad (12.8)$$

are the first moments of such an expansion. For neutral (uncharged) molecules, the total charge of the molecule equals zero and then the dipole–dipole type of interaction becomes the dominant term in the series expansion of the intermolecular interaction:

$$\hat{V}_{nm}^{(d-d)} = \eta \left(\frac{(\mu_n \cdot \mu_m)}{|r_{nm}|^3} - 3\frac{(r_{nm} \cdot \mu_n)(r_{nm} \cdot \mu_m)}{|r_{nm}|^5} \right), \qquad (12.9)$$

where $r_{nm} = r_n - r_m$ is the radius vector determining the intermolecular distance. If this distance is much larger than the size of both molecules, any higher-order moment can be disregarded, leading to the point dipole (or dipole–dipole) approximation $\hat{V}_{nm} \approx \hat{V}_{nm}^{(d-d)}$.

12.3 EXCITONICALLY COUPLED DIMER

12.3.1 Excitonic states

The simplest system for which exciton effects can be demonstrated is a pair of interacting molecules. Therefore, we will first consider such a system by largely following the relevant description presented in [2]. We call a pair of such interacting molecules an excitonically coupled dimer, whereas it should be noted that the word "dimer" does not imply that the molecules are chemically bound to each other or that they are in the van der Waals contact; in fact, they can even be spatially separated. Therefore, a physical dimer, which we are speaking about, should be distinguished from a chemical dimer, where chemical bonds are present between the two monomers. First, we consider two *identical* molecules (the so-called *homodimer*) in vacuum at a fixed distance r_{12} with some fixed mutual orientation and each having only two energy levels. For a given Hamiltonian, $\hat{H} = \hat{H}_1 + \hat{H}_2 + \hat{V}$, the isolated (noninteracting) molecules have their own two eigenstates $\psi^{(i)}$ determined by the corresponding Schrödinger equation:

$$\hat{H}_n \psi_n^{(i)} = \varepsilon_n^{(i)} \psi_n^{(i)}, \qquad (12.10)$$

where the subscript n identifies the pigment (either pigment 1 or 2) and the superscript i refers to the ground (g) or excited (e) state of the corresponding molecule. Since the eigenenergies of both pigments are identical, we may omit the subscript. In the following, we take the ground-state energy to be $\varepsilon_n^{(g)} = 0$.

The intermolecular interaction gives rise to a perturbation of the energy spectrum of the individual molecules. Since the intermolecular interactions are weak as compared to the intra-molecular energy terms, the perturbation theory for degenerate states can be used by applying the Heitler–London approximation, stating that the eigenfunctions of the dimer are equal to linear combinations of the product of the molecular eigenfunctions. We describe the electronic ground state as $\Phi^{(g)} = \psi_1^{(g)}\psi_2^{(g)}$. Note that we do not properly antisymmetrize $\Phi^{(g)}$, thereby implicitly assuming that exchange of electrons between the participating molecules 1 and 2 does not occur (for example, molecules are well separated in space). The corresponding ground-state energy of the dimer is now expressed by

$$E_g = \left\langle \psi_1^{(g)}\psi_2^{(g)} \middle| \hat{H}_1 + \hat{H}_2 + \hat{V} \middle| \psi_1^{(g)}\psi_2^{(g)} \right\rangle$$

$$= \varepsilon_1^{(g)} + \varepsilon_2^{(g)} + \left\langle \psi_1^{(g)}\psi_2^{(g)} \middle| \hat{V} \middle| \psi_1^{(g)}\psi_2^{(g)} \right\rangle$$

$$= \varepsilon_1^{(g)} + \varepsilon_2^{(g)} + V^{(gg)} = V^{(gg)}. \tag{12.11}$$

This indicates that coupling between molecules leads to a displacement of the ground-state energy by $V^{(gg)}$. The excited states can formally be written as

$$\Phi_{ex}^{(j)} = c_{j1}\psi_1^{(e)}\psi_2^{(g)} + c_{j2}\psi_1^{(g)}\psi_2^{(e)}. \tag{12.12}$$

The coefficients c_{j1} and c_{j2} are normalized and thus fulfill the following equation:

$$\left|c_{j1}\right|^2 + \left|c_{j2}\right|^2 = 1. \tag{12.13}$$

The excited state $\Phi_{ex}^{(j)}$ of the coupled system is a linear combination of two terms, in which either one or the other molecule is excited. The relative contributions of these two terms are determined by the coefficients $c_{j1,2}$. These new eigenstates are requested to be the stationary solutions of the Schrödinger equation of the dimer; thus

$$\left(\hat{H}_1 + \hat{H}_2 + \hat{V}\right)\Phi_{ex}^{(j)} = E_j\Phi_{ex}^{(j)}. \tag{12.14}$$

Multiplying both sides from the left with either $\left(\psi_1^{(e)}\psi_2^{(g)}\right)^*$ or $\left(\psi_1^{(g)}\psi_2^{(e)}\right)^*$ and integrating over space gives the two equations (recall that we have set the monomeric ground-state energies $\varepsilon_n^{(g)}$ to zero)

$$c_{j1}\left[\varepsilon_1^{(e)} + \left\langle \psi_1^{(e)}\psi_2^{(g)} \middle| \hat{V} \middle| \psi_1^{(e)}\psi_2^{(g)} \right\rangle\right]$$

$$+ c_{j2}\left\langle \psi_1^{(e)}\psi_2^{(g)} \middle| \hat{V} \middle| \psi_1^{(g)}\psi_2^{(e)} \right\rangle = c_{j1}E_j, \tag{12.15}$$

$$c_{j1}\left\langle \psi_1^{(g)}\psi_2^{(e)} \middle| \hat{V} \middle| \psi_1^{(e)}\psi_2^{(g)} \right\rangle$$

$$+ c_{j2}\left[\varepsilon_2^{(e)} + \left\langle \psi_1^{(g)}\psi_2^{(e)} \middle| \hat{V} \middle| \psi_1^{(g)}\psi_2^{(e)} \right\rangle\right] = c_{j2}E_j, \tag{12.16}$$

which can be written in a shorthand notation as

$$c_{j1}\left(\varepsilon_1^{(e)} + V_{11} - E_j\right) + c_{j2}V_{12} = 0, \tag{12.17}$$

$$c_{j1}V_{21} + c_{j2}\left(\varepsilon_2^{(e)} + V_{22} - E_j\right) = 0. \tag{12.18}$$

Here, V_{12} and V_{21} are resonance interaction terms, which (as will be shown as follows) are also important when considering excitation energy transfer. Evidently, the nontrivial solutions (when c_{j1} and c_{j2} are not simultaneously equal to zero) are obtained only if the determinant

$$\begin{vmatrix} \varepsilon_1^{(e)} + V_{11} - E_j & V_{12} \\ V_{21} & \varepsilon_2^{(e)} + V_{22} - E_j \end{vmatrix} = 0. \tag{12.19}$$

The matrix elements of the Hermitian interaction operator $\hat{V}$ satisfy $V_{21} = V_{12}^*$. Also, for identical molecules, $\varepsilon_1^{(e)} = \varepsilon_2^{(e)} = \varepsilon^{(e)}$ and $V_{11} = V_{22}$; thus, from Equation 12.19, it follows that

$$\left(\varepsilon^{(e)} + V_{11} - E_j\right)^2 = \left|V_{12}\right|^2, \tag{12.20}$$

leading to two eigenenergies:

$$E_{1,2} = \varepsilon^{(e)} + V_{11} \pm \left|V_{12}\right|. \tag{12.21}$$

We see that the transition energies of the dimer have changed as compared to those of the constituent single molecules (cf. Figure 12.2). These values correspond to the excitonic energies of the dimer. The energy levels are separated by the amount of

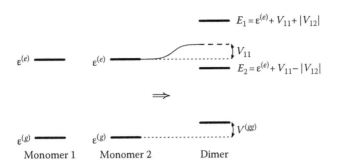

Figure 12.2 The energy scheme for the homodimer, where $\varepsilon^{(g)}$ and $\varepsilon^{(e)}$ determine energies of the ground and excited states of the monomers. Excitonic energy levels of the dimer, $E_{1,2} = \varepsilon^{(e)} + V_{11} \pm |V_{12}|$, are separated by the amount of $2|V_{12}|$; here, V_{12} is the resonance interaction of both pigments.

$2|V_{12}|$ (the so-called Davydov splitting), and the mean energy of these two levels has been shifted with respect to the new ground state E_g (see Equation 12.11) by an amount of $\Delta = V_{11} - V^{(gg)}$, which is usually called the displacement energy. In fact, Δ is comparable to the spectral change that a molecule experiences when it goes from the gas phase into solution.

The value of the resonance interaction V_{12} can then be calculated by using the expression for the intermolecular interaction defined by Equation 12.6, thus giving

$$V_{12} = \eta \iint \frac{\rho_1^{(eg)}(r_1)\rho_2^{(ge)}(r_2)}{|r_1 - r_2|} dr_1 dr_2, \quad (12.22)$$

where $\rho_n^{(eg)}(r_n) = -e\psi_n^{(e)*}(r_n) \cdot \psi_n^{(g)}(r_n)$ is the transition charge density of the nth molecule and $-e$ is the electron charge. Similarly, the displacement energy

$$\Delta = V_{11} - V^{(gg)}$$

$$= \eta \iint \frac{\left[\rho_1^{(ee)}(r_1) - \rho_1^{(gg)}(r_1)\right]\rho_2^{(gg)}(r_2)}{|r_1 - r_2|} dr_1 dr_2, \quad (12.23)$$

where $\rho_n^{(ee)}(r_n) = -e\psi_n^{(e)*}\psi_n^{(e)}$ and $\rho_n^{(gg)}(r_n) = -e\psi_n^{(g)*}\psi_n^{(g)}$ are the charge densities of the nth molecule in the excited and ground states, respectively. When the dipole–dipole approximation is applicable, it follows from Equation 12.9 that

$$V_{12} = \eta \left(\frac{\left(\mu_1^{(eg)} \cdot \mu_2^{(ge)}\right)}{|r_{12}|^3} - 3 \frac{\left(r_{12} \cdot \mu_1^{(eg)}\right)\left(r_{12} \cdot \mu_2^{(ge)}\right)}{|r_{12}|^5} \right), \quad (12.24)$$

where $\mu_n^{(eg)} = \left\langle \psi_n^{(e)} | \hat{\mu}_n | \psi_n^{(g)} \right\rangle$ is the transition dipole moment of the nth molecule. Similarly, the displacement energy in the dipole–dipole approximation is given by

$$\Delta = \eta \left(\frac{\left(d_1^{(ee)} \cdot \mu_2^{(gg)}\right)}{|r_{12}|^3} - 3 \frac{\left(r_{12} \cdot d_1^{(ee)}\right)\left(r_{12} \cdot \mu_2^{(gg)}\right)}{|r_{12}|^5} \right), \quad (12.25)$$

where $d_n^{(ee)} = \mu_n^{(ee)} - \mu_n^{(gg)}$ is the difference of the dipole moments of the nth molecule in the excited and ground states, $\mu_n^{(ee)} = \left\langle \psi_n^{(e)} | \hat{\mu}_n | \psi_n^{(e)} \right\rangle$ and $\mu_n^{(gg)} = \left\langle \psi_n^{(g)} | \hat{\mu}_n | \psi_n^{(g)} \right\rangle$, respectively.

In the case of nonidentical transition energies of the monomers (the so-called *heterodimer*), the excitonic eigenenergies and eigenstates can also be obtained by solving Equation 12.19:

$$E_{1,2} = \frac{\varepsilon_1^{(e)} + V_{11} + \varepsilon_2^{(e)} + V_{22}}{2}$$

$$\pm \frac{1}{2}\sqrt{\left[\left(\varepsilon_1^{(e)} + V_{11}\right) - \left(\varepsilon_2^{(e)} + V_{22}\right)\right]^2 + 4|V_{12}|^2}. \quad (12.26)$$

This expression can be simplified by defining the new zero-energy level as the average of the $\varepsilon_1^{(e)} + V_{11}$ and $\varepsilon_2^{(e)} + V_{22}$ energies of the two molecules and introducing the difference δ between these two values (see Figure 12.3). By using this notation, the excitonic energies become $E_{1,2} = \pm\frac{1}{2}\sqrt{\delta^2 + 4|V_{12}|^2}$. The dependence of this solution on the magnitude of the energy difference δ is shown in Figure 12.4a.

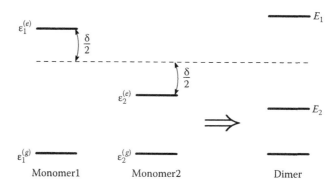

Figure 12.3 The energy scheme for the heterodimer, where $\varepsilon_n^{(g)}$ and $\varepsilon_n^{(e)}$ determine the energies of the ground and excited states of the nth monomer and $\delta = \varepsilon_1^{(e)} - \varepsilon_2^{(e)}$ indicates the difference between the excitation energies of monomers. E_1 and E_2 are the excitonic states of the dimer (for simplicity, it is assumed that $V^{(gg)} = V_{11} = V_{12} = 0$).

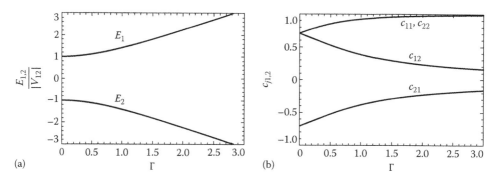

Figure 12.4 Dependencies of the excitonic eigenenergies $E_{1,2}$ (a) and wavefunction expansion coefficients $c_{j1,2}$ (b) of the heterodimer on $\Gamma = \delta/|2V_{12}|$.

The coefficients $c_{j1,2}$ in Equations 12.17 and 12.18 can easily be obtained by plugging in the corresponding excitonic energies $E_{1,2}$:

$$c_{11} = \cos\frac{\beta}{2}, \qquad c_{12} = \sin\frac{\beta}{2},$$
$$c_{21} = -\sin\frac{\beta}{2}, \qquad c_{22} = \cos\frac{\beta}{2}, \qquad (12.27)$$

where $\tan\beta = 2V_{12}/\delta$. The dependence of these coefficients on δ is shown in Figure 12.4b. In the case of either identical (when $\delta = 0$) or strongly interacting (when $|V_{12}| \gg \delta$) molecules, the coefficients $c_{j1,2}$ in Equations 12.17 and 12.18 become $c_{11} = c_{12} = c_{22} = 1/\sqrt{2}$ and $c_{21} = -1/\sqrt{2}$, indicating that exciton states are defined by coherently delocalized molecular excitations. The corresponding two excited state energy levels are split by the

doubled "resonance" interaction energy V_{12}, as was already discussed earlier. In the opposite case, when $|V_{12}| \ll \delta$, the different molecules maintain their own identity, and their energy levels are only slightly perturbed due to the intermolecular interaction:

$$E_{1,2} \approx \pm\frac{\delta}{2}\left(1 + \frac{2|V_{12}|^2}{\delta^2}\right). \qquad (12.28)$$

The probabilities to find the excitation on either one of the molecules then become $1/(2\Gamma)^2$ and $1 - 1/(2\Gamma)^2$, respectively, where $\Gamma = \delta/|2V_{12}|$. In the considered limit, Γ is large; thus, the excitations are almost entirely localized on individual molecules.

The excitonic states discussed so far are sometimes called the single-exciton states since their eigenfunctions $\Phi_{ex}^{(j)}$ were constructed by assuming one monomer being in the excited state and the other one in the ground state (*cf.* Equation 12.12). On the other hand, in some applications (e.g., while describing exciton–exciton annihilation or third-order spectroscopy [3,4]), the double-exciton states become relevant. For a dimer, the wavefunction of such a state is $\Phi_{2ex}^{(ee)} = \psi_1^{(e)}\psi_2^{(e)}$; thus, the corresponding energy level is

$$
\begin{aligned}
E_{2ex}^{(ee)} &= \left\langle \psi_1^{(e)}\psi_2^{(e)} \middle| \hat{H}_1 + \hat{H}_2 + \hat{V} \middle| \psi_1^{(e)}\psi_2^{(e)} \right\rangle \\
&= \varepsilon_1^{(e)} + \varepsilon_2^{(e)} + \left\langle \psi_1^{(e)}\psi_2^{(e)} \middle| \hat{V} \middle| \psi_1^{(e)}\psi_2^{(e)} \right\rangle \\
&= \varepsilon_1^{(g)} + \varepsilon_2^{(g)} + V^{(ee)}.
\end{aligned}
\tag{12.29}
$$

Here, $V^{(ee)}$ can be expressed in terms of the charge densities $\rho_n^{(ee)}$ of both molecules being in the excited state:

$$
V^{(ee)} = \eta \iint \frac{\rho_1^{(ee)}(r_1)\rho_2^{(ee)}(r_2)}{|r_1 - r_2|} dr_1 dr_2.
\tag{12.30}
$$

If the dipole–dipole approximation is valid, this interaction energy can be described in terms of the excited-state dipole moments $\boldsymbol{\mu}_n^{(ee)}$:

$$
V^{(ee)} = \eta \left(\frac{\left(\boldsymbol{\mu}_1^{(ee)} \cdot \boldsymbol{\mu}_2^{(ee)}\right)}{|r_{12}|^3} - 3\frac{\left(r_{12} \cdot \boldsymbol{\mu}_1^{(ee)}\right)\left(r_{12} \cdot \boldsymbol{\mu}_2^{(ee)}\right)}{|r_{12}|^5} \right),
\tag{12.31}
$$

similarly to the expression we previously had in Equation 12.24 for the resonance interaction energy.

12.3.2 Optical transitions in the dimer

Excitonic properties of interacting molecules are resolved in their optical spectra. The interaction of an electromagnetic field with the molecular system under consideration stems from the interaction term $\hat{H}_{int} = -E \cdot \hat{\boldsymbol{\mu}}$, where $E = E_0 \cos(k \cdot r - \omega t)$ is the electric part of the electromagnetic field, whereas the direction of the polarization and amplitude of the wave are defined by the constant vector E_0; k is the wavevector of the electromagnetic field; ω is its frequency; and $\hat{\boldsymbol{\mu}}$ is the electric dipole moment operator. In the following, we will use the long-wave approximation ($k \cdot r \ll 1$), which means that the wavelength ($\lambda = 2\pi/|k|$) of the electromagnetic wave is much longer than a typical length scale of interest, such as the size of the molecule or the intermolecular distance between two interacting molecules. For instance, the excitonic effects usually remain relevant only when the intermolecular distances do not exceed 2–3 nm, which is almost negligible compared with the light wavelength in the visible region (400–700 nm). According to Fermi's golden rule, the rate W_{eg} of the transition from the ground to the excited state during light absorption is proportional to the square of the matrix element of the interaction Hamiltonian. Thus, in the case of a separate molecule, we will get

$$
W_{eg} \propto |E_0|^2 \left| e\left\langle \psi^{(e)} \middle| \hat{\boldsymbol{\mu}} \middle| \psi^{(g)} \right\rangle \right|^2 = |E_0|^2 \left| \mu^{(eg)} \right|^2 \cos^2(e \cdot \mu_0),
\tag{12.32}
$$

where $\mu^{(eg)} = \left\langle \psi^{(e)} \middle| \hat{\boldsymbol{\mu}} \middle| \psi^{(g)} \right\rangle$, while e and $\boldsymbol{\mu}_0$ indicate the unit vectors corresponding to the light field and the transition dipole moment, respectively (i.e., $E_0 = E_0 e$ and $\hat{\boldsymbol{\mu}} = \hat{\mu}\mu_0$). For an isotropic sample, all orientations of the molecular transition dipole moment with respect to the direction of e are equally likely to occur; thus, averaging over all orientations leads to

$$
\left\langle \cos^2(e \cdot \mu_0) \right\rangle = 1/3
\tag{12.33}
$$

independent of the linear polarization direction of the light. Thus, the transition dipole moment is a key vector property of the molecule: its size squared $D_{eg} = |\mu^{(eg)}|^2$, which is called the dipole strength, is a characteristic value of the light absorption, in this case, of an individual (monomeric) molecule. Transitions, for which $D_{eg} = 0$, are called optically dark or forbidden, but in principle, they can still be observed (albeit with much smaller probability) due to higher-order terms of perturbation theory.

It is also relatively straightforward to calculate the absorption capacity of an excitonically coupled dimer. As was already mentioned earlier, the ground-state wavefunction of a dimer is given by $\Phi^{(g)} = \psi_1^{(g)}\psi_2^{(g)}$ and the wavefunction of the single-exciton state $\Phi_{ex}^{(j)}$ is defined in Equation 12.12. Thus, the probability of an optical transition in this case is proportional to:

$$W_{jg} \propto |E_0|^2 \left| e \left\langle \Phi_{ex}^{(j)} | \hat{\mu} | \Phi^{(g)} \right\rangle \right|^2$$

$$= |E_0|^2 \left| e \cdot \left(c_{j1} \mu_1^{(eg)} + c_{j2} \mu_2^{(eg)} \right) \right|^2, \quad (12.34)$$

where $\mu_1^{(eg)}$ and $\mu_2^{(eg)}$ are the transition dipole moments of molecules 1 and 2, respectively.

For the singly excited dimer, there are two possible exciton states defined by transformation coefficients c_{j1} and c_{j2}. Thus, the transition strength is defined by these coefficients and by the mutual orientations of both transition dipole moments of the constituent monomers. The dipole strength of the corresponding optical transition is defined accordingly as

$$D_{jg} = \left| c_{j1} \mu_1^{(eg)} + c_{j2} \mu_2^{(eg)} \right|^2 = |c_{j1}|^2 \left| \mu_1^{(eg)} \right|^2$$

$$+ |c_{j2}|^2 \left| \mu_2^{(eg)} \right|^2 + 2 \operatorname{Re} \left(c_{j1} c_{j2}^* \right) \left(\mu_1^{(eg)} \cdot \mu_2^{(eg)} \right).$$

$$(12.35)$$

In the case of a *homodimer* with identical dipole strengths for both monomers, $D_{eg}^{mon} = \left| \mu_n^{(eg)} \right|^2$, we obtain

$$D_{jg} = D_{eg}^{mon} \left[1 + 2 \operatorname{Re} \left(c_{j1} c_{j2}^* \right) \cos\theta \right], \quad (12.36)$$

where θ is the angle between the two transition dipole moments of the individual molecules. Using the expressions for the coefficients c_{j1} and c_{j2} as defined in Equation 12.27, the dipole strength can be obtained for different δ values. For isoenergetic molecules (when $\delta = 0$), we simply have

$$D_{jg} = D_{eg}^{mon} \left(1 \pm \cos\theta \right). \quad (12.37)$$

For instance, if $\cos\theta = -1$, the dipole strength for the transition corresponding to the "+" sign equals 0, that is, there is no absorption to this exciton level. On the other hand, the dipole strength for the other transition equals $2D_{eg}^{mon}$. It is important to notice that the sum of the dipole strengths of both excitonic transitions of the dimer is equal to the sum of the dipole strengths of the isolated molecules. Thus, in the excitonically coupled dimer, only a redistribution of the dipole strength takes place, whereas the total absorption probability remains unchanged.

As follows from Equation 12.9, the transition dipole moments also determine the resonance interaction within the dimer (when the dipole–dipole approximation is valid). The simplest mutual arrangements of the monomers are the so-called "head-to-tail" and "sandwich" arrangements, shown in Figure 12.5. In both cases, the transition dipole moments are parallel; thus, $\cos\theta = 1$ and, according to Equation 12.37, in both cases, one transition is forbidden and the other has a dipole strength of $2D_{eg}^{mon}$. However, which transition is allowed depends on the sign of the value V_{12}, as follows from Equations 12.27 and 12.36. For the "head-to-tail" configuration, the dipole–dipole interaction energy $V_{12} = -2\eta \left(\mu_1^{(eg)} \cdot \mu_2^{(eg)} \right) / r_{12}^3 < 0$;

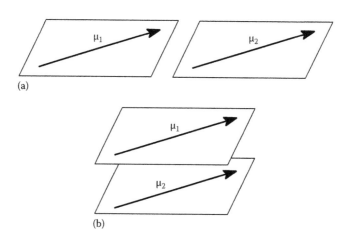

(a)

(b)

Figure 12.5 "Head-to-tail" **(a)** and "sandwich" **(b)** arrangements of monomers in the dimer.

thus, $c_{11}c_{12} = -\dfrac{1}{2}$ and $c_{21}c_{22} = +\dfrac{1}{2}$. As a result, for the energetically higher excitonic transition, we obtain $D_{1g} = D_{eg}^{mon}\left[1 - \cos\theta\right] = 0$, whereas for the lower state $D_{2g} = D_{eg}^{mon}\left[1 + \cos\theta\right] = 2D_{eg}^{mon}$. On the other hand, for the "sandwiched" dimer, we obtain $V_{12} = \eta\left(\mu_1^{(eg)} \cdot \mu_2^{(eg)}\right)/r_{12}^3 > 0$, $c_{11}c_{12} = \dfrac{1}{2}$, and $c_{21}c_{22} = -\dfrac{1}{2}$, which makes the transition to the higher state be the only possible $D_{1g} = 2D_{eg}^{mon}$, whereas $D_{2g} = 0$.

12.4 EXCITONS IN MOLECULAR COMPLEXES

The excitonic properties of a molecular complex containing N molecules can be described in a similar way as those of an excitonically coupled dimer. For simplicity, we assume that each of these molecules has only two electronic states—the ground state and the excited state—with the wavefunctions for the nth monomer denoted as $\psi_n^{(g)}$ and $\psi_n^{(e)}$, respectively. By constructing the basis set for the molecular aggregate within the Heitler–London approximation, we obtain

$$\Phi^{(g)} = \prod_{n=1}^{N}\psi_n^{(g)} \tag{12.38}$$

for the ground state and

$$\Phi_n^{(e)} = \psi_n^{(e)}\prod_{\substack{m=1 \\ m \neq n}}^{N}\psi_m^{(g)} \tag{12.39}$$

for a singly excited state with an excitation residing on the nth molecule.

The Hamiltonian of the molecular aggregate can be given accordingly:

$$\hat{H} = \sum_{n=1}^{N}\hat{H}_n + \sum_{n,m=1}^{N}\hat{V}_{nm}, \tag{12.40}$$

where
  $\hat{H}_n$ is the Hamiltonian of the nth monomer, which satisfies Equation 12.10
  $\hat{V}_{nm}$ is the operator for the intermolecular interaction defined by Equation 12.6

The N single-excitons are related to the molecular excitations described by Equation 12.39 via the unitary transformation matrix c_{jn}:

$$\Phi_{ex}^{(j)} = \sum_{n=1}^{N}c_{jn}\Phi_n^{(e)}. \tag{12.41}$$

The exciton properties can be identified after solving the Schrödinger equation for the Hamiltonian defined by Equation 12.40:

$$\hat{H}\Phi_{ex}^{(j)} = E_j\Phi_{ex}^{(j)}. \tag{12.42}$$

By multiplying this equation from the left side by $\Phi_n^{(e)*}$ and integrating over space, we will get a set of N equations, which allows us to obtain the transformation coefficients c_{jn} that determine the wavefunctions of the exciton states. The exciton energy spectrum (eigenvalues of the Hamiltonian in Equation 12.40) is schematically presented in Figure 12.6 and can be determined by solving

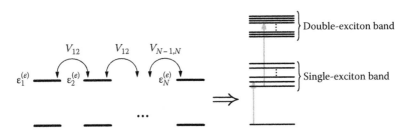

Figure 12.6 Exciton spectrum corresponding to one- and two-exciton states (shown on the right) of a molecular complex with N resonantly coupled monomers (shown on the left). Transitions to the optically allowed one-exciton state and from the one-exciton to two-exciton are indicated by arrows.

the relevant algebraic equation of the Nth order obtained by setting the corresponding determinant to zero (like it was done in Equation 12.19 for a dimer).

Having determined the excitonic states and their eigenfunctions $\Phi_{ex}^{(j)}$, one can calculate the corresponding transition dipole moments of the molecular aggregate, given by

$$\mu_j = \left(\Phi_{ex}^{(j)}|\hat{\mu}|\Phi^{(g)}\right) = \sum_{n=1}^{N} c_{jn}\mu_n^{(eg)}, \quad (12.43)$$

where $\mu_n^{(eg)}$ is the transition dipole moment of molecule n. The dipole strength can then be defined accordingly:

$$D_j = \left|\sum_{n=1}^{N} c_{jn}\mu_n^{(eg)}\right|^2 = \sum_{n,m=1}^{N} \rho_{nm}(j)D_{mn}, \quad (12.44)$$

where $\rho_{nm}(j) = c_{jn}c_{jm}^*$ and $D_{mn} = \mu_n^{(eg)} \cdot \mu_m^{(eg)}$ are the so-called density matrix of state j at time 0 and the dipole strength matrix, respectively. Since the wavefunctions are normalized and orthogonal to each other, the coefficients c_{jn} obey the following equations:

$$\sum_{n=1}^{N} c_{j_1 n}c_{j_2 n}^* = \delta_{j_1 j_2}, \quad \sum_{j=1}^{N} c_{jn}c_{jm}^* = \delta_{nm}, \quad (12.45)$$

so that the matrix of coefficients c_{jn} is unitary. It can then be easily shown that for the aggregate comprised from the identical monomers we will have

$$\sum_{j=1}^{N} D_j = ND_{eg}^{mon}, \quad (12.46)$$

that is, although the absorption strengths are distributed over the different transitions, their total sum is equal to that of the individual (noninteracting) molecules. This relationship is known as the sum rule as was already briefly mentioned for the dimer earlier.

Double-exciton states can be defined in a similar way. Wavefunctions of an aggregate with the mth and nth molecules being excited can be denoted as

$$\Phi_{nm}^{(ee)} = \psi_n^{(e)}\psi_m^{(e)}\prod_{l\neq n,m}^{N}\psi_l^{(g)}. \quad (12.47)$$

Therefore, by using a similar unitary transformation, the eigenfunctions reflecting the two-exciton states can be obtained:

$$\Phi_{2ex}^{(f)} = \sum_{n,m=1}^{N} d_{f,nm}\Phi_{nm}^{(ee)}. \quad (12.48)$$

By using these wavefunctions for the two-exciton states, the eigenvalues of the corresponding Schrödinger equation (schematically presented in Figure 12.6) can be obtained and the transition dipole moments for the one-exciton to two-exciton transitions can be calculated.

12.5 EXCITONS IN MOLECULAR CRYSTALS

The first formulation of an exciton as a nonconductive electronic excitation was given by Frenkel in 1931 [5,6], and afterwards this description was applied to explain the optical spectra of molecular crystals [7–9]. Such an exciton is nowadays called a zero-radius exciton, or *Frenkel exciton*. As an example, let us consider a molecular aggregate containing N equally spaced equivalent molecules (see Figure 12.6). This aggregate is evidently translationally invariant if the appropriate boundary conditions are assumed. Such a system is a relevant model for J-aggregates and one-dimensional molecular crystals [7,9,10]. The translational invariance implies that all molecules are identical. To fulfill this demand, periodic boundary conditions have to be applied, and for a one-dimensional aggregate of N sites, the wavefunction turns out to be an eigenfunction of the translational operator $\hat{T}_n$, where $n = 0, 1, \ldots, N-1$ is an integer number of translations

$$\hat{T}_n\Phi_k = e^{-ikn}\Phi_k, \quad (12.49)$$

where k is the wavevector corresponding to the translational operator. In this case, the wavevector is a quantum number of the system characterizing the exciton band, which is a collection of closely spaced exciton levels (see Figure 12.6). In the case of a single equivalent molecule per unit cell, it follows directly from Equation 12.49 that expansion coefficients in Equation 12.41 are equal to

$$c_{kn} = \frac{1}{\sqrt{N}}e^{ikn}, \quad (12.50)$$

thus single-exciton eigenfunctions are

$$\Phi_k = \frac{1}{\sqrt{N}} \sum_{n=1}^{N} e^{ikn} \Phi_n^{(e)}, \qquad (12.51)$$

which implies that all molecules in the crystal are equally likely to be excited while only the phase factors defined by Equation 12.49 are different. For the eigenenergies of the Hamiltonian defined by Equation 12.40, we then get

$$E_k = \langle \Phi_k | \hat{H} | \Phi_k \rangle - E_g = \varepsilon^{(e)} + \Delta + L(k), \quad (12.52)$$

where

E_g is the ground-state energy
$\varepsilon^{(e)}$ is the molecular excitation energy
Δ is the displacement energy due to other molecules

$$\Delta = \sum_{\substack{m=1 \\ m \neq n}}^{N} \langle \psi_n^{(e)} \psi_m^{(g)} | \hat{V}_{nm} | \psi_n^{(e)} \psi_m^{(g)} \rangle$$

$$- \sum_{\substack{m=1 \\ m \neq n}}^{N} \langle \psi_n^{(g)} \psi_m^{(g)} | \hat{V}_{nm} | \psi_n^{(g)} \psi_m^{(g)} \rangle, \quad (12.53)$$

and the last term is

$$L(k) = \sum_{\substack{m=1 \\ m \neq n}}^{N} e^{ik(m-n)} V_{nm}^{(eg)}, \qquad (12.54)$$

where $V_{nm}^{(eg)} = \langle \psi_n^{(g)} \psi_m^{(e)} | \hat{V}_{nm} | \psi_n^{(e)} \psi_m^{(g)} \rangle$ is the resonance interaction between the nth and mth molecules. Due to translational symmetry, the expressions in Equations 12.53 and 12.54 do not depend on the molecular position, so one can choose n arbitrarily (since $e^{ikN} = 1$, it is more convenient to set $n = N$ in Equation 12.54). By generalizing this approach to two- or three-dimensional crystals, the wavevector is related to the set of crystallographic axes. These orientational properties are reflected in the absorption of polarized light [7,9,10].

In the long-wavelength approximation, which is applicable for systems where the intermolecular distance is much smaller than the wavelength of the light, the absorption properties are well defined by

the transition dipole moments. In case the transition dipole moments of the molecules in the crystal are parallel to each other, $\mu_n^{(eg)}$ no longer depends on n, and then by applying the periodic boundary conditions for the crystal, the selection rule for the optical transitions is as follows:

$$\sum_{n=1}^{N} e^{ikn} = N\delta_{k,0}, \qquad (12.55)$$

where N is the amount of molecules in the crystal. This means that the optical transition is only allowed into the single-exciton state $k = 0$, for which the transition dipole strengths of all molecules in the crystal are summed up. When this state is also the lowest exciton state, the luminescence/fluorescence from this state is enhanced by a factor of N, and therefore this state is called superradiant.

In the case of cyclic molecular aggregates with the transition dipoles oriented in the plane of the aggregate, $\mu_n^{(eg)}$ explicitly depends on n. By introducing the rotating angle $\gamma = 2\pi/N$, we get the following equation [2,4]:

$$\mu_n^{(eg)} = \mu \cos(\gamma n). \qquad (12.56)$$

In this case, the selection rule for the optical transition does not follow Equation 12.55 but changes into

$$\sum_{n=1}^{N} e^{i(k\pm\gamma)n} = N\delta_{k,\pm\gamma}, \qquad (12.57)$$

that is, the optically allowed (superradiant) state now is not the lowest exciton state ($k = 0$) but the one next to the lowest exciton state, which is degenerate due to two different rotational directions available.

12.6 WANNIER–MOTT EXCITONS AND CHARGE-TRANSFER STATES

In atomic crystals or crystals with covalent/ionic bonds, interactions between crystal constituents are not weak, and thus the Heitler–London approximation is not applicable. Due to strong interactions, the electrons can be considered as belonging to the whole crystal; thus, they experience a periodic potential rather than a potential of a single lattice site [9,11]. Therefore, we consider a single electron moving in a periodic potential. The Hamiltonian describing this situation can be written as

$$H = -\frac{\hbar^2}{2m_e}\nabla^2 + V(\mathbf{r}), \qquad (12.58)$$

where

$V(\mathbf{r})$ is a periodic function having the property
$V(\mathbf{r}) = V(\mathbf{r} + \mathbf{a})$

$\mathbf{a}$ is the translational vector of the system—the smallest possible vector, which translates the system into itself

The wavefunction of this Hamiltonian must have the same translational symmetry, but the phase can differ. Therefore, following the Bloch theorem, it will have the form

$$\psi(\mathbf{r}) = u(\mathbf{r})\exp(i\mathbf{k}\cdot\mathbf{a}), \qquad (12.59)$$

with $u(\mathbf{r})$ being a periodic function having translational invariance $u(\mathbf{r}) = u(\mathbf{r} + \mathbf{a})$ and $\mathbf{k}$ being defined as a vector in the reciprocal lattice space, which is usually introduced to determine the exponential phase factor (cf. Equation 12.50). This definition allows one to consider free electrons in the conduction band of the crystal due to the translational property of the system. In this case, the electron in the conduction band is characterized by a specific effective mass, m_e (as well as m_h for the hole in the valence band), reflecting the specificity of the crystal.

The exciton spectrum is now defined by two particles, an electron in the conduction band, and a hole in the valence band, and therefore the Coulomb interaction between them. The Hamiltonian of such an electron–hole system (the so-called Hamiltonian of the *Wannier–Mott exciton*) can be defined as

$$\hat{H}_{WM} = \frac{\hat{p}_e^2}{2m_e} + \frac{\hat{p}_h^2}{2m_h} - \frac{\eta e^2}{|\mathbf{r}_e - \mathbf{r}_h|}, \qquad (12.60)$$

where

$\hat{p}_e$ and $\hat{p}_h$, m_e and m_h, and $\mathbf{r}_e$ and $\mathbf{r}_h$ are momenta, masses, and radius vectors of the electron and hole, respectively

e is the unitary charge of an electron

$\eta = (4\pi\varepsilon_0)^{-1}$

Using the common transformation of this two-particle system into the center of mass motion and the relative motion, we obtain the following equivalent representation of the Hamiltonian:

$$\hat{H}_{WM} = \frac{\hat{P}^2}{2(m_e + m_h)} + \frac{\hat{p}^2}{2\mu} - \frac{\eta e^2}{r}, \qquad (12.61)$$

where

$\hat{P}$ and $\hat{p}$ are the momenta of the center of mass and the relative movement of electron and hole, respectively

$r = |\mathbf{r}_e - \mathbf{r}_h|$

μ is the reduced mass defined as

$$\frac{1}{\mu} = \frac{1}{m_e} + \frac{1}{m_h}. \qquad (12.62)$$

After this transformation of the coordinates, the obtained variables describe the motion of the center of mass (as a free particle) and the relative motion of the electron and the hole separated by r. The total wavefunction can then be factorized into

$$\Psi(\mathbf{R}, \mathbf{r}) = \frac{1}{\sqrt{L^d}}e^{i\mathbf{K}\cdot\mathbf{R}}\varphi(\mathbf{r}), \qquad (12.63)$$

where the initial factor ensures the normalization of the wavefunction (L is the linear size of the crystal and d is its dimension) and $\mathbf{K}$ is the wavevector (or quantum number) describing the translational symmetry of the center of mass of the electron and hole. The relative movement of electron and hole corresponding to the wavefunction $\varphi(\mathbf{r})$ is determined by a hydrogen-type energy spectrum [12]:

$$E_n = E_g - \frac{E_e}{\left(n + \dfrac{d-3}{2}\right)^2}, \qquad (12.64)$$

where

E_g is the energy gap for the electron transition from the valence band to the conduction band

$E_e = (\varepsilon\mu/m_e)\text{Ry}$

$\text{Ry} = m_e e^4/(32\pi^2\varepsilon_0^2\hbar^2)$ is the Rydberg energy of the hydrogen atom [1]

Similarly, the exciton binding energy and the exciton radius corresponding to the lowest excited state ($n = 1$) can also be determined, thus giving

$$E_b = E_g - E_1 = \left(\frac{2}{d-1}\right)^2 E_e \qquad (12.65)$$

$$r_{\text{ex}} = \frac{d-1}{2} a_e, \qquad (12.66)$$

where
$a_e = (m_e/(\varepsilon\mu))a_B$ is the effective Bohr radius
$a_B = 4\pi\varepsilon_0\hbar^2/(m_e e^2)$ is the Bohr radius defined by considering the spectrum of the hydrogen atom

For typical parameters of a semiconducting crystal, E_g-values are of the order of eV, and E_e is of the order of tens or hundreds of meV. In three-dimensional crystals, the exciton binding energy equals E_e and exciton radius $r_{\text{ex}} = a_e$; in the case of $d=2$, the exciton binding energy increases four times and the exciton radius decreases two times; finally, in one-dimensional crystals, the binding energy $E_b \to \infty$ while $r_{\text{ex}} = 0$. The divergence of the exciton binding energy and the δ-type wavefunction in the case of $d=1$ can be understood as follows: in two- or three-dimensional crystals, a charged particle can freely move around the origin of the Coulomb potential, while in the case of $d=1$, it should move through the origin because of its spatial restriction. This causes a substantial enhancement of the exciton binding energy [12]. A typical example of a quasi one-dimensional system, where Wannier–Mott excitons are observed, is carbon nanotubes [13].

By accounting for the kinetic energy of the center of mass, the total eigenvalues of the exciton energy spectrum are given by

$$E_{K,n} = E_n + \frac{\hbar^2 K^2}{2(m_e + m_h)}. \qquad (12.67)$$

Thus, the exciton level diagram shown in Figure 12.7 defines the coupled electron–hole pair freely moving together through the crystal. Their binding states are quantized in a hydrogen atom-like fashion.

In contrast to neutral Frenkel (zero-radius) excitons, where the excited electron and the hole are both located on the same molecular site, excited states exhibiting electron transfer to a

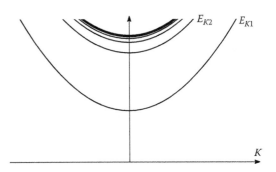

Figure 12.7 The energy levels of the Wannier–Mott exciton.

nearby molecule also exist. Such states, which are intermediate between Frenkel excitons and Wannier–Mott excitons, are called charge-transfer (CT) states [9]. Similar to Wannier–Mott excitons, these CT states are positioned below the lowest conduction band due to the Coulomb interaction between the electron and the hole (or between molecular ions) freely moving together through the crystal. Evidently, the CT state is determined by the ability of the molecules to accept/donate electrons, that is, the excited state energy depends on the molecular ionization potential (I_g) of the electron donor and the molecular electron affinity of the electron acceptor (A_g). Due to the presence of charges in the system, the polarization energy $P_{\text{eh}}(r)$ of an electron–hole pair, separated at some distance r, has also to be taken into account. This polarization energy reflects the adaptation of the system to the presence of the CT state. Thus, the energy of the CT exciton states in the presence of an external electric field is given by [9,10]

$$E_{\text{CT}} = I_g - A_g - P_{\text{eh}}(r) - C(r), \qquad (12.68)$$

where $C(r)$ denotes the Coulomb interaction of the charged pair separated by a distance r.

Evidently, the existence of CT excitons is expected in heteromolecular structures consisting of electron donating and accepting molecular pairs. Then the lowest CT state corresponds to the promotion of an electron from the highest occupied molecular orbital (HOMO) of the donor molecule to the lowest unoccupied molecular orbital (LUMO) of the acceptor molecule. From a theoretical point of view, these ion-pair states must be included in the complete basis set of wavefunctions describing the excited states of the crystal. This produces mixing of the CT states with the Frenkel excitons, thus changing the resonance interaction values of the latter.

The CT excitons are usually involved in photo-induced charge generation, when optically excited Frenkel exciton states are in resonance with the CT states. In this case, the transition from the initial Frenkel-type exciton state into the CT state can occur and further separation of charges as a subsequent step becomes possible.

A combined representation of all excitations can be obtained by using the so-called tight-binding description [4]. In this model, each molecule is represented by two electronic orbitals: its HOMO and LUMO. When the electron in the LUMO and the hole in the HOMO reside on the same molecule, we have the molecular excited state. On the other hand, when the electron and the hole are located on different sites, we obtain a CT state. By taking into account the possibility of the electron and the hole to resonantly move between molecules, a mixed situation between Frenkel excitons and CT states is obtained. This model was already applied to describe the spectral dynamics of the reaction centers of Photosystem II [14].

12.7 DYNAMICS

The exciton states described here are stationary states corresponding to electronic excitations in various molecular aggregates and solid states that can be obtained by diagonalizing the relevant Hamiltonian. This description takes into account only the electronic degrees of freedom while neglecting the intramolecular vibrations and phonons of the crystal lattice or the protein surrounding it. Thus, the corresponding transitions reflected in the absorption or emission spectra are represented as a set of sticks without widths (the so-called stick spectrum), the heights of which are defined by the relevant dipole strengths. In order to understand the spectral bandwidths as well as their temporal evolution, the interaction with intramolecular vibrations and the interaction with phonons of the environment have to be taken into account [4]. To demonstrate that, let us get back to the adiabatic approximation of the electronic spectra described by Equation 12.5. In this approximation, due to the slow motion of the nuclei in comparison with that of the electrons, the wavefunction of

the system is defined by the product of the electron wavefunction, $\psi^{(i)}(r,R)$, and the wavefunction of the nuclear degrees of freedom, $\chi_i(R)$. The transition dipole moment $\mu^{(eg)}$ of a single molecule thus can be expressed explicitly as

$$\mu^{(eg)} = e \iint \chi_e^*(R)\psi^{(e)*}(r,R)r\chi_g(R)\psi^{(g)}(r,R)drdR.$$

$$(12.69)$$

If the pure electronic transition dipole moment

$$\mu_{el}^{(eg)} = e \int \psi^{(e)*}(r,R)r\psi^{(g)}(r,R)dr \quad (12.70)$$

is independent of R (the so-called Franck–Condon approximation), it follows that the total transition dipole moment is given by

$$\mu^{(eg)} = \mu_{el}^{(eg)} \cdot \text{FC}, \qquad (12.71)$$

where $\text{FC} = \int \chi_e^*(R)\chi_g(R)dR$ is the Franck–Condon factor. As a result of the solution of Equation 12.4, the wavefunction is determined as the eigenfunction corresponding to multiple vibrational modes. Thus, the Franck–Condon factor becomes also dependent on these vibrational modes, and therefore it determines the shape of the corresponding electronic transitions of the molecule. Similarly, the same modulating Franck–Condon factors appear while calculating the resonance intermolecular interaction in terms of the dipole–dipole approximation (Equation 12.24). Thus, the strength of the resonance interaction, which is a measure of the exciton coherence, diminishes due to the interaction with molecular vibrations.

This effect, caused by renormalization of the transition dipole moments, does not describe the excitation dynamics, which occurs due to the exciton interaction with intramolecular vibrations and with phonons of the surrounding environment. As was already mentioned while describing the adiabatic approximation, such type of interactions are obtained from the difference of the electronic potential energies, $\varepsilon^{(e)}(R) - \varepsilon^{(g)}(R)$. This means that the intramolecular vibrations and/or phonons are considered in the ground electronic state, whereas the changes of the functional dependence of the excited electronic energy on the nuclear

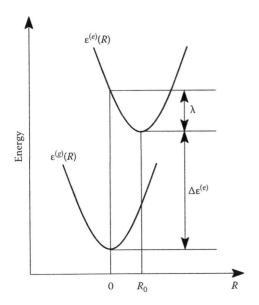

Figure 12.8 Projections of the ground-($\varepsilon^{(g)}$) and excited ($\varepsilon^{(e)}$) state potentials of a single molecule along a single component of nuclear coordinate vector R and definition of the reorganization energy λ.

coordinate R appear due to the interaction of the molecular excitation with the vibrational degrees of freedom. This difference between the energies is usually expanded into a power series of R in the vicinity of the minimum of the $\varepsilon^{(g)}(R)$ potential (see Figure 12.8), thus giving

$$\varepsilon^{(e)}(R) - \varepsilon^{(g)}(R) = \Delta\varepsilon^{(e)} + \lambda + aR + \delta R^2, \quad (12.72)$$

where

$\Delta\varepsilon^{(e)}$ determines the difference between the minima of the potential energies in the excited (e) and ground (g) states

λ is the so-called reorganization energy (or the Franck–Condon energy) that is also related to the Stokes shift via $S = 2\lambda$

a and δ are the expansion coefficients [2]

The first coefficient a is determined by the shift of the potential surface in the excited state in comparison with that in the ground state, and the second coefficient δ is related to changes of the shape of the potential surfaces, that is, it is proportional to the difference of the vibrational frequencies in the excited and ground electronic potentials.

Most frequently, only the linear term is taken into account while the quadratic term is neglected (assuming $\delta = 0$). Evidently, the R-dependence of the potential surfaces results in a modulation of the excitation energy of the molecule. The most convenient form to describe such a modulation effect caused by the vibrational fluctuations is the time-dependent coordinate–coordinate correlation function

$$C(t) = \mathrm{Tr}_B\{R(t)R(0)\rho_B\}, \quad (12.73)$$

where

$R(t)$ is the collective vibrational coordinate at time t

ρ_B is the thermally equilibrated density of vibrations

and the trace $\mathrm{Tr}_B\{\dots\}$ is taken over all vibrational modes

Introducing the correlation function means that the interaction of the electronic excitation with the vibrational degrees of freedom causes relaxation processes, and thus the considered system is treated as an open quantum system [4,15]. This means that the electronic degrees of freedom, which describe a separate molecule or an excitonically coupled molecular aggregate, form the system under consideration, while the vibrations that are taken into account are considered as the external thermally equilibrated bath (hence, the lower index "B" in Equation 12.73).

One of the most elaborate ways to treat the non-stationary dissipative dynamics of open quantum systems is the density operator theory. In principle, one could also solve the Schrödinger equation in a larger Hilbert space by including the environmental degrees of freedom, or use the stochastic Schrödinger equations [4]. However, the density operator (or matrix in a specific representation) characterizes state populations (diagonal elements of the matrix) and interstate phase relation—the so-called coherences that are off-diagonal elements of the matrix—in a unified way and thus makes a direct relation with the dynamics of the observables. The dynamics of the whole system density matrix $\hat{W}$ is governed by the Liouville equation (we take $\hbar = 1$) [4]

$$\frac{\mathrm{d}}{\mathrm{d}t}\hat{W} = -\mathrm{i}\left[\hat{H}, \hat{W}\right], \quad (12.74)$$

where the total Hamiltonian can be split into three terms:

$$\hat{H} = \hat{H}_S + \hat{H}_B(\hat{p}, \hat{q}) + \hat{H}_{SB}(\hat{q}). \quad (12.75)$$

The system (first term) is directly observable and should be attributed to the exciton Hamiltonian described in the previous sections. This system is in contact with the bath (second term), which is not directly observed. The bath is usually considered to be represented by a set of thermally equilibrated harmonic oscillators with generalized coordinates $\hat{q}$ and generalized momenta $\hat{p}$. Finally, the system–bath coupling is represented by the term $\hat{H}_{SB}(\hat{q})$, which is a system operator parametrically dependent on the bath coordinates. Similarly to the interaction term given by Equation 12.72, we generalize $\hat{H}_{SB}(\hat{q})$ by assuming a product form of the system–bath interaction:

$$\hat{H}_{SB}(\hat{q}) = \sum_n \hat{S}_n \hat{q}_n, \quad (12.76)$$

where

$\hat{S}_n$ is the system operator (usually a projector)
$\hat{q}_n$ is the associated bath coordinate (similar to the R-coordinate used to characterize a separate molecule)

To obtain the solution of Equation 12.74, we introduce the interaction representation of the density operator

$$\hat{W}_I(t) = \hat{U}_0^\dagger(t)\hat{W}(t)\hat{U}_0(t), \quad (12.77)$$

where $\hat{U}_0(t)$ denotes the evolution operator generated by the system and bath Hamiltonians $\hat{H}_S + \hat{H}_B(\hat{p}, \hat{q})$:

$$\hat{U}_0(t) = \exp\left[-i\left(\hat{H}_S + \hat{H}_B(\hat{p}, \hat{q})\right)t\right]. \quad (12.78)$$

The system–bath coupling Hamiltonian, Equation 12.76, is likewise defined in the interaction picture and is denoted as $\hat{H}_I(t) = \hat{U}_0^\dagger(t)\hat{H}_{SB}(\hat{q})\hat{U}_0(t)$.

Often, only a few degrees of freedom constitute the observable system, while the others affect the system but cannot be directly monitored. We can derive an approximate closed equation for the *reduced density matrix* of the system $\hat{\rho}$ and consider

the other irrelevant degrees of freedom as the thermal bath. Several approximations have to be used in order to arrive at this result. The first approximation is the factorization of the total density matrix $\hat{W}$ into the system and the bath components (known as the Born approximation). Moreover, considering that the bath has an infinite number of degrees of freedom, it is taken to be in the thermal equilibrium state $\hat{\rho}_B$ at all times. We thus write

$$\hat{W}_I(t) = \hat{\rho}_I(t) \otimes \hat{\rho}_B, \quad (12.79)$$

where $\hat{\rho}_I(t) = \hat{U}_0^\dagger(t)\hat{\rho}\hat{U}_0(t)$ is the reduced density operator of the system in the interaction representation and "$\otimes$" denotes the tensor product. We can formally integrate the Liouville equation for the density operator in the interaction picture (Equation 12.74) and then plug the result back into the right-hand side of the same equation. Then, performing the trace operation over the equilibrium bath variables gives the Quantum Master Equation (QME) for the reduced density matrix in the interaction representation

$$\frac{d}{dt}\hat{\rho}_I(t) = -\int_{t_0}^t d\tau \mathrm{Tr}_B\left(\left[\hat{H}_I(t), \left[\hat{H}_I(\tau), \hat{\rho}_I(\tau) \otimes \hat{\rho}_B\right]\right]\right).$$

$$(12.80)$$

The Born approximation essentially allows one to isolate the reduced density matrix, and it brings irreversibility into the dynamics. We have to understand Equation 12.80 as follows. At time $t < t_0$ the system and the bath are uncoupled. Their dynamics are uncorrelated, and the total density matrix is block diagonal with respect to the system and the bath. At time $t = t_0$, the interaction is switched on and the dynamics become correlated. Equation 12.80 exactly describes the system dynamics since the system–bath correlations are included up to infinite order. The relaxation kernel thus carries the memory effects. If the bath is not in the equilibrium state at time t_0 or if it is correlated with the system, then the system has to be extended to include these correlation effects.

By performing the trace over the bath in Equation 12.80 and taking the initial condition $t_0 \to -\infty$, the rate operator governing the dissipative dynamics is obtained as a function of the interaction delay times $t - \tau$. The evolution is still of infinite order in the system–bath interaction. It is convenient to introduce the delay time explicitly, which yields the following equation in the Schrödinger representation:

$$\frac{d}{dt}\hat{\rho}(t) = -i\left[\hat{H}_S, \hat{\rho}(t)\right] - \int_0^\infty d\tau \hat{\mathcal{R}}(\tau)\rho(t-\tau). \quad (12.81)$$

The obtained QME is an integro-differential equation with a complex expression for the rate superoperator $\hat{\mathcal{R}}(\tau)$ and cannot be easily solved. Therefore, various approximations are used.

12.7.1 Redfield theory for exciton relaxation

The obtained integro-differential form of the QME, Equation 12.81, can be simplified by using the Redfield approximation, that is, assuming the second-order approach by determining the relaxation kernel $\hat{\mathcal{R}}(\tau)$ [4]. Additionally, the Markovian approximation is invoked by assuming that the system–bath interaction is weak, and therefore the system density matrix in the interaction picture, $\hat{\rho}_I(t)$, is a slowly evolving function, as compared to the decay time of the relaxation tensor. Under these conditions, the QME becomes the time-local equation with time-independent rate matrix. The obtained equation is known as the Redfield equation, which is given by

$$\frac{d}{dt}\hat{\rho}(t) = -i\left[\hat{H}_S, \hat{\rho}(t)\right] - \hat{\mathcal{K}}\hat{\rho}(t). \quad (12.82)$$

The Redfield relaxation superoperator $\hat{\mathcal{K}}$ can be simplified considerably if the system operators are expanded into an arbitrary orthogonal basis $|a\rangle$. Then the Hamiltonian reads

$$\hat{H} = \sum_{ab}\left(h_{ab} + \tilde{h}_{ab}\hat{q}_{ab}\right)|a\rangle\langle b| + \hat{H}_B(\hat{p},\hat{q}). \quad (12.83)$$

This form can be directly applied to the single-exciton manifold, where the exciton relaxation and transfer take place. We then have $h_{ab} = \delta_{ab}\varepsilon_a + V_{ab}$. The remaining part can be partitioned out into the term $\hat{H}_B(\hat{p},\hat{q})$, representing the fluctuating environment, with the weak coupling amplitude $\tilde{h}_{ab}$ as a system–bath coupling amplitude. Meanwhile, $\hat{q}_{ab}$ determines the generalized coordinate of the bath,

coupled to the system Hamiltonian element ab (analogous to the $\mathbf{R}$-coordinate in the description of the molecular spectra using the adiabatic approximation). We obtain the relaxation matrix defined by the fluctuation correlation functions:

$$\mathcal{K}_{ab,a'b'} = \sum_{cd}\int_0^\infty d\tau \left[\delta_{bb'}\sum_e \tilde{h}_{ae}\tilde{h}_{cd}C_{ae,dc}(\tau)U_{ed}(\tau)U_{ca'}(-\tau)\right.$$

$$- \tilde{h}_{aa'}\tilde{h}_{cd}C_{cd,aa'}(-\tau)U_{b'c}(\tau)U_{db}(\tau)$$

$$- \tilde{h}_{dc}\tilde{h}_{b'b}C_{b'b,dc}(\tau)U_{ad}(\tau)U_{ca'}(-\tau)$$

$$\left. + \delta_{aa'}\sum_e \tilde{h}_{cd}\tilde{h}_{eb}C_{cd,eb}(-\tau)U_{b'c}(\tau)U_{de}(-\tau)\right]. \quad (12.84)$$

The correlation function $C_{ab,cd}(\tau)$ describes fluctuations of the Hamiltonian elements ab and cd, and $U_{ab}(\tau)$ are the corresponding matrix element of the system evolution operator $\hat{U}(\tau) = \exp(-i\hat{H}_S\tau)$.

A natural choice for the basis set for the Redfield relaxation superoperator is the eigenstate basis of the system Hamiltonian. This choice makes simulations much simpler and allows us to introduce the *secular approximation* and to define the requirements for the long-time limit. Let us assume that states $|a\rangle$ are eigenstates of the system Hamiltonian (the exciton states in the case of the exciton Hamiltonian). In that case, $h_{ab} = \delta_{ab}E_a$ form a diagonal matrix, thus the Redfield equation reduces to

$$\frac{d}{dt}\rho_{ab}(t) = -i\omega_{ab}\rho_{ab}(t) - \mathcal{K}_{ab,cd}\rho_{cd}(t), \quad (12.85)$$

where $\omega_{ab} = E_a - E_b$.

The secular Redfield relaxation equation can then be written in the form

$$\frac{d}{dt}\rho_{ab}(t) = -i(\omega_{ab} - i\gamma_{ab})\rho_{ab}(t) - \delta_{ab}\sum_c k_{ac}\rho_{cc}(t), \quad (12.86)$$

where $\gamma_{aa} \equiv 0$ and γ_{ab} $(a \neq b)$ are complex numbers representing the dephasing rates of coherences, while k_{ac} are real numbers that represent the rates of population transport. All these rates are given by the one-sided Fourier transforms of the coordinate–coordinate correlation function [4]. The Redfield theory provides a commonly used description for excitation dynamics and relaxation in pigment–protein complexes.

12.7.2 Modified Redfield rates for excitons

The Redfield approach assumes that the bath is Markovian and certain types of fluctuations are independent, while all the states are in thermal equilibrium with the bath. However, it is important to realize that the bath is often affected by the system, and thus the bath equilibrium for different system states can be slightly shifted; therefore, within short distances, molecular fluctuations might be highly correlated. The modified Redfield theory includes these effects: it is nonperturbative with respect to diagonal fluctuations and accounts for the correlations between the diagonal and off-diagonal fluctuations [4].

The population transfer rate (from state a to state b) within the modified Redfield scheme reads

$$k_{ba} = 2\,\mathrm{Re}\left|\tilde{h}_{ba}\right|^2 \int_0^\infty \mathrm{d}\tau e^{i\omega_{ab}\tau}\left\{\ddot{g}_{ba,ab}(\tau)\right.$$

$$- \left[\dot{g}_{aa,ba}(\tau) - \dot{g}_{bb,ba}(\tau) + 2i\lambda_{ba,aa}\right]$$

$$\times \left[\dot{g}_{ab,aa}(\tau) - \dot{g}_{ab,bb}(\tau) + 2i\lambda_{ab,aa}\right]\}$$

$$\times \exp\left[-g_{aa,aa}(\tau) - g_{bb,bb}(\tau) + g_{aa,bb}(\tau)\right.$$

$$\left. + g_{bb,aa}(\tau) + 2i\left(\lambda_{aa,bb} - \lambda_{aa,aa}\right)\tau\right], \qquad (12.87)$$

where the so-called excitonic lineshape function $g_{j_4j_3,j_2j_1}(t)$ has been introduced [3,4]. It is given by

$$g_{j_4j_3,j_2j_1}(t) = \sum_m c_{j_4m}c^*_{j_3m}c^*_{j_2m}c_{j_1m}g_m(t), \qquad (12.88)$$

and the monomeric lineshape function $g_m(t)$ is defined as

$$g_m(t) = \int_0^t \mathrm{d}\tau \int_0^\tau \mathrm{d}\tau' C_m(\tau - \tau'). \qquad (12.89)$$

$C_m(t)$ is the correlation function corresponding to the local vibrations of the mth molecule, which is defined by Equation 12.73. The dots and double dots in Equation 12.87 denote the time derivatives, and $\lambda_{ab,cd}$ are the reorganization energies (Stokes shifts) given as the limit

$$\lambda_{ab,cd} = -\lim_{t\to\infty}\dot{g}_{ab,cd}(t). \qquad (12.90)$$

The modified Redfield theory allows one to interpolate the description of the excitation dynamics obtained by using the Redfield approach and Förster resonance energy transfer theory (FRET; see Section 12.7.3) [4]. It also includes correlations of the diagonal and off-diagonal fluctuations. The rates determined within the modified Redfield framework explicitly includes the Stokes shifts.

12.7.3 Förster resonance energy transfer

The description of the excitation dynamics in the excitonically coupled molecular aggregate in terms of the Redfield theory is based on the assumption that the resonance intermolecular interaction dominates over the interaction with the vibrational degrees of freedom. Therefore, the latter were mainly considered perturbatively. In the opposite case of a weak resonance interaction, the excitation energy transfer is usually considered within the framework of the FRET theory [16]. This is a widely employed method, working remarkably well even in situations where the condition of weak chromophore–chromophore coupling might be questionable. However, for some applications, FRET has several important closely related deficiencies. Since FRET is an approach of the Fermi's golden rule type, it provides population transfer rates but no propagation of coherences. In fact, it assumes that all coherences are washed out instantaneously. Moreover, FRET intrinsically assumes that the excitations are localized on individual chromophores despite their mutual interaction, and the intermolecular interaction defined by Equation 12.24 is treated perturbatively, while the interaction between electronic excitations with intramolecular vibrations is taken into account explicitly. This allows one to determine the excitation transfer rate from the molecule a to the molecule b as

$$k_{ba} = k_a^{(\mathrm{r})}\left(\frac{R_0}{r_{ba}}\right)^6, \qquad (12.91)$$

where

$k_a^{(\mathrm{r})}$ is the radiative rate of the excitation donating molecule

r_{ba} is the intermolecular distance between the two molecules

R_0 is the so-called Förster radius, related to the overlap integral of the fluorescence spectrum of the excitation-donating molecule a and the absorption spectrum of the excitation-accepting molecule b [2]:

$$R_0^6 \propto \frac{\kappa^2}{n^4} \int \frac{\varepsilon_b(v) f_a(v)}{v^4} dv. \qquad (12.92)$$

In this expression, the orientation prefactor is $\kappa = \cos\theta - 3\cos\theta_a \cos\theta_b$, where θ is the angle between the transition dipole moments of both molecules, while θ_a and θ_b are the angles between these transition dipole moments and the radius vector r_{ba} determining the distance between the molecules, respectively. Finally, n in Equation 12.92 is the refractive index of the direct environment of donor and acceptor (for a discussion, see Reference 17), $f_a(v)$ is the normalized fluorescence spectrum of the donating monomer a, and $\varepsilon_b(v)$ is the extinction coefficient of the accepting molecule b; v is the frequency.

It is worthwhile to mention that the Förster energy transfer rate expression can be defined directly from the modified Redfield rate expression [4] and even can be generalized by using the reduced density matrix formalism in the weak resonance coupling limit [18]. The derivation is similar to that of the QME, except that this time, the resonance coupling instead of the system–bath interaction is treated as a perturbation.

12.7.4 Lindblad theory

The relaxation process of a quantum system is, in general, not time local, that is, it has some memory. The memory is present since in practice, the energy transfer through the bath has a finite timescale. This is formally described by the bath correlation function. However, when the bath correlation time is short compared to the timescale of the system dynamics, the time-local equation describes the exciton dynamics reasonably well. In such a description, the memory effects may still remain present due to the nonsecular nature of the equation, that is, the effect of excitation being transferred from population to coherence and back, which leads to some effective phase delay.

It is possible to construct the most general form of the equation of motion for the reduced density matrix based on the requirement of complete positivity. We can write such an equation in analogy to Equation 12.81 in the so-called Lindblad form [19]:

$$\frac{d}{dt}\hat{\rho}(t) = -i\left[\hat{H}_S, \hat{\rho}(t)\right]$$
$$+ \sum_k \left(\hat{L}_k \hat{\rho}(t)\hat{L}_k^\dagger - \frac{1}{2}\hat{L}_k^\dagger \hat{L}_k \hat{\rho}(t) - \frac{1}{2}\hat{\rho}(t)\hat{L}_k^\dagger \hat{L}_k \right). \qquad (12.93)$$

Here, $\hat{L}_k$ are the Lindblad operators. While they are obtained as mathematical constructions, they have the meaning of various modes, which couple the system to the bath (collective coordinates). The sum over k here is assumed to run over an uncountable number of independent bath modes.

Such a description leaves a lot of undefined "off-diagonal" parameters, which can be chosen in a specific way to lead to a physically reasonable result [4]. The Lindblad equation determines all other off-diagonal rates responsible for the coherence–coherence transfer and for the population–coherence mixing, which can be calculated microscopically from the Redfield rate expressions.

12.7.5 HEOM theory

Hierarchical equations of motion (HEOM) is a nonperturbative theory describing the exciton dynamics for open quantum systems [20,21]. Because it is a full theory, it contains a hierarchy of coupled kinetic equations for auxiliary density operators defined in the Liouville space. Formally, the hierarchy of equations is infinite since the description is nonperturbative and non-Markovian. However, various truncation schemes are used.

Since the HEOM theory is derived as an operator equation and makes no approximation for the bath, it is independent of the basis chosen to solve the problem. It can thus capture such effects as polaron formation [22,23] and localization of an exciton onto a single chromophore. Therefore, the theory can describe a much broader class of problems. However, it is clear that computationally, this is a very expensive approach.

REFERENCES

1. Atkins, P. and Friedman, R. (2010) *Molecular Quantum Mechanics*, 5 edn. (Oxford University Press, New York).

2. van Amerongen, H., Valkunas, L., and van Grondelle, R. (2000) *Photosynthetic Excitons*. (World Scientific Co., Singapore).

3. Mukamel, S. (1995) *Principles of Nonlinear Optical Spectroscopy*. (Oxford University Press, New York).

4. Valkunas, L., Abramavicius, D., and Mančal, T. (2013) *Molecular Excitation Dynamics and Relaxation*. (Wiley-VCH, Weinheim, Germany).

5. Frenkel, J. (1931) On the transformation of light into heat in solids. II. *Phys. Rev.* **37**, 1276.

6. Frenkel, J. (1931) On the transformation of light into heat in solids. I. *Phys. Rev.* **37**, 17.

7. Davydov, A. (1962) *A Theory of Molecular Excitions*. (McGraw-Hill, New York).

8. Knox, R. S. (1963) *Theory of Excitons*. (Academic Press Inc., New York).

9. Pope, M. and Swenberg, C. E. (1999) *Electronic Processes in Organic Crystals and Polymers*. (Oxford University Press, New York).

10 Silinsh, E. A. and Capek, V. (1994) *Organic Molecular Crystals. Interaction, Localization and Transport Phenomena*. (AIP Press, New York).

11. Haken, H. (1983) *Quantum Field Theory of Solids*. (North-Holland Publishing, Amsterdam, the Netherlands).

12. He, X. F. (1991) Excitons in anisotropic solids: The model of fractional-dimensional space. *Phys. Rev. B* **43**, 2063.

13. Jorio, A., Dresselhous, G., and Dresselhous, M. S., eds. (2008) *Carbon Nanotubes, Topics in Applied Physics*, vol. 111. (Springer Verlag, Berlin, Germany).

14. Gelzinis, A., Valkunas, L., Fuller, F. D., Ogilvie, J. P., Mukamel, S., and Abramavicius, D. (2013) Tight-binding model of the photosystem II reaction center: Application to two-dimensional electronic spectroscopy. *New J. Phys.* **15**, 075013.

15. Butkus, V., Valkunas, L., and Abramavicius, D. (2014) Vibronic phenomena and exciton–vibrational interference in two-dimensional spectra of molecular aggregates. *J. Chem. Phys.* **140**, 034306.

16. Förster, T. (1948) Zwischenmolekulare energiewanderung und fluoreszenz. *Ann. Physik* **6**, 55.

17. Knox, R. S. and van Amerongen, H. (2002) Refractive index dependence of the Förster resonance excitation transfer rate. *J. Phys. Chem. B* **106**, 5289–5293.

18. Mančal, T., Balevičius, V., and Valkunas, L. (2011) Decoherence in weakly coupled excitonic complexes. *J. Phys. Chem. A* **115**, 3845–3858.

19. Lindblad, G. (1976) On the generators of quantum dynamical semigroups. *Commun. Math. Phys.* **48**, 119.

20. Ishizaki, A. and Tanimura, Y. (2005) Quantum dynamics of system strongly coupled to lowtemperature colored noise bath: Reduced hierarchy equations approach. *J. Phys. Soc. Jpn.* **74**, 3131.

21. Tanimura, Y. (2006) Stochastic Liouville, Langevin, Fokker-Planck, and master equation approaches to quantum dissipative systems. *J. Phys. Soc. Jpn.* **75**, 082001.

22. Gelzinis, A., Abramavicius, D., and Valkunas, L. (2011) Non-Markovian effects in time-resolved fluorescence spectrum of molecular aggregates: Tracing polaron formation. *Phys. Rev. B* **84**, 245430.

23. Chorošajev, V., Gelzinis, A., Valkunas, L., and Abramavicius, D. (2014) Dynamics of exciton–polaron transition in molecular assemblies: The variational approach. *J. Chem. Phys.* **140**, 244108.

Modeling of energy transfer in photosynthetic light harvesting

VLADIMIR I. NOVODEREZHKIN AND RIENK VAN GRONDELLE

ABBREVIATIONS

BChl	Bacteriochlorophyll
CD	Circular dichroism spectrum
Chl	Chlorophylls
DBV	Dihydrobiliverdin
EA	Excitation anisotropy spectrum
EET	Excitation energy transfer
FL	Fluorescence spectrum
FLN	Fluorescence line-narrowing
FMO	Fenna–Matthews–Olson
HEOM	Hierarchical equation of motion
LH1	Bacterial light-harvesting complex 1
LH2	Bacterial light-harvesting complex 2
LHCII	Major plant light-harvesting complex II
MR	Modified Redfield theory
OD	Absorption spectrum
PE545	Phycoerythrin 545
PEB	Phycoerythrobilin
PSI	Photosystem I

PSII	Photosystem II
RC	Reaction center
SF	Standard Förster theory
SR	Standard Redfield theory
TA	Transient absorption
TG	Transient grating

13.1 INTRODUCTION

In the primary steps of photosynthesis, solar photons are absorbed by special membrane-associated pigment–protein complexes (light-harvesting antennas) and the electronic excitations are efficiently transferred to a reaction center (RC), where they are used to drive a transmembrane charge separation.[1-8] Antenna complexes consist of ordered arrays of light-harvesting pigments (i.e., chlorophylls (Chls) in higher plants, bacteriochlorophylls (BChls) in photosynthetic bacteria, or bilins in marine photosynthesis) bound to proteins. High-resolution studies revealed the structure of the Fenna–Matthews–Olson (FMO) complex of green bacteria,[9] the LH2[10-12] and LH1[13-15] complexes of purple bacteria, the major light-harvesting complex (LHCII) of higher plants,[16-18] core complexes of photosystem I (PSI)[19-21] and photosystem II (PSII),[22-25] and biliprotein complexes phycoerythrin 545 (PE545)[26,27] and phycocyanin 645[28] from photosynthetic cryptophyte algae. The number of pigments in an elementary antenna subunit is ranging from a few (7–8 BChls in FMO, 8 bilins in PE545, 14 Chls in the LHCII monomer) to 96 Chl in the PSI-core. The average distance between nearest-neighbor pigments in most of these complexes can be as short as 9–12 Å, thus giving rise to strong pigment–pigment interactions (i.e., interactions producing appreciable mixing of the excited states of individual pigments). As a result, the whole antenna is generally characterized by a complicated manifold of excited states, including collective electronic excitations (excitons) with a high degree of delocalization in combination with more localized excitations due to the presence of weakly coupled pigments.

Generally, electronic excitations of the pigments are coupled to collective nuclear modes of the pigment–protein complex. Modulation of the electronic transition energies of individual pigments (denoted below as the "site energies") by slow conformational motion of the protein matrix produces disorder, resulting in more localized exciton wavefunctions as well as inhomogeneous broadening of the electronic transitions due to ensemble averaging. Coupling of excitations to fast nuclear motion (intra- and interpigment vibrations, phonons) results in: (1) homogeneous broadening of the electronic transition spectra, (2) their red shift due to reorganization effects (associated with changes in equilibrium position of the nuclear modes after electronic excitation), (3) a further decrease of the delocalization size due to polaron effects, and (4) the transfer of electronic excitation within the excited state manifold, including fast (fs) relaxation between exciton states within strongly coupled clusters and slower (ps) energy migration between clusters or monomeric sites. Relaxation/migration induced by the coupling of the electronic excited states to a continuum of low-frequency phonons and high-frequency vibrational modes[29-35] is the basic mechanism of photosynthetic light harvesting producing ultrafast cascading from higher- to lower-energy states, effective energy migration in the antenna and the delivery of excitation energy to the RC. Since the biochemical isolation of antenna complexes and the discovery of their structure, these energy transfer events have been studied using a variety of advanced laser spectroscopic methods, including time-resolved (sub-100 fs) nonlinear techniques together with theoretical modeling (see for a review References 2–7, 36–39).

Due to the collective character of an excitation in photosynthetic complexes, the usual theory of resonance energy transfer, that is, Förster theory[40] (treating the interactions between chromophores perturbatively and, therefore, valid only for localized excitations) cannot give an adequate picture of excitation energy transfer (EET). The generalized Förster theory[41-46] considers energy transfer between clusters with arbitrary degree of delocalization but is restricted to weak intercluster interactions.

In the standard Redfield theory[47] (see Chapter 12), all exciton couplings are taken into account explicitly, thus allowing a description of all types of exciton relaxation/migration processes within strongly coupled antenna complexes, including coupled dynamics of the populations and coherences between the exciton states. In this theory, the dynamics is described in the pure exciton basis, where the relaxation between exciton states is accounted for by including exciton-phonon coupling as an off-diagonal perturbation. Such an approach was used to model the energy transfer dynamics within the B850 band of LH2,[48–50] intra- and interband B800-B850 energy transfer in the whole LH2 antenna,[51,52] intra- and interband dynamics in the Chl *b*-Chl *a* LHCII complex,[53] excited state equilibration dynamics in FMO,[54,55] equilibration dynamics in the PSI core,[56] and energy/electron transfer in the PSII-RC.[57–59] Recently, the Redfield theory was successfully applied to model the time evolution of 2D-electronic spectra obtained for the PSII-RC.[60] The standard Redfield approach can be generalized by including strong coupling of excitations to a few vibrational modes. Relaxation in such a system can be described on the basis of electron-vibrational eigenstates. This approach allowed to describe the electron transfer coupled to coherent nuclear motion in the bacterial RC,[61–64] exciton-vibrational relaxation in Chl *a-b* heterodimers from LHCII,[65,66] long-lived vibrational coherences in LH1,[67,68] and coupled exciton-vibrational relaxation in LH1.[38,69] In recent studies, this method has been used for quantitative explanation of a coherent exciton-vibrational dynamics revealed by a 2D electronic spectroscopy of PSII-RC,[60,70,71] including identification of the predominantly excitonic, mixed exciton-vibrational, and pure vibrational coherences in the experimental static[72] and time-resolved[73] 2D frequency maps.

In the modified version of the Redfield theory,[74,75] the diagonal (in the exciton basis) part of the electron-phonon coupling is taken into account nonperturbatively, thus giving more realistic line shapes and relaxation rates due to the inclusion of multiphonon processes.[76] The modified Redfield theory allowed a better quantitative treatment of spectra and dynamics in LH2/LH1,[77] FMO,[78] PE545,[79,80] Lhca4,[81] PSI,[82] LHCII,[83–86] PSII-RC,[87–91] and PSII-core,[92] including the conformational fluctuations of the single-molecule spectra observed for LH1/LH2[77,93–96] and later in LHCII.[4,97] Recently, the theory was used to model the 2D-photon echo spectra in FMO,[98] B800-820 complex,[99] and LHCII.[100] Notice that the present version of the modified Redfield theory is restricted to relaxation dynamics of populations and does not include one-exciton coherences.

In more versatile approaches, both exciton couplings and phonon dynamics are taken into account in a nonperturbative manner, for example, in the recently developed method based on hierarchically coupled equations of motion (HEOM) for the density operator.[101,102] In this approach, the exciton mixing is a function of the phonon reorganization dynamics. When the value of the exciton coupling exceeds the reorganization energy, the mixing is not significantly affected by the phonons, and the dynamics can be described in terms of unperturbed exciton eigenfunctions (Redfield limit). In the case of a big reorganization energy (i.e., strong coupling to phonons) the relaxation of nuclear modes from the region of strong mixing to the bottom of the potential surfaces of the electronic excited states produce "dynamic localization," corresponding to the Förster limit. Thus, the method of hierarchical equations includes the Redfield and Förster pictures as limiting cases. The hierarchical equations have been used to model the coherent dynamics in the FMO complex,[102–106] in the B850 band of the LH2 complex,[107,108] dynamics within the LH1-RC core complex,[109] energy transfer between two B850 rings,[107] and transfers between the B850 and LH1(B875) rings,[110] energy transfer in the B800 ring of LH2,[111] and recently in LHCII monomeric subunit.[112,113]

In this chapter, we present a quantitative comparison of the different energy transfer theories, that is, modified Redfield, standard and generalized Förster theories, as well as the combined Redfield–Förster approach. We demonstrate the physical limitations of these approaches and determine critical values of the key parameters (energy gaps and couplings between the pigments) corresponding to these limits. We show that the modified Redfield theory is able to describe quantitatively the spectroscopy and energy transfer dynamics of a variety of photosynthetic light-harvesting complexes. We also notice that the Redfield-model predicts unrealistically large transfer rates between weakly coupled isoenergetic pigments (resonance artifact).

We consider three examples where these theories are applied to photosynthetic light-harvesting complexes.

1. In the first example, we consider the phycobiliprotein complex PE545, which consists of eight relatively strongly coupled pigments and therefore can be modeled with the modified Redfield theory.

2. The second example is the trimeric LHCII complex that contains clusters of Chls with strong coupling within clusters and weak coupling between quasi-isoenergetic pigments

from adjacent clusters. In this case, a realistic estimation of the energy transfer rates requires the combined Redfield–Förster theory.

3. Finally, we consider the excitation dynamics within the B800 ring that is characterized by relatively weak couplings between neighboring BChls with the same transition energies. The modified Redfield theory does not work in this case. On the other hand, the B800 complex displays sizable excitonic features in linear spectra and coherent components in the nonlinear responses. Thus, the Förster theory is not suitable for modeling as well. We model the spectra and dynamics in this complex using the hierarchical equation method. Interestingly, reasonably good results can also be obtained with the standard Redfield approach, that is, free from the resonance artifact if nonsecular population-to-coherence transfer terms are included.

13.2 PHYSICAL MODELS OF ENERGY TRANSFER

13.2.1 Collective excited states of the light-harvesting antenna

The effective storage of solar energy in photosynthesis is possible due to the presence of light-harvesting pigments (chlorophylls, carotenoids, bilins) in antenna complexes with long-lived excited states (with respect to energy transfer times) and a high cross-section for light absorption. The elementary excitation of the antenna is described by the wavefunction $|n\rangle$, which corresponds to excitation of the n-th pigment. Quantum mechanics dictates that when neighboring pigments are coupled, because they are close by, the excited state of the complex is given by a superposition of such wavefunctions, that is, $c_1|n_1\rangle + c_2|n_2\rangle + \cdots$, where one excitation is shared between a number of molecules. Such a collective excitation (denoted "exciton") is different from independently excited molecules n_1, n_2, ... due to correlations ("coherences") between them, given by $c_1{}^*c_2\ldots$. Such coherences can be produced if the electronic Hamiltonian contains off-diagonal terms, that is, $H \sim |n_2\rangle\langle n_1|$. In this coherent state, one molecule "knows" about the excitation of its neighbors. This dramatically changes the spectrum of a pigment aggregate as well as the energy transfer dynamics. In natural antenna complexes, these

features produce more efficient light absorption, faster conversion from short- to long-wavelength spectral bands, and increase the irreversible trapping of excitations by the RC.

The Hamiltonian of the antenna on the basis of the local excited-state wavefunctions $|n\rangle$ is

$$H = \sum_{n=1}^{N} E_n |n\rangle\langle n| + \sum_{n \neq m}^{N} M_{nm} |n\rangle\langle m| \quad (13.1)$$

where

N is the number of light-harvesting molecules in the antenna

E_n is the electronic transition energy of the n-th molecule (for simplicity we consider two-level molecules)

M_{nm} is the interaction energy between the n-th and m-th molecules

The off-diagonal couplings M_{nm} produce new (collective) eigenstates delocalized over a number of sites instead of excitations of individual pigments. The energies ω_k and wavefunctions $|k\rangle$ of the exciton eigenstates can be obtained by the diagonalization of the Hamiltonian (Equation 13.1):

$$H = \sum_{k=1}^{N} \omega_k |k\rangle\langle k|; \quad |k\rangle = \sum_{n=1}^{N} c_n^k |n\rangle \quad (13.2)$$

where the collective exciton states $|k\rangle$ are composed of a coherent superposition of the individual molecular excitations $|n\rangle$. The wavefunction amplitudes c_n^k reflect the participation of the n-th site in the k-th exciton state.

Generally, the energies ω_k of the exciton states exhibit some shifts from the site energies E_n due to exciton splitting (determined by couplings M_{nm}). The resulting exciton band is then given by a manifold of N discrete exciton transitions. Interaction with phonons induces a homogeneous broadening (in combination with additional shifting) of the corresponding spectral lines (as described in the next section).

13.2.2 Exciton–phonon coupling and spectral line shapes

The line shape corresponding to excitation of the k-th exciton level (from the manifold of N one-exciton levels) is given by the coupling of this level to fast nuclear modes (low-frequency phonons determined by collective modes of the pigment-protein matrix and intramolecular high-frequency vibrations). Absorption (OD) and fluorescence (FL) spectra of the whole complex are then given by the sum of contributions from all exciton components:

$$OD(\omega) = \omega \sum_k d_k^2 \, \mathrm{Re} \int_0^\infty A_k(t) dt$$

$$FL(\omega) = \omega \sum_k P_k d_k^2 \, \mathrm{Re} \int_0^\infty F_k(t) dt \qquad (13.3)$$

$$A_k(t) = e^{i(\omega - \omega_k)t - g_{kkkk}(t)}$$

$$F_k(t) = e^{i(\omega - \omega_k)t + 2i\lambda_{kkkk}t - g_{kkkk}^*(t)}$$

where

P_k denotes the steady-state population

d_k represents the transition dipole moment of the k-th exciton state

The function $g_{kkkk}(t)$ determines the line-broadening of the k-th exciton state due to exciton–phonon coupling

λ_{kkkk} is the corresponding reorganization energy (the Stokes shift of the emission of the k-th level is equal to $2\lambda_{kkkk}$)

Here, we use the simplest form for the OD/FL spectra that can be further generalized by including the phenomenological Markovian term responsible for relaxation-induced broadening or complex non-Markovian term describing the off-diagonal fluctuations inducing broadening and additional shift of the exciton transition energies.[114] The $g_{kkkk}(t)$ function and λ_{kkkk} are related to the exciton–phonon spectral density $C(\omega)$[74,115,116]:

$$g_{kk'k''k'''}(t) = -\int_{-\infty}^\infty \frac{d\omega}{2\pi\omega^2} C_{kk'k''k'''}(\omega)$$

$$\times \left[(\cos\omega t - 1)\coth\frac{\omega}{2k_BT} - i(\sin\omega t - \omega t) \right]$$

$$\lambda_{kk'k''k'''} = \int_{-\infty}^\infty \frac{d\omega}{2\pi\omega} C_{kk'k''k'''}(\omega)$$

$$C_{kk'k''k'''}(\omega) = \sum_n c_n^k c_n^{k'} c_n^{k''} c_n^{k'''} C_n(\omega)$$

$$(13.4)$$

where

T is the temperature

k_B is the Boltzmann constant

$C_n(\omega)$ is the spectral density in the site representation corresponding to phonon-induced modulation of the transition energy of the n-th site (we assume that modulations of different sites are uncorrelated and we do not include modulations of the pigment–pigment interaction energies)

Equation 13.3 yields the homogeneous line shape. In the presence of static disorder (disorder of the site energies E_n and couplings M_{nm}), the homogeneous FL profiles should be averaged over a random distribution of E_n and M_{nm} that will perturb energies and eigenfunctions of the exciton states. Nonlinear spectral responses can be expressed in terms of the line-broadening functions (Equation 13.4) as well, for example, pump-probe,[74,83,84,87,88,117] Stark spectra,[90,118] three-pulse photon echoes,[116] and 2D echoes.[98]

The spectral density needed to evaluate the spectral responses can be obtained from molecular dynamics simulations[35] or can be extracted from experiments, for example, from hole-burning[29] or fluorescence line-narrowing (FLN) spectra.[30,31] In Figure 13.1, we show the $C_n(\omega)$ function for the chlorophyll (Chl) *a* molecules in the LHCII complex obtained from FLN data.[30]

13.2.3 Modified Redfield and generalized Förster theories

When the spectral density $C_n(\omega)$ is specified, the functions (Equation 13.4) provide a unified tool for calculating not only the spectral shapes (Equation 13.3) but also energy transfer rates. Thus, the rate of population transfer from the state k' to the state k expressed in terms of line-broadening functions (Equation 13.4) is[74–76]:

$$R_{kkk'k'} = 2\,\mathrm{Re}\int_0^\infty dt A_k(t) F_{k'}^*(t) V_{kk'}(t)$$

$$V_{kk'}(t) = \exp\left(2g_{k'k'kk}(t) + 2i\lambda_{k'k'kk}t\right)$$

$$\times \left[\ddot{g}_{kk'k'k}(t) - \left\{ \dot{g}_{k'kk'k'}(t) - \dot{g}_{k'kkk}(t) + 2i\lambda_{kk'k'k'} \right\} \right.$$

$$\left. \times \left\{ \dot{g}_{k'k'kk'}(t) - \dot{g}_{kkkk'}(t) + 2i\lambda_{k'k'kk'} \right\} \right]$$

$$(13.5)$$

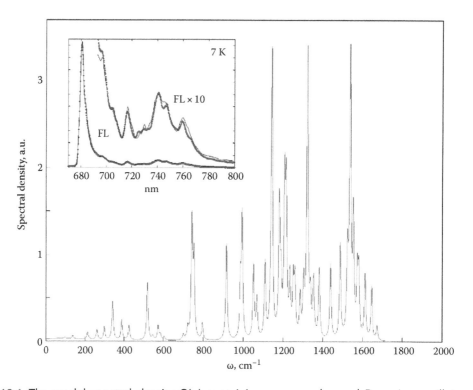

Figure 13.1 The model spectral density C(ω) containing one overdamped Brownian oscillator and 48 high-frequency modes with frequencies and relative Huang–Rhys factors taken from a low-temperature FLN experiment on LHCII.[30] Parameters of the Brownian oscillator and couplings of high-frequency modes have been adjusted from the fit of the low-temperature nonselective 7 K fluorescence spectrum of LHCII complex (see insert).

where F(t) and A(t) are line-shape functions corresponding to fluorescence of the donor state and absorption of the acceptor, respectively, defined by Equation 13.3, while V describes the interaction between donor and acceptor. Equation 13.5 (usually denoted as the modified Redfield equation[76]) is valid for arbitrary delocalization of the donor and acceptor states. In contrast to the standard Redfield equation, here the exciton–phonon coupling is not supposed to be weak, but the corresponding displacements of the equilibrium positions of the nuclear modes are taken to be independent of the exciton wavefunctions, that is, polaron effects are neglected. Equation 13.5 is therefore valid if the exciton delocalization is controlled by the static disorder rather than by a phonon-induced dynamic localization (i.e., the disorder-induced localization size is less than the polaron length).[74,75] However, in the case of transfer between weakly coupled molecules (where any delocalization is completely destroyed by polaron

effects), the application of the modified Redfield equations (Equation 13.5) can give the wrong results. In this case an approach with localization of excitation at the donor and acceptor sites will work better.[5,6,8,76]

If the donor and acceptor states are localized at the m-th and n-th sites (i.e., $c_m^{k'} = 1$ and $c_n^k = 1$), then V is time independent and given by

$$V_{kk'} = |M_{nm}|^2 \qquad (13.6)$$

where M_{nm} is the interaction energy corresponding to a weak coupling between the localized sites n and m. Switching to the Fourier-transforms of F(t) and A(t), we can rewrite the integral in the form of donor-acceptor spectral overlap.[76] Thus, we obtain the Förster formula.[40]

Notice that in the Redfield (13.5) and Förster (13.6) formulas the "V" term describes a perturbation inducing energy transfer. In the Förster picture, the phonons are included explicitly, but the exciton coupling M (time independent!) is taken into account perturbatively. In contrast, in the Redfield picture the exciton coupling M is included explicitly (because we use the exciton basis), and the phonons are treated as a perturbation. As a result the interaction term V in (13.5) is modulated by the fast phonon dynamics, that is, becomes time dependent.

The standard Förster formula can be generalized to the case of energy transfer between two weakly connected excitonic clusters.[41-46] The rate of energy transfer from the k'-th exciton state of one cluster to the k-th state of the other cluster is

$$V_{kk'} = \left| \sum_{n,m} c_n^k M_{nm} c_m^{k'} \right|^2 \quad (13.7)$$

where n and m designate molecules belonging to different clusters. In this generalized Förster formula, the donor and acceptor states k' and k can have an arbitrary degree of delocalization (corresponding to arbitrarily strong excitonic interactions within each cluster), but the intercluster interactions M_{nm} are supposed to be weak, thus producing only a small spatial overlap between the k' and k wavefunctions. It is important that in the case of delocalized donor and/or acceptor states the transfer rate can be increased (depending on the geometry) in proportion to the product of the effective delocalization lengths of donor $(N_{del})^D$ and acceptor $(N_{del})^A$, that is, $(N_{del})^D(N_{del})^A$ as was first pointed out in References 41,42,119. Due to this phenomenon, the rate of excitation trapping by the bacterial reaction center located in the center of the ring-like LH1 antenna is increased in proportion to the degree of delocalization in the antenna.[5,41,42] Similarly, the rate of transfer between the two ring-like LH1 or LH2 complexes can be increased due to delocalisation,[48] as well as transfer within cylindrical aggregates.[119]

In the case of significant spatial overlap of the wavefunctions $c_m^{k'}$ and c_n^k the transfer rate cannot be calculated by treating M_{nm} as a perturbation (as in Equations 13.6 and 13.7). In this case, the energy transfer should be calculated on the basis of the exciton states of the whole system. The rates of relaxation between these states (with arbitrary wavefunction overlap) are then given by the modified Redfield theory (Equation 13.5).

13.3 QUANTITATIVE COMPARISON OF REDFIELD AND FÖRSTER FORMULAS

In this section, we compare the modified Redfield and Förster approaches by calculating energy transfer rates within a Chl aggregate assuming a realistic spectral density. All essential features can be illustrated for the simplest cases of energy transfer between two monomeric Chl sites, or between a Chl dimer and a monomeric Chl.

13.3.1 Transfer between two molecules

In Figure 13.2, we compare the transfer rates within a Chl dimer (as a function of the energy gap $\Delta E = E_1 - E_2$ and interaction energy M_{12} between them) calculated using the Förster (Equation 13.6) and modified Redfield (Equation 13.5) theory with the experimental spectral density (shown in Figure 13.1). We calculate the downhill transfers between the two eigenstates at 77 K. In the Redfield approach, this corresponds to a relaxation rate from a higher to a lower level, whereas in the Förster theory this is a hopping from one localized state to another (because delocalization is just simply ignored in the Förster picture). The difference between the two theories vanishes for large energy gaps and becomes dramatic when the energy gap is reduced. Figures 13.3 through 13.5 illustrate this issue in more detail.

In Figure 13.3, we show transfer rates as a function of the energy gap for different couplings. The difference between the two rates is compared with the delocalization length $N_{del} = \Sigma_n \left(c_n^k \right)^{-4}$ calculated as the inverse participation ratio[120] of the k-th level that varies from 1 (in the localized limit) to 2 (uniform delocalization over two molecules).

In the case of strong coupling ($M_{12} = 255$ cm^{-1}), the Redfield rates are higher than those predicted by the Förster theory. Deviation between the two rates is proportional to the deviation of the delocalization length from the localized limit. For large energy gaps ($\Delta E > 5M_{12}$) corresponding to the localized limit ($1 < N_{del} < 1.1$) Förster and Redfield theories give approximately the same rate.

For lower couplings ($M_{12} = 100$ and 55 cm^{-1}) the two theories give the same results everywhere, but for small gaps ($\Delta E < 5M_{12}$) excitations become delocalized and the Redfield theory starts to give

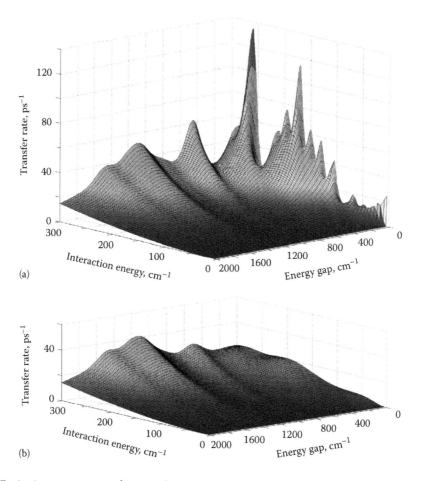

Figure 13.2 Excitation energy transfer rates between two Chl molecules (as a function of the energy gap and the interaction energy between them) calculated according to modified Redfield **(a)** and Förster **(b)** expressions. The specific nonmonotonous dependence of the rates on the energy gap and interaction energy is determined by the shape of exciton–phonon spectral density for Chl.

bigger rates than the Förster theory, where delocalization is not taken into account.

Both for strong and moderate coupling cases ($M_{12} = 25$–255 cm^{-1}), the Redfield theory predicts 2–5 times faster energy transfer in the isoenergetic case ($\Delta E = 0$) due to delocalization. However, this deviation of the Redfield rates from the Förster limit becomes anomalously (and unrealistically) high in the weak coupling limit ($M_{12} < 20$ cm^{-1}). Thus, for $M_{12} = 8$ cm^{-1} we obtain localized excitations for $\Delta E > 30$ cm^{-1} and transfer rates of about 0.1 ps^{-1} (time constant of 10 ps) predicted both by the Förster and Redfield theories. On the other hand, for very small gaps ($\Delta E < 20$ cm^{-1}) the Förster theory gives 0.1–0.25 ps^{-1} (time constants of 4–10 ps), whereas the Redfield theory predicts an abrupt increase in the transfer rate up to 25 ps^{-1} (time constant of 40 fs)

in the $\Delta E = 0$ limit, that is, we can have a more than two orders of magnitude increase in transfer rate! Formally, such fast transfer is possible if the excitation is truly delocalized over two isoenergetic molecules (even if these molecules are well separated in space and weakly interacting). In reality such kind of delocalization will be destroyed by the polaron effects (dynamic localization), a feature that is not included into the Redfield approach. Obviously, the Förster theory (where localization is assumed a priori) gives more correct results in this case.

Figure 13.4 shows the difference between the two theories as a function of interaction energy for fixed values of the energy gap ΔE. In the delocalized limit (for small energy gaps) the curves corresponding to Redfield rates display sharp peaks corresponding to resonance between the vibrational frequencies

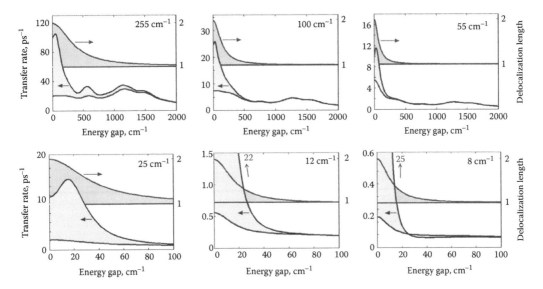

Figure 13.3 Comparison of the energy transfer rate calculated with the Förster (red lines) and modified Redfield (blue lines) theories. The rate of energy transfer between two Chl molecules at 77 K is calculated as a function of the energy gap between them (only downhill rates between the two excited states are shown). The delocalization length (pink lines) is calculated as the inverse participation ratio of the exciton wavefunctions (which is the same for lower and upper eigenstates). We show deviations of the delocalization length of a dimer from the localized limit (where the delocalization length is equal to unity). The top three frames correspond to strong coupling (M_{12} = 255, 100, and 55 cm^{-1}). The bottom frames correspond to moderate (M_{12} = 25 cm^{-1}) and weak coupling cases (M_{12} = 12 and 8 cm^{-1}). (Note that the energy gaps range from 0 to 2000 cm^{-1} for strong coupling and from 0 to 100 cm^{-1} for moderate and weak coupling.) In the weak coupling case, we have expanded the vertical scale to visualize how the two rates start to deviate upon decreasing the energy gap.

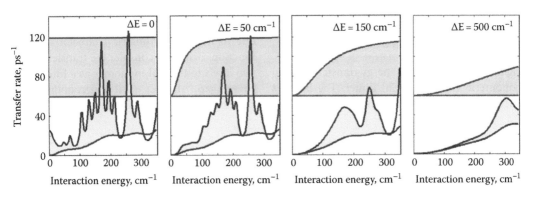

Figure 13.4 The same as in Figure 13.3, but now the rates are shown as a function of interaction energy for fixed values of the energy gap.

(corresponding to high-frequency modes of the spectral density shown in Figure 13.1) and the exciton splitting value. This is not surprising because the Redfield transfer is phonon induced. For larger gaps, this phonon resonance becomes damped due to dephasing induced by the energy difference between the donor and acceptor states

(note that this dephasing is determined by the $\sim \exp(i(\omega_k - \omega_{k'})t)$ term in (13.5) appearing due to the $A_k F_{k'}^*$ factor).

Figure 13.5 shows the ratio between the Redfield and Förster rates in the weak coupling limit. In the region $M_{12} < 30$ cm^{-1}, this ratio is close to unity for large gaps, that is, for $\Delta E > 100$ cm^{-1} (the case

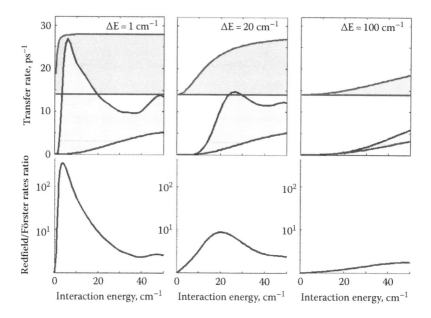

Figure 13.5 *Top frames*: Comparison of the energy transfer rates (same as in Figure 13.4) in the weak interaction limit. *Bottom frames*: ratio of the Förster and Redfield rates shown in the top frames.

$\Delta E = 100$ cm^{-1} is shown in the right panel of Figure 13.5). For smaller gaps (see the panel $\Delta E = 20$ cm^{-1} in Figure 13.5), the Redfield rates can be 10-fold bigger (in the region around $M_{12} = 20$ cm^{-1}). Finally, for the gap values close to zero the Redfield rate can be more than two orders of magnitude higher (in the region $M_{12} < 10$ cm^{-1}).

13.3.2 Transfer between clusters

In Figure 13.6, we consider energy transfer between a monomeric Chl and a Chl dimer. The latter is characterized by equal site energies ($E_1 = E_2$),

parallel configuration of the transition dipoles, and coupling of $M_{12} = 50$ cm^{-1}. Such a dimer is characterized by two exciton levels with symmetric and anti-symmetric exciton wavefunctions. The first one is higher in energy and has a dipole strength that is two times larger than that of a monomer (this level can be denoted as "superradiant"). The second (for this geometry the lower) level is dipole forbidden due to the anti-symmetric character of its wavefunction. Below we consider the transfer from these two levels to the Chl monomer as a function of the energy gap ($\Delta E = E_1 - E_3$) and couplings between the monomeric and two dimeric sites (M_{31} and M_{32}).

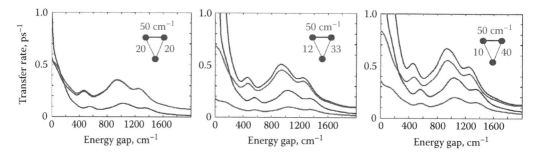

Figure 13.6 The rate of energy transfer between a Chl dimer and a monomeric Chl. Blue lines show transfer rates from the two exciton states of the dimer to the monomer calculated with the modified Redfield theory as a function of the energy gap between the isoenergetic dimeric sites and the monomer. Red lines give the same rates calculated with the generalized Förster theory. Intradimer coupling (50 cm^{-1} for all the examples) and couplings to the monomer are indicated in the inserts.

First, we consider the case of equal couplings ($M_{31} = M_{32} = 20$ cm^{-1}). In this case, the rate of Förster-type transfer from the forbidden state equals to zero (due to absence of dipole–dipole interactions), whereas the Redfield-type transfer from this level yields a finite rate (due to phonon-induced relaxation) (Figure 13.6, left frame). Notice that for the higher superradiant level, both theories give the same result at least for large energy gaps.

For different couplings ($M_{31} \neq M_{32}$), the effective dipole moment responsible for the Förster-type transfer from the forbidden state becomes nonzero (due to nonsymmetric interaction with the dimeric sites, producing a nonequal superposition of the two anti-parallel dipoles). Also, the Förster rate becomes slightly less than the Redfield rate for the superradiant level (Figure 13.6, right and middle frames).

We conclude that modified Redfield is more preferable for energy transfer between the exciton bands separated by a large energy gap (because modified Redfield always works well for large gaps, while the generalized Förster may underestimate rates in this case). In the case of a small gap between the bands, the Förster theory should be used to avoid the resonance artifacts of the Redfield approach, but the generalized Förster rates may lead to an underestimate of the true rates in this case.

13.3.3 Combined Redfield–Förster approach

In the combined Redfield–Förster approach,[8,82,98,99,113] the relaxation dynamics within strongly coupled clusters (with intracluster interactions $M_{nm} > M_{cr}$, where M_{cr} is some critical cutoff value) is calculated with the modified Redfield theory, whereas transfers between these clusters (with intercluster couplings $M_{nm} < M_{cr}$) are modeled by the generalized Förster theory. A typical value for the coupling cutoff M_{cr} is about 15–20 cm^{-1}.[82,86,98] The most difficult is the case when the real couplings are near this cutoff. In this case, the Redfield theory overestimates the rates due to the presence of "resonance" artifacts (2–3 orders of magnitude increase of the rates for realizations of the disorder giving zero gap between donor and acceptor states), whereas the generalized Förster approach underestimates them due to neglecting the delocalization and also due to the appearance of symmetry-forbidden transfers (as illustrated by Figure 13.6).

13.4 MODELING OF BILIPROTEIN COMPLEX PE545

As a first example, we consider the phycobiliprotein antenna complex PE545 isolated from the unicellular photosynthetic cryptophyte algae *Rhodomonas CS24*. The structure of the isolated complex PE545 is known at 0.97 Å resolution[27] (Figure 13.7). Spectroscopic studies performed after the discovery of the structure combine the results from absorption (OD), circular dichroism (CD), fluorescence (FL), and excitation anisotropy (EA) spectroscopies and ultrafast nonlinear techniques, including transient grating (TG) and transient absorption (TA)[27,28,121] and more recently 2D-photon echo.[122] The 8 bilins of the complex are relatively strongly coupled with nearest-neighbor couplings ranging between 17 and 92 cm^{-1}, thus allowing modeling based on the pure modified Redfield approach. In this way the steady-state spectra and TA kinetics have been reproduced at a quantitative level.[79]

13.4.1 The model

PE545 consists of four polypeptide chains, α_1, α_2 plus two β subunits, arranged in a complex known

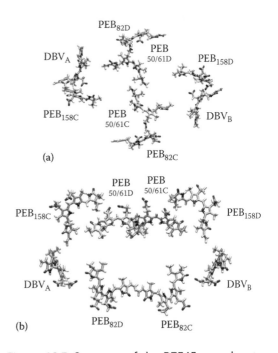

Figure 13.7 Structure of the PE545 complex, top view **(a)** and side view **(b)**. (From Novoderezhkin, V.I. et al., *Biophys. J.*, 99, 344, 2010.)

by convention as a dimer of $\alpha\beta$ monomers. This complex is unusual in that it contains a deep, water-filled slot between the monomers. Each β subunit is covalently linked to three phycoerythrobilin (PEB) chromophores, labeled PEB_{82C}, PEB_{158C}, $PEB_{50/61C}$ (for subunit C) and PEB_{82D}, PEB_{158D}, $PEB_{50/61D}$ (for subunit D), where the numbers indicate the cysteine residues to which the chromophores are bound.[26,27] The $PEB_{50/61}$ chromophores are linked to two cysteine residues, via their A and D pyrroles. Each α subunit contains a covalently linked 15,16-dihydrobiliverdin (DBV) chromophore, DBV_A and DBV_B, respectively, which are spectrally red-shifted compared to the PEBs due to extended conjugation. Transition dipole moments of each bilin have been calculated using the CIS/6-31G method, and the couplings between them have been obtained using the transition density cube method.[79]

As shown in Figure 13.7, the eight pigments of the complex are arranged in two parallel layers. In the first layer ("top" in Figure 13.7) there is strong coupling (92 cm^{-1}) between $PEB_{50/61C}$ and $PEB_{50/61D}$. In the "bottom" layer, there are two pairs with moderate coupling, that is, DBV_B–PEB_{82C} and DBV_A–PEB_{82D} (45 and 46 cm^{-1}). These two pairs are well separated and almost uncoupled, but they have sizable coupling (33–40 cm^{-1}) with the nearest pigments from the top layer, in particular with the $PEB_{50/61C}$–$PEB_{50/61D}$ dimer. As a result, it is impossible to find any pigment (or group of pigments) with weak (<30 cm^{-1}) coupling to the remaining molecules of the complex. This means that there are no channels for Förster-type energy transfer, so that the complex should be considered as a single strongly coupled cluster with the energy transfer rates calculated in the delocalized basis. To this end, we use the modified Redfield approach. The exciton model includes static disorder (accounted for by introducing uncorrelated shifts of the site energies). The unperturbed (unshifted) site energies are free parameters of our model that should be determined from the fit of the spectra using an evolutionary-based search.

13.4.2 Simultaneous fit of spectra and kinetics

Simultaneous fit of the steady-state spectra (OD, FL, circular dichroism [CD], and excitation anisotropy [EA]) obtained for PE545 at 77 and 300 K is shown in Figure 13.8. A similar quality of the fit can be obtained for several sets of the site energies. Fitting of the 77 K TA kinetics (Figure 13.9) rules out most of them (giving poor TA shapes or wrong time scales of the TA spectral evolution), leading us to one model with the best fit of the data (a detailed description of the model was presented in Reference 79).

The origin of the eight exciton states of the complex are revealed by the bar plot in Figure 13.8, displaying the disorder-averaged participation of each of the pigments in each of the exciton states.

The three lowest states k = 1–3, responsible for the 567 nm spectral sub-band, are determined by contributions from DBV_A, DBV_B, and PEB_{82C}. There is a sizable coherence between DBV_B, and PEB_{82C}, which appears due to the relatively strong coupling (45 cm^{-1}) and the small energy difference between these pigments. Coherence between them and DBV_A is negligible. This means that depending on the realization of the disorder an excitation within the 567 nm band is localized either at DBV_A or at the DBV_B-PEB_{82C} dimer.

The higher energy 545 nm sub-band (levels from k = 4 to 8) is determined by PEB_{82D}, two PEB_{158}, and two $PEB_{50/61}$ pigments. The states k = 4 and 8 correspond to the exciton levels of the $PEB_{50/61C}$–$PEB_{50/61D}$ heterodimer. Exciton mixing of the two $PEB_{50/61}$ sites is relatively small due to the big energy gap between them. The k = 5 state is mostly determined by PEB_{158C} with a small contribution of $PEB_{50/61C}$. The moderate coupling between these sites (23 cm^{-1}) still produces sizable coherence due to the small energy gap between them. The levels k = 6 and 7 are determined by PEB_{82D} and PEB_{158D} with some coherence between them because they are also close in energy.

In our model, these excitonic effects produce some redistribution of dipole strength over the levels (Figure 13.8d). The individual PEB and DBV pigments are characterized by a dipole strength of about 150 and 200 debye^2, respectively, whereas some of the exciton states can have a dipole strength of 300 and 400 debye^2 (in the PEB and DBV regions around 567 and 545 nm, respectively). The corresponding delocalization length (calculated as the inverse participation ratio) is varying from 1 to 2–2.5 with an averaged delocalization size no more than 1.2–1.3. Interestingly, this relatively small degree of delocalization is still capable of producing a sizable electronic

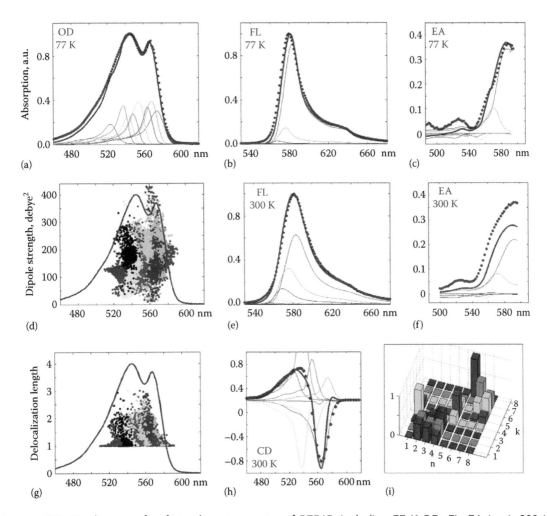

Figure 13.8 Simultaneous fit of steady-state spectra of PE545, including 77 K OD, FL, EA **(a–c)**, 300 K FL, EA, and CD **(e, f, h)**. The spectra are normalized to unity (except EA). Red points show experimental data[18] and blue lines correspond to calculated spectra. Contributions from individual exciton components are shown by thin lines. **(i)** Squared wavefunction amplitudes $\left(c_n^k\right)^2$ (giving the participation of the n-th pigment in the k-th exciton state) are shown for the exciton levels from k = 1 to 8 (shown by the same colors as in the calculated spectra). Distribution of the amplitudes is averaged over disorder. Site numbers from n = 1 to 8 correspond to $PEB_{50/61C}$, DBV_A, DBV_B, PEB_{82C}, PEB_{158C}, $PEB_{50/61D}$, PEB_{82D}, and PEB_{158D}. **(d, g)** Delocalization lengths **(g)** and dipole strengths **(d)** of the exciton states as a function of their peak positions were calculated for 1000 realizations of the disorder. The 77 K OD profile is shown for comparison. Points corresponding to different exciton states have the same colors as the exciton components in the spectra. Delocalization length is calculated as the inverse participation ratio of the exciton wavefunctions. The dipole strength of individual sites n = 1–8 are 148, 196, 204, 155, 148, 141, 152, and 148 debye^2 as obtained by ab initio calculation.[79] (From Novoderezhkin, V.I. and Van Grondelle, R., *Phys. Chem. Chem. Phys.*, 12, 7352, 2010.)

coherence and support quantum-coherent energy transfer as observed recently in 2D-photon echo experiments.[122,123]

Recently, the site energies determined from the fit (shown in Figure 13.8) have been verified by direct molecular dynamics MD-QM/ MM calculation.[80] The MD-QM/MM site energies allowed to model the spectra of PE545 with a similar quality as compared to our original model,[79] which was extracted from the spectral and kinetic fit. The largest difference between the two energy sets is that the MD-QM/MM

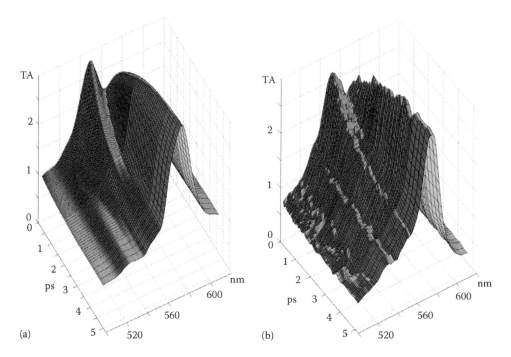

Figure 13.9 3D view (shaded surfaces with lighting) of the calculated **(a)** and measured **(b)** TA spectra at 77 K for delays from 0 to 5 ps (the spectra are inverted, so that the negative bleaching peaks are now positive). (From Novoderezhkin, V.I. and Van Grondelle, R., *Phys. Chem. Chem. Phys.*, 12, 7352, 2010.)

calculations predict a smaller gap between the $PEB_{50/61C}$ and $PEB_{50/61D}$ sites, in better accordance with chemical intuition.

13.4.3 Excited-state population dynamics

The fit of the steady state and TA spectra provides us with a detailed picture of the pathways and timescales of the energy transfer in the PE545 complex.

Short-wavelength excitation (505 nm in the TA experiment) populates the higher states of the 545 nm sub-band (k = 4–8) determined by PEB_{82D}, two PEB_{158}, and two $PEB_{50/61}$. The highest level determined by $PEB_{50/61D}$ exhibits fast (180 fs) decay due to relaxation within the $PEB_{50/61}$ dimer (with a time constant of about 300 fs) in combination with transfer to other sites. Transfer to DBV_A, DBV_B, and PEB_{82C} includes the superposition of many pathways with sub-ps and ps time constants altogether producing energy flow away from the 545 nm region and population of the 567 nm sub-band within 2–4 ps. This is followed by slow redistribution within the 567 nm sub-band within 50 ps and even longer times.

The fastest channels of energy transfer in the complex (faster than 1.5 ps) are shown in Figure 13.10, where they are related to the position of the pigments in real space. According to this scheme, the initial excitation is quickly redistributed within the $PEB_{50/61}$ dimer in the center of the complex. Subsequently, the excitation "slowly" migrates to the outer layer (containing four pigments, that is, DBV_A, DBV_B, and two PEB_{82} bilins). Viewed from the top as in Figure 13.10, this looks like motion from the central dimer to the peripheral sites.

The coherent oscillations of the 2D echoes[122] observed during the first 130 fs with a period of 500 cm^{-1} can be assigned to a quantum coherence created between $PEB_{50/61D}$ and PEB_{82D} (with some admixture of PEB_{158D}). The resulting quantum beats correlate approximately with the expected frequencies and they should disappear with similar times due to pure dephasing in combination with fast relaxation from $PEB_{50/61D}$ to $PEB_{50/61C}$. Notice that at first glance the $PEB_{50/61}$ dimer is a better candidate for a coherent mixing. Indeed, the $PEB_{50/61D}$–$PEB_{50/61C}$ coupling (92 cm^{-1}) exceeds the $PEB_{50/61D}$–PEB_{82D} one (40 cm^{-1}), but also the

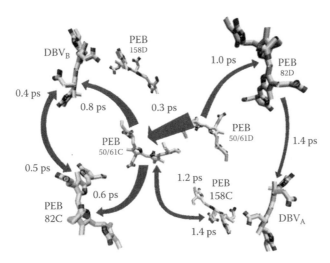

Figure 13.10 Time constants of energy transfer in the site representation (with averaging over disorder). Transfer pathways with the time constant less than 1.5 ps are shown. (From Novoderezhkin, V.I. et al., *Biophys. J.*, 99, 344, 2010.)

$PEB_{50/61D}$–$PEB_{50/61C}$ energy gap is larger, that is, 1000 cm^{-1} according to our model. Therefore, it is much more difficult to coherently excite such a pair of states with a 25 fs laser pulse.[122]

13.5 ENERGY TRANSFER IN THE MAJOR LIGHT-HARVESTING COMPLEX II OF HIGHER PLANTS

The energy transfer in the peripheral light-harvesting complex LHCII from higher plants is an example of transfer that cannot be modeled by pure Redfield theory.

The structure of the trimeric LHCII complex has been obtained at 2.72 Å[17] and later at 2.5 Å resolution.[18] The 14 chlorophylls (Chls) present in each monomeric subunit of LHCII were unambiguously assigned to 8 Chls a and 6 Chls b. Knowledge of the effective dipole moments of Chl a and b[124] together with the spectral density of their exciton-phonon coupling[30,83] allows a quantitative calculation of all the spectral and dynamic properties of the complex using one unified physical picture. Only the site energies of the Chls still remain unspecified and therefore should be determined from the spectroscopic data.

Studies by transient absorption, time-resolved fluorescence, and photon echo techniques revealed major components of the Chl $b \rightarrow$ Chl a transfer of 150–300 and 600 fs[83,125–134] together with minor ps

components. Experiments with selective excitation in the region between the absorption peaks of Chl b (650 nm) and Chl a (675 nm) provided strong evidence for long-lived (10–20 ps) excited states in this spectral region.[83,127–129] In a more recent study,[100] the energy transfer dynamics in LHCII has been explored by the 2D-photon echo technique.

In previous papers,[4,6,84] we have presented an exciton model (with specific site energies for the 14 Chls) that allowed the simultaneous and quantitative fit of all the available spectroscopic data using the modified Redfield approach, including recent studies by single molecule spectroscopy.[6,97] These studies, however, were restricted to modeling of the excitation dynamics within isolated monomeric subunits. In the following sections, we discuss our latest results obtained by modeling of the whole trimeric LHCII complex.[8,86]

13.5.1 Simultaneous fit of linear spectra and transient absorption in LHCII trimer

The arrangement of the Chls within the LHCII trimer according to the crystal structure[17] is shown in Figure 13.11a and b. On the stromal side there are two tightly packed clusters of Chls a (red-encircled for one monomeric subunit), that is, the trimer $a610$–$a611$–$a612$ and the dimer $a602$–$a603$. They are closely connected with three Chls b on the stromal side, Chls $b601$, $b608$, and $b609$. The structure

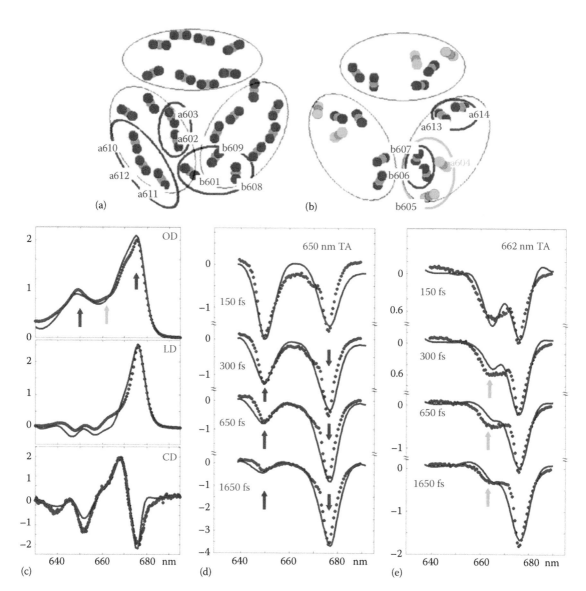

Figure 13.11 Arrangement of chlorophylls within the LHCII trimer at the stromal (a) and lumenal (b) sides (in both frames the top view from the stromal side is shown). Chlorophylls are represented by three atoms: the central magnesium atom and two nitrogen atoms. The connecting line between the two nitrogen atoms defines the direction of the Q_y transition dipole. Red, Chl a nitrogen; blue, Chl b nitrogen; gray, magnesium; green, nitrogen of Chl a604 and Chl b605 (according to the structure reported by Liu et al.[17]). Clusters of Chls a, Chls b, and a mixed group containing long-lived intermediate sites (Chl a604 and Chl b605) are encircled by red, blue, and green, respectively. (c–e) Simultaneous fit of OD, LD, CD, and time-dependent TA spectra using the modified Redfield approach. Experimental OD/LD spectra[135] and CD spectrum[136] have been measured for the LHCII trimer at 77 K (c, red points). Experimental TA spectra[8] are obtained upon 650 nm (d, red points) and 662 nm excitation (e, red points) with 120 fs pulse and pump-probe delays of 150, 300, 650, and 1650 fs. Calculated spectra (c, d, e, blue lines) are obtained with the disordered exciton model for the whole trimer, where the unperturbed site energies within a monomeric subunit have been adjusted in order to obtain the best simultaneous fit of the data. Blue, red, and green arrows show the bands corresponding to Chls b, Chls a, and intermediate sites, respectively. (From Novoderezhkin, V.I., Phys. Chem. Chem. Phys., 12, 7352, 2010. With permission.)

suggests significant coupling between Chls b601' from an adjacent monomeric subunit and b609 (and similarly for the other subunits). Hence, one can consider the b601'–b608–b609 group (encircled in blue in Figure 13.11a) as a b-cluster, which is expected to have a short excited-state lifetime due to excitation energy transfer to the a610–a611–a612 and a602–a603 clusters of the two subunits. In the Chl a-region, one can expect fast exciton relaxation within the a610–a611–a612 and a602–a603 clusters as well as slower hopping between them.

On the lumenal side, there are two groups of pigments, that is, the a613–a614 dimer (encircled by red) and the b605–b606–b607–a604 tetramer (within the green circle). The b605, b606, b607, and a604 sites of the tetramer are weakly connected to the remaining part of the complex. The excited-state dynamics within this latter cluster must be dominated by fast downhill transfer of excitations from the b-sites to the monomeric a604 pigment, which is the best candidate for a long-lived "bottleneck" state, the presence of which was suggested in previous experimental studies.[127–129]

The fits of the room-temperature spectra (OD/CD) of the LHCII monomer and trimer are not so critical with respect to variation of the site energies within the Chl a and b bands,[137] in contrast to the low-temperature fits.[8,84]

A simultaneous fit of the 77 K OD, LD, and CD spectra of the LHCII trimer is shown in Figure 13.11c. Modeling has been done using the modified Redfield theory with the same parameters as in our previous model.[84] However, including the CD into the fit required some (slight) adjustment of the site energies.[8]

This updated model has been further verified by a calculation of the magic angle 77 K TA spectra upon 650 and 662 nm excitation.[8] In contrast to the previous model (where TA spectra have been calculated for the monomeric subunit only),[84] here we have modeled the spectral evolution within the whole LHCII trimer.

13.5.2 Coupling cut-off and combined Redfield–Förster approach

A consistent fit of the data (shown in Figure 13.11) allows to obtain detailed information about pathways and time scales of energy transfer in the LHCII complex. However, the pure Redfield approach is not sufficient to obtain a realistic estimate of the relaxation rates, because the couplings between the monomeric subunits are small, as well as some couplings within the monomeric subunit (i.e., couplings between the stromal and lumenal clusters). Thus, the energy transfer picture in the whole LHCII trimer should be built using the combined Redfield–Förster theory. Such an approach requires the determination of the cutoff coupling M_{cr} showing where the phonon-induced localization of excitations (i.e., dynamic localization and/or polaron effects) becomes dominant. This phenomenon is expected when the reorganization energy exceeds the exciton bandwidth,[138] which is the case for most photosynthetic antenna complexes. However, the phonon-induced localization can be neglected if it is overruled by the disorder-induced localization (determined both by the nonequality of the site energies and their random shifts due to slow conformational dynamics). In the region $M > M_{cr}$, the phonon-induced effects can be present but they are not dominant, as supposed in the modified Redfield theory.[74,75] In contrast, in the $M < M_{cr}$ region the delocalization is completely destroyed by phonons, as supposed a priori in the Förster approach. The M_{cr} value can be roughly estimated by the presence (or absence) of excitonic effects in the spectra. For example, in the bacterial B800 antenna the couplings of 19 cm^{-1} (and varying from 14 to 27 cm^{-1} in earlier calculations[139]) are still capable to produce excitonic linear spectra[140] and signatures of electronic coherence in the nonlinear spectral responses.[52,99] Thus, the M_{cr} should not exceed these values. A value of $M_{cr} = 20$ cm^{-1} was taken by Fleming cs[82,98]. In our recent paper,[86] we have modeled energy transfer within the LHCII trimer by using the Redfield–Förster theory with $M_{cr} = 15$ cm^{-1}.

Verification of the M_{cr} values can be done by comparing the Redfield–Förster picture with the exact solution given by hierarchical equation approach (HEOM).[112,113] It was found that the critical cutoff M_{cr} indicating which exciton couplings should be broken is dependent on the energy gap between the corresponding sites. In the case of LHCII, the most realistic is the Redfield–Förster mode with $M_{cr} = 16$ and 30 cm^{-1} for the stromal- and luminal-side Chls, respectively.[113]

13.5.3 Intramonomeric relaxation rates

The scheme of energy transfer within one monomeric subunit obtained with the Redfield–Förster theory[86] is shown in Figure 13.12.

At the stromal side there is very fast (50 fs) relaxation within the $a610$–611–612 and $b608$–609–$601'$ trimers, slower (425 fs) relaxation within the $a602$–603 dimer, and sub-ps (300–800 fs) transfer between these clusters, including fast $b \rightarrow a$ conversion (with the fastest channel of 290 fs).

At the lumenal side, there is fast (100–200 fs) relaxation within the $a613$–614 and $a604$–$b605$–606–607 clusters. This relaxation within $a604$–$b605$–606–607 produces a quick localization at the "bottleneck" sites, that is, the relatively long-lived $b605$ and the even longer-lived $a604$ site. The relaxation from $b605$ to $a604$ is slow (3.6 ps), but still much faster

than the transfer from $b605$ to other sites. Thus, the real "bottleneck" is $a604$. The transfer from $a604$ to the $a613$–614 cluster is very slow (33 ps).

Migration between the stromal- and lumenal-side clusters is slow. Weak coupling between the stromal $b608$–609–$601'$ and lumenal $a604$–$b605$–606–607 clusters produces slow transfer between them with the fastest channel of about 7 ps (mostly determined by the $b606$–$b608$ coupling). This is much slower than the downhill transfer from the corresponding sites.

There is slow transfer from the lumenal-side "bottleneck" $a604$ to the stromal-side clusters, that is, to $a602$–603 (with a total 43 ps time constant) and to $a610$–611–612 (with a total 40 ps time constant). Thus, the total decay of $a604$ is about 21 ps due to transfer to the stromal side clusters. (In combination with the 33 ps transfer to the lumenal $a613$–614 cluster, we then obtain a 13 ps lifetime).

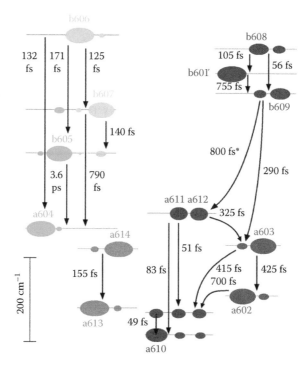

Figure 13.12 Time constants of relaxation between the exciton levels of the Chl a and Chl b clusters within one monomeric subunit of the trimeric LHCII complex at 77 K. Arrows indicate the fastest relaxation channels (slow 10–50 ps channels of intercluster migration are not shown—see the text). The numbers associated with each of the arrows show the time constants. Vertical positions of the levels correspond to the energies of the exciton states averaged over disorder. A vertical bar of 200 cm⁻¹ is shown as a reference. Participation of certain pigments in the exciton states is shown by ellipses with the area proportional to the square of the wavefunction amplitude averaged over disorder. (From Novoderezhkin, V.I. et al., *Phys. Chem. Chem. Phys.*, 13, 17093, 2011.)

Equilibration between the stromal- and lumenal-side a-clusters (i.e., a602–603, a610–611–612, and a613–614) is determined by several components with time constants of 10–50 ps. The overall equilibration rate is determined by population of the donor states, that is, it is temperature dependent.

13.5.4 Intra- and intermonomeric transfers in LHCII trimer

The competition between intra- and intermonomeric energy transfer in the whole LHCII trimer as revealed by the combined Redfield–Förster theory[86] is illustrated in Figures 13.13 and 13.14.

Figure 13.13 shows the energy-transfer scheme within the stromal layer of the LHCII trimer, following the hypothetical excitation of the stromal blue-most b608 site. After excitation of the b608–609–601′ cluster, energy is transferred on a sub-ps time scale to the a602–603 and a610–611–612 clusters within the same subunit that contains b608. The fastest transfer (290 fs) occurs from the lower exciton level of the b608–609 dimer (with b609

having a stronger participation) to the higher exciton level of the a602–603 dimer (which is localized at a603). This transfer pathway is expected, considering the small distance between b609 and a603. Furthermore, from b601′ being located in the second subunit, energy transfer takes place to the two neighboring clusters, that is, a602′–603′ and a610′–611′–612′. The total effective time constant of the latter transfer is ~700 fs. Thus, the strong coupling within the b601′–b608′–b609 cluster results in fast sub-ps conversion to the stromal-side Chl a clusters in the two monomeric subunits connected to this Chl b cluster.

The initial, fast b → a transfer is followed by slower intermonomeric transfer from the a602–603 cluster to the same clusters in the other two subunits, that is, a602′–603′ and a602″–603″. The higher and lower exciton states of these clusters are localized at the a603 and a602 pigments, respectively. Excitation of the higher level (localized at a603) gives rise to a relatively fast (6.3 ps) transfer to the nearest a602′ site, but the largest fraction of energy from a603 flows to the lower

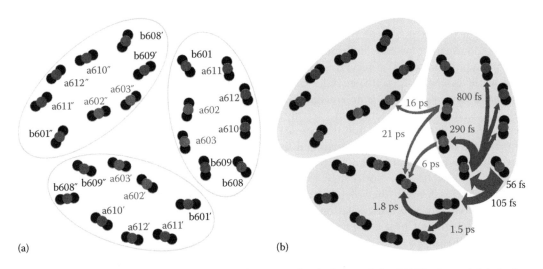

(a)   (b)

Figure 13.13 Intermonomeric transfer within the stromal-side layer of the LHCII trimer. (a) Labeling of the pigments is the same as in Liu et al.[17] but we use notations like a603, a603′, a603″ instead of a603(1), a603(2), a603(3) to distinguish between pigments from different monomeric subunits (each of the three subunits is encircled). Chls are represented by three atoms: the central magnesium atom (red, pink, or blue; notice that the colors are the same as in Figure 13.12) and two nitrogen atoms N_B and N_D (black) that define the directions of the Q_y transition dipole. (b) Main pathways of energy transfer from the b608–609–601′ cluster among the stromal pigments (implying initial excitation of the highest level with a predominant contribution of b608), including intracluster relaxation (blue arrows), transfer to the Chl a clusters of the two nearest subunits (red arrows), and inter-monomeric transfers in the Chl a-region starting from the a602–603 dimer (magenta arrows). (From Novoderezhkin, V.I. et al., *Phys. Chem. Chem. Phys.*, 13, 17093, 2011.)

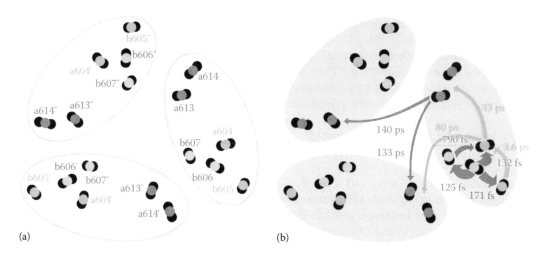

Figure 13.14 Inter- and intramonomeric transfer within the lumenal-side layer of the LHCII trimer. Arrangement of pigments corresponds to the view from the stromal side (as in Figure 13.13), but the stromal-side layer (which is above the lumenal one) is not shown. Colors of the magnesium atom (cyan, green, or orange) correspond to the colors in Figure 13.12. **(a)** Labeling of the pigments (each of the three LHCII monomers is encircled). **(b)** Main intermonomeric pathways of energy transfer from the a604–b605–606–607 cluster (implying initial excitation of the blue-most state with predominant contribution of b606), including fast intracluster relaxation (light blue arrows), slower migration between the intermediate long-lived a604 and b605 sites (green arrow), migration from the "bottleneck" a604 site (also shown by green arrows), and transfer between the Chl a clusters (orange arrows). (From Novoderezhkin, V.I. et al., *Phys. Chem. Chem. Phys.*, 13, 17093, 2011.)

a602 site or to the a610–611–612 trimer. From a602, the dominant (fastest) intermonomeric pathway is uphill to a603″ (occurring within 16 ps), while isoenergetic transfers take place to both a602′ and a602″ with time constants of 21 and 26 ps, respectively. Hence, an exchange between the subunits in quasi-equilibrium must be temperature dependent, being determined by isoenergetic and uphill energy transfer from the high energy a602/a603 site. Notice that experimental studies[141] showed that depolarization upon long-wavelength 680 nm excitation decays at 5 ps over the whole Chl a Q$_y$–b and at room temperature and slows down to 8–30 ps at 13 K (for 675–685 nm detection, respectively).

On the lumenal side of LHCII (see Figure 13.14), b606 is the blue-most pigment. This pigment is part of the a604–b605–606–607 cluster, which contains the long-living, "bottleneck" a604 site. After rapid equilibration within this cluster, only slow migration takes place from a604, including 33- and 80-ps transfer steps to the a613–614 and a613′–614′ dimers, respectively.

Energy transfer obviously also occurs between the lumenal and stromal layers of the LHCII

trimer. The intermonomeric stromal-lumenal transfer in the Chl a region is mostly determined by migration between the a602–603 and a613–614 clusters. There is a 12-ps transfer from the higher energy state of a613–614 (localized at a614) to the a602″–603″ dimer, and 18- and 62-ps transfers from the higher and lower states of a602–603 (localized at a603 and a602, respectively) to a613′–614′.

A combination of the intra- and intermonomeric pathways produces complicated dynamics within the LHCII trimer. The complete energy-transfer dynamics in the LHCII trimer includes the following:

Fast (sub-ps) conversion from Chls b to Chls a at the stromal side, followed by fast (sub-ps) equilibration between the Chl a clusters on the stromal side

Slower (10 ps or more) equilibration within each monomeric subunit between the Chl a clusters at the stromal side and those at the lumenal side

Even slower (20 ps and more) migration between the Chl a clusters located on different monomeric subunits

The slow energy-transfer channels through the "bottleneck" $a604$ site at the lumenal side give rise to additional dynamics, including

Fast (sub-ps to 4 ps) migration from the Chls b at the lumenal side to the "bottleneck" $a604$ site

Slow (33 ps) transfer from $a604$ to the lumenal-side Chls a

Reasonably slow (21 ps) transfer from $a604$ to the stromal-side Chl a clusters

Interestingly, the total decay of the $a604$ "bottleneck" (with time constants of 13 ps and more) is still faster than the intermonomeric equilibration between the Chl a clusters.

13.6 EXCITATION DYNAMICS WITHIN THE B800 ANTENNA

The peripheral light-harvesting complex LH2 from purple photosynthetic bacteria consists of two highly symmetric rings of bacteriochlorophyll (BChl) molecules (see Figure 13.15): the inner ring of 18(16) tightly packed BChls absorbing around 850 nm (B850 band), and the outer ring of 9(8) weakly interacting BChls with an absorption maximum near 800 nm (B800 band).[10-12] The energy transfer dynamics includes migration of excitations around the B800 ring, superimposed on the transfer to exciton states of the B850 ring (for a review see References 3–5). The possible interplay

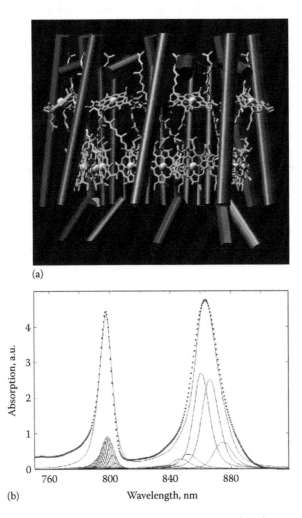

(a)

(b)

Figure 13.15 **(a)** Structure of LH2 antenna of *Rs. molischianum*.[11] **(b)** The absorption spectrum of the LH2 antenna of *Rs. molischianum* measured at 77 K[149] (points) and calculated using the standard Redfield theory[52] (solid lines). The calculated absorption is shown together with contributions of the individual exciton components. (From van Grondelle, R. and Novoderezhkin, V.I., *Phys. Chem. Chem. Phys.*, 8, 793, 2006.)

of intraband B800 → B800 and interband B800 → B850 energy transfer has been the subject of intense studies by hole burning,[142-147] pump-probe,[148-158] three-pulse photon echo,[159,160] and two-dimensional (2D) photon echo techniques[99] together with theoretical modeling.[51,52,99,161]

The isotropic transient absorption (TA) kinetics in the 800 nm region exhibits a biexponential decay.[149-158] According to earlier interpretations, the slow (1.2–1.9 ps) component has been assigned to B800 → B850 transfer, whereas the fast (0.3–0.8 ps) component has been taken to reflect the hopping of localized excitations around the B800 ring. The TA anisotropy near 800 nm decays with approximately the same (0.3–0.5 ps) time constant, apparently supporting the idea of fast hopping-type migration around the B800 ring giving rise to depolarization. However, this intuitive picture (neglecting excitonic effects and implying incoherent migration within the B800 band) is not supported by quantitative modeling. Due to the isoenergetic character of the BChls excited states, even weak coupling between them creates sizable excitonic coherences (not destroyed by the disorder) that should be taken into account in order to reproduce the absorption (OD),[140] circular dichroism (CD),[162] polarized pump-probe kinetics,[52] and the shape of the 2D photon echo spectra[99] observed within the B800 band.

Modeling performed with the standard Redfield theory has led to a detailed picture of the energy transfers in the B800 ring (as well as in the whole B800–B850 antenna) based on a quantitative fit of the measured TA kinetics.[52] It has been shown that impulsive excitation creates coherences between the B800 states that decay with a time constant of 0.3–0.5 ps, producing a fast decay of the TA anisotropy from 0.4 to 0.2. This is followed by a slow migration around the B800 ring with a time constant of several ps producing further depolarization from 0.2 to 0.1. Thus, it has been demonstrated that the fast sub-ps component of the anisotropy decay does not reflect any population migration at all, but is connected with the coherence dynamics. The population equilibration within the B800 ring occurs on the ps time scale being even slower than the interband B800 → B850 energy transfer.[52]

A recent 2D photon echo study of the B800-820 complex[99] has revealed the presence of excitonic effects within the B800 band, thus confirming earlier suggestions.[52,140] The dynamics within the B800 band indicates that there is no significant intraband B800 → B800 transfer on a time scale preceding the interband B800 → B850 transfer, as suggested by polarized TA modeling.[52]

In a more recent study,[111] we have revisited the excitation dynamics within the B800 ring using the hierarchically coupled equations of motion for the density operator.[101,102] In the following sections, we discuss some of these results, including the findings that unveil the origins of the resonance artifact of the modified Redfield theory (discussed in Section 13.3).

13.6.1 Linear spectra: Homogenous B800 ring

To understand the origin of the excitonic spectra of the LH2 B800 ring (in the case of LH2 of *Rs. molischianum* consisting of 8 B800 BChls), it is useful to analyze the OD spectra for a homogeneous ring calculated with the standard Redfield (SR), modified Redfield (MR), hierarchical equations of motion (HEOM), and standard Förster model (SF).

In the HEOM approach, the dynamics is given by the density matrix in the site representation p_{nm} ($j_1, j_2, ..., j_N$), where n, m number the sites (from 1 to N), and integers j_n describe the state of the phonon bath of the n-th site. Each density matrix element (for each pair of states) feels the state of phonons for all other sites, meaning that the coherent superposition of the excited molecules depends now on the phonon reorganization dynamics in the whole system. Thus, there are no time-independent collective (exciton) eigenstates anymore. Initially some of the sites can be excited coherently (near the crossing point of their potential surfaces), but the following phonon relaxation to the bottom of the potentials (where they are not so strongly mixed with the potential surfaces of other sites) can break the intersite coherences, producing more localized excitations (so-called dynamic localization). This dynamics is given by the hierarchy of coupled equations for p_{nm} ($j_1, j_2, ..., j_N$) with all possible sets of n.m and {j_n} values. The hierarchy is terminated if $j_1 + j_2 + \cdots + j_N$ exceeds some cutoff value.[101,102,106] The linear optical responses can be calculated by solving HEOM for the density matrices elements $p_{0n}(j_1, j_2, ..., j_N)$, corresponding to the coherences between the ground "0" state and the excited state of the n-th site.

In Figure 13.16, we calculate the OD spectra of a homogeneous B800 ring for nearest-neighbor

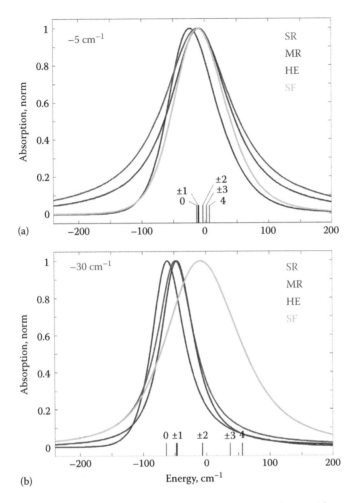

Figure 13.16 Comparing OD spectra for a homogeneous B800 ring (i.e., without energetic disorder) calculated with the standard Redfield (SR, magenta), modified Redfield (MR, red), hierarchical equations (HEOM, blue), and localized (Förster) model (SF, green). The spectra are calculated for two different pigment–pigment couplings, that is, M = −5 cm⁻¹ **(a)** and −30 cm⁻¹ **(b)** at 77 K. The energy is counted from the energy of the pure electronic transition (that is the same for all the sites). Vertical bars show the positions of the eight exciton levels (k = 0, ±1, ±2, ±3, and 4), including their reorganization shift. The dipole allowed k = ±1 degenerate pair is shown by a thick bar (other levels shown by thin bars are forbidden). All the spectra are normalized to unity. (From Novoderezhkin, V.I. and van Grondelle, R., *J. Phys. Chem. B*, 117(38), 11076, 2013.)

coupling M = −5 and −30 cm⁻¹ (note that the real coupling value is 19 cm⁻¹).[139] In the absence of static disorder and without dynamic localization (as supposed in SR/MR) the excited state manifold consists of delocalized exciton eigenstates, that is, k = 0, ±1, ±2, ±3, and 4 levels shift from the monomeric transition energy by 2M, M, 0, −M, and −2M, respectively, where M is the nearest-neighbor coupling. For an in-plane orientation of the transition dipoles, all the dipole strength is concentrated in the two k = ±1 levels broadened due to relaxation to other levels.

This exciton relaxation between delocalized states (with a big spatial overlap of their wavefunctions) is rather fast, producing a significant broadening. The largest broadening is observed for small couplings M, when all the levels are almost isoenergetic.

In our example, the SR/MR theories predict too broad an OD spectrum at M = −5 cm⁻¹ (as shown in Figure 13.16a). The real spectrum is narrower as shown by the HEOM calculation. This narrowing is caused by dynamic localization destroying the exciton states and producing more localized excitations with slower hopping-type energy transfer. Notice that for $M = -5$ cm^{-1} the HEOM approach almost coincides with the fully localized SF model. The only difference is the more asymmetric line shape predicted by HEOM.

Increase of coupling to $M = -30$ cm^{-1} produces bigger splitting of the levels, thus blocking the up-hill transfer from the $k = \pm 1$ states. The allowed $k = \pm 1$ levels become more red-shifted, more separated from the higher levels, and are not so much lifetime broadened. The SR/MR models give the OD spectrum with a characteristic narrowing and red-shifting, in contrast to SF, where the exciton splitting is not included (see Figure 13.16b, where the SF spectrum is peaking near the zero energy with some red-shifting due to reorganization). The HEOM spectrum is closer to the SR/MR picture, but the HEOM line shape is more asymmetric.

13.6.2 Excited-state dynamics in a homogenous ring: Secular versus nonsecular Redfield

The excited-state dynamics for a homogeneous ring is shown in Figure 13.17. Using the HEOM method, we calculate the dynamics of the one-exciton density matrix in the site representation (supposing that the site $n = 1$ is initially populated, whereas all other initial populations and coherences are set to zero).We show populations p_{11} and imaginary parts of the coherences p_{12}'' (where p_{nm} denote elements of the density matrix in the site representation). Notice that the imaginary parts of the coherences in the site representation p_{nm}'' are responsible for the creation of coherences between the exciton states. If they are oscillating, they will produce beats in the third-order spectral responses (pump-probe anisotropy, 2D photon echo). If they exhibit aperiodic decay, they will give nonoscillatory contributions to the spectral responses, for example, additional anisotropy decaying on the same time scale.

The results obtained with HEOM are compared with the other approaches, that is, full (nonsecular) SR$_f$ and secular SR$_s$ versions of the SR theory.

For strong coupling (-60 cm^{-1}), the difference between the three theories is not so big. The full Redfield SR$_f$ gives the same oscillatory amplitudes as the HEOM-model, but oscillations exhibit faster decay. In the secular Redfield SR$_s$ the oscillations decay even faster, and also the oscillatory amplitudes

are significantly reduced. Decrease of coupling to -40 and -30 cm^{-1} results in more pronounced differences between the secular SR$_s$ and nonsecular approaches (SR$_f$ and HEOM). The amplitude of the first maximum of the coherence p_{12}'' developing during 50 fs is significantly less in SR$_s$, and also the population p_{11} decay becomes faster in the secular version displaying an increasingly larger deviation from both SR$_f$ and HEOM.

In the weak coupling case ($-20, -12,$ and -7 cm^{-1}), the initial population decay occurs on the ps time scale. Nonsecular theories predict also a very long-lived coherence. In contrast, the coherence in the secular case decays quickly (i.e., during 0.25 ps). Population equilibration in the SR$_s$ picture is unrealistically fast, showing about 100 fs time constant.

On the other hand, for very weak coupling (-12 and -7 cm^{-1}) there exists some deviation of the full Redfield from the HEOM. According to the SR$_f$ model, the amplitude of the coherence p_{12}'' developing in 100–200 fs is significantly smaller than for the HEOM approach. In addition, the population p_{11} decay is slower as compared to HEOM (due to the one-phonon character of the relaxation in the standard Redfield).

Comparing the difference between the secular and nonsecular kinetics in the examples discussed, we conclude that nonsecular terms (coupling between one-exciton populations and coherences between the exciton states) give only a minor contribution to the dynamics in the case of strong exciton coupling, but can change dramatically the apparent kinetics in the weakly coupled antenna.

The strongly coupled system is characterized by a big exciton splitting, so that the coherences between exciton states p_{k1k2} are oscillating at high frequencies ω_{k1k2} that are very different for different pairs of the eigenstates ($k1, k2$). Thus, the coherence to coherence ($p_{k1k2} \leftrightarrow p_{k3k4}$) and population to coherence transfers ($p_{k1k2} \leftrightarrow p_{k1k1}, p_{k2k2}$) are essentially nonresonant, and therefore not effective (at least on the timescale exceeding $(\omega_{k1k2} - \omega_{k3k4})^{-1}$ or $(\omega_{k1k2})^{-1}$, respectively).

In the case of weak coupling, the splitting ω_{k1k2} is small and there is no significant difference between the pure population ($p_{k1k1} \leftrightarrow p_{k2k2}$) and population to coherence transfer ($p_{k1k2} \leftrightarrow p_{k2k2}, p_{k1k1}$) even on the ps time scale. The coherences are continuously repumped from the populations, and the density matrix determined by a coherent superposition of $p_{k1k1}, p_{k2k2},$ and p_{k1k2} remains localized in the site

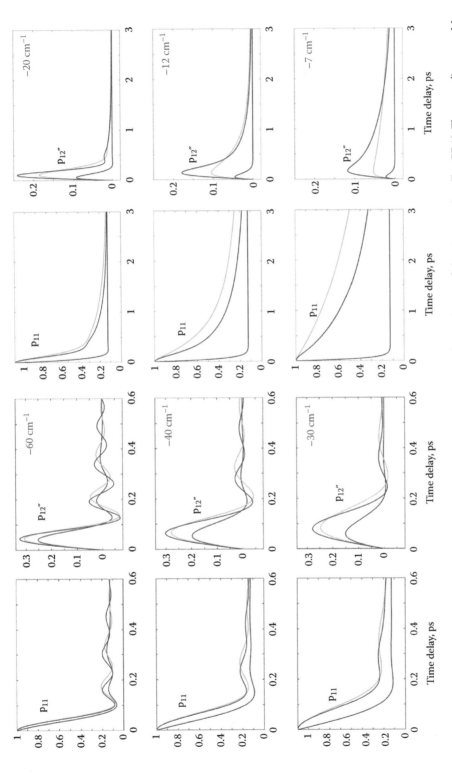

Figure 13.17 Excitation dynamics in a homogeneous ring of eight BChls with initial population of the site n = 1 at T = 77 K. The couplings are M = −60, −40, −30, −20, −12, and −7 cm⁻¹, as indicated in the frames. For each coupling value, we calculate the population of the initially excited site p_{11} and the imaginary part of the coherence p_{12}'' using the HEOM (blue), SR_f (green), and SR_s (red) theories. Note the different time scales for strong and weak couplings, that is, 0.6 ps (M = −60, −40, and −30 cm⁻¹) and 3 ps (M = −20, −12, and −7 cm⁻¹). (From Novoderezhkin, V.I. and van Grondelle, R., J. Phys. Chem. B, 117(38), 11076, 2013.)

representation. In other words, in the presence of coherences the excitations keep memory about their initial localization, and this can significantly slow down the transfer to other sites. Without re-pumping from the populations (as supposed in the SR_s model), the coherences decay very quickly (in 100–200 fs for all couplings, as shown in Figure 13.17) rapidly resulting in the formation of a density matrix determined by the noncoherent contribution from delocalized exciton eigenstates. The relaxation between delocalized exciton states is very fast (about 100 fs in our examples) and does not depend on the exciton coupling M. Thus, we obtain the 100 fs equilibration even in very weakly coupled systems (see the SR_s kinetics at M = −12 and −7 cm^{-1}), whereas more realistic approaches (SR_f and HEOM) give time constants of several ps in the case of (very) weak coupling.

Similar unrealistically fast transfer has been obtained with the modified Redfield theory, where the coherences are not included at all (see Section 13.3, Figures 13.2 through 13.5). Considering two weakly coupled isoenergetic sites, we formally obtain delocalization between them. The MR theory (where the coherences between exciton states are neglected) gives a sub-100 fs population transfer in such systems for M < 15–20 cm^{-1}. To exclude this "resonant artifact" a combined Förster–Redfield theory can be used. As nicely shown in Figure 13.17, this resonance artifact is absent in the SR_f approach and in the more general HEOM approach.

13.7 CONCLUSIONS

A quantitative comparison of the energy transfer theories and their application to photosynthetic antenna complexes reveals possibilities and limitations of these approaches. In the case of compact and relatively strongly coupled antenna subunits, the Redfield relaxation theory in the exciton basis is a powerful tool to explore the spectral and equilibration dynamics at a quantitative level. However, in the presence of weakly coupled and isoenergetic pigments (or clusters) the Redfield theory can strongly overestimate the transfer rates. This is connected to the formal use of delocalized wavefunctions in the region where phonon-induced dynamic localization occurs. The corresponding energy transfer channels can be better modeled by the Förster or generalized Förster theories assuming localization a priori. The corresponding rates are more realistic (especially for

transfer between isoenergetic levels), but they are generally underestimated (e.g., in the case of dipole forbidden levels connected via phonon-induced transfer). Combination of the modified Redfield and generalized Förster theories is the best solution for heterogeneous systems with different couplings between the pigments (covering both weak and strong coupling limits).

Application of the theory to light-harvesting complexes of algae (PE545) and higher plants (LHCII) (as illustrated in this review) and other complexes allowed to build a consistent and quantitative picture of the pattern of energy transfer, thus leading to deeper and more detailed understanding of how photosynthetic antenna complexes perform their function of light harvesting.

The B800 ring is an excellent example of a system where the interpigment coupling (−19 cm^{-1}) is exactly in between the strong and weak coupling limit. In this region, the excitations are predominantly localized, but there are still some coherences giving sizable contributions to the spectral responses. When using the localized (Förster) theory, one can obtain rather realistic population kinetics (for a more detailed discussion of this feature see Reference 111), but the coherences are omitted. On the other hand, in the Redfield picture it is quite difficult to model the almost localized excitations in terms of the excitonic (delocalized) basis. To obtain the correct localized dynamics, one should describe very precisely a superposition of the exciton populations and the coherences between the exciton states (including nonsecular transfer between them). This is not included into the secular version of the standard Redfield and in the modified Redfield theory. In the full version of the standard Redfield, the nonsecular transfer is taken into account but only in terms of single-phonon transitions. The hierarchical equations (HEOM) method would be the best solution to the problem, but this method is numerically expensive.

ACKNOWLEDGMENTS

VN was supported by a visitor's grant from the Netherlands Organisation for Scientific Research (NWO) and by the Russian Foundation for Basic Research, Grant No. 15-04-02136.

RvG was supported by the Royal Dutch Academy of Sciences (KNAW), the VU University Amsterdam, TOP grant (700.58.305) from the

Foundation of Chemical Sciences part of NWO, the advanced investigator grant (267333, PHOTPROT) from the European Research Council and the support from the EU FP7project PAPETS (GA 323901).

REFERENCES

1. R. van Grondelle, J. P. Dekker, T. Gillbro, and V. Sundström, Energy-transfer and trapping in photosynthesis. *Biochim. Biophys. Acta*, 1994, **1187**, 1.

2. H. van Amerongen, L. Valkunas, and R. van Grondelle, *Photosynthetic Excitons*, World Scientific Publishers, Singapore, 2000.

3. X. Hu, T. Ritz, A. Damjanović, F. Autenrieth, and K. Schulten, Photosynthetic apparatus of purple bacteria. *Q. Rev. Biophys.*, 2002, **35**, 1.

4. R. van Grondelle and V. I. Novoderezhkin, Energy transfer in photosynthesis: Experimental insights and quantitative models. *Phys. Chem. Chem. Phys.*, 2006, **8**, 793.

5. R. van Grondelle and V. I. Novoderezhkin, Spectroscopy and dynamics of excitation transfer and trapping in purple bacteria. in: *The Purple Phototrophic Bacteria, Advances in Photosynthesis and Respiration*, Vol. 28, eds., C. N. Hunter, F. Daldal, M. C. Thurnauer, and J. T. Beatty, Springer, Dordrecht, the Netherlands, 2008, Ch. 13, pp. 231–252.

6. R. van Grondelle, V. I. Novoderezhkin, and J. P. Dekker, Modeling light-harvesting and primary charge separation in photosystem I and photosystem II. in: *Photosynthesis In Silico: Understanding Complexity from Molecules to Ecosystems*, eds., A. Laisk, L. Nedbal, and Govindjee, Springer, Dordrecht, the Netherlands, 2009, Ch. 3, pp. 33–53.

7. Y. C. Cheng and G. R. Fleming, Dynamics of Light Harvesting in Photosynthesis. *Annu. Rev. Phys. Chem.*, 2009, **60**, 241.

8. V. I. Novoderezhkin and R. Van Grondelle, Physical origins and models of energy transfer in photosynthetic light harvesting, *Phys. Chem. Chem. Phys.*, 2010, **12**, 7352–7365.

9. R. E. Fenna and B. W. Matthews, Chlorophyll arrangement in a bacteriochlorophyll protein from *Chlorobium-Limicola*. *Nature*, 1975, **258**, 573.

10. G. McDermott, S. M. Prince, A. A. Freer, A. M. Hawthornthwaite-Lawless, M. Z. Papiz, R. J. Cogdell, and N. W. Isaacs, Crystal structure of an integral membrane light-harvesting complex from photosynthetic bacteria. *Nature*, 1995, **374**, 517.

11. J. Koepke, X. Hu, C. Muenke, K. Schulten, and H. Michel, The crystal structure of the light-harvesting complex II (B800-850) from *Rhodospirillum molischianum*. *Structure*, 1996, **4**, 581.

12. M. Z. Papiz, S. M. Prince, T. Howard, R. J. Cogdell, and N. W. Isaacs, The structure and thermal motion of the B800-850 LH2 complex from *Rps. acidophila* at 2.0 Å resolution and 100 K: New structural features and functionally relevant motions. *J. Mol. Biol.*, 2003, **326**, 1523.

13. S. Karrasch, P. A. Bullough, and R. Ghosh, The 8.5 A projection map of the light-harvesting complex I from *Rhodospirillum rubrum* reveals a ring composed of 16 subunits. *EMBO J.*, 1995, **14**, 631.

14. S. Scheuring, J. Seguin, S. Marco, D. Levy, B. Robert, and J. L. Rigaud, Nanodissection and high-resolution imaging of the *Rhodopseudomonas viridis* photosynthetic core complex in native membranes by AFM. *Proc. Natl. Acad. Sci. U. S. A.*, 2003, **100**, 1690.

15. A. W. Roszak, T. D. Howard, J. Southall, A. T. Gardiner, C. J. Law, N. W. Isaacs, and R. J. Cogdell, Crystal structure of the RC-LH1 core complex from *Rhodopseudomonas palustris*. *Science*, 2003, **302**, 1969.

16. W. Kühlbrandt, D. N. Wang, and Y. Fujiyoshi, Atomic model of plant light-harvesting complex by electron crystallography. *Nature*, 1994, **367**, 614.

17. Z. Liu, H. Yan, K. Wang, T. Kuang, J. Zhang, L. Gui, X. An, and W. Chang, Crystal structure of spinach major light-harvesting complex at 2.72 A resolution. *Nature*, 2004, **428**, 287.

18. J. Standfuss, A. C. T. Van Scheltinga, M. Lamborghini, and W. Kühlbrandt, Mechanisms of photoprotection and nonphotochemical quenching in pea light-harvesting complex at 2.5 A resolution. *EMBO J.*, 2005, **24**, 919.

19. P. Jordan, P. Fromme, H. T. Witt, O. Klukas, W. Saenger, and N. Krauss, Three-dimensional structure of cyanobacterial photosystem I at 2.5 angstrom resolution. *Nature*, 2001, **411**, 909.

20. X. Qin, M. Suga, T. Kuang, and J.-R. Shen, Structural basis for energy transfer pathways in the plant PSI-LHCI supercomplex. *Science*, 2015, **348**, 989–995.

21. Y. Mazor, A. Borovikova, and N. Nelson, The structure of plant photosystem I supercomplex at 2.8 Å resolution. *elife*, 2015, **4**, e07433.

22. A. Zouni, H. T. Witt, J. Kern, P. Fromme, N. KrauX, W. Saenger, and P. Orth, Crystal structure of photosystem II from Synechococcus elongatus at 3.8 Angström resolution. *Nature*, 2001, **409**, 739.

23. N. Kamiya and J.-R. Shen, Crystal structure of oxygen-evolving photosystem II from *Thermosynechococcus vulcanus* at 3.7-angstrom resolution. *Proc. Natl. Acad. Sci. U. S. A.*, 2003, **100**, 98.

24. K. N. Ferreira, T. M. Iverson, K. Maghlaoui, J. Barber, and S. Iwata, Architecture of the photosynthetic oxygen-evolving center. *Science*, 2004, **303**, 1831.

25. Y. Umena, K. Kawakami, J.-R. Shen, and N. Kamiya. Crystal structure of oxygen-evolvingphotosystem II at a resolution of 1.9A. *Nature*, 2011, **473**, 55–61.

26. K. E. Wilk, S. J. Harrop, L. Jankova, D. Edler, G. Keenan, F. Sharples, R. G. Hiller, and P. M. G. Curmi, Evolution of a light-harvesting protein by addition of new subunits and rearrangement of conserved elements: Crystal structure of a cryptophyte phycoerythrin at 1.63-Å resolution. *Proc. Natl. Acad. Sci. U. S. A.*, 1999, **96**, 8901.

27. A. B. Doust, C. N. J. Marai, S. J. Harrop, K. E. Wilk, P. M. G. Curmi, and G. D. Scholes, Developing a structure-function model for the cryptophyte phycoerythrin 545 using ultrahigh resolution crystallography and ultrafast laser spectroscopy. *J. Mol. Biol.*, 2004, **344**, 135.

28. A. B. Doust, K. E. Wilk, P. M. G. Curmi, and G. D. Scholes, The photophysics of cryptophyte light-harvesting. *J. Photochem. Photobiol. A Chem.*, 2006, **184**, 1.

29. J. Pieper, J. Voigt, and G. J. Small, Chlorophyll a Franck-Condon factors and excitation energy transfer. *J. Phys. Chem. B*, 1999, **103**, 2319.

30. E. J. G. Peterman, T. Pullerits, R. van Grondelle, and H. van Amerongen, Electron-phonon coupling and vibronic fine structure of light-harvesting complex II of green plants: Temperature dependent absorption and high-resolution fluorescence spectroscopy. *J. Phys. Chem. B*, 1997, **101**, 4448.

31. E. J. G. Peterman, H. van Amerongen, R. van Grondelle, and J. P. Dekker. The nature of the excited state of the reaction center of photosystem II of green plants: A high-resolution fluorescence spectroscopy study. *Proc. Natl. Acad. Sci. U. S. A.*, 1998, **95**, 6128.

32. J. Pieper, R. Scholdel, K.-D. Irrgang, J. Voigt, and G. Renger, Electron-phonon coupling in solubilized LHC II complexes of green plants investigated by line-narrowing and temperature-dependent fluorescence spectroscopy. *J. Phys. Chem. B*, 2001, **105**, 7115.

33. M. Rätsep and A. Freiberg, Resonant emission from the B870 exciton state and electron–phonon coupling in the LH2 antenna chromoprotein. *Chem. Phys. Lett.*, 2003, **377**, 371.

34. J. Pieper, K.-D. Irrgang, G. Renger, and R. E. Lechner, Density of vibrational states of the light-harvesting complex II of green plants studies by inelastic neutron scattering. *J. Phys. Chem. B*, 2004, **108**, 10556.

35. A. Damjanović, I. Kosztin, U. Kleinekathöfer, and K. Schulten, Excitons in a photosynthetic light-harvesting system: A combined molecular dynamics, quantum chemistry, and polaron model study. *Phys. Rev. E*, 2002, **65**, 031919.

36. G. R. Fleming and R. van Grondelle, Femtosecond spectroscopy of photosynthetic light-harvesting systems. *Curr. Opin. Struct. Biol.*, 1997, **7**, 738.

37. V. Sundström, T. Pullerits, and R. van Grondelle, Photosynthetic light-harvesting: Reconciling dynamics and structure of purple bacterial LH2 reveals function of photosynthetic unit. *J. Phys. Chem. B*, 1999, **103**, 2327.

38. R. van Grondelle and V. Novoderezhkin, The dynamics of excitation energy transfer in the LH1 and LH2 light-harvesting complexes of photosynthetic bacteria. *Biochemistry*, 2001, **40**, 15057.

39. T. Renger, V. May, and O. Kühn, Ultrafast excitation energy transfer dynamics in photosynthetic pigment-protein complexes. *Phys. Rep.*, 2001, **343**, 137.

40. T. Förster, in: *Modern Quantum Chemistry, Part III. B. Action of Light and Organic Crystals*, ed., O. Sinanoğlu, Academic Press, New York, 1965, pp. 93–137.

41. V. I. Novoderezhkin and A. P. Razjivin, Exciton states of the antenna and energy trapping by the reaction center. *Photosynth. Res.*, 1994, **42**, 9.

42. V. I. Novoderezhkin and A. P. Razjivin, The theory of Förster-type migration between clusters of strongly interacting molecules: Application to light-harvesting complexes of purple bacteria. *Chem. Phys.*, 1996, **211**, 203.

43. H. Sumi, Theory on rates of excitation-energy transfer between molecular aggregates through distributed transition dipoles with application to the antenna system in bacterial photosynthesis. *J. Phys. Chem. B*, 1999, **103**, 252.

44. K. Mukai, S. Abe, and H. Sumi, Theory of rapid excitation-energy transfer from B800 to optically-forbidden exciton states of B850 in the antenna system LH2 of photosynthetic purple bacteria. *J. Phys. Chem. B*, 1999, **103**, 6069.

45. G. D. Scholes and G. R. Fleming, On the mechanism of light harvesting in purple bacteria: B800 to B850 energy transfer. *J. Phys. Chem. B*, 2000, **104**, 1854.

46. S. Jang, M. D. Newton, and R. J. Silbey, Multichromophoric Förster resonance energy transfer. *Phys. Rev. Lett.*, 2004, **92**, 218301-1.

47. A. G. Redfield, The theory of relaxation processes. *Adv. Mag. Res.*, 1965, **1**, 1.

48. O. Kühn and V. Sundström, Pump-probe spectroscopy of dissipative energy transfer dynamics in photosynthetic antenna complexes: A density matrix approach. *J. Chem. Phys.*, 1997, **107**, 4154.

49. O. Kühn, V. Sundström, and T. Pullerits, Fluorescence depolarization dynamics in the B850 complex of purple bacteria. *Chem. Phys.*, 2002, **275**, 15.

50. B. Brüggemann and V. May, Exciton exciton annihilation dynamics in chromophore complexes. II. Intensity dependent transient absorption of the LH2 antenna system. *J. Chem. Phys.*, 2004, **120**, 2325.

51. O. Kühn and V. Sundström, Energy transfer and relaxation dynamics in light-harvesting antenna complexes of photosynthetic bacteria. *J. Phys. Chem. B*, 1997, **101**, 3432.

52. V. Novoderezkhin, M. Wendling, and R. van Grondelle, Intra- and interband transfers in the B800-B850 antenna of *Rhodospirillum molischianum*: Redfield theory modeling of polarized pump-probe kinetics. *J. Phys. Chem. B*, 2003, **107**, 11534.

53. V. Novoderezkhin, J. M. Salverda, H. van Amerongen, and R. van Grondelle, Exciton modeling of energy-transfer dynamics in the LHCII complex of higher plants: A Redfield theory approach. *J. Phys. Chem. B*, 2003, **107**, 1893.

54. T. Renger and V. May, Ultrafast exciton motion in photosynthetic antenna systems: The FMO-complex. *J. Phys. Chem. A*, 1998, **102**, 4381.

55. S. I. E. Vulto, M. A. de Baat, S. Neerken, F. R. Nowak, H. van Amerongen, J. Amesz, and T. J. Aartsma, Excited-state dynamics in FMO antenna complexes from photosynthetic green sulfur bacteria: A kinetic model. *J. Phys. Chem. B*, 1999, **103**, 8153.

56. B. Brüggemann, K. Sznee, V. Novoderezhkin, R. van Grondelle, and V. May, From structure to dynamics: Modeling exciton dynamics in the photosynthetic antenna PS1. *J. Phys. Chem. B*, 2004, **108**, 13536.

57. J. A. Leegwater, J. R. Durrant, and D. R. Klug, Exciton equilibration induced by phonons: Theory and application to PS II reaction centers. *J. Phys. Chem. B*, 1997, **101**, 7205.

58. V. I. Prokhorenko and A. R. Holzwarth, Primary processes and structure of the photosystem II reaction center: A) Photon echo study. *J. Phys. Chem. B*, 2000, **104**, 11563.

59. L. M. C. Barter, J. R. Durrant, and D. R. Klug, A quantitative structure-function relationship for the Photosystern II reaction center: Supermolecular behavior in natural photosynthesis. *Proc. Natl. Acad. Sci. U. S. A.*, 2003, **100**, 946.

60. E. Romero, R. Augulis, V. I. Novoderezhkin, M. Ferretti, J. Thieme, D. Zigmantas, and R. van Grondelle. Quantum coherence in photosynthesis for efficient solar-energy conversion. *Nature Physics*, 2014, **10**, 676–682.

61. J. M. Jean, R. A. Friesner, and G. R. Fleming, Application of a multilevel Redfield theory to electron transfer In condensed phases. *J. Chem. Phys.*, 1992, **96**, 5827.

62. J. M. Jean, Time- and frequency-resolved spontaneous emission as a probe of coherence effects in ultrafast electron transfer reactions. *J. Chem. Phys.*, 1994, **101**, 10464.

63. J. M. Jean and G. R. Fleming, Competition between energy and phase relaxation in electronic curve crossing processes. *J. Chem. Phys.*, 1995, **103**, 2092.

64. V. I. Novoderezhkin, A. G. Yakovlev, R. van Grondelle, and V. A. Shuvalov, Coherent nuclear and electronic dynamics in primary charge separation in photosynthetic reaction centers: A Redfield theory approach. *J. Phys. Chem. B*, 2004, **108**, 7445.

65. T. Renger and V. May, Theory of multiple exciton effects in the photosynthetic antenna complex LHC-II. *J. Phys. Chem. B*, 1997, **101**, 7232.

66. T. Renger and V. May, Influence of higher excited singlet states on ultrafast exciton motion in pigment-protein complexes. *Photochem. Photobiol.*, 1997, **66**(5), 618.

67. M. Chachisvilis and V. Sundström, Femtosecond vibrational dynamics and relaxation in the core light-harvesting complex of photosynthetic purple bacteria. *Chem. Phys. Lett.*, 1996, **261**, 165.

68. V. Novoderezhkin, R. Monshouwer, and R. van Grondelle, Electronic and vibrational coherence in the core light-harvesting antenna *Rhodopseudomonas viridis*. *J. Phys. Chem. B*, 2000, **104**, 12056.

69. V. Novoderezhkin and R. van Grondelle, Exciton-vibrational relaxation and transient absorption dynamics in LH1 of *Rhodopseudomonas viridis*: A Redfield theory approach. *J. Phys. Chem. B*, 2002, **106**, 6025.

70. E. Romero, V. I. Novoderezhkin, and R. van Grondelle, Quantum design of photosynthesis for bio-inspired solar-energy conversion, *Nature*, 2017, **543**, 355–365.

71. E. Romero, J. Prior, A. W. Chin, S. E. Morgan, V. I. Novoderezhkin, M. B. Plenio, and R. van Grondelle, Quantum-coherent dynamics in photosynthetic charge separation revealed by wavelet analysis, *Sci. Rep.*, 2017, **7**, 2890.

72. I. Vladimir, E. R. Novoderezhkin, and R. van Grondelle, How exciton-vibrational coherences control charge separation in the photosystem II reaction center, *Phys. Chem. Chem. Phys.*, 2015, **17**, 30828–30841.

73. I. Vladimir E. R. Novoderezhkin, J. Prior, and R. van Grondelle, Exciton-vibrational resonance and dynamics of charge separation in the photosystem II reaction center, *Phys. Chem. Chem. Phys.*, 2017, **19**, 5195–5208.

74. W. M. Zhang, T. Meier, V. Chernyak, and S. Mukamel, Exciton-migration and three-pulse femtosecond optical spectroscopies of photosynthetic antenna complexes. *J. Chem. Phys.*, 1998, **108**, 7763.

75. W. M. Zhang, T. Meier, V. Chernyak, and S. Mukamel, Simulation of three-pulse echo and fluorescence depolarization in photosynthetic aggregates. *Phil. Trans. R. Soc. Lond. A*, 1998, **356**, 405.

76. M. Yang and G. R. Fleming, Influence of phonons on exciton transfer dynamics: Comparison of the Redfield, Forster, and modified Redfield equations. *Chem. Phys.*, 2002, **275**, 355.

77. V. I. Novoderezhkin, D. Rutkauskas, and R. van Grondelle, Dynamics of the emission spectrum from single LH2 complex: Interplay of slow and fast nuclear motions. *Biophys. J.*, 2006, **90**, 2890.

78. J. Adolphs and T. Renger, How proteins trigger excitation energy transfer in the FMO complex of green sulfur bacteria. *Biophys. J.*, 2006, **91**, 2778.

79. V. I. Novoderezhkin, A. B. Doust, C. Curutchet, G. D. Scholes, and R. van Grondelle, Excitation dynamics in phycoerythrin 545: Modeling of steady-state spectra and transient absorption with modified Redfield theory. *Biophys. J.*, 2010, **99**, 344–352.

80. C. Curutchet, V. I. Novoderezhkin, J. Kongsted, A. Muñoz-Losa, R. van Grondelle, G. D. Scholes, and B. Mennucci, Energy flow in the cryptophyte PE545 antenna is directed by bilin pigment conformation, *J. Phys. Chem. B*, 2013, **117**(16), 4263–4273.

81. V. I. Novoderezhkin, R. Croce, M. Wahadoszamen, I. Polukhina, E. Romero, and R. van Grondelle, Mixing of exciton and charge-transfer states in light-harvesting complex Lhca4, *Phys. Chem. Chem. Phys.*, 2016, **18**, 19368–19377.

82. M. Yang, A. Damjanović, H. M. Vaswani, and G. R. Fleming, Energy transfer in photosystem I of cyanobacteria *Synechococcus elongatus*: Model study with structure-based semi-empirical hamiltonian and experimental spectral density. *Biophys. J.*, 2003, **85**, 140.

83. V. Novoderezhkin, M. Palacios, H. van Amerongen, and R. van Grondelle, *J. Phys. Chem. B*, 2004, **108**, Energy-transfer dynamics in the LHCII complex of higher plants: Modified Redfield approach. 10363.

84. V. Novoderezhkin, M. Palacios, H. van Amerongen, and R. van Grondelle, Excitation dynamics in the LHCII complex of higher plants: Modeling based on 2.72 A crystal structure. *J. Phys. Chem. B*, 2005, **109**, 10493.

85. T. Renger, M. Madjet, A. Knorr, and F. Muh, How the molecular structure determines the flow of excitation energy in plant light-harvesting complex II. *J. Plant Physiol.*, 2011, **168**, 1497–1509.

86. V. I. Novoderezhkin, A. Marin and R. van Grondelle, Intra- and inter-monomeric transfers in the light harvesting LHCII complex: The Redfield-Förster picture. *Phys. Chem. Chem. Phys.*, 2011, **13**, 17093–17103.

87. G. Raszewski, W. Saenger, and T. Renger, Theory of optical spectra of photosystem II reaction centers: Location of the triplet state and the identity of the primary electron donor. *Biophys. J.*, 2005, **88**, 986.

88. V. I. Novoderezhkin, E. G. Andrizhiyevskaya, J. P. Dekker, and R. van Grondelle, Pathways and timescales of primary charge separation in the photosystem II reaction center as revealed by a simultaneous fit of time-resolved fluorescence and transient absorption. *Biophys. J.*, 2005, **89**, 1464.

89. G. Raszewski, B. A. Diner, E. Schlodder, and T. Renger, Spectroscopic properties of reaction center pigments in Photosystem II core complexes: Revision of the multimer model. *Biophys. J.*, 2008, **95**, 105.

90. V. I. Novoderezhkin, J. P. Dekker and R. van Grondelle, Mixing of exciton and charge-transfer states in Photosystem II reaction centers: Modeling of Stark spectra with modified Redfield theory. *Biophys. J.*, 2007, **93**, 1293.

91. I. Vladimir E. R. Novoderezhkin, J. P. Dekker, and R. van Grondelle, Multiple charge separation pathways in photosystem II: Modeling of transient absorption kinetics, *Chem. Phys. Chem.*, 2011, **12**, 681–688.

92. G. Raszewski and T. Renger, Light harvesting in photosystem II core complexes is limited by the transfer to the trap: Can the core complex turn into a photoprotective mode? *J. Am. Chem. Soc.*, 2008, **130**, 4431.

93. D. Rutkauskas, V. Novoderezhkin, R. J. Cogdell, and R. van Grondelle, Fluorescence spectral fluctuations of single LH2 complexes from *Rhodopseudomonas acidophila* strain 10050. *Biochemistry*, 2004, **43**, 4431.

94. D. Rutkauskas, V. Novoderezhkin, R. J. Cogdell, and R. van Grondelle, Fluorescence spectroscopy of conformational changes of single LH2 complexes. *Biophys. J.*, 2005, **88**, 422.

95. D. Rutkauskas, V. Novoderezhkin, A. Gall, J. Olsen, R. J. Cogdell, C. N. Hunter, and R. van Grondelle, Spectral trends in the fluorescence of single bacterial light-harvesting complexes: Experiments and modified Redfield simulations. *Biophys. J.*, 2006, **90**, 2475.

96. V. I. Novoderezhkin, D. Rutkauskas, and R. van Grondelle, Multistate conformational model of a single LH2 complex: Quantitative picture of time-dependent spectral fluctuations. *Chem. Phys.*, 2007, **341**, 45.

97. T. P. J. Krüger, V. I. Novoderezhkin, C. Ilioaia, and R. van Grondelle, Fluorescence spectral dynamics of single LHCII trimers. *Biophys. J.*, 2010, **98**, 3093–3101.

98. M. Cho, H. M. Vaswani, T. Brixner, J. Stenger, and G. R. Fleming, Exciton analysis in 2D electronic spectroscopy. *J. Phys. Chem. B*, 2005, **109**, 10542.

99. D. Zigmantas, E. L. Read, T. Mančal, T. Brixner, A. T. Gardiner, R. J. Cogdell, and G. R. Fleming, Two-dimensional electronic spectroscopy of the B800-820 light-harvesting complex. *Proc. Natl. Acad. Sci. U. S. A.*, 2006, **103**, 12672.

100. G. S. Schlau-Cohen, T. R. Calhoun, N. S. Ginsberg, E. L. Read, M. Ballottari, R. Bassi, R. van Grondelle, and G. R. Fleming, Pathways of energy Flow in LHCII from two-dimensional electronic spectroscopy. *J. Phys. Chem. B*, 2009, **113**, 15352.

101. A. Ishizaki and G. R. Fleming, Unified treatment of quantum coherent and incoherent hopping dynamics in electronic energy transfer: Reduced hierarchy equation approach, *J. Chem. Phys.*, 2009, **130**, 234111(1–10).

102. A. Ishizaki and G. R. Fleming, Theoretical examination of quantum coherence in a photosynthetic system at physiological temperature, *Proc. Natl. Acad. Sci. U. S. A.*, 2009, **106**, 17255–17260.

103. C. Kreisbeck, T. Kramer, M. Rodriguez, and B. Hein, High-performance solution of hierarchical equations of motions for studying energy-transfer in light-harvesting complexes, *J. Chem. Theor. Comput.*, 2011, **7**, 2166–2174.

104. B. Hein, C. Kreisbeck, T. Kramer, and M. Rodrıguez Modelling of oscillations in two-dimensional echo-spectra of the Fenna–Matthews–Olson complex, *New J. Phys.*, 2012, **14**, 023018(1–20).

105. C. Kreisbeck and T. Kramer, Long-lived electronic coherence in dissipative exciton dynamics of light-harvesting complexes, *J. Phys. Chem. Lett.*, 2012, **3**, 2828–2833.

106. J. Zhu, S. Kais, P. Rebentrost, and A. Aspuru-Guzik, Modified scaled hierarchical equation of motion approach for the study of quantum coherence in photosynthetic complexes, *J. Phys. Chem. B*, 2011, **115**, 1531–1537.

107. J. Strumpfer and K. Schulten, Light harvesting complex II B850 excitation dynamics, *J. Chem. Phys.*, 2009, **131**, 225101(1–9).

108. S.-H. Yeh, J. Zhu, and S. Kais, Population and coherence dynamics in light harvesting complex II (LH2), *J. Chem. Phys.*, 2012, **137**, 084110.

109. J. Strumpfer and K. Schulten, Excited state dynamics in photosynthetic reaction center and light harvesting complex 1, *J. Chem. Phys.*, 2012, **137**, 065101(1–8).

110. J. Strumpfer and K. Schulten, Open quantum dynamics calculations with the hierarchy equations of motion on parallel computers, *J. Chem. Theor. Comput.*, 2012, **8**, 2808–2816.

111. V. I. Novoderezhkin and R. van Grondelle, Spectra and dynamics in the B800 antenna: Comparing hierarchical equations, redfield and Förster theories, *J. Phys. Chem. B*, 2013, **117**(38), 11076–11090.

112. C. Kreisbeck, T. Kramer, and A. Aspuru-Guzik, Scalable high-performance algorithm for the simulation of exciton dynamics. Application to

the light-harvesting complex II in the presence of resonant vibrational modes. *J. Chem. Theory Comput.*, 2014, **10**, 4045–4054.

113. V. I. Novoderezhkin and R. van Grondelle, Modeling of excitation dynamics in photosynthetic light-harvesting complexes: Exact vs perturbative approaches, *J. Phys. B*, 2017, **50**, 124003.

114. A. Gelzinis, D. Abramavicius, and L. Valkunas, Absorption lineshapes of molecular aggregates revisited. *J. Chem. Phys.*, 2015, **142**, 154107.

115. S. Mukamel, *Principles of Nonlinear Optical Spectroscopy*, Oxford University Press, New York, 1995.

116. T. Meier, V. Chernyak, and S. Mukamel, Femtosecond photon echoes in molecular aggregates. *J. Chem. Phys.*, 1997, **107**, 8759.

117. T. Renger and R. Marcus, Photophysical properties of PS-2 reaction centers and a discrepancy in exciton relaxation times. *J. Phys. Chem. B*, 2002, **106**, 1809.

118. S. Mukamel, P. Rott, and V. Chernyak, Optical Stark spectroscopy of molecular aggregates. *J. Chem. Phys.*, 1996, **104**, 5415.

119. Z. G. Fetisova, A. M. Freiberg, K. Mauring, V. I. Novoderezhkin, A. S. Taisova, and K. E. Timpmann, Excitation energy transfer in chlorosomes of green bacteria: Theoretical and experimental studies. *Biophys. J.*, 1996, **71**, 995–1010.

120. V. Novoderezhkin, R. Monshouwer, and R. Van Grondelle, Exciton (de)localization in the LH2 antenna of *Rhodobacter sphaeroides* as revealed by relative difference absorption measurements of the LH2 antenna and the B820 subunit, *J. Phys. Chem. B*, 1999, **103**, 10540–10548.

121. A. B. Doust, I. H. M. van Stokkum, D. S. Larsen, K. E. Wilk, P. M. G. Curmi, R. van Grondelle, and G. D. Scholes, Mediation of ultrafast light-harvesting by a central dimer in phycoerythrin 545. Studied by transient absorption and global analysis. *J. Phys. Chem. B*, 2005, **109**, 14219.

122. E. Collini, C. Y. Wong, K. E. Wilk, P. M. G. Curmi, P. Brumer, and G. D. Scholes, Coherently wired light-harvesting in photosynthetic marine algae at ambient temperature. *Nature*, 2010, **463**, 644.

123. R. van Grondelle and V. I. Novoderezhkin, Quantum design for a light trap. *Nature*, 2010, **463**, 614.

124. R. S. Knox and B. Q. Spring, Dipole strengths in the chlorophylls. *Photochem. Photobiol.*, 2003, **77**, 497.

125. R. Agarval, B. P. Krueger, G. D. Scholes, M. Yang, J. Yom, L. Mets, and G. R. Fleming, Ultrafast energy transfer in LHC-II revealed by three-pulse photon echo peak shift measurements. *J. Phys. Chem. B*, 2000, **104**, 2908.

126. T. Bittner, G. P. Wiederrecht, K.-D. Irrgang, G. Renger, and M. R. Wasielewski, Femtosecond transient absorption spectroscopy on the light-harvesting Chl *a/b* protein complex of photosystem II at room temperature and 12K. *Chem. Phys.*, 1995, **194**, 311.

127. H. M. Visser, F. J. Kleima, I. H. M. van Stokkum, R. van Grondelle, and H. van Amerongen, Probing of many energy-transfer processes in the photosynthetic light-harvesting complex II at 77K by energy-selective sub-picosecond transient absorption spectroscopy. *J. Chem. Phys.*, 1996, **210**, 297.

128. F. J. Kleima, C. C. Gradinaru, F. Calkoen, I. H. M. van Stokkum, R. van Grondelle, and H. van Amerongen, Energy transfer in LHCII monomers at 77K studied by sub-picosecond transient absorption spectroscopy. *Biochemistry*, 1997, **36**, 15262.

129. C. C. Gradinaru, S. Özdemir, D. Gülen, I. H. M. van Stokkum, R. van Grondelle, and H. van Amerongen, The flow of excitation energy in LHCII monomers. Implications for the structural model of the major plant antenna. *Biophys. J.*, 1998, **75**, 3064.

130. C. C. Gradinaru, I. H. M. van Stokkum, A. A. Pascal, R. van Grondelle, and H. van Amerongen, Identifying the pathways of energy transfer between carotenoids and chlorophylls in LHCII and CP29. A multicolor, femtosecond pump-probe study. *J. Phys. Chem. B*, 2000, **104**, 9330.

131. M. Du, X. Xie, L. Mets, and G. R. Fleming, Direct observation of ultrafast energy transfer processes in light-harvesting complex II. *J. Phys. Chem.*, 1994, **98**, 4736.

132. T. Bittner, K.-D. Irrgang, G. Renger, and M. R. Wasielewski, Ultrafast excitation energy transfer and exciton-exciton annihilation processes in isolated light harvesting complexes of photosystem II (LHC II) from spinach. *J. Phys. Chem.*, 1994, **98**, 11821.

133. J. P. Connelly, M. G. Müller, M. Hucke, G. Gatzen, C. W. Mullineaux, A. V. Ruban, P. Horton, and A. R. Holzwarth, Ultrafast spectroscopy of trimeric light-harvesting complex II from higher plants. *J. Phys. Chem. B*, 1997, **101**, 1902.

134. J. M. Salverda, M. Vengris, B. P. Krueger, G. D. Scholes, A. R. Czarnoleski, V. Novoderezhkin, H. van Amerongen, and R. van Grondelle, Energy transfer in light-harvesting complexes LHCII and CP29 of spinach studied with three-pulse echo peakshift and transient grating. *Biophys. J.*, 2003, **84**, 450.

135. H. van Amerongen, S. L. S. Kwa, B. M. van Bolhuis, and R. van Grondelle, Polarized fluorescence and absorption of macroscopically aligned light harvesting complex II. *Biophys. J.*, 1994, **67**, 837.

136. P. W. Hemelrijk, S. L. S. Kwa, R. van Grondelle, and J. P. Dekker, Spectroscopic properties of LHC-II, the main light-harvesting chlorophyll a/b protein complex from chloroplast membranes. *Biochim. Biophys. Acta*, 1992, **1098**, 159.

137. S. Georgakopoulou, G. van der Zwan, R. Bassi, R. van Grondelle, H. van Amerongen, and R. Croce, Understanding the changes in the circular dichroism of light harvesting complex II upon varying its pigment composition and organization, *Biochemistry*, 2007, **46**(16), 4745–4754.

138. T. Meier, Y. Zhao, V. Chernyak, and S. Mukamel, Polarons, localization, and excitonic coherence in superradiance of biological antenna complexes. *J. Chem. Phys.*, 1997, **107**, 3876.

139. S. Tretiak, C. Middleton, V. Chernyak, and S. Mukamel, Bacteriochlorophyll and carotenoid excitonic couplings in the LH2 system of purple bacteria. *J. Phys. Chem. B*, 2000, **104**, 9540.

140. Y. C. Cheng and R. J. Silbey, Coherence in the B800 Ring of Purple Bacteria LH2, *Phys. Rev. Lett.*, 2006, **96**, 028103.

141. S. Savikhin, H. Van Amerongen, S. L. Kwa, R. Van Grondelle, and W. S. Struve, Low-temperature energy transfer in LHC-II trimers from the Chl *a/b* light-harvesting antenna of photosystem II, *Biophys. J.*, 1994, **66**(5), 1597–1603.

142. H. van der Laan, T. Schmidt, R. W. Visschers, K. J. Visscher, R. van Grondelle, and S. Völker, Energy transfer in the B800–850 antenna complex of purple bacteria *Rhodobacter sphaeroides*: A study by spectral hole-burning, *Chem. Phys. Lett.*, 1990, **170**, 231–238.

143. N. R. S. Reddy, G. J. Small, M. Seibert, and R. Picorel, Energy-transfer dynamics of the B800-B850 antenna complex of *Rhodobacter sphaeroides*—A hole burning study, *Chem. Phys. Lett.*, 1991, **181**, 391–399.

144. C. D. de Caro, R. W. Visschers, R. van Grondelle, and S. Völker, Inter- and intraband energy transfer in LH2-antenna complexes of purple bacteria. A fluorescence line-narrowing and hole-burning study, *J. Phys. Chem.*, 1994, **98**, 10584–10590.

145. H.-M. Wu, S. Savikhin, N. R. S. Reddy, R. Jankowiak, R. J. Cogdell, and G. J. Small, Femtosecond and hole-burning studies of B800's excitation energy relaxation dynamics in the LH2 antenna complex of *Rhodopseudomonas acidophila* (strain 10050), *J. Phys. Chem.*, 1996, **100**, 12022–12033.

146. S. Matsuzaki, V. Zazubovich, N. J. Fraser, R. J. Cogdell, and G. J. Small, Energy transfer dynamics in LH2 complexes of *Rhodopseudomonas acidophila* containing only one B800 molecule, *J. Phys. Chem. B*, 2001, **105**, 7049–7056.

147. V. Zazubovich, R. Jankowiak, and G. J. Small, On B800-B800 energy transfer in the LH2 complex of purple bacteria, *J. Lumin.*, 2002, **98**, 123–129.

148. T. Joo, Y. Jia, J.-Y. Yu, D. M. Jonas, and G. R. Fleming, Dynamics in isolated bacterial light harvesting antenna (LH2) of *Rhodobacter sphaeroides* at room temperature, *J. Phys. Chem.*, 1996, **100**, 2399–2409.

149. M. Wendling, F. Van Mourik, I. H. M. Van Stokkum, J. M. Salverda, H. Michel, and R. Van Grondelle, Low-intensity pump-probe measurements on the B800 band of *Rhodospirillum molischianum*, *Biophys. J.*, 2003, **84**, 440–449.

150. S. Hess, F. Feldchtein, A. Babin, I. Nurgaleev, T. Pullerits, A. Sergeev, and V. Sundström, Femtosecond energy transfer within the LH2 peripheral antenna of the photosynthetic purple bacteria *Rhodobacter sphaeroides* and *Rhodopseudomonas palustris* LL, *Chem. Phys. Lett.*, 1993, **216**, 247–257.

151. S. Hess, E. Åkesson, R. J. Cogdell, T. Pullerits, and V. Sundström, Energy transfer in spectrally inhomogeneous light-harvesting pigmentprotein complexes of purple bacteria, *Biophys. J.*, 1995, **69**, 2211–2225.

152. R. Monshouwer, I. Ortiz De Zarate, F. Van Mourik, and R. Van Grondelle, Low-intensity pump-probe spectroscopy on the B800 to B850 transfer in the light harvesting 2 complex of *Rhodobacter sphaeroides*, *Chem. Phys. Lett.*, 1995, **246**, 341–346.

153. Y. Z. Ma, R. J. Cogdell, and T. Gillbro, Energy transfer and exciton annihilation in the B800–850 antenna complex of the photosynthetic purple bacterium *Rhodopseudomonas acidophila* (strain 10050). A femtosecond transient absorption study, *J. Phys. Chem. B*, 1997, **101**, 1087–1095.

154. Y. Z. Ma, R. J. Cogdell, and T. Gillbro, Femtosecond energy-transfer dynamics between bacteriochlorophylls in the B800-820 antenna complex of the photosynthetic purple bacterium *Rhodopseudomonas acidophila* (strain 7750), *J. Phys. Chem. B*, 1998, **102**, 881–887.

155. J. T. M. Kennis, A. M. Streltsov, T. J. Aartsma, T. Nozava, and J. Amesz. Energy transfer and exciton coupling in isolated B800-850 complexes of the photosynthetic purple sulfur bacterium *Chromatium Tepidum*. The effect of structural symmetry on bacteriochlorophyll excited states, *J. Phys. Chem.*, 1996, **100**, 2438–2442.

156. J. T. M. Kennis, A. M. Streltsov, S. I. E. Vulto, T. J. Aartsma, T. Nozava, and J. Amesz, Femtosecond dynamics in isolated LH2 complexes of various species of purple bacteria, *J. Phys. Chem. B*, 1997, **101**, 7827–7834.

157. T. Pullerits, S. Hess, J. L. Herek, and V. Sundström, Temperature dependence of excitation transfer in LH2 of *Rhodobacter sphaeroides*, *J. Phys. Chem. B*, 1997, **101**, 10560–10567.

158. J. A. Ihalainen, J. Linnanto, P. Myllyperkiö, I. H. M. van Stokkum, B. Ücker, H. Scheer, and J. E. I. Korppi-Tommola, Energy transfer in LH2 of *Rhodospirillum molischianum*,

studied by subpicosecond spectroscopy and configuration interaction exciton calculations, *J. Phys. Chem. B*, 2001, **105**, 9849–9856.

159. J. M. Salverda, F. Van Mourik, G. Van Der Zwan, and R. Van Grondelle, Energy transfer in the B800 rings of the peripheral bacterial light harvesting complexes of *Rhodopseudomonas acidophila* and *Rhodospirillum molischianum* studied with photon echo techniques, *J. Phys. Chem. B.*, 2000, **104**, 11395–11408.

160. R. Agarwal, M. Yang, Q. H. Xu, and G. R. Fleming, Three pulse photon echo peak shift study of the B800 band of the LH2 complex of *Rps. acidophila* at room temperature: A coupled master equation and nonlinear optical response function approach, *J. Phys. Chem. B*, 2001, **105**, 1887–1894.

161. T. Renger, V. May, and O. Kühn, Ultrafast excitation energy transfer dynamics in photosynthetic pigment-protein complexes, *Phys. Rep.*, 2001, **343**, 137–254.

162. S. Georgakopoulou, R. N. Frese, E. Johnson, M. H. C. Koolhaas, R. J. Cogdell, R. Van Grondelle, and G. Van Der Zwan, Absorption and CD spectroscopy and modeling of various LH2 complexes from purple bacteria, *Biophys. J.*, 2002, **82**, 2184–2197.

Quantum aspects of photosynthetic energy transfer

SUSANA F. HUELGA AND MARTIN B. PLENIO

14.1 EXPERIMENTAL EVIDENCE OF QUANTUM TRANSPORT IN PHOTOSYNTHESIS

In recent years, quantum biology has been benefitting considerably from the refinement of experimental tools that are beginning to provide direct access to the observation of quantum dynamics in biological systems [1–5] thanks to their increasing sensitivity to quantum phenomena at short length and timescales. The experiments that especially triggered an increasing theoretical effort in understanding the origin and the meaning of possible coherent effects in photosynthesis employ ultrafast 2D electronic spectroscopy to probe the dynamics of light-harvesting aggregates present in plants and bacteria. The resulting spectral response, as analyzed extensively in other chapters of this book, show unambiguous oscillatory features

that demonstrate on one hand the existence of delocalized excitons, and on the other hand the existence of excitonic beating and therefore a clear indication of a nontrivial system–environment interaction, where excitons do not simply relax monotonically in an energy funnel toward the reaction center. We will discuss a microscopic model that provides quantitative understanding of the mechanisms that allow for coherence preservation in the noisy environment of protein vibrations, together with a spectral response that is consistent with the observations. The key ingredient of the model is the interplay of electronic and specific vibrational degrees of freedom that facilitates the generation of robust quantum behavior [6,7].

At this point it seems pertinent to make some short remarks concerning the relevance of the observation of long-lived coherence in photosynthetic systems in experiments that use laser excitation as opposed to the natural sunlight that these objects are normally subjected to. In fact, it has been argued that under incoherent excitation by sunlight, no coherence will be observable and that coherence plays no role in the dynamics of the system. Hence, it was concluded that these laser experiments are of little relevance for the understanding of the possible quantum effects in photosynthetic systems in vivo. Indeed, it is correct that the coherence properties of the states of a system that are observable in an experiment may depend very much on the specific experimental setup and the specific excitation regime that the system is subjected to. Excitation by incoherent sunlight as well as ensemble averages may suppress the observed coherence and have tempted some researchers to reach the conclusion that under natural conditions coherence may be of no relevance in these systems. This, however, is not necessarily correct as the crucial point is the coherence properties of the underlying dynamics and not of signatures of coherence of states in an experimental signal that is possibly averaged over a large disordered ensemble. It is the equations of motion that determine the performance of each individual system and are unaffected by the nature of the initial preparation or ensemble averages that may obscure the observed coherence (Consider a set of pendula, each of which oscillates independently from the others at a fixed frequency and phase. If the phase for each pendulum is chosen at random, then the global signal appears incoherent, while clearly each pendulum is oscillating coherently). While natural conditions do not resemble laser light, laser spectroscopic experiments on individual specimens provide the sharpest tools for the identification of the dynamical equations that govern the system evolution and thus have a crucial role to play in the determination of quantum effects in biological systems. Hence, we argue that these experiments do have a direct bearing on our understanding of natural photosynthetic units.

14.2 THEORY OF QUANTUM DYNAMICS IN PHOTOSYNTHETIC TRANSPORT

Before we start to discuss a set of general and generalizable principles that govern the quantum dynamics of biological systems, we would like to consider

first an example that presents us with a "smoking gun" for the importance of quantum dynamics in the presence of structured environments as opposed to purely broadband noise.

14.2.1 Setting the scene

Let us consider the Fenna–Matthews–Olson (FMO) complex [8,9], which forms an integral part of the photosynthetic light-harvesting complexes of green sulfur bacteria [10]. It is composed of a set of bacteriochlorophyll molecules, each of which may support an electronic excitation (a Frenkel exciton). These molecules will be arranged in space by a protein scaffold and together they form a pigment–protein complex. Such complexes serve to transport electronic excitations, excitons [11], in the presence of a vibrational environment. At the hand of some example parameters, let us consider the transport dynamics through the FMO complex to show that it actually benefits from the presence of an environment. Then we will move on to discuss in more detail how vibrational environments may assist transport. The full Hamiltonian describing the exciton–vibrational interaction as well as the exciton–exciton interaction is given by $H = H_e + H_I + H_B$ where

$$H_e = \sum_{n=1}^{N} E_n |n\rangle\langle n| + \frac{1}{2}\sum_{m\neq n}\left[J_{mn}|m\rangle\langle n| + h.c. \right], \quad (14.1)$$

$$H_B = \sum_{i,k} \hbar\omega_k a_{ik}^\dagger a_{ik}, \quad (14.2)$$

$$H_I = \frac{1}{2}\sum_n\left[\sum_k \sqrt{S_{nk}}\,\omega_k\left(a_{nk} + a_{nk}^\dagger\right)|n\rangle\langle n| + h.c.\right]. \quad (14.3)$$

Here

$|n\rangle$ describes an excitation on site n

J_{mn} describes the dipolar interaction between excitation on sites m and n

The operators $a_{nk}, a_{nk}^\dagger$ denote bosonic destruction and creation operators for the kth independent vibrational mode coupled to site n [8]

The exciton–mode interaction is determined by the strength of their Huang-Rhys factors S_{nk} [12].

Note that H_I describes a purely dephasing interaction because the vibrational degrees of freedom have energies that are at least 10–100 times smaller than the excitation energies of the excitons, thus suppressing direct exciton–phonon interconversion.

The dephasing can be understood to originate from the fact that vibrations will change the local environment of each site (e.g., the local charge distribution) and thus affect the excitation energy of the relevant site [12]. We assume that the dynamics is dominated by contributions to the spectral density where each site interacts with its own independent environment, an assumption that is corroborated by first-principles numerical studies of photosynthetic complexes [13–17] and normal-mode analysis combined with quantum chemical methods [18,19]. The linear interaction between site n and its environment is characterized by the spectral density $J_n(\omega) = \sum_k S_{nk}\omega_k^2\delta(\omega - \omega_k)$, which is a joint property of the environment and the system combining the strength of interaction of modes with the mode density.

The full dynamical equation also needs to include further contributions: first, one describing the spontaneous annihilation of an exciton and the concomitant loss of energy into the general environment at a rate γ_{loss}, second, the transfer of the excitation from a specific site, in the FMO complex, that is the site labeled 3, into the reaction center, which is again described by an irreversible decay at rate γ_{RC}, motivated by the fact that in the reaction center charge separation is achieved to irreversibly stabilize the excitonic energy; and third, the rate γ_{in} at which excitations enter the complex, which we assume to be much smaller than γ_{RC} to model low-light conditions. In order to obtain our first observation, we follow the approximate treatments presented in [20–23], which treat these effects in simple master equation treatments [24]. These methods can be improved upon, but there is no need for this at this point.

The typical, but at first sight perhaps surprising, result is presented in Figure 14.1. Contrary to what one might have expected, the conductivity of the electronic transport network in the FMO complex (quantified as the rate at which the reaction center is populated in steady state divided by the rate at which excitations enter the transport network) exhibits a maximum at a finite dephasing rate, that is, dephasing noise can actually assist the electronic transport [22,23].

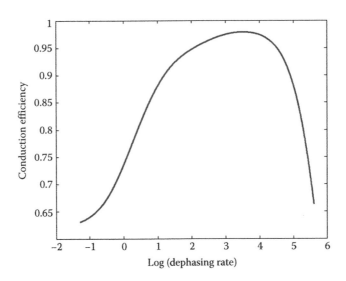

Figure 14.1 Plot of the conduction efficiency of the FMO complex where excitations enter the FMO complex at site 1 and exit at site 3 versus the overall strength of the dephasing noise due to the interaction between electronic and vibrational degrees of freedom. The key observation is that optimal performance is found for an intermediate level of dephasing noise. No dephasing noise or very strong dephasing noise is counterproductive for the performance of the transport network. See [20] for the Hamiltonian parameters and the spontaneous emission rate, the noise is modeled as local dephasing Lindblad master equation in the site basis.

This immediately raises the question as to whether the regime that results in optimal transport performance is found to be essentially classical in the sense that the dynamics is well represented by a rate equation model, or whether, despite the dephasing noise, it remains firmly in the quantum mechanical regime in which quantum coherent dynamics is only weakly perturbed by the dephasing noise. Questions of this type can be answered both in a qualitative and a more quantitative manner. First, an examination of the parameters that are typically entering the dynamical equations when describing photosynthetic complexes reveals that the strength of the intrasystem coupling, for example, the dipolar interaction, is comparable in strength to the system–environment coupling. Hence, one already expects that the dynamics is taking place in a regime in which neither dephasing noise nor quantum coherent dynamics clearly dominate. This is further corroborated by the examination of the coherence and entanglement [25,26] properties of states [20,27] and, more importantly, the dynamics of the system [28], which demonstrate quite clearly that on shorter length and timescales quantum coherence is present in the systems, while for longer distances and times classical properties dominate. This suggests that indeed, the optimal operating regime in this setting is found to be "halfway" between the classical and the quantum world. These observations raise the questions as to why optimal performance is achieved in this intermediate regime. Answering this question will lead us to identify the dynamical and structural principles that are underlying optimal performance of quantum transport networks. It will drive us toward uncovering a rich interplay between electronic degrees of freedom and their vibrational environment and point toward the possibility that nature has optimized both electronic networks and vibrational environment in an evolutionary process. This will be the subject of the next section.

14.2.2 Design principles

Here we will elucidate basic principles that have been found to underlie the fruitful interplay between vibrational environments and coherent quantum dynamics [6,7,20–23,28–40]. Identifying

and understanding these principles at a deep, intuitive level and seeing how nature may have used them to optimize performance does provide additional value, even if individual processes in specific circumstances have been known in different physical situations.

14.2.2.1 CONTROLLING RESONANCES: THE PHONON ANTENNA

We will begin by elucidating a first principle, which will provide an understanding why optimal transport performance in the FMO complex can be achieved at intermediate noise levels. More importantly, this principle is sufficiently general to provide a mechanism that can support the surprisingly long-lasting oscillatory features observed in recent ultrafast laser spectroscopy experiments and explain key aspects of the dynamics that may underlie the process olfaction (see [41] for a more detailed exposition of this aspect).

We will approach this topic by means of a simple but instructive question concerning the optimization of a simple transport network (see Figure 14.2 for a schematic representation of the following). Consider a network made up of only three sites, namely, site 1, which accepts excitations from the antenna, and site 3, which is connected to the reaction center. Both site 1 and site 3 are fixed in their properties (position, orientation, and excitation energy). The system is completed by site 2, whose excitation energy, position, and orientation, and hence dipolar interaction strength with sites 1 and 3, we are free to choose. We assume that site 3 provides the zero of excitation energy, while site 1 has an excitation energy that is $300 \, \text{cm}^{-1}$ higher. The question that we would like to answer concerns the optimal choice of excitation energy, position, and orientation of site 2, or in other words, the optimal choice of the excitation energy of site 2 and its dipolar coupling strengths to sites 1 and 3. As such, this question cannot be answered unambiguously as we are missing a crucial piece of information, namely, that of the structure of the spectral density of the environmental fluctuations. Typical spectral densities in pigment–protein complexes possess considerable structure with sharp peaks originating from long-lived vibrational modes as well as a broad background whose maximum tends to be in the range of around $200 \, \text{cm}^{-1}$, which we will now assume for the subsequent optimization.

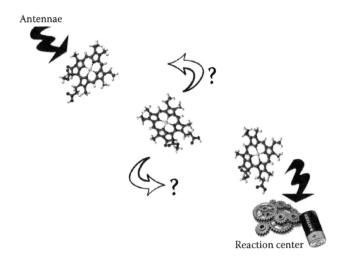

Figure 14.2 Network made up of only three sites: site 1, which accepts excitation from the antenna and site 3, which is connected to the reaction center, both of which are fixed in their properties (position, orientation, and excitation energy), as well as another site 2, whose excitation energy, position, and orientation, and hence dipolar interaction strength with sites 1 and 2, we are free to choose. What is the optimal choice of the excitation energy of site 2 and its dipolar coupling strengths to sites 1 and 3?

For the sake of clarity, let us assume that the environmental spectral density has a single maximum and thus takes roughly the shape depicted in Figure 14.3. A numerical optimization employing Redfield equations to take account of the spectral structure of the environment (see [6,21] for theoretical and numerical details) now finds that the optimal position of site 2 is close to site 1 such that it exhibits a strong coherent dipolar interaction and close in excitation energy. Having found these numerical results, we now would like to rationalize its origin and thereby arrive at a very useful design principle—the phonon antenna.

Indeed, the strong coherent dipolar interaction between sites 1 and 2 suggests that we move to a new basis made up of the eigenstates of the coherent part of the dynamics of these two sites, that is the excitonic states of that system, or, for quantum opticians, the dressed state picture. This change of picture leads us to rewrite the Hamiltonian equation (14.3) that describes the system–environment interaction in the excitonic basis of eigenstates $\{|e_n\rangle\}$ of Equation 14.1, so that $|i\rangle = \sum_n C_n^i |e_n\rangle$, and the coupling terms

$$H_I = \frac{1}{2}\sum_{n,m}\left(Q_{nm}|e_n\rangle\langle e_m| + h.c.\right), \quad (14.4)$$

where

$$Q_{nm} = \sum_{ik}\sqrt{S_k}\,\omega_k C_n^i C_m^i\left(a_{ik} + a_{ik}^\dagger\right). \quad (14.5)$$

This provides us with two insights. First, in the exciton (dressed-state) basis, the action of the dephasing noise now leads to transitions between excitons, that is amplitude noise, which facilitates transport toward the lower of the two exciton states. Second, the two excitons (dressed states) are separated by an energy difference that is related to the coherent dipolar coupling strength and the energy difference of sites 1 and 2. The dominant contribution to the transition between these excitons (dressed states) arises from those environmental modes whose frequency closely matches the energy difference between dressed states. Indeed, the optimal solution is such that the energy separation of the dressed states matches the maximum of the environmental spectral density, that is, where the environmental fluctuations are strongest, so that the environment may bring about transitions between the dressed states most effectively. In this sense, we can argue that the two eigenstates of the coupled Hamiltonian are tuned to harvest environmental fluctuations to achieve optimal excitation energy

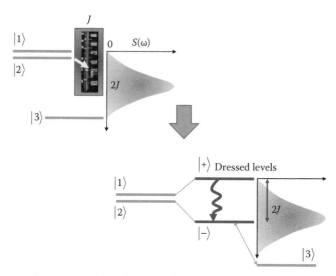

Figure 14.3 In the upper figure, two closely spaced energy levels are separated from a third level to which excitations should be delivered. They are subject to a dephasing noise from an environment, with a finite bandwidth that exhibits a maximum. A coherent interaction between the upper two energy levels leads to dressed states $|\pm\rangle$ with an energy splitting, which, if matched to the maximum of the environment spectral density, will optimize transport from the upper to the lower level. Hence, the dressed states act as an antenna to harvest environmental fluctuations.

transport through the formation of a tunable "phonon antenna."

Observing these two points alone, that is inducing strong coherent coupling to move to a dressed basis and tuning the coupling such that it matches the maximum of the spectral density of the environmental fluctuations, one already comes close to the numerically obtained solution of the above optimization problem and can thus optimize excitation energy transport. It should be noted that the phonon antenna principle is also capable of making predictions about more complex transport networks, such as that of the FMO complex (see [34] for a study of transport in cryptophyte antenna protein phycoerythrin 545). Indeed, it was found that the physically important relaxation pathway between sites 1 and 3 is mediated by pigments that are spectrally and spatially positioned by the protein to efficiently sample the spectral function of the proteins' fluctuations [6,21]. Whether this optimality is a determinant in the emergence of this structure in nature is another matter altogether, but it is striking how well the phonon antenna concept can be used to rationalize the site energies and the couplings of the pigments participating in this pathway.

It now becomes transparent as to why the optimal operating regime for excitation energy transport may actually be found to be where coherent dipolar interactions and system–environment interactions, that is, dephasing noise, are of broadly comparable strength. Indeed, if the environmental noise is too weak, then the formation of dressed states will present little benefit for transport. On the other hand, if dephasing noise is very strong then it will suppress the formation of the dressed states and thus of the phonon antenna effect in the first place. As a consequence, an intermediate regime where dephasing noise of intermediate strength is present naturally appears as the optimal operating regime according to the phonon antenna. We will later see that there are a number of additional mechanisms in which noise and coherent dynamics coexist to lead to similar conclusions [20,42].

With this in mind, we now move on to bring out two other conclusions and connections that we can draw from the phonon antenna concept, in the case, when the spectral density of the environment possesses very sharply peaked features, that is, well-defined long-lived vibrational modes. Indeed, this will provide both a mechanism explaining the origin of long-lived coherences observed in recent ultrafast spectroscopy experiments and a connection to biological sensors.

14.2.2.2 LONG-LIVED COHERENCES AS A NONEQUILIBRIUM PROCESS

Experimental observations employing ultrafast 2D spectroscopy on various photosynthetic complexes exhibited long-lived oscillatory features, which were interpreted as evidence for long-lived electronic coherence in the systems under investigation [1,2,4]. Under this hypothesis, electronic coherence appears to exhibit lifetimes that can reach the picosecond range, thus exceeding expectations from condensed matter systems at least tenfold. This interesting observation gave rise to a variety of attempts for explanations of the long-lived coherences, including (1) the overall reduction of dephasing [16,43], which is however not compatible with the observed very short lifetimes of optical exciton coherences in the system; (2) correlations in the noise sources between different sites [17,44,45] that are however not supported by first principles calculations of spectral densities [13–16] and normal-mode analysis combined with quantum chemical methods [18,19]; and (3) variations of the electronic structure of the FMO complex [46]; which are not sufficient however to explain the observed durations. In the following, we show that the inclusion of significant coupling of electronic motion to long-lived vibrational modes as first proposed in [32,39,47] and further developed in [6,7] are capable of explaining the observations [48–50] and even more so to give support to the idea that vibrational motion plays an important role for electronic transport, quantum or classical—a principle that is now increasingly recognized for being of broader importance in biology [41].

The basic idea is an application of the phonon antenna principle to a system–environment interaction in which the broad features of the spectral density are supplemented by sharp features due to the presence of some long-lived and well-defined vibrational mode, as proposed in [7]. As we have learnt in the previous section, tuning the dipolar interaction in the electronic system such that the energy difference of the excitonic states matches the maximum of the spectral density will maximize the rate at which transitions between these states occur. For a broad and smooth spectral density these transitions will be dominantly incoherent (following essentially a Fermi's golden rule argument). In the presence of a long-lived vibrational mode, the phonon antenna principle remains the same but, crucially, the nature of the interaction changes as the interaction between a single vibrational mode and the electronic degrees of freedom is coherent and leads to an oscillatory response, at least for as long as the mode itself remains coherent.

Let us examine this mechanism in more detail by considering an exciton transport network in which excitons enter in one site and exit in another. Initially the network is in its ground state and no excitons are present. This situation is depicted schematically in Figure 14.4a where a pendulum represents the long-lived vibrational mode that we assume to be in a thermal state with small excitation number and where the black dot represents the population of the ground state of the electronic system. The higher lying electronic levels, representing the various exciton eigenstates of the electronic system, are not excited even at room temperature as the excitation energy is in the range of eV. The initial (fast) injection of an exciton, *either* coherently or incoherently, populates one of the exciton states of the system (the raised black dot in Figure 14.4b) and creates a sudden force on the electrons and nuclei and thus changes their equilibrium positions (compressed spring in Figure 14.4b). Now the environment will start to react to these forces that initiate transient oscillations of the modes at approximately their natural frequency ω_k. The continuous background of the spectral density will relax very rapidly into the new equilibrium state as it contains a broad range of frequencies and thus possesses a very short correlation time. The well-defined long-lived vibrational mode will oscillate for a considerable time (which can be up to several picoseconds) and will interact with the electronic system (in Figure 14.4c we see that the spring connecting the vibrational mode to the electronic motion is periodically expanded and compressed). This in turn leads to oscillations between different exciton states. We make

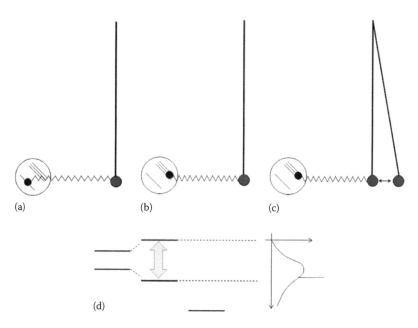

Figure 14.4 Simplified mechanical illustration of a possible principle behind long-lived coherence in biological systems. **(a)** The excitonic system (represented by the circle) and the vibrational mode (represented by the pendulum) are both in the ground state. They are linearly coupled as represented by the spring. **(b)** The excitonic system is excited and exerts a force on the vibrational mode, setting it in motion. **(c)** The coherently oscillating vibrational mode acts back on the excitonic system, forcing it into a coherently oscillating response. **(d)** The oscillatory coupling between excitonic and vibrational motion is most effective when excitonic transition energies are matched to the energy of a single vibrational quantum.

two observations: first, these oscillations will have the largest amplitude between those exciton states whose energy difference is nearly resonant with the frequency of the vibrational mode (see Figure 14.4d).

Second, we note that the sudden displacement of this mode implies that it is now found to be in a displaced thermal state, which will be close to a coherent state if the frequency of the mode is such that its thermal occupation number is low. As is well-known in quantum optics, a mode in a coherent state acts on a two-level system, essentially like a time-dependent classical driving field resulting in coherent Rabi-like oscillations following approximately the Hamiltonian

$$H_{driving} \approx \frac{1}{2} \sum_{n \neq m} \left(\langle Q_{nm} \rangle (t) |e_n\rangle\langle e_m| + h.c. \right), \quad (14.6)$$

where $\langle Q_{nm} \rangle (t) \propto \sum_{ik} \sqrt{S_k}\, \omega_k C_n^i C_m^i \sin(\omega_k t)$ in the mentioned approximation of the initial, transient, and coherent response of the modes to exciton injection. As a result, we will observe coherent transitions between

dissipative excitonic states, and thus coherences that will last for the coherence time of the vibrational mode. The conclusions of this semiclassical picture is confirmed by more sophisticated numerical methods that treat the system–environment interaction numerically exactly for arbitrary spectral densities [7,47,51]. The physical picture presented here illustrates a key point: electronic coherence may emerge from transiently exciting robust, weakly dephasing vibrational coherences, which then transfer back coherence to those exciton transitions that are well matched to the mode [7,49,50]. The importance of vibrational modes for interpreting experimental observations in multidimensional spectroscopy has only recently begun to be appreciated [7,47–50] and, from the discussion above, is seen to be intimately related to the phonon antenna concept (see Figure 14.4d).

14.2.2.3 SPECTRAL RESPONSE

As we have shown, the coupling of excitons and spectrally sharp local vibrational modes in the environment (most likely due to intramolecular motion of the chromophores) can provide mechanisms to explain the different lifetime of excitonic superposition

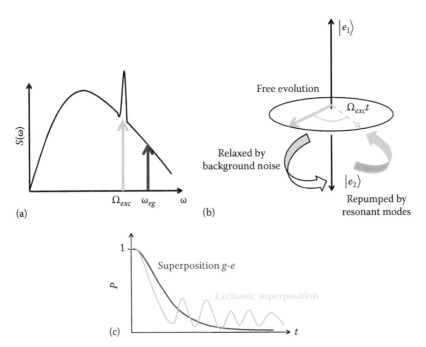

Figure 14.5 As pictorially illustrated in **(a)**, typical spectral densities for light-harvesting aggregates contain sharp spectral features embedded in a broadband background. When an excitonic resonance matches one such spectral peak, the excitonic lifetime of a coherent superposition, which we depict in **(b)** using a Bloch vector representation for the symbolic two-level system describing the dimeric time evolution in the one-exciton sector, can be extended by the recoherence effect that results from the quasi-resonant mode effective driving of excitonic transitions, as described in the main text. As a result, **(c)** excitonic superposition can significantly outlive local (site) superpositions of ground and excited state.

versus superposition involving the ground state, which are typically damped out in a fs timescale (see Figure 14.5). To link this microscopic, nonequilibrium mechanism to support long-lived coherence at ambient temperatures with actual experimental observations of oscillatory behavior using 2D photon echo techniques, we need to evaluate the nonlinear response of the prototypical dimer system.

We show that, for realistic parameter regimes, the resulting overall spectral signal includes contributions that have a genuine excitonic component and whose weight is comparable to those resulting from dominantly vibrational coherence within the electronic ground state manifold. Moreover, it should be noted that the manifestation of the latter still requires the existence of significant excitonic coupling, as remarked also in [7,50]. These results are also compatible with those obtained within the framework of discussing the exciton–phonon interaction in terms of intensity borrowing of dipolar strength [48] and confirm the relevance of coherent excitonic coupling to explain current spectral observations (Figure 14.6).

The Hamiltonian of Equation 14.1 for $n = 2$ provides us with the simplest model system that can give rise to delocalized eigenstates, the so-called exitonically coupled dimer (ecd). Following conventional notation, let us relabel the exciton states as $|A\rangle$ and $|B\rangle$. The exact form of these eigenvectors can be expressed in terms of the mixing angle θ, defined implicitly as $\tan(2\theta) = 2J/\mathcal{E}_{ab}$, where $\mathcal{E}_{ab} \equiv \mathcal{E}_a - \mathcal{E}_b$ is the difference in site energy. With this definition, the eigenstates $|A\rangle$ and $|B\rangle$ in the single excitation sector of the Hamiltonian equation (14.1) can be written as

$$\begin{pmatrix} |A\rangle \\ |B\rangle \end{pmatrix} = \begin{pmatrix} \cos(\theta) & \sin(\theta) \\ -\sin(\theta) & \cos(\theta) \end{pmatrix} \begin{pmatrix} |a\rangle \\ |b\rangle \end{pmatrix}. \quad (14.7)$$

These expressions make explicit the degree of delocalization of the new eigenstates (excitons) as a function of the mixing angle θ. The exciton energies are split by $\Delta_{ex} = E_B^{ecd} - E_A^{ecd}$, where $E_{A,B}^{ecd} = \mp (1/2)\sqrt{\mathcal{E}_{ab}^2 + 4J^2}$.

As discussed before, the vibrational environment of real protein–pigment complexes is characterized by a

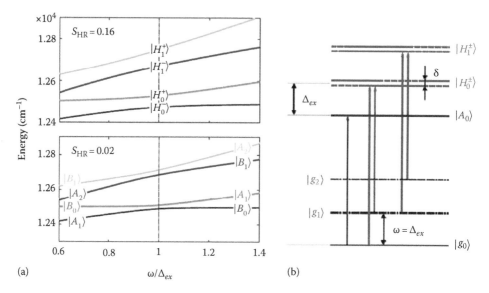

Figure 14.6 (a) Hybridization of electronic–vibrational energy eigenstates under vibronic coupling. **(b)** In the presence of a harmonic mode resonant with the excitonic splitting results in hybrid states in the one-exciton manifold $|H_0^\pm\rangle$ and $|H_1^\pm\rangle$ that can act, even in the limit of weak electronic–vibrational coupling, as a bridge between states in the ground state manifold with different vibrational numbers, therefore behaving as a *catalyzer* for sustaining (purely vibrational) coherent oscillations within the electronic ground state manifold.

highly structured spectral function that encompasses both smooth and sharp features. The coupling of excitonic degrees of freedom to vibrational modes with energies in the vicinity of the excitonic splitting Δ_{ex} can have a significant impact on the electronic dynamics [7]. Here we are interested in studying explicitly the impact of these vibrational modes in nonlinear photon echo signals and, more specifically, in analyzing the effect that these modes have on the dynamics of the peak beatings in the population time. To this effect we will extend the electronic Hamiltonian to include the degrees of freedom of two identical underdamped modes coupled to each chromophore as

$$H_{e-v} = H_e + \sum_{k=a,b} \omega a_k^\dagger a_k + \frac{\omega}{2}\sqrt{S_{HR}}\sum_{k=a,b}\sigma_k^z\left(a_k^\dagger + a_k\right),$$

$$(14.8)$$

where

ω stands for the frequency of the normal mode

$a_k^\dagger$, a_k, $(k=a,b)$ denote the phonon creation and annihilation operators on the corresponding underdamped vibrational modes

In the regime where the interaction strength with the harmonic mode is moderate, the structure of

eigenstates of the full Hamiltonian can still be understood as different vibrational progressions of each electronic state of the purely electronic dimer. We will denote the vibrational progression of states corresponding to the ground state electronic manifold with $|g_n\rangle$ and the first and second excitonic states with $|A_n\rangle$, and $|B_n\rangle$ respectively. Here, the index n increases with increasing energy states within the vibrational progression. The interaction with a vibrational mode with energy close to the electronic splitting $\omega \simeq \Delta_{ex}$ produces a hybridization of the states $|A_1\rangle$ and $|B_0\rangle$, resulting in a new pair of states $|H_0^\pm\rangle = \cos(\phi)|A_1\rangle \pm \sin(\phi)|B_0\rangle$, with the angle ϕ increasing from 0 to $\pi/2$ as we cross the resonance point from lower to higher values of ω/Δ_{ex}. Mixing between $|A_1\rangle$ and $|B_0\rangle$ is therefore quite significant close to the resonant point $\omega/\Delta_{ex}=1$. An analogous reasoning can be extended to the hybrid states $|H_1^\pm\rangle$ resulting from the coupling between states $|A_2\rangle$ and $|B_1\rangle$. The energy splitting between these hybrid states grows with increasing interaction strength S_{HR}.

We will now compute the nonlinear optical response of this vibronic model. The effect of the coupling between vibrational and electronic degrees of freedom will immediately lead to two phenomena that will leave long-lived beating traces in the population time. One of these vibrationally induced effects is expressed in the one-exciton sector of our model while the other concerns entirely the ground state electronic manifold. Our core interest is the study of the population time beatings in the various peaks of the 2D spectrum of the vibronic dimer as well as the mechanisms underlying their unexpectedly long lifetimes. With this in mind, rather than considering a detailed microscopic description of the dephasing processes involved in a real photosynthetic complex, we will consider a simplified model of the environment that, nevertheless, contains those features of the full description that are necessary to understand the origin of long-lasting population time beatings.

In order to describe the dynamical evolution of our system we will therefore use a Markovian master equation where the effect of the electronic dephasing will be explicitly included with the appropriate rate γ_{deph}, and we include vibrational modes that couple both to the electronic degrees of freedom and are additionally coupled dissipatively to a Markovian environment at a finite temperature. This results in equations of motion given by

$$\frac{d\rho}{dt} = \mathcal{L}(\rho) \qquad (14.9)$$

with

$$\mathcal{L}(\rho) \equiv -i/\hbar [H_{e-v}, \rho] + \gamma_{deph} (\sigma_a^z \rho \sigma_a^z + \sigma_b^z \rho \sigma_b^z - 2\rho)$$
$$+ \gamma_{mod}(n_T + 1) \sum_{k=a,b} \left[-a_k^\dagger a_k \rho - \rho a_k^\dagger a_k + 2a_k \rho a_k^\dagger \right]$$
$$+ \gamma_{mod} n_T \sum_{k=a,b} \left[-a_k a_k^\dagger \rho - \rho a_k a_k^\dagger + 2a_k^\dagger \rho a_k \right], \qquad (14.10)$$

where

n_T is the mean thermal occupation number of the vibrational mode

γ_{mod} is the damping rate into the Markovian thermal reservoir to which the vibrational modes are coupled

Using tetradic notation,

$$\frac{d|\rho\rangle\rangle}{dt} = \mathcal{L}|\rho\rangle\rangle, \qquad (14.11)$$

the equation of motion reads

$$|\rho(t)\rangle\rangle = \mathcal{G}(t)|\rho(0)\rangle\rangle, \qquad (14.12)$$

with the propagator

$$\mathcal{G}(t) \equiv e^{\mathcal{L}t}. \qquad (14.13)$$

Imposing rotating wave approximation and in the impulsive limit together with strict time ordering of the pulses, the form of the 2D photon echo signal measured in the spatial direction $\mathbf{k}_s = -\mathbf{k}_1 + \mathbf{k}_2 + \mathbf{k}_3$, the rephasing component, involves only three contributions:

$$P^{(3)}(t_1, t_2, t_3) \simeq E_0^3 \left(-R_{1f}^*(t_1, t_2, t_3) + R_{2g}(t_1, t_2, t_3) \right.$$
$$\left. + R_{3g}(t_1, t_2, t_3) \right), \qquad (14.14)$$

with

$$R_{1f}(t_1, t_2, t_3) = \left\langle\left\langle V | \mathcal{G}(t_3) \vec{V}_{ef} \mathcal{G}(t_2) \vec{V}_{ge} \mathcal{G}(t_1) \vec{V}_{ge} | \rho(-\infty) \right\rangle\right\rangle$$
$$R_{2g}(t_1, t_2, t_3) = \left\langle\left\langle V | \mathcal{G}(t_3) \vec{V}_{ge} \mathcal{G}(t_2) \vec{V}_{ge} \mathcal{G}(t_1) \vec{V}_{ge} | \rho(-\infty) \right\rangle\right\rangle$$
$$R_{3g}(t_1, t_2, t_3) = \left\langle\left\langle V | \mathcal{G}(t_3) \vec{V}_{ge} \mathcal{G}(t_2) \vec{V}_{ge} \mathcal{G}(t_1) \vec{V}_{ge} | \rho(-\infty) \right\rangle\right\rangle. \qquad (14.15)$$

Note that the total electric-dipole operator is given by $V = \mu_a \cdot \mathcal{E}\sigma_a^x + \mu_b \cdot \mathcal{E}\sigma_b^x$ so that the super operators $\vec{V}_{ge}$, $\vec{V}_{ef}$ (and $\overleftarrow{V}_{ge}$, $\overleftarrow{V}_{ef}$) defined from the truncated dipolar operators V_{ge} and V_{ef}, whose form is the same as the complete operator V but retaining only the matrix elements connecting the ground state and the one-exciton manifolds (for V_{ge}) and the elements connecting the one-exciton and doubly excited manifolds (for V_{ef}) (Figure 14.7).

In the presence of a resonant electron–phonon interaction, the distribution of dipolar transition becomes significant. In this situation a description via electronic–vibrational product states is no longer appropriate, and it is advantageous to adopt a description in terms of the dressed states $|H_0^\pm\rangle$. The dipolar transition strength between the states $|H_0^\pm\rangle$ and the states $|g_0\rangle$ and $|g_1\rangle$ in the ground state manifold may indeed have comparable values, even though the transitions involve different quantum vibrational numbers. This is easy to see from the definition of the dipolar transition strength

$$\langle H_0^\pm | \hat{\mu} | g_0 \rangle = \cos(\phi) \langle A_1 | \hat{\mu} | g_0 \rangle \pm \sin(\phi) \langle B_0 | \hat{\mu} | g_0 \rangle$$
$$\simeq \sin(\phi) \mu_B \cdot \mathcal{E} \qquad (14.16)$$

and

$$\langle H_0^\pm | \hat{\mu} | g_1 \rangle = \cos(\phi) \langle A_1 | \hat{\mu} | g_1 \rangle \pm \sin(\phi) \langle B_0 | \hat{\mu} | g_1 \rangle$$
$$\simeq \cos(\phi) \mu_A \cdot \mathcal{E}, \qquad (14.17)$$

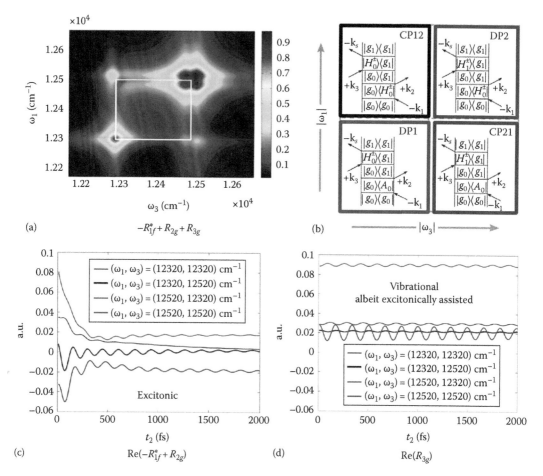

Figure 14.7 **(a)** Absolute value of the electronic 2D spectra $-R_{1f}^* + R_{2g} + R_{3g}$ evaluated for a population time of $t_2 = 1.6\Delta_{ex}^{-1}$ and the vibrational modes on-resonance with the excitonic splitting Δ_{ex}. **(b)** Subdiagrams result in vibrational superpositions of states within the ground state manifold during the population time t_2 of the experiment. The position of the diagrams in the figure is in correspondence with the cross-diagonal (CP12, CP21) and diagonal (DP1, DP2) peak amplitudes to which they contribute. **(c)** Partial contribution $\text{Re}\left(-R_{1f}^* + R_{2g}\right)$ for the peaks (DP1 (blue), CP21 (black), CP12 (red), DP2 (magenta)) at $(\omega_1, \omega_3) = (12320, 12320)$, $(12320, 12520)$, $(12520, 12320)$, and $(12520, 12520)$ cm^{-1}. This contribution produces population time peak beating given by coherences on the one-exciton sector. **(d)** The partial contribution $\text{Re}(R_{3g})$ for the same points. This contribution does not result in population time beating on any peak in a purely electronic dimer. However, the mixing between electronic and vibrational degrees of freedom allows the *catalyzed* transitions described in the text. The resulting beating in the population time is a signature of quantum coherence between vibrational states within the ground state electronic manifold. Simulations include an average over the orientations of the dimer and over the static disorder with a probability density $p(\omega) \sim e^{-\omega^2/2\sigma^2}$ with $\sigma = 0.17\Delta_{ex}$.

where the mixing angle ϕ has some value close to $\pi/4$ near the resonance. This situation is illustrated in Figure 14.8b, where now the state $\langle H_0^\pm |$ can act as a *catalyzer* for vibrational transitions within the electronic ground state manifold. This particular redistribution allows the pathways described in panel (b) to contribute to the total 2D spectra signal with strong amplitudes in the spirit described

in [50]. Contrary to the nonresonant case, all these diagrams result in population time peak beating due to coherent superposition of states in the ground state electronic manifold. In particular, only the diagram corresponding to the cross-diagonal peak CP12 contains four electronic vibrationally enhanced transitions. The diagrams corresponding to peaks DP2 and DP1 contain three enhanced

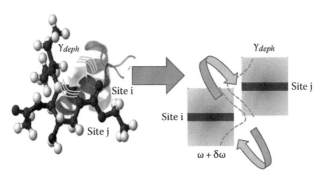

Figure 14.8 Local dephasing, for example, due to random fluctuations of the energy levels generated from random vibrational motion, leads to line broadening and hence, increased overlap between sites. Viewing these fluctuations dynamically, as illustrated by the double arrows, one finds that the energy gap between levels varies in time. The resulting nonlinear dependence of the transfer rate on the energy gap may therefore lead to an enhancement of the average transfer rate in the presence of dephasing noise.

transitions while the diagram corresponding to CP21 contains only two enhanced transitions. This intrinsic asymmetry between the different spectroscopic pathways has been postulated to be at the root of the differences found in actual experiments [50].

It is then apparent that the effect of the interaction between vibrational and electronic degrees of freedom is twofold. First, it relates to "lifetime borrowing" due to dynamics in the one-exciton sector and, secondly, it allows for the generation of coherent superpositions of vibrational states in the electronic ground state manifold through the excited state hybridization. Both phenomena contribute to the long-lasting population time beating signals in 2D spectra (full signal depicted in panel (a)). However, the intrinsic difference between the electronic sectors in which the relevant dynamics occurs in each case allows for a simple and transparent way to study their effects separately. According to the Feynman paths specified in panel (b), the spectroscopic pathways whose dynamics is within the ground state sector during the population time are those described by $R_{3g}(\omega_1, t_2, \omega_3)$. On the other hand, the diagrams whose population time dynamics takes place within the one-exciton sector are $R_{2g}(\omega_1, t_2, \omega_3)$ and $R^*_{1f}(\omega_1, t_2, \omega_3)$. In panel (c) we have plotted separately the contributions from these diagrams to the 2D spectrum in order to demonstrate that both are of comparable size. The environment temperature was set to $T = 77$ K, and the mode, initiated in a thermal state, suffers an inverse damping rate of $\gamma_{mode} = 0.005\Delta_{ex}$ into a reservoir at temperature T. The inverse dephasing

rate for each site has been chosen as $\gamma_{deph} = 0.025\Delta_{ex}$. The population time has been set to $t_2 = 32\Delta^{-1}_{ex}$ and the Huang–Rhys factor is $S_{HR} = 0.02$. The dipole moments of the two sites were chosen as $\mu_a \cdot \mathcal{E} = (1, 0.5, 0)$ and $\mu_b \cdot \mathcal{E} = (0, 1, 0)$ and the spectra include averaging over the orientation of the chromophores. All plots are normalized to the same maximal peak amplitude.

The spectral response of the analyzed model system therefore exhibits signatures of excitonic, vibrational, and vibronic coherence. While vibrational coherence in the ground state is not expected to exhibit functional relevance directly, the fact that it is excitonically mediated emphasizes the importance of the coherent electron–phonic coupling in explaining actual experiments. More importantly, however, the excited state contributions to the measured signal can be of equal size and it is this vibronic component that can have a very direct impact on function, for example, by enhancing excitonic transport. Previous results show that the spectral response of the considered vibronic model includes significant contributions resulting from excited state dynamics. We expect that current numerical techniques may soon allow for the computation of the exact 2D response for small aggregates, like perhaps the full FMO, together with spectral densities involving sufficiently narrow peaks. This would allow for direct comparison with experimental data. At present, results obtained by probing different types of artificial light harvesters support the feasibility of the vibronic model.

Wang et al. [52] packed structurally flexible synthetic heterodimers on single-walled carbon nanotubes as a way to restrict the motions of chromophores. Using 2D electronic spectroscopy they found that by limiting the relative rotation of chromophores and tuning the energy difference between the two electronic transitions in the dimer to match a vibrational mode of the lower-energy monomer it was possible to enhance the observed quantum-beating signals. These observations in controlled experiments neatly illustrate the theoretical results we discussed in previous sections and, in particular, emphasize that the resonance condition needs to be complemented with a reduced excitonic dephasing for the mechanism of coherence regeneration to be effective. J-aggregates of cyanine dyes provide a particularly suitable system to observe this effect [53]. This is facilitated by a relatively simple excitonic structure yielding excitonic bands with almost orthogonal transition dipole moments. As a result, the system exhibits a decongested spectral 2D response, with only two peaks with oscillatory components in specific regions of the 2D-maps, that is, one on the diagonal and one as a cross-peak for nonrephasing and rephasing signal components, respectively. In this case, it is possible to formulate a vibronic model and derive analytical expressions that show how system parameters such as electronic decoherence rates and exciton–vibrational resonance determine the amplitude and lifetime of oscillatory signals. By fitting the analytical expressions to measured data, the vibronic model was shown to be in quantitative agreement with experimental observations [53]. Importantly, long-lived oscillatory signals were shown to be dominated by excited-state coherence rather than ground-state coherence, and the constraints imposed by the observed asymmetry of the excitonic decoherence rates and the fast relaxation of exciton population allowed us to rule out explanations in terms of incoherent models, where long-lived oscillations are sustained by Markovian correlated fluctuations, for this experiment. Vibronic interactions have also been shown to underpin ultrafast polaron pair formation, on a sub-20-fs timescale, in a prototypical polymer thin film [54]. Here coherent vibronic coupling is shown to accelerate the process of charge separation and making it insensitive to disorder. These results open up a new perspective for the optimization of charge transport in organic semiconductors by controlling vibronic coherence. The efficiency of organic photovoltaics is often limited by the short diffusion lengths in conjugated polymers and therefore achieving sizeable coherent transport lengths may significantly boost the performance. This could be achieved by tuning the electronic and vibronic couplings via chemical synthesis as a way to control the yield of photo-induced charge formation, a crucial step for optimizing the performance of organic-based optoelectronic devices.

14.2.2.4 NOISE ASSISTED PROCESSES

Needless to say, not all biological systems will possess long-lived vibrational modes that are strongly coupled to electronic degrees of freedom and can therefore play a role in the dynamics of the system. What is present in all biological systems, however, is thermal fluctuations of the molecular and protein structures, as well as the surrounding solvents, water, etc. These may lead to a broad noisy background, mainly resulting in dephasing noise on the electronic degrees of freedom, which may enter an interplay with electronic quantum dynamics.

14.2.2.5 BRIDGING ENERGY GAPS AND BLOCKING PATHS

Pigment–protein complexes consist of a number of sites whose excitation energies generally exhibit a certain degree of static disorder, that is, their on-site energies will differ from site to site and also from one pigment-protein complexes to another. If the energy difference between sites that exchange excitation is larger than the intersite coupling matrix element in the relevant Hamiltonian, then transitions will be severely reduced because of energy conservation, unless of course there are quasi-resonant vibrational modes present that, as was explained in the phonon–antenna mechanism, can take up the energy difference. The presence of a well-matched mode is not necessary though. Broadband dephasing noise alone may already come to the rescue in a manner that can be understood from two different viewpoints. On the one hand one notices that dephasing noise will lead to a broadening of the excitation energy of each site and thus to an increased overlap between the two energy levels, while it does not cause loss of excitations from the system (see Figure 14.8). Alternatively, one may take a dynamical viewpoint of the same phenomenon by realizing that dephasing noise arises from the random fluctuations of the excitation energies of each site. As a consequence, these fluctuating energy levels will occasionally come sufficiently close in energy to allow for excitation energy transfer between the sites as the energy difference has been reduced to a value smaller than the direct coupling matrix element (see Figure 14.8). Again, we observe that this mechanism will lead to an optimal operating regime at intermediate levels of environmental noise. Indeed, a low level of fluctuations will not bring the site energies sufficiently close and transport remains suppressed, while excessive fluctuations of the site energies will reduce the time intervals in which the sites are energetically sufficiently close to allow for efficient energy transfer. This can be estimated easily by computing the overlap between Lorentzian lines of width γ that are displaced by an amount ω_0, in which case we find

$$\frac{1}{\pi^2}\int_{-\infty}^{\infty}\frac{\gamma}{\gamma^2+\omega^2}\frac{\gamma}{\gamma^2+(\omega-\omega_0)^2}d\omega=\frac{2\gamma}{\pi(4\gamma^2+\omega_0^2)}, \tag{14.18}$$

which takes on a maximum at $\gamma=\omega_0/2$.

It is worth noting that while the application of an excessive amount of dephasing noise suppresses transport, this may in itself serve a useful function if, for example, a transport network contains sites, for example, for structural reasons, that may lead to leakage of excitations into domains from where further transport may be slow. In such a case, dephasing may be useful to reduce the effective transition rate to such sites and thus block unfavorable transfer paths from being followed [32,55].

14.2.2.6 DESTRUCTIVE INTERFERENCE, SYMMETRY, AND NOISE

While linear networks can already exhibit interesting noise-assisted transport phenomena [23,56–59], multisite networks may exhibit more complex behavior, which arises due to the interplay between a wealth of constructive and destructive interference effects in a quantum dynamical system on the one hand and environmental noise on the other hand [20,23,30,36,60–63].

A basic example that exhibits the essential nature of this type of effect consists of a simple three-site network depicted in Figure 14.9. Here, two sites 1 and 2 are coupled to a third site, which in turn leaks excitations irreversibly into a reaction center. The coherent interaction is described by a Hamiltonian

$$H=\sum_{k=1}^{3}E_i|i\rangle\langle i|+\sum_{k=1}^{2}J_{k3}\left(|k\rangle\langle 3|+h.c.\right), \tag{14.19}$$

where $|i\rangle$ corresponds to an excitation in site i and we assume $J_{13}=J_{23}$. Let us begin by considering an excitation initially prepared in the antisymmetric state

$$|\psi\rangle=\frac{1}{\sqrt{2}}(|1\rangle-|2\rangle), \tag{14.20}$$

which forms an eigenstate of this Hamiltonian, whose overlap with site 3, which we assume to be coupled dissipatively to a reaction center, vanishes.

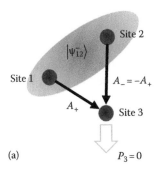 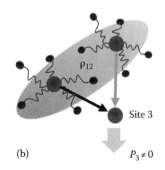

Figure 14.9 Three-site network in which two sites 1 and 2 are each coupled to a third site 3 via an exchange interaction of the same strength. Site 3 is irreversibly connected to a sink. In **(a)** the excitation is delocalized over two sites (red and green) with equal probability of being found at either site but with a wave function that is antisymmetric with respect to the interchange of red and green. This state will not evolve due to destructive interference and hence no excitation will ever reach the reaction center. In **(b)** pure dephasing causes the loss of phase coherence and the two tunneling amplitudes no longer cancel, eventually leading to a complete excitation transfer to the sink.

As a consequence, this excitation remains localized and does not propagate through the system. Eventually, the finite lifetime of the excitation implies that it will be lost to the general environment due to a spontaneous annihilation process, an event that is not in the interest of a transport network. Hence, coherence effects may lead to a strongly reduced or even vanishing transport rate.

One may argue, however, that under natural conditions a pigment-protein complex is not excited in such an antisymmetric state, but will tend to receive a single excitation locally, for example, on site 1 (this is, for example, the case of the FMO complex [8]). Nevertheless, it is easy to see that the subsequent dynamics has a propensity for leaving the system in an antisymmetric state or, more generally, a state that propagates slowly through the network due to quantum interference [20]. To this end, note that we can write the initial state localized on site 1 as an equally weighted coherent superposition of the symmetric and the anti-symmetric states, that is,

$$|1\rangle = \frac{1}{\sqrt{2}}\left(\frac{|1\rangle - |2\rangle}{\sqrt{2}} + \frac{|1\rangle + |2\rangle}{\sqrt{2}}\right). \quad (14.21)$$

Thanks to constructive interference, a symmetric superposition experiences a coherently enhanced coupling to site 3 to which it will then propagate rapidly, and from there into the reaction center, while the antisymmetric part will not evolve at all. Hence, in 50% of the cases the system will remain in the antisymmetric state, while in the other 50% of the cases the excitation reaches the reaction center. Therefore, the transfer efficiency is limited to 50% in this setting.

Now it becomes evident that a dephasing noise whose strength is correctly tuned can have a beneficial effect in such situations. Indeed, uncorrelated dephasing noise acting locally on each site will randomly flip the relative phase between $|1\rangle$ and $|2\rangle$ and thus lead to transitions between the symmetric and the antisymmetric state. Hence, the presence of dephasing noise inhibits both constructive and destructive interference and therefore slows the propagation of an excitation in a system initiated in the symmetric state and accelerates the propagation of an excitation in a system initiated in the anti-symmetric state. As a consequence, we expect again that an intermediate noise level will be optimal, too low it will not suppress destructive

interference efficiently and too high it will suppress all transport. Noise will be beneficial if the overall propagation under the noisy environment is still sufficiently rapid to be completed within the natural lifetime of the excitation. Similar considerations have also been conjectured independently to play a role in biological electron transport in photosynthetic reaction centers [64].

14.2.2.7 A THERMODYNAMICAL APPROACH

In the previous sections we have discussed how a careful interplay between coherent and incoherent interactions seems to be required to explain the experimental results obtained when light-harvesting complexes are probed using linear and multidimensional spectroscopy. In particular, the coherent coupling to intramolecular modes whose frequency is close to an excitonic splitting has been argued to provide a rather natural explanation for the existence of sustained oscillatory features in the 2D response. At this stage it is however unclear whether those coherent signatures may have any relevance for the actual function of the photosynthetic machinery. A thermodynamical approach to EET provides a convenient theoretical framework for addressing this question and examining the constructive role of quantum coherence [65]. See Figure 14.10a for a pictorial illustration. In this context, we analyze the possible functional relevance of the vibronic coupling by formulating a quantitative measure of efficiency, which depends explicitly on a (vibronic) coherent measure [66]. We should emphasize that recent developments in the field of quantum information have led to the definition of precise coherent measures so that we can now unambiguously characterize the amount of coherence of both states and evolutions [67].

Imagine, first, the situation where the light-harvesting network involves only the (electronic) coherent interaction between chromophores governed by a Hamiltonian H_e, while the effect of the rest of the complex is modeled in terms of suitable dissipative contributions. For simplicity, those can be assumed to be of Lindblad form. As illustrated in Figure 14.10b, the resulting two subsets of electronic states (ground/excitonic subspaces) are separated by a large energy gap. The incoming solar energy creates an excitation that is then transported through a network of excited states to the edge of the gap (lowest energy exciton). At this stage it is useful

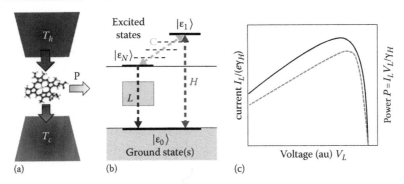

Figure 14.10 Quantum heat engine (QHE) modelization of excitation energy transfer (EET). **(a)** The chromophoric network is represented by a working system operating between two reservoirs at different temperatures, which play the role of the (hot) radiation field (T_h) and the (cold) vibrational reservoir provided by the fluctuating protein environment at T_c. **(b)** The efficiency of the process can be characterized in terms of the output power P delivered to a hypothetical load placed between the lowest excitonic state and the ground state of the network. Excitation energy is transferred to the load, which may mimic a reaction center, at a rate Γ_L. **(c)** The output power of the thermal machine is different, depending on whether or not the system involves a coherent exciton–phonon coupling, which facilitates non-thermal paths for energy transfer. Typically, in the presence of coherent vibronic interactions, the output power (solid line) can exceed the values provided by a thermal machine with a purely incoherent excitonic dynamics (dashed line).

to consider an abstract load bridging the separated electronic "terminals," very much as in an electric circuit, which might represent a LHC's reaction center. By giving its excess energy to the load, the electron can return to the low-energy subspace and recombine with the remaining low-energy hole, completing the cycle. QHE performance is determined by the rate at which useful energy is transferred to the load, that is, by the power $P = I_L V_L$, where I_L and V_L are the current and voltage across the load. The current is simply $I_L = e\rho_{NN}\Gamma_L$, with ρ_{NN} the population of the lowest excited state (e is the electron charge). The voltage is given by $eV_L = E_g + k_B T_C \ln(\rho_{NN}/\rho_{00})$, where $E_g = \varepsilon_N - \varepsilon_0$ and k_B is Boltzmann's constant. This quantifies the energy that could be released by equilibrating a given population ratio at temperature T_C. Using detailed balance arguments, it can be shown that the maximum achievable voltage is $eV_L \leq \eta_C E_g$, where $\eta_C := 1 - T_C/T_H$ is the Carnot efficiency (Shockley–Queisser bound). Since V_L is fundamentally limited, the best strategy to increase power may therefore be to increase I_L, which is directly related to the population ρ_{NN}. Thus, the larger the population in this lowest excited state during operation, the more power the heat engine can deliver. Remarkably, this is the key advantage offered by coherently coupling electron and vibration systems [66]. Heuristically, strong vibrational interactions open up alternate, coherence-mediated, pathways

for excitation transfer, allowing the total system to deliver energy faster than possible by incoherent thermal processes alone. Quantitatively, and for a prototype 3 site model, the flow of energy in our heat engine contains two contributions, one is due to the coherent (Hamiltonian) H_S, which includes both H_e and the terms accounting for the vibrational modes. Specifically, $H_S := H_e + H_m + H_I$, where

$$H_m = \hbar\omega_i \sum_{i=1}^{2} a_i^\dagger a_i, \quad H_I = \sum_{i=1}^{2} \hbar g_i \left\langle s_i | \right| \left| \langle s_i | \otimes \left(a_i + a_i^\dagger \right). \right.$$

$$(14.22)$$

We take $\omega_i = \omega_m$ and $g_i = g$ for each mode. The second contribution results from the incoherent (Lindbladian) parts of the master equation:

$$\frac{d}{dt}\langle H_e \rangle = i\mathrm{Tr}\left(\left[\rho_S(t), H_S \right] H_e \right) + \mathrm{Tr}\left(\mathcal{L}(\rho_S(t)) H_e \right).$$

$$(14.23)$$

The incoherent part takes the form

$$\mathrm{Tr}\left(\mathcal{L}(\rho_S) H_e \right) = \underbrace{\varepsilon_{10} I_H}_{\dot{Q}_H} - \underbrace{\sum_{\{ij\}} \varepsilon_{ij} I_C^{i \to j}}_{\dot{Q}_C} - \underbrace{\varepsilon_{30} I_L}_{\dot{Q}_L}, \quad (14.24)$$

where $\dot{Q}_\alpha$ are net energy flows and $\varepsilon_{ij} := \varepsilon_i - \varepsilon_j$ are energy differences between the various states.

Without coherent interactions, the electronic populations $\rho_{kk}^e := \text{Tr}\left(\rho_S|\varepsilon_k\rangle\langle\varepsilon_k|\otimes 1_m\right)$ obey detailed balance conditions, which dictate the value of the load current I_L in steady state. If the interaction H_I is present, then $[H_S, H_e] \neq 0$, and the vibronic coupling leads to an additional coherent energy exchange between the electron and the mode, $\dot{Q}_{e-m} := i\text{Tr}\left[\rho_S(t), H_S\right]H_e$. Simplifying, we find

$$\dot{Q}_{e-m} = \varepsilon_{12}\left[\sqrt{2}g\sum_{n=0}^{\infty}\sqrt{n+1}\,\text{Im}\left(\rho_{2n+1,1n}\right)\right] =: \varepsilon_{12}I_{coh},$$

$$(14.25)$$

with $\rho_{2n+1,1n} := \langle\varepsilon_2; n+1|\rho_S(t)|\varepsilon_1; n\rangle$. The *coherent current* I_{coh} has been defined with the convention that $-I_{coh} > 0$ when a net current flows from electron to mode. Clearly this is a nonclassical flow of energy, taking place exclusively via coherence between the uncoupled basis vectors $|\varepsilon_i; n\rangle$. The steady state energy currents can be determined by setting Equation 14.23 to zero, giving

$$I_L = \frac{1}{\varepsilon_{30}}\left[\varepsilon_{10}I_H - \varepsilon_{12}I_{coh} - \sum_{\varepsilon_i > \varepsilon_j}\varepsilon_{ij}I_C^{i\rightarrow j}\right].\quad (14.26)$$

Thus, the coherent current pushes the steady state away from the rate equation solution (involving only the incoherent terms I_H and $I_C^{i\rightarrow j}$) expected by purely thermal relaxation, thereby allowing the load current to be increased overall. This simple model illustrates how a thermodynamical approach can be useful in providing a quantitative link coherence $\leftrightarrow$ function. When considering actual parameters for model systems of natural light harvesters, parameter regimes leading to a possible increase of efficiency were identified [66]. However, conclusions concerning natural systems should be taken with caution and further investigations are needed. As an additional note, and following arguments put forward by Lindblad himself [68], a relevant conceptual issue emerges in this scenario, and that is the good thermodynamical behavior of the considered master equations in relation to the second law [69]. In this respect, the correct quantification of the entropy production within a vibronic model and the accurate modeling of the transfer to the load can be considered as open problems.

14.3 SUMMARY AND CONCLUSIONS

We have discussed quantum aspects of photosynthetic energy transport and its interaction. In this context, we have identified as a central theme—the interplay of this quantum dynamics of the system with its highly structured vibrational environment, often simply perceived and disregarded as noise. This theme is indeed fundamental to the study of quantum effects in biology [41].

It is this unavoidable lack of isolation of the electronic components that provides the boundary conditions under which natural evolution had to operate. Therefore it is, in hindsight, not surprising that nature has found solutions in which optimal biological quantum dynamics tends to be achieved in a regime where the interaction within the quantum system is of the order of its interaction with the environment and that both contributions do not merely coexist but enter a fruitful interplay. That this regime exists is no accident either. It can be understood from simple and generalizable principles that we have identified in our discussions to explain that too much or too little coherence can be detrimental and that it is in fact natural to expect that there is an intermediate regime that is optimal. Moreover, we argue that biological systems are able to tune their structures to achieve these optimal regimes in a controlled fashion. Indeed, by the use of protein structure to arrange molecules and their local environment, they are capable of tuning the properties of transport or sensory networks and, at the same time, adjust the environment of these networks by providing isolation or, for example, by inserting specific molecules with desirable vibrational or spin properties to fashion an environment.

It is this toolbox that allows for mutual tuning through evolutionary adaptation, which can then be used to achieve optimal performance whose origin we understand from generalizable design principles. The importance of these design principles goes beyond merely understanding what has been created already, but also paves the way by which these principles, when spelled out and made quantitative, can allow for the rational design of optimal structures as well as the execution of optimized experiments by which to amplify and verify quantum effects in biology.

REFERENCES

1. Collini, E., Wong, C., Wilk, K., Curmi, P., Brumer, P., and Scholes, G. (2010), Coherently wired light-harvesting in photosynthetic marine algae at ambient temperature, *Nature* 463, 644–649.

2. Engel, G.S., Calhoun, T.R., Read, E.L., Ahn, T.K., Mancal, T., Cheng, Y.C., Blankenship, R.E., and Fleming, G.R. (2007), Evidence for wavelike energy transfer through quantum coherence in photosynthetic systems, *Nature* 446, 782–786.

3. Fuller, F.D., Pan, J., Gelzinis, A., Butkus, V., Seckin Senlik, S., Wilcox, D.E., Yocum, C.F., Valkunas, L., Abramavicius, D., and Ogilvie, J.P. (2014), Vibronic coherence in oxygenic photosynthesis, *Nature Chemistry 6,* 706–711.

4. Panitchayangkoon, G., Hayes, D., Fransted, K.A., Caram, J.R., Harel, E., Wen, J.Z., Blankenship, R.E., and Engel, G.S. (2010), Long-lived quantum coherence in photosynthetic complexes at physiological temperature, *Proceedings of the National Academic Sciences of America* 107, 12766–12770.

5. Romero, E., Augulis, R., Novoderezhkin, V.I., Ferretti, M., Thieme, J., Zigmantas, D., and van Grondelle, R. (2014), Quantum coherence in photosynthesis for efficient solar-energy conversion, *Nature Physics* 10, 676–682.

6. Chin, A.W., Huelga, S.F., and Plenio, M.B. (2012), Coherence and decoherence in biological system: Principles of noise assisted transport and the origin of long-lived coherences, *Philosophical Transactions of the Royal Society A* 370, 3638–3657.

7. Chin, A.W., Prior, J., Rosenbach, R., Caycedo-Soler, F., Huelga, S.F., and Plenio, M.B. (2013), Vibrational structures and long-lasting electronic coherence, *Nature Physics* 9, 113–118.

8. Adolphs, J. and Renger, T. (2006), How proteins trigger excitation energy transfer in the FMO complex of green sulfur bacteria, *Biophysics Journal* 91, 2778–2797.

9. Busch, M.S.A., Müh, F., Madjet, M.E., and Renger, T. (2011), The eighth bacteriochlorophyll completes the excitation energy funnel in the FMO protein, *Journal of Physical Chemistry Letters* 2, 93–98.

10. van Amerongen, H., Valkunas, L., and van Grondelle, R. (2000), Photosynthetic excitons, World Scientific, Singapore.

11. Davydov, A.S. (1964), The theory of molecular excitons, *Soviet Physics Uspeki* 7, 145–178.

12. May, V. and Kühn, O. (2004), *Charge and Energy Transfer Dynamics in Molecular Systems,* Wiley-VCH Verlag, Weinheim.

13. Olbrich, C., Jansen, T.L.C., Liebers, J., Aghtar, M., Strumpfer, J., Schulten, K., Knoester, J., and Kleinekathöfer, U. (2011), From atomistic modeling to excitation transfer and two-dimensional spectra of the FMO light-harvesting complex, *Journal of Physical Chemistry B* 115, 8609–8621.

14. Olbrich, C., Strumpfer, J., Schulten, K., and Kleinekathöfer, U. (2011), Theory and simulation of the environmental effects on FMO electronic transitions, *Journal of Physical Chemistry Letters* 2, 1771–1776.

15. Olbrich, C., Strumpfer, J., Schulten, K., and Kleinekathöfer, U. (2011), Quest for spatially correlated fluctuations in the FMO light-harvesting complex, *Journal of Physical Chemistry B* 115, 758–764.

16. Shim, S., Rebentrost, P., Valleau, S., and Aspuru-Guzik, A. (2012), Atomistic study of the long-lived quantum coherences in the fenna-matthews-olson complex, *Biophysics Journal* 102, 649–660.

17. Strumpfer, J. and Schulten, K. (2011), The effect of correlated bath fluctuations on exciton transfer, *Journal of Chemical Physics* 134, 095102.

18. Renger, T., Klinger, A., Steinecker, F., Schmidt am Busch, M., Numata, J., and Müh, F. (2012), Normal mode analysis of the spectral density of the fenna-matthews-olson light-harvesting protein: How the protein dissipates the excess energy of excitons, *Journal of Physical Chemistry B* 116, 14565–14580.

19. Renger, T. and Müh, F. (2013), Understanding photosynthetic light-harvesting: A bottom up theoretical approach, *Physical Chemistry Chemical Physics* 15, 3348–3371.

20. Caruso, F., Chin, A.W., Datta, A., Huelga, S.F., and Plenio, M.B. (2009), Highly efficient energy excitation transfer in light-harvesting complexes: The fundamental role of noise-assisted transport, *Journal of Chemical Physics* 131, 105106.

21. del Rey, M., Chin, A.W., Huelga, S.F., and Plenio, M.B. (2013), Exploiting structured environments for efficient energy transfer: The phonon antenna mechanism, *Journal of Physical Chemistry Letters* 4, 903–907.

22. Mohseni, M., Rebentrost, P., Lloyd, S., and Aspuru-Guzik, A. (2008), Environment-assisted quantum walks in photosynthetic energy transfer, *Journal of Chemical Physics* 129, 174106.

23. Plenio, M.B. and Huelga, S.F. (2008), Dephasing assisted transport: Quantum networks and biomolecules, *New Journal of Physics* 10, 113019.

24. Rivas, A. and Huelga, S.F. (2012), *Open Quantum Systems—An Introduction*, Springer Briefs in Physics, Springer Verlag, Heidelberg.

25. Plenio, M.B. and Vedral, V. (1998), Teleportation, entanglement and thermodynamics in the quantum world, *Contemporary Physics* 39, 431–446.

26. Plenio, M.B. and Virmani, S. (2007), An introduction to entanglement measures, *Quantum Information and Computation* 7, 1–71.

27. Fassioli, F. and Olaya-Castro, A. (2010), Distribution of entanglement in light-harvesting complexes and their quantum efficiency, *New Journal of Physics* 12, 085006.

28. Caruso, F., Chin, A.W., Datta, A., Huelga, S.F., and Plenio, M.B. (2010), Entanglement and entangling power of the dynamics in light-harvesting complexes, *Physical Review A* 81, 062346.

29. Campos Venuti, L. and Zanardi, P. (2011), Excitation transfer through open quantum networks: A few basic mechanisms, *Physical Review B* 84, 134206.

30. Cao, J.S. and Silbey, R.J. (2009), Optimization of exciton trapping in energy transfer processes, *Journal of Physical Chemistry A* 113, 13825–13838.

31. Caycedo-Soler, F., Chin, A.W., Almeida, J., Huelga, S.F., and Plenio, M.B., The nature of the low energy band of the Fenna-Matthews-Olson complex: Vibronic signatures, *Journal of Chemical Physics* 136, 155102.

32. Chin, A.W., Datta, A., Caruso, F., Huelga, S.F., and Plenio, M.B. (2010), Noise-assisted energy transfer in quantum networks and light-harvesting complexes, *New Journal of Physics* 12, 065002.

33. Hoyer, S., Sarovar, M., and Whaley, K.B. (2010), Limits of quantum speedup in photosynthetic light harvesting, *New Journal of Physics* 12, 065041.

34. Kolli, A., O'Reilly, E.J., Scholes, G.D., and Olaya-Castro, A. (2012), The fundamental role of quantized vibrations in coherent light harvesting by cryptophyte algae, *Journal of Chemical Physics* 137, 174109.

35. Moix, J., Wu, J.L., Huo, P.F., Coker, D., and Cao, J.S. (2011), Efficient energy transfer in light-harvesting systems, III: The influence of the eighth bacteriochlorophyll on the dynamics and efficiency in FMO, *Journal of Physical Chemistry Letters* 2, 3045–3052.

36. Olaya-Castro, A., Lee, C.F., Fassioli Olsen, F., and Johnson, N.F. (2008), Efficiency of energy transfer in a light-harvesting system under quantum coherence, *Physical Review B* 78, 085115.

37. Plenio, M.B. and Huelga, S.F. (2011), Quantum dynamics of bio-molecular systems in noisy environments, *Proceedings of the 22nd Solvay Conference on Chemistry 2010*, Brussels, Belgium on "Quantum Effects in Chemistry and Biology," *Procedia Chemistry* 3, 248–255.

38. Rebentrost, P., Mohseni, M., Kassal, I., Lloyd, S., and Aspuru-Guzik, A. (2009), Environment-assisted quantum transport, *New Journal of Physics* 11, 033003.

39. Womick, J.M. and Moran, A.M. (2011), Vibronic enhancement of exciton sizes and energy transport in photosynthetic complexes, *Journal of Physical Chemistry B* 115, 1347–1356.

40. Wu, J.L., Liu, F., Shen, Y., Cao, J.S., and Silbey, R.J. (2010), Efficient energy transfer in light-harvesting systems, I: Optimal temperature, reorganization energy and spatial-temporal correlations, *New Journal of Physics* 12, 105012.

41. Huelga, S.F. and Plenio, M.B. (2013), Vibrations, quanta and biology, *Contemporary Physics* 54, 181–207.

42. Mohseni, M., Shabani, A., Lloyd, S., and Rabitz, H. (2014), Energy-scales convergence for optimal and robust quantum transport in photosynthetic complexes, *Journal of Chemical Physics* 140, 035102.

43. Pachon, L.A. and Brumer, P. (2011), Physical basis for long-lived electronic coherence in photosynthetic light-harvesting systems, *Journal of Physical Chemistry Letters* 2, 2728–2732.

44. Fassioli, F., Nazir, A., and Olaya-Castro, A. (2010), Quantum state tuning of energy transfer in a correlated environment, *Journal of Physical Chemistry Letters* 1, 2139–2143.

45. Lim, J., Tame, M., Yee, K.H., Lee, J.-S., and Lee, J. (2014), Phonon-induced dynamic resonance energy transfer, *New Journal of Physics* 16, 053018.

46. Ritschel, G., Roden, J., Strunz, W.T., Aspuru-Guzik, A., and Eisfeld, A. (2011), Absence of quantum oscillations and dependence on site energies in electronic excitation transfer in the fenna-matthews-olson trimer, *Journal of Physical Chemistry Letters* 2, 2912–2917.

47. Prior, J., Chin, A.W., Huelga, S.F., and Plenio, M.B. (2010), Efficient simulation of strong system-environment interactions, *Physical Review Letters* 105, 050404.

48. Christensson, N., Kauffmann, H.F., Pullerits, T., and Mancal, T. (2012), Origin of long-lived coherences in light harvesting complexes, *Journal of Physical Chemistry B* 116. 7449–7454.

49. Plenio, M.B., Almeida, J., and Huelga, S.F. (2013), Long-lived oscillatory signals in electronic 2-D spectra from exciton-vibrational interaction, *Journal of Chemical Physics* 139, 235102.

50. Tiwari, V., Peters, W.K., and Jonas, D.M. (2013), Electronic resonance with anticorrelated pigment vibrations drives photosynthetic energy transfer outside the adiabatic framework, *Proceedings of the National Academy of Sciences USA* 110, 1203–1208.

51. Chin, A.W., Rivas, A., Huelga, S.F., and Plenio, M.B. (2010), Exact mapping between system-reservoir quantum models and semi-infinite discrete chains using orthogonal polynomials, *Journal of Mathematical Physics* 51, 092109.

52. Wang, L., Griffin, G.B., Zhang, A., Zhai, F., Williams, N.E., Jordan, R.F., and Engel, G.S. (2017), Controlling quantum-beating signals in 2D electronic spectra by packing synthetic heterodimers on single-walled carbon nanotubes, *Nature Chemistry* 9, 219–225.

53. Lim, J., Paleček, D., Caycedo-Soler, F., Lincoln, C.N., Prior, J., von Berlepsch, H., Huelga, S.F., Plenio, M.B., Zigmantas, D., and Hauer, J. (2015), Vibronic origin of long-lived coherence in an artificial molecular light harvester, *Nature Communications* 6, 7755.

54. De Sio, A. et al. (2017), Watching the coherent birth of polaron pairs in conjugated polymers, *Nature Communications* 7, 13742.

55. Chen, G.-Y., Lambert, N., Li, C.-M., Chen, Y.-N., and Nori, F. (2013), Rerouting excitation transfer in the Fenna-Matthews-Olson complex, *Physical Reviews E* 88, 032120.

56. Gaab, K.M. and Bardeen, C.J. (2004), The effects of connectivity, coherence, and trapping on energy transfer in simple light-harvesting systems studied using the Haken-Strobl model with diagonal disorder, *Journal of Chemical Physics* 121, 7813–7820.

57. Kassal, I. and Aspuru-Guzik, A. (2012), Environment-assisted quantum transport in ordered systems, *New Journal of Physics* 14, 053041.

58. Semião, F.L., Furuya, K., and Milburn, G.J. (2010), Vibration-enhanced quantum transport, *New Journal of Physics* 12, 083033.

59. Vaziri, A. and Plenio, M.B. (2010), Quantum coherence in ion channels: Resonances, transport and verification, *New Journal of Physics* 12, 085001.

60. Caruso, F., Huelga, S.F., and Plenio, M.B. (2010), Noise-enhanced classical and quantum capacities in communication networks, *Physical Review Letters* 105, 190501.

61. Giorda, P., Garnerone, S., Zanardi, P., and Lloyd, S. (2011), Interplay between coherence and decoherence in LHCII photosynthetic complexes, E-print arXiv:1106.1986.

62. Mohseni, M., Shabani, A., Lloyd, S., Omar, Y., and Rabitz, H. (2013), Geometrical effects on energy transfer in disordered open quantum systems, *The Journal of Chemical Physics* 138, 204309, E-print arXiv:1212.6804.

63. Wu, J., Silbey, R.J., and Cao, J. (2013), Generic mechanism of optimal energy transfer efficiency: A scaling theory of the mean first-passage time in exciton systems, *Physical Review Letters* 110, 200402.

64. Balabin, I.A. and Onuchic, J.N. (2000), Dynamically controlled protein tunneling paths in photosynthetic reaction centers, *Science* 290, 114–117.

65. Scully, M.O. (2010), Quantum photocell: Using quantum coherence to reduce radiative recombination and increase efficiency, *Physical Review Letters* 104, 207701.

66. Killoran, N., Huelga, S.F., and Plenio, M.B. (2015), Enhancing light-harvesting power with coherent vibrational interactions: A quantum heat engine picture, *The Journal of Chemical Physics* 143, 155102.

67. Baumgratz, T., Cramer, M., and Plenio, M.B. (2014), Quantifying coherence, *Physical Review Letters* 113, 140401.

68. Lindblad, G. (1975), Completely positive maps and entropy inequalities, *Communications in Mathematical Physics* 40, 147.

69. Gelbwaser-Klimovsky, D. and Aspuru-Guzik, A. (2017), On thermodynamic inconsistencies in several photosynthetic and solar cell models and how to fix them, *Chemical Science* 8, 1008.

15

Photoinduced electron transfer in the reaction centers

THOMAS RENGER

15.1 INTRODUCTION

Photosynthesis forms the basis of our life on Earth. It delivers essentially all the carbon compounds and the oxygen that make aerobic metabolism possible. During the light reaction of photosynthesis, solar energy is used to generate the high energy compound ATP and the reduction equivalent NADPH, which are utilized in the dark reaction to convert atmospheric CO_2 into carbohydrates and other forms of organic matter.

At the heart of the light reaction of photosynthesis is the transmembrane ET driven by solar energy. In higher plants and cyanobacteria, two photosystems (PSII and PSI) work in series. Both PSs contain a reaction center (RC) that is composed of several pigments (Figure 15.1, middle and lower parts), which are arranged in two nearly symmetric branches that span the photosynthetic membrane. The RCs are surrounded by light-harvesting complexes that collect the light energy and transfer it with high efficiency to the RC (see Chapters 10 through 14). Once the excitation energy has arrived at the RC, a stable charge-separated state is created by subsequent ET reactions along the chain of cofactors (from bottom to top in Figure 15.1)

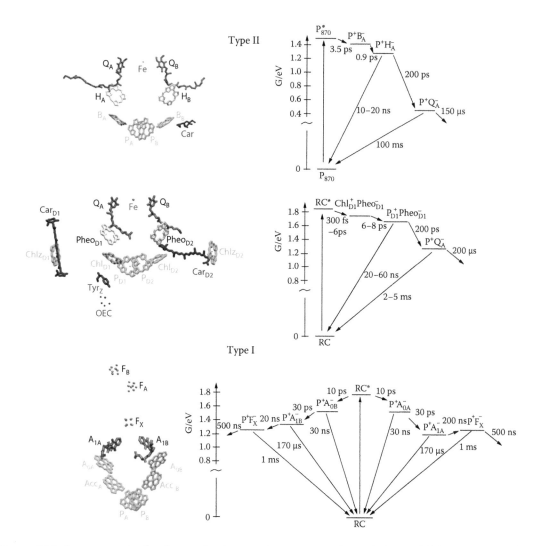

Figure 15.1 Arrangement of cofactors in the reaction centers of purple bacteria[1] (upper part), photosystem II[2,3] (PSII, middle part), and photosystem I[4] (PSI, lower part) of cyanobacteria and higher plants. Bacteriochlorophylls and chlorophylls are shown in green, bacteriopheophytins and pheophytins in yellow, carotenoids in red, quinones in black, and nonheme irons and iron sulfur clusters in orange. The substituents of chlorine pigments have been removed for clarity. On the right of the respective structures, the free-energy levels of the functional states of the transmembrane electron transfer (ET) chain and the time constants for forward ET and recombination reactions are shown. References for these values are given in the text.

through the membrane. Whereas in type II RCs, only one branch is used for ET (for review, see References 5 and 6), in type I RCs both branches are ET active.[7–12]

Denoting the excited state of the RC by RC* and the radical pair states as RP_1, RP_2, RP_3, ..., we may describe the transmembrane ET schematically as

$$RC^* \xrightarrow{k_1} RP_1 \xrightarrow{k_2} RP_2 \xrightarrow{k_3} RP_3 \cdots \quad (15.1)$$

Due to the small distances between pigments, the excited states of the RC are delocalized over a certain number of pigments, and exciton states $|M\rangle = \sum_m c_m^{(M)} |m\rangle$ are formed, where $|m\rangle$ is an excited state that is localized on pigment m. Since the ET couplings are much weaker than the excitonic couplings (except for the central dimers, the so-called special pairs, as will be discussed in more detail further below), it is reasonable to assume that exciton relaxation in the RC is fast compared to primary ET. In this case primary ET starts from an excitonically equilibrated state of the RC and the rate constant for primary ET is given as[13]

$$k_1 = P_p^{(eq)} k_{intr},\qquad (15.2)$$

where

k_{intr} is the intrinsic rate constant for primary ET
$P_p^{(eq)}$ is the probability to find the primary electron donor excited in a system where the population of exciton states is thermally equilibrated.

Hence, $P_p^{(eq)}$ is given as

$$P_p^{(eq)} = \left\langle \sum_M \left| c_p^{(M)} \right|^2 f(M) \right\rangle_{dis},\qquad (15.3)$$

where $\left| c_p^{(M)} \right|^2$ is the probability of the primary electron donor p to be excited in exciton state M and the Boltzmann factor $f(M) = \exp(-E_M/k_B T)/\sum_N \exp(-E_N/k_B T)$ gives the thermal population of exciton state M with energy E_M. Since the contribution of the primary donor to the low-energy exciton states varies due to static disorder in local optical transition energies (site energies) of the pigments, we have included a disorder average in Equation 15.3, denoted as $\langle ...\rangle_{dis}$. From Equations 15.2 and 15.3, it becomes clear that only those exciton states $|M\rangle$ of the RC contribute to k_1 which have a sufficiently low energy $E_M = \hbar\omega_{M0}$ and a large contribution $\left| c_p^{(M)} \right|^2$ from the primary electron donor.

15.1.1 Reaction center of photosystem II

From an extensive analysis of optical spectra, in particular optical difference spectra, it became possible to infer an exciton Hamiltonian for the PSII RC[13–15] that provided evidence also for the molecular identity of several functional states of the RC related to ET and charge recombination (as reviewed in refs.[16–18]). Recently, this Hamiltonian was also confirmed by direct electrostatic/quantum chemical calculations.[19] By comparing the exciton state pigment distribution function[13]

$$d_m(\omega) = \left\langle \sum_M \left| c_m^{(M)} \right|^2 \delta(\omega - \omega_{M0}) \right\rangle_{dis}\qquad (15.4)$$

of the six core pigments of the RC ($m = 1...6$) with the density of exciton states

$$d_M(\omega) = \left\langle \delta(\omega - \omega_{M0}) \right\rangle_{dis}\qquad (15.5)$$

(Figure 15.2), it is seen that the lowest exciton state has the largest contribution from Chl_{D1} and that the special pair pigments P_{D1} and P_{D2} contribute mostly to the third and sixth exciton states. As before, the $\langle ...\rangle_{dis}$ in Equations 15.4 and 15.5 denotes an average over disorder in site energies. Due to the large contribution of Chl_{D1} in the lowest exciton state, the $P_m^{(eq)}$ (Equation 15.3) of this pigment is largest, as can be seen in Figure 15.3 for $T = 5$ K and room temperature. If ET would start from any other pigment than Chl_{D1}, a large temperature dependence of ET would result (Equations 15.2 and 15.3). In the experiment, however, only a very weak dependence of ET on temperature was found.[20,21]

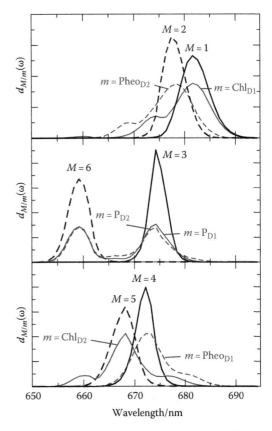

Figure 15.2 Exciton states pigment distribution function $d_m(\omega)$ (Equation 15.4) and density of exciton states $d_M(\omega)$ (Equation 15.5) for the six core pigments of the RC of PSII.

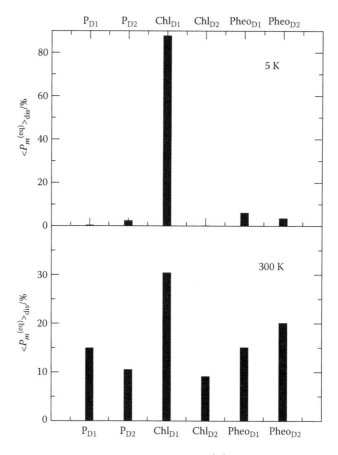

Figure 15.3 Thermal population of excited pigment states $P_m^{(eq)}$ (Equation 15.3) of the six core pigments in the RC of PSII at $T = 5\,K$ (upper part) and $T = 300\,K$ (lower part).

Therefore, it can be concluded that primary ET in PSII starts at Chl_{D1},[13] at least at low temperatures. At higher temperatures, in principle, additional pathways could open, but the simplest assumption is that Chl_{D1} is the main primary donor at all T. Hence, the reaction scheme for primary ET in PSII reads

$$RC^* \xrightarrow{k_1} P_{D1}Chl_{D1}^+Pheo_{D1}^-P_{D2}Chl_{D2}Pheo_{D2} \quad (15.6)$$

$$\xrightarrow{k_2} P_{D1}^+Chl_{D1}Pheo_{D1}^-P_{D2}Chl_{D2}Pheo_{D2}. \quad (15.7)$$

Independent evidence for the primary electron donor Chl_{D1} was reported from femtosecond VIS pump/IR probe and VIS pump/VIS probe experiments by van Grondelle and coworkers[22] and by Holzwarth and coworkers,[23] respectively. Concerning the inferred time scales of primary ET, the value for $1/k_1$ varies between 300 fs, obtained

from a fit of time-resolved fluorescence decay data of PSII core particles,[24] and 600–800 fs,[22] estimated from VIS pump/IR probe data, to 6 ps obtained from VIS pump/VIS probe data.[23] The value estimated so far for $1/k_2$ are in the 6 ps[22] to 8 ps[23] range.

From femtosecond pump–probe spectra of isolated RC (D1D2cyt$b$559) complexes and PSII core complexes measured at room temperature, Shelaev et al.[25,26] concluded that primary charge transfer (CT) starts at the special pair with a time constant of 0.9 ps. However, their interpretation is based on the assumption that Chl_{D1} absorbs at 670 nm, in contradiction to the exciton models described in the beginning of this section.

Based on their time-resolved spectroscopic data, van Grondelle and coworkers[14,27–29] suggested that at room temperature, depending on the realization of disorder, governing the contributions of different excited states to RC*, also P_{D1} might act as primary electron donor in PSII. Although it is tempting to assume that in this way the RC is more flexible to choose a fast pathway for every realization of disorder, the upper part of Figure 15.3 suggests that for all realizations of disorder the contribution of Chl_{D1} in the lowest exciton state is dominant. Therefore, an additional pathway seems to be not really necessary from a functional point of view, and so far it has not been possible to unambiguously resolve the time constants of the reactions in the two pathways. A target analysis of transient data revealed a 3 ps time constant for ET starting at Chl_{D1} and a 1 ps time constant for ET initiated at P_{D1},[27] whereas a microscopic modeling of these data suggests an inverse assignment, that is, the fast ET pathway starting at Chl_{D1}.[28]

From the work of Krausz and coworkers,[30,31] it is known that charge separation in PSII can be initiated by long-wavelength excitation of a homogeneously broadened CT state, which absorbs in the 690–730 nm spectral region. It is, therefore, quite likely that multiple ET reaction pathways exist. What remains to be investigated is how much these additional pathways contribute under physiological conditions. Novoderezhkin et al.[14] reported the first attempt to include CT states in the simulations of optical spectra of PSII RCs and found them to be important in the description of the amplitude of the low energy wing of the Stark spectrum (for a more detailed review, see Reference 17). A tight-binding Hamiltonian, which also included CT states, was applied to model time-resolved 2D spectra.[32] From the results, reported for the simulations using one or two ET pathways, it is, however, difficult to decide whether other approximations have to be checked first, before it can be decided whether both ET pathways are needed to describe the experimental data. 2D experiments[33] and analyses[29] have concentrated on the role of various types of coherences and have so far not led to a refinement of the kinetic scheme of the proposed multiple ET pathways.[34]

In order to obtain a fit of the time-resolved fluorescence that takes into account microscopically calculated energy transfer rates, it had to be assumed that the free energy difference $\Delta G^{(0)}$ for the primary ET is not smaller than 175 meV.[24] From measurements of the recombination fluorescence, $\Delta G^{(0)}$ values of 110 meV[35] for D1–D2–cyt$b$559 complexes and 170 meV[36] for PSII core complexes were obtained. A similar value $\Delta G^{(0)} = 170 - 190$ meV was reported for the bacterial RC.[37] We note that much smaller $\Delta G^{(0)}$ values have been inferred

in the exciton–radical pair equilibrium (ERPE) model,[23,38] which is, however, based on the assumption of unrealistically fast excitation energy transfer between the core antennae and the RC, as will be discussed in detail in Section 15.2.3.2.

After formation of the state $P_{D1}^+Pheo_{D1}^-$, the electron is transferred further to plastoquinone Q_A and finally to the secondary plastoquinone Q_B[23,39]

$$P_{D1}^+Pheo_{D1}^-Q_AQ_B \xrightarrow{200\,ps} P_{D1}^+Pheo_{D1}Q_A^-Q_B$$
$$\xrightarrow{100\,\mu s} P_{D1}^+Pheo_{D1}Q_AQ_B^-. \tag{15.8}$$

Subsequently, Q_B^- gets protonated and accepts a second electron from the next excitation and, after it got doubly protonated, leaves the binding pocket as plastoquinol QH_2. The latter diffuses to the cyt$b6f$ complex, where the reducing power of QH_2 is used to pump protons through the photosynthetic membrane, generating a proton motive force (pmf) that is utilized by the ATPase to generate ATP. The electrons from QH_2 are shuttled further via a soluble plastocyanin to PSI, to be discussed further below in Section 15.1.3.

The electron hole P_{D1}^+ in PSII is filled by ET from the oxygen-evolving center (OEC) via an intervening Tyr_Z residue. The OEC uses water as an external electron source. The water is split into electrons; molecular oxygen, the basis of oxygenic life on earth; and protons, which contribute to the pmf utilized by the ATPase.

We note the fact that at room temperature the $P^+Pheo_{D1}^- - PPheo$ and the $P^+Q_A^- - PQ_A$ optical difference spectra show that a bleaching at 680 nm (for review, see, e.g., Reference 17) has led to the terminus "P_{680}" for the special pair in PSII. On the other hand, the excited state, from where ET starts, is sometimes termed "P_{680}^*," although it is dominated by the excited state of Chl_{D1}, as discussed earlier. Historically, the term "P^*" was introduced to denote the excited state of the special pair of purple bacteria, where indeed the primary electron donor and the location of hole stabilization is the special pair.

To avoid confusion, we recommend to avoid the term "P_{680}" for PSII and to use instead the molecular identities of the cofactors, for example, "P_{D1}" for the special pair chlorophyll of the D1-branch of the RC and "Chl_{D1}" for the accessory chlorophyll of this branch. Since the excited states, from where ET starts, are partially delocalized, we will use "RC*" for those states.

Besides forward ET, there are competing reactions that limit the photochemical efficiency of the RC. The side reactions, involving charge recombination, are at least an order of magnitude slower than the forward ET reactions, and there is a successive drop in free energy that suppresses ET back reactions (Figure 15.1, middle right part). Experimental information about charge recombination reactions can be obtained at low temperature, at which ET from Q_A to Q_B and from Tyr_Z to P^+ is blocked. Charge recombination between Q_A^- and P^+ occurs with a time constant between 2 and 5 ms.[40,41] When ET between $Pheo_{D1}$ and Q_A is blocked (either by prereduction or extraction of Q_A), charge recombination between $Pheo_{D1}^-$ and P^+ was detected to occur with time constants of 20–60 ns.[42] Singlet–triplet mixing in the radical-pair state prior to charge recombination leads to the formation of a triplet state, in addition to the singlet ground state of the RC. EPR studies[43] with oriented PSII samples have shown that the plane of the Chl carrying the triplet state is oriented like that of the accessory Chls Chl_{D1} and Chl_{D2}. From calculations of triplet-minus-singlet spectra[13] and from mutagenesis studies,[44] it was found that the triplet in PSII is localized on Chl_{D1}. Interestingly, the lifetime of the triplet state was found to reduce by two orders of magnitude if Q_A is in the reduced state Q_A^-.[42,45]

15.1.2 Reaction center of purple bacteria

Whereas the interpretation of optical experiments of PSII RCs is difficult due to the overlap of optical bands, in its ancestor the type II RC from purple bacteria (bRC) (upper part in Figure 15.1), the optical bands are much better separated and the identification of functional states is easier.[46]

The overall arrangement of cofactors is very similar to that in PSII, discussed earlier. As will be discussed later, there is a slight tilt in mutual geometry of the pigments in the special pair with respect to the special pair in PSII that leads to a larger π-electron overlap and a larger coupling of exciton states to charge-transfer states in the special pair of bRC. This coupling gives rise to a large redshift of the low-energy excited state of the special pair[47,48] such that it is well separated from the other excited states of the RC. Even at physiological temperatures, the $P_m^{(eq)}$ (Equation 15.3) is only nonzero for the special pair pigments

P_A and P_B, from where, therefore, primary ET has to start (Equations 15.2 and 15.3).

Despite the C2 symmetry of the two branches of cofactors, such as in PSII, only the A-branch is ET active, as revealed from absorbance difference spectra involving the oxidized special pair P and a reduced pheophytin H, which was identified as H_A, that is, the one of the A-branch (for review, see Reference 5). Site-directed mutagenesis studies suggest that the differences in free energies of charge-separated states are responsible for the exclusive use of the A-branch in ET. In particular, the state $P^+B_A^-$ is lower in free energy than the low-energy excited-state P^* of the special pair, whereas the free energy of the state $P^+B_B^-$ is higher (Reference 49 and references therein).

The accessory bacteriochlorophyll B_A serves as an intermediate electron carrier, that is, the initial ET scheme reads (for review, see Reference 46)

$$P^*B_AH_A \xrightarrow{3ps} P^+B_A^-H_A \xrightarrow{0.9ps} P^+B_AH_A^-. \quad (15.9)$$

The short lifetime of the intermediate state $P^+B_A^-H_A$ has provided a challenge for its spectroscopic detection, and therefore, alternative models assuming a direct ET from P to H_A where the intermediate B_A contributes only by mediating the electronic coupling between P and H_A in a superexchange reaction were discussed initially, but were finally shown to be invalid.[46,50] Quantum chemical calculations identified the strong electronic coupling between B_A and H_A as an important factor for a fast reaction.[51,52]

From H_A the electron is transferred in 200 ps[53] to the quinone Q_A and from there in about 150 μs[54] to the second quinone Q_B much like in PSII of cyanobacteria and higher plants, discussed earlier.

$$P^+H_A^-Q_AQ_B \xrightarrow{200ps} P^+H_AQ_A^-Q_B \xrightarrow{150μs} P^+H_AQ_AQ_B^-. \quad (15.10)$$

Whereas the primary ET reactions do not depend on temperature or even accelerate somewhat at lower temperatures,[55] ET between Q_A and Q_B is an activated process that was shown, however, to be not very sensitive to the free energy difference between the states $Q_A^-Q_B$ and $Q_AQ_B^-$.[54] A thermally activated gating mechanism, for example, a proton transfer reaction or a conformational change, seems to form the bottleneck of the reaction. As in PSII, Q_B is doubly reduced and protonated. The resulting ubiquinol QH_2 diffuses to the cytbc1 complex where it drives transmembrane proton transfer generating a pmf that is utilized by the ATP*ase*. In contrast to PSII, in purple bacteria the electrons extracted from QH_2 are shuttled back to the RC, via ET from the cytbc1 protein to a soluble cytochrome c1 complex, thus completing the ET cycle.

In biochemically isolated RCs, the final state of the ET chain is $P^+Q_A^-$, which recombines with a time constant of about 100 ms.[37,56] Charge recombination reactions between P^+ and H_A^- were measured, after removing or reducing Q_A, to occur with a time constant of about 10–20 ns.[57,58] Singlet–triplet mixing in the radical-pair states in combination with these charge recombination reactions creates triplet states in the RC that localize on the special pair $P_A - P_B$.[59]

Finally, we note that, as in the case of PSII, the forward ET reactions outcompete the side and back ET reactions by orders of magnitude, thus enabling the RC to work with a quantum efficiency close to unity (Figure 15.1, upper part).

15.1.3 Reaction center of photosystem I

As noted in the introduction, in PSI RCs both branches are ET active, both chlorophyll A_0s act as primary electron acceptors. However, the A-branch is used more frequently.[8–12,60,61] Primary ET creates a charge-separated state, where the electron is localized at the primary acceptor A_0 and the hole stabilizes at one of the two special pair chlorophylls.[62] From mutagenesis studies, removing a hydrogen bond to the special pair chlorophyll P_A, and measuring absorbance difference spectra (P^+–P), it became clear that the lowest excited state has strong contributions from the special pair.[63] Most likely, the energy sink is determined by short-range effects in the special pair, which were calculated[48] to redshift the site energies of the special pair by about 500 cm^{-1}. This assignment is indirectly supported by electrostatic calculations of site energy shifts of the RC pigments,[64] which are significantly smaller than the short-range shifts of the site energies of the special pair.

From a structure-based analysis of time-resolved spectroscopic data on the excited state decay of PSI core complexes, it was concluded that primary charge separation in the RC occurs with an intrinsic rate constant of $(500 \text{ fs})^{-1}$.[65,66] From ultrafast pump–probe spectroscopy, using 25 fs pulses, Semenov and coworkers concluded that primary CT starts at the special pair and occurs in less than 100 fs.[67,68] These time constants are considerably faster than the one for primary ET in purple bacteria and similar to that in PSII, discussed earlier. From mutagenesis studies involving transient absorbance spectra, Holzwarth and coworkers[23,69] concluded that the primary electron donor in PSI is not the special pair but the two accessory chlorophylls Acc_A and Acc_B. This suggestion implies either that there is considerable mixing of excited states between the special pair and the accessory chlorophylls such that the primary electron donors still have a large enough population in the lowest exciton state of the complex for efficient ET or that ET starts from a nonequilibrated excited state of the RC. Structure-based calculations of site energies of PSI core complexes[64] found that the concentration of low-energy exciton states in the core antenna is higher on the A-branch site of the RC, which is used more frequently for ET. Although highly hypothetical, it is tempting to assume that the asymmetry in the light-harvesting process together with primary ET starting at the accessory chlorophylls leads to a more frequent ET in the A-branch.

Ultimately, the hole stabilizes at the special pair (which is rereduced by the electrons from PSII). From the two chlorophyll A_0s, the electrons are transferred to the two phylloquinone A_1s with a time constant of 30 ps.[70] Concerning the oxidation of phylloquinone A_1^- by ET to the iron sulfur cluster F_X, a slow (200 ns) and a fast (20 ns) phase were measured and assigned to ET in the A- and B-branch, respectively.[7,71] The origin of the difference in ET rate constants along the two branches most likely lies in the different redox potentials of phylloquinones A_{1A} and A_{1B}. According to electrostatic calculations of Ishikita and Knapp,[72] the redox potential of A_{1A} is 150 mV more negative than that of A_{1B} leading to an uphill ET reaction to F_X. Rutherford and coworkers[73] suggested that due to the low free energy of the state $P^+A_{1A}^-$, triplet formation by charge recombination of the state $P^+A_{0A}^-$ is reduced due to the slow back transfer between $P^+A_{1A}^-$ and $P^+A_{0A}^-$. As Rutherford et al.[73] pointed out, this reduction also holds for the overall triplet yield, since electrons initially transferred along the B-branch will fall back to $P^+A_{1A}^-$ if forward ET from F_X is blocked.

Under normal conditions, the electron is transferred further from F_X via iron-sulfur clusters

F_A and F_B to the soluble ferredoxin protein, and ultimately $NADP^+$ is reduced. In this way light energy is stored in a chemical form as reduction equivalent. Experiments on ferredoxin reduction by PSI (a review and references can be found in Reference 70) revealed a time constant of 500 ns for this process. Hence, it was concluded that both ET reactions involved in $F_X \to F_A \to F_B$ occur on a faster time scale.[70]

Due to the uphill ET step from A_{1A} to F_X, reduction of the iron sulfur clusters along the A-branch can be frozen out at low T, and charge recombination of the state $P^+A_1^-$ can be measured. A time constant of 170 μs was reported for this process.[74] When A_1 is prereduced, charge recombination of the state $P^+A_0^-$ was measured to occur with a time constant of 30 ns, and the charge recombination time constant of $P^+F_X^-$ is about 1 ms (reviews are given in refs.[70,75]). As in bRC and PSII, also in PSI, the forward ET reactions under physiological conditions are orders of magnitude faster than the side reactions, thus leading to a near unity quantum efficiency of the charge separation process (Figure 15.1, lower part). Similar to PSII, also in the RC of PSI, a low energy charge-transfer state was identified[76] that can be optically excited with long wavelengths ($\lambda > 800$ nm) and leads to the formation of the charge-separated state $P^+A_0^-$, however, with lower yield than for high-energy excitation. The molecular identity of this charge-transfer state is still unknown. In general, low-energy excited states in photosynthesis are an active field of research.[77]

15.2 BASIC PHOTOPHYSICAL PRINCIPLES

The rather similar arrangement of cofactors in all known RCs (Figure 15.1) suggests that nature has found an optimal solution for a high photochemical efficiency of a photosynthetic RC. Two important factors are an efficient transfer of excitation energy from the antenna complexes to the RC and the generation of a stable charge-separated state in the latter. In the following, we will focus first on the second aspect and discuss what theory tells us about the basic principles that can be used to suppress ET back and side reactions, such as recombination, in order to reach the near unity quantum efficiency of charge separation in photosynthesis.

For this purpose, we discuss the theoretical basis for ET in the weak electronic coupling regime

(nonadiabatic ET), where second-order perturbation theory in the electronic coupling is used to derive a rate constant. We will focus first on the role of nuclear dynamics in suppressing back ET, and afterwards introduce the electronic energy gap between the reactant and product state as a generalized reaction coordinate and derive the Marcus rate constant in the framework of a classical description of the nuclear degrees of freedom. This rate constant will explain why recombination reactions can be suppressed by a large free energy difference. Next, we derive corrections to the classical Marcus rate constant of ET that are due to nuclear tunneling effects and require a quantum mechanical treatment of the nuclear degrees of freedom. It will become clear that, in addition to a large free energy difference, a relatively weak electron–vibrational coupling, as realized in the RCs by the nonpolar protein interior and by using chlorin pigments exhibiting a very weak intramolecular vibronic coupling, is important to suppress charge recombination. Next, we will discuss limitations of nonadiabatic ET theory that concern the assumption of fast equilibration of nuclear degrees of freedom prior to ET and briefly discuss experimental evidence for such nonequilibrium effects in the RCs and theories that are capable to describe these effects. Next, structure-based methods are reviewed for the determination of parameters (electronic couplings, free energies, spectral densities) entering the ET theories, and we will discuss how these simulations have uncovered structure–function relationships of the RCs.

15.2.1 Theory of electron transfer

From Fermi's golden rule, the rate constant $k_{1 \to 2}$ between two weakly coupled electronic states, which may represent the excited donor state D^*A and the charge-separated state D^+A^- in an ET reaction, is given as

$$k_{1 \to 2} = \frac{2\pi}{\hbar} |V_{12}|^2 \delta(E_1 - E_2), \quad (15.11)$$

where V_{12} is the matrix element of the electronic coupling between the two states with energies E_1 and E_2. The delta function $\delta(E_1 - E_2)$ ensures energy conservation during the ET event. The following question arises: How can an electron be transferred in a certain direction, for example, across the photosynthetic membrane, if forward and backward rate constants are equal, as in Equation 15.11? The answer to this question lies in the nuclear degrees of freedom of the complex that stabilize the charge-separated state, for example, by orientational polarization effects. One way to include the effect of nuclear dynamics on the rate constant is by combining the quantum mechanical description of the ET in Equation 15.11 with a classical description of the nuclear degrees of freedom and their coupling to the electrons.

15.2.1.1 MARCUS THEORY OF NONADIABATIC ELECTRON TRANSFER

Let us assume that we have identified a reaction coordinate X that takes certain average values X_1 and X_2 in the initial and the final electronic state, respectively. In thermal equilibrium, the probability $P_i(X)$ to find a certain value of the reaction coordinate X for electronic state i is related to the free energy $G_i(X)$ of this state by

$$P_i(X) \propto e^{-\frac{G_i(X)}{k_B T}}. \tag{15.12}$$

An illustration of the free energy surfaces of the two electronic states is given in Figure 15.4. The free energy surfaces of the two electronic states cross at the transition state with reaction coordinate value $X = X_t$, where the corresponding electronic energies $E_i(Q_1 \ldots Q_N)$ of the electron, which depend on the vibrational degrees of freedom Q_ξ of the complex, in the two electronic states are equal. The rate constant in Equation 15.11 can now be expressed as a product of the probability $P_1(X_t)$ to reach the transition state and the electronic factor $2\pi|V_{12}|^2/\hbar$ from Fermi's golden rule (Equation 15.11).

$$k_{1 \to 2} \propto |V_{12}|^2 e^{-\frac{G_1^\#}{k_B T}} \tag{15.13}$$

with the activation free energy $G_1^\# = G_1(X_t) - G_1^{(0)}$. Energy conservation during ET is achieved by thermal fluctuations of the nuclear degrees of freedom, which bring the system into the transition state, where the energies of both electronic states are equal.

After the electronic transition, nuclei relax in the free energy surface $G_2(X)$ of the charge-separated state. The ratio between forward and backward transfer rate constants is obtained as $k_{1 \to 2}/k_{2 \to 1} = \exp(-G_1^\#/k_B T)/\exp(-G_2^\#/k_B T) = \exp(-\Delta G_{12}^{(0)}/k_B T)$ with the standard free-energy difference $\Delta G_{12}^{(0)} = G_1^{(0)} - G_2^{(0)}$. In this way directional ET becomes possible by tuning $\Delta G_{12}^{(0)}$. The standard free-energy differences along the ET chain across the photosynthetic membrane are lowered in (almost) every step such as to guide the electrons across the membrane and to avoid back transfer reactions that could lead to unwanted recombination of the electron–hole pair (Figure 15.1).

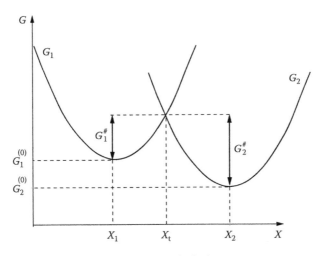

Figure 15.4 Free energy surfaces of two weakly coupled electronic states as a function of a reaction coordinate X, $G_1^\#$ denotes the free energy of activation for the transition between electronic states 1 and 2 and $G_2^\#$ that of the back reaction.

For the kinetics of the ET reaction, it is important to evaluate the activation free energy $G_1^{\#}$, as a function of a reaction coordinate X. The latter might be, for example, a critical bond length that changes upon ET. It can, however, also be a more complicated function of the nuclear coordinates. A safe choice is the difference between the potential energy of nuclei in the two electronic states, that is, the electronic energy difference between the two electronic states at a given configuration of nuclei $Q_1 \ldots Q_N$,

$$X(Q_1 \ldots Q_N) = E_2(Q_1 \ldots Q_N) - E_1(Q_1 \ldots Q_N). \quad (15.14)$$

In the following, we will derive free-energy curves for this reaction coordinate assuming a linear electron–vibrational coupling, which is valid in a strict sense if any anharmonic dependencies of $E_i(i = 1, 2)$ on the nuclear coordinates can be neglected. We will later comment on a generalization of this procedure. In the spirit of a normal mode analysis, we may write the electronic energies $E_i(Q_1 \ldots Q_N)$ as

$$E_i(Q_1 \ldots Q_N) = E_i^{(0)} + \sum_\xi \frac{\hbar \omega_\xi}{4} \left(Q_\xi - Q_\xi^{(i)} \right)^2 \quad (15.15)$$

with dimensionless normal coordinates Q_ξ and their different equilibrium positions $Q_\xi^{(i)} = -2g_\xi^{(i)}$ in the two electronic states $i = 1, 2$.[78] Hence, the reaction coordinate X (Equation 15.14) reads

$$X = \Delta G^{(0)} + \lambda + \sum_\xi \hbar \omega_\xi \Delta g_\xi Q_\xi \quad (15.16)$$

where the standard free-energy difference

$$\Delta G^{(0)} = E_2^{(0)} - E_1^{(0)} = \hbar \omega_{21} \quad (15.17)$$

corresponds to the energy difference between the minima of the two high-dimensional potential energy surfaces. The reorganization energy λ in Equation 15.16 was introduced as

$$\lambda = \sum_\xi \hbar \omega_\xi \Delta g_\xi^2 \quad (15.18)$$

and $\Delta g_\xi = g_\xi^{(2)} - g_\xi^{(1)}$. In order to evaluate the free energy $G_1(X)$ of the initial electronic state, we

need to determine the probability $P_1(X)$ to find a certain value X of the energy gap to the final electronic state. From Equation 15.16 we notice that X, up to a constant, is given as a sum of independent variables $\hbar \omega_\xi, \Delta g_\xi, Q_\xi$ containing normal coordinates Q_ξ. In thermal equilibrium the Q_ξ, according to Boltzmann, are distributed with the probability $P(Q_\xi) \propto \exp(-E_1(Q_1 \ldots Q_N)/k_B T)$. By choosing the origin of the Q_ξ-axis equal to $Q_\xi^{(1)}$ (i.e., setting $Q_\xi^{(1)} = 0$), we obtain (using Equation 15.16)

$$P(Q_\xi) = \left(\frac{4\pi k_B T}{\hbar \omega_\xi} \right)^{-1/2} e^{-\frac{\hbar \omega_\xi}{4 k_B T} Q_\xi^2} \quad (15.19)$$

where the prefactor ensures normalization, $\int dQ_\xi P(Q_\xi) = 1$. The variance of this distribution function is obtained as

$$\sigma_\xi^2 = \langle Q_\xi^2 \rangle - \langle Q_\xi \rangle^2 = \langle Q_\xi^2 \rangle = \frac{2 k_B T}{\hbar \omega_\xi}. \quad (15.20)$$

The central limit theorem of statistical mechanics tells us that X (Equation 15.16), as a sum over independent variables, is distributed Gaussian $P(X) \propto \exp\left(-(X - \overline{X})^2 / 2\sigma_X^2 \right)$ with variance $\sigma_X^2 = \langle X^2 \rangle - \langle X \rangle^2$ that is simply given as a sum over the variances of the independent contributions

$$\sigma_X^2 = \sum_\xi \hbar^2 \omega_\xi^2 \Delta g_\xi^2 \sigma_\xi^2 = 2\lambda k_B T \quad (15.21)$$

and mean value

$$\langle X \rangle = \Delta G^{(0)} + \lambda + \hbar \omega_\xi \Delta g_\xi \langle Q_\xi \rangle = \Delta G^{(0)} + \lambda. \quad (15.22)$$

With a proper normalization, $P(X)$ is thus obtained as

$$P(X) = \frac{1}{\sqrt{4\pi \lambda k_B T}} e^{-\frac{\left(X - \Delta G^{(0)} - \lambda \right)^2}{4 \lambda k_B T}}. \quad (15.23)$$

For $X = 0$ the electron has the same energy in the initial and final states, and an electronic transition between the states can occur. Hence, we may write the rate constant for ET as a product of the probability $P(X = 0)$ to reach the transition state and the electronic transition factor $\frac{2\pi}{\hbar} |V_{12}|^2$ from Fermi's golden rule (Equation 15.11),

$$k_{1\to 2} = \frac{2\pi}{\hbar}\left|V_{12}\right|^2 P(0) = \sqrt{\frac{\pi}{\hbar^2\lambda k_B T}}\left|V_{12}\right|^2 e^{-\frac{\left(\Delta G^{(0)}+\lambda\right)^2}{4\lambda k_B T}},$$

$$(15.24)$$

which is the Marcus rate constant for nonadiabatic ET.[79] It is illuminating to discuss this rate constant in the framework of free energy surfaces of the initial and final states, introduced in the following.

From Equation 15.12 we have $G_i(X) = -k_B T \ln P_i(X) + G_i^{(0)}$ where the index i indicates the electronic state i from where the reaction starts, and $G_i^{(0)}$ is the minimum free energy of state i. So far, we have considered the forward reaction, that is, we may identify $P_1(X)$ with the $P(X)$ in Equation 15.23, and we obtain the free energy surface of state 1 as

$$G_1(X) = G_1^{(0)} + \frac{\left(X - \left(\Delta G^{(0)}+\lambda\right)\right)^2}{4\lambda}. \qquad (15.25)$$

In complete analogy (using now $Q_\xi^{(2)}$ as the origin of the Q_ξ axis), the free-energy surface of the charge-separated state 2 is obtained as

$$G_2(X) = G_2^{(0)} + \frac{\left(X - \left(\Delta G^{(0)}-\lambda\right)\right)^2}{4\lambda}. \qquad (15.26)$$

An illustration of these free-energy surfaces is given in Figure 15.5. By noting that $G_2(\Delta G^{(0)} + \lambda) - G_2(\Delta G^{(0)} - \lambda) = \lambda$, it is seen that the reorganization energy λ corresponds to the difference between the minimum of the free-energy surface of a given state (here state 2) and the value of the free energy of the same state but at the equilibrium position of the reaction coordinate of the other state, as illustrated in the middle part of Figure 15.5.

In addition, we see that the free energy of activation is given as

$$G^\# = G_1(0) - G_1\left(\Delta G^{(0)}+\lambda\right) = \frac{\left(\Delta G^{(0)}+\lambda\right)^2}{4\lambda}. \qquad (15.27)$$

Based on the relative positions of free-energy surfaces G_1 and G_2, three regions of ET can be distinguished[79]: (1) normal, (2) activationless, and

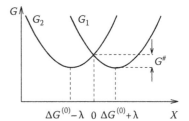

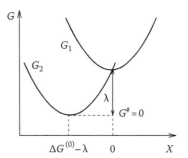

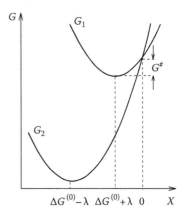

Figure 15.5 Free-energy surfaces of two electronic states involved in ET, according to Equations 15.25 and 15.26. Three situations are illustrated that differ in the relative size of $\Delta G^{(0)}$ (Equation 15.17) and λ (Equation 15.18): (1) normal region of ET, $-\Delta G^{(0)} < \lambda$ (upper part); (2) activationless region of ET, $-\Delta G^{(0)} = \lambda$ (middle part); and (3) inverted region of ET, $-\Delta G^{(0)} > \lambda$ (lower part). The middle part contains an illustration of the reorganization energy λ, which is assumed to be the same in all three situations. $G^\#$ is the free energy of activation of the reaction from state 1 to state 2 (Equation 15.27).

(3) inverted, as illustrated in Figure 15.5. Whereas in the normal region thermal fluctuations of nuclei are needed to bring the system into the transition state, in the activationless region such a thermal activation is not needed and the rate constant takes its maximum value.

$$k_{1\to2}^{(\max)} = \sqrt{\frac{\pi}{\hbar^2\lambda k_BT}}\,|V_{12}|^2. \qquad (15.28)$$

Upon further decrease of $\Delta G^{(0)}$, a barrier between the initial and final states of the reaction is formed again (Figure 15.5, lower part), and the rate constant decreases. In this inverted region of ET, however, corrections due to nuclear tunneling are expected. In order to include the latter, a quantum treatment of the vibrations will be discussed in the following.

15.2.1.2 QUANTUM CORRECTIONS

If the nuclear motion is treated quantum mechanically, Fermi's golden rule gives for the rate constant

$$k_{1\to2} = \frac{2\pi}{\hbar}|V_{12}|^2\sum_{\alpha,\beta}b(1\alpha)|<1\alpha\,|\,2\beta>|^2\delta\big(E_{1\alpha}-E_{2\beta}\big). \qquad (15.29)$$

The high-dimensional states $|1\alpha> = \prod_\xi|1\alpha_\xi>$ and $|2\beta\ge\prod_\xi|2\beta_\xi>$ contain a product of vibrational eigenstates of the initial and final electronic states, respectively. The factor $b(1\alpha)$ describes a Boltzmann distribution for the vibrational states of the initial electronic state. The Hamiltonian of the initial electronic state $|1>$ is defined via $H_1|1\alpha\ge E_{1\alpha}|1\alpha>$, with the energy $E_{1\alpha}=\sum_\xi E_{1\alpha_\xi}$. Similarly, for the final electronic state, we have $H_2|2\beta\ge E_{2\beta}|2\beta>$.

By writing the δ-function in Equation 15.29 as $\delta\big(E_{1\alpha}-E_{2\beta}\big)=\dfrac{1}{2\pi\hbar}\displaystyle\int_{-\infty}^{\infty}dt\,\exp\big\{i\big(E_{1\alpha}-E_{2\beta}\big)t/\hbar\big\}$ and using the completeness relation, that is, $\sum_\beta|2\beta><2\beta|=1$, Fermi's golden rule becomes

$$k_{1\to2} = \frac{|V_{12}|^2}{\hbar^2}\int_{-\infty}^{\infty}dt\,\big\langle e^{iH_1t/\hbar}e^{-iH_2t/\hbar}\big\rangle, \qquad (15.30)$$

where $<\cdots>$ denotes an average with respect to the equilibrium statistical operator of the vibrational degrees of freedom of the initial state $|1>$, $W_{eq}=e^{-H_1/k_BT}/\mathrm{Tr}_{vib}\{e^{-H_1/k_BT}\}$. In complete analogy with the semiclassical case (see Equation 15.14), the energy difference

$$\hat{X} = H_2 - H_1 \qquad (15.31)$$

is introduced as a reaction coordinate.[80] In the following, use is made of the identity[81]

$$e^{iH_1t/\hbar}e^{-iH_2t/\hbar} = Te^{-i\int_0^t d\tau\,\hat{X}(\tau)/\hbar}, \qquad (15.32)$$

which contains the time-ordering operator $\hat{T}$ that acts on a product of operators at different times $\hat{O}(t_1)\hat{O}(t_2)$ by time ordering of all the possible permutations $\hat{T}\hat{O}(t_1)\hat{O}(t_2)=\theta_{t_1-t_2}\hat{O}(t_1)\hat{O}(t_2)+\theta_{t_2-t_1}\hat{O}(t_2)\hat{O}(t_1)$, where θ_t is the Heaviside step function, that is, zero for $t<0$ and one for $t>0$. The time evolution of the reaction coordinate

$$\hat{X}(\tau) = e^{iH_1\tau/\hbar}\hat{X}(0)e^{-iH_1\tau/\hbar} \qquad (15.33)$$

is defined by the Hamiltonian H_1 of the nuclei in the initial electronic state.

Using a second-order cumulant expansion,[82] which is exact for harmonic oscillators, the rate constant is obtained as[80]

$$k_{1\to2} = \frac{|V_{12}|^2}{\hbar^2}\int_{-\infty}^{\infty}dt\,e^{-i<\hat{X}>t/\hbar+\gamma(t)}, \qquad (15.34)$$

with

$$\gamma(t) = -\frac{1}{\hbar^2}\int_0^t d\tau(t-\tau)C(\tau). \qquad (15.35)$$

where $C(t)$ is the correlation function of the energy gap

$$C(t) = <\delta\hat{X}(t)\delta\hat{X}(0)> = <\hat{X}(t)\hat{X}(0)> - <X>^2. \qquad (15.36)$$

In the case of harmonic oscillators, the Hamiltonian H_i reads $H_i=E_i+T_{\mathrm{nucl}}$ with the kinetic energy of nuclei T_{nucl} and the state-specific potential energy E_i (Equation 15.15), where $Q_\xi=C_\xi^++C_\xi$ are given now in terms of creation $\big(C_\xi^+\big)$ and annihilation (C_ξ) operators of the vibrational quanta of protein mode ξ.[78] Choosing the origin of the Q_ξ axis at the minimum position of the PES of the initial state, results in the Hamiltonian are $H_1=\sum_\xi\hbar\omega_\xi\big(C_\xi^+C_\xi+1/2\big)$, and the $\delta\hat{X}(t)=X(t)-\langle X\rangle$ in Equation 15.36 is obtained as[83]

$$\begin{aligned}
\delta\hat{X}(t) &= \sum_\xi\Delta g_\xi^2\hbar\omega_\xi Q_\xi(t)\\
&= \sum_\xi\Delta g_\xi^2\hbar\omega_\xi e^{i\omega_\xi tC_\xi^\dagger C_\xi}\big(C_\xi+C_\xi^\dagger\big)e^{-i\omega_\xi tC_\xi^\dagger C_\xi}\\
&= \sum_\xi\Delta g_\xi^2\hbar\omega_\xi\big(C_\xi e^{i\omega_\xi t}+C_\xi^\dagger e^{-i\omega_\xi t}\big). \qquad (15.37)
\end{aligned}$$

Introducing this $\delta\hat{X}(t)$ into Equation 15.36, and performing the Boltzmann average, results in the correlation function

$$C(t) = \sum_\xi \hbar^2 \omega_\xi^2 \Delta g_\xi^2 \left\{ \left(1 + n(\omega_\xi)\right) e^{-i\omega_\xi t} + n(\omega_\xi) e^{i\omega_\xi t} \right\}$$

$$= \int d\omega J(\omega) \hbar^2 \omega^2 \left(\left(1 + n(\omega)\right) e^{-i\omega t} + n(\omega) e^{i\omega t} \right),$$

(15.38)

where the spectral density $J(\omega)$ was introduced as

$$J(\omega) = \sum_\xi \Delta g_\xi^2 \delta(\omega - \omega_\xi),$$

(15.39)

and the Bose–Einstein distribution function $n(\omega_\xi)$ describes the mean number of vibrational quanta with energy $\hbar\omega_\xi$ that are excited at a given temperature T

$$n(\omega_\xi) = \mathrm{Tr}_{\mathrm{vib}}\left\{ W_{\mathrm{eq}} C_\xi^\dagger C_\xi \right\} = \frac{1}{e^{\hbar\omega_\xi/k_{\mathrm{B}}T} - 1}.$$

(15.40)

With the above correlation function $C(t)$, the function $\gamma(t)$ in Equation 15.35 becomes[81]

$$\gamma(t) = \int d\omega J(\omega) \left[\left(2n(\omega) + 1\right) \omega^2 \cos\omega t \right.$$

$$\left. - i\left(\sin\omega t - \omega t\right) \right].$$

(15.41)

Assuming that the spectral density $J(\omega)$ is a continuous function that is broad enough in energy to allow for a fast decay of $\gamma(t)$, we may approximate the latter by using a short time approximation for the sin and cos functions, that is, $\cos(\omega_\xi t) \approx 1 - (\omega_\xi t)^2/2$, $\sin(\omega_\xi t) \approx \omega_\xi t$, and $\gamma(t)$, is obtained as

$$\gamma(t) \approx \frac{k_{\mathrm{B}}T}{\hbar} \sum_\xi g_\xi^2 \omega_\xi^2 t^2$$

$$= -\frac{k_{\mathrm{B}}T}{\hbar^2} \lambda t^2,$$

(15.42)

where also the high-temperature assumption $k_{\mathrm{B}}T \gg \hbar\omega_\xi$ was used to approximate $n(\omega_\xi)$ by $k_{\mathrm{B}}T/\hbar\omega_\xi$ in Equation 15.41. The integral in Equation 15.34 then yields the classical Marcus-type rate constant[79] (Equation 15.24).

If one vibrational mode $\xi = \nu$ is treated quantum mechanically, and its energy is sufficiently high,

that is, $\hbar\omega_\nu \gg k_{\mathrm{B}}T$, whereas the remaining modes are treated classically as above in Section 15.2.1.1, the function $\gamma(t)$ in Equation 15.41 becomes

$$\gamma(t) \approx -\frac{k_{\mathrm{B}}T}{\hbar^2} \lambda t^2$$

$$+ g_\nu^2 \left(\cos(\omega_\nu t) - i\sin(\omega_\nu t)\right) + i g_\nu^2 \omega_\nu t,$$

(15.43)

where λ now includes only the classical modes. The mean energy difference $<X>$ reads

$$<X> = \Delta G^{(0)} + \lambda + S_\nu \hbar\omega_\nu,$$

(15.44)

where $S_\nu = g_\nu^2$ is the Huang–Rhys factor of the quantum mode ν. Using the last two equations, the rate constant is obtained as

$$k = \frac{|V_{12}|^2}{\hbar^2} e^{-S_\nu} \int_{-\infty}^{\infty} dt\, e^{-\frac{i}{\hbar}\left(\Delta G^{(0)} + \lambda\right)t - \frac{k_{\mathrm{B}}T}{\hbar^2}\lambda t^2} e^{S_\nu e^{-i\omega_\nu t}}.$$

(15.45)

After expanding the last term in this equation into an exponential series, the integral for the terms of that series can be performed, and the final expression for the rate constant reads[78,84]

$$k = 2\pi \frac{|V_{12}|^2}{\hbar} \frac{1}{\left(4\pi\lambda kT\right)^{1/2}} e^{-S_\nu} \sum_{N=0}^{\infty} \frac{\left(S_\nu\right)^N}{N!} e^{-\frac{\left(\Delta G^{(0)} + \lambda + N\hbar\omega_\nu\right)^2}{4\lambda k_{\mathrm{B}}T}}.$$

(15.46)

Comparison of this result with the classical Marcus result in Equation 15.24 suggests to introduce free-energy surfaces $G_{i,N}$ of vibronic states $|i, N>$, where i denotes the electronic state and N the vibrational quantum number. The rate constant then follows as

$$k = e^{-S_\nu} \sum_N \frac{\left(S_\nu\right)^N}{N!} k_N,$$

(15.47)

where

$$k_N = \sqrt{\frac{\pi}{\hbar^2 \lambda k_{\mathrm{B}}T}} |V_{12}|^2 e^{-\frac{\left(\Delta G_N^{(0)} + \lambda\right)^2}{4\lambda k_{\mathrm{B}}T}}$$

(15.48)

is the classical rate constant for transfer between the vibronic states |1,0> and |2,N>, with the standard free-energy difference

$$\Delta G_N^{(0)} = \Delta G^{(0)} + N\hbar\omega_v. \qquad (15.49)$$

An illustration of the nuclear tunneling in the inverted region of ET is given in Figure 15.6. As is seen there, by excitation of the high-frequency vibrational quanta in the final electronic state, the regime of ET changes from inverted for $N=0$ to activationless for larger N ($N = 14$ in Figure 15.6) excited quanta. The relative weight of the different transitions is given by the Franck–Condon factor $e^{-S_v}\dfrac{(S_v)^N}{N!}$ in Equation 15.47, which describes the overlap of the vibrational ground-state wavefunction of the initial state with the different vibrational wavefunctions of the final electronic state. It is seen thereby that besides a required large $\Delta G^{(0)}$, a suppression of charge recombination reactions also requires a weak electron–vibrational coupling, in order to keep the prefactor $e^{-S_v}\dfrac{(S_v)^N}{N!}$ in

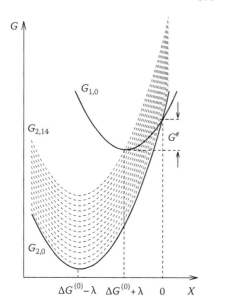

Figure 15.6 Nuclear tunneling between two electronic states in the inverted region of ET. The free-energy surfaces of vibronic states of the effective high-frequency mode treated quantum mechanically are shown for the two electronic states. The character of the electronic transition changes from normal ET ($G_{1,0} \rightarrow G_{2,0}$) to activationless ($G_{1,0} \rightarrow G_{2,14}$).

Equation 15.47 small for large quantum numbers N for which the activationless transfer regime between vibronic states is reached.

Finally, we note that although expression (15.45) shows that nuclear tunneling is particularly important in the inverted region of ET, we have to be aware also of the distinction between classical and quantum degrees of freedom made in this model. At very low T, where it holds that $\hbar\omega_\xi \gg k_B T$ for all ξ, all vibrational modes need to be described quantum mechanically. Therefore, the classical reorganization energy λ in Equation 15.45 becomes zero and nuclear tunneling remains the only possible mechanism for ET in the normal and inverted regions.

15.2.1.3 LIMITATIONS OF STANDARD THEORIES AND EXTENSIONS

A critical assumption in the derivations of ET rate constants seen so far was a fast equilibration of nuclear degrees of freedom. Interestingly, the relaxation of some of the nuclear degrees of freedom was found to be slow compared to the CT (see References 85–89 and references therein). In practice this slow relaxation often is taken into account by introducing time- and temperature-dependent parameters (free-energy differences and reorganization energies) in the analysis of experimental data. A more rigorous theoretical treatment is given by Sumi–Marcus theory,[90,91] which was originally created to include slow solvent diffusion processes in the description of ET. A theory that is related in spirit is the one that Agmon and Hopfield[92] derived for the description of chemical reaction kinetics in the presence of slow conformational motion. Parson and Warshel[55] investigated the temperature dependence of primary ET with a density matrix model that explicitly included one vibrational mode per molecule and found quantitative agreement with the experimental data showing a slight increase of the rate constant with decreasing temperature.

Another assumption in our derivation was a linear electron–vibrational coupling, giving rise to the normal mode picture used to construct the free-energy surfaces. As noted by Georgievski et al.,[80] the cumulant expansion in the derivation of the rate constant in Equation 15.34 may be valid also for strongly anharmonic systems. In other words, anharmonic potentials do not necessarily lead to anharmonic free energies. In the classical derivation, this is seen by noting that the Gaussian shape of the free-energy surface results from the central limit theorem of statistical mechanics. As long as the energy difference can be approximately written as a sum over independent contributions, the resulting free-energy surface will be Gaussian. Indeed, microscopic simulations of free energies by Warshel and Parson[93] that use a full molecular mechanics force field (including anharmonic terms) show that the resulting free energies are indeed Gaussian. Independent evidence was reported by Simonson,[94] who investigated ET in cytochrome c in a similar way.

15.2.2 Determination of parameters

In order to understand the building principles of the RCs, it is important to determine the parameter values that enter the ET theories discussed earlier. In the following sections, we want to discuss how these parameters can be obtained either from independent experiments or from structure-based simulations.

15.2.2.1 DUTTON RULER

From Equation 15.24 it is seen that the semiclassical rate constant takes its maximum value if the reorganization energy λ equals $-\Delta G^{(0)}$. In experiments, $\Delta G^{(0)}$ can be varied by exchanging the pigments with chemically related pigments that exhibit a different redox potential. By measuring the ET rate for different $\Delta G^{(0)}$, from the maximum rate constant $k_{max} = |V|^2 / \sqrt{4\pi\lambda kT}$ for a known $\lambda = \Delta G^{(0)}$, the ET coupling matrix element V can be inferred. Dutton and coworkers[95,96] performed an extensive analysis of ET reactions in different proteins and found that the coupling V roughly varies as $\exp(-\beta R)$ with the donor–acceptor (edge to edge) distance R, and a $\beta \approx 1.4$ Å^{-1}.[95] They converted the logarithm of the Marcus rate constant into a simple empirical expression[95]

$$\log_{10} k = 15 - 0.6R - 3.1\left(\Delta G^{(0)} + \lambda\right)^2 / \lambda \quad (15.50)$$

where k is given in units of s^{-1}, R in Å, and $\Delta G^{(0)}$ and λ in eV.

15.2.2.2 STRUCTURE-BASED SIMULATIONS

Free-energy surfaces can be obtained by running classical molecular dynamics (MD) simulations in the initial electronic state and monitoring the time-dependent energy gap $X(t)$ between this state and the final electronic state.[93,97] From the histogram of probabilities $p(X)$ of this generalized reaction

coordinate, the free-energy surface is obtained as described earlier (see Equation 15.12). Free-energy perturbation techniques are used to explore also such regions of the free-energy surface, which reflect small probabilities of X. Parson and Warshel obtained the free-energy surfaces of P*, P$^+$B$_A^-$, and P$^+$B$_B^-$ of bRC in this way. Interestingly, the resulting $\Delta G^{(0)}$ and λ provide a convincing explanation of why ET proceeds exclusively along the A-branch of the RC. Whereas A-branch ET is found to occur in the activationless region of ET; in the case of the B-branch, a positive $\Delta G^{(0)}$ was obtained.[93,97]

Alternatively, standard free-energy differences $\Delta G^{(0)}$ can be obtained from electrostatic calculations[98] of the redox potentials of protein-bound cofactors. A thermodynamic cycle is constructed that relates the measured redox potential of a cofactor in solution to that in a specific binding site in the protein. The electrostatic free-energy difference between the oxidized (reduced) state of the cofactor in the solvent and in the protein is obtained from a solution of a Poisson–Boltzmann equation, where the flexibility of the protein and solvent environments are approximated by that of a continuous dielectric with dielectric constant $\varepsilon_{prot} = 4$ for the protein and $\varepsilon_{prot} = 80$ for the aqueous solvent. From the calculations of Ishikita et al.[99] on the redox potential of the special pairs in bRC, PSI, and PSII, it was concluded that the electrostatic coupling between the protein and the special pair is responsible for the large difference between the redox potentials of the special pairs of PSI (500 meV) and PSII (1200 meV). A major contributor is the difference in interaction with transmembrane helices. There is also a 700 meV difference in redox potential between the special pair of PSII and that of bRC (500 meV). Ishikita et al.[99] estimate that about 160 meV are due to intrinsic differences between BChl a and Chl a, 140 meV are due to stronger electron exchange in bRC, and 200 meV are due to differences in electrostatic pigment–protein coupling. Using the same method, Ishikita and Knapp[72] also provided a microscopic explanation for the difference in redox potentials of the phylloquinones of the A- and B-branch of PSI A$_{1A}$ and A$_{1B}$, respectively, which is thought to be responsible for the biphasic ET kinetics of F_X reduction, as discussed above.

In order to evaluate the quantum corrections in ET reactions according to Equation 15.34, the spectral density $J(\omega)$ in Equation 15.39 is needed. Warshel and coworkers developed the dispersed polaron model,[100] where first classical MD simulations in the initial electronic state are run to calculate the classical correlation function $C(t)$ of the energy gap between the initial and final electronic states. The system is then described by a set of effective harmonic oscillators by setting the $C(t)$ equal to the high-temperature limit $(n(\omega) \approx k_B T / \hbar \omega_\xi)$ of Equation 15.38. The spectral density $J(\omega)$ is then obtained from this equation by applying an inverse Fourier transform.

The extracted $J(\omega)$ can afterwards be applied to calculate the quantum rate constant (Equations 15.34, 15.35, and 15.38) for arbitrary temperatures. This model was applied to study ET between bacteriopheophytin and quinone and charge recombination between quinone and the special pair in bRC. Quantum corrections were found to be important for the latter reaction, which occurs in the inverted region of ET.[100]

ET coupling matrix elements have been obtained by semiempirical[52,101] and ab initio quantum chemical methods,[102] for the cofactors of bRC. Interestingly, the ET coupling between the states $P^+B_A^-$ and $P^+H_A^-$ was found to be considerably stronger than the coupling between P^*B_A and $P^+B_A^-$. This result provided a molecular explanation of the faster transfer time of ET measured[46] between the latter states. Interestingly, van Brederode et al.[103,104] found that optical excitation of the accessory bacteriochlorophyll B_A in bRC can trigger ultrafast ET to the bacteriopheophytin H_A, with a time constant of 400 fs, which is close to the time constant estimated for the related reaction in PSII of higher plants, discussed earlier. The perfect overlay of the $B_A - H_A$ dimer of bRC and the $Chl_{D1} - Pheo_{D1}$ dimer of the PSII RC (see, e.g., Reference 15) suggests that there is a similar difference in ET coupling matrix elements in PSII as calculated for bRC, providing further support for Chl_{D1} as the primary electron donor in PSII. In order to understand why bRC cannot utilize B_A as a primary electron donor and PSII can, we will discuss the trapping of excitation energy in the following.

15.2.3 Trapping of excitation energy

15.2.3.1 MIXING BETWEEN CHARGE TRANSFER AND EXCITON STATES IN THE SPECIAL PAIRS

For the trapping of excitation energy, it is important to tune the optical properties of the RC pigments in such a way as to reach near resonance with the low-energy excited states of the core antenna, as will be discussed in the next subsection. Here, we want to discuss how electron exchange[47,48,107,108] between pigments changes the optical properties of the RCs. The strongest wavefunction overlap occurs between the pigments of the special pairs (compare Figure 15.1). In bRC this wavefunction overlap is particularly strong (Figure 15.7). It gives rise to a strong mixing of excited states with intermolecular CT states of the special pair.[47,48] This mixing leads to a strong redshift of the low-energy absorbance band of the special pair, which is clearly separated in the spectrum from the remaining optical bands of the RC.[46,109] In the case of PSII, such a separation is not observed, instead all optical bands of the RC strongly overlap. An explanation of this behavior is found by evaluating the long- and short-range effects of intermolecular interactions in the special pairs (Figure 15.7). Whereas the short-range effects rely on wavefunction overlap, a prerequisite for electron exchange, the long-range effects occur due to the Coulomb coupling between charge and transition densities, without electron exchange. By relating quantum chemical calculations of transition energies and transition dipole moments of isolated monomers and whole dimers, it is possible to decipher the different contributions to the optical properties.[48] As seen in Figure 15.7, the strong overlap of π-electron wavefunction in the special pair of purple bacteria gives rise to a large redshift of the optical transition energies of the monomers. In addition, there is a dominating influence of short-range excitonic coupling between the optical transitions that gives rise to a further redshift of the low-energy exciton state of the special pair. About 80% of the excitonic coupling in the special pair of bRC is due to electron exchange.[48] In PSI and PSII, the mutual geometry of special pair pigments is changed by introducing inward (PSI) and outward (PSII) tilts of the two pigments. The resulting disruption of ; π-electron wavefunction diminishes the short-range effects considerably (Figure 15.7).[48,106] The long-range site energy shifts in PSI and PSII special pairs partially compensate the short range shifts because of different signs (Figure 15.7). Despite the close distance between the two special pair pigments in PSI and PSII, the long-range excitonic couplings are very small and still the short-range excitonic couplings dominate.[48,106]

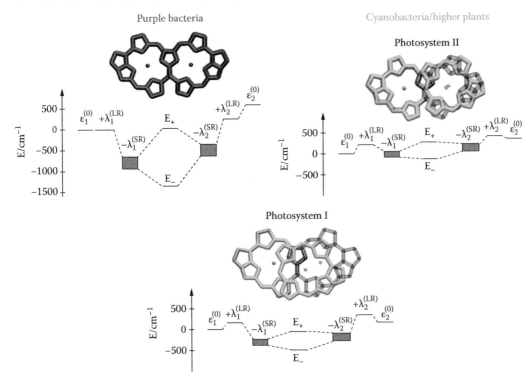

Figure 15.7 Upper part: Macrocycles of the two special pair bacteriochlorophylls P_A and P_B of the RC from the purple bacterium *Rhodobacter* (*Rb. sphaeroides*).[105] The scheme below the structure relates the optical transition energies $\varepsilon_i^{(0)}$ of the two isolated monomers to the transition energies (E_- and E_+) of the dimer. These quantities were obtained from quantum chemical calculations using TDDFT with different hybrid functionals.[48,106] An effective two-state Hamiltonian was used to relate these transition energies and the respective transition dipole moments and to extract long-range and short-range site energy shifts and excitonic couplings. The gray areas reflect uncertainties due to limitations of the quantum chemical method and the effective two-state Hamiltonian used in the analysis. Middle and lower parts: same as the upper part but for the special pairs of PSII and PSI, respectively. The structures of the special pairs $P_{D1}-P_{D2}$ of PSII and P_A-P_B of PSI (shown in green) are overlaid with the structure of P_A-P_B of purple bacteria (shown in transparent blue).

15.2.3.2 EXCITATION ENERGY TRANSFER BETWEEN THE CORE ANTENNAE AND THE REACTION CENTERS

To prevent the oxidation of antenna pigments by pigments in the RC, the interpigment distances between pigments in the antenna and those in the RC are large (Figure 15.8). Hence, the electronic coupling between pigments in different subunits is much weaker than that between pigments in the same complex and the electron–vibrational coupling. The latter destroys any coherence (delocalization) between excited states in different subunits and restricts the delocalization of excited states to pigments within certain domains in the RC and the antenna. (Also within the antenna of PSI and PSII, different exciton domains exist.[15,64])

Excitation energy transfer between the antenna and the RC occurs between excited states that are delocalized in different domains. The transfer of excitation energy between two such states, say, $|M_a\rangle$ and $|N_b\rangle$, is described by generalized Förster theory, giving for the rate constant[110,111]

$$k_{M_a \to N_b} = 2\pi \frac{\left|V_{M_a N_b}\right|^2}{\hbar^2} \int_{-\infty}^{\infty} d\omega D'_{M_a}(\omega) D_{N_b}(\omega), \quad (15.51)$$

with the interdomain excitonic coupling

$$V_{M_a N_b} = \sum_{m_a, n_b} c_{m_a}^{(M_a)} c_{n_b}^{(N_b)} V_{m_a n_b}, \quad (15.52)$$

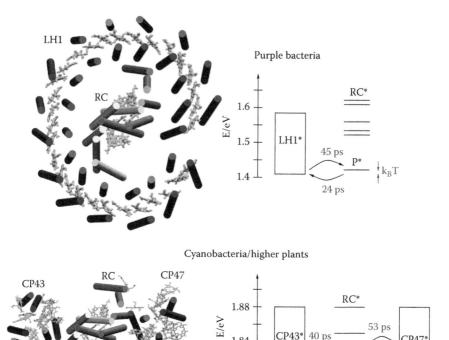

Figure 15.8 Upper part: RC and core antenna complex LH1 of the purple bacterium *Rhodopseudomonas palustris*.[115] Transmembrane helices (red and yellow cylinders) are depicted together with the macrocycles of bacteriochlorophyll a pigments (green). The energy levels of the excited states of LH1 and RC and the transfer times between them, as obtained from an analysis of time-resolved spectra,[116] are shown in the right hand side. Lower part: RC and core antenna complexes CP43 and CP47 of PSII of the cyanobacterium *Thermosynechococcus vulcanus*.[3] Transmembrane helices (purple, yellow, and red cylinders) are depicted together with the macrocycles of chlorophyll a (and pheophytin a) pigments (green). The energy levels of excited states of CP43, RC, and CP47 and the transfer times CP43 ↔ RC and CP47 ↔ RC, as calculated,[15] are shown in the right hand side. k_BT denotes the thermal energy at room temperature. The transfer times were obtained from structure-based modeling of excitation energy transfer and trapping in PSII core complexes.[24]

which contains the coefficients of the intradomain exciton states $c_{k_c}^{(K_c)}$ and the interdomain excitonic couplings $V_{m_a n_b}$ between pigments m_a in domain a and n_b in domain b. If the distance between the centers of the domains is large compared to their extensions, the coupling $V_{M_a N_b}$ equals the dipole–dipole coupling between the exciton transition dipole moments $\vec{d}_{M_a}$ and $\vec{d}_{N_b}$, where $\vec{d}_{M_c} = \sum_{m_c} c_{m_c}^{(M_c)} \vec{d}_{m_c}$ and $c = a, b$. Hence, for large interdomain distances, the original Förster result[112,113] is recovered since each domain can be considered as a supermolecule. If the domains are closer, also optically dark exciton states can contribute to the transfer.[114]

The integral in Equation 15.51 contains an overlap of the lineshape function $D'_{M_a}(\omega)$ for fluorescence of the donor domain a in the antenna and the lineshape function $D_{N_b}(\omega)$ of the absorbance of the acceptor domain in the RC. Roughly, these lineshape functions contain the density of exciton states in the different domains. Since exciton relaxation between the different delocalized states of a domain is much faster than interdomain exciton transfer, the latter starts from an equilibrated manifold of exciton states of the donor domain in the antenna, and the rate constant for transfer of excitation energy from domain a in the antenna to domain b in the RC reads

$$k_{a \to b} = \sum_{M_a, N_b} f(M_a) k_{M_a \to N_b} \qquad (15.53)$$

with the Boltzmann factor $f(M_a) = \exp(-E_{M_a}/k_\mathrm{B}T)/\sum_{K_a}\exp(-E_{K_a}/k_\mathrm{B}T)$. From Equations 15.51 and 15.53, it follows that an efficient trapping of excitation energy by the RC requires excited states $|N_b\rangle$ of the latter that are in resonance with the low-energy exciton states $|M_a\rangle$ in the antenna. We note that exciton transfer to the RC may compete with exciton transfer between different domains in the antenna. If the latter is fast, an effective rate constant

$$k_{\mathrm{Ant}\to\mathrm{RC}} = \sum_a^{a\neq\mathrm{RC}}\sum_{M_a}\sum_{N_\mathrm{RC}}\tilde{f}(M_a)k_{M_a\to N_\mathrm{RC}} \quad \text{for}$$

exciton transfer between the antenna and the RC can be defined, where the Boltzmann factor $\tilde{f}(M_a) = \exp(-E_{M_a}/k_\mathrm{B}T)/\sum_c\sum_{K_c}\exp(-E_{K_c}/k_\mathrm{B}T)$ now takes into account exciton equilibration between all exciton domains c in the antenna prior to exciton transfer to the RC.[24]

In purple bacteria, ring-shaped pigment–protein complexes serve as peripheral (LH2) and core (LH1) light-harvesting complexes (for review, see Reference 117). The LH1 complex, surrounding the RC in purple bacteria (Figure 15.8, upper part) absorbs at around 875 nm but also contains dark exciton states for efficient harvesting of excitation energy from the peripheral LH2 complex. After the excitation energy has arrived in the LH1 complex, it relaxes within the exciton manifold and afterwards is transferred to the low-energy exciton state of the special pair in the RC absorbing at 870 nm. This low-energy excited state of the special pair in the RC is the only one in resonance with the low-energy exciton states of LH1 (Figure 15.8), because of the short-range effects discussed earlier (Figure 15.7). Obviously, the coupling between pigments is utilized in the antenna system of purple bacteria to create an excitation energy funnel. Whereas in the peripheral LH2 antenna and the core antenna LH1 variations in the long-range excitonic coupling are used, the energy sink in the RC is caused by electron exchange effects in the special pair, as discussed earlier. The mixing with CT states, in addition, leads to a large homogeneous broadening of the low-energy exciton state[109] thereby helping to provide resonance with the low-energy exciton states of the LH1 antenna for efficient energy transfer.[118]

In cyanobacteria and higher plants, a different strategy is used for collecting light. The pigments in the core antenna are not coupled as strongly as in the LH2 and LH1 antenna, and their mutual orientation is much less regular. In this way the domains of strongly coupled pigments are smaller, and not

the excitonic coupling but the pigment–protein coupling determines the flow of excitation energy. Site energies of the pigments have been fitted to reproduce linear optical spectra of the CP43, CP47, and D1D2cyt$b$559 (RC) subunits, and these parameters have been applied afterwards in the description of nonlinear optical spectra revealing good agreement with experimental data.[24,119] Independent support for the site energies comes from quantum chemical/electrostatic calculations.[19,120] Interestingly, not all energy sinks of the core antenna are located close to the RC pigments; in particular, the lowest energy state of the CP47 complex is localized at a chlorophyll far away from the RC.[24,121] The exciton Hamiltonian of the CP43 and CP47 subunits has been evaluated by measuring circularly polarized fluorescence at low temperature.[121,122] Whereas full support was obtained for the Hamiltonian of CP43 as suggested[24,119,120]; in case of CP47, a new lowest-energy site was identified,[121] which is, however, still located at a large distance to the RC. Overall the RC is a very shallow trap for the excitation energy, and the ratio between effective rate constants for transfer between the RC and the core complexes and back approximately equals the inverse ratio of the number of pigments.[24] Hence, the entropic factor determines the free-energy difference. According to structure-based calculations of energy transfer and trapping by primary ET, the transfer from the antenna to the RC takes about 50 ps, whereas the back transfer is about a factor of three faster.[24] Energy transfer times up to one order of magnitude faster had been inferred earlier in the ERPE model by Holzwarth and coworkers.[23,123] In the ERPE model, the slow time constants in the fluorescence decay arise from back ET processes. As noted by van der Weij–de Wit et al.,[124] both the transfer-to-the-trap limited model and the ERPE model can fit the fluorescence decay equally well, but the transfer-to-the-trap limited model, in the light of the crystal structure revealing large distances between the RC and the antenna pigments, is more realistic. The first direct experimental proof of the slow transfer between the CP43 and CP47 core antennae and the RC, as predicted by the transfer-to-the-trap limited model,[24] was obtained from polarized VIS pump/ IR probe experiments on oriented single crystals of PSII core complexes.[125]

Despite the completely different organization of the core antennae in purple bacteria and higher plants/cyanobacteria very similar time constants were found for excitation energy transfer between the core antenna and the RC (Figure 15.8). Of course, the question arises, why do purple bacteria and cyanobacteria/higher plants possess a very similar RC and a completely different antenna. Obviously, there exist more strategies for efficient light harvesting than for transmembrane ET. Of course, the relevant interactions are much less distance dependent in excitation energy transfer as compared to ET, since no wavefunction overlap is required.

We will argue in Section 15.3.2 below that PSII had to reorganize the special pair in order to achieve a high enough redox potential for the splitting of water. This change made it impossible to use a LH1-type core antenna for efficient light harvesting, since the low-energy exciton states of the antenna would have been off resonant to the excited states of the RC.

15.3 DESIGN PRINCIPLES

15.3.1 Common design principles

The quantum efficiency of the primary charge separation and excitation energy transfer in photosynthesis under low light conditions is close to unity, that is, almost every photon absorbed by the PSs is converted into chemical energy in the RC. At high light intensities, antenna complexes are able to quench excitation energy in order to avoid photochemical damage of the RC. We will not discuss this aspect further in the present chapter but refer to Reference 126 and the references therein.

The key for high quantum efficiency, of course, is to suppress any competing side reaction. One limit is given by the fluorescence/intersystem crossing lifetime of the chlorin pigments that is in the 5 ns range. By limiting the antenna size to a couple of hundred pigments and by arranging them in optimal distances for fast excitation energy transfer, it became possible to realize an overall ≈ 100–200 ps[127,128] decay of excited states in practically all PSs, that is, there is a factor of 25–50 difference between the rate constants for excitation energy transfer/trapping and fluorescence/intersystem crossing.

Once the excitation energy has arrived in the RC, it is efficiently trapped by a fast primary CT reaction. To avoid oxidation of antenna pigments, the distances between antenna and RC pigments are relatively large giving rise to a slow transfer to the trap. Due to energetic and entropic (number of pigments) differences, the back transfer from the RC to the antenna is faster by a factor of 2–5 than the forward transfer (Figure 15.1). However, the primary ET reactions in the RC of PSII and purple bacteria are still a factor of 30 and 7, respectively, faster than the back excitation energy transfer to the antenna.

A fast primary CT puts limits on the distance between the primary electron donor and acceptor. According to the simple Dutton ruler (Equation 15.50), the maximum inverse rate constant (i.e., assuming $-\Delta G^{(0)} = \lambda$) for an edge-to-edge distance of 5 Å is 1 ps; whereas for twice the distance, it is already 100 ns, that is, much slower than even the fluorescence/intersystem crossing lifetime of the pigments. The exponential distance dependence of the ET matrix element gives rise to this dramatic decrease of the rate constant with distance. Therefore, multiple ET steps are required for transmembrane ET, that is, over a distance of a few nm.

As seen in Figure 15.1, the differences in free energy for most of the ET steps are large compared to the thermal energy, thus suppressing a back ET reaction, and the charge recombination reactions are at least a factor of 100 slower than the respective forward ET reactions. An interesting question is: What is the mechanism behind the large difference between forward ET and charge recombination rates? With increasing distance (i.e., number of ET steps) between the electron and the hole, the electronic coupling factor for the recombination reaction becomes of course smaller than the electronic factor for shifting the electron to the neighboring pigment. For the first two ET steps, however, a different explanation has to be found. The primary ET reaction has to compete with exciton back transfer from the RC to the antennae. Hence, the relatively large separation between RC and antenna pigments not only serves to prevent an unwanted oxidation of antenna pigments but also to keep the exciton back transfer rate small enough, such that every exciton that reaches the RC can be trapped by primary ET.

After primary ET, electron and hole are just located on neighboring pigments, and therefore, the electronic coupling for the next ET step can be assumed to be similar to the coupling that is responsible for recombination of the electron–hole pair recovering the electronic ground state. If we use the simple Dutton ruler, which is based on the Marcus expression, and assume that the forward rate is activationless, we obtain a ratio $r = 10^{3.1(\Delta G^0)^2/(4\lambda k_B T)}$ of rate constants of forward and recombination reactions (assuming the electronic couplings of the two reactions are the same). Since charge recombination reactions occur in the inverted region of ET, we have $|\Delta G^0| \gg \lambda$ and hence a large ratio r. The ratio is particularly large for small λ, which is realized by the small polarity of the protein environment inside the membrane.[129] Taking into account the nuclear tunneling effects, of course, diminishes the inverted effect. However, the coupling of electrons to high-frequency intramolecular vibrations of chlorophylls and bacteriochlorophylls is small,[130] and hence the Huang–Rhys factor S_v in Equation 15.47 is small. The weak temperature dependence of some recombination reactions[131] suggests that, indeed, nuclear tunneling (Figure 15.6) is the dominating mechanism.

In summary, we may conclude that the common design principle of all RCs are (1) a large spatial separation between antenna and RC pigments to avoid oxidation of the antenna and to ensure the primary ET is fast compared to exciton back transfer to the antenna; (2) a stepwise ET, where the separation between electron and hole is increasing with every ET step and the free energy of involved states is decreasing, thus making recombination reactions more and more unlikely, and suppressing back ET, respectively; and (3) a nonpolar environment and a weak intramolecular vibronic coupling of the pigments giving rise to small Franck–Condon factors between the vibronic states of the charge-separated states and the electronic ground state, thus suppressing charge recombination and nuclear tunneling in the inverted regime of ET.

15.3.2 Special adaptations of photosystem II for water splitting

The water-splitting reaction requires a high redox potential of the RC chlorophylls in PSII.[18] Since pigment dimerization, in general, increases the energy of the highest occupied molecular orbital (HOMO), it diminishes the redox potential of the oxidized molecular system for attracting electrons. Therefore, in the course of evolution from purple bacteria to cyanobacteria, it was necessary to disrupt the wavefunction overlap between the two special pair chlorophylls and to introduce an asymmetry in HOMO energies of the two special pair chlorophylls. The latter should be larger than the remaining short-range coupling, such that the hole becomes localized on the special pair pigment with the higher HOMO (lower redox potential).

Indeed, there is a significant tilt of P_{D1} relative to P_{D2} with respect to their purple bacterial counterpart $P_A - P_B$ leading to a more staggered arrangement of π-electron wavefunction than the eclipsed one seen in purple bacteria (Figure 15.7), as has been discussed earlier. If the short-range coupling becomes much weaker than the difference in local redox potentials, the hole will localize. From QM/MM calculations, a P_{D1}^+ / P_{D2}^+ charge ratio of 80/20 was obtained, where asymmetric electrostatic interaction of the two special pair chlorophylls with the protein environment were found responsible for the redox potential difference of about 100 meV between P_{D1} and P_{D2}.[132]

Site directed mutagenesis experiments[133] were performed on bRC, where one of the two special pair bacteriochlorophylls was replaced by bacteriopheophytin, which has a sufficiently higher redox potential to localize the hole on the remaining bacteriochlorophyll. The redox potential of the latter was found to increase by 140 meV in this mutant. This value is in nice agreement with DFT calculations on the special pair of PSII by Noguchi and coworkers,[134] who used a polarizable continuum model for the protein and found a decrease of the redox potential of the dimer by about 140 meV with respect to that of the monomers. Since no explicit protein charge density was included, the local redox potentials of the isolated monomers were so close that the hole was found to be delocalized in the dimer. From this result, however, it could be concluded that charge localization by switching on asymmetric pigment–protein coupling would increase the redox potential of the special pair by up to 140 meV (or to a lower value, if charge localization is not complete).

Another increase of the redox potential between purple bacteria and cyanobacteria/higher plants occurred by the replacement of bacteriochlorophyll by chlorophyll (and bacteriopheophytin by pheophytin). The redox potential of chlorophyll measured in solution is higher by about 160 meV than that of bacteriochlorophyll.[135,136] An additional 200 meV increase in redox potential was achieved by changing the electrostatic coupling with the protein environment,[99] as has been described earlier. The latter electrostatic coupling also explains a large part of the remarkable 0.8 V difference in oxidation potential between the special pairs of PSI and PSII.[99] The different influences on the redox potentials of the special pairs, discussed earlier, are summarized in Figure 15.9.

The decreased wavefunction overlap in the "special pair" diminishes the mixing of excited states with CT states to such an extent that the excitation energy sink is not located at the special pair but has large contributions from the accessory chlorophyll of the D1-branch Chl_{D1}[13,19] (Figure 15.2). The almost perfect overlay of macrocycles of the B_A–H_A dimer of bRC and the Chl_{D1}–$Pheo_{D1}$ dimer of PSII (e.g., Reference 24) suggests that the respective ET matrix

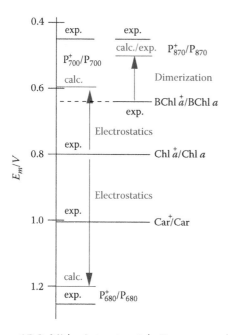

Figure 15.9 Midpoint potentials E_m measured for the oxidation of Car (1.1 V)[137] in aqueous micellar solution; Chl a (0.80 V) and BChl a (0.64 V) in CH_2Cl_2[135,136] are compared with oxidation potentials inferred from experimental data on the state P_{680}^+ of PSII (1.26 V),[138] P_{700}^+ (0.45 V)[139] of PSI, and P_{870}^+ (0.45 eV)[140] of the bacterial RC. The blue arrows represent electrostatic calculations of the shift in oxidation potential between Chl a in solution and the special pairs of the RCs of PSI and PSII[99] and estimates of the effect of dimerization (hole delocalization) on the oxidation potential of the special pair in bacterial RCs from mutant studies[133] and quantum chemical calculations.[134] Please note that the electrostatic contribution to the shift in oxidation potential of the special pair in bacterial RCs was calculated to be very small.[99]

elements[101,102] are similar, and therefore Chl_{D1} seems to be the most efficient primary electron donor in PSII. Whereas in purple bacteria, the different free energies of the two primary radical pair states $P^+B_A^-$ and $P^+B_B^-$ most likely are responsible for unidirectional Et along the A-branch; in PSII the difference in excitation energies between the two accessory chlorophylls Chl_{D1} and Chl_{D2} may be an important factor.[13,15] Recently, it was shown that this asymmetry is caused by the difference in electrostatic interactions of the pigments with their protein environments.[19]

The high redox power of the central pigments in PSII RC and the side reactions with molecular oxygen represent a major challenge for the stability of the system.[18,73,141–143] Under high light intensity the mean *in vivo* lifetime of the D1-protein is only about half an hour.[141,142] During this time the D1-polypeptide is irreversibly damaged and replaced by a new polypeptide. Different molecular mechanisms are discussed for the damage.[142,144] One possibility is an oxidation of pigments or parts of the protein by the highly oxidizing state P_{680}^+ and a subsequent degradation of the unstable cationic states. Another possibility is the formation of Chl triplet states (Chl) by intersystem crossing/charge recombination reactions and the subsequent reaction with triplet oxygen leading to the formation of the highly reactive singlet oxygen, which damages the D1-protein. Experimental studies and calculations have revealed that these triplet states accumulate at Chl_{D1} in PSII (for review, see Reference 16).

In the antenna the triplet states ^3Chl (^3BChl) are quenched by exchange coupling with Cars. There are no Cars found in van der Waals contact of any of the six core pigments of the RC in PSII. The high redox power of these pigments, necessary for the water oxidation, would lead to unwanted oxidation of the Cars (for oxidation potentials, see Figure 15.9). A difference between PSII and the purple bacteria is that in the former the triplet state localizes at the accessory Chl_{D1},[13,145] whereas in the latter at the special pair P_A–P_B.[59] In purple bacteria the RC triplet energy is quenched via energy transfer to B_B, which is in van der Waals contact with a carotenoid[1,146] (Figure 15.1). The lifetime of the triplet state in the RC of PSII was found to decrease by two orders of magnitude if Q_A is in the reduced state.[42] This finding led Noguchi to conclude that triplet localization on Chl_{D1} is functionally important for efficient triplet quenching by Q_A^-.[147] The molecular mechanism of this quenching is still unknown.

One possibility to avoid damage by P_{D1}^+ is a controlled secondary ET that most likely involves Cars[148] (for review, see Reference 149). From calculations of Q_A^- Car$^+$ - Q_A Car optical difference spectra and comparison with experimental data, it was concluded that only Car$_{D2}$ and not Car$_{D1}$ is involved in secondary ET in PSII.[44] Secondary ET involving Car$_{D2}$ leads to a controlled reduction of P_{680}^+ by cyt$b$559, which may be completed by a subsequent charge recombination with the reduced Q_B.[150] A second photoprotective function of Car$_{D2}$ discussed[144,149] is the quenching of singlet oxygen, which is generated by reaction of triplet oxygen with triplet states of Chl.

Another protection mechanism concerns the slowing down of primary CT by the reduced Q_A, and thereby the escape of excitation energy from the RC to the antenna.[24] This effect is enhanced in PSII because of the change in primary electron donor from the special pair in purple bacteria to Chl$_{D1}$ in PSII, due to the close distance between the latter and Q_A (as discussed earlier for the triplet quenching). Once the excitation energy has escaped the RC, it most likely is converted into a triplet state in the antenna that can be quenched by a nearby carotenoid. Alternatively, the excitation energy of antenna Chls could be quenched directly in the single state by internal conversion. In this respect, it is interesting to note that the PSII core antenna does not represent a simple excitation energy funnel but contains a low energy trap state far away from the RC that may be involved in photoprotection.[119,121]

15.4 SUMMARY

In this chapter, we understood that there are certain structural requirements that are important for efficient transmembrane ET, namely, (1) a chain of cofactors, (2) a coupling to vibrational degrees of freedom that is large enough to allow for activation less ET with sufficiently large free-energy gaps to suppress back ET, and (3) a small enough coupling to vibrational degrees of freedom to suppress nuclear tunneling effects in recombination reactions, occurring in the inverted region of ET. In addition, we saw life with molecular oxygen required special adaptations in the evolution of PSs: (1) tuning of the electronic properties of the special pair, (2) use of a different light-harvesting strategy, and (3) different photoprotective mechanisms.

ACKNOWLEDGMENT

It is a pleasure to acknowledge the support by the Austrian Science Fund (FWF): P 24774-N27.

REFERENCES

1. T. O. Yeates, H. Komiya, A. Chirino, D. C. Rees, J. P. Allen, and G. Feher, Structure of the reaction center from *Rhodobacter sphaeroides* R-26 and 2.4.1: Protein-cofactor (bacteriochlorophyll, bacteriopheophytin, and carotenoid) interactions, *Proc. Natl. Acad. Sci. USA*, 1988, **85**, 7993–7997.

2. A. Guskov, J. Kern, A. Gabdulkhakov, M. Broser, A. Zouni, and W. Saenger, Cyanobacterial photosystem II at 2.9 Å resolution and the role of quinones, lipids, channels and chloride, *Nat. Struct. Mol. Biol.*, 2009, **16**, 334–342.

3. Y. Umena, K. Kawakami, J.-R. Shen, and N. Kamiya, Crystal structure of oxygenevolving photosystem II at a resolution of 1.9 Å, *Nature*, 2011, **473**, 55–60.

4. P. Jordan, P. Fromme, O. Klukas, H. T. Witt, W. Saenger, and N. Krauß, Three dimensional structure of cyanobacterial photosystem I at 2.5 Å resolution, *Nature*, 2001, **411**, 909–917.

5. W. W. Parson, Electron donors and acceptors in the initial steps of photosynthesis in purple bacteria: a personal account, *Photosynth. Res.*, 2003, **76**, 81–92.

6. F. Rappaport and B. A. Diner, Primary photochemistry and energetics leading to the oxidation of the (Mn$_4$)Ca cluster and to the evolution of molecular oxygen in photosystem II, *Coord. Chem. Rev.*, 2008, **252**, 259–272.

7. P. Joliot and A. Joliot, In *vivo* analysis of the electron transfer within photosystem I: Are the two phylloquinones involved?, *Biochemistry*, 1999, **38**, 11130–11136.

8. M. Guergova-Kuras, B. Boudreaux, A. Joliot, P. Joliot, and K. Redding, Evidence for two active branches for electron transfer in photosystem I, *Proc. Natl. Acad. Sci. USA*, 2001, **98**, 4437–4442.

9. J. A. Bautista, F. Rappaport, M. Guergova-Kuras, R. O. Cohen, J. H. Golbeck, J. Y. Wang, D. Beal, and B. A. Diner, Biochemical and biophysical characterization of photosystem I

from phytoene desaturase and zeta-carotene desaturase deletion mutants of *Synechocystis Sp.* PCC 6803, *J. Biol. Chem.*, 2005, **280**, 20030–20041.

10. Y. Li, A. van der Est, M. G. Lucas, V. M. Ramesh, F. Gu, A. Petrenko, S. Lin, A. N. Webber, F. Rappaport, and K. Redding, Directing electron transfer within photosystem I by breaking H-bonds in the cofactor branches, *Proc. Natl. Acad. Sci. USA*, 2006, **103**, 2144–2149.

11. F. Rappaport, B. A. Diner, and K. Redding, Optical measurements of secondary electron transfer in photosystem I, in: *The Light-Driven Water-Plastocyanin Ferredoxin Oxidoreductase* (Golbeck, J. H. Ed.), Springer, Dordrecht, the Netherlands, 2006, p. 223–244.

12. K. Redding and A. van der Est, The directionality of electron transport in photo-system I, in: *The Light-Driven Water-Plastocyanin Ferredoxin Oxidoreductase* (Golbeck, J. H. Ed.), Springer, Dordrecht, the Netherlands, 2006, p. 413–437.

13. G. Raszewski, W. Saenger, and T. Renger, Theory of optical spectra of photosystem II reaction centers: Location of the triplet state and the identity of the primary electron donor, *Biophys. J.*, 2005, **88**, 986–998.

14. V. I. Novoderezhkin, J. P. Dekker, and R. Van Grondelle, Mixing of exciton and chargetransfer states in photosystem II reaction centers: Modeling of Stark spectra with Modified Redfield theory, *Biophys. J.*, 2007, **93**, 1293–1311.

15. G. Raszewski, B. A. Diner, E. Schlodder, and T. Renger, Spectroscopic properties of reaction center pigments in photosystem II core complexes: Revision of the multimer model, *Biophys. J.*, 2008, **95**, 105–119.

16. T. Renger and E. Schlodder, Primary photophysical processes in photosystem II: Bridging the gap between crystal structure and optical spectra, *ChemPhysChem*, 2010, **11**, 1141–1153.

17. T. Renger and E. Schlodder, Optical properties, excitation energy and primary charge transfer in photosystem II: Theory meets experiment, *J. Photochem. Photobiol. B*, 2011, **104**, 126–141.

18. T. Cardona, A. Sedoud, N. Cox, and A. W. Rutherford, Charge separation in photosystem II: A comparative and evolutionary overview, *Biochim. Biophys. Acta*, 2012, **1817**, 26–43.

19. F. Müh, M. Plöckinger, and T. Renger, Electrostatic asymmetry in the reaction center of photosystem II, *J. Phys. Chem. Lett.*, 2017, **8**, 850–858.

20. S. R. Greenfield, M. Seibert, Govindjee, and M. R. Wasielewski, Direct measurments of the effective rate constant for primary charge separation in isolated photosystem II reaction centers, *J. Phys. Chem. B*, 1997, **101**, 2251–2255.

21. S. R. Greenfield, M. Seibert, Govindjee, and M. R. Wasielewski, Time-resolved absorption changes of the pheophytin Q_x band in isolated photosystem II reaction centers at 7 K: Energy transfer and charge separation, *J. Phys. Chem. B*, 1999, **103**, 8364–8374.

22. M. L. Groot, N. P. Pawlowicz, L. J. G. W. Van Wilderen, J. Breton, I. H. M. van Stokkum, and R. van Grondelle, Initial electron donor and acceptor in isolated photosystem II reaction centers identified with femtosecond mid-IR spectroscopy, *Proc. Natl. Acad. Sci. USA*, 2005, **102**, 13087–13092.

23. A. R. Holzwarth, M. G. Müller, M. Reus, M. Nowaczyk, J. Sander, and M. Rögner, Kinetics and mechanism of electron transfer in intact photosystem II and in the isolated reaction center: Pheophytin is the primary electron donor, *Proc. Natl. Acad. Sci. USA*, 2006, **103**, 6895–6900.

24. G. Raszewski and T. Renger, Light harvesting in photosystem II core complexes is limited by the transfer to the trap: Can the core complex turn into a photoprotective mode?, *J. Am. Chem. Soc.*, 2008, **130**, 4431–4446.

25. I. Shelaev, F. Gostev, V. A. Nadtochenko, A. Shkuropatov, A. A. Zabelin, M. Mamedov, A. Semenov, O. Sarkisov, and V. Shuvalov, Primary light-energy conversion in tetrameric chlorophyll structure of photosystem II and bacterial reaction centers: II. Femto- and picosecond charge separation in PSII D1/D2/Cyt b559 complex, *Photosynth. Res.*, 2008, **98**, 95–103.

26. I. Shelaev, F. Gostev, M. Vishnev, A. Shkuropatov, V. Ptushenko, M. Mamedov, O. Sarkisov, V. Nadtochenko, A. Semenov, and V. Shuvalov, P_{680} ($P_{D1}P_{D2}$) and Chl_{D1} as alternative electron donors in photosystem II core complexes and isolated reaction centers, *J. Photochem. Photobiol. B*, 2011, **104**, 44–50.

27. E. Romero, I. H. M. van Stokkum, V. I. Novoderezhkin, J. P. Dekker, and R. van Grondelle, Two different charge separation pathways in photosystem II, *Biochemistry*, 2010, **49**, 4300–4307.

28. V. I. Novoderezhkin, E. Romero, J. P. Dekker, and R. van Grondelle, Multiple charge separation pathways in photosystem II: Modeling of transient absorption kinetics, *ChemPhysChem*, 2011, **12**, 681–688.

29. V. I. Novoderezhkin, E. Romero, and R. van Grondelle, How exciton-vibrational coherences control charge separation in the photosystem II reaction center, *Phys. Chem. Chem. Phys.*, 2015, **17**, 30828–30841.

30. E. Krausz, J. L. Hughes, P. Smith, R. Pace, and S. P. Årsköld, Oxygen-evolving photosystem II core complexes: A new paradigm based on the spectral identification of the charge separating state, the primary acceptor and assignment of low-temperature fluorescence, *Photochem. Photobiol. Sci.*, 2005, **4**, 744–753.

31. J. L. Hughes, P. Smith, R. Pace, and E. Krausz, Charge separation in photosystem II core complexes induced by 690–730 nm excitation at 1.7 K, *Biochim. Biophys. Acta*, 2006, **1757**, 841–851.

32. A. Gelzinis, L. Valkunas, F. D. Fuller, J. P. Ogilvie, S. Mukamel, and D. Abramavicius, Tight-binding model of the photosystem II reaction center: Application to two-dimensional electronic spectroscopy, *New J. Phys.*, 2013, **15**, 075013.

33. E. Romero, R. Augulis, V. I. Novoderezhkin, M. Ferreti, J. Thieme, D. Zigmantas, and R. van Grondelle, Quantum coherence in photosynthesis for efficient solar-energy conversion, *Nat. Phys.*, 2014, **10**, 676–682.

34. E. Romero, V. I. Novoderezhkin, and R. van Grondelle, Quantum design of photosynthesis for bio-inspired solar energy conversion, *Nature*, 2017, **543**, 355–365.

35. P. J. Booth, B. Crystall, L. B. Giorgi, J. Barber, D. R. Klug, and G. Porter, Thermodynamic properties of D1/D2/Cytochrome-B-559 reaction centers investigated by time-resolved fluorescence measurements, *Biochim. Biophys. Acta*, 1990, **1016**, 141–152.

36. B. Hillmann, I. Moya, and E. Schlodder, Temperature dependence of the recombination fluorescence from PS II core complexes, in: *Photosynthesis: From Light to Biosphere; Proceedings of the 10th International Congress on Photosynthesis* (P. Mathis Ed.), Kluwer Academic Publishers, Dordrecht, the Netherlands, 1995, pp. 603–606.

37. N. W. Woodbury and W. W. Parson, Nanosecond fluorescence from chromatophores of *Rhodopseudomonas sphaeroides* and *Rhodospirillum rubrum*, *Biochim. Biophys. Acta*, 1986, **850**, 197–210.

38. G. H. Schatz, H. Brock, and A. R. Holzwarth, Kinetic and energetic model for the primary processes in photosystem II, *Biophys. J.*, 1988, **54**, 397–405.

39. A. M. Nuijs, H. J. van Gorkom, J. J. Plijter, and L. N. M. Duysens, Primary charge separation and excitation of chlorophyll a in photosystem II particles from spinach as studied by picosecond absorbance difference spectroscopy, *Biochim. Biophys. Acta*, 1986, **848**, 167–175.

40. P. Mathis and A. Vermeglio, Chlorophyll radical cation in photosystem II of chloroplasts-Millisecond decay at low temperatures, *Biochim. Biophys. Acta*, 1975, **396**, 371–381.

41. B. Hillmann and E. Schlodder, Electron transfer reactions in Photosystem II core complexes from *Synechococcus* at low temperature-difference spectrum of $P_{680}^+Q_A^-/P_{680}Q_A$ at 77 K, *Biochim. Biophys. Acta*, 1995, **1231**, 76–88.

42. F. Van Mieghem, K. Brettel, B. Hillmann, A. Kamlowski, A. W. Rutherford, and E. Schlodder, Charge recombination reactions in photosystem II. 1. Yields, recombination pathways, and kinetics of the primary pair, *Biochemistry*, 1995, **34**, 4789–4813.

43. F. Van Mieghem, K. Satoh, and A. W. Rutherford, A chlorophyll tilted 30° relative to the membrane in the photosystem II reaction center, *Biochim. Biophys. Acta*, 1991, **1058**, 379–385.

44. E. Schlodder, T. Renger, G. Raszewski, W. J. Coleman, P. J. Nixon, R. O. Cohen, and B. A. Diner, Site directed mutations at D1-Thr179 of photosystem II in *Synechocystis* PCC 6803 modify the spectroscopic properties of the accessory chlorophyll in the D1-branch of the reaction center, *Biochemistry*, 2008, **47**, 3143–3154.

45. W. O. Feikema, P. Gast, I. B. Klenina, and I. I. Proskuryatov, EPR characterization of the triplet state in photosystem II reaction centers with singly reduced primary acceptor Q_A, Biochim. Biophys. Acta, 2005, **1709**, 105–112.

46. W. Zinth and J. Wachtveitl, The first picoseconds in bacterial photosynthesis-ultrafast electron transfer for efficient conversion of light energy, ChemPhysChem, 2005, **6**, 871–880.

47. A. Warshel and W. W. Parson, Spectroscopic properties of photosynthetic reaction centers. 1. Theory, J. Am. Chem. Soc. USA, 1987, **109**, 6143–6151.

48. M. E. Madjet, F. Müh, and T. Renger, Deciphering the influence of short-range electronic couplings on optical properties of molecular dimers: Application to special pairs in photosynthesis, J. Phys. Chem. B, 2009, **113**, 12603–12614.

49. C. Kirmaier, C. He, and D. Holton, Manipulating the direction of electron transfer in the bacterial reaction center by swapping Phe for Tyr near $BChl_M$ and Tyr for Phe near $BChl_L$ (M208), Biochemistry, 2001, **40**, 12132–12139.

50. W. Holzapfel, U. Finkele, W. Kaiser, D. Oesterhelt, H. Scheer, H. U. Stilz, and W. Zinth, Initial electron-transfer in the reaction center from Rhodobacter sphaeroides, Proc. Natl. Acad. Sci. USA, 1990, **87**, 5168–5172.

51. S. F. Fischer and P. O. J. Scherer, On the early charge separation and recombination processes in bacterial reaction centers, Chem. Phys., 1987, **115**, 151–158.

52. A. Warshel, S. Creighton, and W. W. Parson, Electron-transfer pathways in the primary event of bacterial photosynthesis, J. Phys. Chem., 1988, **92**, 2696–2701.

53. D. Holten, M. W. Windsor, W. W. Parson and J. P. Thornber, Primary photochemical processes in isolated reaction centers of Rhodopseudomonas viridis, Biochim. Biophys. Acta, 1978, **501**, 112–126.

54. M. S. Graige, G. Feher, and M. Y. Okamura, Conformational gating of the electron transfer reaction $Q_A^- Q_B \rightarrow Q_A Q_B^-$ in bacterial reaction centers of Rhodobacter sphaeroides determined by a driving force assay, Proc. Natl. Acad. Sci. USA, 1998, **95**, 11679–11684.

55. W. W. Parson and A. Warshel, Dependence of photosynthetic electron transfer kinetics on temperature and energy in a density matrix model, J. Phys. Chem. B, 2004, **108**, 10474–10483.

56. M. R. Gunner and P. L. Dutton, Temperature and $-\Delta G_0$ dependence of the electron transfer from BPh^- to Q_A in reaction centers from Rhodobacter sphaeroides with different quinones as Q_A, J. Am. Chem. Soc., 1989, **111**, 3400–3412.

57. V. A. Shuvalov and W. W. Parson, Energies and kinetics of radical pairs involving bacteriochlorophyll and bacteriopheophytin in bacterial reaction centers, Proc. Natl. Acad. Sci. USA, 1981, **78**, 957–961.

58. A. Ogrodnik, H. W. Kruger, H. Orthuber, R. Haberkorn, M. E. Michel-Beyerle, and H. Scheer, Recombination dynamics in bacterial photosynthetic reaction centers, Biophys. J., 1982, **39**, 91–95.

59. H. Frank, J. D. Bolt, S. M. B. Costa, and K. Sauer, Electron paramagnetic resonance detection of carotenoid triplet states, J. Am. Chem. Soc., 1980, **102**, 4893–4898.

60. R. O. Cohen, G. Shen, J. H. Golbeck, W. Xu, P. R. Chitnis, A. I. Valieva, A. van der Est, Y. Pushkar, and D. Stehlik, Evidence for asymmetric electron transfer in cyanobacterial photosystem I: Analysis of a methionine-to-leucine mutation of the ligand to the primary electron acceptor A_0, Biochemistry, 2004, **43**, 4741–4754.

61. N. Dashdorj, W. Xu, R. O. Cohen, J. H. Golbeck, and S. Savikhin, Asymmetric electron transfer in cyanobacterial photosystem I. Charge separation and secondary electron transfer dynamics of mutations near the primary electron acceptor A_0, Biophys. J., 2005, **88**, 1238–1249.

62. H. Kaes, P. Fromme, H. T. Witt, and W. Lubitz, Orientation and electronic structure of the primary donor radical cation P_{700^+} in photosystem I: A single crystals EPR and ENDOR study, J. Phys. Chem. B, 2001, **105**, 1225–1239.

63. H. Witt, E. Schlodder, C. Teutloff, J. Niklas, E. Bordignon, D. Carbonera, S. Kohler, A. Labahn, and W. Lubitz, Hydrogen bonding to P_{700}: Site-directed mutagenesis of threonine A739 of Photosystem I in Chlamydomonas reinhardtii, Biochemistry, 2002, **41**, 8557–8569.

64. J. Adolphs, F. Müh, M. E. Madjet, M. Schmidt am Busch, and T. Renger, Structure-based calculations of optical spectra of photosystem I suggest an asymmetric lightharvesting process, *J. Am. Chem. Soc.*, 2010, **132**, 3331–3343.

65. M. Byrdin, I. Rimke, E. Schlodder, D. Stehlik, and T. A. Roelofs, Decay kinetics and quantum yields of fluorescence in photosystem I from *Synechococcus elongatus* with P_{700} in the reduced and oxidized state: Are the kinetics of excited state decay traplimited or transferlimited?, *Biophys. J.*, 2000, **79**, 992–1007.

66. B. Gobets and R. van Grondelle, Energy transfer and trapping in photosystem I, *Biochim. Biophys. Acta*, 2001, **1507**, 80–99.

67. I. Shelaev, F. Gostev, M. Mamedov, O. Sarkisov, V. A. Nadtochenko, A. A. Zabelin, V. Shuvalov, and A. Semenov, Femtosecond primary charge separation in *Synechocystis* sp. PCC 6803 photosystem I, *Biochim. Biophys. Acta*, 2010, **1797**, 1410–1420.

68. A. Semenov, I. Shelaev, F. Gostev, M. Mamedov, V. Shuvalov, O. Sarkisov, and V. A. Nadtochenko, Primary steps of electron and energy transfer in photosystem I: Effect of excitation pulse wavelength, *Biochem. Mosc.*, 2012, **77**, 1011–1020.

69. M. G. Müller, C. Slavov, R. Luthra, K. E. Redding, and A. R. Holzwarth, Independent initiation of primary electron transfer in the two branches of the photosystem I reaction center, *Proc. Natl. Acad. Sci. USA*, 2010, **107**, 4123–4128.

70. K. Brettel and W. Leibl, Electron transfer in photosystem I, *Biochim. Biophys. Acta*, 2001, **1507**, 100–114.

71. I. Karyagina, Y. Pushkar, D. Stehlik, A. van der Est, H. Ishikita, E.-W. Knapp, B. Jagannathan, R. Agalarov, and J. H. Golbeck, Contributions of the protein environment to the midpoint potentials of the A_1 phylloquinones and the F_x iron-sulfur cluster in photosystem I, *Biochemistry*, 2007, **46**, 10804–10816.

72. H. Ishikita and E.-W. Knapp, Redox potentials of quinones in both electron transfer branches of photosystem I, *J. Biol. Chem.*, 2003, **278**, 52002–52011.

73. A. W. Rutherford, A. Osyczka, and F. Rappaport, Back-reactions, short-circuits, leaks and other energy wasteful reactions in biological electron transfer: Redox tuning to survive life in O_2, *FEBS Lett.*, 2012, **586**, 603–616.

74. E. Schlodder, K. Falkenberg, M. Gergeleit, and K. Brettel, Temperature dependence of forward and reverse electron transfer from A_1^-, the reduced secondary electron acceptor in photosystem I, *Biochemistry*, 1998, **37**, 9466–9476.

75. K. Brettel, Electron transfer and arrangement of the redox cofactors in photosystem I, *Biochim. Biophys. Acta*, 1997, **1318**, 322–373.

76. E. Schlodder, F. Lendzian, J. Meyer, M. Cetin, M. Brecht, T. Renger, and N. V. Karapetyan, Long-wavelength limit of photochemical energy conversion in photosystem I, *J. Am. Chem. Soc.*, 2014, **136**, 3904–3918.

77. J. R. Reimers, M. Biczysko, D. Bruce, D. F. Coker, T. J. Frankcombe, H. Hashimoto, J. Hauer, et al., Challenges facing an understanding of the nature of low-energy excited states in photosynthesis, *Biochim. Biophys. Acta*, 2016, **1857**, 1627–1640.

78. V. May and O. Kühn, *Charge and Energy Transfer Dynamics in Molecular Systems: A Theoretical Introduction*, Wiley–VCH, Berlin, Germany, 2000.

79. R. A. Marcus and N. Sutin, Electron transfer in chemistry and biology, *Biochim. Biophys. Acta*, 1985, **811**, 265–322.

80. Y. Georgievskii, C.-P. Hsu, and R. A. Marcus, Linear response theory of electron transfer reactions as an alternative to the molecular harmonic oscillator model, *J. Chem. Phys.*, 1999, **110**, 5307–5317.

81. M. Lax, The Franck-Condon principle and its application to crystals, *J. Chem. Phys.*, 1952, **20**, 1752–1760.

82. R. Kubo, Generalized cumulant expansion method, *J. Phys. Soc. Jpn.*, 1962, **17**, 1100–1120.

83. H. Haken, *Quantum Field Theory of Solids: An Introduction*, North Holland Pub. Co., Amsterdam, the Netherlands, 1976.

84. J. Jortner, Temperature-dependent activation energy for electron transfer between biological molecules, *J. Chem. Phys.*, 1976, **64**, 4860–4867.

85. J. M. Peloquin, J. C. Williams, X. Lin, R. G. Alden, A. K. W. Taguchi, J. P. Allen, and N. W. Woodbury, Time-dependent thermodynamics during early electron transfer in reaction centers from *Rhodobacter sphaeroides*, *Biochemistry*, 1994, **33**, 8089–8100.

86. W. W. Parson, Z. T. Chu, and A. Warshel, Reorganization energy of the initial electron-transfer step in photosynthetic reaction centers, *Biophys. J.*, 1998, **74**, 182–191.

87. J. M. Kriegl and G. U. Nienhaus, Structural, dynamic, and energetic aspects of long-range electron transfer in photosynthetic reaction centers, *Proc. Natl. Acad. Sci. USA*, 2004, **101**, 123–128.

88. H. Wang, S. Lin, J. P. Allen, J. C. Williams, S. Blankert, C. laser, and N. W. Woodbury, Protein dynamics control the kinetics of initial electron transfer in photosynthesis, *Science*, 2007, **316**, 747–750.

89. A. B. Wohri, G. Katona, L. C. Johansson, E. Fritz, E. Malmerberg, M. Andersson, J. Vincent et al., Light-induced structural changes in a photosynthetic reaction center caught by Laue diffraction, *Science*, 2010, **328**, 630–633.

90. H. Sumi and R. A. Marcus, Dynamical effects in electron transfer reactions, *J. Chem. Phys.*, 1986, **84**, 4894–4914.

91. H. Wang, S. Lin, E. Katilius, C. Laser, J. P. Allen, J. C. Williams, and N. W. Woodbury, Unusual temperature dependence of photosynthetic electron transfer due to protein dynamics, *J. Phys. Chem. B*, 2009, **113**, 818–824.

92. N. Agmon and J. J. Hopfield, Transient kinetics of chemical reactions with bounded diffusion perpendicular to the reaction coordinate: Intramolecular processes with slow conformational changes, *J. Chem. Phys.*, 1983, **78**, 6947–6959.

93. A. Warshel and W. W. Parson, Dynamics of of biochemical and biophysical reactions: Insights from computer simulations, *Q. Rev. Biophys.*, 2001, **34**, 563–670.

94. T. Simonson, Gaussian fluctuations and linear response in an electron transfer protein, *Proc. Natl. Acad. Sci. USA*, 2002, **99**, 6544–6549.

95. C. C. Moser, J. M. Keske, K. Warncke, R. S. Farid, and P. L. Dutton, Nature of biological electron–transfer, *Nature*, 1992, **355**, 796–802.

96. C. P. Page, C. C. Moser, X. Chen, and P. L. Dutton, Natural engineering principles of electron tunneling in biological oxidation-reduction, *Nature*, 1999, **402**, 47–52.

97. W. W. Parson and A. Warshel, Calculations of electrostatic energies in proteins using microscopic, semimicroscopic and macroscopic models and free energy perturbation

approaches in: *Biophysical Techniques in Photosynthesis* (T. J. Aartsna and J. Matysik, Eds.), Springer, Dordrecht, the Netherlands, 2006, pp. 401–420.

98. G. M. Ullmann and E. W. Knapp, Electrostatic models for computing protonation and redox equilibria in proteins, *Eur. Biophys. J.*, 1999, **28**, 533–550.

99. H. Ishikita, W. Saenger, J. Biesiadka, B. Loll, and E.-W. Knapp, How photosynthetic reaction centers control oxidation power in chlorophyll pairs P_{680}, P_{700}, and P_{870}, *Proc. Natl. Acad. Sci. USA*, 2006, **103**, 9855–9860.

100. A. Warshel, Z. T. Chu, and W. W. Parson, Dispersed polaron simulations of electron transfer in photosynthetic reaction centers, *Science*, 1989, **246**, 112–116.

101. P. O. J. Scherer and S. Fischer, Quantum treatment of the optical spectra and the initial electron transfer process within the reaction center of *Rhodopseudomonas viridis*, *Chem. Phys.*, 1989, **131**, 115–127.

102. L. Y. Zhang and R. A. Friesner, *Ab initio* calculation of electronic coupling in the photosynthetic reaction center, *Proc. Natl. Acad. Sci. USA*, 1998, **95**, 13603–13605.

103. M. E. Van Brederode, M. R. Jones, F. Van Mourik, I. H. M. Van Stokkum, and R. Van Grondelle, A new pathway for transmembrane electron transfer in photosynthetic reaction centers of *Rhodobacter sphaeroides* not involving the excited special pair, *Biochemistry*, 1997, **36**, 6855–6861.

104. M. E. Van Brederode, F. Van Mourik, I. H. M. Van Stokkum, M. R. Jones, and R. Van Grondelle, Multiple pathways for ultrafast transduction of light energy in the photosynthetic reaction center of *Rhodobacter sphaeroides*, *Proc. Natl. Acad. Sci. USA*, 1999, **96**, 2054–2059.

105. M. H. B. Stowell, T. M. McPhillips, S. M. Soltis, D. C. Rees, E. Abresch, and G. Feher, Light-induced structural changes in photosynthetic reaction center: Implications for mechanism of electron-proton transfer, *Science*, 1997, **276**, 812–816.

106. T. Renger, E. M. Madjet, M. Schmidt am Busch, J. Adolphs, and F. Müh, Structure-based modeling of energy transfer in photosynthesis, *Photosynth. Res.*, 2013, **116**, 367–388.

107. D. L. Dexter, A theory of sensitized luminescence in solids, *J. Chem. Phys.*, 1953, **21**, 836–850.

108. R. D. Harcourt, K. P. Ghiggino, G. D. Scholes, and S. Speiser, On the origin of matrix elements for electronic excitation (energy) transfer, *J. Chem. Phys.*, 1996, **105**, 1897–1901.

109. T. Renger, Theory of optical spectra involving charge transfer states: Dynamic localization predicts a temperature dependent optical band shift, *Phys. Rev. Lett.*, 2004, **93**, 188101.

110. K. Mukai, S. Abe, and H. Sumi, Theory of rapid excitation energy transfer from B800 to optically forbidden exciton states of B850 in the antenna system LH2 of photosynthetic purple bacteria, *J. Phys. Chem. B*, 1999, **103**, 6096–6102.

111. T. Renger, Theory of excitation energy transfer: From structure to function, *Photosynth. Res.*, 2009, **102**, 471–485.

112. T. Förster, Zwischenmolekulare Energiewanderung und Fluoreszenz, *Ann. Phys. Leipzig*, 1948, **2**, 55–75.

113. T. Förster, Delocalized excitation and excitation transfer in: *Modern Quantum Chemistry, Part III, Action of Light and Organic Crystals* (O. Sinanoglu, Ed.), Academic Press, New York, 1965, pp. 93–137.

114. H. Sumi, Theory on rates of excitation energy transfer between molecular aggregates through distributed transition dipoles with application to the antenna system in bacterial photosynthesis, *J. Phys. Chem. B*, 1999, **103**, 252–260.

115. W. Roszak, T. D. Howard, J. Southall, A. T. Gardiner, C. J. Law, N. W. Isaacs, and R. J. Cogdell, Crystal structure of the RC-LH1 core complex from *Rhodopseudomonas palustri*, *Science*, 2003, **302**, 1969–1972.

116. Z. Katiliene, E. Katilius, and N. W. Woodbury, Energy trapping and detrapping in reaction center mutants from *Rhodobacter sphaeroides*, *Biophys. J.*, 2003, **84**, 3240–3251.

117. R. J. Cogdell, A. T. Gardiner, A. W. Roszak, C. J. Law, J. Southall, and N. W. Isaacs, Rings, ellipses and horseshoes: How purple bacteria harvest solar energy, *Photosynth. Res.*, 2004, **81**, 207–214.

118. H. Sumi, Uphill energy trapping by reaction center in bacterial photosynthesis. 2. Unistep charge separation, virtually mediated by special pair, by photoexcitation in place of excitation transfer from the antenna system, *J. Phys. Chem. B*, 2004, **108**, 11792–11801.

119. Y. Shibata, S. Nishi, K. Kawakami, J.-R. Shen, and T. Renger, Photosystem II does not possess a simple excitation energy funnel: Time-resolved fluorescence spectroscopy meets theory, *J. Am. Chem. Soc.*, 2013, **135**, 6903–6914.

120. F. Müh, M. Plöckinger, H. Ortmayer, M. Schmidt am Busch, D. Lindorfer, J. Adolphs, and T. Renger, The quest for energy traps in the CP43 antenna of photosystem II, *J. Photochem. Photobiol. B*, 2015, **152**, 286–300.

121. J. Hall, T. Renger, F. Müh, R. Picorel, and E. Krausz, The lowest-energy chlorophyll of photosystem II is adjacent to the peripheral antenna: Emitting states of CP47 assigned via circularly polarized luminescence, *Biochim. Biophys. Acta*, 2016, **1857**, 1580–1593.

122. J. Hall, T. Renger, R. Picorel, and E. Krausz, Circularly polarized luminescence spectroscopy reveals low-energy excited states and dynamic localization of vibronic transitions in CP43, *Biochim. Biophys. Acta*, 2016, **1857**, 115–128.

123. Y. Miloslavina, M. Szczepaniak, M. G. Müller, J. Sander, M. Nowaczyk, M. Rogner, and A. R. Holzwarth, Charge separation kinetics in intact photosystem II core particles is trap-limited. A picosecond fluorescence study, *Biochemistry*, 2006, **45**, 2436–2442.

124. C. D. van der Weij-de Wit, J. P. Dekker, R. van Grondelle, and I. H. M. van Stokkum, Charge separation is virtually irreversible in photosystem II core complexes with oxidized primary quinone acceptor, *J. Phys. Chem. A*, 2010, **115**, 3947–3956.

125. M. Kaucikas, K. Maghlaoui, J. Barber, T. Renger, and J. van Thor, Ultrafast infrared observation of exciton equilibration from oriented single crystals of photosystem II, *Nat. Commun.*, 2016, **7**, 13977.

126. P. Horton, M. P. Johnson, M. L. Perez-Bueno, A. Z. Kiss, and A. V. Ruban, Photosynthetic acclimation: Does the dynamic structure and macro-organisation of photosystem II in higher plant grana membranes regulate light harvesting states?, *FEBS Lett.*, 2008, **275**, 1069–1079.

127. R. Croce and H. van Amerongen, Light-harvesting in photosystem I, *Photosynth. Res.*, 2013, **116**, 153–166.

128. H. van Amerongen and R. Croce, Light-harvesting in photosystem II, *Photosynth. Res.*, 2013, **116**, 251–263.

129. A. Warshel and D. W. Schlosser, Electrostatic control of the efficiency of light-induced electron transfer across membranes, *Proc. Natl. Acad. Sci. USA*, 1981, **78**, 5564–5568.

130. M. Rätsep, J. Pieper, K. D. Irrgang, and A. Freiberg, Excitation wavelength-dependence electron-phonon and electron-vibrational coupling in the CP29 antenna complex of green plants, *J. Phys. Chem.*, 2008, **112**, 110–118.

131. J. M. Ortega, P. Mathis, J. C. Williams, and J. P. Allen, Temperature dependence of the reorganization energy for charge recombination in the reaction center from *Rhodobacter sphaeroides*, *Biochemistry*, 1996, **35**, 3354–3361.

132. K. Saito, T. Ishida, M. Sugiura, K. Kawakami, Y. Umena, N. Kamiya, J.-R. Shen, and H. Ishikita, Distribution of the cationic state over the chlorophyll pair of the photosystem II reaction center, *J. Am. Chem. Soc.*, 2011, **133**, 14379–14388.

133. J. P. Allen, K. Artz, X. Lin, J. C. Williams, A. Ivancich, D. Albouy, T. A. Mattioli, A. Fetsch, M. Kuhn, and W. B. Lubitz, Effects of hydrogen bonding to a bacteriochlorophyll bacteriopheophytin dimer in reaction centers from *Rhodobacter sphaeroides*, *Biochemistry*, 1996, **35**, 6612–6619.

134. R. Takahashi, K. Hasegawa, and T. Noguchi, Effect of charge distribution over a chlorophyll dimer on the redox potential of P680 in photosystem II as studied by density functional theory calculations, *Biochemistry*, 2008, **47**, 6289–6291.

135. J. Fajer, D. C. Brune, M. S. Davis, A. Forman, and L. D. Spaulding, Two photochemical systems in photosynthesis, *Proc. Natl. Acad. Sci. USA*, 1975, **72**, 4956–4960.

136. L. L. Maggiora, J. D. Petke, D. Gopal, R. T. Iwamoto, and G. M. Maggiora, Experimental and theoretical studies of Schiff-base chlorophylls, *Photochem. Photobiol.*, 1985, **42**, 69–75.

137. R. Edge, E. J. Land, D. J. McGarvey, M. Burke, and T. G. Truscotti, The reduction potential of the β-carotene⁺/β-carotene couple in an aqueous micro-heterogeneous environment, *FEBS Lett.*, 2000, **471**, 125–127.

138. F. Rappaport, M. Guergova-Kuras, P. J. Nixon, B. A. Diner, and J. Lavergne, Kinetics and pathways of charge recombination in photosystem II, *Biochemistry*, 2002, **41**, 8518–8527.

139. A. N. Webber, H. Su, S. E. Bingham, H. Käss, L. Krabben, M. Kuhn, R. Jordan, E. Schlodder, and W. Lubitz, Site-directed mutations affecting the spectroscopic characteristics and midpoint potential of the primary donor in photosystem I, *Biochemistry*, 1996, **35**, 12857–12863.

140. W. W. Parson, and R. J. Cogdell, Primary photochemical reaction of bacterial photosynthesis, *Biochim. Biophys. Acta*, 1975, **416**, 105–149.

141. J. H. Kim, J. A. Nemson, and A. Melis, Photosystem II reaction center damage and repair in *Dunaliella salina* (green algae), *Plant Physiol.*, 1993, **103**, 181–189.

142. J. Barber, Molecular basis of the vulnerability of photosystem II to damage by light, *Aust. J. Plant Physiol.*, 1994, **22**, 201–208.

143. I. Vass, Molecular mechanisms of photodamage in the Photosystem II complex, *Biochim. Biophys. Acta*, 2012, **1817**, 209–217.

144. A. Telfer, Too much light? How β-carotene protects the photosystem II reaction centre, *Photochem. Photobiol. Sci.*, 2005, **4**, 950–956.

145. E. Schlodder, M. Cetin, H. J. Eckert, F.-J. Schmitt, J. Barber, and A. Telfer, Both chlorophylls *a* and *d* are essential for the photochemistry in photosystem II of the cyanobacteria, *Acaryochloris marina*, *Biochim. Biophys. Acta*, 2007, **1767**, 589–595.

146. C. C. Schenck, P. Mathis, and M. Lutz, Triplet formation and triplet decay in reaction centers from the photosynthetic bacterium *Rhodopseudomonas sphaeroides*, *Photochem. Photobiol.*, 1984, **39**, 407–417.

147. T. Noguchi, Dual role of triplet localization on the accessory chlorophyll in the photosystem II reaction center: photoprotection and photodamage of the D1 protein, *Plant Cell Physiol.*, 2002, **43**, 1112–1116.

148. C. C. Schenck, B. Diner, P. Mathis, and K. Satoh, Flash-induced carotenoid radical cation formation in photosystem II, *Biochim. Biophys. Acta*, 1982, **680**, 216–227.

149. P. Faller, C. Fufezan, and A. W. Rutherford, Side-path electron donors: Cytochrome b559, chlorophyll Z and β-carotene in photosystem II, in: *The Light-Driven Water-Plastoquinone Oxidoreductase* (Wydrzynski, K. and Satoh, K. Eds.), Springer, Dordrecht, the Netherlands, 2005, pp. 347–365.

150. C. A. Buser, B. A. Diner, and G. W. Brudvig, Photooxidation of cytochrome b559 in oxygen-evolving photosystem II, *Biochemistry*, 1992, **31**, 11449–11459.

16

Modulation of the redox potentials

FABRICE RAPPAPORT

Among the numerous possible definitions of life, the ability to multiply and spread genetic material is one of the most commonly used. Yet, this wide usage makes its implicit consequences often overlooked. Multiplying and spreading is synonymous to ordering and hierarchizing matter. The first and second laws of thermodynamics imply that, to be everlasting, this process must be constantly supplied with energy. This statement readily follows from the relationship that links the free energy, enthalpy, and entropy changes associated to a given chemical reaction: $\Delta G = \Delta H - T\Delta S$. Let us consider the case of such a process whereby order is maintained or even built up, or, in other terms, which results in a decrease in entropy. For the reaction to produce a net flux of matter leading to the transient ordering of matter, ΔG must be negative. Thus, for the product to be more ordered than the reactants, that is, $\Delta S < 0$ and thus $-T\Delta S > 0$, the reaction must be accompanied by a decrease in enthalpy that compensates the change in entropy. This calls for a "reservoir of enthalpy" that can be consumed to sustain the maintenance of order within a living system (see Schoepp-Cothenet et al. [1] for a recent review and a comprehensive overview). Life on Earth uses two different types of such reservoirs, the phosphate potential and the redox potential, the two being intimately linked. The former involves adenosine triphosphate (ATP) and adenosine diphosphate (ADP) as, respectively, the reactant and product of a hydrolysis reaction: ATP → ADP + P_i, where P_i stands short for PO_4^{2-}. The standard free energy change associated with the reaction, that is, the free energy difference between the right-hand and left-hand terms of the reaction, assuming that the concentrations of the reactants and products are 1 M, is negative ($\Delta G_0 = -30.5$ kJ mol^{-1}). This reflects the propensity of the ATP molecule to transfer one phosphoryl group or, rather, the fact that both ADP and P_i have larger resonance stabilization than ATP. As pointed by Lubert Stryer in his biochemistry textbook [2], while the phosphoanhydride bonds in ATP are often referred to as high-energy bonds, there is nothing special about these bonds except that much energy is released when they are hydrolyzed because the products require less energy to be stabilized than the reactants. In addition, the steady-state intracellular concentrations of ATP, ADP, and P_i are such (ADP/ATP ≤ 10 and P_i ~ 20 mM in a resting nonphotosynthetic cell) that the free energy released during the hydrolysis of ATP, $\Delta G = \Delta G_0 + R \cdot T \ln \dfrac{ADP \cdot P_i}{ATP}$, is larger than ΔG_0.

The second reservoir involves electrons. Every single atom or molecule bears at least one electron and, depending on its electronic structure, may accept or donate one or more electrons. Electron transfer between two molecules, one acting as the electron donor and the other one as the electron acceptor, thus constitutes, with proton transfer, the simplest conceivable chemistry. This mere statement highlights the potential and spontaneous flow of matter which may exist between molecules and thereby supply the cellular request for enthalpy. Here, we will describe the basic principles that determine the reactivity of molecules in terms of electron transfer and put these reactions in the perspective of the biological processes in which they are mainly involved, bioenergetic electron transfer chains.

16.1 THERMODYNAMICS OF ELECTRON TRANSFER REACTION AND ITS DEPENDENCE UPON THE MEDIUM IN WHICH IT TAKES PLACE

16.1.1 Thermodynamic aspects of electron transfer reaction

From a theoretical standpoint, since any atom or molecule bears electron(s), it can participate in electron transfer according to Red → Ox + e$^-$, where Red and Ox are the reduced and oxidized states of the same molecule, respectively. As any chemical species, either Red or Ox may be characterized by its chemical potential: $\mu_i = \mu_i^{std} + RT \ln a_i$, where R and T are, respectively, the gas constant and temperature (in K), μ_i^{std} is the chemical potential of species i in standard conditions, and a_i is its activity, which, for the sake of simplicity, we will hereafter approximate by concentration. This formal reaction, whereby *Red* provides one electron and is converted into Ox, is thus associated with a variation in chemical potential, $\Delta\mu = \Delta\mu_i^{std} + RT \ln \dfrac{Ox}{Red}$. The well-known, and commonly used (by electrochemist), Nernst equation, $E = E^0 + \dfrac{RT}{n \cdot F} \ln \dfrac{Ox}{Red}$, where F is the Faraday constant, n is the number of electrons involved in the redox process (considered as being equal to 1 in the preceding text), E is the ambient electrical potential, and E^0 is the standard redox potential (often called the midpoint potential because Ox = Red when E = E^0), is nothing else than a variant of the preceding one when equilibrium is achieved. The standard redox potential thus characterizes the propensity of a given chemical entity to gain (be reduced) or lose (be oxidized) one electron. Such a formulation implies the interplay of another chemical entity that will either lose or gain one electron, because electron transfer to the bulk is thermodynamically highly unfavorable. If we, thus, now consider two redox couples, Red$_1$/Ox$_1$ and Red$_2$/Ox$_2$, the free energy change associated with the oxidation of Red$_1$ by Ox$_2$ is $\Delta G = n F \left(E^1 - E^2 \right) + RT \ln \dfrac{Ox_1 Red_2}{Red_1 Ox_2}$.

Thus, depending on the value of the $\dfrac{Ox_1 Red_2}{Red_1 Ox_2}$ term, either Ox$_1$ or Ox$_2$ can act as the oxidizing species.

Yet, this role will more often fall to the oxidant of the redox couple having the more positive standard redox potential. Despite the practicality of this rule of thumb, one should not overlook the $\frac{Ox_1 Red_2}{Red_1 Ox_2}$ term, which, depending on the relative concentrations of all the actors, may make the opposite direction more favorable from a thermodynamic standpoint. As regards the kinetic aspects of electron transfer reaction, we refer the reader to Chapter 15, which provides a comprehensive and thorough description of electron transfer theory.

16.1.2 Influence of the surrounding medium on standard redox potential

In a solution with an ambient potential E^h, the free energy for a reduced species to lose an electron is $\Delta G = n\, F\left(E^h - E^0\right) + RT \ln \frac{Ox}{Red}$ with $G = \int \mu \cdot d\eta$, where η is the number of molecules. These considerations highlight two points that are particularly important when dealing with redox potentials and electron transfer reactions: (1) the propensity of a given molecule to attract or lose one electron is defined with respect to standard conditions that need to be described, and (2) it depends on the relative

abundance of the Ox and Red states in the medium where the reaction takes place. The direct and most important consequence of the first point is that E^0 depends on the physicochemical properties of the medium where the reaction occurs. This is because as the molecule loses or gains one electron, its coulombic charge changes, and this causes the medium to reorganize in response to this change in the local charge distribution. Since in cells electron transfer reactions take place within proteins, describing and discussing the way redox properties contribute to biological energy conversion first requires the understanding of why and how the protein environment modifies the redox properties of a given molecule. To this aim, we will first consider the situation where there is absolutely no interaction between the molecule of interest, in either of its redox state, and its surrounding medium, that is, when the reaction occurs in vacuum. We will thereafter note ΔG_{vac}^0, the free energy change associated with the redox change in vacuum. Such a figure has been made accessible by the development of quantum mechanics/molecular mechanics (QM/MM) approaches [3–7]. The calculations provide a reference system that can be experimentally assessed by the measure of the redox properties of the molecule of interest in a solvent. As illustrated in Figure 16.1, this involves the transfer of the molecule from

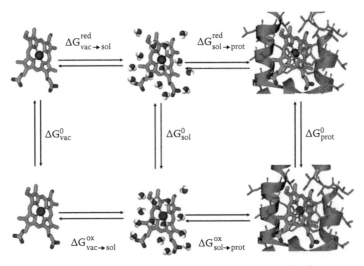

Figure 16.1 A scheme depicting the influence of the surrounding medium on the free energy change associated with the redox changes of a redox-active species. The left-hand side of the scheme shows the redox changes in vacuum, the blue and red color symbolizing the reduced and oxidized states, respectively. In the middle are shown the redox changes in water and the reorganization of the water molecules upon the redox change of the redox species (the configurations shown here have only illustration purposes and are not the result of any particular calculation). The right-hand side of the scheme depicts the redox changes within a protein.

vacuum to solvent, a process that is associated with a free energy change, $\Delta G_{vac \to sol}$, which determines the interaction energy between the solute and the solvent (often referred to as the solvation or Born energy; see Bashford and Karplus [8], Gunner and Honig [9], Warshel and Aqvist [10]).

The important point here is that $\Delta G_{vac \to sol}^{Red}$ and $\Delta G_{vac \to sol}^{Ox}$ differ because the oxidized and reduced states have different electrostatic charge and thus different interaction energies with the solvent molecules. The interaction energy has different components: one is intrinsic to the solute, the change in electronic polarization resulting from the transfer to the solvent; another is intrinsic to the solvent, the reorganization of the solvent induced by the transfer of the solute; and, finally, the third one, the pairwise interaction between the solute and the reorganized solvent, involves both actors. The thermodynamic square in the left-hand side of the scheme in Figure 16.1, which links the four different states of our system, illustrates that, as a result of these different interaction energies, the free energy change associated with the redox changes in solution differs from that in vacuum and one has $\Delta G_{sol}^0 = \Delta G_{vac}^0 + \Delta G_{vac \to sol}^{0} + n \cdot \Delta G_{vac \to sol}^{e^-} - \Delta G_{vac \to sol}^{Red}$. The midpoint potential of the redox couple in solution is $E_{sol}^0 = \dfrac{\Delta G_{sol}^0}{nF}$.

This discussion highlights the role played by the solvent surrounding the redox-active species in determining the free energy change associated with the redox changes of this species (see [11–14]). Notably, this statement does not rely on any implicit assumption regarding the nature of the considered solvent. Thus, transferring the redox species of interest from one solvent to another translates into different solvation energies and thus changes in the redox-induced free energy change. In other words, transferring the redox species within the protein has an energy cost, $\Delta G_{sol \to prot}^{Red}$ and $\Delta G_{sol \to prot}^{Ox}$ for the reduced and oxidized states, respectively, and this modifies the midpoint potential of the redox species with respect to its value in solution: $\Delta G_{prot}^0 = \Delta G_{vac}^0 + \Delta G_{vac \to sol}^{0x} + \Delta G_{sol \to prot}^{0x} + n \cdot \Delta G_{vac \to sol}^{e^-} - \Delta G_{vac \to sol}^{Red} - \Delta G_{sol \to prot}^{Red}$. The high dielectric constant of water means that it efficiently interacts with and thus stabilizes any charged species. These favorable interactions are, at least partly, lost upon the transfer to the protein so that $\Delta G_{sol \to prot}^{Red}$ and $\Delta G_{sol \to prot}^{Ox}$ are positive figures. Yet, the difference between these two, $\Delta\Delta G_{vac \to sol} = \Delta G_{vac \to sol}^{0x} - \Delta G_{vac \to sol}^{Red}$, can be either positive or negative. On the one hand, one

expects the free energy change associated with the transfer to increase with the charge of the considered state. But, on the other hand, one cannot predict with certainty, which is, between the two redox states, the one that bears the larger charge. Indeed, although the oxidation unerringly corresponds to the increase of coulombic charge by one unit, the net charge of the oxidized state may be either larger or smaller than that of the reduced state, depending on the electronic configuration or the possible involvement of other charge transfer, such as proton transfer, in the process. For example, the oxidized state may bear one negative charge (in coulomb unit) and the reduced state will be neutral; conversely, the oxidized state may be neutral and the reduced state will bear one negative charge.

16.2 MIDPOINT POTENTIALS WITHIN PROTEINS

16.2.1 Most common redox cofactors found in biology: A limited number of chemical compounds for a large span of midpoint potentials

As discussed earlier, proteins exert a fine electrochemical tuning of the redox potential of the cofactors they bind, and this likely contributes to the amazing diversity of redox reactions that they can catalyze. A glance at Figure 2 in Schoepp-Cothenet et al. [1] highlights the large span of redox potentials covered by biological processes that sometimes take place within the same organism. As an example, oxygenic photosynthesis extracts electrons from water, which requires producing oxidizing equivalents with a midpoint potential of ~1.25 V [15–17], and transfers them to CO_2 to make carbohydrates or, in some instances, to H^+ to make H_2, and this involves reducing equivalent with midpoint potentials of −0.4 and −0.5 V (see Antal et al. [18] and references therein). This impressive diversity in terms of midpoint potentials contrasts with the relative paucity of redox-active prosthetic groups, which sums up to a handful of compounds. Here is a nonexhaustive list of the most commonly used compounds:

1. Iron–sulfur clusters, which can be either Fe_2S_2 or Fe_4S_4 or Fe_3S_4 [19,20]
2. Hemes, made of a Fe^{2+}/Fe^{3+} ion contained in the center of a porphyrin ring—the most common variations being a-type, b-type, and o-type hemes,

which are not covalently bound to the protein in which they act, and c-type hemes, which are covalently bound via two cysteine-S-yl bonds [21] (with the exception of heme c_i in the cytochrome b_6f complex of the photosynthetic chain that is bonded by a single bridge; Stroebel et al. [22])

3. In the photosynthetic electron transfer chain, pigments (see Chapter 1) such as chlorophyll, a porphyrin ring with a Mg^{2+}, or pheophytin, a porphyrin ring

4. Flavins, such as flavin adenine dinucleotide or flavin mononucleotide

5. Metals, such as Cu^+/Cu^{2+} and molybdenum [23–26]

In addition to these various prosthetic groups, which all, except pheophytin, bear a metal as the redox-active atom, organic compounds are found to be involved in electron transport as well. Prominent among these are quinones. These play a special and central role in bioenergetics processes because they not only act as intraprotein electron carriers, similarly to the redox cofactors mentioned earlier, but being neutral in both their oxidized and reduced forms (at physiological pH's) they can diffuse in the lipid phase and act as an electron carrier between membrane-bound complexes.

16.2.2 Redox potential in proteins

16.2.2.1 THE DIELECTRIC MEDIUM PROVIDED BY THE PROTEIN MOIETY

We have seen earlier that the redox properties of a given redox species strongly depend on the physicochemical properties of the medium in which the redox reaction takes place. In simple terms, this basically reflects the fact that the change in the net charge of the redox entity, which occurs upon its oxidation or reduction, results in a change in its energy level, which is, for a point charge q, $E = q \cdot \psi(\vec{r})$, where $\psi(\vec{r})$ is the electrostatic potential at the position $\vec{r}$. This electrostatic potential $\psi(\vec{r})$ is the result of the electrostatic charge distribution within the surrounding medium and, in the case of a continuous medium, is given by the Poisson–Boltzmann equation: $\nabla \cdot \left[\varepsilon(\vec{r}) \nabla \psi(\vec{r}) \right] = -4\pi\rho(\vec{r})$, where $\varepsilon(\vec{r})$ and $\rho(\vec{r})$ are, respectively, the dielectric constant and the charge density at position $\vec{r}$. Although proteins are not a continuous medium, $\psi(\vec{r})$ can be computed following a similar approach as the

one defined by the Poisson–Boltzmann equation. Applications that calculate electrostatic potentials and the corresponding electrostatic energies in and around macromolecules have been developed and made freely available (http://www.ibridgenetwork. org/columbia/ir_1790). In addition, as detailed in the references given earlier, the combination of QM/MM and computation of electrostatic potential relying on the Poisson–Boltzmann equation allows the calculation of the distribution of charges within the protein, when knowing the structure of a protein at atomic resolution, and thus of the redox potential of the redox species borne by this protein.

16.2.2.2 ROLE OF AMINO ACID SIDE CHAINS IN TUNING THE REDOX POTENTIAL

The diversity of charge distribution, and thus of electrostatic landscape, and interaction energies mediated for examples by ligation or H-bonds, that the protein moiety may provide translates into a remarkable span of redox potentials. As a famous example, in the tetraheme cytochrome of the *Blastochloris viridis*, the redox potentials (E_m) of c-type cytochromes are tuned by more than 500 mV (see Gunner et al. [27] and references therein). But this is only one among many examples, and the reader is referred to [28–30] for additional cases such as the multiheme cytochromes of *Desulfovibrio desulfuricans* or *Shewanella* and to Zheng and Gunner for a survey of 141 heme-binding proteins in which the redox span exceeds 1000 mV [31]. The physicochemical basis of such modulations has been rationalized by various authors who agree that the protein medium has the unique property of providing a dielectric environment in which the redox cofactors are embedded with an intricate charge or dipole distribution [10,27,32]. Prominent among the structural determinants of the midpoint potential are the different possible binding interactions such as hydrogen bonds or ligation. By providing additional electropositivity to the redox-active entity, H-bonds stabilize the reduced state. In other terms, the reduction/oxidation is, respectively, favored/disfavored by the presence of H-bonds. A remarkable example of the consequences of the addition/removal of H-bonds on the midpoint potential of a redox-active species is the delicate engineering of the redox potential of the chlorophyll dimer in the reaction center of purple bacteria, which allowed Allen and colleagues to increase its oxidizing power by 400 mV [33]. Photosystem I provides another illustration of the role of H-bonds in tuning the midpoint potentials of redox cofactors. It bears two phylloquinones involved in two converging electron transfer branches [34,35]. Despite the almost identical structural environment of the two quinones in their oxidized state, the reoxidation rates of the semiquinones differ by one order of magnitude [36,37]. The current framework accounting for such a difference is that the interaction(s) between the two semiquinones and the protein moiety, which is not accessible to X-ray crystallography, differ [38–40], and this yields distinct midpoint potentials for the two semiquinone/quinone couples [41–43]. For additional discussion, the reader is referred to studies that present the functional analysis of the consequences of site-directed mutations of amino acids predicted to be involved in H-bond formation based on structural data in, for example, Photosystem II [15,44,45] or Photosystem I [46,47]. Typically, the changes in redox potential resulting from the breakage or the strengthening of an H-bond are in the 50–200 mV range [47]. As regards ligands, the most exhaustive study to our knowledge is the survey of 141 proteins bearing hemes with midpoint potentials spanning over 1000 mV [31]. Even though this study is restricted to a single type of redox cofactor, in the present case heme, it highlights the importance of the axial ligation system with bis-His ligation lowering the E_m by 200 mV relative to His-Met ligands (see also, e.g., Raphael and Gray [48] for Cys ligand). But above all, the survey shows that no single type of interaction can be identified as the most important in setting heme electrochemistry in proteins. For example, as discussed earlier, the loss of solvation energy has been suggested to be a major factor in tuning *in situ* E_ms. However, the analyses resulting from Multi-Conformation Continuum Electrostatics show that the correlation between calculated and experimental E_ms is poor when considering a single type of interaction, but very satisfying when considering the sum of these individual terms. This highlights that there are multiple solutions to the problem posed by the tuning of heme electrochemistry in proteins and that proteins rely on different structural features to make the redox reactivity of their cofactors compatible with the catalysis they perform.

16.2.2.3 EQUILIBRIUM MIDPOINT POTENTIALS VERSUS OPERATING POTENTIALS

We have seen earlier that estimating the respective values of the different interaction energy terms that determine the electrochemical properties of a redox species is not always straightforward. Yet, their sum is readily accessible experimentally and can be obtained by comparing the absolute values of the E_m in solution and in protein. The latter may be obtained by two different methods. The most commonly used one is equilibrium redox titration. The other relies on the determination, by a functional analysis, of the free energy change associated with an electron transfer reaction between two cofactors. Such a change in free energy is considered as equal to the difference in E_ms between the electron donor and acceptor. Thus, provided that one of these two E_ms is known, the other is readily inferred from such a calculation. However, these two methods sometimes yield different results, and this has led to the distinction between "equilibrium redox potential" and "operating redox potential." The differences arise because the two methods do not probe the same state of the redox cofactors. Two types of phenomena may account for these differences. One comes from the different time domains involved in equilibrium redox titration and functional analysis. Whereas redox titrations require thermodynamic equilibrium between the sample and the solution poised at a given electrochemical potential, the functional analysis allows one to probe transient states whose free energy may differ significantly from that of the equilibrated states. Indeed, in response to the change in the redox state of a given cofactor, the protein environment may undergo energetic relaxation (such as proton transfer, conformational changes), which may be slower than the lifetime of the transient oxidized (or reduced) cofactor. Armstrong and colleagues have nicely illustrated this point with the "fast-scan electrovoltammetric" technique [49,50]. Other examples are found in photosynthetic reaction centers (see, e.g., Woodbury et al. [51], Sebban and Wraight [52]). An alternative but nonexclusive explanation of the different E_ms yielded by redox titration and functional analysis relies on the fact that most of the membrane proteins involved in electron transfer reactions bind several cofactors, which are usually located at less than 15 Å from one another. Such short distances may result in significant electrostatic interactions between the different cofactors. Thus, the free energy level of a given cofactor includes the electrostatic contributions of the nearby electron carriers. We refer the reader to Gunner and Honig [9], Nitschke and Rutherford [53], Alric et al. [54], and Leger et al. [55] for examples of such contributions. It is noteworthy, however, that throughout a redox titration, all the cofactors undergo an identical charge change in terms of sign (i.e., all are either reduced or oxidized), whereas in an electron transfer chain, two nearby cofactors involved in an electron transfer reaction will undergo charge changes of opposite sign (one will be oxidized at the expense of the other). Consequently, if the electrostatic interaction between the two is significant, the difference between their equilibrium E_ms will be greater than the free energy change associated with the electron transfer between them.

16.3 ELECTRON TRANSFER WITHIN AND BETWEEN PROTEINS

The theoretical aspects of electron transfer are addressed in detail in this book in Chapter 15 by Thomas Renger. We will thus limit ourselves here to remind the reader that the electron transfer rate between a donor and an acceptor is empirically given by the following equation (which is Equation 15.50): $\log_{10}(k_{et}) = 15 - 0.6R - 3.1(\Delta G_i + \lambda)^2$, where R is the distance (in Å) between the electron donor and acceptor, ΔG the free energy charge associated with the reaction, and λ the reorganization energy (see Chapter 15 for a definition of this term). Notably, this equation provides an empirical but reliable estimate of the electron transfer rate when the reaction is downhill in energy, that is, when ΔG_0 is negative. For endergonic reactions (uphill electron transfer (ET)), the rate constant is obtained by applying a Boltzmann factor to the downhill expression: $k_{uphill} = k_{downhill} e^{\frac{\Delta G_i}{kT}}$. The need to introduce this second equation results in a distortion of the curve yielding k_{et} as a function of ΔG. This problem does not arise when using the classical [56] or fully quantum mechanical description [57] (see Crofts and Rose [58] for discussion and Winkler and Gray [59] for a recent review). Nevertheless, the distortion implied by the approximate formula remains acceptable and hardly impacts its practical merit. From the preceding text, it follows that electron transfer rates can be readily estimated with reasonable

accuracy once the parameters ΔG and R are known. Importantly, this equation applies to intra- as well as to interprotein electron transfer reactions as long as the complex comprising the two protein partners is stable enough to remain immobile during the electron transfer reaction. In photosynthesis, this will be the case for the plastocyanin–PSI complex [60], for the ferredoxin–PSI complex [61,62], and for the plastocyanin–cytochrome f complex (see Cruz-Gallardo et al. [63] for a review).

16.3.1 Interprotein electron transfer

Most biological electron transfer chains involve electron carriers, such as cytochrome c, c_6, plastocyanin, and iron–sulfur proteins, which are found in the aqueous phase (the lumen in the chloroplast, the intermembrane space in mitochondria, the periplasmic space in gram-negative photosynthetic bacteria). This contributes to the necessary compartmentalization of bioenergetics processes as it prevents the free encounter of these electron carriers with the other diffusing electron shuttles, which are soluble in the lipid phase. These soluble and diffusing electron carriers usually interact with membrane-bound enzymes, and the electron transfer reaction involves the transient formation of a tight complex. The most direct evidence for the formation of such tight complexes is the occurrence of the electron transfer reaction in frozen conditions (see, e.g., Venturoli et al. [64]). The formation of these complexes involves the recognition of the binding site and the proper docking of the soluble partner to its site. The structural motifs that contribute to determining the recognition pattern are often charged residues that enable the formation of salt bridges between the two partners (see, e.g., Fischer et al. [65], Hippler et al. [66]). Other types of interaction may also be involved, and listing them all goes beyond the scope of this chapter. What matters here are the possible consequences of this binding interaction on the free energy change associated with the electron transfer reaction. Indeed, the involvement of a reaction complex inevitably introduces an additional state in the thermodynamic picture that describes the entire sequence of events. This includes binding of the soluble carrier to its site, electron transfer within the transient reaction complex, and release of the soluble carrier after it has undergone a change in its redox state. If we neglect the kinetic aspects for the time being and restrict

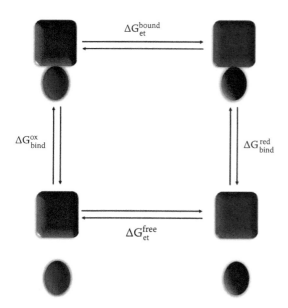

Figure 16.2 The thermodynamic square highlighting the possible influence on the driving force of an electron transfer reaction of the binding of the two electron-transferring partners.

the discussion to the thermodynamic aspects, one may draw a thermodynamic square in which two parallel sides correspond to the electron transfer reaction either within the complex or between the freely diffusing electron carrier and its membrane-bound partner, and the two other parallel sides depict the binding of the reduced or oxidized states of the soluble carrier (see Figure 16.2).

This thermodynamic square highlights the fact that if the free energy change associated with the binding of the soluble partner to its docking site depends on its redox state, then the free energy change associated with the electron transfer reaction within the complex differs from that occurring between unbound electron carrier and the membrane-bound enzyme. With the nomenclature of Figure 16.2, one has $\Delta G_{bind}^{ox} - \Delta G_{bind}^{red} = \Delta G_{et}^{free} - \Delta G_{et}^{bound}$. If adopting an engineering standpoint, one would intuitively consider that a smart way to optimize the electron transfer reaction between the soluble electron carrier and its membrane-bound partner is to promote the formation of the reacting complex and the dissociation of the product. In other words, and in the present example of the electron transfer reaction involving plastocyanin and Photosystem I, one would think that increasing the binding constant of the reduced form of plastocyanin and decreasing that of the oxidized form can contribute to the optimization of

this electron transfer step. Saliently, the photosynthetic electron transfer chain has evolved following this engineering principle since the binding constant of the reduced plastocyanin to Photosystem I is sixfold larger than that of the oxidized form [67]. This statement calls, however, for two qualifications. First, this tuning of the respective binding constant of the oxidized and reduced forms has only kinetic consequences and does not modify the overall driving force of the reaction. Indeed, irrespective of the pathway followed, the starting point of the reaction remains $P_{700}^{+} + PC_{unbound}$ and the end point $P_{700} + PC_{unbound}^{ox}$ so that the free energy change associated with the overall reaction remains ΔG_{et}^{free}. The second, and maybe more important qualification, is the price of this purported kinetic optimization: a decrease in the driving force of the electron transfer step within the complex. Indeed, according to the equality $\Delta G_{bind}^{ox} - \Delta G_{bind}^{red} = \Delta G_{et}^{bound} - \Delta G_{et}^{free}$, if the binding constant of the reduced form is larger than that of the oxidized one, then $\left(\Delta G_{bind}^{ox} - \Delta G_{bind}^{red} \right) \geq 0$ (we consider that all the individual reactions are exergonic) so that $\Delta G_{et}^{bound} \geq \Delta G_{et}^{free}$ and thus $k_{et}^{bound} \leq k_{et}^{free}$. In brief, promoting the formation of the state that is competent for electron transfer, that is, the PC–PSI complex in the present example, is at the expense of the equilibrium constant and of the rate of the intrinsic electron transfer reaction within the complex.

16.3.2 Intraprotein electron transfer

This topic is dealt with in this book by Thomas Renger and Mike Jones who, respectively, review the theoretical aspects of electron transfer theory and its application to the photosynthetic energy conversion performed by reaction centers. We refer the reader to Chapters 9 and 15 in which the issue of individual electron transfer steps is treated in detail. Here, we will discuss the kinetic characteristics of the enzymes comprising a chain of electron carriers. From the description given earlier of the thermodynamic properties of electron transfer reactions, one might intuitively consider that most, if not all, intraprotein electron transfer chains are made of redox centers with decreasing midpoint potentials. Such an energetic picture would provide a monotonic driving force to each of the individual electron transfer steps, which would all benefit from being downhill in energy. However, there are many exceptions to this view. The first example of such a counterintuitive arrangement is the tetraheme cytochrome of some photosynthetic reaction centers. Particularly demonstrative cases are [NiFe]-hydrogenase [68] and fumarate reductase [69], which exhibit chains of multiple iron–sulfur clusters in which neutral [3Fe–4S] centers are inserted between highly electronegative [4Fe–4S] clusters, introducing very endergonic ET steps [70]. Other well-documented examples are nitrate reductase [71], formate dehydrogenase [71], Complex I [72], and additional cases such as the multiheme cytochromes of *Desulfovibrio desulfuricans* or *Shewanella* [29,30]. Several experimental and theoretical studies of such "roller-coaster" electron transfer chains have shown that they can perfectly well accomplish the kinetic function of driving quickly and efficiently the transfer of an electron from one end of the chain to the other. As discussed in detail by others [55,73], a critical parameter at each link in the electron transfer chain is the ratio between forward over backward rate constants out of a redox center (e.g., for the second redox center in a chain made of three, this ratio would be k_{23}/k_{21}, where k_{23} is the forward electron transfer rate constant between centers 2 and 3 and k_{21} is the backward electron transfer rate constant between centers 2 and 1). This ratio enters in the steady-state rate constant as the "transmission coefficient" $1/(1+(k_{23}/k_{21}))$, which defines the probability that forward electron transfer outcompetes backward electron transfer. In *B. viridis*, this transmission coefficient is close to one for every single redox center in the tetraheme electron transfer chain. Why have such arrangements been maintained throughout evolution? Even though they do not compromise the overall catalytic efficiency of the enzyme, one may wonder what has been the nature of the evolutionary pressure that counterselected the more intuitively smooth and energetically downhill arrangement. As far as biological electron transfers are concerned, it has become clear that the usual picture, where the reaction pathway involves a series of irreversible steps, is oversimplified. Because every single step in the process is reversible, backward rates must be considered to generate a complete picture that takes into account the thermodynamic as well as the kinetic features of the entire redox chain, where the behavior of the enzyme as a whole depends on every relay. In their theoretical analysis of this issue, Leger et al. distinguish two extreme cases on the basis of how

fast intramolecular ET is with respect to the chemical transformation at the active site [55]. They conclude that when intramolecular ET does not limit turnover, the system behaves as if there were no individual steps and the only effect of the redox chain is to transmit the driving force across the enzyme. In contrast, when intramolecular ET is not very fast with respect to the catalytic turnover of the enzyme, reaction intermediates may build up as catalysis goes on and this will affect the driving force of the reactions because of the $\ln \dfrac{Ox_1 Red_2}{Red_1 Ox_2}$ term in the equation $\Delta G = nF(E^1 - E^2) + RT \ln \dfrac{Ox_1 Red_2}{Red_1 Ox_2}$. Under such conditions, the catalytic driving force thus depends both on the reduction potentials of the relays, which sets the $nF(E^1 - E^2)$ term, and on the kinetics of intramolecular electron transfer, which sets the $RT \ln \dfrac{Ox_1 Red_2}{Red_1 Ox_2}$ term by determining the concentrations of the oxidized and reduced states at steady state. In other words, everything happens as if the active site had redox properties that are controlled, or tuned, by the ET chain. The authors convincingly advocate that optimization of the electron transfer chain need not be considered in terms of speed: the properties of the relays, and therefore the electron transfer kinetics, may well have been optimized, in some cases, to tune the apparent reduction potential of the active site.

16.4 TUNING REDOX POTENTIALS TO ADAPT BIOLOGICAL FUNCTIONS TO CHANGING ENVIRONMENT

We have described earlier the relationship between the midpoint potentials of redox cofactors within a protein and the thermodynamic and kinetic characteristics of intra- or interprotein electron transfer reactions, which determine the biological function of the enzyme. Yet, the biological realm offers an amazingly wide spectrum of environmental conditions, or ecological niches, and these biological functions have been shaped accordingly throughout evolution (see Schoepp-Cothenet et al. [1] for a recent review). Obviously, these are long-term adaptations. Other striking examples of the necessary functional plasticity in response to short-term changes in environmental conditions come from the photosynthetic realm and the changes in gene expression induced by changing illumination conditions (see Sicora et al. [74], Vass and Cser [75], Rutherford et al. [76] and references therein).

16.4.1 Quinones, the redox cornerstone of bioenergetic electron transfer chain

If one would want to design an electron transfer chain that would rely upon an energy-converting process such as those described by the chemiosmosis theory [77], one would pay particular attention to the free energy change associated with each step in the series. Even though this statement must be qualified because we have seen earlier that steps being uphill in energy do not necessarily constitute an impassable barrier, what really matters is the difference in midpoint potential rather than their absolute values. In other words, the kinetic and thermodynamic properties of a given reaction will remain fully unchanged if the midpoint potentials of the electron donor and acceptor undergo exactly the same shift. Thus, when designing a chemiosmotic electron transfer chain, the first step would be to define a redox potential that would serve as a reference, and the following ones would be to tune the midpoint potential of the various redox cofactors by subtle changes in their protein environment, using the principles described earlier, so as to conserve the free energy change associated with all the individual electron transfer steps. This design principle is remarkably illustrated by comparing the (available) individual midpoint potentials of the redox cofactors that make the electron transfer chain of distinct but related species. With regard to the midpoint potentials of the iron–sulfur cluster of its Rieske protein and of the photo-oxidizable pigment of its photosynthetic reaction center, the purple bacteria, *Halorhodospira halophila (H. halophita)*, more closely resembles the Heliobacteria *Heliobacterium chlorum (H. chlorum)* than the related purple bacteria *Rodocyclus gelatinosus (R. gelatinosus)*. Indeed, as shown by Schoepp-Cothenet et al., the midpoint potential of the FeS cluster is close to 100 mV versus 250 mV in *R. gelatinosus*, and that of the photo-oxidizable pigment P is 270 mV versus 420 mV in *R. gelatinosus* [78]. As discussed by Schoepp-Cothenet et al. [78], the point in common between *H. chlorum* and *H. halophita* is that, at variance with *R. gelatinosus* and most of its sister photosynthetic purple bacteria, they use menaquinone and not ubiquinone as a redox shuttle between

their reaction center and their quinol:cytochrome oxidoreductase. Because the midpoint potential of the menaquinol/menaquinone couple is about 170 mV more negative than that of the ubiquinol/ubiquinone couple, it is tempting to conclude with Nitschke and colleagues that the midpoint potential of the quinone, which acts as a lipophilic soluble electron carrier, serves as the reference upon which the entire redox chain is built to perform chemiosmosis (see Nitschke and Russell [79] for a discussion). This conclusion begs the question: What is so special about quinones? One possible answer to this question is that they may be the only link that is universally found in bioenergetic electron transfer chains that is not carried by a protein (with the exception of those of acetogens and methanogens). Quinones thus constitute a link that cannot undergo the fine redox tuning that a protein environment can provide, making them a poor target for the evolutionary tinkering. A second, and nonexclusive, possible answer is that the redox chemistry of quinones has a remarkable specificity that may have constituted a strong evolutionary pressure. Because they are two-electron carriers, quinones can be found in three different redox states: the fully reduced state, which bears two additional electrons, the quinol QH_2; the singly reduced state, which bears one electron, the semiquinone SQ; and the fully oxidized state, the quinone Q. There are thus three redox couples to be considered: QH_2/SQ (hereafter noted E_{QH2}), SQ/Q (E_{SQ}), and QH_2/Q ($E_{QH2/Q}$). Expectedly, the midpoint potentials of these three couples are related and, as illustrated in Figure 16.3, one has $E_{QH_2/Q} = \dfrac{E_{SQ} + E_{QH2}}{2}$ and $E_{QH_2} - E_{SQ} = \dfrac{RT}{F} \ln\left(\dfrac{Q \cdot QH_2}{SQ^2}\right)$. The $\dfrac{Q \cdot QH_2}{SQ^2}$ term is the equilibrium constant of the disproportionation reaction by which two semiquinones react and yield a quinol and a quinone. The larger this equilibrium constant, the lower the probability that the semiquinone state is encountered, so that this equilibrium constant is often noted K_{stab}, for stability constant. As discussed in detail by others [79–83], the second equation shows that, as depicted in Figure 16.3, two extreme cases can be considered:

1. $K_{stab} \gg 1$: in this case, $E_{QH2} \gg E_{SQ}$, and the semiquinone is both a very good electron donor and electron acceptor (hence its low stability).
2. $K_{stab} \ll 1$: in this case, $E_{QH2} \ll E_{SQ}$, and the semiquinone is a poor electron acceptor and a poor electron donor (hence its large stability).

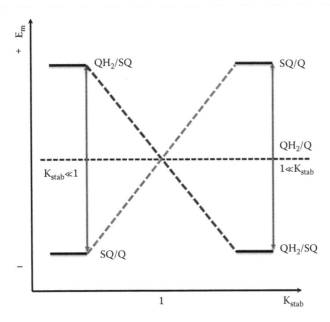

Figure 16.3 A schematic illustration of the relationship between the stability of a semiquinone and the midpoint potentials of the two redox couples in which it is involves.

We now consider the first case, $K_{stab} \gg 1$. According to the equation that relates $E_{QH2/Q}$, E_{QH2}, and E_{SQ}, the semiquinone is a better electron donor than the quinol, and this allows an electron transfer reaction that would be uphill in energy if involving the quinol as the electron-donating species. In other words, the energy change associated with the transfer of two electrons from the quinol to a soluble protein, such as plastocyanin or cytochrome c, can be split into two one-electron reactions associated with a smaller free energy change, and the remaining free energy can be converted into work by, for example, pumping a H^+ across the membrane against its electrochemical potential. As proposed by Mitchell [84] and others [80], such a bifurcation is at work in the Q cycle catalyzed by Rieske/bc complexes (and their related b_6f complex; see for a recent review ten Brink et al. [85]) and contributes to the bioenergetics energy conversion (but see Malnoe et al. [86] for a discussion). This is presumably another reason that makes the quinones so special in bioenergetic electron transfer chain and why the entire redox edifice seems to be constructed upon the E_{QH2} reference (see Bailleul et al. [87], Schoepp-Cothenet et al. [78] for discussions).

16.4.2 Redox tuning to avoid detrimental reactions such as those potentially leading to the production of reactive oxygen species

As discussed by Rutherford et al., enzymes involved in energy conversion are no exception to other enzymes and act as catalysts [76]. Thus, irrespective of the reaction pathway and the possibly high number of intermediates it involves, completion of the enzyme cycle brings the enzyme back to its ground state. However, one specificity of energy-converting devices is that their reaction pathways necessarily involve high-energy intermediates. Their decay is downhill in energy and provides the work required to promote the accumulation of the electrochemical potential.

These two basic characteristics are viewed as being so self-evident that their consequences may be overlooked. Indeed, these enzymes have to undertake a real challenge to obey the preceding principle: how to favor the energy-productive processes over competing reactions in which the high-energy intermediates can decay without going through the energy-converting step(s). Because all the forward and productive reactions are reversible in essence, intermediates may simply decay by back-reacting to the ground state of the enzyme via the thermally activated repopulation of higher energy states. In this case, the energy that has been transiently stored is thus simply released, and no work is extracted from the overall process. These backward reactions are the most obvious source of potential loss. Yet, as detailed by Rutherford et al., there are other pathways that can also result in unproductive reactions or short circuits [76]. Such processes could involve recombination from the high energy states directly to the ground state or decay to lower energy states of the same components (such as formation of a triplet excited state at the expense of a singlet). An additional route of energy loss worth mentioning more specifically involves molecular actors that are not bound by the enzyme. The most famous examples are all the reactions involving oxygen as an "exogenous" reactant—for instance, the one-electron reduction of O_2, which produces superoxide radical, $O_2{}^{\cdot-}$, or the energy transfer between chlorophyll triplet and oxygen (which is in the triplet state in its ground state), which yield singlet oxygen. These various oxygen species are known as "reactive oxygen species" (ROS) and are responsible for severe damage to biological material, prominent among which are the light-induced damages to Photosystem II, which has a lifetime as short as 5 minutes under high light conditions (see Krieger-Liszkay et al. [88], Murata et al. [89], Vass [90] for recent reviews). The consequences of these unwanted processes go far beyond these damages since they can impact the overall photosynthetic productivity to a critical extent [91–93].

Since oxygen has shaped the biosphere we live in (see, e.g., Kutschera and Niklas [94]), ROS are involved in a large variety of biological processes, such as aging (see, e.g., Breitenbach et al. [95] and references therein), signaling (Kim et al. [96] and references therein), and cellular disorder ([97,98] and references therein). There are many reviews on the subject of ROS production in the biological electron transfer chain, and discussing these goes beyond the scope of this chapter. However, we may briefly recast here the energetic aspects of these issues. Before anything, among the vast family of ROS, one must distinguish singlet oxygen from the other species, which are oxygen derivatives in various redox states (oxygen radical species, such as superoxide, hydroxyl radical, and peroxide). Singlet oxygen differs from triplet oxygen (the ground state of molecular oxygen) in that the two unpaired electrons have antiparallel, rather than parallel, spins. The two states of the molecule have the same number of electrons and thus the same redox state. Singlet oxygen production involves energy transfer from another molecule in its excited state and not electron transfer. In photosynthetic electron transfer chains, singlet oxygen is thought to be produced in reaction centers as the result of energy transfer from a chlorophyll in its triplet state. This stems from the simple fact that triplet chlorophylls have the two properties required to interact with triplet oxygen and $^1O_2$, they lie high enough in energy to allow this reaction (which requires about 1 eV; Schweitzer and Schmidt [99]), and, above all, (and at variance with the singlet chlorophyll counterparts, which also bears enough energy to promote the formation of $^1O_2$), they are relatively long-lived (or at least longer lived than the singlet state, which is quenched by charge separation in a few picoseconds; see the relevant chapters in this book). But how is the triplet state of the chlorophyll formed, and why is it not avoided if it is so potentially deleterious? These questions have been addressed by many authors [76,100–103], and we will only review the mainstream answers here. In principle, a chlorophyll triplet can be formed via intersystem crossing (or spin dephasing) of a singlet excited state. When the charge separating trap is active and the singlet excited state is connected to the photochemical trap, the probability of occurrence of triplet formation is very low because trapping of the singlet excited state is much faster. Yet, in some instances, the reduced state of the secondary or tertiary electron acceptor may build up, thereby preventing forward electron transfer from proceeding normally. The formed radical pair then decays via charge recombination to repopulate either the ground state or a chlorophyll triplet, depending on the spin of the transferred electron and the energy level of the recombining radical pair. As shown in Figure 16.4, the matter thus sums up to a competition between different decay pathways, one of which has a 2/3 probability to generate ^{3}Chl (we note here that the ^{3}Chl can be quenched by nearby pigments, prominent among which are

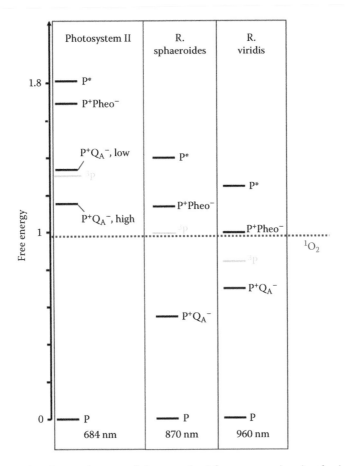

Figure 16.4 A scheme showing estimates of the standard free energy levels of relevant excited states and radical pairs in type II photosynthetic reaction centers. In blue is shown the chlorophyll triplet (^{3}P) formed by charge recombination of the ^{3}P$^+$Pheo$^{\circ-}$ states in each of the illustrated cases, which differ by the energy input of the photon. In red is shown the energy required to form singlet oxygen. For each type of reaction center, the energy available in the first excited singlet state (P*) corresponds to the photon absorbed and that is indicated at the foot of each column (see Chapter 9).

carotenoids; see Chapter 3 and Telfer [104] and references therein).

In type II reaction centers, the main determinant of the ^{3}Chl production yield is the energy gap between the P$^+$Q$_A$$^{\circ-}$ and P$^+$Pheo$^{\circ-}$ states: the larger this gap, the smaller the yield of the indirect charge recombination pathway and hence that of ^{3}Chl formation. Yet, the larger this gap, the less the free energy borne by the P$^+$Q$_A$$^{\circ-}$ state. As discussed by Rutherford et al., there is a trade-off between two antagonistic requirements, the preservation of the energetic yield upon a single turnover on one hand and the minimization of the triplet yield formation on the other [76]. This trade-off manifests itself by the redox tuning of the free energy difference between P$^+$Q$_A$$^{\circ-}$ and P$^+$Pheo$^{\circ-}$, either by tuning the midpoint potential of the quinone by changing its chemical nature, as happens in the reactions of purple bacteria such as Rhodobacter sphaeroides (R. sphaeroides) or B. viridis, or by tuning the midpoint potential through point mutations, such as at position 130 of the D$_1$ subunit of Photosystem II, which is involved in an H-bond with the Pheo. Whereas the former case likely reflects the evolutionary pressure exerted on R. sphaeroides reaction centers in which the free energy of the P$^+$Pheo$^{\circ-}$ state is larger than that of the corresponding state in B. viridis, the latter case involves the regulation of gene variants coding for the D$_1$ subunit (see [76,105,106]).

As mentioned earlier, the other oxygen-derived species that may cause proteolytic damages are superoxide, hydroxyl radical, hydrogen peroxide, etc. At variance with singlet oxygen, their production involves electron transfer from a reducing species to oxygen, so that it competes directly with the forward and productive electron transfer reactions. As such, these can be found in any biological process involving electron transfer reactions and intermediate species having enough reducing power to react with oxygen (the redox potential of the $O_2^{\cdot-}/O_2$ couple relative to 1 mol L^{-1} of O_2 is -160 mV [107], so that under atmospheric oxygen pressure, which corresponds to $[O_2] = 210$ μM, it lies in the -380 mV range*). But, as in the previous case, the reducing power cannot account on its own for the yield of ROS production, which also strongly depends on the lifetime of the reduced state. Under optimal conditions, normal electron flow thus outcompetes these bypass reactions involving oxygen owing to the short lifetime of the reducing intermediates. There is, however, one particularly interesting case where decreasing the lifetime of the potential electron donor is synonymous to increasing its reactivity: the semiquinone case. Indeed, as detailed earlier, the lower the stability of the semiquinone, the larger its propensity to act as an efficient electron donor. Conversely, the larger its stability, the smaller its electron-donating capability. As discussed in detail by Cape et al., one way that the cyt bc/b6f complexes can prevent superoxide production and other bypass reactions is by either stabilizing (decreasing the reducing power) or destabilizing (decreasing the lifetime) the intermediate SQ [83].

To conclude, it seems fair to claim that the basic requirements for electron transfer in biology have been established and that a few basic and user-friendly rules have been defined [108–110]. They highlight the requirement for proximity between cofactors and appropriate overall driving forces. Once these are in place, the promotion of productive electron transfer hardly demands any additional fine-tuning, as demonstrated by the fact that large variations in the energy levels of intermediates have little effect on the final (quantum) yield of the forward reactions (see, e.g., Alric et al. [73]). However, there are some notorious cases where the fine-tuning of energy levels does occur such as circumstances in which the desired energy-useful outcome of catalysis becomes less important than saving the system from damaging reactions, particularly with O_2, that put the viability of the organism in danger.

ACKNOWLEDGMENTS

This work was supported by the Centre National de la Recherche Supérieure and the "Initiative d'Excellence" program from the French state (Grant "DYNAMO," ANR-11-LABX-0011-01).

REFERENCES

1. Schoepp-Cothenet, B., R. van Lis, A. Atteia, F. Baymann, L. Capowiez, A. L. Ducluzeau, S. Duval, F. ten Brink, M. J. Russell, and W. Nitschke. 2013. On the universal core of bioenergetics. *Biochim Biophys Acta* 1827:79–93.
2. Berg, J. M., J. L. Tymoczko, and L. Stryer. *Biochemistry*. W. H. Freeman, New York, 2002.
3. Zwanzig, R. W. 1954. High-temperature equation of state by a perturbation method. I. Nonpolar gases. *J Chem Phys* 22:1420–1426.
4. Brooks, C. L., M. Karplus, and M. B. Pettitt. 1987. *Advances in Chemical Physics: A Theoretical Perspective of Dynamics, Structure, and Thermodynamics*. Wiley, New York.
5. Field, M. J., P. A. Bash, and M. Karplus. 1990. A combined quantum mechanical and molecular mechanical potential for molecular dynamics simulations. *J Comput Chem* 11:700–733.
6. Kollman, P. 1993. Free energy calculations: Applications to chemical and biochemical phenomena. *Chem Rev* 93:2395–2417.
7. Blomberg, M. R. and P. E. Siegbahn. 2010. Quantum chemistry as a tool in bioenergetics. *Biochim Biophys Acta* 1797:129–142.

---

* This rather negative value may give the false impression that only few reduced intermediates may bear enough reducing power to produce superoxide. This impression is false because it confuses redox and midpoint potential. Whereas the latter corresponds to the potential reached when the concentration of the reduced and oxidized forms are equal, the former characterizes the reactivity of a redox couple in all the other cases. As an example, the redox potential of a couple with a midpoint potential of -240 mV is -360 mV when the ratio Red/Ox = 100.

8. Bashford, D. and M. Karplus. 1990. pKa's of ionizable groups in proteins: Atomic detail from a continuum electrostatic model. *Biochemistry* 29:10219–10225.

9. Gunner, M. R. and B. Honig. 1991. Electrostatic control of midpoint potentials in the cytochrome subunit of the *Rhodopseudomonas viridis* reaction center. *Proc Natl Acad Sci USA* 88:9151–9155.

10. Warshel, A. and J. Aqvist. 1991. Electrostatic energy and macromolecular function. *Annu Rev Biophys Biophys Chem* 20:267–298.

11. Warshel, A. and A. Papazyan. 1998. Electrostatic effects in macromolecules: Fundamental concepts and practical modeling. *Curr Opin Struct Biol* 8:211–217.

12. Gunner, M. R., J. Mao, Y. Song, and J. Kim. 2006. Factors influencing the energetics of electron and proton transfers in proteins. What can be learned from calculations. *Biochim Biophys Acta* 1757:942–968.

13. Warshel, A., P. K. Sharma, M. Kato, and W. W. Parson. 2006. Modeling electrostatic effects in proteins. *Biochim Biophys Acta* 1764:1647–1676.

14. Ullmann, G. M. and E. Bombarda. 2013. pK(a) values and redox potentials of proteins. What do they mean? *Biol Chem* 394:611–619.

15. Rappaport, F., M. Guergova-Kuras, P. J. Nixon, B. A. Diner, and J. Lavergne. 2002. Kinetics and pathways of charge recombination in photosystem II. *Biochemistry* 41:8518–8527.

16. Grabolle, M. and H. Dau. 2005. Energetics of primary and secondary electron transfer in Photosystem II membrane particles of spinach revisited on basis of recombination-fluorescence measurements. *Biochim Biophys Acta* 1708:209–218.

17. Rappaport, F. and B. A. Diner. 2008. Primary photochemistry and energetics leading to the oxidation of the Mn$_4$Ca cluster and to the evolution of molecular oxygen in Photosystem II. *Coord Chem Rev* 252:259–272.

18. Antal, T. K., I. B. Kovalenko, A. B. Rubin, and E. Tyystjarvi. 2013. Photosynthesis-related quantities for education and modeling. *Photosynth Res* 117:1–30.

19. Peters, J. W. and J. B. Broderick. 2012. Emerging paradigms for complex iron-sulfur cofactor assembly and insertion. *Annu Rev Biochem* 81:429–450.

20. Roche, B., L. Aussel, B. Ezraty, P. Mandin, B. Py, and F. Barras. 2013. Iron/sulfur proteins biogenesis in prokaryotes: Formation, regulation and diversity. *Biochim Biophys Acta* 1827:455–469.

21. Pettigrew, G. W. and G. R. Moore. 1987. *Cytochromes c: Biological Aspects*. Springer-Verlag, Berlin, Germany.

22. Stroebel, D., Y. Choquet, J. L. Popot, and D. Picot. 2003. An atypical haem in the cytochrome b(6)f complex. *Nature* 426:413–418.

23. Merchant, S. S. and A. Sagasti. 2006. Precious metal economy. *Cell Metab* 4:99–101.

24. Schoepp-Cothenet, B., R. van Lis, P. Philippot, A. Magalon, M. J. Russell, and W. Nitschke. 2012. The ineluctable requirement for the trans-iron elements molybdenum and/or tungsten in the origin of life. *Sci Rep* 2:263.

25. Grimaldi, S., B. Schoepp-Cothenet, P. Ceccaldi, B. Guigliarelli, and A. Magalon. 2013. The prokaryotic Mo/W-bisPGD enzymes family: A catalytic workhorse in bioenergetic. *Biochim Biophys Acta* 1827:1048–1085.

26. Mendel, R. R. 2013. The molybdenum cofactor. *J Biol Chem* 288:13165–13172.

27. Gunner, M. R., E. Alexov, E. Torres, and S. Lipovaca. 1997. The importance of the protein in controlling the electrochemistry of heme proteins: Methods of calculations and analysis. *J Biol Inorg Chem* 2:126–134.

28. Pessanha, M., L. Morgado, R. O. Louro, Y. Y. Londer, P. R. Pokkuluri, M. Schiffer, and C. A. Salgueiro. 2006. Thermodynamic characterization of triheme cytochrome PpcA from Geobacter sulfurreducens: Evidence for a role played in e–/H+ energy transduction. *Biochemistry* 45:13910–13917.

29. Bewley, K. D., K. E. Ellis, M. A. Firer-Sherwood, and S. J. Elliott. 2013. Multi-heme proteins: Nature's electronic multi-purpose tool. *Biochim Biophys Acta* 1827:938–948.

30. Paquete, C. M. and R. O. Louro. 2014. Unveiling the details of electron transfer in multicenter redox proteins. *Acc Chem Res* 47:56–65.

31. Zheng, Z. and M. R. Gunner. 2009. Analysis of the electrochemistry of hemes with E(m)s spanning 800 mV. *Proteins* 75:719–734.

32. Parson, W. W., Z.-T. Chu, and A. Warshel. 1990. Electrostatic control of charge separation in bacterial photosynthesis. *Biochim Biophys Acta* 1017:251–272.

33. Lin, X., H. A. Murchison, V. Nagarajan, W. W. Parson, J. P. Allen, and J. C. Williams. 1994. Specific alteration of the oxidation potential of the electron donor in reaction centers from *Rhodobacter sphaeroides*. *Proc Natl Acad Sci USA* 91:10265–10269.

34. Brettel, K. 1997. Electron transfer and arrangement of the redox cofactors in photosystem I. *Biochim Biophys Acta* 1318:322–373.

35. Fromme, P., P. Jordan, and N. Krauss. 2001. Structure of photosystem I. *Biochim Biophys Acta* 1507:5–31.

36. Joliot, P. and A. Joliot. 1999. In vivo analysis of the electron transfer within photosystem I: Are the two phylloquinones involved? *Biochemistry* 38:11130–11136.

37. Guergova-Kuras, M., B. Boudreaux, A. Joliot, P. Joliot, and K. Redding. 2001. Evidence for two active branches for electron transfer in photosystem I. *Proc Natl Acad Sci USA* 98:4437–4442.

38. Agalarov, R. and K. Brettel. 2003. Temperature dependence of biphasic forward electron transfer from the phylloquinone(s) A1 in photosystem I: Only the slower phase is activated. *Biochim Biophys Acta* 1604:7–12.

39. Santabarbara, S., I. Kuprov, P. J. Hore, A. Casal, P. Heathcote, and M. C. Evans. 2006. Analysis of the spin-polarized electron spin echo of the [P700+ A1-] radical pair of photosystem I indicates that both reaction center subunits are competent in electron transfer in cyanobacteria, green algae, and higher plants. *Biochemistry* 45:7389–7403.

40. Karyagina, I., Y. Pushkar, D. Stehlik, A. van der Est, H. Ishikita, E. W. Knapp, B. Jagannathan, R. Agalarov, and J. H. Golbeck. 2007. Contributions of the protein environment to the midpoint potentials of the A1 phylloquinones and the Fx iron-sulfur cluster in photosystem I. *Biochemistry* 46:10804–10816.

41. Srinivasan, N. and J. H. Golbeck. 2009. Protein-cofactor interactions in bioenergetic complexes: The role of the A1A and A1B phylloquinones in Photosystem I. *Biochim Biophys Acta* 1787:1057–1088.

42. Srinivasan, N., R. Chatterjee, S. Milikisiyants, J. H. Golbeck, and K. V. Lakshmi. 2011a. Effect of hydrogen bond strength on the redox properties of phylloquinones:

43. Srinivasan, N., S. Santabarbara, F. Rappaport, D. Carbonera, K. Redding, A. van der Est, and J. H. Golbeck. 2011b. Alteration of the H-bond to the A(1A) phylloquinone in Photosystem I: Influence on the kinetics and energetics of electron transfer. *J Phys Chem B* 115:1751–1759.

44. Merry, S. A. P., P. J. Nixon, L. M. C. Barter, M. J. Schilstra, G. Porter, J. Barber, J. R. Durrant, and D. Klug. 1998. Modulation of quantum yield of primary radical pair formation in photosystem II by site directed mutagenesis affecting radical cations and anions. *Biochemistry* 37:17439–17447.

45. Cuni, A., L. Xiong, R. T. Sayre, F. Rappaport, and J. Lavergne. 2004. Modification of the pheophytin midpoint potential in Photosystem II: Modulation of the quantum yield of charge separation and of charge recombination pathways. *Phys Chem Chem Phys* 6:4825–4831.

46. Li, Y., M. G. Lucas, T. Konovalova, B. Abbott, F. MacMillan, A. Petrenko, V. Sivakumar et al. 2004. Mutation of the putative hydrogen-bond donor to P700 of photosystem I. *Biochemistry* 43:12634–12647.

47. Olson, T. L., J. C. Williams, and J. P. Allen. 2013. Influence of protein interactions on oxidation/reduction midpoint potentials of cofactors in natural and de novo metalloproteins. *Biochim Biophys Acta* 1827:914–922.

48. Raphael, A. and H. Gray. 1991. Semisynthesis of axial-ligand (position 80) mutants of cytochrome c. *J Am Chem Soc* 113:1038–1040.

49. Armstrong, F. A., R. Camba, H. A. Heering, J. Hirst, L. J. Jeuken, A. K. Jones, C. Leger, and J. P. McEvoy. 2000. Fast voltammetric studies of the kinetics and energetics of coupled electron-transfer reactions in proteins. *Faraday Discuss*:191–203.

50. Baymann, F., N. L. Barlow, C. Aubert, B. Schoepp-Cothenet, G. Leroy, and F. A. Armstrong. 2003. Voltammetry of a "protein on a rope". *FEBS Lett* 539:91–94.

51. Woodbury, N. W., W. W. Parson, M. R. Gunner, R. C. Prince, and P. L. Dutton. 1986. Radical-pair energetics and decay mechanisms in reaction

centers containing anthraquinones, naphtho-quinones of benzoquinones in place of ubiquinone. *Biochim Biophys Acta* 851:6–22.

52. Sebban, P. and C. A. Wraight. 1989. Heterogeneity of the $P^+Q_A^-$ recombination kinetics in reaction centers from *Rhodopseudomonas viridis*: The effect of pH and temperature. *Biochim Biophys Acta* 974:54–65.

53. Nitschke, W. and A. W. Rutherford. 1994. The tetrahaem cytochromes associated with photosynthetic reaction centres: A model system for intraprotein redox centre interactions. *Biochem Soc Trans* 22:694–699.

54. Alric, J., A. Cuni, H. Maki, K. V. Nagashima, A. Vermeglio, and F. Rappaport. 2004. Electrostatic interaction between redox cofactors in photosynthetic reaction centers. *J Biol Chem* 279:47849–47855.

55. Leger, C., F. Lederer, B. Guigliarelli, and P. Bertrand. 2006. Electron flow in multicenter enzymes: Theory, applications, and consequences on the natural design of redox chains. *J Am Chem Soc* 128:180–187.

56. Marcus, R. A. 1956. On the theory of oxidation-reduction reactions involving electron transfer. *J Chem Phys* 24:966–978.

57. Jortner, J. 1980. Dynamics of electron transfer in bacterial photosynthesis. *Biochim Biophys Acta* 594:193–230.

58. Crofts, A. R. and S. Rose. 2007. Marcus treatment of endergonic reactions: A commentary. *Biochim Biophys Acta* 1767:1228–1232.

59. Winkler, J. R. and H. B. Gray. 2014. Long-range electron tunneling. *J Am Chem Soc* 136(8):2930–2939.

60. Santabarbara, S., K. E. Redding, and F. Rappaport. 2009. Temperature dependence of the reduction of P(700)(+) by tightly bound plastocyanin in vivo. *Biochemistry* 48:10457–10466.

61. Barth, P., B. Lagoutte, and P. Setif. 1998. Ferredoxin reduction by photosystem I from *Synechocystis* sp. PCC 6803: Toward an understanding of the respective roles of subunits PsaD and PsaE in ferredoxin binding. *Biochemistry* 37:16233–16241.

62. Fromme, P., H. Bottin, N. Krauss, and P. Setif. 2002. Crystallization and electron paramagnetic resonance characterization of the complex of photosystem I with its natural electron acceptor ferredoxin. *Biophys J* 83:1760–1773.

63. Cruz-Gallardo, I., I. Diaz-Moreno, A. Diaz-Quintana, and M. A. De la Rosa. 2012. The cytochrome f-plastocyanin complex as a model to study transient interactions between redox proteins. *FEBS Lett* 586:646–652.

64. Venturoli, G., F. Drepper, J. C. Williams, J. P. Allen, X. Lin, and P. Mathis. 1998. Effects of temperature and deltaGo on electron transfer from cytochrome c2 to the photosynthetic reaction center of the purple bacterium *Rhodobacter sphaeroides*. *Biophys J* 74:3226–3240.

65. Fischer, N., M. Hippler, P. Setif, J. P. Jacquot, and J. D. Rochaix. 1998. The PsaC subunit of photosystem I provides an essential lysine residue for fast electron transfer to ferredoxin. *EMBO J* 17:849–858.

66. Hippler, M., F. Drepper, W. Haehnel, and J. D. Rochaix. 1998. The N-terminal domain of PsaF: Precise recognition site for binding and fast electron transfer from cytochrome c6 and plastocyanin to photosystem I of *Chlamydomonas reinhardtii*. *Proc Natl Acad Sci USA* 95:7339–7344.

67. Drepper, F., M. Hippler, W. Nitschke, and W. Haehnel. 1996. Binding dynamics and electron transfer between plastocyanin and photosystem I. *Biochemistry* 35:1282–1295.

68. Volbeda, A., M. H. Charon, C. Piras, E. C. Hatchikian, M. Frey, and J. C. Fontecilla-Camps. 1995. Crystal structure of the nickel-iron hydrogenase from *Desulfovibrio gigas*. *Nature* 373:580–587.

69. Haas, A. H. and C. R. Lancaster. 2004. Calculated coupling of transmembrane electron and proton transfer in dihemic quinol:fumarate reductase. *Biophys J* 87:4298–4315.

70. Rousset, M., Y. Montet, B. Guigliarelli, N. Forget, M. Asso, P. Bertrand, J. C. Fontecilla-Camps, and E. C. Hatchikian. 1998. [3Fe-4S] to [4Fe-4S] cluster conversion in Desulfovibrio fructosovorans [NiFe] hydrogenase by site-directed mutagenesis. *Proc Natl Acad Sci USA* 95:11625–11630.

71. Bertero, M. G., R. A. Rothery, M. Palak, C. Hou, D. Lim, F. Blasco, J. H. Weiner, and N. C. Strynadka. 2003. Insights into the respiratory electron transfer pathway from the structure of nitrate reductase A. *Nat Struct Biol* 10:681–687.

72. Hirst, J. 2013. Mitochondrial complex I. *Annu Rev Biochem* 82:551–575.

73. Alric, J., J. Lavergne, F. Rappaport, A. Vermeglio, K. Matsuura, K. Shimada, and K. V. Nagashima. 2006. Kinetic performance and energy profile in a roller coaster electron transfer chain: A study of modified tetra-heme-reaction center constructs. *J Am Chem Soc* 128:4136–4145.

74. Sicora, C. I., C. M. Brown, O. Cheregi, I. Vass, and D. A. Campbell. 2008. The psbA gene family responds differentially to light and UVB stress in *Gloeobacter violaceus* PCC 7421, a deeply divergent cyanobacterium. *Biochim Biophys Acta* 1777:130–139.

75. Vass, I. and K. Cser. 2009. Janus-faced charge recombinations in photosystem II photoinhibition. *Trends Plant Sci* 14:200–205.

76. Rutherford, A. W., A. Osyczka, and F. Rappaport. 2012. Back-reactions, short-circuits, leaks and other energy wasteful reactions in biological electron transfer: Redox tuning to survive life in O(2). *FEBS Lett* 586:603–616.

77. Mitchell, P. 1961. Coupling of phosphorylation to electron and hydrogen transfer by a chemi-osmotic type of mechanism. *Nature* 191:144–148.

78. Schoepp-Cothenet, B., C. Lieutaud, F. Baymann, A. Vermeglio, T. Friedrich, D. M. Kramer, and W. Nitschke. 2009. Menaquinone as pool quinone in a purple bacterium. *Proc Natl Acad Sci USA* 106:8549–8554.

79. Nitschke, W. and M. J. Russell. 2012. Redox bifurcations: Mechanisms and importance to life now, and at its origin—A widespread means of energy conversion in biology unfolds. *BioEssays* 34:106–109.

80. Crofts, A. R. and S. W. Meinhardt. 1982. A Q-cycle mechanism for the cyclic electron-transfer chain of *Rhodopseudomonas sphaeroides*. *Biochem Soc Trans* 10:201–203.

81. Osyczka, A., C. C. Moser, F. Daldal, and P. L. Dutton. 2004. Reversible redox energy coupling in electron transfer chains. *Nature* 427:607–612.

82. Osyczka, A., C. C. Moser, and P. L. Dutton. 2005. Fixing the Q cycle. *Trends Biochem Sci* 30:176–182.

83. Cape, J. L., M. K. Bowman, and D. M. Kramer. 2006. Understanding the cytochrome bc complexes by what they don't do. The Q-cycle at 30. *Trends Plant Sci* 11:46–55.

84. Mitchell, P. 1975. The protonmotive Q cycle: A general formulation. *FEBS Lett* 59:137–199.

85. ten Brink, F., B. Schoepp-Cothenet, R. van Lis, W. Nitschke, and F. Baymann. 2013. Multiple Rieske/cytb complexes in a single organism. *Biochim Biophys Acta* 1827:1392–1406.

86. Malnoe, A., F. A. Wollman, C. de Vitry, and F. Rappaport. 2011. Photosynthetic growth despite a broken Q-cycle. *Nat Commun* 2:301.

87. Bailleul, B., X. Johnson, G. Finazzi, J. Barber, F. Rappaport, and A. Telfer. 2008. The thermodynamics and kinetics of electron transfer between cytochrome b6f and photosystem I in the chlorophyll d-dominated cyanobacterium, *Acaryochloris marina*. *J Biol Chem* 283:25218–25226.

88. Krieger-Liszkay, A., C. Fufezan, and A. Trebst. 2008. Singlet oxygen production in photosystem II and related protection mechanism. *Photosynth Res* 98:551–564.

89. Murata, N., S. I. Allakhverdiev, and Y. Nishiyama. 2012. The mechanism of photoinhibition in vivo: Re-evaluation of the roles of catalase, alpha-tocopherol, non-photochemical quenching, and electron transport. *Biochim Biophys Acta* 1817:1127–1133.

90. Vass, I. 2012. Molecular mechanisms of photodamage in the Photosystem II complex. *Biochim Biophys Acta* 1817:209–217.

91. Marosvolgyi, M. A. and H. J. van Gorkom. 2010. Cost and color of photosynthesis. *Photosynth Res* 103:105–109.

92. Raven, J. A. 2011. The cost of photoinhibition. *Physiol Plant* 142:87–104.

93. Vinyard, D. J., G. M. Ananyev, and G. C. Dismukes. 2013. Photosystem II: The reaction center of oxygenic photosynthesis. *Annu Rev Biochem* 82:577–606.

94. Kutschera, U. and K. J. Niklas. 2013. Metabolic scaling theory in plant biology and the three oxygen paradoxa of aerobic life. *Theory Biosci (Theorie in den Biowissenschaften)* 132:277–288.

95. Breitenbach, M., M. Rinnerthaler, J. Hartl, A. Stincone, J. Vowinckel, H. Breitenbach-Koller, and M. Ralser. 2014. Mitochondria in ageing: There is metabolism beyond the ROS. *FEMS Yeast Res* 14:198–212.

96. Kim, C., R. Meskauskiene, K. Apel, and C. Laloi. 2008. No single way to understand singlet oxygen signalling in plants. *EMBO Rep* 9:435–439.

97. Bleier, L. and S. Drose. 2013. Superoxide generation by complex III: From mechanistic rationales to functional consequences. *Biochim Biophys Acta* 1827:1320–1331.

98. Gutowski, M. and S. Kowalczyk. 2013. A study of free radical chemistry: Their role and pathophysiological significance. *Acta Biochim Pol* 60:1–16.

99. Schweitzer, C. and R. Schmidt. 2003. Physical mechanisms of generation and deactivation of singlet oxygen. *Chem Rev* 103:1685–1757.

100. van Gorkom, H. J. 1985. Electron transfer in photosystem II. *Photosynth Res* 6:97–112.

101. van Gorkom, H. J. and J. P. M. Schelvis. 1993. Kok's oxygen clock: What makes it tick? The structure of P680 and consequences of its oxidizing power. *Photosynth Res* 38:297–301.

102. Fufezan, C., A. W. Rutherford, and A. Krieger-Liszkay. 2002. Singlet oxygen production in herbicide-treated photosystem II. *FEBS Lett* 532:407–410.

103. Vass, I., K. Cser, and O. Cheregi. 2007. Molecular mechanisms of light stress of photosynthesis. *Ann N Y Acad Sci* 1113:114–122.

104. Telfer, A. 2014. Singlet oxygen production by photosystem II under light stress: Mechanism, detection and the protective role of beta-carotene. *Plant Cell Physiol* 55(7):1216–1223.

105. Sander, J., M. Nowaczyk, J. Buchta, H. Dau, I. Vass, Z. Deak, M. Dorogi, M. Iwai, and M. Rogner. 2010. Functional characterization and quantification of the alternative PsbA copies in *Thermosynechococcus elongatus* and their role in photoprotection. *J Biol Chem* 285:29851–29856.

106. Sugiura, M. and A. Boussac. 2014. Some Photosystem II properties depending on the D1 protein variants in *Thermosynechococcus elongatus*. *Biochim Biophys Acta* 1837(9):1427–1434.

107. Wood, P. M. 1988. The potential diagram for oxygen at pH 7. *Biochem J* 253:287–289.

108. Moser, C. C., J. M. Keske, K. Warncke, R. S. Farid, and P. L. Dutton. 1992. Nature of biological electron transfer. *Nature* 355:796–802.

109. Moser, C. C., C. C. Page, R. Farid, and P. L. Dutton. 1995. Biological electron-transfer. *J Bioenerg Biomembr* 27:263–274.

110. Moser, C. C. and P. L. Dutton. 1996. Outline of theory of protein electron transfer. In *Protein Electron Transfer*. D. S. Bendall, ed. BIOS Scientific Publishers, Oxford, U.K., pp. 1–21.

PART 4

Light-Harvesting Systems in Action: Spectroscopy

Basic optical spectroscopy for light harvesting

ARVI FREIBERG AND GYŐZŐ GARAB

ABBREVIATIONS

17.1 INTRODUCTION

Photosynthesis starts with the absorption of light by photosynthetic pigment molecules and the migration of the excitation energy in light-harvesting antenna systems toward the reaction centers for primary charge separation, initiating a series of electron and proton transport processes in the photosynthetic redox systems. These are the light reactions of photosynthesis, that is, reactions elicited by electromagnetic radiation in the visible spectral range (with some organisms extending also to the ultraviolet and near-infrared ranges). In view of these basic processes, it is not surprising, and probably not an overstatement, that historically, steady-state and time-resolved optical spectroscopic methods have contributed significantly more to our knowledge about photosynthesis than any other technique. The "traditional" methods, electronic absorption and fluorescence spectroscopy, as well as the basic tools of polarization spectroscopic techniques, such as linear dichroism (LD), circular dichroism (CD), and fluorescence anisotropy, have become part of the everyday laboratory practice. These "classical" optical spectroscopic techniques still continue to provide important information on the molecular architecture, function, and structural–functional plasticity of systems at different levels of complexity—especially if combined with other techniques, from tools of structural and molecular biology to methods for monitoring physiological processes. In parallel, more advanced optical spectroscopic techniques such as fluorescence line narrowing (FLN), spectral hole burning (SHB), single molecule microscopy and spectroscopy (SMS), as well as anisotropic circular dichroism (ACD) and differential polarization (DP) microscopy techniques have been developed during the last decades.

This chapter, after outlining the physical bases of the relevant photophysical processes and spectroscopic techniques, will survey both the classical and advanced techniques, along with examples of the application of these techniques on photosynthetic antenna pigments, complexes, and in vivo systems. The functional significance of the information provided by the spectroscopic techniques is discussed.

The purpose of this chapter is to provide an introduction, both on the physical background and the biological significance—with no attempt for a comprehensive discussion. For further details, interested readers are referred to excellent textbooks and monographs (and references therein) published in the past years. The textbook of Blankenship [1] surveys photosynthesis and the molecular mechanisms of the light reactions. The monographs by Green and Parson [2], by Hunter et al. [3], and by Laisk et al. [4] provide a comprehensive treatment on light-harvesting antennas in plants and bacteria, from physical mechanisms to ecological significance and from basic synthetic pathways to evolution. For photosynthetic excitons, the monograph by Van Amerongen et al. [5] is a primary source. Many relevant chapters are contained in the two books on *Biophysical Techniques in Photosynthesis* edited by Amesz and Hoff [6] and Aartsma and Matysik [7], and in the books by Lakowicz [8], Serdyuk et al. [9], and Parson [10]. For basic properties of the main pigment molecules, chlorophylls, bacteriochlorophylls, and carotenoids, from quantum chemistry to biological functions, see the monographs edited by Grimm et al. [11] and Frank et al. [12]. On in vivo chlorophyll fluorescence and on nonradiative dissipation processes in light-harvesting antennas, respectively, one should see Papageorgiu and Govindjee [13] and Demmig-Adams et al. [14]. On general questions of photochemistry and photophysics of organic molecules, one should visit Turro et al. [15]. Last but not least, the reader should consult different chapters of this book.

17.2 ABSORPTION AND EMISSION OF LIGHT BY MOLECULES AND MOLECULAR ASSEMBLIES

Biological matter consists of molecules. In molecules, the negatively charged electrons are spatially confined by Coulomb interaction around the positively charged atomic nuclei. Quantum mechanics teaches that only certain spatial configurations where the electrons can be found around the nuclei are available. These so-called atomic (in atoms) or molecular (in molecules) orbitals have a specific spatial "shape" (distribution of electrons within the atom or molecule) and fixed energy. That is, when the electron is moved between the states, which involves change in the distribution of electrons, its

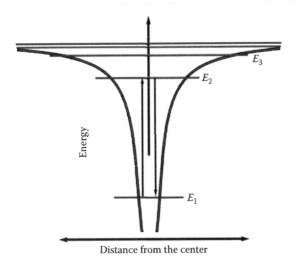

Figure 17.1 The potential energy of interaction between a proton at the coordinate origin and an electron as a function of the distance between them. Horizontal red lines represent the energy levels available for such a system (hydrogen atom) in vacuum. When energy from the electromagnetic field is absorbed (upward arrow), the electron initially in the ground state E_1 is moved to the excited state E_2 (or to any other available state energetically above the ground state). The downward arrow corresponds to emission, release of energy back to the electromagnetic field.

energy must change. For example, if the electron in a hydrogen atom (Figure 17.1) makes a transition from the higher energy state E_2 to the lower energy state E_1, its energy decreases. The energy difference $\Delta E = E_2 - E_1$ is released by emitting a photon, a quantum of electromagnetic field, of energy $\Delta E = h\nu$, where h is the Planck constant and ν is the frequency. The opposite process, from E_1 to E_2, is possible if the energy deficiency between the states is covered by an external source of photons. In this case, the photon is absorbed. The photon absorption and emission in these transitions is a prime consequence of the energy conservation law. Since the spatial image of atoms and molecules is related to their electronic structure, the "shape" of the atoms and molecules changes along with excitation and/or de-excitation.

Measuring the energy of the absorbed or emitted photons upon transitions between different electronic states constitutes the basis of electronic absorption and emission spectroscopies, respectively. This is an important point to notice—in

spectroscopy always energy differences between the quantum states are recorded, not the energies of the states themselves. The set of different transition energies available in the system, plotted as a function of wavelength or frequency, makes up a spectrum of this system. The power of various spectroscopic methods relies on the fact that the observed spectra uniquely characterize not only individual atoms and molecules but also aggregates of molecules and their condensed phases, liquids and solids. Optical spectroscopy, with a history that can be traced back to Newton's seventeenth century optics experiments, has by now become one of the major tools of qualitative as well as quantitative analysis of materials.

Due to the additional degrees of freedom available in molecules, the spectra of molecules are in general more complex than the spectra of atoms. In contrast to the clear-cut line spectra of atoms, in the spectra of isolated (e.g., in low-pressure gas phase), small molecules, closely spaced groups of many different spectral lines are frequently observed. These groups, known as spectral bands, are due to rotational, rotational–vibrational, and rovibronic (combined rotational, vibrational, and electronic) transitions. The rotational spectral structure is due to free rotations of the molecule. The vibration of nuclei relative to their stable equilibrium positions causes the vibrational structure of the spectra. Usually, the following hierarchy is followed between the electronic transition energy E_n, the vibrational energy E_v, and the rotational energy E_J: $E_n \gg E_v \gg E_J$; E_n is often in the visible range, E_v in the infrared range, and E_J in the microwave range. The rotational fine structure is typically not revealed in the vibronic spectra of large biomolecules. This is why in the further discussion we do not pay attention to the rotational degrees of freedom.

In large biological macromolecules, absorption and/or emission of a photon can frequently be related to the excitation of electrons that belong to just a limited group of atoms within the macromolecule. A well-known example is the carbonyl group ($>C=O$) in the peptide bond of proteins, which is liable for the characteristic absorption of proteins in the ultraviolet region of the optical spectrum, at about 290 nm. Parts of a molecule that are responsible for their color are called chromophores. The chromophores are usually orientationally well fixed and do not rotate with respect to their macromolecular frame.

In the hydrogen atom referred to in Figure 17.1, the oscillating electromagnetic field induces oscillations of the negatively charged electron about the heavy positively charged proton (nucleus). If two equal and opposite electric charges e are displaced by a distance r, an electric dipole moment of magnitude er is created. The induced dipole moment is greatly enhanced when the resonance condition $h\nu = \Delta E_{fi}$ is met between the energy of the electrical field oscillations of light, $h\nu$, and the energy gap $\Delta E_{fi} = E_f - E_i$ between the initial (subscript i) and final (f) states. The interaction between the electromagnetic field and the electron thus produces a transition dipole moment $\boldsymbol{\mu}_{fi}$ as it oscillates between the two states of the atom, causing an exchange of energy between the electromagnetic field and the material system. In biological macromolecules, the transition dipole moment associated with the two states is a complex vector quantity related to the distribution of charge within the system. The square of its magnitude $|\boldsymbol{\mu}_{fi}|^2$ gives the strength of the transition. Its spatial orientation determines how the system will interact with an electromagnetic wave of a given polarization. Light is plane or linearly polarized if the oscillating electric field vector $\boldsymbol{E}$ maintains its orientation as the wave propagates. It is circularly (generally elliptically) polarized if the

electric field vector describes a circle (ellipse) in any fixed plane intersecting and normal to the direction of propagation. The well-defined orientation of the transition dipole moments with respect to the molecular coordinates serves the basis of structural information that can be derived from LD or polarized fluorescence emission measurements (see Section 17.3).

Chlorophyll molecules in photosynthetic chlorophyll–protein complexes that are responsible for the green color of flora are one of the best known examples of chromophores. The absorption spectra of chlorophylls can be understood based on the properties of cyclic tetrapyrroles, which exhibit four transitions, each of which usually is accompanied by one or more vibrational bands (Gouterman model [16]). The orientation of different transition dipoles of chlorophylls is given with respect to the x–y molecular framework of the tetrapyrrole plane. By convention, the x and y molecular axes are defined as passing through the N atoms in pyrrole rings B and D, and A and C, respectively (see Figure 17.2). Correspondingly, the main electronic transitions are labeled with indices x and y; these are, in the order of increasing energy, Q_y and Q_x, and B_x and B_y transitions, for the main and minor red and the two Soret bands, respectively. The indices indicate that the

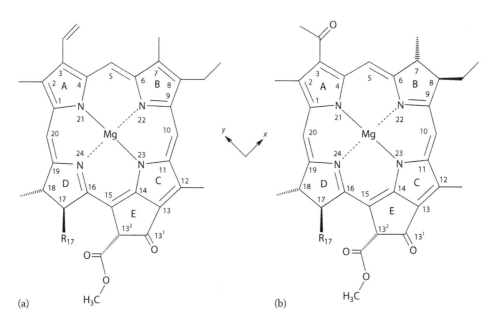

(a) (b)

Figure 17.2 The structures of chlorophyll-a **(a)** and bacteriochlorophyll-a **(b)**. Directions of the x- and y-axes in both molecules are shown by arrows; C-atoms are numbered according to the IUPAC nomenclature. See Figure 17.10 for the optical absorption and fluorescence emission spectra of these molecules.

transition dipoles are polarized along or close to the corresponding axes, albeit for some chlorophyll species and transitions there might be mixed cases (see [17]) and Section 17.2.4). The polarization of the Q_y transition of many chlorophylls is well defined; this has important consequences on the polarization of the fluorescence emission, which occurs from the lowest energy singlet excited state (see following text), and thus it is polarized along the y-axis. For the spectroscopic properties of other major photosynthetic molecules, carotenoids, and phycobilisome (bilin) pigments, see Chapters 3 and 5.

The wavelength of absorption and emission as well as the strength of optical transitions depend not only on the chemical nature of chromophores but also on their interactions with surroundings. Both the chromophore and the surrounding parts of photosynthetic light-harvesting proteins are complex. First, functional proteins are only defined together with their solvent environment. Therefore, the host matrix of a chromophore is the protein together with its immediate solvent surroundings. Second, in most cases, there are multiple chromophores in the light-harvesting proteins, which by interacting with each other form chromophore assemblies. The assembled chromophores may be of similar or different molecular origins (see Chapters 1, 2, and 5 of this book for more details of photosynthetic pigment assemblies). If a single chromophore in the closely packed assembly of chromophores is photoexcited, the excitation may transfer from one chromophore to another. Such a mobile excitation is called exciton. The exciton as a propagating excitation was introduced in the context of condensed matter physics in 1931 [18]. Although the first attempts to apply this concept to photosynthesis were unsuccessful [19], nowadays, it has become a cornerstone of the understanding of spectroscopic and excitation energy transfer properties of photosynthetic antenna complexes [5] (see also Chapter 11). The exciton moves relatively slowly between widely spaced chromophores but becomes "delocalized" between the two or more chromophores when they are close together (yet still avoiding significant overlapping of their molecular orbitals) and interacts strongly [5,20]. Since exciton interactions between the chromophores determine the energy level structure of photosynthetic antenna complexes, absorption and emission spectroscopies are able to provide very useful insights into the electronic structure of the chromophore assemblies. As it will be shown in Part 3 of this chapter, polarization

spectroscopic techniques offer especially detailed information on the architecture of the molecular assemblies and also carry information on intermolecular interactions. Interactions between the chromophores and their surroundings are the main cause of transitions between exciton levels and dissipation of the photosynthetic exciton energy. Various aspects of these interactions can be revealed by broadening of the spectra using the chromophores or their assemblies as local probes. These probes, which may be of native origin (as chlorophylls in photosynthetic protein complexes) or specially designed by genetic engineering methods or even the "label molecules" added externally, have spatial resolution in a nanometer scale [21].

The various aspects of the interaction between the chromophore and its matrix can be revealed by broadening and shift of the spectra [22,23]. Distinction is frequently made between the so-called inhomogeneous and heterogeneous spectral broadenings from one side and the homogeneous spectral broadening from another side. It is considered that the former two broadenings are due to slow (compared with the experimental time scale) fluctuations in the system, while the latter broadening is caused by much faster dynamic mechanisms. The so-called lifetime broadening and the pure dephasing broadening are both examples of homogeneous broadening along with the complex lineshape due to electron–vibrational and electron–phonon coupling, which in some detail will be discussed in the following. The difference between inhomogeneous and heterogeneous broadening, however, is rather subtle [24]. Let us take a protein as an example. It is said that the quasi-continuous conformational variations of the protein matrix lead to inhomogeneous broadening of the spectra. On the other hand, proteins may provide multiple binding pockets for the chromophores of the same chemical identity resulting in clearly distinguishable spectral bands. This is called heterogeneous spectral broadening and may be considered as an analog of the Shpol'skii effect in the spectroscopy of organic molecules in solid crystalline matrices [25]. A well-known example of this kind of spectra provides the light-harvesting 2 (LH2) complex from purple photosynthetic bacteria. The absorption spectrum of the LH2 complex in the near-infrared spectral range shows two distinct bands peaking at 800 and 850 nm. Both of these bands are related to the same bacteriochlorophyll-a chromophores, albeit in very different multichromophore arrangement and surrounding (see [26] and Chapter 6 for structural aspects of LH2). The demonstrated sensitivity of the chlorophyll spectra in response to the surrounding protein matrix is an evolutionary advantage, which allows the light-harvesting systems of photosynthetic organisms (plants, algae, and bacteria) not only to optimize collection and trapping of solar energy but also to flexibly adapt to varying environmental conditions by changing their spectral properties. Physical mechanisms of such regulation have attracted considerable interest [27–34].

17.2.1 Absorption and excitation spectra

When a beam of light is incident on a sample of chromophores dissolved in a transparent solvent, part of the light may be absorbed. The absorption spectrum measures wavelengths (or frequencies) at which the molecules absorb light. A quantitative measure of the absorbed light is called absorbance, which is defined as

$$A = -\log\left(\frac{I_T}{I_0}\right) = -\log(T), \qquad (17.1)$$

where

I_0 is the intensity of the incident light
I_T is that of the transmitted light
$T = I_T/I_0$ is called the transmittance

Multiple concepts like absorption coefficient, oscillator strength, cross section, transition dipole moment, and Einstein's A and B coefficients are used to describe the strength of molecular optical transitions. The relationships among these descriptions of the light–matter interaction are in the case of two-level atomic systems rigorously reviewed in [35]. Here, we briefly explain the relationship between the most often used characteristics of the spectra.

The phenomenological molar absorption (or extinction) coefficient $\varepsilon'(v)$ is defined by Beer's (also known as Beer–Lambert) law

$$\frac{dI}{dl} = -\varepsilon'(v)IC, \qquad (17.2)$$

where

I is the light intensity
l is the path length of light usually measured in centimeters (cm)
C is the molar concentration of the sample measured in mole per liter units, mol/L (or M)

There is a common practice to solve Equation 17.2 using a decade (base 10) logarithm instead of natural logarithm. One then obtains

$$I(l) = I_0 \cdot 10^{-\varepsilon(v)lC} = I_0 \cdot 10^{-A(v)}. \qquad (17.3)$$

In Equation 17.3, $I(l)$ is the transmitted light at path length l, $\varepsilon(v) = \varepsilon'(v)/\ln 10$ is the molar absorptivity (or molar decadic extinction coefficient), measured in M/cm, and $A(v) = \varepsilon lC$ is the absorbance or optical density (OD) already defined in Equation 17.1.

Since $A \equiv OD$ depends linearly on the solute concentration and is additive for a mixture of chromophores, it is well suited for quantitative analysis of substances. The reader should be aware that OD is sometimes also defined per unit path length—in this case, $A = \varepsilon C$.

Absorption cross section $\sigma(v)$ is measured in cm^2 and is related to the absorption coefficient as

$$\sigma(v) = \varepsilon(v)\ln 10/N_A, \qquad (17.4)$$

where N_A is the Avogadro number.

As it was already indicated, molecules undergo a transition from an initial to a final electronic state of energy through the coupling of the electromagnetic field vector E to the transition dipole moment vector μ. The transition probability is proportional to the square of the scalar product of these two vectors (α is the angle between E and μ):

$$|E \cdot \mu|^2 = E^2\mu^2 \cos^2\alpha \propto A. \qquad (17.5)$$

The latter relation results from the fact that μ^2 is proportional to the integrated absorption coefficient $\int \frac{\varepsilon(v)}{v}dv$. Equation 17.5 serves as the basis of LD measurements (see Section 17.3).

Although the measurement of absorption spectra looks straightforward in a spectrophotometer, in samples containing particles with sizes commensurate with the wavelength of the visible light, such as granal thylakoid membrane ultrastructures (reviewed in [36]), the determination of "true" absorption spectra is not an easy task and requires a cautious approach. Light-harvesting complex II (LHCII), the main chlorophyll-a/b complex of plants readily forms extended multilamellar arrays or random aggregates. The investigation of such self-aggregated systems is of great interest because they can be used to mimic some of the in vivo properties of the thylakoid membranes, including the nonphotochemical quenching of chlorophyll fluorescence [37]. Aggregates of this kind in vitro or in vivo lead to substantial changes in the measured absorbance spectra. First, a long tail appears outside the principal absorption bands, which originates from elastic light scattering. Via deflection of the measuring beam from a straight path, the light scattering leads to a weakening of the transmitted light. This phenomenon occurs over the entire spectral range, irrespective of whether the particles absorb light or not. The scattering cross section is a complex function, depending on the geometry and optical parameters of the particle, on the properties of the interface between the particle and the medium, and on the wavelength of the measuring light relative to the size of the particle. The general theory of the scattering of electromagnetic radiation is based on Maxwell's equations. Their solution for scattering by spheres of arbitrary size is known as Mie theory [38], which in the case of very small particles (compared with the incident light wavelength) reduces to Rayleigh scattering. In the Rayleigh scattering limit, the scattering cross section increases proportional to ν^4 (or λ^{-4}) [39]. Since the scattering spectrum is physically related, via the Kramers–Kronig relationship, to the absorption spectrum, which depends on the imaginary part of the refractive index, in case of absorbing particles, there also is spectrally selective light scattering [40,41]. The wavelength and angular dependence of light scattering can be quite complex, but in simplified cases, it can be expressed as a function of a size parameter and the complex relative refractive index of the particle; see [42–45] for related examples on photosynthetic materials.

The analysis of the scattering spectra as well as the more refined measurements of the so-called static and dynamic light scattering, techniques widely used in colloids, yield important information on the size and shape of scattering particles [46]. However, since light scattering significantly distorts the absorption spectrum, one frequently wishes to minimize its effect. The scattering contribution to the absorption spectra can be diminished by placing the cuvette close to the detector, in front of which there is an opal glass to increase its acceptance angle. Scattering can also be taken into account by measuring two transmission spectra recorded by placing the sample cuvette at two extreme positions in the sample compartment: one position close to and another far from the detector. This allows offsetting the scattering contribution to the absorption spectrum [47]. Further corrections might be needed with large clusters of small units, such as LHCII trimers. The presence of large clusters can lead to flattening (or sieving) of the main absorption bands [48]. When the measuring beam passes through a suspension of absorbing particles, all particles do not get equal exposure to the beam but those "in front" with strong absorbance can screen some of those that lie behind. This distortion can be corrected by appropriate modeling [47]. In order to record essentially distortion-free absorption spectra, integrating spheres are to be used, which warrant both high sensitivity and the removal of the scattering artifacts of the absorption spectra. In the integrating cavity absorption meters, the liquid sample (e.g., a dilute suspension) fills in the entire sphere with diffusive or specular reflecting walls, prolonging the optical path length. In these systems, however, deviations from Beer's law occur, which must be taken into account [49,50]. Scattering and flattening artifacts in the absorption spectra can also be corrected via measuring excitation spectra—provided that the fluorescence intensity of the sample is proportional to the absorption.

The excitation spectrum is determined by those wavelengths/frequencies of light that produce emission (luminescence) from the molecule. In large organic molecules including chlorophylls, the luminescence occurs in appreciable yield only from the lowest excited electronic state of a given multiplicity (Kasha rule). This implies very fast vibrational relaxation of excitation energy compared with the lifetime of the lowest excited state (see Section 17.2.2). Depending on the task, the luminescence can be monitored selectively, at a particular wavelength or integrally, over the whole emission spectrum. The intensity of the peaks in the excitation spectrum I_{ex} is directly proportional to the number of photons absorbed, $I_0 - I_T$, as well as to the quantum yield, φ_{lu}, of the luminescence at a selected wavelength/frequency:

$$I_{ex} = \varphi_{lu}\left(I_0 - I_T\right) = \varphi_{lu}I_0\left(1 - T\right). \quad (17.6)$$

Speaking specifically about the fluorescence excitation spectrum, the emission quantum yield φ_{lu} in Equation 17.6 is replaced by the fluorescence quantum yield φ_{flu}. As will be in some detail explained in Section 17.2.2, the excited state of a molecule may dissipate (get de-excited) by producing fluorescence with the rate k_{flu} and phosphorescence with the rate k_{pho}, or nonradiatively via internal conversion and intersystem crossing with the common rate of k_{nr}. The fluorescence quantum yield is then evaluated as

$$\varphi_{flu} = \frac{k_{flu}}{k_{flu} + k_{pho} + k_{nr}}. \qquad (17.7)$$

Since all the rates generally depend on the wavelength/frequency of the exciting light, so does φ_{flu}. One can now recognize that the information content of the fluorescence excitation spectrum even in the case of simple fluorescing molecules (fluorophores) is much richer than the absorption spectrum. It specifically reports about the relative quantum yield of transitions producing fluorescence.

Equation 17.7 can easily be generalized by accounting processes that take place in complex systems and compete with the fluorescence. Relevant processes in light-harvesting proteins and photosynthetic membranes include energy transfer and trapping, described by the rates k_{et} and k_{tr}, respectively. Then the modified equation (Equation 17.7) reads as

$$\varphi_{flu} = \frac{k_{flu}}{k_{flu} + k_{pho} + k_{nr} + k_{et} + k_{tr}}. \qquad (17.8)$$

Superior sensitivity of fluorescence detection as compared with absorption and the high selectivity of identifying different fluorophores renders fluorescence excitation spectroscopy an invaluable tool for spectroscopic studies of complex molecular assemblies such as light-harvesting antennas (see [51] for specific examples).

It should finally be noticed that even when the quantum yield of luminescence is independent of the wavelength of the exciting radiation (Kasha–Vavilov rule), the peaks in the excitation spectrum will not have the same relative intensity as the peaks in the absorption spectrum, though they appear exactly at the same wavelength/frequency. This is because the absorption $A = -\log(T)$ and the excitation $I_{ex} = \varphi_{lu}I_0(1 - T) = \varphi_{lu}I_0(1 - 10^{-A})$ spectra are not directly proportional to each other. Only in a very dilute sample (where $I_{ex} \approx \varphi_{lu}I_0 A$ approximately holds), the relative intensities of the peaks in the two types of spectra approach the same value.

17.2.2 Emission: Fluorescence and phosphorescence

Fluorescence is a form of photoluminescence, which generally describes the phenomenon of light emission from matter after the absorption of a photon and not resulting from thermal excitations [8]. Another form is phosphorescence. The distinction between fluorescence and phosphorescence is related to the electron spin. According to the Pauli exclusion principle, a pair of electrons occupying the same electronic ground (or valence) state has opposite spins and is said to be in a singlet spin state. Absorbing a photon promotes one of the electrons to an excited state. Excited states are generally unstable, spontaneously decaying to the lowest electronic (ground) state via photon emission or via radiationless transition, releasing the excess energy as heat. The radiative decay of the singlet excited state into the singlet ground state results in fluorescence. That is, in the fluorescence process, the electron does not change its spin direction. Quantum mechanics sets formal constraints (selection rules) for possible transitions of a system from one state to another. It follows that transitions in which the spin direction is conserved are allowed, while the transitions accompanied with the change of the spin are forbidden. However, in reality, various mechanisms exist that break the spin selection rules. As a consequence, there is a small but finite probability that during the decay the electron undergoes a spin-flip, ending up in a triplet excited state. The radiation emitted along the triplet state decay into the singlet ground state (more generally, between any two levels that differ by their spin direction) is called phosphorescence. The Jablonski diagram [52] presented in Figure 17.3 visualizes these processes.

Very different lifetimes of singlet and triplet states reflect their variant origin. The fluorescence of molecules involves emitting light in 10^{-9}–10^{-6} s, while the release of phosphorescence is relatively delayed, typically lasting 10^{-4}–10^{2} s. The terms vibrational relaxation, internal conversion, and intersystem crossing are used to distinguish the radiationless deactivation processes taking place, respectively, in the same electronic state, between the vibrational energy levels of different electronic states of the same spin multiplicity, and between the different electronic states having different spin multiplicity. Vibrational energy relaxation, the process by which molecules in high energy quantum

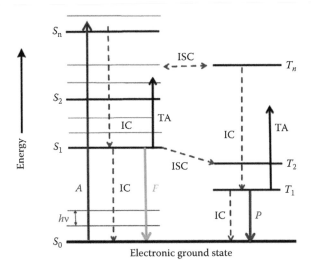

Figure 17.3 Modified Jablonski energy diagram recounting the routes by which an excited molecule can return to the ground state upon absorbing energy in the form of electromagnetic radiation. Thin horizontal lines indicate vibrational states—rotational states are not shown; S and T, respectively, denote singlet and triplet electronic excited states; A, F, and P relate to the photon absorption, fluorescence, and phosphorescence; IC and ISC refer to internal conversion and intersystem crossing processes. Photoexcitation and radiative decays are marked by solid lines, while the radiationless processes are shown with dashed lines. ISC from triplet to singlet excited states leads to the phenomenon named delayed fluorescence. Transient absorption (TA) is also indicated for the first singlet and triplet excited states.

states return to thermal equilibrium, typically takes 10^{-14}–10^{-11} s. The weak light emitted during the vibrational energy relaxation is called hot luminescence [53,54]. Although observed for organic impurity molecules, hot luminescence has so far not been recorded in photosynthetic systems. Fast vibrational relaxation is an important condition for the irreversible internal conversion and intersystem crossing decay pathways, which usually take place in the 10^{-9}–10^{-6} s time range.

The observed fluorescence (phosphorescence) decay time is determined by the intrinsic lifetime of the singlet (triplet) state and the competing decay mechanisms, probing the processes these states are involved in. Light-harvesting, energy transfer, and trapping processes in photosynthesis are all greatly concerned with the singlet excited states of (bacterio)chlorophyll and carotenoid chromophores. The photochemical energy conversion taking place in the reaction center chlorophyll–protein complex effectively competes with the fluorescence of the chromophores in light-harvesting complexes. Therefore, fluorescence is considered as one of the best probes of the primary photosynthesis processes.

Many other virtues of fluorescence in biophysical research are reviewed in [8,9,51]. Fluorescence-lifetime imaging microscopy, or FLIM, uses the differences in the exponential decay rate of fluorescing species for producing an image of a sample [8]. Combining high time resolution with high spatial resolution in FLIM has the advantage of minimizing the effect of photon scattering in thick layers of sample. This determines it as a very prospective technique in photosynthesis cellular research. For the wide range of applications of chlorophyll fluorescence transitions, probing the energy conversion processes as well as their regulatory mechanisms and metabolic and physiological variations in oxygenic photosynthetic organisms, see [13]). Weak, and therefore difficult to access, phosphorescence of chlorophylls has been less useful in the context of studies of photosynthesis' primary processes. However, triplet states are often involved in dark photochemical and photobiological reactions, which require long accumulation times. Also, hazardous photooxidation of photosynthetic pigments by singlet oxygen is an example of processes mediated by triplet states [55,56].

17.2.3 The shapes of absorption and emission spectra of molecules

As already noted, one of the reasons for homogeneous spectral broadening of an electronic transition is coupling with nuclear movements. These movements include vibrations of nuclei of both the chromophore and the matrix. The vibrations of the matrix are generally called phonons, the term originally coined for quantum nuclear excitations in regular lattices (crystals). Naming the vibrations of disordered biological matrices as phonons is justified by their (partly) delocalized nature, while the chromophore vibrations are rather localized on the chromophore. Phonons are responsible for energy relaxation processes as well as for damping of the electronic coherence. In photosynthetic complexes, they serve as acceptor modes in nonadiabatic excitation energy transfer between energetically inequivalent electronic or excitonic energy states, assuring energetically as well as spatially directed flow of excitation energy. In the following, we will demonstrate that whenever the transition from one electronic state to another takes place in a chromophore, it causes changes in the intrachromophore vibrational states and the matrix phonon states. These changes result from interactions between the chromophore electrons and the nuclei of both the chromophore and the matrix called electron–vibrational and electron–phonon coupling, respectively. In practice, the phonon and intrachromophore vibrational parts of optical spectra can easily be distinguished from each other. The spectrum of phonons looks like a continuum spanning from zero frequency to a maximum frequency, whereas the vibrational features are clear-cut and usually lie at higher energies than the delocalized protein vibrations, because of the generally smaller effective mass of the moving entities. These distinctions allow the effects of electron–vibrational coupling to be observed even in ambient-temperature spectra (as in Figure 17.10), differently from the electron–phonon coupling, which is best studied at low temperatures (Figure 17.11).

Exact wave functions to evaluate the homogeneously broadened spectra of molecules are not available. However, the treatment gets much simplified by applying suitable approximations such as the Born–Oppenheimer and harmonic approximations

and the Franck–Condon principle [22,23]. The Born–Oppenheimer approximation rests on the fact that the electrons, by virtue of their small mass, move much faster than the vibrating nuclei. It is thus assumed that the movements of electrons and nuclei are (almost) independent from each other. The total wave function of the state in the Born–Oppenheimer approximation is written as a product of an electronic part and a nuclear (vibrational) part. The parametric dependence of the electronic state energy on nuclear coordinates forms the so-called potential energy well, in which the nuclei can vibrate. The form of the potential energy surface sufficiently close to its minimum can always be considered parabolic, resulting in harmonic nuclear modes. The motion of the nuclei along one conformational coordinate is equivalent to that of a harmonic oscillator moving around an equilibrium position. The various oscillators in harmonic approximation are independent from each other and therefore can all be assessed separately.

The Franck–Condon principle concerns the intensity of vibronic transitions. The principle states that because the nuclei are much more massive than the electron, the most probable transition occurs between the states with the same nuclear position. The classical Franck–Condon principle thus essentially relies on reasoning similar to the Born–Oppenheimer approximation. Since according to the classical Franck–Condon principle the nuclear coordinates remain unchanged during the transition between the electronic states, such transitions are represented by vertical arrows in common energy level diagrams like shown in Figure 17.4. The more precise quantum-mechanical version of the Franck–Condon principle states that the transition probability $P_{fm,in}$ between the vibrational levels n and m in the initial and the final electronic states, respectively, is given by the square of the product of the electronic transition probability $\mu_{i,f}$ and the overlap integral of the respective vibrational wave functions of each oscillator k:

$$P_{fm,in} = \left| \mu_{i,f} \prod_k \langle m_k | n_k \rangle \right|^2. \qquad (17.9)$$

A comprehensive derivation of Equation 17.9 can be found in several excellent textbooks (see, e.g., [22,23]). We only notice that when the electronic transition probability amplitude $\mu_{i,f}$ does not change

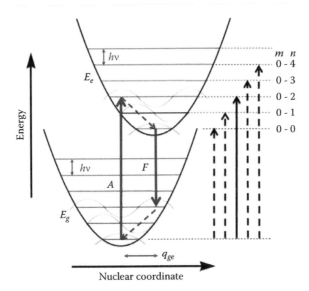

Figure 17.4 Potential energy diagrams of a two-atomic molecule with a single vibrational frequency plotted as a function of a nuclear conformational coordinate q in the electronic ground E_0 and excited E_e states. Schematic vibronic wave functions are given by gray lines for selected vibrational energy levels to illustrate the principle of the overlap integral in Equation 17.9. q_{ge} denotes the displacement of the excited electronic state potential energy curve relative to the ground state. The displacement in this specific example is selected so that the most probable transitions at thermal energies $k_B T \ll h\nu$ occur to the $n = 2$ level in absorption A and to the $m = 2$ level in fluorescence F. These transitions are specified by red solid arrows. Red dashed arrows indicate fast vibrational relaxation following the electronic transition. (Adapted from Pieper, J. and Freiberg, A., Electron-phonon and Exciton-phonon coupling in light harvesting: Insights from line-narrowing techniques, in *The Biophysics of Photosynthesis*, Golbeck, J. and van der Est, A. eds., Springer, New York, 2014.)

with the nuclear coordinates, the overlap integral term $\prod_k \langle m_k | n_k \rangle$ carries all the nuclear dependence in the transition dipole matrix element; the imposed insensitivity of $\mu_{i,f}$ on nuclear coordinates is called the Condon approximation.

As follows, we will discuss the spectral lineshape of an isolated chromophore whose electronic transition is for simplicity coupled to a single localized (intrachromophore) vibration. Later on, coupling to a distribution of delocalized protein phonons is introduced as well. It should be noticed that the theory presented here is strictly valid for localized electronic excitations only. The well separated from each other chlorophyll-a molecules in the CP29 light-harvesting complex of green plants and the ring of bacteriochlorophyll-a in the B800 compartment of the LH2 complex of photosynthetic bacteria constitute such systems. Understanding of the electron–vibrational (vibronic for short) and electron–phonon coupling effects in the spectra of strongly excitonically coupled chromophore systems such

as bacteriochlorophyll-a in the B850 compartment of the LH2 complex requires more involved theory and is deliberately left out from this basic chapter. In this respect, an interested reader may want to consult with references [5,58,59].

In Figure 17.4, two potential energy wells representing the ground E_g and excited E_e electronic states are drawn. The vibrational states of a single nuclear oscillator form a ladder of energy levels indicated by $n(m) = 0, 1, 2, \ldots$ in the excited (ground) state. The potential energy curves describe that upon the electronic excitation due to absorption of a light quantum, the system of nuclei undergoes a rearrangement, which leads to a shift of the equilibrium position of the excited electronic state potential with respect to the ground electronic state potential (displaced harmonic oscillator model). In general, also the change of the curvature of the excited state potential is expected, related to the change of the vibrational frequency at the electronic transition. These modifications determine the Franck–Condon type of vibronic couplings [60]. For simplicity, here, the parabolic potentials representing the ground and the excited electronic states are taken identical, corresponding to an identical oscillator frequency ν in both states.

Assuming the average thermal energy $k_B T \ll h\nu$, the transitions almost exclusively occur from the lowest $m = 0$ vibrational level of the ground state to $n \geq 0$ levels of the excited state. The progression of these vibronic transitions along with the pure electronic (0–0) transition composes the absorption/fluorescence spectrum. The 0–0 transition energy, noted as E_{00}, equals the energy difference between the zero vibrational levels in the ground and excited electronic states. It can be shown (see, e.g., [23]) that the peak intensities in the absorption/fluorescence spectrum, which are energetically separated by $h\nu$, follow the Poisson distribution:

$$|\langle 0|n\rangle|^2 = e^{-S} \frac{S^n}{n!}. \qquad (17.10)$$

In Equation 17.10, the term $|\langle 0|n\rangle|^2$ carries the name Franck–Condon factor for the particular transition $0 \to n$ and S is the so-called Huang–Rhys factor (sometimes also called the S-factor). The Huang–Rhys factor is a measure of the linear coupling strength between the nuclear and electronic degrees of freedom. It is related to the horizontal displacement q_{ge} of the ground and excited electronic state potential energy curves relative to each other: $S = q_{ge}^2/2$. The shift q_{ge} is dimensionless, counted relative to the classical amplitude, $\sqrt{h/\mu\nu}$, of the oscillator on the ground vibrational level (μ denotes the reduced mass of the oscillator).

The shape of the Poisson distribution (Equation 17.10), thus also the vibrational structure of the molecular electronic spectra, strongly varies with S. This is schematically demonstrated in Figure 17.5. In case of very weak coupling ($S \ll 1$, two left graphs of Figure 17.5), there is just a single 0–0 peak in the

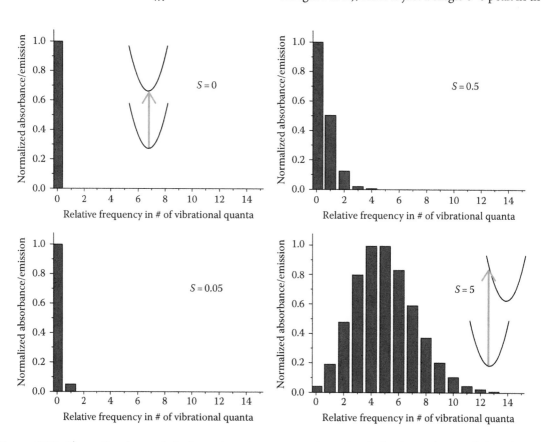

Figure 17.5 Absorption (or emission) spectra of a two-atomic molecule in dependence of the vibronic coupling strength S. The hypothetical electronic transition is coupled to a single vibrational mode; the frequency is counted relative to the 0–0 electronic transition energy. Insets show the ground and excited state potential energy diagrams for the limiting cases of zero ($S = 0$) and strong ($S = 5$) vibronic coupling.

spectrum; in case $S < 1$ (weak coupling, top right graph), there is a dominant 0–0 peak, followed by a short progression of vibronic lines falling off as S^n; and in case $S > 1$ (strong coupling, bottom right), there is a long progression of peaks with very weak 0–0 line and maximum intensity found at $n \approx S$. We will subsequently observe (see Figures 17.7 and 17.9) that the chlorophyll-type chromophores in photosynthetic complexes belong to the very weak coupling class. The strongest coupled vibrations both in chlorophyll-a and bacteriochlorophyll-a are characterized by $S < 0.05$ [61,62], that is, detectable in the vibronic progression is just the strong origin 0–0 peak; the first 0–1 replica of the origin peak is already very weak, as shown in the bottom left graph of Figure 17.5.

One might recognize that in the case of very strong coupling, $S \gg 1$, the envelope of the line progression takes a Gaussian profile with the peak energy at $E = E_{00} \pm Sh\nu$. The positive sign in this expression is related to the absorption and negative sign to the emission process. The width of the profile approximately equals $\sqrt{S}h\nu$. In the chromophores coupled to the protein environment, the energy increment $Sh\nu$ with respect to E_{00} is usually very fast dissipated by vibrational relaxation into heat in the excited (absorption process) or ground (emission process) potential energy surface. During this relaxation, the nuclear coordinate changes by q_{ge}, as shown by Figure 17.4. This is why the mean vibrational energy $Sh\nu$ is also known as the reorganization energy λ:

$$\lambda = Sh\nu. \qquad (17.11)$$

We see that the light absorbed at a particular frequency is generally reemitted at a lower frequency (i.e., at a longer wavelength). Equation 17.11 reveals a simple physical meaning of S, which is a mean number of vibrational quanta $h\nu$ created or lost upon the electronic transition: $\lambda/h\nu = S$. According to Equation 17.10, the coupling strength S for a discreet intramolecular mode can be calculated as

$$S = \frac{I_{0-1}}{I_{0-0}}, \qquad (17.12)$$

where I_{0-1} and I_{0-0} are the integral intensities of the lines attributable to the 0–1 and 0–0 vibronic transitions, respectively.

Within the given model approximations, the absorption and emission spectral shapes appear mirror symmetric with respect to the 0–0 transition energy E_{00}. The energy shift between the absorption and emission spectral maxima thus equals 2λ; it is customarily called the Stokes shift.

The preceding displaced harmonic oscillator model, where for simplicity only a single vibrational mode in harmonic potential energy surface at $T \sim 0$ K was considered, is directly applicable only for describing the spectra of individual two-atomic molecules. In isolated biomolecules, a large number of intramolecular vibrations are available. However, similar considerations generally hold, as long as the vibrations are harmonic (i.e., independent from each other) and the temperature is sufficiently low. The total vibronic coupling strength is then obtained as $S_{vib} = \sum_j S_j$, where j counts different vibrational modes. For example, in a chlorophyll molecule with a molecular formula $C_{55}H_{72}O_5N_4Mg$ (137 atoms), this number amounts to 405 (counted as $3N-6$, where N is the number of atoms in the molecule). Only about 50 vibrations are optically active, though, that is, visible in the absorption or fluorescence spectra (see Figures 17.7 and 17.9).

There also is basic conceptual similarity between the preceding model of countable number of displaced harmonic oscillators and a more general problem of coupling of electronic states of a chromophore to a continuum of phonons. The problem was first recognized and solved in solid-state physics, where the coupling of electronic states to phonons is the main issue [22,23]. It has been shown that the coupling of the electronic states to the harmonic phonon oscillators can be considered using a concept of a weighted density of states or a spectral density function $J(\nu)$ determined at 0 K. The spectral density function is a product of spectral density of phonons $W(\nu)$ and a frequency-dependent coupling $S(\nu)$ [22,23,60,63–65]. It is defined such that the integral of $J(\nu)$ over ν gives S_{ph}, the total Huang–Rhys factor for phonon coupling, while the integral of $\nu J(\nu)$ gives the total reorganization energy related to phonons. The detailed frequency-dependent coupling $S(\nu)$ can be obtained from studying the phonon sidebands if the density of phonon states $W(\nu)$ is available, for example, from neutron scattering. Unfortunately, such studies on chromoproteins are still very rare [66–68]. Alternatively, the shape of $W(\nu)$ can be evaluated theoretically, by molecular dynamics simulations based on the crystal structure of proteins. This approach is becoming increasingly more popular due to the advancement of computational resources [69]. The shape of the phonon spectral density in photosynthetic complexes and its temperature dependence is, for example, analyzed in [65,66,70–73].

The lineshape for an electronic transition coupled to the phonon bath is shown in the inset of Figure 17.6. It consists of a narrow and intense zero-phonon line (ZPL) and a broad phonon sideband (PSB), which in the absorption spectrum is attached to the ZPL from the high energy side. S_{ph} can be estimated from the experimental spectra at low temperatures as

$$\exp\left(-S_{ph}\right) = \frac{I_{ZPL}}{I_{ZPL} + I_{PSB}}, \qquad (17.13)$$

where I_{ZPL} and I_{PSB} are the integral intensities of the ZPL and the PSB. The relative ZPL intensity is also known as the Debye–Waller factor, $\alpha = \exp(-S_{ph})$. Analogous to the localized vibrational modes, one distinguishes weak electron–phonon coupling with S-factors smaller than 1 and strong coupling with S-factors larger than 1.

Given the established qualitative similarity between the localized vibrations of the chromophore and the delocalized phonon oscillators of the matrix, a general spectral density can be defined representing the distribution of all vibrational modes and their coupling strength to the electronic transition. It is then clear that at low temperatures each and every vibronic spectral line in Figure 17.5 has a shape similar to that shown in the inset of Figure 17.6. In the low order of the theory comprising the adiabatic, harmonic, Condon, and linear electron–phonon/vibrational coupling approximations (basic model [23]), the phonon sidebands of the vibronic lines are all similar, being replicas of the pure electronic (0–0) transition sideband. One of the very early demonstrations of this similarity principle was provided by the emission spectrum of the O_2^- impurity center in alkali halide crystals [74].

Although the Huang–Rhys factor has a rigorous meaning only within the Franck–Condon interaction model [60,64], and important deviations from this simplified model have been observed even in monomeric chlorophyll spectra (see following text), this concept is widely used in a much broader context as a general indicator of the coupling strength between the electronic and vibrational degrees of freedom. To discuss exciton–phonon coupling in the photosynthetic complexes that bind the excitonically coupled chromophores [5], however, a number of complicating aspects should be taken into account [22,58,70,75–80]. According to Davydov [58], within the adiabatic approximation, the exciton–bath interaction can be separated into the diagonal and nondiagonal parts. The diagonal part describes the modulation of the transition energies of the chromophores by their spatial and

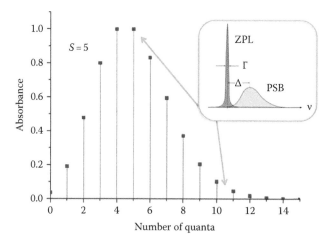

Figure 17.6 Absorption spectrum of a two-atomic molecule coupled to a continuum of phonons. The shapes of the lines in the vibronic progression, for simplicity drawn as vertical drop lines, are all similar and shown in the inset. The vibronic progression is characterized by the vibronic coupling strength $S = 5$.

orientational coordinates. Its physical origin is the change in van der Waals interactions between the chromophores when one of the chromophores is electronically excited. It can be shown that this diagonal interaction is primarily responsible for the phonon as well as vibronic sidebands of the exciton spectra [70,76]. The nondiagonal part of the interaction is due to variation of the exciton coupling energies by nuclear motions. This term is mostly accountable for the phonon-assisted exciton relaxation [70]. Assuming for simplicity only diagonal coupling (the dispersion force coupling limit [76]), a picture of mutually shifted excitonic potential energy surfaces arises, qualitatively similar to the one for individual chromophores depicted in Figure 17.4. However, due to the delocalized nature of excitons and phonons, the Huang–Rhys factor as well as the corresponding reorganization energy, which is related to the specific exciton state, is modified compared with those found in individual chromophores. The coupling for excitons appears generally reduced, and this reduction is increased with delocalization of the exciton (with the strength of the exciton coupling) [76,78]. Furthermore, since delocalization of various exciton states may be different, the reorganization energy and the Huang–Rhys factor both may vary between the exciton states [22].

The value of the Huang–Rhys factor for a selected exciton state can be accessed from the experimental phonon sideband using the same procedure (see Equation 17.12) as for localized excitations. Also, similar criteria of weak ($S < 1$) and strong ($S > 1$) coupling apply. Still, their physical meaning is different. In the case of excitons, the strength of the electron–phonon coupling involves a propagation aspect. At $S \ll 1$, excitons are almost free, only slightly disturbed by lattice vibrations. In the opposite case of $S \gg 1$, the exciton may be considered basically localized. It is then said that the exciton is self-trapped by the electron–phonon coupling [81]. The various situations in between these limiting cases may be described as excitons propagating in the form of an exciton polaron, that is, of an electronic excitation accompanied by phonons [79,82]. Vibrational and phonon sidebands of exciton transitions in dye aggregates, polymer materials, and biological light-harvesting complexes are still a matter of active study. The physical insights and computational methodologies of this research area can be found in [59,70,72,83–94].

17.2.4 Advanced techniques of absorption and fluorescence spectroscopy of light-harvesting systems

Ideally, one is determined to measure the homogeneously broadened spectra to get maximal information about intramolecular and intermolecular couplings in the system that are reflected in the vibronic structure of the molecular states. Conventional optical absorption and fluorescence/phosphorescence spectra of bulk samples, however, provide only limited information. This is because the spectra are generally inhomogeneously broadened. The inhomogeneous broadening relates not only to a distribution of transition frequencies but also to a distribution of electron–phonon and vibronic couplings, and other parameters, because the specific environment of a chromophore generally varies from binding site to binding site in the same protein as well as between the proteins in a macroscopic sample. In most cases, the inhomogeneous spectral broadening of photosynthetic pigment–protein complexes measured at cryogenic temperatures is well characterized by a Gaussian distribution with full width at half maximum, Γ_{inh}, between 80 and 200 cm^{-1} [95,96]. This distribution, referred to as an inhomogeneous distribution function (IDF), is an inherent measure for static heterogeneity of the protein matrix. Experimentally, the IDF can be determined by SHB spectroscopy (see following text). The observed absorption (or fluorescence) spectrum of an ensemble of chromophores in the protein matrix is obtained through a convolution of the homogeneously broadened spectrum with the IDF. Provided the ZPL in the homogeneously broadened spectrum is much narrower than the IDF, the latter also provides an estimate for the spectral resolution feasible by conventional spectroscopy.

To take full advantage of the information contained in the optical spectra, it is necessary to get rid of the inhomogeneous broadening. As long as the latter is an ensemble averaging effect, the obvious way to achieve this goal is to reduce the size of the ensemble of simultaneously recorded molecules, ultimately to just a single molecule. This approach became technically feasible 40–50 years ago, due to the advent of lasers as extremely bright and narrowband excitation sources, and because of advanced theoretical understanding of the purely electronic ZPL as the foundation for the high-resolution matrix spectroscopy and single impurity molecule spectroscopy, reviewed in [23,97–99]. It first resulted in the frequency-selective line-narrowing methods such as the FLN [100,101], persistent SHB [102,103], and differential fluorescence line narrowing (ΔFLN) [104–106] spectroscopies, followed by single molecule spectroscopy (SMS) [107–109]. The line-narrowing techniques are practical only at low temperatures, since they rely on the very high quality factor, E/Γ, of the zero-phonon singlet transitions, which can reach 10^7–10^{10}. In the quality factor, E is the transition energy and Γ is the homogeneous width of the ZPL (see Figure 17.6). Smaller Huang–Rhys factors provide higher selectivity of the line-narrowing methods. Since the Huang–Rhys factors grow with temperature and the ZPL usually disappears from the spectra of photosynthetic chromophores already at temperatures >50 K [73], the line-narrowing techniques are efficient only at cryogenic temperatures. In contrast, single molecule/single protein complex spectra can be studied from cryogenic to ambient temperatures, despite their significant thermal (homogeneous) broadening.

At this point, the reader might wonder, why should we learn low-temperature techniques, while photosynthesis is operating at ambient temperature? The line-narrowing and SMS spectroscopies allow detailed measurements of the homogeneously broadened spectrum of molecules/chromophores. As it was discussed in the previous section, the analysis of such spectra yields the Huang–Rhys factor S for intramolecular as well as for phonon modes. From this information, one can in principle predict the shape of the sample spectrum at any feasible temperature. The reverse process is generally not practical. Important dynamical information is also available from the shape of ZPL. Let us consider as an example the width Γ of the ZPL, which is determined by the excited state lifetime τ and the so-called pure dephasing time T_2:

$$\Gamma = \frac{1}{\pi c}\left(\frac{1}{2\tau} + \frac{1}{T_2}\right). \qquad (17.14)$$

In Equation 17.14, Γ is the measured in wavenumber units, times τ and T_2 in seconds, and c is the speed of light, in cm/s. At temperatures approaching 0 K, T_2 closes zero too, and the measurements of the Γ

yield τ: $\tau = (2\pi c\Gamma)^{-1}$. Due to the high sensitivity of dephasing processes on the temperature, the rising temperature causes significant broadening of the ZPL (along with rapid increase of the PSB intensity at the expense of the ZPL intensity). So much so that above ~10 K the second term in Equation 17.14 becomes dominating, allowing writing $\Gamma \approx (\pi c T_2)^{-1}$. Studies of the ZPL width as a function of temperature are thus potent to uncover the mechanisms of dephasing in photosynthetic systems [110,111] (see also [95] for a review). Dephasing mechanisms, however, are intimately related to all basic energy relaxation, transfer, and trapping processes in photosynthesis.

A major difference between the frequency-selective line-narrowing techniques (i.e., FLN, SHB, and ΔFLN) and SMS should be noticed. The former methods all rely on picking up a subset of chromophores (which still is a large number) according to a single parameter, the 0–0 transition frequency. The obtained spectra along other possible selection criteria, such as vibrational frequency, strength of electron–phonon coupling, or energy gap between the singlet and triplet transitions, still remain averaged, albeit generally less than in bulk spectra. Excitation energy–dependent electric field and pressure effects on spectral line shifts and width [112,113], on strength of electron–phonon coupling [114–116], and on fluorescence blinking [117] are all different aspects of this fact. SMS, by contrast, provides information, which is uniquely associated with an individual chromophore or, in the case of multichromophoric systems, with their assembly. Parallel measurements of large numbers of individual chromophores or their assemblies allow building distributions of physical observables of interest, devoid of ensemble averaging [24]. So the two methods provide complementary rather than identical information about spectral properties of chromophores/assemblies [99,118].

We shall next briefly introduce the different selective spectroscopy approaches. For details, the interested reader is referred to [24,57,95–97,99,108,119–129] and the references given therein. Selective spectroscopy methods for studying high-resolution vibronic spectra of biomolecules, among them chlorophyll and bacteriochlorophyll molecules, were first picked up by Avarmaa and Rebane [122]. The many follow-up works are well reviewed [96,123,130–134].

Conceptually, the FLN method is straightforward and easy to grasp. It is based on the excitation of fluorescence spectra at low temperatures with a spectrally narrow laser beam tuned to the very red edge of the absorption spectrum of the sample. Since in the absorption spectrum the broad PSBs are at higher energy with respect to the ZPL, just this subset of chromophores in the inhomogeneous manifold is selectively excited, which has an intense ZPL in resonance with the laser frequency. As is reviewed in [122,123], in the case of photosynthetic pigments, significant enhancement of vibronic structure and improvement of spectral resolution is achieved in the FLN spectrum compared with the conventional fluorescence spectrum. Figure 17.7 demonstrates this nicely in the case of pheophytin-a molecules. Pheophytins are chemical compounds that serve as the first electron carrier intermediate in the electron transfer pathway of the photosynthetic reaction centers found in purple bacteria and photosystem II. The strong scattered laser line in Figure 17.7 is cut off for clarity. Scanning the excitation light over the absorption spectrum allows in favorable cases also the excited electronic state vibronic structure to be studied by FLN. The main shortcoming of FLN is its inability to resolve the ZPL (and the low-frequency part of the PSB) due to strong

excitation light scattering spectrally overlapping with ZPLs. In addition, the PSB is contaminated with the so-called pseudo-PSB contribution into the spectrum. These aspects strongly prohibit quantitative evaluation of the Huang–Rhys factors from the FLN spectra.

Chromophores tend to photobleach under intense laser excitation. This commonly unfortunate reality of "burning a hole in the spectra" is turned into an advantage in SHB spectroscopy. SHB spectroscopy was discovered in 1974, independently in Tartu [102] and Moscow [103]. In SHB experiments, the transition frequencies of those molecules selected by the excitation laser are altered, so that the difference between the pre- and post-illuminated absorption spectra reveals a hole at the excitation frequency as well as concomitant vibronic structure holes shifted toward the blue (the so-called real-PSB hole) and red (pseudo-PSB hole) from the resonant hole. The difference spectrum, that is, the SHB spectrum (see Figure 17.8), corresponds to the bleached out subset of chromophores. This explains the name of this technique. A distribution of the hole depths, measured at constant burn fluence and plotted over the frequency range, the so-called action spectrum [96], yields the IDF. A nonlinear

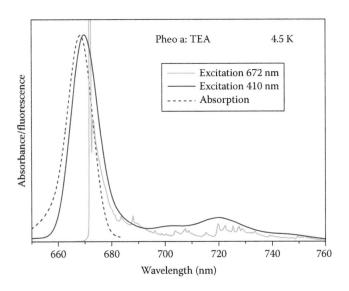

Figure 17.7 Comparison of the conventional fluorescence spectrum (red line) of pheophytin-a molecules in triethylamine matrix, nonselectively excited at 410 nm, and the FLN spectrum (green line), selectively excited at 674 nm with a narrowband laser (bandwidth, 2 cm^{-1}) into the Q_y absorption band. The absorption spectrum is also shown for reference (drawn with blue dashed line). All spectra are measured at 4.5 K.

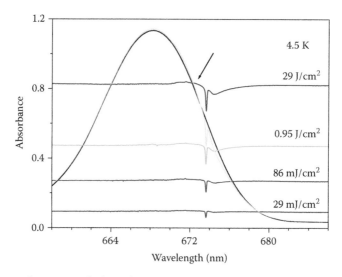

Figure 17.8 Hole-burned spectra of pheophytin-a in triethylamine at different burn fluencies applied at 673.7 nm. Seen toward red from the resonant hole is the pseudo-PSB hole. Arrow indicates the so-called antihole (increased absorbance), which in this sample is completely shadowing the real-PSB hole. The antihole is due to absorption of the hole-burning products. Shown on the background are the initial absorption spectrum (drawn with black line) and the modified absorption spectrum upon hole burning (turquoise).

dependence on the burning light fluence, high sensitivity to the hole-burning mechanism, and limited width of IDF complicate the interpretation of SHB data (for a detailed discussion of these issues, see [96,135–137]). The first obstacle can in principle be overcome by working at the low-fluence limit, which is, however, not always feasible. From the positive side, SHB is very useful for investigating dynamic broadening mechanisms of ZPLs. The technique can also be applied for studying nonfluorescing samples, differently from FLN and ΔFLN.

The ΔFLN technique combines elements from both FLN and SHB. Earlier realizations of this method [104–106] have been applied to enhance spectral selectivity of FLN mainly in the high-frequency range of vibronic lines. The ZPL region has become accessible only relatively recently due to the simultaneous recording of the whole spectrum using modern charge-coupled device (widely known as CCD) cameras [61,62,136,138]. This novel experimental approach for the first time yielded the resonantly excited ZPL together with the phonon and vibrational structure building on it, so that electron–phonon and electron–vibrational coupling strengths of photosynthetic pigments and protein complexes could be uniquely determined.

The ΔFLN spectrum (Figure 17.9) is obtained as a difference between the two FLN spectra recorded before and after an intermediate hole-burning step. This technique provides two important advantages over traditional SHB and FLN spectroscopy: first, the scattered laser light, which otherwise obscures the ZPL in FLN spectra, can be effectively eliminated in the difference spectrum, allowing direct measurement of the ZPL; second, the double selection via both SHB and FLN results in the ΔFLN spectrum, which in the low-fluence limit becomes identical to the homogeneously broadened spectrum [96,104,105,139].

Single molecule/single complex spectroscopy is achieved using very small (picomolar) concentrations of molecules or chromophore complexes and microspectroscopy setups. As already mentioned, differently from the selective spectroscopy techniques, SMS is applicable at both cryogenic and ambient temperatures. Most of the single molecule/single complex work has been focused either on the fluorescence emission spectroscopy or fluorescence excitation spectroscopy of a limited number of "well-behaving" chromophores with strong absorption, high fluorescence quantum efficiency, and exceptional photostability. Photostability is critical, since excitation

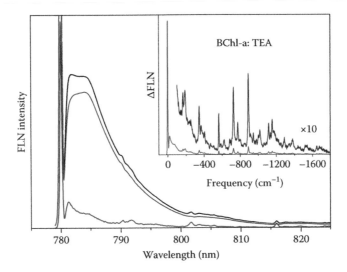

Figure 17.9 The ΔFLN spectrum (red curve) of bacteriochlorophyll-a in triethylamine, recorded at 4.5 K using the 780.2 nm excitation. The black and blue curves show the pre- and postburn FLN spectra recorded with a low burn fluence of 0.20 mJ/cm² and a spectral resolution of 8 cm⁻¹. The intermediate hole-burning fluence was 4.6 mJ/cm². The inset shows the full ΔFLN spectrum in relative wavenumber scale.

light intensities used in SMS frequently reach 1 kW/cm² as an order of magnitude. Therefore, with ambient temperature measurements, great care should be taken to avoid contact with atmospheric oxygen and resulting photooxidation of the sample (for details, see discussion in [140]). Recently, many photosynthetic chromophore–protein complexes from plants and bacteria have been assessed by SMS [141–144], as reviewed in [26,145–147]. These studies, revealing an environmental control of the intrinsic protein disorder, shed light on the structural/functional plasticity of light-harvesting antenna complexes [148].

17.2.5 Selected examples of current research

In this paragraph, a couple of examples are provided, which reflect ongoing research on optical spectroscopy for light harvesting. The examples are meant to demonstrate that basic models introduced earlier describe reasonably well the main features of the spectra of light-harvesting pigments and complexes. At the same time, they also point to certain limitations of these models. The questions raised by the current experimental research thus clearly call for further theoretical developments.

17.2.5.1 BEYOND THE BASIC MODEL: BREAKDOWN OF MIRROR SYMMETRY AND NEW ASSIGNMENT OF THE CHLOROPHYLL SPECTRA

Mirror symmetry between the conjugate absorption and fluorescence spectra is theoretically expected within the basic model [23] and in the presence of fast vibronic relaxation (Kasha–Vavilov rule) [8]. The breakdown of any of these assumptions could lead to asymmetry. Quantitative quantum chemical modeling of the absorption and emission spectra should in principle reveal the nature of the electronic transitions and the origin of any spectral asymmetry. However, without proper experimental guidance, this is a challenging task for the large molecules with extended conjugation, as the chlorophyllides are.

Recent experimental results indicated considerably asymmetry between the Q_y absorption and fluorescence spectra of chlorophyll [61] and bacteriochlorophyll [62] molecules, indicating that the preceding simple model framework may not be useful for studying the intricate photosynthetic exciton dynamics. This asymmetry is demonstrated in Figure 17.10, which shows the low-resolution absorption and fluorescence spectra of pheophytin-a, chlorophyll-a, and bacteriochlorophyll-a in triethylamine solution at 295 K. Taking the bacteriochlorophyll as an example, it is obvious

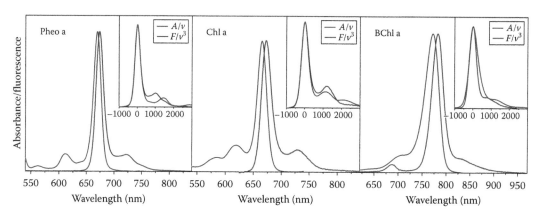

Figure 17.10 Absorption (blue line) and nonresonantly excited fluorescence (red line) spectra of pheo-phytin-a, chlorophyll-a, and bacteriochlorophyll-a in triethylamine at room temperature. The fluorescence was excited in the region of 405–410 nm. A hump visible at 670 nm in the fluorescence spectrum of bacteriochlorophyll-a is due to an oxidation product. The insets show the overlap of the absorption and fluorescence (reflected about its origin) spectra in frequency scale and in transition dipole moment representation. For that, the fluorescence (F) spectra measured in units of photons/nm are first converted to units of photons/cm^{-1} by the relationship $F(\nu) = F(\lambda)/\nu^2$, where λ is the wavelength. Then the absorption (A) and fluorescence spectra are weighted by ν^{-1} and ν^{-3}, respectively.

that the vibronic sidebands in emission and absorption are different; in fluorescence it is centered on modes around 900 cm^{-1}, while in absorption this structure moves out to 1100–1200 cm^{-1}. The long-range corrected density functional CAM-B3LYP simulations reproduced detailed experimental results for the bacteriochlorophyll-a, also predicting asymmetry of the correct magnitude [62]. It was furthermore revealed that the spectral asymmetry arises primarily through Duschinsky rotation [149].

In the case of the chlorophyll-a molecule, the deviation from mirror symmetry was originally assigned to a combined effect of Franck–Condon and Herzberg–Teller interactions [61]. However, the most recent analyses [17,150] exposed a fundamentally new mechanism for the spectral asymmetry, this being the strong coupling between the energetically closely spaced Q_y and Q_x excited electronic states, mediated by high-frequency intramolecular vibrations. This interpretation, which has been checked on more than 30 chlorophyllides in various environments, provides a definitive spectral assignment not only for their absorption and fluorescence spectra but also for the LD and magnetic circular dichroism (MCD) spectra. While most chlorophylls conform to the Gouterman model [16] and display two independent transitions Q_x and Q_y, strong vibronic coupling inseparably mixes these states in chlorophyll-a. This spreads the x-polarized

absorption intensity over the entire Q-band system to influence all exciton transport, relaxation, and coherence properties of chlorophyll-based photosystems. The fraction of the total absorption intensity attributed to the x-polarized absorption is between 10% and 25% for chlorophyll-a. CAM-B3LYP density-functional-theory calculations of the band origins, relative intensities, and vibronic coupling strengths fully support this new assignment.

17.2.5.2 SPECTRAL EFFECTS ACCOMPANYING THE ASSEMBLY OF BACTERIOCHLOROPHYLL-a INTO BACTERIAL LIGHT-HARVESTING COMPLEXES

One might notice in Figure 17.10 that the lowest singlet Q_y electronic transition of a bacteriochlorophyll-a molecule dissolved in normal (nonreactive) organic solvents is around 770 nm. Significant red shift (move toward longer wavelengths) of this transition is observed in light-harvesting systems. As seen in Figure 17.11, the fluorescence spectrum of FMO and LH2 protein complexes peaks at 827 and 890 nm, respectively. Energetically, the FMO spectrum is almost halfway between the spectra of individual pigments and LH2.

Why are the red shifts in the two bacterial light-harvesting complexes so different? To answer this

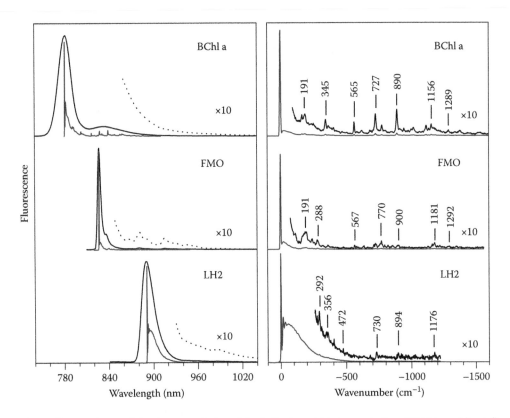

Figure 17.11 Comparison of the low-resolution (black line) and high-resolution (ΔFLN, red line) fluorescence spectra of bacteriochlorophyll-a in solid solution of triethylamine and in FMO and LH2 light-harvesting protein complexes from green and purple photosynthetic bacteria, respectively. The peak-normalized spectra are recorded at 4.5 K. Presented with dashed lines is the amplified tail of the low-resolution spectra. The right panel shows the peak-normalized ΔFLN spectra in relative wavenumber scale (vertical lines label selected vibrational mode frequencies).

question, we have to recall that there are two major mechanisms, which influence the optical spectra of self-assembled chromophores in proteins, these being exciton couplings between the transition densities of the chromophores and interactions between each chromophore and the surrounding matrix. The latter mechanism, which implicitly involves structural modifications of the chromophores, is a source of the solvent/protein spectral shift [151]. The same shift is the origin of static disorder discussed earlier.

Exciton interactions between the loosely packed bacteriochlorophyll-a chromophores in FMO are known to be relatively weak [65,152]. Therefore, the red shift observed in FMO can be considered mostly due to the protein shift. The situation in LH2, where responsible for the 890 nm fluorescence are eighteen tightly coupled bacteriochlorophyll-a chromophores in a ring-like B850 structure, is

quite different. There is a prevailing consensus that the further red shift can be explained by the strong exciton coupling mechanism [26,33,153–155].

As already noticed, the fine vibrational/phonon structure of large biomolecules is rather hidden by inhomogeneous broadening in the conventional absorption/fluorescence spectra, even if recorded at cryogenic temperatures (see left panel of Figure 17.11). From the three systems presented in Figure 17.11, FMO appears the least disordered, appreciated by the narrowest fluorescence origin band. This conclusion is corroborated by the fact that in this exceptional case the relatively clear-cut vibronic structure is observed already in the low-resolution spectrum. The right-hand panel of Figure 17.11 compares the high-resolution ΔFLN spectra of the samples in relative wavenumber scale. In accordance with the basic theory discussed earlier, all the spectra enclose a narrow pure electronic ZPL accompanied by a broad PSB and series of their vibronic (first-order) replicas. Yet a closer look reveals two striking differences in these spectra. First, the vibronic structure appears the strongest in the spectrum of the individual molecules, while it is the weakest in LH2, where the emitting excitons are presumably delocalized over several chromophores [59]. In fact, vibrational lines in LH2 are so weak that they have mostly been ignored. At the same time, in the spectrum of LH2, the PSB is much more powerful than it is in the spectrum for individual bacteriochlorophyll-a molecules. In terms of both electron–phonon and electron–vibrational coupling strengths, the FMO complex keeps an intermediate position between individual molecules and LH2.

We thus notice that, apart from the major spectral shift, the assembly of bacteriochlorophyll-a molecules with localized electronic excitations into cyclic bacterial light-harvesting complexes, where the exciton is delocalized, is accompanied by significant enhancement of the electron–phonon coupling and simultaneous suppression of the electron–vibrational coupling. The suppression of the vibronic structure in strongly excitonically coupled molecular assemblies is rather expected due to the delocalized exciton states. However, the same reasoning should apply for electron–phonon coupling, which it apparently does not. It appears that there are issues of exciton couplings to the dynamic environment which we still do not fully comprehend. It is proposed [59] (see also [94]) that self-trapping of excitons may explain the observed enhancement of the electron–phonon coupling in cyclic light-harvesting complexes might.

17.3 INTERACTION OF PHOTOSYNTHETIC PIGMENT SYSTEMS WITH POLARIZED LIGHT

In the previous sections, our attention was focused on the energy level structure and did not account for polarization properties of light and matter. Accordingly, in these isotropic cases, the absorbance does not contain structural information, which would otherwise be possible to obtain based on Equation 17.5. Also, to obtain information on the chirality of a molecule or an excitonically coupled molecular assembly, circular polarization spectroscopic techniques must be applied, such as CD. Similar considerations can be made on the polarization of fluorescence emission.

In general, polarization optical spectroscopy comprises a set of techniques that analyzes the polarization properties of light transmitted, scattered, or emitted by the sample. The polarization state of electromagnetic radiation can conveniently be characterized by the so-called Stokes parameters. These parameters are often combined into a vector, the Stokes vector: (I,Q,U,V) (with a different, also frequently used notation, the (S_0, S_1, S_2, S_3) vector). For a monochromatic light beam propagating along the z-axis, they describe its intensity, linear polarization contents (in the xz/yz directions, also called as horizontal vs vertical or 0° vs 90°, and at +45° vs −45°), as well as the amount of right- or left-handed circular polarization contained within the beam. Evidently, for any state of (fully or) partially polarized light, the square of the total intensity parameter (I or S_0) is (equal or) larger than the sum of the squares of the polarization components. For illustration, a nonpolarized beam, a horizontally (H) and a vertically (V) linearly polarized and a right-hand circularly polarized (R) beams are represented by the following vectors: (1,0,0,0), (1,1,0,0), (1,−1,0,0), and (1,0,0,1), respectively. For a more comprehensive treatment, also for the so-called Jones' formalism, see, for example, [156].

When such a polarized beam, with initial states (I_0, Q_0, U_0, V_0), interacts with a sample (which absorbs or scatters the beam), its polarization content will almost always be changed. The emerging beam can again be characterized by a (different) Stokes' vector, (I, Q, U, V), while the interaction is characterized by a 4×4 matrix, the Mueller matrix:

$$\begin{Bmatrix} I \\ Q \\ U \\ V \end{Bmatrix} = \begin{pmatrix} M_{11} & M_{12} & M_{13} & M_{14} \\ M_{21} & M_{22} & M_{23} & M_{24} \\ M_{31} & M_{32} & M_{33} & M_{34} \\ M_{41} & M_{42} & M_{43} & M_{44} \end{pmatrix} \begin{Bmatrix} I_0 \\ Q_0 \\ U_0 \\ V_0 \end{Bmatrix}. \quad (17.15)$$

In highly organized systems, all 16 elements of this matrix can be independent and yield useful structural information on the sample, but in many cases there are correlations between different elements [157]. In usual laboratory practice, only the changes in the total intensity, related to the absorption of the sample (element M_{11}), and those related to LD (M_{12} and M_{13}) and CD (M_{14}) are determined—mainly because of the absence of theories exactly describing the content of all elements and also because of the lack of available instrumentation [157].

In the following sections, we confine our discussion to LD and CD, that is, to the two most frequently used in photosynthesis research DP spectroscopic methods. Since by measuring principle the two methods are comparable, and in modern practice, LD and CD measurements are often performed using the same instrument, it is useful to treat the two measurement schemes together, as shown in Figure 17.12. Both LD and CD are defined as the absorption difference between two orthogonally polarized light beams, A_1 and A_2:

$$LD(CD) = A_1 - A_2 = \Delta A \propto \frac{\Delta I}{I}. \quad (17.16)$$

The last part of Equation 17.16 refers to measuring LD or CD; the meaning of ΔI and I can be understood from Figure 17.12.

In LD measurements, two linearly polarized (e.g., vertically (V) and horizontally (H) polarized) beams are used; in CD, the two beams are left (L) and right (R) circularly polarized. As already pointed out in a previous section, in linearly polarized light, the electric field vector E oscillates sinusoidally in the so-called polarization direction (or plane—this is why it is often called plane-polarized light). For circularly polarized light, the magnitude of E remains constant, but it traces out a helix as a function of time. In accordance with the convention used in CD spectroscopy, for R and L circularly polarized light beams, when viewed by an observer looking toward the light source, the end point of the electric vector rotates clockwise and counterclockwise, respectively.

In modern dichrographs (see Figure 17.12), the two orthogonally polarized beams are alternating at

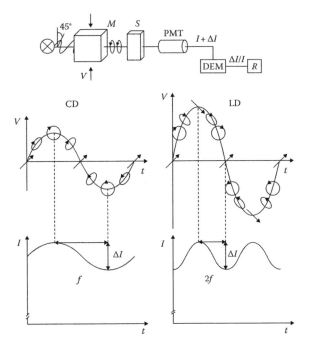

Figure 17.12 Simplified scheme for CD and LD measurements showing that by using a linearly polarized incident light at 45° with respect to the vertical or horizontal plane and a photoelastic modulator (M), on which a sinusoidally changing voltage (v) is applied, the polarization state of the measuring beam modulates between L and R or between +45° and −45°, depending on the maximum voltage applied. After the beam passes through the sample (S), the emerging light intensity difference ΔI is determined using a photomultiplier tube (PMT) and a demodulator circuit (DEM, usually a lock-in amplifier). The same principle is applied when measuring fluorescence-detected CD or LD (FDCD or FDLD, respectively; see following text)—in these cases, the intensity difference of the fluorescence emission elicited by the two orthogonally polarized states is measured. (Reproduced from Garab, G., Linear and circular dichroism, in *Biophysical Techniques in Photosynthesis*, Amesz, J. and Hoff, A.J. eds., Kluwer Academic Publishers, Dordrecht, the Netherlands, 1996, pp. 11–40.)

high frequency (usually at f = 50 kHz for CD and at $2f$ = 100 kHz for LD), and LD and CD are measured via determining ΔI and I (see following text). The key optical element, added to a conventional single beam spectrophotometer, is a photoelastic modulator (PEM), which is capable of inducing a sinusoidally changing phase shift of the vertical linearly polarized beam, when it passes through the transparent slab, with its long axis aligned vertically. (In practice, the long axis is set at 45° to allow $LD = A_V - A_H$.) This is achieved via pressing and drawing periodically the slab using a piezoelectric transducer; the phase shift is proportional to the applied voltage. At the same time, the H polarization remains unaffected. Let us now consider an incident beam with linear polarization at 45°, and let us "decompose" it (using the superposition principle) to V and H components of equal amplitude and phase, and vary the phase shift of the V component; it can easily be seen that at a phase shift of $+\lambda/4$ and is $-\lambda/4$ circularly polarized beams are obtained (see Figure 17.1 in [158] and also the movie supplemented to [159]). This way, a PEM can be adjusted to produce periodically at f the L and R circularly polarized beams. Modulation of the linear polarization state of the beam at $2f$ between + 45° and −45° is obtained via adjusting the maximum phase shifts to $+\lambda/2$ and $-\lambda/2$. (The frequency is doubled because at zero phase shift the +45° polarization state is unaffected.) Recently, a spectropolarimetric instrument, using ferro-liquid-crystal was constructed, which was designed to measure the circular polarization of a sample in both transmittance and reflectance; its performance in transmission was demonstrated on whole leaves [160].

17.3.1 Linear dichroism and related techniques

By today, LD has become a "standard" technique in photosynthesis research, which is of interest for its noninvasive nature; with some restrictions, it can be used both for qualitative and quantitative information in macroscopic and microscopic samples and in vitro and in vivo. Systematic studies in the 1970s and 1980s (reviewed in [158,161]) have led to the recognition of an apparently universal property of photosynthetic pigment systems: all chromophores in the light-harvesting and reaction center complexes display nonrandom orientation not only with respect to each other but to the protein axes and to

the membrane plane as well. Nowadays, this is of course also evident from crystallography data, which can readily be correlated with LD [159]. It must be pointed out, however, that crystallographic data alone are not sufficient for interpreting LD data because, for example, excitonic interactions can basically modify the magnitude and orientation of the molecular transition dipoles (see Section 17.3.2); conversely, LD can aid the identification of these interactions.

In the following, a few theoretical and practical aspects will be discussed in relation to LD measurements. In order to obtain a nonzero LD signal in a macroscopic sample, the protein complexes, particles, or membranes must be aligned. Evidently, the magnitude of the LD depends on the efficiency of the alignment. There is a number of orientation techniques overviewed in [158,159]. Here, the discussion will be confined to the two most frequently used orientation techniques in the photosynthesis literature: by magnetic field and by unidirectional or two-directional squeezing in a polyacrylamide gel matrix.

For whole chloroplasts and entire thylakoid membranes, as shown in Figure 17.13, a magnetic field of about 0.5 T provides a very good, nearly saturating degree of alignment. The same technique can also be used for lamellar aggregates of LHCII, but they require somewhat larger field strengths (about 1–2 T). Upon magnetic orientation, the membranes are aligned with their planes preferentially perpendicular to the field. If trapped at low temperature or in a gel, the same sample can be studied in edge-aligned or face-aligned orientation of the membranes.

For practical reasons, LD is often evaluated in the form $LD = A_{||} - A_\perp$, where the symbols $||$ and $\perp$ refer to directions parallel with and perpendicular to an axis of interest, for example, a protein axis or a membrane normal. In an idealized case, when all the membranes represented by flat planar sheets are completely aligned, one obtains after averaging over the angle between the projection of $\boldsymbol{\mu}$ and the $||$ direction:

$$LD = A_{||} - A_\perp = \frac{3}{2}A\left(1 - \cos^2\Theta\right), \quad (17.17)$$

where

Θ is the orientation angle between the membrane normal and $\boldsymbol{\mu}$

A is the isotropic absorbance

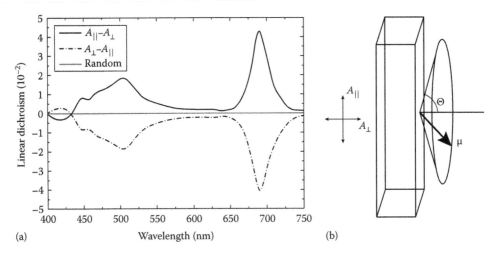

(a)   Wavelength (nm)   (b)

Figure 17.13 LD spectra of edge-aligned thylakoid membranes oriented in a magnetic field of ~0.5 T and trapped in a gel in two different positions of the sample (a). It can be seen that upon rotating the sample by 90° along the axis of the beam, the LD inverts sign but its shape does not change. This test is useful when checking the magnitude of the LD contribution in the CD (see Section 17.3.2). The orientation angle Θ between the membrane normal and μ is shown in the scheme of (b). (Reproduced from Garab, G. and Van Amerongen, H., *Photosynth. Res.*, 101, 135, 2009.)

The quantities that are directly related to the orientation angle are S, the orientation parameter, and DR, the dichroic ratio:

$$S = \frac{LD}{3A}; \quad DR = \frac{A_{II}}{A_{\perp}}. \tag{17.18}$$

Θ can be readily determined from the dichroic ratio of absorbance or, alternatively, from polarized fluorescence ($FP = I_{II}/I_{\perp}$) obtained after nonpolarized excitation to avoid photoselection:

$$DR(FP) = \frac{1}{2}\tan^2\Theta. \tag{17.19}$$

It follows that if a transition dipole is oriented at Θ = 54.7°, that is, at the magic angle, we obtain DR = FP = 1 and LD = 0 as for random samples (random orientations of molecular dipole moments).

Thylakoid membranes and lamellar sheets of LHCII and in general disc-shaped particles can also be aligned by embedding them in polyacrylamide gel, which is then squeezed unidirectionally. This technique was introduced by Abdourakhmanov et al. in 1979 [162]. Let us take M (>1) as the compression or squeezing parameter. Then upon squeezing, for example, a 1 cm³ cube to 1/M cm in one direction and, at the same

time, allowing its expansion to the other two directions by √M cm, so that the initial volume is preserved, the particles will tend to align with their planes perpendicular to the direction of squeezing. A good alignment can be obtained with M between 1.5 and 2; the particles will be distributed around the plane of ideal orientation, achieved when M→∞. The distribution function (T) depends only on M. In turn, S (or DR and FP) can be expressed as a function of T and Θ; and thus via measuring S, and calculating T, Θ can be determined with good precision. Applying M ≤ 2 usually does not distort the structure—this should be tested by measuring S with smaller M and see if the calculated Θ remains unchanged. For further details, see [158] and references therein.

The gel-squeezing technique can also be used for rod-shaped particles. With unidirectional squeezing of these particles, the rods tend to be aligned in the plane perpendicular to the direction of squeezing; however, their long axis will be randomly oriented in this plane. Thus, two-directional squeezing should be applied, deforming the 1 cm³ cube to a block with two edges of 1/√M cm lengths and one with M cm length. Again, Θ, the angle between the transition dipole and the long axis of the particle, can be determined via measuring S (or DR and FP) and calculating the distribution function that belongs to the case of two-directional squeezing. In the first approximation, Θ can be determined by Equation 17.20:

$$LD = \frac{3}{2}A\left(3\cos^2\Theta - 1\right). \qquad (17.20)$$

It is finally interesting to note that even spherical membrane vesicles (such as blebs, stroma thylakoid membrane vesicles, or a chromatophore) can be "oriented" with unidirectional squeezing and deforming them into ellipsoids. Such measurements can again be treated quantitatively in order to determine Θ [163]. These techniques have been thoroughly discussed in the photosynthesis literature; exhaustive examples are provided in the reviews [158,159].

LD can be especially useful for in vivo studies, when combined with information from high-resolution structural models, for example, in identifying transition dipole moments. The orientations of the transition dipole moments are functionally very important: they strongly influence the rates and the routes of excitation energy transfer in the light-harvesting antenna system. The rate of the Förster energy transfer is proportional to the so-called κ^2 factor [80] with κ equal to

$$\kappa = \mu_1 \cdot \mu_2 - 3\left(\mu_1 \cdot r\right)\left(\mu_2 \cdot r\right). \qquad (17.21)$$

In Equation 17.21, μ_1 and μ_2 are the transition dipoles of the acceptor and donor molecules, and $r = R/|R|$ is the unit distance vector between them. Evidently, κ^2 depends strongly on the mutual orientation of the transition dipoles, varying between 0 and 4, the values for perpendicularly and collinearly oriented transition dipoles, respectively. For randomly oriented samples, for example, in the dynamic case, its value is 2/3. As already pointed out earlier and will be further specified in Section 17.3.2, in systems with excitonically coupled molecules, the absorbance and fluorescence properties are modified—since they as well depend on the mutual orientation of the transition dipoles of the interacting molecules.

LD can also be determined in microscopic samples. Under these conditions, the sample does not need to be aligned: the full information on the orientation of the transition dipoles can be obtained just from two measurements, determining the difference between the transmitted intensities in the H and V directions and between the +45° and −45°. When, however, in the microscope we want to measure the orientation of the transition dipoles with respect to a given axis, for example, the membrane normal or the long axis of a rod of microscopic size, one measurement that yields information on parallel and perpendicular orientations is sufficient. In a microscope that digitally records the intensity values in each pixel (e.g., in a laser scanning confocal microscope), in principle, this difference can be obtained by subtracting the two images recorded using orthogonally polarized excitation light beams. This, however, could lead to artifacts, because of possible changes in the biological material between the two consecutive scans. Much higher precision and confidence can be obtained when using the high-frequency modulation/demodulation principle, as in dichrographs (see Figure 17.12), and measuring $\Delta I/I$ pixel by pixel. The intensity difference can be, for example, the difference between the light intensities transmitted with periodically changing orthogonal polarizations, giving information on LD. Measuring the fluorescence intensity difference due to differential absorption of the excitation (FDLD) or the intensity difference between the two orthogonally polarized fluorescence emissions elicited by nonpolarized excitation gives information on FP and $r = (I_{II} - I_\perp)/I$. In the latter equation, I_{II} and $I_\perp$, respectively, refer to the polarization directions parallel and perpendicular to, for example, a membrane plane, and $I = I_{II} + 2I_\perp$.

The theoretical background of DP imaging has been elaborated by Bustamante and coworkers [164,165], showing that different microscopic DP quantities carry important and unique physical information on the anisotropic molecular architecture of the samples. The first DP microscope was constructed in 1985 by Mickols and coworkers [166]. It was based on a scanning-stage microscope and could record, in confocal mode, LD images. A similar setup, a scanning-stage microscope with DP attachment, was used in [167]. In the latter work, high-resolution LD and CD images were recorded on isolated intact plant thylakoid membranes.

The first laser scanning microscope equipped with the modulation/demodulation unit was constructed by Gupta and coworkers [168] and was designed for the determination of linear birefringence and LD, in the nonconfocal regime, on ultrathin Langmuir–Blodgett films. The DP laser scanning microscope designed and constructed by Garab and coworkers [169–172] was extended to the fluorescence (and reflected beam) light paths. In addition to linear birefringence and LD, it is capable of confocally determining FDLD, r, and the degree of fluorescence polarization defined as $P = (I_{\parallel} - I_{\perp})/(I_{\parallel} + I_{\perp})$. For the birefringence of granal chloroplast thylakoid membranes, see [173].

Recently, this microscope was used, in combination with macroscopic FP and LD measurements on magnetically aligned samples, to unravel the anisotropic molecular architecture of microscopic rods of synthetic Zn-porphyrin molecules. In organic solvents, these molecules self-assemble into large structures of rods with typically 5–10 μm length and 1–2 μm width, that is, to "artificial chlorosomes" (see Figure 17.14). Using a DP laser

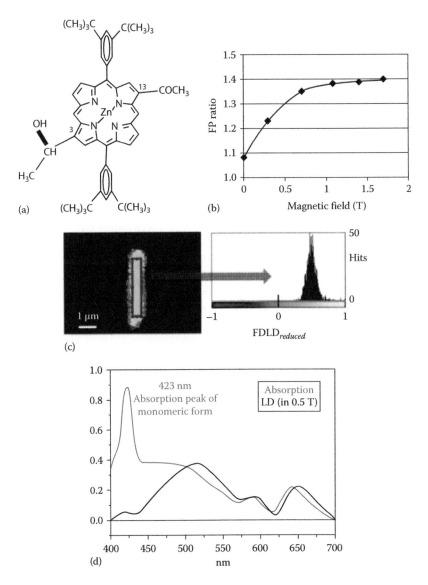

Figure 17.14 Orientation of self-assembled rods from synthetic Zn-porphyrin molecules **(a)** in an external magnetic field and their optical anisotropy properties **(b–d)**. Dependence of FP on the magnetic field strength upon excitation with depolarized 440 nm laser beam, fluorescence recorded at >650 nm **(b)**. FDLD normalized to the total intensity, $FDLD_{reduced} = (I_V - I_H)/(I_V + I_H)$, of a single rod oriented vertically **(c)**. The histogram at the right shows the distribution of local FDLD ($=I_V - I_H$) values in the framed area comprising 948 pixels. Absorption (red trace) and LD (black trace, scaled arbitrarily) measured in a 0.5 T magnetic field **(d)**. The monomer absorption peaks at 423 nm; it practically does not contribute to the LD, which originates only from the aggregates' transitions. (Adapted from Chappaz-Gillot, C. et al., *J. Am. Chem. Soc.*, 134, 944, 2012.)

scanning microscope, it was possible to map the spatial distribution of the emission and absorbance dipoles, via r and fluorescence-detected LD (FDLD) imaging, respectively [174]. These microscopic measurements revealed that the nanorods are highly anisotropic and homogeneous entities and gave crucial information to determine their 3D molecular architecture.

The same setup, with some restrictions and careful corrections of the possible polarization distortions by the dichroic mirrors, based on Mueller matrix analyses, can also be used for CD, for fluorescence-detected CD (FDCD), and for circularly polarized luminescence (CPL) microscopic measurements. These polarization spectroscopy methods will be in some detail introduced next. For a general review on CPL, physical mechanisms, instrumentation, and results, see [175]. On the emission transition dipoles, CPL provides analogous information as CD on the absorption dipoles. Recently, CPL provided sensitive and detailed information on the low-energy exciton states (trap states) of the isolated plant pigment–protein complex CP43, revealing exciton delocalization of the emitting states [176].

17.3.2 Circular dichroism

As it was already noted (see Equation 17.15), CD refers to the phenomenon in optically active materials where the left- and right-handed circularly polarized lights are absorbed to a different extent:

$$CD = A_L - A_R = \Delta A. \qquad (17.22)$$

To illustrate the origin of CD of isolated molecules, the easiest is to consider enantiomers, mirror image isomers. They, similarly to our left and right hands, are not superimposable onto each other. They have handedness or chirality (a term derived from the Greek word $\chi\epsilon\iota\rho$, meaning hand). Since circularly polarized light also has this property, it is easy to imagine that one can discriminate between the two isomers using circularly polarized light. Indeed, for a pair of enantiomers (with identical physical and chemical properties), the CD spectra are of identical shape but sign-inverted relative to each other. While this allows grasping the idea, it is better to look at CD as reflecting the lack of certain symmetries.

CD can most conveniently be measured in a dichrograph (see Figure 17.12) in units of ΔA, or when difference between the molar extinction coefficients

$(\Delta\epsilon = \epsilon_L - \epsilon_R)$ for optically active molecules is recorded, in units of M/cm. Often CD is expressed in ellipticity units, in millidegrees or degrees M/cm. This is because in earlier times, CD measurements were performed using linearly polarized light. A linearly polarized light, according to the superposition principle, contains the two circularly polarized components, L and R, at equal amplitudes. Upon differential absorbance, this equality does not hold any more and the beam becomes elliptic. Since many dichrographs display CD in millidegrees, it is useful to have the conversion factor: 100 mdeg $\approx 3 \times 10^{-3}$ ΔA units.

Although the LD and CD measuring principles are very similar and the parameters derived from the two types of DP measurements are related to each other (as will be demonstrated in the following in the case of an excitonic dimer), the complexities of the two phenomena are quite different. In contrast to LD, CD cannot be understood based on a simple equation such as Equation 17.5. The underlying physical mechanisms can be complex and quite different from each other for various types of molecular interactions. Therefore, as follows, only a brief overview can be given about the physical origin of CD signals in photosynthetic pigment systems in vitro and in vivo. For general applications of CD and for CD spectroscopy of proteins, DNA, and other systems, the interested reader may wish to consult [177].

In photosynthetic systems, three types of CD signals have been identified, which, due to different physical mechanisms, differ in their basic characters, spectral features, and magnitudes. The three types of spectra in the visible region, where only chromophores absorb, are shown in Figure 17.15a and are indicated as intrinsic, excitonic, and psi type. The pigment composition (in this case chlorophylls a and b, and carotenoid molecules) and concentrations are the same in the three measured samples, but their molecular organizations are different. The intrinsic CD, the weakest type of signal, comes from the isolated monomeric chromophores; the excitonic CD, typically an order of magnitude stronger signal, is caused by short-range dipole–dipole interactions between identical or different chromophores; the psi-type CD (psi is for polymer or salt-induced) displays a giant CD, which is attributed to the long-range chiral order of the chromophores in intact granal thylakoid membranes (due to the self-assembly of their protein complexes in macro-arrays with long-range order). The psi-type CD appears evidently greater than the sum of its

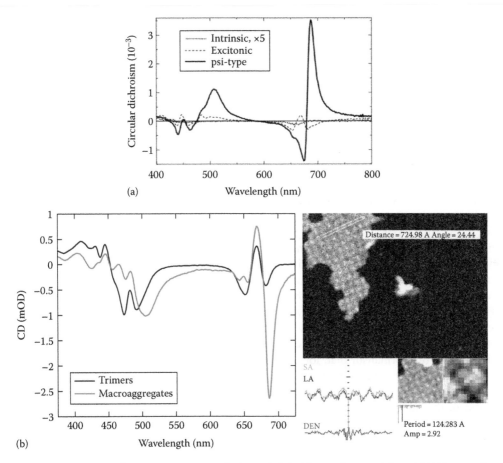

Figure 17.15 CD spectra of pigment systems of plant thylakoid membranes **(a)** and LHCII complexes **(b)** at different levels of their structural complexity. **(a)** Isolated intact thylakoid membranes suspended in an isotonic reaction medium containing 5 mM $MgCl_2$ (a medium that preserves the multilamellar granal ultra-structure allowing for psi-type CD) and the same membranes upon suspending them in hypotonic, low-salt medium (in this case, the long-range order of the protein complexes in otherwise functional membranes is lost [36] and, consequently, the CD signal is essentially the sum of the excitonic CD bands of the constituent protein complexes). Also shown for reference is the pigment extract of the membranes. **(b)** CD spectra of isolated LHCII complexes in their detergent-solubilized trimeric state (containing excitonic bands) and of their tightly stacked (microcrystalline) lamellar aggregates, which exhibit psi-type bands, superimposed on the excitonic CD. (**a:** Reproduced from Garab, G. and Van Amerongen, H., *Photosynth. Res.*, 101, 135, 2009; Figure courtesy of P. Lambrev, based on data published in [178]; scanning transmission electron micrograph and analysis, courtesy of G. Hind and J.S. Wall.)

component parts. This in general also holds for the excitonic CD, indicating in both cases the appearance of a higher degree of the structural hierarchy compared to the pigments. Figure 17.15b demonstrates the pair of spectra of purified LHCII trimer complexes, displaying excitonic CD, and that of the lamellar aggregates of the same complexes, which are dominated by the psi-type features. As pointed out earlier, the underlying physical mechanisms generating these CD signals are different from each other, and in hierarchically organized systems, such as the thylakoid membranes and LHCII lamellar aggregates, the CD of the higher level structures also contains the CD of the lower level units. For instance, the CD of intact thylakoid membranes contains the intrinsic and excitonic CD bands of the components, in addition to the psi-type CD bands.

In isolated molecules, CD arises from intrinsic asymmetry or the asymmetric perturbation of a molecule. The molecular electronic transitions interact with the electric (**E**) and magnetic (**B**) components of the electromagnetic field, respectively, via their electric transition dipole **μ** and the magnetic transition dipole **m**. The quantum-mechanical theory yields for the dependence of the rotational (or rotatory) strength ($\mathcal{R}$) of a transition on **μ** and **m**:

$$\mathcal{R} = -Im(\mathbf{m}\boldsymbol{\mu}). \qquad (17.23)$$

Im indicates that the rotational strength corresponds to the imaginary part of the scalar product (note that **m** is a purely imaginary vector). According to this equation, CD vanishes if either **m** or **μ** is zero (e.g., outside the absorbance band, where **μ** is zero). In order to have nonzero **m**, the molecule must contain a circulation of charge upon excitation (similar to the generation of magnetic field by current in a coil), which can be satisfied in molecules in which the light-induced charge displacement includes a circular motion of the electron. Further, the scalar product of the two must also be nonzero, which can be satisfied by a helical motion of the electron, in a helical molecule. It also follows from this formalism that, for a single electronic transition, CD has the same band shape as the absorption, and its sign is determined by the handedness of the molecule (often referred to as positive or negative Cotton effect). For details, see, for example, references [179,180].

In Figure 17.15a, we have seen that in monomeric solutions chlorophylls and carotenoids exhibit very weak CD signals: for 1 absorbance unit, in the range of some 10^{-5} intensities. This is because chlorophylls, with their planar and rather symmetric structures, and carotenoids, with their rod shapes, exhibit very weak rotational strengths. It is not quite clear how much the protein environment induces some twisting in these chromophores [159]. In the light-harvesting antennas, the interactions between molecules produce moving excitations or excitons, which give rise to excitonic CD [181,182]. The absorption, fluorescence, and LD spectra can also be significantly different from the spectra of the individual chromophores. Taken together, these spectroscopic fingerprints are of special value— they carry unique information on the molecular architecture of the light-harvesting complexes and on the nature of intermolecular excitonic interactions in them. These architectures and interactions can be very complex, rendering interpretation of the experimentally obtained spectroscopic data an enormous task [183–188]. With regard to the assignment of different bands to different structural

entities in near atomic resolution structures, such as that of LHCII [189], further complications arise from heterogeneity of the complexes (e.g., [190]) and spectral variations in different physicochemical environments, perturbed by detergent solubilization [191,192].

For illustration, let us consider the simplest case, an excitonic dimer, formed by two identical pigment molecules that are separated by a distance vector $\boldsymbol{R}$ and characterized by transition dipole vectors $\boldsymbol{\mu}_1$ and $\boldsymbol{\mu}_2$ (see Figure 17.16 and also reference [5]).

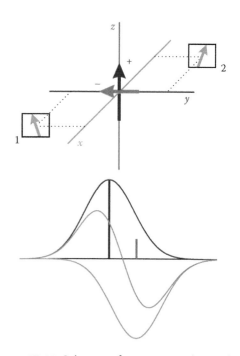

Figure 17.16 Scheme of an exciton homodimer and the calculated absorption (shown by sticks) and CD (yellow line) spectra. The transition dipole moments $\boldsymbol{\mu}_1$ and $\boldsymbol{\mu}_2$ of the individual components (sites) of the dimer are shown with yellow arrows. Exciton coupling between the sites leads to two new transition dipoles, $\boldsymbol{\mu}^+$ and $\boldsymbol{\mu}^-$, marked with red and green color, respectively. The $\boldsymbol{\mu}^+$ and $\boldsymbol{\mu}^-$ vectors are perpendicular to each other but have different lengths (dipole strengths). The calculated positions and relative amplitudes of the exciton spectra are indicated by sticks. The + and − exciton transitions have equal amounts of rotational strengths but possess opposite signs; the opposite sign red and green CD bands are dressed with Gaussians for illustration: (Model calculations, courtesy of P. Lambrev). For an almost perfect experimental separation of the two components of the split bands of a dimer, see [193].

In such a dimer, the energetically degenerate absorption band corresponding to the two noninteracting molecules splits into two bands; the degree of splitting depends on the interaction energy V_{12}:

$$V_{12} \propto \frac{\kappa}{R^3} \mu^2, \tag{17.24}$$

where κ is provided by Equation 17.21. The relative intensities (exciton dipole strengths $D_{\pm}$, where + and − designate the lower and higher energy transitions) of the resulting exciton absorption bands are

$$D_{\pm} = \frac{1}{2}\left(\mu_1^2 + \mu_2^2\right) \pm \mu_1 \cdot \mu_2. \tag{17.25}$$

Equations 17.24 and 17.25 indicate that both the interaction energy and the intensity of the exciton transitions depend not only on the dipole strength of the molecules but also on their mutual orientation.

The rotational strength (in units of Debye–Bohr magnetons) is given as

$$\mathcal{R}_{\pm CD} = \pm 1.7 \cdot 10^{-5} v_0 \mu^2 |R|\left(r \cdot d_2 \times d_1\right), \tag{17.26}$$

where v_0 is the band center energy and the vectors in the triple product are unit vectors of the distance and the transition dipoles. It is clear that the rotational strength for the dimer is independent of that of the monomers—the excitonic CD does not require the optical activity of the participating chromophore. It is also clear that the absolute size of the positive CD is equal to that of the negative CD, that is, the excitonic CD spectrum, when plotted on an energy scale, is conservative. It is to be noted that the equations can readily be generalized to systems with more excitonically interacting pigments [194]. Nevertheless, the conservative nature of the spectrum is retained—and in small complexes, deviations indicate contributions from intrinsic CD bands, which should then be subtracted.

This simple model shows that exciton coupling generally leads to changes in the absorption spectrum, which, in turn, will affect the CD (as well as the LD) spectrum. The magnitude of this effect evidently depends on the orientation of the transition dipoles of the individual or "monomeric" components of the molecular assembly (in exciton parlance they are frequently called (local) sites). Information about the "monomeric" components may be available from high-resolution structural data, like in the case of the LHCII complex [189,195]. Upon identifying the excitonically coupled assemblies, one can perform model calculations on the orientation angles of the exciton transition dipoles with respect to the membrane normal or protein axis. By this means, CD and LD spectroscopies can aid the identification of excitonic interactions in light-harvesting complexes.

Conventionally, CD is measured in an isotropic phase, that is, on randomly oriented complexes and membranes. As shown by Kuball and coworkers [196–199], the ACD of oriented molecules can yield additional information, information that is not available in the isotropic CD, on the molecular organization of pigment–protein complexes. This is well demonstrated in Figure 17.17 by ACD measurements on isolated washed thylakoid membranes oriented by gel squeezing. It can be seen that certain transitions are amplified while others vanish in one or the other ACD component compared with CD [178]. Information from isotropic and anisotropic circular dichroism and LD has been demonstrated to aid attaining the high-resolution structure of the baseplate antenna complex in *Chlorobaculum tepidum* [193]. In general, ACD can also be a powerful tool toward the better understanding of the excitonic interactions in light-harvesting antenna systems via linking structural data with the spectral signatures. Despite these unique features of ACD, to our knowledge, before 2012, this technique on photosynthetic systems was applied only once [200,201]. It was noticed that the CD spectrum of randomly oriented and magnetically (face-) aligned isolated intact thylakoid membranes displays markedly different CD spectra. This area of research is yet to be explored. It is also to be noted that in microscopic measurements [202] and/or on crystals, CD spectroscopy data may easily contain substantial ACD contributions.

In protein complexes and in their small aggregates (e.g., in supercomplexes), the distances are small compared to the wavelength of the visible light; CD is induced by short-range excitonic couplings between chromophores. The same holds in suspensions of small aggregates, in the absence of any sizeable long-range interactions between the chromophores. The situation is different in large 3D assemblies, which contain densely packed (~1 chromophore/nm^3) and intensely interacting chromophores, which are assembled in large

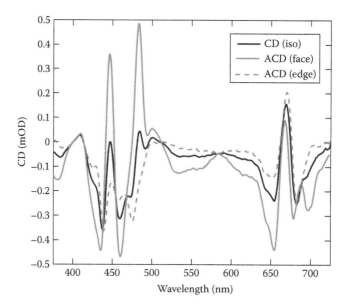

Figure 17.17 Comparison of the conventional CD spectrum of randomly oriented washed thylakoid membranes (brown solid curve) and ACD spectra of the same thylakoid membranes in face-aligned and edge-aligned orientation (solid green and dashed yellow curves, respectively). The samples were oriented by gel squeezing and measured in face-aligned position. The edge-aligned spectra were calculated as $CD_{edge} = 1/2(3CD_{random} - CD_{face})$. (Reproduced from Miloslavina, Y. et al., *Photosynth. Res.*, 111, 29, 2012.)

arrays with long-range chiral order and sizes commensurate with the wavelength of the visible light. This kind of assemblies, the so-called psi-type aggregates, such as DNA condensates, viruses, etc., usually exhibits intense anomalously shaped CD bands, accompanied by long tails outside the absorbance bands. These tails originate from differential scattering of L and R circularly polarized light [157,200,203]. Theory predicts that in psi-type aggregates the magnitude of the anomalous CD, at a constant density of chromophores and a constant pitch of the macrohelix, is controlled by the volume of the aggregate [203]. In general, the CD signal in these highly organized systems is correlated with the macro-organization of the chromophore system and the long-range chiral order of the components.

Our understanding of the psi-type CD is far from being complete [158]. Although there is a general agreement that it can only be observed at a high level of structural hierarchy, for example, in supercoiled DNA [157] or chirally organized macrodomains of protein complexes in vivo or in vitro (see following text), a refined theory, based on a structural model with realistic parameters, is still missing. Such a model might be based, for instance, on LHCII complexes, the structure of which is known at near atomic resolution and which readily self-assembles into lamellar aggregates with psi-type features. The better understanding of the psi-type CD would be essential because in many systems the regulation-related reorganizations occur at a high level of structural complexity, rather than at the level of building blocks (e.g., at the level of the array of protein complexes rather than at the level of every single protein complex). The present state of understanding of the psi-type CD has been reviewed by Garab and van Amerongen [159]. Here, we briefly repeat this latter treatment and then describe some of the key observations on different thylakoid membrane systems and on lamellar aggregates of LHCII.

The CD theory of psi-type aggregates [157,203] is based on the classical theory of coupled oscillators. This theory considers that light induces oscillating transition dipoles in the polarizable groups of the object, and the induced dipoles interact as static dipoles. In small aggregates, it is sufficient to consider the short-range dipole–dipole interactions, with R^{-3} dependence (R is the distance between the dipoles; see Equation 17.23). In contrast, in psi-type aggregates, the full electrodynamic interaction between the dipoles must be taken into account. At distant points of observation, the oscillating dipole can be regarded as a radiating spherical wave. Thus, the chromophores at large distances can be coupled via radiation and intermediate coupling mechanisms between the dipoles (with R^{-1} and R^{-2} dependencies, respectively). In contrast to a suspension of small aggregates, where these interactions can be neglected, because of the large distances between the particles, in psi-type aggregates, the radiation and intermediate couplings between the chromophores play an important role in determining the shape and magnitude of the psi-type CD spectrum. In large aggregates that possess no long-range order, the weak CD signals, arising from these relatively weak interactions, cancel each other. In contrast, in psi-type aggregates, they can sum up due to the long-range chiral order of the chromophores. This explains that the magnitude of the psi-type CD spectrum is controlled by the size of the particle, as indeed shown for the LHCII-containing macrodomains [204]. It is equally important to point out that while in small aggregates the entire aggregate at any instant is essentially at the same phase of the wave upon interaction with the light, in large aggregates, this is not true, and retardation effects can play an important role.

Thylakoid membranes, with their high protein content [205] and LHCII [189], are known to possess a high density of chromophores, as required for psi-type aggregates. As shown in Figure 17.15, granal thylakoid membranes, extended arrays of LHCII, and large semicrystalline domains [206] indeed display features of the psi-type aggregates. The psi-type CD of LHCII is associated with ordered arrays of the complexes [36]. It has been shown that although these particles, for their large sizes, inevitably exhibit intense light scattering (see Section 17.2.1), turbidity alone does not explain the emergence of the giant CD and the long tails outside the principal absorbance bands. It has been demonstrated [207] on isolated thylakoid membranes that upon decreasing the ionic strength and tonicity of the reaction medium, the psi-type CD can essentially be eliminated, while "normal" light scattering is hardly affected. Also, disordered LHCII aggregates display intense light scattering, but they exhibit no psi-type CD, while revealing the essentially unperturbed excitonic CD of small

aggregates with substantially smaller turbidity [208]. These data have indicated that scattering alone affects only marginally the excitonic CD and cannot be held responsible for the emergence of psi-type CD in highly organized macroassemblies. Differential scattering, instead of turbidity, however, is involved in the generation of psi-type CD [157], as it has also been shown for granal thylakoid membranes [200].

It must also be stressed that the multilamellar thylakoid membrane system of grana *per se* cannot be held responsible for the psi-type CD, as shown by the absence of psi-type signal in some minor-antenna mutant plants [209]. The fact that in this mutant the semicrystalline order of the complexes is also changed strongly indicates (in full accordance with the expectations) that psi-type CD carries information on the long-range order in the chromophore system. Also, probably strong variations in the long-range order of the protein complexes in different algal thylakoid membrane systems explain the variations, presence and absence of psi-type CD bands, in different algae [210–212]. Via capitalizing on the noninvasive feature of this technique, CD spectroscopy has been demonstrated to be capable of detecting different macrodomains with different outer antenna compositions in intact leaves [213].

Nevertheless, given the high complexity of the system, psi-type CD had been suspected to contain substantial optical artifacts. In order to settle this important issue, MCD experiments were recently performed [178] on intact and washed thylakoid membranes (i.e., on samples with CD spectra as presented in Figure 17.15a). MCD in this case is used as an internal standard: the external magnetic field generates MCD in every chlorophyll molecule; its band shape is identical with the absorbance of the molecule exhibiting a positive or a negative MCD band (in some cases, it can take the shape of the first derivative of the absorbance band) [214,215]. As it can be seen in Figure 17.18, intact and washed thylakoid membranes display essentially the same MCD spectra, despite the fact that they exhibit strikingly different CD spectra, with about an order of magnitude difference between the two amplitudes in the Q_y region [178]. These data, in accordance with the earlier conclusion, based among others on MCD experiments on randomly oriented and face-aligned intact thylakoid membranes, where the signs of the major CD bands were different in the Q_y region [201], show

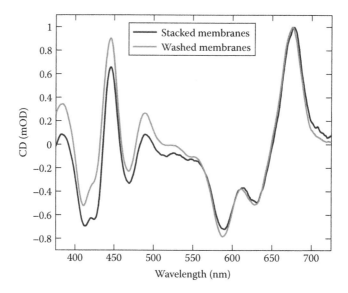

Figure 17.18 MCD spectra of intact and washed thylakoid membranes, induced by an external magnetic field of 1.4 T. MCD spectra were obtained via subtracting two CD spectra (similar to those in Figure 17.15a) measured in magnetic field with reverse direction. In this way, the inherent CD, even though it might be much larger in intensity than the MCD, is eliminated, while all other parameters remain unchanged. (Reproduced from Miloslavina, Y. et al., *Photosynth. Res.*, 111, 29, 2012.)

that the psi-type signals cannot be accounted for by distortions, for example, by scattering. Instead, the psi-type CD, superimposed onto the signals arising from structures at lower levels of the structural hierarchy, carries physical information on the molecular organization of chiral macrodomains.

The existence of chiral macrodomains in intact granal thylakoid membranes has been confirmed by CPL. It was shown that both large CPL and psi-type CD signals appeared in the presence of granal macrostructure and were sensitive to structural changes of the grana [216].

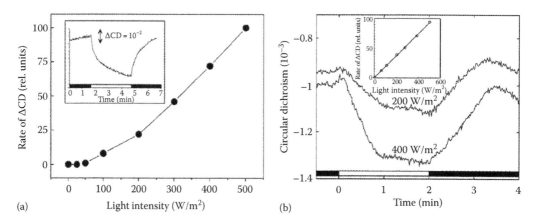

Figure 17.19 Light-induced, dark-reversible changes of the psi-type CD signal of isolated granal thylakoid membranes at 505 nm (a) and loosely stacked lamellar aggregates of isolated LHCII at 495 nm (b). Light and dark periods are indicated by empty and full horizontal bars, respectively. Insets, the kinetics of ΔCD (a) and light intensity dependence (b). Please note that the rate of ΔCD in thylakoid membranes is approximately linearly proportional to the excess light intensity above the saturation of photosynthesis, at around 100–150 W/m². (Based on data published in Barzda, V. et al., *Biochemistry*, 33, 10837, 1994; Simidjiev, I. et al., *Anal. Biochem.*, 250, 169, 1997; Cseh, Z. et al., *Photosynth. Res.*, 86, 263, 2005 [219].)

To finalize this part, it was emphasized in [36,159] that CD as a noninvasive tool allows to observe highly organized molecular assemblies under physiologically relevant conditions, including their variations in the time range of seconds and minutes due to changes in the environmental conditions. It has been documented in a number of studies that the photosynthetic lamellar structures are highly flexible. Reversible reorganizations detected by psi-type CD signals have been shown to occur both in thylakoid membranes and in loosely stacked lamellar aggregates of LHCII (see Figure 17.19 [217,218]). In thylakoid membranes, these light-induced reorganizations in LHCII-containing chiral macrodomains are largely independent of the photochemical activity of the membranes. Therefore, they have been proposed to be driven by a thermo-optic mechanism, according to which fast thermal transients due to dissipated excitation energy can lead to elementary structural transitions in the close vicinity of the site of dissipation. For an overview on the thermo-optic mechanism of the light-induced and dissipation-assisted reversible structural changes in light-harvesting antennas, see [36].

ACKNOWLEDGMENTS

The Estonian Research Council (Grant IUT02-28 to AF), the National Research, Development and Innovation Office of Hungary (Grant OTKA K 112688 to GG) and the Moravian-Silesian Region (Grant 01211/2016/RRC to GG) supported this work. The authors are grateful to P. H. Lambrev, J. Linnanto, and M. Rätsep for kindly preparing selected figures and for critical reading of the manuscript.

REFERENCES

1. Blankenship, R. E. 2014. *Molecular Mechanisms of Photosynthesis.* Wiley-Blackwell, Chichester, U.K.
2. Green, B. R. and W. W. Parson eds. 2003. *Light-Harvesting Antennas in Photosynthesis.* Kluwer Academic Publishers, Dordrecht, the Netherlands.
3. Hunter, C. N., F. Daldal, M. C. Thurnauer, and J. T. Beatty eds. 2008. *The Purple Phototrophic Bacteria.* Springer, Dordrecht, the Netherlands.
4. Laisk, A., L. Nedbal, and Govindjee eds. 2009. *Photosynthesis in Silico. Understanding Complexity From Molecules to Ecosystems.* Springer, Heidelberg, Germany.
5. Van Amerongen, H., L. Valkunas, and R. Van Grondelle. 2000. *Photosynthetic Excitons.* World Scientific, Singapore.
6. Amesz, J. and A. J. Hoff eds. 1996. *Biophysical Techniques in Photosynthesis.* Kluwer Academic Publishers, Dordrecht, the Netherlands.
7. Aartsma, T. J. and J. Matysik eds. 2008. *Biophysical Techniques in Photosynthesis.* Springer, Dordrecht, the Netherlands.
8. Lakowicz, J. R. 2006. *Principles of Fluorescence Spectroscopy.* Springer, New York.
9. Serdyuk, I. N., N. R. Zaccai, and J. Zaccai. 2007. *Methods in Biophysics: Structure, Dynamics, Function.* Cambridge University Press, Cambridge, U.K.
10. Parson, W. W. 2007. *Modern Optical Spectroscopy with Examples from Biophysics and Biochemistry.* Springer, Berlin, Germany.
11. Grimm, B., R. J. Porra, W. Rüdiger, and H. Scheer eds. 2006. *Chlorophylls and Bacteriochlorophylls.* Springer, Dordrecht, the Netherlands.
12. Frank, H. A., A. Young, G. Britton, and R. J. Cogdell eds. 1999. *The Photochemistry of Carotenoids.* Springer, the Netherlands.
13. Papageorgiu, G. C. and Govindjee eds. 2004. *Chlorophyll a Fluorescence: A Signature of Photosynthesis.* Kluwer Academic Publishers, Dordrecht, the Netherlands.
14. Demmig-Adams, B., G. Garab, W. W. Adams III, and Govindjee eds. 2014. *Non-Photochemical Quenching and Thermal Energy Dissipation in Plants, Algae and Cyanobacteria.* Springer, Dordrecht, the Netherlands.
15. Turro, N. J., V. Ramamurthy, and J. C. Scaiano. 2010. *Modern Molecular Photochemisty of Organic Molecules.* University Science Books, Sausalito, CA.
16. Gouterman, M. 1978–1979. Optical spectra and electronic structure of porphyrins and related rings. In *The Porphyrins, V. III,* D. Dolphin ed., Academic Press, New York, pp. 1–166.
17. Reimers, J. R., Z.-L. Cai, K. Kobayashi, M. Rätsep, A. Freiberg, and E. Krausz. 2013. Assignment of the Q-bands of the chlorophylls: Coherence loss via Q_x-Q_y mixing. *Sci. Rep.* 3:2761.

18. Frenkel, J. 1931. On the transformation on light into heat in solids. II. *Phys. Rev.* 37:1276–1294.

19. Franck, J. and E. Teller. 1938. Migration and photochemical action of excitation energy in crystals. *J. Chem. Phys.* 6:861–872.

20. Parson, W. W. and V. Nagarajan. 2003. Optical spectroscopy in photosynthetic antennas. In *Light-Harvesting Antennas in Photosynthesis*, B. R. Green and W. W. Parson, eds. Kluwer Academic Publishers, Dordrecht, the Netherlands, pp. 83–127.

21. Lesch, H., J. Schlichter, J. Friedrich, and J. M. Vanderkooi. 2004. Molecular probes: What is the range of their interaction with the environment? *Biophys. J.* 86:467–472.

22. May, V. and O. Kühn. 2000. *Charge and Energy Transfer Dynamics in Molecular Systems.* Wiley-VCH, Berlin, Germany.

23. Rebane, K. K. 1970. *Impurity Spectra of Solids.* Plenum Press, New York.

24. Naumov, A. V., A. A. Gorshelev, Y. G. Vainer, L. Kador, and J. Köhler. 2011. Impurity spectroscopy at its ultimate limit: Relation between bulk spectrum, inhomogneous broadening, and local disorder by spectroscopy of (nearly) all individual dopant molecules in solids. *Phys. Chem. Chem. Phys.* 13:1734–1742.

25. Naumov, A. V. 2013. Low-temperature spectroscopy of organic molecules in solid matrices: From the Shpol'skii effect to laser luminescent spectromicroscopy for all effectively emitting single molecules. *Physics-Uspekhi* 56:605.

26. Cogdell, R. J., A. Gall, and J. Köhler. 2006. The architecture and function of the light-harvesting apparatus of purple bacteria: From single molecules to in vivo membranes. *Q. Rev. Biophys.* 39:227–324.

27. Wu, H.-M., N. R. S. Reddy, and G. J. Small. 1997. Direct observation and hole burning of the lowest exciton level (B870) of the LH2 antenna complex of *Rhodopseudomonas acidophila* (strain 10050). *J. Phys. Chem. B* 101:651–656.

28. McLuskey, K., S. M. Prince, R. J. Cogdell, and N. W. Isaacs. 2001. The crystallographic structure of the B800-820 LH3 light-harvesting complex from the purple bacteria *Rhodopseudomonas acidophila* stain 7050. *Biochemistry* 40:8783–8789.

29. Gerken, U., F. Jelezko, B. Götze, M. Branschädel, C. Tietz, R. Ghosh, and J. Wrachtrup. 2003. Membrane environment reduces the accessible conformational space available to an integral membrane protein. *J. Phys. Chem. B* 107:338–343.

30. Zerlauskiene, O., G. Trinkunas, A. Gall, B. Robert, V. Urboniene, and L. Valkunas. 2008. Static and dynamic protein impact on electronic properties of light-harvesting complex LH2. *J. Phys. Chem. B* 112:15883–15892.

31. Freiberg, A., K. Timpmann, and G. Trinkunas. 2010. Spectral fine-tuning in excitonically coupled cyclic photosynthetic antennas. *Chem. Phys. Lett.* 500:111–115.

32. Hussels, M. and M. Brecht. 2011. Effect of glycerol and PVA on the conformation of photosystem I. *Biochemistry* 50:3628–3637.

33. Freiberg, A., M. Rätsep, and K. Timpmann. 2012. A comparative spectroscopic and kinetic study of photoexcitations in detergent-isolated and membrane-embedded LH2 light-harvesting complexes. *Biochim. Biophys. Acta* 1817:1471–1482.

34. Kunz, R., K. Timpmann, J. Southall, R. J. Cogdell, J. Köhler, and A. Freiberg. 2013. Fluorescence-excitation and emission spectra from LH2 antenna complexes of *Rhodopseudomonas acidophila* as a function of the sample preparation conditions. *J. Phys. Chem. B* 117:12020–12029.

35. Hilborn, R. C. 1982. Einstein coefficients, cross sections, f values, dipole moments, and all that. *Am. J. Phys.* 50:11982–11986.

36. Garab, G. 2014. Hierarchical organization and structural flexibility of thylakoid membranes. *Biochim. Biophys. Acta* 1837:481–494.

37. Horton, P., A. V. Ruban, and R. G. Walters. 1996. Regulation of light harvesting in green plants. *Annu. Rev. Plant Physiol. Plant Mol. Biol.* 47:655–684.

38. Bohren, C. F. and D. Huffman. 1983. *Absorption and Scattering of Light by Small Particles.* John Wiley & Sons, New York.

39. Cox, A. J., A. J. DeWeerd, and J. Linden. 2002. An experiment to measure Mie and Rayleigh total scattering cross section. *Am. J. Phys.* 70:620–625.

40. Van de Hulst, H. C. 1957. *Light Scattering by Small Particles.* John Wiley & Sons, Inc., New York.

41. Kerker, M. 1969. *The Scattering of Light.* Academic Press, New York.

42. Latimer, P. and E. Rabinowitch. 1959. Selective scattering of light by pigments *in vivo. Arch. Biochem. Biophys.* 84:428–441.

43. Duniec, J. T. and S. W. Thorne. 1977. The relation of light-induced slow absorbancy and scattering changes about 520 nm and structure of chloroplast thylakoids—A theoretical investigation. *J. Bioenerg. Biomembr.* 9:223–235.

44. Garab, G., G. Paillotin, and P. Joliot. 1979. Flash-induced scattering transients in the 10 gs-5 s time range between 450 and 540 nm with *chlorella* cells. *Biochim. Biophys. Acta* 545:445–453.

45. Merzlyak, M. N. and K. R. Naqvi. 2000. On recording the true absorption spectrum and the scattering spectrum of a turbid sample: Application to cell suspensions of the cyanobacterium *Anabaena variabilis. J. Photochem. Photobiol. B Biol.* 58:123–129.

46. Brown, W. ed. 1993. *Dynamic Light Scattering, the Method and Some Applications.* Oxford University Press, Oxford, U.K.

47. Naqvi, K. R., T. B. Melo, B. B. Raju, T. Javorfi, and G. Garab. 1997. Comparison of the absorption spectra of trimers and aggregates of chlorophyll a/b light-harvesting complex LHC II. *Spectrochim. Acta A* 53:1925–1936.

48. Duysens, L. N. M. 1956. The flattening of the absorption spectrum of suspensions, as compared to that of solutions. *Biochim. Biophys. Acta* 19:1–12.

49. Javorfi, T., J. Erostyak, J. Gal, A. Buzady, L. Menczel, G. Garab, and K. R. Naqvi. 2006. Quantitative spectrophotometry using integrating cavities. *J. Photochem. Photobiol. B Biol.* 82:127–131.

50. Erostyak, J., T. Javorfi, A. Buzady, K. R. Naqvi, and G. Garab. 2006. Comparative study of integrating cavity absorption meters. *J. Biochem. Biophys. Methods* 69:189–196.

51. Sauer, K. and M. Debreczeny. 1996. Fluorescence. In *Biophysical Techniques in Photosynthesis*, J. Amesz and A. J. Hoff eds. Kluwer Academic Publishers, Dordrecht, the Netherlands, pp. 41–61.

52. Jablonski, A. 1933. Efficiency of anti-Stokes fluorescence in dyes. *Nature* 131:839–840.

53. Hizhnyakov, V. and I. Tehver. 1970. On the theory of hot luminescence and resonant Raman effect of impurity centres. *Phys. Status Solidi B* 39:67–78.

54. Rebane, K. and P. Saari. 1978. Hot luminescence and relaxation processes in resonant secondary emission of solid matter. *J. Lumin.* 16:223–243.

55. Krieger-Liszkay, A. 2005. Singlet oxygen production in photosynthesis. *J. Exp. Bot.* 56:337–346.

56. Krasnovsky, A. A. 2007. Primary mechanisms of photoactivation of molecular oxygen. History of development and the modern status of research. *Biochem. Mosc.* 72:1065–1080.

57. Pieper, J. and A. Freiberg. 2014. Electron-phonon and exciton-phonon coupling in light harvesting: Insights from line-narrowing techniques. In *The Biophysics of Photosynthesis*, A. Van der Est and J. Golbeck eds. Springer, New York.

58. Davydov, A. S. 1971. *Theory of Molecular Excitons.* Plenum Press, New York.

59. Freiberg, A. and G. Trinkunas. 2009. Unraveling the hidden nature of antenna excitations. In *Photosynthesis in Silico: Understanding Complexity from Molecules to Ecosystems*, A. Laisk, L. Nedbal, and Govindjee eds. Springer, Heidelberg, Germany, pp. 55–82.

60. Osad'ko, I. S. 1983. Theory of light absorption and emission by organic impurity centers. In *Spectroscopy and Excitation Dynamics of Condensed Molecular Systems*, V. M. Agranovich and Hochstrasser, R. M. eds. North-Holland, Amsterdam, the Netherlands, pp. 438–514.

61. Rätsep, M., J. Linnanto, and A. Freiberg. 2009. Mirror symmetry and vibrational structure in optical spectra of chlorophyll a. *J. Chem. Phys.* 130:194501.

62. Rätsep, M., Z.-L. Cai, J. R. Reimers, and A. Freiberg. 2011. Demonstration and interpretation of significant asymmetry in the low-resolution and high-resolution Qy fluorescence and absorption spectra of bacteriochlorophyll a. *J. Chem. Phys.* 134:024506.

63. Keil, T. H. 1965. Shapes of Impurity Absorption Bands in Solids. *Phys. Rev.* 140:601–616.

64. Osad'ko, I. S. 1979. Determination of electron-phonon coupling from structured optical spectra of impurity centers. *Sov. Phys. Usp.* 22:311–329.

65. Pullerits, T., R. Monshouwer, F. van Mourik, and R. van Grondelle. 1995. Temperature dependence of electron-vibronic spectra of photosynthetic systems. Computer simulations and comparison with experiment. *Chem. Phys.* 194:395–407.

66. Pieper, J., K.-D. Irrgang, G. Renger, and R. E. Lechner. 2004. Density of vibrational states of the light-harvesting complex II of green plants studied by inelastic neutron scattering. *J. Phys. Chem. B* 108:10556–10565.

67. Pieper, J. and G. Renger. 2009. Protein dynamics investigated by neutron scattering. *Photosynth. Res.* 102:281–293.

68. Nagy, G., G. Garab, and J. Pieper. 2014. Neutron scattering in photosynthesis research. In *Photosynthesis: Open Questions and What We Know Today*, S. I. Allakhverdiev, A. B. Rubin, and V. A. Schuvalov, eds. Institute of Computer Science, Moscow, Russia.

69. Kurkal-Siebert, V. and J. C. Smith. 2006. Low-temperature protein dynamics: A simulation analysis of interprotein vibrations and the Boson Peak at 150 K. *J. Am. Chem. Soc.* 128:2356–2364.

70. Renger, T. and R. A. Marcus. 2002. On the relation of protein dynamics and exciton relaxation in pigment-protein complexes: An estimation of the spectral density and a theory for the calculation of optical spectra. *J. Chem. Phys.* 116:9997–10019.

71. Timpmann, K., M. Rätsep, C. N. Hunter, and A. Freiberg. 2004. Emitting excitonic polaron states in core LH1 and peripheral LH2 bacterial light-harvesting complexes. *J. Phys. Chem. B* 108:10581–10588.

72. Kell, A., X. Feng, M. Reppert, and R. Jankowiak. 2013. On the shape of the phonon spectral density in photosynthetic complexes. *J. Phys. Chem. B* 117:7317–7323.

73. Pajusalu, M., M. Rätsep, and A. Freiberg. 2014. Temperature dependent electron-phonon coupling in chlorin-doped impurity glass and in photosynthetic FMO protein containing bacteriochlorophyll a. *J. Lumin.* 152:79–83. doi: 10.1016/j.jlumin.2014.12.011.

74. Rolfe, J., F. R. Lipsett, and W. J. King. 1961. Optical absorption and fluorescence of oxygen in alkali halide crystals. *Phys. Rev.* 123:447–454.

75. Fulton, R. L. and M. Gouterman. 1964. Vibronic coupling II spectra of dimers. *J. Chem. Phys.* 41:2280–2286.

76. Hochstrasser, R. M. and P. N. Prasad. 1972. Phonon sidebands of electronic transitions in molecular crystals and mixed crystals. *J. Chem. Phys.* 56:2814–2823.

77. Reynolds, P. A. 1973. The phonon spectrum of molecular crystals from exciton sidebands. *Chem. Phys. Lett.* 22:177–179.

78. Hanson, D. M. 1976. Effect of the exciton bandwidth on electron-phonon coupling in molecular crystals. *Chem. Phys. Lett.* 43:217–220.

79. Toyozawa, Y. 2003. *Optical Processes in Solids.* Cambridge University Press, Cambridge, U.K.

80. Förster, T. 1965. *Delocalized Excitation and Excitation Transfer.* Academic Press, New York.

81. Rashba, I. E. 1982. Self-trapping of excitons. In *Excitons*, M. D. Sturge ed. North-Holland, Amsterdam, the Netherlands, pp. 543–602.

82. Holstein, T. 1959. Studies of polaron motion. Part I. The molecular-crystal model. *Ann. Phys.* 8:325–342.

83. Lu, N. and S. Mukamel. 1991. Polaron and size effects in optical lineshapes of molecular aggregates. *J. Chem. Phys. B* 95:1588–1607.

84. Damjanovic, A., I. Kosztin, U. Kleinekathöfer, and K. Schulten. 2002. Excitons in a photosynthetic light-harvesting system: A combined molecular dynamics, quantum chemistry, and polaron model study. *Phys. Rev. E* 65:031919–031943.

85. Jang, S., J. Cao, and R. J. Silbey. 2002. On the temperature dependence of molecular line shapes due to linearly coupled phonon bands. *J. Phys. Chem. B* 106:8313–8317.

86. Myers Kelley, A. 2003. A multimode vibronic treatment of absorption, resonance Raman, and hyper-Rayleigh scattering of excitonically coupled molecular dimers. *J. Chem. Phys.* 119:3320–3331.

87. Olbrich, C., T. L. C. Jansen, J. R. Liebers, M. Aghtar, J. Strumpfer, K. Schulten, J. Knoester, and U. KleinekathoGfer. 2011. From atomistic modeling to excitation transfer and two-dimensional spectra of the FMO light-harvesting complex. *J. Phys. Chem. B* 115:8609–8621.

88. Spano, F. C. 2006. Excitons in conjugated oligomer aggregates, films, and crystals. *Annu. Rev. Phys. Chem.* 37:217–243.

89. Spano, F. C. 2010. The spectral signatures of Frenkel polarons in H- and J-aggregates. *Acc. Chem. Res.* 43:429–439.

90. Curutchet, C. and B. Mennucci. 2017. Quantum chemical studies of light harvesting. *Chem. Rev.* 117:294–343.

91. Lee, M. K., P. Huo, and D. F. Coker. 2016. Semiclassical path integral dynamics: Photosynthetic energy transfer with realistic environment interactions. *Annu. Rev. Phys. Chem.* 67:639–668.

92. Tempelaar, R., A. Stradomska, J. Knoester, and F. C. Spano. 2013. Anatomy of an exciton: Vibrational distortion and exciton coherence in H- and J-aggregates. *J. Phys. Chem. B* 117:457–466.

93. Kobrak, M. N. and E. R. Bittner. 2000. A quantum molecular dynamics study of exciton self-trapping in conjugated polymers: Temperature dependence and spectroscopy. *J. Chem. Phys.* 112:7684–7692.

94. Pajusalu, M., R. Kunz, M. Rätsep, K. Timpmann, J. Köhler, and A. Freiberg. 2015. Unified analysis of ensemble and single-complex optical spectral data from light-harvesting complex-2 chromoproteins for gaining deeper insight into bacterial photosynthesis. *Phys. Rev. E* 92:052709.

95. Purchase, R. and S. Völker. 2009. Spectral hole burning: Examples from photosynthesis. *Photosynth. Res.* 101:245–266.

96. Jankowiak, R., M. Reppert, V. Zazubovich, J. Pieper, and T. Reinot. 2011. Site selective and single complex laser-based spectroscopies: A window on excited state electronic structure, excitation energy transfer, and electron-phonon coupling of selected photosynthetic complexes. *Chem. Rev.* 111:4546–4598.

97. Sild, O. and K. Haller eds. 1988. *Zero-phonon Lines and Spectral Hole Burning in Spectroscopy and Photochemistry*. Springer Verlag, Berlin, Germany.

98. Rebane, K. K. 2002. Purely electronic zero-phonon line as the foundation for high resolution matrix spectroscopy, single impurity molecule spectroscopy, persistent spectral hole burning. *J. Lumin.* 100:219–232.

99. Osad'ko, I. S. 2003. *Selective Spectroscopy of Single Molecules*. Springer, Berlin, Germany.

100. Denisov, Y. V. and V. A. Kizel. 1967. Energy migration in Europium-activated borate glasses and relative location of energy levels. *Opt. Spektrosk.* 23:251.

101. Szabo, A. 1970. Laser-induced fluorescence-line narrowing in ruby. *Phys. Rev. Lett.* 25:924–926.

102. Gorokhovskii, A. A., R. K. Kaarli, and L. A. Rebane. 1974. Hole burning in the contour of a pure electronic line in a Shpol'skii system. *JETP Lett.* 20:216–218.

103. Kharlamov, B. M., R. I. Personov, and L. A. Bykovskaya. 1974. Stable "gap" in absorption spectra of solid solutions of organic molecules by laser irradiation. *Opt. Commun.* 12:191–193.

104. Jaaniso, R. V. and R. A. Avarmaa. 1986. Measurement of the inhomogeneous distribution and homogeneous spectra of an impurity molecule in a glassy matrix. *Sov. J. Appl. Spectrosc. (USA)* 44:365–370.

105. Fünfschilling, J., D. Glatz, and I. Zschokke-Gränacher. 1986. Hole-burning spectroscopy as a tool to eliminate inhomogeneous broadening. *J. Lumin.* 36:85–92.

106. Wolf, J., K.-Y. Law, and A. B. Myers. 1996. Hole-burning subtracted fluorescence line-narrowing spectroscopy of squaraines in polymer matrixes. *J. Phys. Chem.* 100:11870–11882.

107. Moerner, W. E. and L. Kador. 1989. Optical detection and spectroscopy of single molecules in a solid. *Phys. Rev. Lett.* 62:2535–2538.

108. Moerner, W. E. and M. Orrit. 1999. Illuminating single molecules in condensed matter. *Science* 283:1670–1676.

109. Orrit, M. and J. Bernard. 1990. Single pentacene molecules detected by fluorescence excitation in a p-terphenyl crystal. *Phys. Rev. Lett.* 65:2716–2719.

110. Wu, H.-M., M. Rätsep, R. Jankowiak, R. J. Cogdell, and G. J. Small. 1997. Comparison of the LH2 antenna complexes of *Rhodopseudomonas acidophila* (strain 10050) and *Rhodobacter sphaeroides* by high-pressure absorption, high-pressure hole burning, and temperature-dependent absorption spectroscopies. *J. Phys. Chem. B* 101:7641–7653.

111. Wu, H.-M., M. Rätsep, R. Jankowiak, R. J. Cogdell, and G. J. Small. 1998. Hole-burning and absorption studies of the LH1 antenna complex of purple bacteria: Effects of pressure and temperature. *J. Phys. Chem. B* 102:4023–4034.

112. Reul, S., W. Richter, and D. Haarer. 1991. Spectroscopic studies of the solute-solvent interaction in dye-doped polymers: Frequency-dependent pressure shifts. *Chem. Phys. Lett.* 180:1–5.

113. Kohler, M., J. Friedrich, and J. Fidy. 1998. Proteins in electric fields and pressure fields: Basic aspects. *Biochim. Biophys. Acta* 1386:255–288.

114. Rätsep, M., J. Pieper, K.-D. Irrgang, and A. Freiberg. 2008. Excitation wavelength-dependent electron-phonon and electron-vibrational coupling in the CP29 antenna complex of green plants. *J. Phys. Chem. B* 112:110–118.

115. Rätsep, M., M. Pajusalu, and A. Freiberg. 2009. Wavelength-dependent electron–phonon coupling in impurity glasses. *Chem. Phys. Lett.* 479:140–143.

116. Renge, I., M. Rätsep, and A. Freiberg. 2011. Intermolecular repulsive-dispersive potentials explain properties of impurity spectra in soft solids. *J. Lumin.* 131:262–265.

117. Orlov, S. V., A. V. Naumov, Y. G. Vainer, and L. Kador. 2012. Spectrally resolved analysis of fluorescence blinking of single dye molecules in polymers at low temperatures. *J. Chem. Phys.* 137:194903.

118. Barkai, E., Y. J. Yung, and R. Silbey. 2004. Theory of single-molecule spectroscopy: Beyond the ensemble average. *Annu. Rev. Phys. Chem.* 55:457–507.

119. Rebane, L. A., A. A. Gorokhovskii, and J. Kikas. 1982. Low-temperature spectroscopy of organic molecules in solids by photochemical hole burning. *Appl. Phys. B* 29:235–250.

120. Friedrich, J. and D. Haarer. 1984. Photochemical hole burning: A spectroscopic study of relaxation processes in polymers and glasses. *Angew. Chem. Int. Ed.* 23:113–140.

121. Moerner, W. E. ed. 1988. *Persistent Spectral Hole Burning: Science and Applications.* Springer-Verlag, Berlin, Germany.

122. Avarmaa, R. A. and K. K. Rebane. 1985. High-resolution optical spectra of chlorophyll molecules. *Spectrochim. Acta A* 41:1365–1380.

123. Fidy, J., M. Laberge, A. D. Kaposi, and J. M. Vanderkooi. 1998. Fluorescence line narrowing applied to the study of proteins. *Biochim. Biophys. Acta* 1386:331–351.

124. Weiss, S. 1999. Fluorescence spectroscopy of single biomolecules. *Science* 283:1676–1683.

125. Reinot, T., V. Zazubovich, and J. M. Hayes. 2001. New insights on persistent nonphotochemical hole burning and its application to photosynthetic complexes. *J. Phys. Chem. B* 105:5083–5098.

126. Orrit, M., J. Bernard, and R. I. Personov. 1993. High-resolution spectroscopy of organic molecules in solids: From fluorescence line narrowing and hole burning to single molecule spectroscopy. *J. Phys. Chem.* 97:10256–10268.

127. Berlin, Y., A. Burin, J. Friedrich, and J. Köhler. 2007. Low temperature spectroscopy of proteins. Part II: Experiments with single protein complexes. *Phys. Life Rev.* 4:64–89.

128. Kulzer, F., T. Xia, and M. Orrit. 2010. Single molecules as optical nanoprobes for soft and complex matter. *Angew. Chem. Int. Ed.* 49:854–866.

129. Wagie, H. E. and P. Geissinger. 2012. Hole-burning spectroscopy as a probe of nanoenvironments and processes in biomolecules: A review. *Appl. Spectrosc.* 66:609–626.

130. Jankowiak, R. 2000. Fundamental aspects of fluorescence line-narrowing. In *Shpol'skii Spectroscopy and Other Site-Selective Methods*, C. Gooijer, F. Ariese, and J. W. Hofstraat eds. John Wiley & Sons, New York, pp. 235–271.

131. Chang, H.-C., R. Jankowiak, N. R. S. Reddy, and G. J. Small. 1995. Pressure dependence of primary charge separation in a photosynthetic reaction center. *Chem. Phys.* 197:307–321.

132. Jankowiak, R. and G. J. Small. 1993. Spectral hole burning: A window on excited state electronic structure, heterogeneity, electron-phonon coupling, and transport dynamics of photosynthetic units. In *Photosynthetic Reaction Centers*, J. Deisenhofer ed. Academic Press, Boston, MA, pp. 133–177.

133. Jankowiak, R., J. M. Hayes, and G. J. Small. 1993. Spectral hole-burning spectroscopy in amorphous molecular solids and proteins. *Chem. Rev.* 93:1471–1502.

134. Krausz, E. 2013. Selective and differential optical spectroscopies in photosynthesis. *Photosynth. Res.* 116:411–426.

135. Pieper, J., J. Voigt, G. Renger, and G. J. Small. 1999. Analysis of phonon structure in line-narrowed optical spectra. *Chem. Phys. Lett.* 310:296–302.

136. Rätsep, M. and A. Freiberg. 2003. Resonant emission from the B870 exciton state and electron-phonon coupling in the LH2 antenna chromoprotein. *Chem. Phys. Lett.* 377:371–376.

137. Volker, S. 1989. Hole-burning spectroscopy. *Annu. Rev. Phys. Chem.* 40:499–530.

138. Rätsep, M. and A. Freiberg. 2007. Electron-phonon and vibronic couplings in the FMO bacteriochlorophyll a antenna complex studied by difference fluorescence line narrowing. *J. Lumin.* 127:251–259.

139. Reppert, M., V. Naibo, and R. Jankowiak. 2010. Accurate modeling of fluorescence line narrowing difference spectra: Direct measurement of the single-site fluorescence spectrum. *J. Chem. Phys.* 133:014506.

140. Leiger, K., L. Reisberg, and A. Freiberg. 2013. Fluorescence micro-spectroscopy study of individual photosynthetic membrane vesicles and light-harvesting complexes. *J. Phys. Chem. B* 117:9315–9326.

141. Bopp, M. A., Y. Jia, L. Li, R. J. Cogdell, and R. M. Hochstrasser. 1997. Fluorescence and photobleaching dynamics of single light-harvesting complexes. *Proc. Natl. Acad. Sci. U. S. A.* 94:10630–10635.

142. Van Oijen, A. M., M. Ketelaars, J. Koehler, T. J. Aartsma, and J. Schmidt. 1998. Spectroscopy of single light-harvesting complexes from purple photosynthetic bacteria at 1.2 K. *J. Phys. Chem. B* 102:9363–9366.

143. Tietz, C., O. Chekhlov, A. Dräbenstedt, J. Schuster, and J. Wrachtrup. 1999. Spectroscopy on single light-harvesting complexes at low temperature. *J. Phys. Chem. B* 103:6328–6333.

144. Rutkauskas, D., J. Olsen, A. Gall, R. J. Cogdell, and C. N. Hunter. 2006. Comparative study of spectral flexibilities of bacterial light-harvesting complexes: Structural implications. *Biophys. J.* 90:2463–2475.

145. Vacha, F., L. Bumba, D. Kaftan, and M. Vacha. 2005. Microscopy and single molecule detection in bacterial photosynthesis. *Micron* 36:483–502.

146. Saga, Y., Y. Shibata, and H. Tamiaki. 2010. Spectral properties of single light-harvesting complexes in bacterial photosynthesis. *J. Photochem. Photobiol. C* 11:15–24.

147. Aartsma, T. J. and J. Köhler. 2008. Optical spectroscopy of individual light-harvesting complexes. In *Biophysical Techniques in Photosynthesis*, Vol. II, V. T. J. Aartsma and J. Matysik eds. Academic Publishers, Dordrecht, the Netherlands, pp. 241–266.

148. Krüger, T. P. J., C. Ilioaia, M. P. Johnson, E. Belgio, P. Horton, A. V. Ruban, and R. Van Grodelle. 2013. The specificity of controlled protein disorder in the photoprotection of plants. *Biophys. J.* 105:1018–1026.

149. Duschinsky, F. 1937. On the interpretation of electronic spectra of polyatomic molecules. *Acta Physicochim. URSS* 7:551.

150. Belen Oviedo, M., C. F. A. Negre, and C. Sanchez. 2010. Dynamical simulation of the optical response of photosynthetic pigments. *Phys. Chem. Chem. Phys.* 12:6706–6711.

151. Cogdell, R. J., T. D. Howard, N. W. Isaacs, K. McLuskey, and A. T. Gardiner. 2002. Structural factors which control the position of the Qy absorption band of bacteriochlorophyll a in purple bacterial antenna complexes. *Photosynth. Res.* 74:135–141.

152. Adolphs, J. and T. Renger. 2006. How proteins trigger energy transfer in the FMO complex of green sulfur bacteria. *Biophys. J.* 91:2778–2797.

153. Sundström, V., T. Pullerits, and R. van Grondelle. 1999. Photosynthetic light-harvesting: Reconciling dynamics and structure of purple bacterial LH2 reveals function of photosynthetic unit. *J. Phys. Chem. B* 103:2327–2346.

154. Sauer, K., R. J. Cogdell, S. M. Prince, A. Freer, N. W. Isaacs, and H. Scheer. 1996. Structure-based calculations of the optical spectra of the LH2 bacteriochlorophyll-protein complex from *Rhodopseudomonas acidophila*. *Photochem. Photobiol.* 64:564–576.

155. Alden, R. G., E. Johnson, V. Nagarajan, W. W. Parson, C. J. Law, and R. G. Cogdell. 1997. Calculations of spectroscopic properties of the LH2 bacteriochlorophyll - Protein antenna complex from *Rhodopseudomonas acidophila*. *J. Phys. Chem. B* 101:4667–4680.

156. Goldstein, D. 2003. *Polarized Light*, 2nd edn., Revised and Expanded. Marcel Dekker, New York.

157. Tinoco, I. J., W. Mickols, M. F. Maerstr, and C. Bustamante. 1987. Absorption, scattering, and imaging of biomolecular structures with polarized light. *Annu. Rev. Biophys. Biomol. Struct.* 16:319–349.

158. Garab, G. 1996. Linear and circular dichroism. In *Biophysical Techniques in Photosynthesis*, J. Amesz and A. J. Hoff eds. Kluwer Academic Publishers, Dordrecht, the Netherlands, pp. 11–40.

159. Garab, G. and H. Van Amerongen. 2009. Linear and circular dichroism in photosynthesis research. *Photosynth. Res.* 101:135–146.

160. Patty, C. H. L., L. J. J. Visser, F. Ariese, W. J. Buma, W. B. Sparks, R. J. M. van Spanning, W. F. M. Roling, and F. Snik. 2017. Circular spectropolarimetric sensing of chiral photosystems in decaying leaves. *J. Quant. Spectrosc. RA* 189:303–311.

161. Breton, J. and A. Vermeglio. 1982. Orientation of photosynthetic pigments in vivo. In *Photosynthesis*, Govindjee ed. Academic Press, New York, pp. 153–194.

162. Abdourakhmanov, I. A., A. O. Ganago, Y. E. Erokhin, A. A. Solov'ev, and V. A. Chugunov. 1979. Orientation and linear dichroism of the reaction centers from *Rhodopseudomonas sphaeroides* R-26. *Biochim. Biophys. Acta* 546:183–186.

163. Kiss, L., A. O. Ganago, and G. Garab. 1985. Quantitative method for studying orientation of transition dipoles in membrane vesicles of spherical symmetry. *J. Biochim. Biophys. Methods* 11:213–225.

164. Kim, M., D. Keller, and C. Bustamante. 1987. Differential polarization imaging, Part 1: Theory. *Biophys. J.* 52:911–927.

165. Kim, M. and C. Bustamante. 1991. Differential polarization imaging. 4. Images in higher born approximations. *Biophys. J.* 59:1171–1182.

166. Mickols, W., I. J. Tinoco, J. E. Katz, M. F. Maestre, and C. Bustamante. 1985. Imaging differential polarization microscope with electrical readout. *Rev. Sci. Instrum.* 56:2228–2236.

167. Finzi, L., C. Bustamante, G. Garab, and C. B. Juang. 1989. Direct observation of large domains in chloroplast thylakoid membranes by differential polarization microscopy. *Proc. Natl. Acad. Sci. U. S. A.* 86:8748–8752.

168. Gupta, V. K., J. A. Kornfield, and G. Wenger. 1994. Controlling molecular order in "hairy-rod" Langmuir-Blodgett films: A polarization-modulation microscopy study. *Science* 265:940–942.

169. Garab, G., R. Pomozi, R. Jörgens, and G. Weiss. 2005. Method and apparatus for determining the polarization properties of light emitted, reflected or transmitted by a material using a laser scanning microscope. US Patent 6:391.

170. Steinbach, G., F. Besson, R. Pomozi, and G. Garab. 2005. Differential polarization laser scanning microscope: Biological applications. *Proc. SPIE* 5969:566–575.

171. Steinbach, G., R. Pomozi, O. Zsiros, A. Pay, G. V. Horvath, and G. Garab. 2008. Imaging fluorescence detected linear dichroism of plant cell walls in laser scanning confocal microscope. *Cytometry A* 73:202–208.

172. Steinbach, G., R. Pomozi, O. Zsiros, L. Menczel, and G. Garab. 2009. Imaging anisotropy using differential polarization laser scanning confocal microscopy. *Acta Histochem.* 111:316–325.

173. Garab, G., P. Galajda, R. Pomozi, L. Finzi, T. Praznovszky, P. Ormos, and H. Van Amerongen. 2005. Alignment of biological microparticles by a polarized laser beam. *Eur. Biophys. J.* 34:335–343.

174. Chappaz-Gillot, C., P. L. Marek, B. J. Blaive, G. Canard, J. Brück, G. Garab, H. Hahn et al. 2012. Anisotropic organization and microscopic manipulation of self-assembling synthetic porphyrin microrods that mimic chlorosomes: Bacterial light-harvesting systems. *J. Am. Chem. Soc.* 134:944–954.

175. Gussakovsky, E. E. 2008. Circularly polarized luminescence (CPL) of proteins and protein complexes. In *Reviews in Fluorescence*, C. D. Geddes ed. Springer, New York, pp. 425–460.

176. Hall, J., T. Renger, R. Picorel, and E. Krausz. 2016. Circularly polarized luminescence spectroscopy reveals low-energy excited

states and dynamic localization of vibronic transitions in CP43. *Biochim. Biophys. Acta* 1857:115–128.

177. Berova, N., K. Nakanishi, and R. W. Woody eds. 2000. *Circular Dichroism: Principles and Applications*. Wiley, New York.

178. Miloslavina, Y., P. H. Lambrev, T. Javorfi, Z. Varkonyi, V. Karlicky, J. S. Wall, G. Hind, and G. Garab. 2012. Anisotropic circular dichroism signatures of oriented thylakoid membranes and lamellar aggregates of LHCII. *Photosynth. Res.* 111:29–39.

179. Cantor, C. R. and P. R. Schimmel. 1980. *Biophysical Chemistry, Part II: Techniques for the Study of Biological Structure and Function*. H. Freeman & Co., San Francisco, CA.

180. Warshel, A. and W. W. Parson. 1987. Spectroscopic properties of photosynthetic reaction centers I. Theory. *J. Am. Chem. Soc.* 109:6143–6152.

181. Tinoco, I. J. 1962. Theoretical aspects of optical activity. Part two: Polymers. *Adv. Chem. Phys.* 4:113–160.

182. DeVoe, H. 1965. Optical properties of molecular aggregates. II. Classical theory of the refraction, absorption, and optical activity of solutions and crystals. *J. Chem. Phys.* 43:3199–3208.

183. Ke, Y. L. and Y. Zhao. 2017. Hierarchy of stochastic Schrodinger equation towards the calculation of absorption and circular dichroism spectra. *J. Chem. Phys.* 146:174105.

184. Reinot, T., J. H. Chen, A. Kell, M. Jassas, K. C. Robben, V. Zazubovich, and R. Jankowiak. 2016. On the conflicting estimations of pigment site energies in photosynthetic complexes: A case study of the CP47 complex. *Anal. Chem. Insights* 11:35–48.

185. Reimers, J. R., M. Biczysko, D. Bruce, D. F. Coker, T. J. Frankcombe, H. Hashimoto, J. Hauer et al. 2016. Challenges facing an understanding of the nature of low-energy excited states in photosynthesis. *Biochim. Biophys. Acta* 1857:1627–1640.

186. Renger, T. and F. Müh. 2013. Understanding photosynthetic light-harvesting: A bottom up theoretical approach. *Phys. Chem. Chem. Phys.* 15:3348–3371.

187. Müh, F. and T. Renger. 2012. Refined structure-based simulation of plant light-harvesting complex II: Linear optical spectra of trimers and aggregates. *Biochim. Biophys. Acta* 1817:1446–1460.

188. Müh, F., M. E. Madjet, and T. Renger. 2012. Structure-based simulation of linear optical spectra of the CP43 core antenna of photosystem II. *Photosynth. Res.* 111:87–101.

189. Liu, Z., H. Yan, K. Wang, T. Kuang, J. Zhang, L. Gui, X. An, and W. Chang. 2004. Crystal structure of spinach major light-harvesting complex at 2.72 Å resolution. *Nature* 428:287–292.

190. Natali, A. and R. Croce. 2015. Characterization of the major light-harvesting complexes (LHCBM) of the green alga *Chlamydomonas reinhardtii*. *PLoS One* 10:e0119211.

191. Akhtar, P., M. Dorogi, K. Pawlak, L. Kovacs, A. Bota, T. Kiss, G. Garab, and P. H. Lambrev. 2015. Pigment interactions in light-harvesting complex II in different molecular environments. *J. Biol. Chem.* 290:4877–4886.

192. Lambrev, P. H., Z. Varkonyi, S. Krumova, L. Kovacs, Y. Miloslavina, A. R. Holzwarth, and G. Garab. 2007. Importance of trimer-trimer interactions for the native state of the plant light-harvesting complex II. *Biochim. Biophys. Acta* 1767:847–853.

193. Nielsen, J. T., N. V. Kulminskaya, M. Bjerring, J. M. Linnanto, M. Ratsep, M. O. Pedersen, P. H. Lambrev et al. 2016. In situ high-resolution structure of the baseplate antenna complex in *Chlorobaculum tepidum*. *Nat. Commun.* 7:12454.

194. Somsen, O. J. G., L. Valkunas, and R. van Grondelle. 1996. A perturbed two-level model for exciton trapping in small photosynthetic systems. *Biophys. J.* 70:669–683.

195. Standfuss, J., M. Lamborghini, W. Kühlbrandt, and A. C. T. van Scheltinga. 2005. Mechanisms of photoprotection and nonphotochemical quenching in pea light-harvesting complex at 2.5 Å resolution. *EMBO J.* 24:919–928.

196. Kuball, H. G. 2002. CD and ACD spectroscopy on anisotropic samples: Chirality of oriented molecules and anisotropic phases—A critical analysis. *Enantiomer* 7:197–205.

197. Kuball, H. G., G. Sieber, S. Neubrecht, B. Schultheis, and A. Schonhofer. 1993. Circular-dichroism of oriented molecules—Magnetic dipole and electric quadrupole contribution to the ACD of chirally substituted diaminoanthraquinones. *Chem. Phys.* 169:335–350.

198. Kuball, H. G., E. Dorr, T. Höfer, and O. Turk. 2005. Exciton chirality method. Oriented molecules—Anisotropic phases. *Monatsh. Chem.* 136:289–324.

199. Kuball, H. G. and T. Höfer. 2000. Chirality and circular dichroism of oriented molecules and anisotropic phases. *Chirality* 12:278–286.

200. Garab, G., K. S. Wells, L. Finzi, and C. Bustamante. 1988. Helically organized macroaggregates of pigment-protein complexes in chloroplasts: Evidence for circular intensity differential scattering. *Biochemistry* 27:5839–5843.

201. Garab, G., A. Faludi-Daniel, J. C. Sutherland, and G. Hind. 1988. Macroorganization of chlorophyll a/b light-harvesting complex in thylakoids and aggregates: Information from circular differential scattering. *Biochemistry* 27:2425–2430.

202. Furumaki, S., Y. Yabiku, S. Habuchi, Y. Tsukatani, D. A. Bryant, and M. Vacha. 2012. Circular dichroism measured on single chlorosomal light-harvesting complexes of green photosynthetic bacteria. *J. Phys. Chem. Lett.* 3:3545–3549.

203. Keller, D. and C. Bustamante. 1986. Theory of the interaction of light with large inhomogeneous molecular aggregates. *J. Chem. Phys.* 84:2972–2980.

204. Barzda, V., L. Mustardy, and G. Garab. 1994. Size dependency of circular dichroism in macroaggregates of photosynthetic pigment-protein complexes. *Biochemistry* 33:10837–10841.

205. Garab, G., K. Lohner, P. Laggner, and T. Farkas. 2000. Self-regulation of the lipid content of membranes by non-bilayer lipids: A hypothesis. *Trends Plant Sci.* 5:489–494.

206. Kouril, R., J. P. Dekker, and E. J. Boekema. 2012. Supramolecular organization of photosystem II in green plants. *Biochim. Biophys. Acta* 1817:2–12.

207. Garab, G., J. Kieleczawa, J. C. Sutherland, C. Bustamante, and G. Hind. 1991. Organization of pigment-protein complexes into macrodomains in the thylakoid membranes of wild-type and chlorophyll b-less mutant of barely as revealed by circular dichroism. *Photochem. Photobiol.* 54:273–281.

208. Simidjiev, I., V. Barzda, L. Mustardy, and G. Garab. 1997. Isolation of lamellar aggregates of the light-harvesting chlorophyll a/b protein complex of photosystem II with long-range chiral order and structural flexibility. *Anal. Biochem.* 250:169–175.

209. Kovacs, L., J. Damkjaer, S. Kereiche, C. Ilioaia, A. V. Ruban, E. J. Boekema, S. Jansson, and P. Horton. 2006. Lack of the light-harvesting complex CP24 affects the structure and function of the grana membranes of higher plant chloroplasts. *Plant Cell* 18:3106–3120.

210. Goss, R., C. Wilhelm, and G. Garab. 2000. Organization of the pigment molecules in the chlorophyll a/b/c containing alga Mantoniella aquamata studied by means of absorption, circular and linear dichroism spectroscopy. *Biochim. Biophys. Acta* 1457:190–199.

211. Büchel, C. and G. Garab. 1998. Molecular organization of the chlorophyll a/c light-harvesting complex of *Pleurochloris meiringensis* (Xanthophyceae). Pigment binding and secondary structure of the protein. *J. Photochem. Photobiol. B* 42:191–194.

212. Szabo, M., B. Lepetit, R. Goss, C. Wilhelm, L. Mustardy, and G. Garab. 2008. Structurally flexible macroorganization of the pigment-protein complexes of the diatom Phaeodactylum tricornutum. *Photosynth. Res.* 95:237–245.

213. Toth, T. N., N. Rai, K. Solymosi, O. Zsiros, W. P. Schroder, G. Garab, H. van Amerongen, P. Horton, and L. Kovacs. 2016. Fingerprinting the macro-organisation of pigment-protein complexes in plant thylakoid membranes in vivo by circular-dichroism spectroscopy. *Biochim. Biophys. Acta* 1857:1479–1489.

214. Sutherland, J. C. 1978. The magnetic optical activity of porphyrins. In *The Porphyrins*, D. Dolphin ed. Academic Press, Inc. Ltd, London, U.K., pp. 225–248.

215. Sutherland, J. C. and B. Holmquist. 1980. Magnetic circular dichroism of biological molecules. *Annu. Rev. Biophys. Bioeng.* 9:293–326.

216. Gussakovsky, E. E., Y. Shahak, H. Van Amerongen, and V. Barzda. 2000. Circularly polarized chlorophyll luminescence reflects the macro-organization of grana in pea chloroplasts. *Photosynth. Res.* 65:83–92.

217. Garab, G., R. C. Leegood, D. A. Walker, J. C. Sutherland, and G. Hind. 1988. Reversible changes in macroorganization of the light-harvesting chlorophyll *a/b* pigment protein complex detected by circular-dichroism. *Biochemistry* 27:2430–2434.

218. Barzda, V., A. Istokovics, I. Simidjiev, and G. Garab. 1996. Structural flexibility of chiral macroaggregates of light-harvesting chlorophyll *a/b* pigment-protein complexes. Light-induced reversible structural changes associated with energy dissipation. *Biochemistry* 35:8981–8985.

219. Cseh, Z., A. Vianelli, S. Rajagopal, S. Krumova, L. Kovacs, E. Papp, V. Barzda, R. C. Jennings, and G. Garab. 2005. Thermo-optically induced reorganizations in the main light harvesting antenna of plants. I. Non-Arrhenius type of temperature dependence and linear light-intensity dependencies. *Photosynth. Res.* 86:263–273.

18

Advanced Ultrafast Spectroscopy Methods

TOMÁŠ POLÍVKA AND DONATAS ZIGMANTAS

18.1 INTRODUCTION

Photosynthesis is a complex process that involves multiple phenomena occurring on time scales ranging from femtoseconds to years. Following and characterizing the slow processes of photosynthesis has been a relatively accessible task since the beginning of photosynthesis research. On the other hand, resolving the fast processes has been crucially dependent on the availability of the light sources, producing pulses that are shorter than the investigated processes. The millisecond and nanosecond domains became accessible in the 1960s, but entering the femtosecond domain awaited until the dawn of the femtosecond lasers in the late 1980s (Zewail, 1998). The possibility to follow the processes occurring on picosecond and femtosecond time scales was certainly a breakthrough in the photosynthesis research. It opened the time window into the key initial photosynthetic processes, such as energy transfer in light-harvesting proteins and fast electron transfer in reaction centers. Since the 1990s, femtosecond spectroscopy has become a standard technique for studying the initial processes in photosynthesis. Today in the femtosecond spectroscopy laboratories, time resolution better than 50 fs is routinely reached, with a broad variety of pulse energies and repetition rates, and a spectral range that covers from the deep UV to the infrared region.

18.2 TIME-RESOLVED SPECTROSCOPIES PROBING POPULATION DYNAMICS

The first part of this chapter deals with "standard" time-resolved spectroscopy methods, which enable to follow population dynamics of the studied system in a convenient way without the need for sophisticated analysis. All these methods are based on the creation of excited-state population by an excitation (pump) pulse, and subsequent population dynamics is followed by monitoring time-dependent changes in either absorption or emission of the sample by means of a probe (absorption) or a gate (emission) pulse that interacts with the sample at a defined time following the excitation pulse.

One key parameter of any ultrafast experiment is time resolution. In a time-resolved experiment described in this section, time resolution is given by the cross correlation (or convolution) between pump and probe (gate) pulses. A natural way toward higher time resolution is thus generation and compression of the shorter pulses, but this comes with a price of losing spectral selectivity in the excitation, which is crucial in photosynthetic systems to selectively excite pigments of interest. Thus, even though pulses shorter than 5 fs are now available for the ultrafast spectroscopy, they are hardly ever used to monitor population dynamics in photosynthetic systems. The relation between time duration and spectral bandwidth of a short laser pulse that is described by the Gaussian function envelope is determined by the time–bandwidth product

$$\tau_p \Delta v \geq 0.441 \tag{18.1}$$

where

τ_p is the pulse duration
Δv is the frequency bandwidth of the pulse

In terms of wavelengths for $\Delta v \ll v$, this relation is written as

$$\frac{\tau_p c \Delta v}{\lambda^2} \geq 0.441 \tag{18.2}$$

The equality sign holds for the so-called transform-limited pulses, which are the shortest pulses achievable for a given spectral bandwidth. It must also be noted that the equation is valid only for the pulses with Gaussian profiles that are most commonly used in ultrafast spectroscopy. For other profiles (or envelopes), the constant on the right side of the equation is different and can be obtained from the Fourier transform relationship.

Thus, for the 5 fs pulse with the central wavelength of 800 nm, the spectral bandwidth is at least 180 nm, which makes any selective excitation of pigments impossible. On the other hand, a 100 fs pulse at the same central wavelength has the bandwidth of ~8 nm that certainly allows for the selective excitation. It is therefore obvious that the high spectral selectivity and high time resolution cannot be achieved simultaneously in a standard time-resolved experiment.

18.2.1 Transient absorption spectroscopy

The basic ultrafast spectroscopy method is transient absorption (or pump–probe) spectroscopy that measures the time dependence of absorption changes induced by excitation of the studied system. One could consider transient absorption spectroscopy as the "mother" of all time-resolved spectroscopies, and therefore understanding how it works is prerequisite for discussion of more complex spectroscopy techniques. A simplified scheme of this method is depicted in Figure 18.1. The primary beam (train of pulses) generated by a femtosecond laser system is divided into two beams, serving as excitation and probe pulses. The excitation beam excites the sample at the desired wavelength, while the probe beam, which arrives at the sample after a certain delay that is realized by a mechanical delay line that simply changes the optical path length of one of the beams, monitors the absorption changes in the sample induced by excitation. By repeating the experiment for various delays between excitation and probe, a time-evolution picture of the induced absorption changes can be constructed.

Since current femtosecond lasers are mostly based on the Ti/sapphire amplifiers generating pulses with a central wavelength at around 800 nm, both excitation and probe beams are usually further modified to achieve a certain desired spectral properties. As mentioned for the excitation beam, a spectral selectivity is required, and various types of wavelength converters, such as second (third) harmonic generators or optical parametric amplifiers, are used to convert the excitation wavelength to the desired spectral range. Nowadays, a spectral range from 200 to 20,000 nm can be obtained by commercially available systems. Whereas the excitation beam should be spectrally narrow to achieve the desired spectral selectivity (absorption bands of the photosynthetic pigments are relatively narrow), the probe beam for most applications should

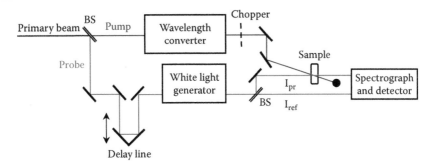

Figure 18.1 Basic scheme of a transient absorption experiment. The primary beam is divided into two beams, pump (red) and probe (blue). A delay line, introducing a time delay between excitation and probe pulse, is placed into one of the beams. The probe beam is usually further divided into reference and probe beams, which are both dispersed by a spectrograph onto a diode-array or CCD detector. See text for details. BS beam splitter.

be as spectrally broad as possible, allowing to record absorption changes at various wavelengths simultaneously. The broadband probe beam is usually generated by focusing a small fraction of the spectrally narrow primary beam into a certain medium (typically sapphire, quartz, or calcium fluoride), in which due to a nonlinear process called self-phase modulation, a broadband beam, known as white-light continuum (WLC), is generated. Depending on the properties of the medium, the spectral width of the WLC pulses reaches hundreds of nanometers and can cover the spectral region from 300 to 1600 nm.

After passing the sample, the excitation beam is blocked, whereas the probe beam is led to a detection system. If the WLC is used as a probe beam, the beam must first pass through a spectrograph where it is dispersed onto a detector. For measurements in the 350–1000 nm spectral region (that is crucial for studies of photosynthetic systems), diode-array or CCD detectors are almost exclusively used because they allow the detection of the whole probe spectrum in one measurement, thereby greatly improving the information content of the measurements. Yet, for experiments outside this spectral region, narrow-band probe beams are still usually used. The reason why this is done in the UV region is because extending the WLC spectra range below 300 nm is rather complicated, in the spectral range beyond 1000 nm; it is mainly because of the decreased sensitivity of the diode-array or CCD detectors. Yet, the diode-array detectors sensitive in the near-infrared region up to 2 μm are becoming increasingly available.

Transient absorption experiment measures the difference between absorption of the excited and

nonexcited sample. The most common realization of the experiment is shown in Figure 18.1. In this scheme the probe beam is split into two beams before the sample. One of the beams, the probe, overlaps with the excitation beam at the sample, while the second one, denoted as a reference beam, either passes through the sample without overlapping with the excitation or passes outside the sample. Both beams are detected; the intensity of the probe beam is denoted I_{pr}, and the intensity of the reference beam is I_{ref}. A chopper, whose frequency is set to block every second pulse, is placed into the excitation beam. The signal is then calculated as

$$\Delta A(t, \lambda) = A_{ON}(t, \lambda) - A_{OFF}(t, \lambda)$$

$$= \log\left[\frac{I_{ref}}{I_{pr}}\right]_{ON} - \log\left[\frac{I_{ref}}{I_{pr}}\right]_{OFF} \quad (18.3)$$

where the "ON" term refers to the situation when the excitation beam is present while the "OFF" term is detected with the excitation beam blocked. For the experimental setup with the reference beam passing outside the sample, the second term corresponds to a standard ground-state absorption spectrum; while for the reference beam passing through the sample, the second term should be ideally zero. In both cases, the second term represents a background signal that is subtracted and helps to improve signal-to-noise ratio of the experiment.

Because the transient absorption experiment measures difference signal, ΔA, in contrast to a standard absorption spectrum, the transient absorption spectrum may contain both positive and negative contributions. Generally, three types of signals contribute to the transient absorption spectrum:

1. Ground-state bleaching (GSB)
2. Stimulated emission (SE)
3. Excited-state absorption

First, in the spectral region of the ground-state absorption, a negative transient signal, denoted as *ground-state bleaching*, is observed. This contribution occurs because excitation of the sample transfers part of the ground-state population of the molecules to an excited state; thus consequently, the sample absorbs less in this spectral region, resulting in a negative signal (in Equation 18.1, $I_{pr} > I_{ref}$ when the excitation beam is ON). Similarly, when a molecule is promoted to the excited state, an incident probe photon can induce a process called *stimulated emission* (SE), resulting in return of the molecule to the ground state via the emission of a photon. The SE is usually red-shifted to the GSB. The third signal, which is positive, is denoted as *excited-state absorption* (ESA). It appears due to the fact that molecules in the excited state can be promoted to even higher excited states via the absorption of photons from the probe beam, resulting in $I_{pr} < I_{ref}$ when the excitation beam is ON, generating a positive signal.

In addition to these three signals that are always present, the fourth possible signal is again positive and appears as a result of the formation of new *products* generated by the excitation beam. In photosynthesis studies, these products can typically be radicals, ions, or isomers of photosynthetic pigments. Since these products have usually different absorption properties than the parent molecule, they appear in the transient absorption spectrum as new absorption bands.

A typical result of a transient absorption experiment is shown in Figure 18.2. The water-soluble peridinin-chlorophyll-a-protein (PCP) from photosynthetic dinoflagellates, containing the carotenoid peridinin as energy donor and chlorophyll (Chl)-a as the acceptor, is taken as an example. Peridinin is excited at 500 nm, thus in the spectral region where Chl-a has essentially zero absorption. Figure 18.2a shows the full spectrotemporal matrix data resulting from a transient absorption experiment using a diode-array detection system. To visualize the spectral evolution, cuts along the wavelength axis at a few different time delays are depicted in Figure 18.2b. These cuts clearly show that at 0.1 ps after excitation, the spectrum is dominated by GSB (below 550 nm) and ESA (the broad band extending from 550 to 750 nm) of peridinin. Yet, a dip at 672 nm is superimposed on the broad peridinin ESA band. This dip is GSB of Chl-a, indicating that some energy transfer occurs at a very short time scale. At longer delays, the Chl-a GSB deepens, while both ESA and GSB of peridinin disappear as the energy flows from the initially excited peridinin to Chl-a.

Cuts along the time axis allows following of the time evolution of the individual spectral bands. These cuts are denoted as transient absorption kinetics (or kinetic traces). Figure 18.2c shows three kinetics monitoring the time evolution of the peridinin GSB (490 nm), ESA (630 nm), and Chl-a GSB (672 nm). The kinetics demonstrate that the appearance of the Chl-a signal has the same, but inverted, dynamics as the disappearance of the peridinin signals, indicating the excitation energy flow from peridinin to Chl-a. Fitting these kinetics by a multiexponential decay (a sum of several exponential functions) enables extracting exponential time constants associated with energy transfer. For the particular case of PCP, most of the energy transfer occurs with a time constant of 3 ps. Global fitting procedure (see Chapter 20), which fits a large number of kinetics across a wide spectral region simultaneously, is often used to obtain species-specific information on the evolution of the studied system.

Transient absorption spectroscopy has become a standard tool in photosynthetic research. Besides the energy transfer dynamics described in the example, it is routinely used to reveal dynamics of electron transfer in reaction centers (Holzwarth et al. 2006) and fast photoprotective processes (Ruban et al. 2006). Importantly, transient absorption spectroscopy also provides crucial information about excited-state dynamics of isolated photosynthetic pigments (carotenoids, chlorophylls) (Polívka and Sundström, 2004), which are crucial input parameters to assess energy and electron transfer processes in photosynthetic proteins.

18.2.2 Pump–dump–probe

The traditional pump–probe scheme used in transient absorption spectroscopy described in the previous section can be extended by adding one

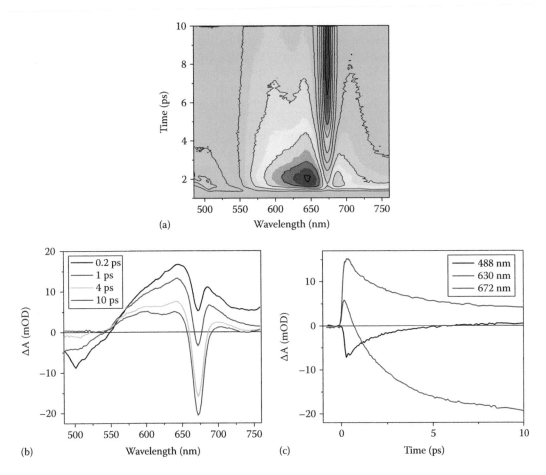

Figure 18.2 Transient absorption spectroscopy data recorded for the peridinin–chlorophyll protein excited into the carotenoid peridinin at 500 nm. **(a)** Full spectrotemporal matrix showing negative (bleaching and SE) signals in blue and positive (ESA) signals in red. **(b)** Transient absorption spectra taken at four delay times after excitation. The positive signal is due to peridinin ESA peaking around 635 nm, negative signal in the first two spectra (black and red) corresponds to peridinin GSB. **(c)** Kinetic traces monitoring decay of peridinin GSB (488 nm), peridinin ESA (630 nm), and GSB of Chl-a (672 nm). It is obvious that peridinin decay matches nicely the rise of the Chl-a signal, indicating energy transfer from peridinin to Chl-a. Global fitting of the kinetic traces reveals time constant of energy transfer of 2.7 ps.

(or more) excitation pulses. This multipulse method is usually denoted as pump–dump–probe (PDP) spectroscopy, though, depending on the action of the second excitation pulse, various names can be found in the literature (pump–repump–probe, pre-pump–pump–probe). PDP methods allow direct manipulation of populations in the excited states.

The limitation of the standard transient absorption experiment described in the previous section is that in a complex system consisting of a few pigments (each having a few excited states), the observed spectra sometimes suffer from the overlap of various signals originating from GSB, ESA, SE, and possible reaction products. This congestion

of the measured signals may obscure the dynamics one is looking for. Adding an extra excitation pulse alters the relaxation or reaction pathways and thereby disentangles the signals, which overlap in the standard transient absorption experiment. In combination with sophisticated global fitting methods (see Chapter 20), the PDP approach enables revealing dynamic processes hidden in the pump–probe experiment.

A basic scheme of a PDP experiment is the same as that of the transient absorption experiment shown in Figure 18.1 except the PDP experiment requires the second excitation beam with an extra delay line controlling the delay between the two excitation pulses. Depending on whether the second pulse is resonant with transitions corresponding to SE or ESA, the second pulse is called either dump or repump pulse. Figure 18.3 shows a possible application of the PDP technique. As indicated in Figure 18.3a, application of a dump pulse is useful in systems containing a short-lived excited-state intermediate. If the dynamics of a system containing three excited states A, B, and C is such that the process A → B is slower than the subsequent process B → C (lifetime of B is shorter than that of A), the population of B will be virtually zero, preventing observation of the B intermediate in a standard transient absorption experiment. If, however, the dump pulse is tuned to the A → B transition, it will increase the population of B by actively dumping the population of A to B. This way the state B can be revealed in the PDP experiment and the lifetime of B can be measured (Papagiannakis et al., 2006, Redeckas et al., 2016).

The practical application of this approach is depicted in Figure 18.4, which shows two sets of PDP data measured after excitation of the carotenoid peridinin at 530 nm. Two probing wavelengths, 435 and 590 nm, monitor the population of the ground state (GSB at 435 nm) and the intramolecular charge-transfer (ICT) state (ESA at 590 nm), respectively. The 800 nm dump pulse is resonant with ICT SE; thus when the dump pulse is applied, the population of the ICT state monitored at 590 nm is instantaneously depleted (Figure 18.4b). Yet, the population of the ground state is not affected

instantaneously by the dump pulse (Figure 18.4a), implying that an intermediate state must be involved. By fitting the rise of the ground-state recovery after the dump pulse, the lifetime of the intermediate of 900 fs was determined (Papagiannakis et al., 2006). The peridinin ground-state intermediate was revealed by the same experimental approach also in PCP (Redeckas et al., 2017). Dumping the ICT-state emission in the PDP experiment has also been used to explore the relation between the S_1 and ICT states of the carotenoid fucoxanthin in a solution (Redeckas et al., 2016).

Using a repump pulse, which is tuned to some ESA transitions, allows populating high-energy states that may not be accessible in a standard pump–probe experiment. A subsequent probe pulse then monitors relaxation pathways from these high-energy states. A typical example is the formation of carotenoid radicals by using the repump pulse to excite higher excited states (Papagiannakis et al., 2006, Amarie et al., 2009). A promising application of PDP spectroscopy for photosynthetic systems is to use the extra pump pulse to reveal properties of the energy donor in various light-harvesting systems (Figure 18.3b). One of the major obstacles in determining the efficiency of a particular energy transfer channel is lack of knowledge of the excited-state properties (mainly the relaxation rate) of the energy donor in the absence of energy transfer. This is a typical situation for the carotenoid–(Bacterio)chlorophyll (B)Chl pairs in which the carotenoid acts as the

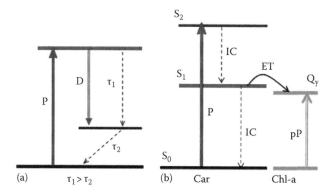

Figure 18.3 Energy level schemes demonstrating the use of the pump–dump spectroscopy. **(a)** If a system contains an intermediate state (black) whose lifetime is shorter than the lifetime of a precursor (red), in a standard pump–probe experiment, the population of the intermediate is virtually zero. Application of a dump pulse (D) following the excitation (P) steers some population into the intermediate state whose properties can be measured. **(b)** A prepump (pP) excitation pulse applied prior to excitation (P) can be used to saturate the acceptor state and prevent energy transfer.

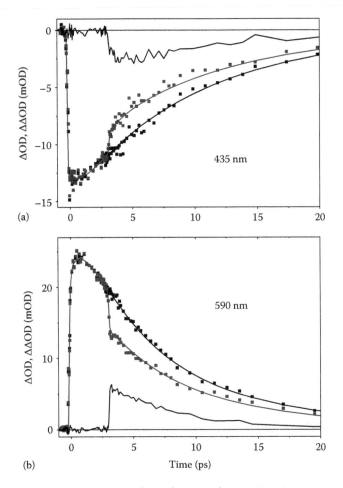

Figure 18.4 PDP experiment on the carotenoid peridinin in solution. Kinetic traces monitor the population of the ground state (**a**, probe in the GSB region at 435 nm) and excited state (**b**, probe in the ESA region at 590 nm). The black traces represent results of a standard transient absorption experiment; red traces are recorded after excitation of the sample followed by application of a dump pulse 3 ps after excitation. See text for details. (Based on data from Papagiannakis, E. et al., *J. Phys. Chem. B*, 110, 512, 2006.)

energy donor. Although we can obtain the excited-state properties of the carotenoid in solution, when it is embedded in an antenna protein these intrinsic excited-state properties may be altered. Tuning the extra pump pulse to the resonance with the Q_y band of (B)Chl and then applying it prior to the excitation of carotenoid (the extra pulse is then called prepump pulse) may saturate the (B)Chl transition. The energy transfer channel is therefore blocked, and the excited-state properties of the carotenoid donor in the antenna protein without energy transfer can be determined.

Experimentally, the PDP experiment in any of the previously mentioned versions is usually realized by measuring both pump–dump–probe

(with dump) and pump–probe (without dump) signals, and finally constructing the double difference spectra $\Delta\Delta A = PDP - PP$. This approach defines the standard PP signals as a background and allows extracting solely the effect of the dump pulse (Papagiannakis et al., 2006, Redeckas et al., 2016). Delay lines in the excitation and dump beams enable precise setting of the timing between excitation and dump pulses, and between excitation and probe pulses. Two experimental modes are used: (1) the time delay between the excitation and dump pulse is fixed, and by varying the delay between the excitation and probe pulses, the "standard" dynamics affected by the dump pulse is monitored; and (2) the time delay between excitation and probe is fixed, and by varying the delay between the excitation and dump pulse, the "action kinetic" of the dump pulse is obtained (Papagiannakis et al., 2006).

18.2.3 Fluorescence up-conversion

The advantage of the methods described in the previous sections is that they explore time-dependent changes in both ground and excited states and can therefore monitor the entire reaction or relaxation pathway. Yet, a drawback of transient absorption methods lies in the fact that signals consist of overlapping positive and negative bands. If the goal of the measurement is to determine the properties of a particular excited state, extraction of desired information from complex transient spectra may sometimes be hindered by overlapping bands in such spectra. Thus, time-resolved methods based on the detection of the emission from the excited state whose properties are to be revealed are especially useful as they represent essentially a background-free experiment.

To achieve a time resolution of ~100 fs in fluorescence measurement, it is not possible to apply detection techniques based on fast electronics such as, for example, in a time-correlated single-photon counting experiment (Chapter 17). Instead, fluorescence induced by an excitation pulse is mixed in a nonlinear crystal with another femtosecond laser pulse denoted as a gate pulse (see experimental scheme in Figure 18.5). If the gate pulse has wavelength λ_1 and we detect an emission at wavelength λ_2, then, by a nonlinear process called sum–frequency generation, these two wavelengths are mixed in the nonlinear crystal to produce a signal at wavelength λ

$$\frac{1}{\lambda} = \frac{1}{\lambda_1} + \frac{1}{\lambda_2} \tag{18.4}$$

Thus, if the emission is at 680 nm (Chl-a) and the gate pulse is at 800 nm, the resulting sum–frequency signal is detected in the UV region at 367 nm. The key advantage of this method is that the sum–frequency signal at wavelength λ is detected only when the gate pulse is present. The evolution of the system is monitored by controlling the time delay between the excitation and gate pulses (Shah, 1988), providing an elegant way of measuring very short fluorescence lifetimes. Here, the time resolution is determined not only by the duration of the excitation and gate pulses but also by the thickness of the nonlinear crystal. Thin crystals (100–200 μm) must be used to push the time resolution significantly below 100 fs.

In photosynthesis research, the fluorescence up-conversion technique was successfully used, for example, to determine the lifetimes of the S_2 state of carotenoids. This state has a lifetime of 50–250 fs in solution, depending on carotenoid, exhibits weak fluorescence (quantum yield of about 10^{-5}); however, the up-conversion of this weak emission was the first reliable method to determine the S_2 lifetime (Macpherson and Gillbro, 1998). The carotenoid S_2 state is known to transfer energy to (B)Chl in many photosynthetic systems; thus comparison of the S_2 lifetimes in solution and in

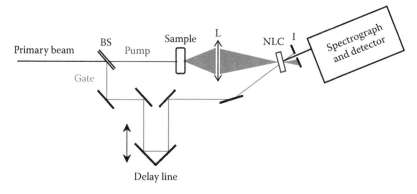

Figure 18.5 Basic scheme of a fluorescence up-conversion experiment. The primary beam is divided by a beam splitter (BS) to the excitation (red) and gate (blue) beams. Emission from the sample is focused onto a nonlinear crystal (NLC) in which the emission is mixed with the gate pulse. The resulting up-converted signal is filtered out by iris (I) and spectrograph, and eventually detected.

protein provided a basis for determining the efficiency of the S_2-mediated energy transfer channel (Ricci et al., 1996, Krueger et al., 1998). A typical example of using the fluorescence up-conversion to follow dynamics of excited states in photosynthetic systems is shown in Figure 18.6. After excitation of the carotenoid absorbing at 500 nm in the LHCII antenna, it is possible to resolve the carotenoid S_2-state emission at 575 nm, of Chl-b at 655 nm, and of Chl-a at 685 nm. By fitting the resulting kinetic traces, dynamics of the energy transfer from carotenoid to chlorophylls and from Chl-b to Chl-a have been determined (Walla et al., 2000a).

Although there is no doubt that fluorescence up-conversion is an ideal experiment to determine short lifetimes of emissive states, it has certain drawbacks. First, the experimental setup of the fluorescence up-conversion experiment is more demanding compared to transient absorption experiment. It is based on mixing weak fluorescence signal with a gate pulse via a nonlinear process, resulting in very weak sum–frequency signals, usually in the UV region. To collect as much emission as possible and to focus it properly into the nonlinear crystal, a detection scheme using ellipsoidal mirror was successfully used (Krueger et al., 1998). If a sample is placed in one focus of the elliptical mirror and the nonlinear crystal is in the other focus, this scheme assures an efficient conversion of photons emitted from the sample to the sum–frequency signal.

The major limitation of fluorescence up-conversion is due to the fact that the sum–frequency generation requires a phase matching condition that depends on the light frequency and crystal orientation. This prevents detection of full up-converted fluorescence spectra, and usually only narrow spectral slices of the fluorescence emission are up-converted simultaneously. The up-conversion bandwidth may be broadened by using low-dispersion and very thin nonlinear crystals and/or moving the gate pulse to the infrared region, but the resulting spectral bandwidth cannot compete with those obtained in transient absorption experiments using white-light continuum (Section 18.2.1). A possible solution of this problem is to scan the crystal angle during the measurement of each time point (Haacke et al., 1998, Cannizzo et al., 2007).

To circumvent the problem with narrowband detection in fluorescence up-conversion experiment, modification of the experiment, exploiting a phenomenon known as the optical Kerr effect, was proposed (Schmidt et al., 2003). The optical Kerr effect is a nonlinear process in which the electromagnetic field (light) modifies the optical properties of the Kerr medium. In the Kerr gate fluorescence experiment, the fluorescence generated by the excitation pulse passes two crossed polarizers, and an isotropic Kerr medium is placed between the polarizers. Without the gate pulse, no fluorescence can pass through two

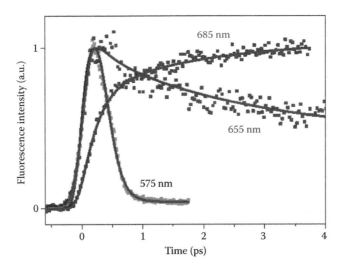

Figure 18.6 Fluorescence up-conversion data measured after excitation of carotenoid in LHCII. Detection at 575 nm (black) monitors emission from the S_2 state of the carotenoid, detection at 655 nm (red) corresponds to Chl-b emission, and detection at 685 nm (blue) nm follows arrival of excitations at Chl-a. (Based on data from Walla, P.J. et al., *J. Phys. Chem. B*, 104, 4799, 2000a.)

crossed polarizers. But when the gate pulse hits the Kerr medium, it induces birefringence in the medium that changes the polarization state of the fluorescence light from linear to elliptical. Since a fraction of the elliptically polarized fluorescence can pass through the second polarizer, the gate pulse via Kerr medium acts as a shutter that opens the second polarizer for a time corresponding to the duration of the gate pulse. Thus, as for the fluorescence up-conversion experiment, the signal is detected only when the gate pulse is present, and the time-resolved experiment is realized by controlling the delay between the excitation and gate pulses. Since no phase-matching conditions are required for the opening of the Kerr shutter, broadband fluorescence can be detected. Yet, a disadvantage of the Kerr shutter is that background emission from the Kerr medium is often present and needs to be suppressed by suitable filters. In photosynthetic systems, optical Kerr gate spectroscopy was used to study the properties of the S_2 state of some carotenoids (Nakamura and Kanematsu, 2004).

18.2.4 Two-photon excitation

While the techniques described in the previous sections provide a wealth of information about population dynamics in various photosynthetic systems, to reveal some specific details about a particular system, more sophisticated time-resolved techniques are sometimes needed. For example, common constituents of photosynthetic systems, carotenoids, feature lowest excited state, the S_1 state, which is forbidden for a one-photon transition, meaning that this state is invisible in the absorption spectrum. Yet this state is involved in energy transfer processes. Thus, to reveal and characterize the energy transfer pathways involving the S_1 state by the standard methods, the S_1 state is populated by internal conversion after excitation of the allowed higher-energy S_2 state. Since the S_2 state, besides relaxing to the S_1 state via internal conversion, also transfers energy to (B)Chl, excitation of the S_2 state results in a complicated network of relaxation pathways (Polívka and Sundström, 2009).

This complexity can be simplified if two-photon excitation is used to initiate excited-state processes, because the transition from the ground state to the S_1 state is allowed for two-photon processes.

Two-photon absorption is a nonlinear process involving simultaneous absorption of two photons whose energies $\hbar\omega_1$ and $\hbar\omega_2$ are added to populate a state with transition energy $\hbar(\omega_1 + \omega_2)$. Thus, to directly populate the S_1 state of carotenoids, which, depending on the carotenoid, has an energy in the range of 11,000–16,000 cm^{-1} (900–625 nm), it is necessary to use near-infrared (IR) excitation at wavelengths in the 1250–1800 nm range.

In photosynthetic systems, the two-photon absorption has been used to confirm rates of energy transfer between the carotenoid S_1 state and (B)Chl Q_y state. Although these rates are routinely obtained also from standard pump–probe experiments using one-photon excitation of carotenoid into the S_2 state, data analysis of such one-photon excitation experiments is always more complicated, as other relaxation channels are inevitably present in data. Thus, exciting the carotenoid S_1 state directly via two-photon excitation allows for separation of the S_1-mediated channel from other relaxation pathways. The two-photon excitation experiments aiming to determine the S_1-mediated energy transfer rate have been so far carried out in two detection modes. Either the two-photon excitation is incorporated into a standard transient absorption setup (Section 18.2.1), in which both carotenoid ESA and (B)Chl GSB signals can be detected, or it is coupled to a fluorescence up-conversion experiment (Section 18.2.3) detecting solely the (B)Chl fluorescence.

The fluorescence up-conversion experiment with two-photon excitation at 1300 nm was successfully used to determine time constants of energy transfer from the S_1 state of carotenoids in the LHCII complex (Walla et al., 2000a). Similarly, a pump–probe experiment with two-photon excitation was employed to determine the rates of S_1-mediated energy transfer of carotenoid-to-BChl energy transfer in purple bacterial light-harvesting complexes (Walla et al., 2000b). It has been also demonstrated that the ratio of (B)Chl fluorescence after one- and two-photon excitations correlates with the coupling of the carotenoid S_1 state and (B)Chl Q_y state (Wehling et al., 2006). This technique was later used to estimate the carotenoid–Chl coupling in the LHCII complexes, including mutants with enhanced nonphotochemical quenching (Bode et al., 2009). It must be however noted that for determining the S_1 spectral profiles and S_1–Q_y coupling from a two-photon excitation experiment in any system that contains (B)Chl, the contribution of (B)Chl to the two-photon absorption (how much of the near-IR photons is absorbed by the two-photon process directly by (B)Chl) is a crucial parameter that is still under debate (Leupold et al., 2002).

18.3 MULTIDIMENSIONAL SPECTROSCOPIES

18.3.1 Femtosecond Raman-type spectroscopies

Various time-resolved spectroscopy methods have been employed through the years for accessing dynamic structural information about molecules and molecular ensembles (Fayer, 2001). Here we outline techniques that use UV to near-IR femtosecond pulses to probe vibrational transitions via the Raman scattering process. The wavenumber resolution achieved in these techniques approaches 10 cm^{-1}, which is comparable to the natural linewidths of the vibrational transitions of molecules in the solution. Three groups of such techniques can be distinguished: (1) femtosecond stimulated Raman spectroscopy (FSRS), (2) pump–degenerate four-wave mixing (pump–DFWM), and (3) pump–impulsive vibrational spectroscopy (pump–IVS). The latter two share the basic principles and provide very similar information, and thus in this chapter, we will describe only the pump–DFWM technique as it offers more flexibility in experiments. All of these methods are fifth-order nonlinear spectroscopy techniques, signifying that during the measurements, there are five interactions of incoming laser fields with the molecules. Note that one laser pulse can interact with the sample more than once.

One general remark on the time resolution of these techniques is worth noting here. To measure Raman lines with the $\sim 10 \text{ cm}^{-1}$ resolution, it is necessary to collect the spectrum over the full course of the vibrational free induction decay (FID), which in condensed phases is typically ~ 1 ps. Thus, strictly speaking, these techniques do not record an instantaneous vibrational spectrum (Mukamel and Biggs, 2011). Instead, a transient state of the molecule is captured, from which a particular Raman transition is initiated, and consequently an FID recorded. For clarity of the presentation, we will not make this distinction throughout this section and will attribute the time resolution of the outlined methods to the duration of the femtosecond pulses that are used in these experiments.

18.3.1.1 FEMTOSECOND STIMULATED RAMAN SPECTROSCOPY

Transient Raman spectroscopy techniques have a long history and started with nanosecond pulses that were followed by experiments on a picosecond time scale. However, many of the photochemical and photophysical processes in molecules occur on the femtosecond time scale. At the end of the 1990s, by an ingenious trick of separating Raman pump and probe pulses, FSRS was implemented, and femtosecond time resolution was finally achieved (Yoshizawa and Kurosawa, 2000). FSRS can obtain instantaneous Raman spectra over a 2000 cm^{-1} window, which is basically free of fluorescence background (McCamant et al., 2004, Kukura et al., 2007). The basic principles of FSRS are illustrated in Figure 18.7 (McCamant et al., 2004). First, the femtosecond *actinic pump* pulse resonant with an electronic transition (usually in the UV or visible region) excites the sample to the excited electronic state. To obtain a substantial signal from the vibrational transitions in the excited state, a large part of the molecular population has to be promoted to the excited electronic state. This is achieved by a rather intense actinic pump pulse. Then, after time Δt, two more pulses interact with the sample: *picosecond Raman pump* and *femtosecond Raman probe* (see Figure 18.7b). Together they initiate Raman transitions and Stokes Raman lines appear as a gain signal on the Raman probe spectrum (Figure 18.7a). In the majority of experiments, a nonresonant Raman pump at 800 nm is used. This pulse dictates the spectral resolution of the experiment and therefore usually has a duration of 1–4 ps providing the 15–4 cm^{-1} wavenumber resolution. The time resolution in the experiment is determined by the duration of the actinic pump and Raman probe pulses, which can be shorter than 30 fs. Quite often a white-light continuum, generated in some materials like sapphire or quartz, is used as the broadband Raman probe pulse. When the continuum is generated from a widely utilized Ti/sapphire laser with central wavelength at ~ 800 nm, a Raman probe spanning 830–950 nm range is routinely used. All the beams can be arranged in an either collinear or noncollinear geometry. It follows from the phase-matching condition that the FSRS signal always propagates in the same direction as the Raman probe. In other words, the signal is heterodyned with the remaining Raman probe that did not interact with the sample.

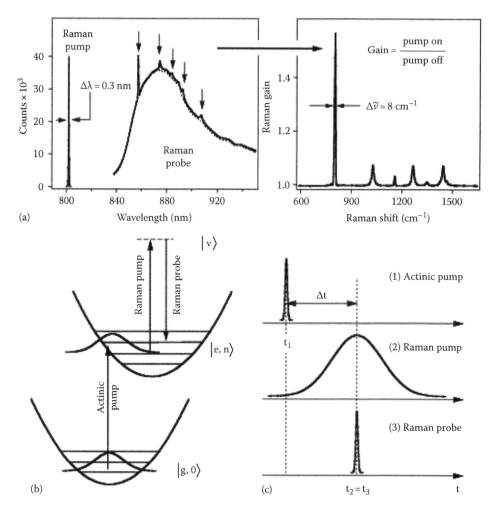

Figure 18.7 Illustrations for FSRS. **(a)** The spectra of the Raman pump and probe pulses with stimulated Raman lines visible on the top. On the right: resulting Raman spectrum. **(b)** Electronic/vibrational energy levels with arrows indicating energies of the laser pulses used in the experiments. **(c)** Timing of the laser pulses in FSRS. Time resolution is obtained by delaying the femtosecond actinic pulse by Δt relative to the Raman probe. (Reprinted with permission from McCamant, D.W., Kukura, P., Yoon, S. et al., Femtosecond broadband stimulated Raman spectroscopy: Apparatus and methods, *Rev. Sci. Inst.*, 75, 4971–4980. Copyright 2004, American Institute of Physics.)

The time-resolved Raman gain spectrum is calculated by dividing the Raman pump-on and pump-off spectra after correcting for the background and fluctuations in the continuum using the reference:

Raman gain

$$= \frac{\left[\dfrac{(\text{probe} + \text{actinic} - \text{bkgnd})}{(\text{ref} - \text{bkgnd})}\right]_{\text{RamanPumpOn}}}{\left[\dfrac{(\text{probe} + \text{actinic} - \text{bkgnd})}{(\text{ref} - \text{bkgnd})}\right]_{\text{RamanPumpOff}}} \quad (18.5)$$

To obtain excited-state vibrational Raman spectra, first the Raman probe background and contribution from the solvent have to be measured and subtracted. A contribution from the ground state of the studied molecules is most commonly subtracted (McCamant et al., 2004, Shim and Mathies, 2008a) or alternatively is considered as a part of the total signal as a transient bleaching (Kloz et al., 2011).

Most of the experiments to date use nonresonant Raman pump pulses, but resonant Raman pump FSRS, which offers strong enhancement of the Stokes Raman lines, was also developed (Shim and Mathies, 2008a). However, this approach at the same time suffers from the presence of multipulse nonlinear signals generating a complex background.

The FSRS technique has been mostly applied to biological molecules in which vibrations are strongly coupled to the electronic transitions. These pigments are usually polyene-type molecules, like carotenoids and retinal. Recently, FSRS was used to study carotenoids in the light-harvesting complex from photosynthetic bacteria (Yoshizawa et al., 2012). There are promising developments in the field, including femtosecond Raman-induced Kerr effect spectroscopy (Shim and Mathies, 2008b) and wavelength-modulated FSRS (Kloz et al., 2011, 2016), which will hopefully broaden the spectrum of molecules that can be studied with time-resolved Raman techniques.

18.3.1.2 TIME-RESOLVED RAMAN STUDY OF β-CAROTENE

Excitation dynamics in the manifold of the β-carotene excited states was monitored by the resonant FSRS technique (Shim and Mathies, 2008a). In these experiments, <100 fs temporal and <35 cm^{-1} spectral resolution was achieved. β-Carotene molecules in solution were pumped by the <50 fs pulses at 480 nm to resonantly excite the S_2 transition. To achieve selective resonance with the S_1–S_n transition, a tunable narrow bandwidth optical parametric amplifier was used to generate the ~0.7 ps Raman pump pulses at 560 nm. For the Raman probe, a white-light continuum was generated with the 800 nm laser pulses in a sapphire plate and the spectrum in the range of 810–960 nm was used.

The ground-state FSRS spectrum and a sequence of time-resolved FSRS spectra of β-carotene are shown in Figure 18.8. The characteristic Raman-active vibrations in the ground spectrum are found at 1004, 1156, and 1524 cm^{-1}, corresponding to the C–CH$_3$ methyl rock, C–C stretch, and C=C stretch vibrations, respectively. Right after the excitation (in the time window of 0–300 fs), dispersive features appear at the location of the ground-state vibrational lines. These features are attributed to Raman initiated by nonlinear emission (RINE), in which vibrational coherence is created in the ground state by Raman pump and probe interaction. RINE decay is determined by the convolution of the excited-state lifetime (S_2 in this case) with the dephasing time of the ground-state vibrational coherence. Later, as population is transferred from the S_2 to S_1 state, new vibrational features assigned to the S_1 state appear: 1010 (C–CH$_3$), 1196 and 1250 (C–C), as well as 1540 and 1790 cm^{-1} (C=C). At even longer delay times (>1 ps), the S_1 state relaxes to the vibrationally hot ground state (hot S_0), which is associated with the appearance of additional lines at 1145 and 1512 cm^{-1}. After 50 ps, all β-carotene molecules relax to the bottom of the S_0 state and all vibrational lines disappear.

Evolution of the excited states can be studied in detail by monitoring the kinetics of distinctive vibrational lines. For example, decay of the RINE lines and rise of the "signature" S_1 line at 1790 cm^{-1} allows investigation of the S_2–S_1 relaxation process. Since for both processes the lifetime of ~180 fs was observed, it was concluded that there are no intermediate states, such as 1B$_u^-$ symmetry state, involved in the relaxation process, in contrast to the earlier proposals (see, e.g., Yoshizawa et al., 2001). The 1790 cm^{-1} line decays with the time constant of 9.5 ps, which corresponds to the well-known S_1–S_0 relaxation. In addition, processes of vibrational cooling can be monitored by FSRS as shifting and narrowing of the time-resolved Raman lines.

18.3.1.3 PUMP–DEGENERATE FOUR-WAVE MIXING SPECTROSCOPY

Another technique for studying fast processes of excited molecules by observing specific molecular vibrations is the pump–degenerate four-wave mixing (pump-DFWM) spectroscopy (Motzkus et al., 1996, Kraack et al., 2013). This is a full time–domain technique, where time-resolved vibrational signatures are measured as signals oscillating with specific frequencies. Just like FSRS, this is a fifth-order nonlinear spectroscopy technique, which scales steeply with transition dipole strength of the involved electronic transitions. Likely, this is the reason why only molecules possessing strong transition dipole moments and strong coupling to vibrations, such as carotenoids, have been investigated so far (Hauer et al., 2007, Marek et al., 2011 Kraack et al., 2013).

A sequence of pump–DFWM excitation pulses and an illustration of the processes taking place in the manifold of electronic/vibrational states in a molecule (carotenoid is taken as an example) are shown in Figure 18.9. The sequence consists of the *initial pump* (IP) pulse tuned into resonance with

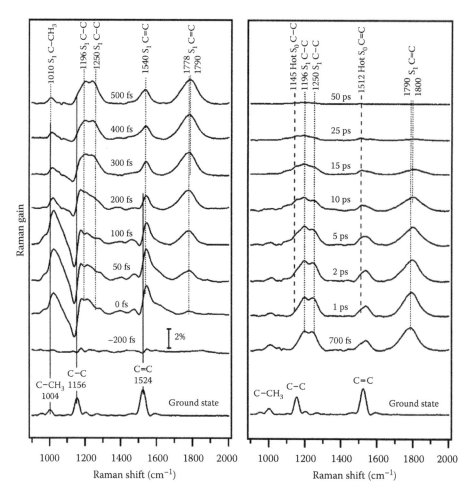

Figure 18.8 Resonance FSRS spectra of β-carotene at selected delay times after the actinic excitation at 480 nm. The Raman pump pulse at 560 nm was used to match the resonance of the carotenoid S_1–S_n transition. Dispersive peaks at early delay times are attributed to RINE process. The peaks from the S_1 state (dotted line) and the hot S_0 state are observed. Evolution of the S_1 "signature" vibration at ~1790 cm^{-1} allows to follow the S_1-state dynamics. (Reprinted with permission from Shim, S. and Mathies, R.A., Development of a tunable femtosecond stimulated Raman apparatus and its application to beta-carotene, *J. Phys. Chem. B*, 112, 4826–4832, 2008a. Copyright 2008 American Chemical Society.)

absorption, which is followed by a train of three pulses, denoted as *pump*, *Stokes*, and *probe*. The latter three have usually the same frequency, which is different from IP, and are tuned to address a specific electronic resonance in the molecule, like S_1–S_n absorption in the carotenoids (Figure 18.9a). Usually all pulses are shorter than 30 fs, ensuring excellent time resolution of the experiments. The sequence of pulses is arranged in the following way. The initial IP pulse excites the molecules (to the S_2 state in this example); after delay T (population is transferred to the S_1 state), the IP pulse is followed by a pair of

pump and Stokes pulses, which can be separated by τ_{12} (usually set to 0). By the impulsive Raman process, these two pulses create vibrational wave packets on the excited state (the S_1 state). The Raman wave packets are consequently probed by the arrival of the last pulse (probe), which is delayed from the Stokes pulse by the probe delay τ_{23}. Finally, the signal is emitted immediately after the probe pulse (Figure 18.9b). Whereas delay T tracks the evolution of the population in the excited states after IP, probe delay τ_{23} monitors several processes: the population of the excited states (the S_1 state) and the vibrational FID

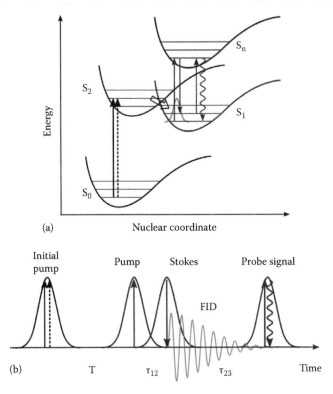

Figure 18.9 Schematics of the excitation processes in a pump–DFWM experiment. **(a)** Arrows indicate all interactions with the laser pulses on the carotenoid energy level diagram. **(b)** Time ordering of the pulses in the experiment; T is the initial population delay, τ_{12} is the pump-to-Stokes delay usually set to 0, τ_{23} is the probe delay. FID, free induction decay containing characteristic vibrational frequencies. Dashed lines indicate the second interaction with a single pulse and wave arrows denote the signal.

of the vibrational wave packets (coherences) created on these states. By first subtracting the monotonous decay components and then Fourier transforming oscillating residues along τ_{23} the specific vibrational frequencies of the electronic states are obtained.

In a pump–DFWM experiment, the laser beams are arranged in the so-called folded boxcar arrangement and the signal is detected in the background-free phase-matched direction. Standard homodyne detection is usually implemented either by integrating the whole signal on a single detector or by detecting a spectrally dispersed signal on a CCD. Homodyne detection of the oscillating signals requires some monotonous signal, which is usually readily available as a population dynamics signal. However, this signal could decay faster than the dephasing time of the vibrational coherence, and thus vibration dephasing would appear shorter in the measurement (Kraack et al., 2013).

One advantage of the pump–DFWM technique as compared to FSRS is that FID is fully characterized in the time domain and potentially provides complete information about vibrational dephasing dynamics. In addition, besides probing the structural dynamics via detecting vibrational frequencies, the pump–DFWM technique tracks electronic population dynamics as well. Yet, this large amount of information could become overwhelming, because in pump–DFWM a lot of different signals are probed at the same time (in the case of carotenoids there are about 80 different signals that can contribute to the total signal) (Marek et al., 2011). Disentangling of all these contributions can be very challenging unless only a few of them dominate the total signal.

18.3.1.4 SEARCH FOR A DARK STATE IN THE LYCOPENE'S PUMP–DFWM EXPERIMENTS

Pump–DWWM was used to study intermediate dark states in the carotenoid lycopene (Marek et al., 2011). The IP pulse with central wavelength at

510 nm was used to effectively excite lycopene into the S_2 absorption band. The DFWM spectrum was centered at 600 nm to match the S_1–S_n absorption. When scanning the IP delay, a very sharp DFWM feature with a large amplitude was observed at T = 40 fs. To separate vibrational signatures of various states that could be populated during relaxation, probe delay τ_{23} transients were recorded at 40 and 800 fs IP delay times (Figure 18.10a). At T = 800 fs, most of the population of lycopene has relaxed from the S_2 state and occupies the S_1 state; whereas

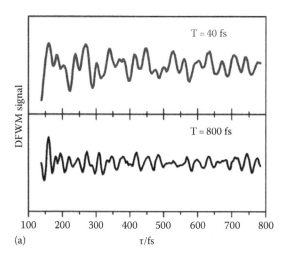

(a)

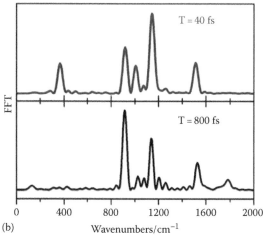

(b)

Figure 18.10 **(a)** Carotenoid lycopene oscillatory signals in the DFWM transients and **(b)** their FFT spectra at 570 nm detection wavelength at initial delay T = 40 fs (red) and T = 800 fs (black). (Reprinted with permission from Marek, M.S., Buckup, T., and Motzkus, M., Direct observation of a dark state in lycopene using pump-DFWM, *J. Phys. Chem. B*, 115, 8328–8337. Copyright 2011 American Chemical Society.)

at T = 40 fs, most molecules can be found in the transition from the S_2 to S_1 state. The Fourier transform spectra of the oscillating residuals of the two traces look remarkably different (Figure 18.10b). At T = 800 fs, two solvent modes at 915 and 1080 cm^{-1} are clearly visible as well as the frequencies that can be attributed to the S_1 state: methyl deformation at 1030 cm^{-1}; the C–C and C=C stretching modes at 1140, 1205, and 1530 cm^{-1}; and a C=C mode at 1800 cm^{-1}. In the time window of the first 100 fs, illustrated by the Fourier spectrum at T = 40 fs, several distinctive differences from the T =800 fs spectrum can be found: a component at 370 cm^{-1} is observed, the 1800 cm^{-1} mode is absent, and other modes have shifted frequencies. All these observations indicated that a distinctive state (transition) is observed at 40 fs.

The 40 fs signal with characteristic vibrational features that appear and disappear within the 100 fs time window was attributed to the dark state S_x (other than the S_1 state) of lycopene. According to the proposed model, S_2-population of the lycopene molecules leaves the S_2 state for the dark S_x state with a 20 fs time constant and then is transferred from the S_x to S_1 state with the 110 fs time constant. After the initial relaxation, temporal evolution of the lycopene pump–DFWM signal is determined by the vibrationally hot species of S_1 and S_0. Thus, this experiment demonstrated that transient species with very short lifetime can be monitored in the pump–DFWM experiment.

18.3.2 Two-dimensional electronic spectroscopy

Encouraged by the huge success of multidimensional magnetic resonance spectroscopies, the end of twentieth century saw the emergence of the optical multidimensional techniques, most often tagged as two-dimensional (2D) spectroscopies (Lepetit and Joffre, 1996, Hamm et al., 1998, Hybl et al., 1998). Foundations for the theoretical framework of vibrational and electronic 2D spectroscopies were laid by Mukamel and co-workers (Tanimura and Mukamel, 1993, Mukamel, 1995, 2000). The basic idea of 2D spectroscopy lies in finding the time-dependent correlations between different vibrational or electronic transitions in molecules, molecular complexes, or materials, and thus unraveling couplings. Correlations are plotted as 2D maps on two so-called "excitation" and "detection" frequency axes. Whereas 2D infrared spectroscopy immediately became a powerful and popular tool for studying structural dynamics of amino acids, proteins and other molecules (Hamm and Zanni, 2001), the 2D spectroscopy in visible range came of age together with the maturing of laser and optical technologies (Brixner et al., 2004).

18.3.2.1 PRINCIPLES OF TWO-DIMENSIONAL SPECTROSCOPY

2D electronic spectroscopy (2DES) is a third-order nonlinear spectroscopy technique. In the widely used implementation of 2DES, three equivalent laser pulses are used to excite the sample, and the fourth, so-called local oscillator (LO) pulse, is used to "read" the signal. A typical 2DES setup is shown in Figure 18.11. All four pulses are arranged in the boxcar geometry, where the three excitation pulses can be found in the three corners of the square. These pulses interact with the sample and induce the third-order polarization. Consequently, the signal is emitted by the sample as an electromagnetic wave collinear with the attenuated LO pulse in the direction of the fourth corner of the square. Signal emission direction is dictated by the phase-matching condition $k_S = -k_1 + k_2 + k_3$, where k_i are wave vectors of the signal and the three excitation pulses. The signal is mixed with the LO and heterodyne

detected in a spectral interferometry scheme. Since the amplitude and phase of LO are known, the full information about the signal amplitude and phase is determined, which can be also represented as real and imaginary parts of the signal.

The timing between the pulses employed in the 2DES experiment is depicted in the inset of Figure 18.11. Time delay between the first two pulses t_1 is called coherence delay, which tracks the evolution of the optical coherence between the ground and electronic states created by the first pulse. Time delay between the second and third pulses t_2 is called population delay, which tracks the population on one of the electronic states and coherence (quantum superposition) between vibrational, electronic or vibronic states. Delay between the third excitation pulse and emitted signal t_3 is called detection delay. Excitation and detection frequencies of the 2D plot are obtained by Fourier transform of the coherence and detection delays, correspondingly. When spectral interferometry detection is used in 2DES, the latter Fourier transform "is performed" by the spectrometer. The first two pulses in 2DES serve the analogous function as a single excitation pulse in the pump–probe spectroscopy (Section 18.2.1). To obtain an undistorted signal on the excitation frequency axis, the coherence delay has to be scanned with an

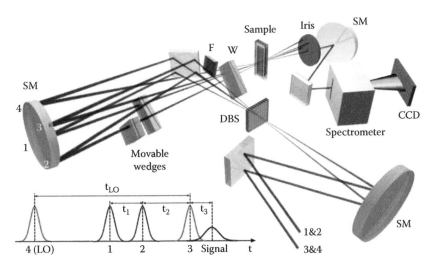

Figure 18.11 Schematic 2DES setup and the sequence of pulses used in the experiments with indicated time delays. SM spherical mirror, DBS diffractive beam splitter, W compensation wedges, F neutral density filter in LO. The first two excitation pulses are separated by the coherence time t_1, and the second and third pulses by the population time t_2. Third-order nonlinear signal is observed in the phase-matched direction collinear with the LO pulse at arrival time t_3. The LO pulse arrives first at time t_{LO} before pulse three.

interferometric precision. Most easily, this is achieved by moving glass wedges that are placed in the beams (Brixner et al., 2004) or by using pulse shapers (Tian et al., 2003, Vaughan et al., 2007). To obtain the total 2D spectrum, negative and positive coherence times have to be scanned, corresponding to the so-called rephasing and nonrephasing parts of the spectrum. Phase stability between the first two pulses as well as between the third and LO pulses has to be maintained during the experiments. This implies stringent requirements for the experimental setup. In the experimental scheme depicted in Figure 18.11, the overall phase stability is achieved in a passive way by a chosen geometry of the setup. However, the absolute phase of the signal is often undetermined. There are several methods to recover the absolute phase; among the most widely used procedures is "phasing" of the 2D spectrum to the transient absorption spectrum obtained from a standard pump–probe experiment (see Section 18.2.1). The projection-slice theorem implies that the 2D spectrum integrated along the excitation frequency axis should be identical to the transient absorption spectrum measured at the same population time (Jonas, 2003). Thus "phasing" is done by adjusting the phase of the 2D spectrum to match the projection to the transient absorption spectrum.

The information content obtained from 2DES experiments depends on the spectral coverage of the laser spectrum; whatever is not covered by the LO spectrum is not detected. Therefore, quite often very short (10–20 fs) spectrally broad (750–1500 cm^{-1} FWHM) pulses are used in 2DES. The key advantage of 2DES is that simultaneous time and spectral resolution of the experiments is not limited by the Fourier transform relationship of the pulses, but rather by the dynamic processes in the sample. Simultaneous time resolution of 20 fs and spectral resolution of 40 cm^{-1} can be routinely achieved in 2DES experiments. In general, the total 2D spectrum (after conversion to a unitless quantity) can be understood as a stacked sequence of frequency-resolved transient absorption spectra with the sharp resolution on the excitation frequency. Note that all this data is obtained in a single experiment.

2D spectra contain full information about the emitted electric field, which is directly related to the polarization created in the sample (Jonas, 2003). This means that information on the nonlinear response of the sample that is obtained in the other third-order time-resolved techniques can be extracted from the sequence of 2D spectra. For example, transient grating, three pulse photon echo peak shift, and transient absorption data can be recovered. Note, however, that for particular information on photophysical or photochemical properties of the sample, the technique optimized for the particular task will often provide the best result.

A 2D spectrum is a complex quantity and contains phase information about the nonlinear response. Whereas the real part of the 2D spectrum provides information about the absorptive response of the sample, the imaginary part corresponds to the refractive response. Usually only the real part of 2D spectra is shown, since we are most often interested in the absorptive features. Note that correct "phasing" of the 2D spectrum described earlier is crucial in separating the real and imaginary parts of the signal.

Depending on the interaction of the laser fields with the electronic states, three types of signals are recorded (as explained in Section 18.2.1): GSB, SE, and ESA (Mukamel, 1995). The convention in the 2DES community is to display 2D spectra in such a way that positive features correspond to the positive signal stemming from SE and GSB, and negative features to the negative signal due to ESA. Note that this is the opposite from what is normally used for displaying transient absorption spectra from pump–probe experiments. It should be also mentioned that different research groups working in the 2DES field use different presentations of the 2DES spectra; the excitation and detection frequency axes are often interchanged. Here we use the notation in which the excitation frequency corresponds to the abscissa axis, and detection frequency to the ordinate axis.

Most of the 2DES experiments are carried out by setting polarization of all beams parallel to each other. However, manipulation of the polarization of the beams can provide new insights into coherent and incoherent molecular dynamics. One informative polarization scheme is where polarization of the four pulses used in the experiments are set to (−45°, 45°, 90°, 0°), which allows investigating exclusively coherences with the electronic character and coherences excited via vibronically coupled states, whereas population and "standard" vibrational coherence dynamics are heavily suppressed (Schlau-Cohen et al., 2012, Westenhoff et al., 2012).

18.3.2.2 TRACKING THE ENERGY TRANSFER THROUGH THE LIGHT-HARVESTING COMPLEXES

2DES is especially well suited for studying multichromophore systems, as couplings between the molecules can clearly be revealed in the 2D maps. At the same time, energy transfer processes can be directly monitored in the evolution of the 2D spectra as disappearance of the diagonal peaks (and some cross peaks) and appearance of the cross peaks below the diagonal. 2DES was successfully applied in investigations of energy transfer processes in a variety of light-harvesting complexes. A few examples include studies of Fenna–Matthews–Olson (FMO) (Brixner et al., 2005, Thyrhaug et al., 2016) and chlorosome (Dostál et al., 2012) from green sulfur bacteria, LH3 and LH2 complexes from purple bacteria (Zigmantas et al., 2006, Ostroumov et al., 2013), and LHC II complex from

higher plants (Schlau-Cohen et al., 2009). Even the energy flow through the light-harvesting apparatus of the intact green sulfur bacteria cells has been characterized (Dostál et al., 2016). In addition, it has been demonstrated that 2DES provides a new level of insight when studying coherent dynamics in light-harvesting complexes. These experimental advances are summarized in Chapter 19.

As an illustration of the information that can be gained on the intermolecular connectivity and energy transfer processes in the light-harvesting complexes, we show four representative 2D spectra of FMO complex from the green sulfur bacterium *Chlorobaculum tepidum* measured at 77 K (Figure 18.12). Each FMO complex contains 7 or 8 BChl *a* molecules and serves as the energy conduit from the light-harvesting antenna chlorosome to the reaction centers in the photosynthetic apparatus of green sulfur bacteria. The 2D spectra depicted

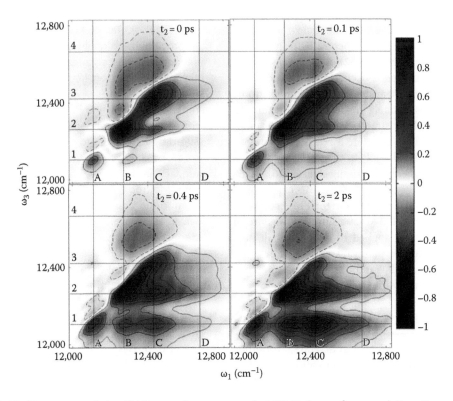

Figure 18.12 2D spectra of the FMO complex measured at 77 K shown for population times $t_2 = 0$ fs, 100 fs, 400 fs, and 2 ps. Contour lines are drawn in 10% intervals of the peak amplitude, with solid lines representing positive features (red) and dashed lines negative features (blue). Each spectrum is normalized to the maximum signal. Horizontal and vertical grid lines indicate the four most prominent excitonic levels. Coupling between transitions can be observed as cross peaks in the 2D correlation spectrum at $t_2 = 0$ fs. Energy transfer is monitored as a decay of the diagonal peaks (and higher cross peaks) as well as a rise of the lower-lying cross peaks.

here are similar to the 2D spectra presented earlier (Brixner et al., 2005). The grid lines in the spectra are drawn to indicate the four most intense electronic transitions, which correspond to the excitonic states of the complex. Laser pulses of 15 fs duration with spectrum covering all FMO absorption bands were used in these 2DES experiment.

The 2D correlation spectrum (at $t_2 = 0$ fs) is dominated by the diagonal features corresponding to the FMO absorption bands. The ESA contributions (transitions from the single to double exciton band) appear as negative signals above the diagonal. Some cross peaks at B1, C2, and D3 positions can be seen as well. This is a clear indication that excitonic transitions are observed, which are correlated through the common ground state. The cross peaks appear in a correlation spectrum, because the different excitonic states share one or more (B)Chl molecules. Correlation cross peaks should also appear above the diagonal, but they are masked by the ESA signals.

The energy flow through FMO can be monitored as rising amplitudes of cross peaks below the diagonal, as observed in the 2D spectra with increasing population time (Figure 18.12). It can be easily seen that energy relaxation is occurring on several time scales. Each energy transfer step between excitonic states can be separately inspected. For example, judging from the rising amplitudes of the cross peaks, energy from the state 3 relaxes not only to the state 2 but also directly to the state 1.

In a recent paper by Thyrhaug et al. (2016), 2DES with polarization control was used to experimentally determine all eight states in the FMO complex. In addition, the global kinetic fitting procedure of the 2D spectra was applied to determine all possible energy transfer pathways and rates between the FMO states, and a "multibranching," interconnected transfer network was unraveled.

REFERENCES

Amarie, S., L. Wilk, T. Barros et al. 2009. Properties of zeaxanthin and its radical cation bound to the minor light-harvesting complexes CP24, CP26 and CP29. *Biochim. Biophys. Acta* 1787: 747–752.

Bode, S., C. C. Quentmeier, P. N. Liao et al. 2009. On the regulation of photosynthesis by excitonic interactions between carotenoids and chlorophylls. *Proc. Natl. Acad. Sci. U. S. A.* 106: 12311–12316.

Brixner, T., T. Mančal, I. V. Stiopkin et al. 2004. Phase-stabilized two-dimensional electronic spectroscopy. *J. Chem. Phys.* 121: 4221–4236.

Brixner, T., J. Stenger, H. M. Vaswani et al. 2005. Two-dimensional spectroscopy of electronic couplings in photosynthesis. *Nature* 434: 625–628.

Cannizzo, A., O. Braem, G. Zgrablic et al. 2007. Femtosecond fluorescence upconversion setup with broadband detection in the ultraviolet. *Opt. Lett.* 32: 3555–3557.

Dostál, J., T. Mančal, R. Augulis et al. 2012. Two-dimensional electronic spectroscopy reveals ultrafast energy diffusion in chlorosomes. *J. Am. Chem. Soc.* 134: 11611–11617.

Dostál, J., J. Pšenčík, and D. Zigmantas. 2016. In situ mapping of the energy flow through the entire photosynthetic apparatus. *Nat. Chem.* 8: 705–710.

Fayer, M. D. 2001. *Ultrafast Infrared and Raman Spectroscopy: Practical Spectroscopy.* New York: Marcel Dekker, p. 709.

Haacke, S., R. A. Taylor, I. Bar-Joseph et al. 1998. Improving the signal-to-noise ratio of femtosecond luminescence up-conversion by multichannel detection. *J. Opt. Soc. Am. B* 15: 1410–1417.

Hamm, P., M. H. Lim, and R. M. Hochstrasser. 1998. Structure of the amide I band of peptides measured by femtosecond nonlinear-infrared spectroscopy. *J. Phys. Chem. B* 102: 6123–6138.

Hamm, P. and M. T. Zanni. 2001. *Concepts and Methods of 2D Infrared Spectroscopy.* Cambridge University Press, London, U.K.

Hauer, J., T. Buckup, and M. Motzkus. 2007. Pump-degenerate four wave mixing as a technique for analyzing structural and electronic evolution: Multidimensional time-resolved dynamics near a conical intersection. *J. Phys. Chem. A* 111: 10517–10529.

Holzwarth, A. R., M. G. Müller, M. Reus et al. 2006. Kinetics and mechanism of electron transfer in intact photosystem II and in the isolated reaction center: Pheophytin is the primary electron acceptor. *Proc. Natl. Acad. Sci. U. S. A.* 103: 6895–6900.

Hybl, J. D., A. W. Albrecht, S. M. G. Faeder et al. 1998. Two-dimensional electronic spectroscopy. *Chem. Phys. Lett.* 297: 307–313.

Jonas, D. M. 2003. Two-dimensional femtosecond spectroscopy. *Annu. Rev. Phys. Chem.* 54: 425–463.

Kloz, M., R. van Grondelle, and J. T. M. Kennis. 2011. Wavelength-modulated femtosecond stimulated Raman spectroscopy—Approach towards automatic data processing. *Phys. Chem. Chem. Phys.* 13: 18123–18133.

Kloz, M., J. Weissenborn, T. Polívka, H. A. Frank, and J. T. M. Kennis. 2016. Spectral watermarking in Raman spectroscopy. *Phys. Chem. Chem. Phys.* 18: 14619–14628.

Kraack, J. P., A. Wand, T. Buckup et al. 2013. Mapping multidimensional excited state dynamics using pump-impulsive-vibrational-spectroscopy and pump-degenerate-four-wave-mixing. *Phys. Chem. Chem. Phys.* 15: 14487–14501.

Krueger, B. P., G. D. Scholes, R. Jimenez et al. 1998. Electronic excitation transfer from carotenoid to bacteriochlorophyll in the purple bacterium Rhodopseudomonas acidophila. *J. Phys. Chem. B* 102: 2284–2292.

Kukura, P., D. W. McCamant, and R. A. Mathies. 2007. Femtosecond stimulated Raman spectroscopy. *Annu. Rev. Phys. Chem.* 58: 461–488.

Lepetit, L. and M. Joffre. 1996. Two-dimensional nonlinear optics using Fourier-transform spectral interferometry. *Opt. Lett.* 21: 564–566.

Leupold, D., K. Teuchner, J. Ehlert et al. 2002. Two-photon excited fluorescence from higher electronic states of chlorophylls in photosynthetic antenna complexes: A new approach to detect strong excitonic chlorophyll a/b coupling. *Biophys. J.* 82: 1580–1585.

Macpherson, A. N. and T. Gillbro. 1998. Solvent dependence of the ultrafast S_2-S_1 internal conversion rate of beta-carotene. *J. Phys. Chem. A* 102: 5049–5058.

Marek, M. S., T. Buckup, and M. Motzkus. 2011. Direct observation of a dark state in lycopene using pump-DFWM. *J. Phys. Chem. B* 115: 8328–8337.

McCamant, D. W., P. Kukura, S. Yoon et al. 2004. Femtosecond broadband stimulated Raman spectroscopy: Apparatus and methods. *Rev. Sci. Instrum.* 75: 4971–4980.

Motzkus, M., S. Pedersen, and A. H. Zewail. 1996. Femtosecond real-time probing of reactions .19. Nonlinear (DFWM) techniques for probing transition states of uni- and bimolecular reactions. *J. Phys. Chem.* 100: 5620–5633.

Mukamel, S. 1995. *Principles of Nonlinear Optical Spectroscopy.* New York: Oxford University Press.

Mukamel, S. 2000. Multidimensional femtosecond correlation spectroscopies of electronic and vibrational excitations. *Annu. Rev. Phys. Chem.* 51: 691–729.

Mukamel, S. and J. D. Biggs. 2011. Communication: Comment on the effective temporal and spectral resolution of impulsive stimulated Raman signals. *J. Chem. Phys.* 134: 161101.

Nakamura, R. and Kanematsu, Y. 2004. Femtosecond spectral snapshots based on electronic optical Kerr effect. *Rev. Sci. Instrum.* 75: 636–644.

Ostroumov, E. E., R. M. Mulvaney, R. J. Cogdell et al. 2013. Broadband 2D electronic spectroscopy reveals a carotenoid dark state in purple bacteria. *Science* 340: 52–56.

Papagiannakis, E., M. Vengris, D. S. Larsen et al. 2006. Use of ultrafast dispersed pump-dump-probe and pump-repump-probe spectroscopies to explore the light-induced dynamics of peridinin in solution. *J. Phys. Chem. B* 110: 512–521.

Polívka, T. and V. Sundström. 2004. Ultrafast dynamics of carotenoid excited states—From solution to natural and artificial systems. *Chem. Rev.* 104: 2021–2071.

Polívka, T. and V. Sundström. 2009. Dark excited states of carotenoids: Consensus and controversy. *Chem. Phys. Lett.* 477: 1–11.

Redeckas, K., V. Voiciuk, and M. Vengris. 2016. Investigation of the S_1/ICT equilibrium in fucoxanthin by ultrafast pump–dump–probe and femtosecond stimulated Raman scattering spectroscopy. *Photosynth. Res.* 128: 169–181.

Redeckas, K., Voiciuk, V., Zigmantas, D. et al. 2017. Unveiling the excited state energy transfer pathways in peridinin-chlorophyll a-protein by ultrafast multi-pulse transient absorption spectroscopy. *BBA Bioenergetics* 1858: 197–307.

Ricci, M., S. E. Bradforth, R. Jimenez et al. 1996. Internal conversion and energy transfer dynamics of spheroidene in solution and in the LH1 and LH2 light-harvesting complexes. *Chem. Phys. Lett.* 259: 381–390.

Ruban, A. V., R. Berera, C. Ilioaia et al. 2006. Identification of a mechanism of photoprotective energy dissipation in higher plants. *Nature* 450: 575–578.

Schlau-Cohen, G. S., T. R. Calhoun, N. S. Ginsberg et al. 2009. Pathways of energy flow in LHCII from two-dimensional electronic spectroscopy. *J. Phys. Chem. B* 113: 15352–15363.

Schlau-Cohen, G. S., A. Ishizaki, T. R. Calhoun et al. 2012. Elucidation of the timescales and origins of quantum electronic coherence in LHCII. *Nat. Chem.* 4: 389–395.

Schmidt, B. S. Laimgruber, W. Zinth et al. 2003. A broadband Kerr shutter for femtosecond fluorescence spectroscopy. *Appl. Phys. B Lasers Opt.* 76: 809–814.

Shah, J. 1988. Ultrafast luminescence spectroscopy using sum frequency generation. *IEEE J. Quant. Electron.* 24: 276–288.

Shim, S. and R. A. Mathies. 2008a. Development of a tunable femtosecond stimulated Raman apparatus and its application to beta-carotene. *J. Phys. Chem. B* 112: 4826–4832.

Shim, S. and R. A. Mathies. 2008b. Femtosecond Raman-induced Kerr effect spectroscopy. *J. Raman Spectrosc.* 39: 1526–1530.

Tanimura, Y. and S. Mukamel. 1993. 2-Dimensional femtosecond vibrational spectroscopy of liquids. *J. Chem. Phys.* 99: 9496–9511.

Thyrhaug, E., K. Zidek, J. Dostál et al. 2016. Exciton structure and energy transfer in the Fenna–Matthews–Olson complex. *J. Chem. Phys. Lett.* 7: 1653–1660.

Tian, P. F., D. Keusters, Y. Suzaki et al. 2003. Femtosecond phase-coherent two-dimensional spectroscopy. *Science* 300: 1553–1555.

Vaughan, J. C., T. Hornung, K. W. Stone et al. 2007. Coherently controlled ultrafast four-wave mixing spectroscopy. *J. Phys. Chem. A* 111: 4873–4883.

Walla, P. J., P. A. Linden, C. P. Hsu et al. 2000b. Femtosecond dynamics of the forbidden carotenoid S_1 state in light-harvesting complexes of purple bacteria observed after two-photon excitation. *Proc. Natl. Acad. Sci. U. S. A.* 97: 10808–10813.

Walla, P. J., J. Yom, B. P. Krueger et al. 2000a. Two-photon excitation spectrum of light-harvesting complex II and fluorescence upconversion after one- and two-photon excitation of the carotenoids. *J. Phys. Chem. B* 104: 4799–4806.

Wehling, A. and P. Walla. 2006. A two-photon excitation study on the role of carotenoid dark states in the regulation of plant photosynthesis. *Photosynth. Res.* 90: 101–110.

Westenhoff, S., D. Paleček, P. Edlund et al. 2012. Coherent picosecond exciton dynamics in a photosynthetic reaction center. *J. Am. Chem. Soc.* 134: 16484–16487.

Yoshizawa, M., H. Aoki, and H. Hashimoto. 2001. Vibrational relaxation of the $2A_g^-$ excited state in all-trans-β-carotene obtained by femtosecond time-resolved Raman spectroscopy. *Phys. Rev. B* 63: 180301.

Yoshizawa, M. and M. Kurosawa. 2000. Femtosecond time-resolved Raman spectroscopy using stimulated Raman scattering. *Phys. Rev. A* 61: 013808.

Yoshizawa, M., R. Nakamura, O. Yoshimatsu et al. 2012. Femtosecond stimulated Raman spectroscopy of the dark S-1 excited state of carotenoid in photosynthetic light harvesting complex. *Acta Biochim. Pol.* 59: 49–52.

Zewail A. H. 1998. Laser femtochemistry. *Science* 242: 1645–1653.

Zigmantas, D., E. L. Read, T. Mančal et al. 2006. Two-dimensional electronic spectroscopy of the B800-B820 light-harvesting complex. *Proc. Natl. Acad. Sci. U. S. A.* 103: 12672–12677.

19

Experimental evidence of quantum coherence in photosynthetic light harvesting*

GREGORY D. SCHOLES, JACOB C. DEAN, JESSICA M. ANNA,
GREGORY S. ENGEL, AND RIENK VAN GRONDELLE

19.1 INTRODUCTION

A biological system is incredibly complex, and the complexity hardly diminishes even as we focus on processes occurring in the cell such as respiration or the action of enzymes. We therefore employ an average picture of the process rather than mapping out every possibility that transpires in each cell, each time the process occurs. As a consequence of this statistical approach, quantum mechanical (or coherent) phenomena are hidden in the average response of a very complex system. Schrödinger pioneered this idea in his book *What Is Life?* (Schrödinger 1944). Schrödinger hypothesized that

quantum mechanical effects in living systems may be evident, and indeed active, if the number of "particles" in the system of interest is very small.

To understand what Schrödinger's statement means, first consider a process being performed by the particles, where "particle" could represent an enzyme, for instance. Let's assume we need quantum mechanics to describe the mechanism of this process. However, the "behavior" of the particle, for example, its position or energy levels along a reaction coordinate, fluctuates randomly and is therefore unpredictable. The average over many particles washes away the evidence of quantum mechanics because we lose the ability to distinguish

---

* This chapter was adapted from Anna, J.M., A little coherence in photosynthetic light harvesting, *Bioscience*, 2014, 64, 1, 14–25 by permission of Oxford University Press.

the signature of wavelike behavior: its phase. Schrödinger reasoned that if the number of particles involved in the process is small enough, then the quantum mechanical effects could be preserved from start to end, and therefore play a role in the mechanism. Nowadays, we know this viewpoint is a bit too simplistic, but it's a useful way to start thinking about why we don't usually consider quantum mechanics in biology, even though it must be lurking beneath the surface.

Schrödinger hypothesized that as the *size* of a statistical system diminishes, quantum effects start to show up. We should define this *size* not as the physical size of a molecule or enzyme but instead as the size of configuration space that is sampled as we "watch" a process happen over and over again. Configuration space can include the positions and orientations of reactants and their relative energies, all of which fluctuate randomly owing to thermal effects (Frauenfelder and Wolynes 1985, Voth and Hochstrasser 1996). Schrödinger's idea is evident; then, when we reduce configuration space by confining molecules to isolated environments (Figure 19.1).

But what about in a biological environment? Spectroscopic experimental results have suggested that quantum mechanics—or at least coherence— can be observed in light-harvesting complexes and may play a role in the initial process of photosynthesis (for review, see Cheng and Fleming 2009, Novoderezhkin and van Grondelle 2010, Renger 2009, Schlau-Cohen et al. 2012, Scholes 2010, Scholes et al. 2012, Sundström 2000, van Grondelle and Novoderezhkin 2006, Yang et al. 2003). These results suggest that while incoherent hopping of excitation energy is an excellent first approximation, it is not the whole story and we need to introduce concepts from quantum mechanics to explain the mechanism satisfactorily. In this chapter, we describe the nature of these quantum effects, how these effects can be detected, and the role coherence could play in photosynthesis.

19.2 ELECTRONIC ENERGY TRANSFER IN LIGHT-HARVESTING COMPLEXES

Light harvesting, the initial step in photosynthesis, refers to the process by which antenna complexes absorb photons and funnel this excitation energy to reaction centers where charge separation and subsequent electron transfer reactions take place (Blankenship 2002). These electron transfer reactions eventually drive photosynthesis. This process of light harvesting is common to all photosynthetic organisms. However, on the molecular level, there is a huge amount of diversity among the different light-harvesting complexes, for example, in

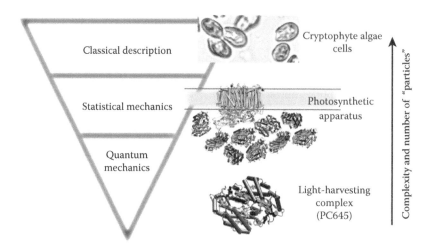

Figure 19.1 Depiction of the idea of the scales of biological length and complexity, as described in the text. Atomic scale, and possibly quantum mechanical, effects govern the function of a protein. These details are lost as the system considered becomes more complex. In that case, certain "emergent" phenomena, or functions, are the primary observables. (Reprinted from Anna, J.M. et al., *BioScience*, 64, 15, 2014.)

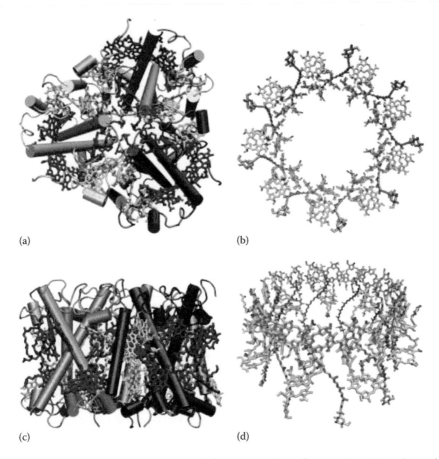

(a)　　　　　　　　　　　　(b)

(c)　　　　　　　　　　　　(d)

Figure 19.2 Structural models of **(a and c)** LHCII from peas (Standfuss et al. 2005), where the red molecules are chlorophyll-*a* and the blue molecules are chlorophyll-*b*. **(b and d)** Chromophores of LH2 from *Rhodopseudomonas acidophila* strain 10050 (Prince et al. 1997). Here the green bacteriochlorophyll-*a* molecules comprise the B850 ring and the light blue ones comprise the B800 ring. The carotenoids are drawn in grey. (Reprinted from Anna, J.M. et al., *BioScience*, 64, 15, 2014.)

Figure 19.2 the antenna light-harvesting complexes of purple bacteria, LH2, and plants (LHCII) are shown. As can be seen from Figure 19.2, the structures of these light-harvesting complexes are quite different. The different structural arrangement of the chromophores within the protein environment leads to different chromophore–chromophore interactions, electronic couplings (*J*), different chromophore–proteins interactions, and system-bath couplings (λ). The relative magnitude of the λ and *J* dictates the mechanisms by which electronic energy transfer occurs. To further explore this concept and to understand the role that quantum mechanics may play in light harvesting, it is helpful to review how to think about electronic energy transfer in these complex biological systems (Cheng and Fleming 2009, Fassioli et al. 2014, Hu et al. 1997, Novoderezhkin

and van Grondelle 2010, Renger 2009, Schlau-Cohen et al. 2012, van Grondelle and Novoderezhkin 2006, van Grondelle et al. 1994).

19.2.1 Weak electronic coupling regime: $\lambda > J$

When the system-bath coupling is much stronger than the electronic coupling between the molecules, energy transfer can be described using Förster theory (Andrews 1989, Förster 1965, Scholes 2003). From the perspective of Förster theory, electronic energy transfer requires two factors. First, there needs to be an electronic interaction between pairs of molecules. Such an interaction is turned on when one molecule is electronically excited. Then the "transition dipole" for de-excitation of the photo-excited molecule couples to the transition dipole for excitation of a nearby ground-state molecule, promoting a "radiationless" jump of electronic excitation from one molecule to another. This is a quantum mechanical phenomenon, but the interaction and its distance dependence resemble the classical electrostatic interaction between two dipoles—in this case, interaction between the transition dipoles of "donor" and "acceptor" (Andrews 1989, Krueger et al. 1998). The second requirement for energy transfer is that the energy is conserved. In practice, that means that the fluorescence spectrum of the photo-excited molecule should overlap in frequency with the absorption spectrum of the energy acceptor.

Förster theory handles system-bath coupling well but treats electronic coupling perturbatively to second order and is most accurate when states are weakly coupled. Within this Förster picture, excitation transfer occurs when electronic states in close spatial proximity experience fluctuations that permit the two states to have identical energy. The fluctuations of each state are presumed be to be statistically independent from one another and also across an array of complexes. In this regard, Förster theory gives rise to incoherent energy transfer that manifests experimentally as smooth, monotonic, exponential dynamics.

19.2.2 Strong electronic coupling regime: $\lambda < J$

The quantum mechanical aspect of light harvesting can be traced to the assumption in Förster theory that electronic coupling between the molecules is extremely weak (compared to line broadening). When the system-bath coupling is much weaker than the electronic coupling, Redfield theory can be used to describe the dynamics of the system (Cheng and Fleming 2009, Redfield 1957, 1965, Renger 2009).

Redfield theory represents a rather different picture from Förster theory. The theory treats system-bath coupling perturbatively to second order but treats electronic coupling among excited states exactly. Thus, typical Redfield models depend strongly on spatial overlap of electronic wavefunctions. Because the model assumes weak system-bath coupling, the bath promotes energy transfer between states that overlap on the same site, that is, a fluctuation localizes the excitation on a site and then the probability of finding that excitation in a given state will depend on the participation of the site in that state.

19.2.3 Intermediate electronic coupling regime: $\lambda \approx J$

Of course, there is no reason a priori that either electronic coupling or system-bath coupling needs to be weak. In fact, in most photosynthetic systems, electronic coupling, system-bath coupling, dephasing, thermal energy, and energy gaps among states are all of approximately the same magnitude (Mohseni et al. 2014). In this intermediate regime, both Redfield theory and Förster theory fail to represent accurately the dynamics of the energy transfer. While numerous approximate theories have been developed to understand dynamics in this intermediate regime, an exact theory, the hierarchical equations of motion, has recently been developed (Ishizaki and Fleming 2009b, Ishizaki and Tanimura 2005, Tanimura 2014, Tanimura and Kubo 1989). This method offers exact dynamics in the intermediate regime for specific (but broadly applicable) bath structures.

19.2.4 Nature of the quantum effect

As Schrödinger described, quantum effects may matter when not overwhelmed by random noise in the system. What is not so obvious is how nature might avoid such noise within a biological system, what the advantages might be, and how we might detect such effects experimentally. The key to understanding these questions is to look at the noise. What is the magnitude of the system-bath coupling? What is the nature of the bath? Is this "noise" truly random? How does it compare to the electronic coupling?

Perhaps not surprisingly, the simplest framework for understanding this noise and the quantum effects that it produces or masks is through the lens of Förster and Redfield models. First, in a canonical Förster model, consider a simple dimer consisting of two weakly coupled ($J \ll \lambda$) monomers, A and B. If one monomer (A) is excited, we would expect to see energy transfer to the second monomer (B) over time. This energy transfer would manifest as a simple exponential growth of excited state population on the second monomer (B). This energy transfer dynamic is mathematically fully equivalent to a classical hopping process whereby the excitation randomly hops to the other monomer with some time constant.

The second example of a strongly coupled dimer is subtly more complicated. In this strongly coupled ($\lambda \ll J$) dimer, Redfield theory is the proper lens through which to view this system. In this case, if monomer A is excited, the excitation will coherently oscillate between monomers A and B with a frequency proportional to the energy gap between the electronic eigenstates. In the absence of any system-bath interaction, this coherent oscillation will continue indefinitely, but in the presence of some random noise from the bath, the oscillation will decay with time.

In the intermediate regime, dephasing (rapid damping of oscillatory dynamics) will be observed for the same reasons. However, in most measurements, it is impossible to excite just monomer A impulsively within an ensemble. For example, an optical excitation will excite eigenstates that span both monomers (recall that A and B are strongly coupled resulting in two mixed eigenstates). Energy relaxation from one eigenstate to the next will occur due to system-bath fluctuations and will appear as exponential dynamics within the system, assuming that the system-bath motions are incoherent. The reason that we would not observe oscillations hinges on the nature of the original excitation. If the excitation were spatially localized rather than energetically localized, oscillatory dynamics would be observed. Thus, we need to craft an excitation strategy that will allow us to see such dynamics and to do so across an ensemble so that macroscopic signals can be observed. These macroscopic signals, though induced by our experimental configuration, will report of the microscopic coherent dynamics.

Within light-harvesting complexes, excitations are absorbed and then shuttled from one complex to the next. Coherent dynamics can be important to drive fast, ballistic transport across hundreds of nanometers. Of course, no coherent oscillations will be observed when driven by incoherent light, but the ballistic transport dynamics can still impact overall transport efficiency. In all cases, whether oscillations can be observed or not, the important quantity giving rise to the coherent, ballistic transport is the system-bath coupling, which drives dephasing. This quantity simply reports on how the electronic states "feel" the world around them.

The chromophores in light-harvesting complexes are packed at high density. As a consequence, the average center-to-center separation of neighboring molecules is typically only ~10 Å, and therefore the electronic couplings are moderately large.

One ramification of this is that several chromophores can act cooperatively in the absorption and transfer of electronic excitation (Renger 2009, Scholes 2003, van Amerongen et al. 2000 and references therein). When chromophores carry electronic excitation cooperatively, it means not only that there is a good chance of finding the excitation on more than one chromophore simultaneously but that the excitation is spread with a defined *amplitude* across those molecules. This amplitude factor carries information on the relative sign of the excitation wave on each molecule.

19.2.5 What does quantum mechanics do for energy transfer?

In this section, we briefly describe why coherent superpositions of states and particularly how long they survive are interesting in these systems. The short answer is that it changes the way we think about the energy "jumps" in Pearlstein's random walk model. Let's say there are two pathways for transferring excitation from molecule A to B in a light-harvesting complex: directly from A to B (P_{AB}) and via a third molecule C (P_{ACB}). Classically, the probability of the energy transfer is simply the sum of the probability of taking each path:

$$P_{total} = P_{AB} + P_{ACB} \qquad (19.1)$$

In quantum mechanics, however, that probability law is modified, resulting in the common explanation that both paths are taken simultaneously. What happens is that the probability of energy transfer from A to B is calculated differently: we assign a probability "amplitude" to each path, sum those amplitudes, then convert the sum to a probability by taking the modulus squared:

$$\Pi_{total} = \left| A_{AB} + A_{ACB} \right|^2 \qquad (19.2)$$

The result of that procedure is that the pathways can interfere, as if we represented each as a wave, then we added those waves. If the crests of the waves are lined up for the two paths, then constructive interference boosts the energy transfer rate compared to the classical calculation. The quantum law reduces to the familiar probability law when we lose the ability to discriminate the waves, a process called "decoherence," or when a system is so intrinsically complex that all the constructive and destructive interferences cancel on average—an idea exploited in semiclassical simulations of dynamics (Miller 2012). It should be noted that while Miller expertly represented these dynamics by classical degrees of freedom within his simulations, Stock and Thoss have shown that this representation is homomorphic to a full quantum calculation (Stock and Thoss 1997).

19.3 HOW TO DETECT QUANTUM COHERENCE

With the advent of femtosecond spectroscopy, the possibility of exploring quantum mechanical effects (i.e., coherences) in natural photosynthetic complexes became an experimental reality (Fleming and van Grondelle 1997, Martin and Vos 1992, Zewail 2000). Some of the earlier work in this area included the observation of oscillatory intensity modulations in pump-probe measurements on natural photosynthetic complexes (Arnett et al. 1999, Chachisvilis et al. 1994, Martin and Vos 1992, Savikhin et al. 1997a, Vos et al. 1991). Though there has been much previous work in this area, there has been a recent revival in the field since the striking report by Engel et al. describing evidence for remarkably long-lived electronic coherences after excitation of the Fenna–Matthews–Olson (FMO) complex (Engel et al. 2007). Those observations were facilitated by the experimental spectroscopic technique of two-dimensional electronic spectroscopy (2DES), a method that yields a clearer measurement of dynamic coherences than other femtosecond spectroscopic experiments. These experiments stimulated a rapid increase in the number of theoretical (Caruso et al. 2009, Cheng and Fleming 2009, Chin et al. 2013, Hoyer et al. 2010, Ishizaki and Fleming 2009a, Mohseni et al. 2008, Novoderezhkin et al. 2015, Tiwari et al. 2013) and experimental (Collini et al. 2010, Harel and Engel 2012, Hildner et al. 2013, Lee et al. 2007, Panitchayangkoon et al. 2010, Turner et al. 2012) studies exploring the possible role that coherences may play in photosynthetic complexes. Recently, coherent dynamics have even been observed in single LH2 complexes (Hildner et al. 2013). In the following section, we give a detailed account of how 2DES reveals coherences.

19.3.1 Two-dimensional electronic spectroscopy to reveal coherence

To demonstrate how 2DES has been important in providing new insights, we first describe how this information is manifested in a two-dimensional spectrum for a model system. The 2DES experiment uses femtosecond laser pulses to excite electronic absorption bands of the system being studied (e.g., a protein dispersed in solution) and monitors how these excited transitions change over time. This allows for information on how energy has been redistributed among the excited states to be obtained, for example, by energy transfer causing excitation energy to flow from one molecule to another. In fact, 2DES is very similar to pump-probe spectroscopy where a probe pulse arrives at various times after the pump and examines how the excited state population has redistributed among the excited states. One main way that 2DES differs from pump-probe spectroscopy is that the pump pulse is spectrally resolved in a 2D electronic spectrum. For a 2D spectrum, one can think of the pump pulse as effectively "labeling" the system according to the electronic absorption bands and reporting this labeled state along the excitation axis ω_{excite}. Then after some time, the waiting time (t_2), in which the system is free to evolve, the probe pulse interacts to effectively "read out" the current state of the previously labeled system, and that current state is recorded along the detection axis ω_{detect}. In this manner, a frequency–frequency correlation map is generated where each excitation frequency is correlated to each detection frequency (Cho 2008, Jonas 2003).

The experimental details for obtaining this frequency–frequency correlation map are quite involved and go beyond the scope of this chapter. The following books and review articles offer a comprehensive account of the experimental details (Cho 2008, 2009, Hamm and Zanni 2011, Jonas 2003, Mukamel 2000). Though we do not discuss the experimental techniques here, we would like to comment on some of the limitations associated with 2DES before moving on to the model system. 2DES is based on Fourier transform techniques and therefore requires attosecond timing precision and mechanisms for phase stabilization (for a detailed account on experimental techniques for overcoming these issues, see Ogilvie and Kubarych 2009).

So though 2DES can be used to obtain more direct, unambiguous information regarding the system, one of the drawbacks to employing this technique lies in the fact that it is harder to implement experimentally compared to pump-probe spectroscopies. Another limitation lies in the difficulty in extracting kinetics from the wealth of information contained in the 2D spectra. However, progress has been made in this area with the Ogilvie group (Myers et al. 2010) and Scholes group (Ostroumov et al. 2013), extending well-established techniques for extracting kinetics from 1D pump-probe spectra to 2D electronic spectra.

To demonstrate how information can be extracted from the 2D spectra, we first consider a model system. The model system is shown in Figure 19.3, which consists of three identical chromophores, A, B, and C, in a protein scaffold (e.g., the chlorin ring of chlorophyll molecules in light-harvesting chromophore–protein complexes). The energy level diagram together with the linear electronic absorption spectrum is also shown in Figure 19.3. Owing to chromophore–protein and chromophore-chromophore interactions, the three identical chromophores give rise to three resolvable peaks in the electronic absorption spectrum. The chromophores lie in different local protein environments, leading to shifts in the transition frequencies associated with the individual chromophores (the site energies). In the model system, chromophore C is shifted to a higher energy with respect to chromophores A and B. The electronic structure is further perturbed due to the chromophore-chromophore interactions. Chromophores A and B lie within close proximity to one another, and when electronically excited, the wavefunction is shared coherently over A and B like a wave that has a specific arrangement of peaks and troughs at the position of the molecules. This kind of delocalized excited electronic state is known as a molecular exciton (Scholes and Rumbles 2006, Spano 2010, van Amerongen et al. 2000).

Schematic representations of two 2D spectra at different waiting times are also shown in Figure 19.3. The peaks lying along the diagonal correspond to the peaks in the linear spectrum. At early

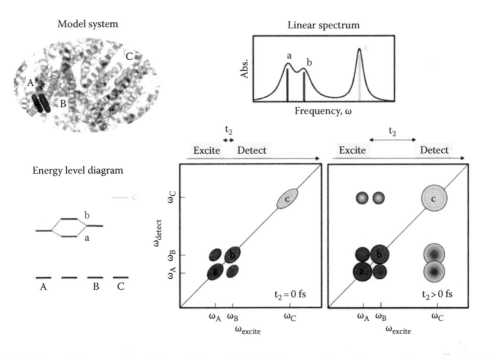

Figure 19.3 A model system consisting of three identical chromophores in a protein scaffold is shown along with the corresponding energy level diagram and linear spectrum. Schematic 2D spectra at $t_2 = 0$ fs and a later waiting time are shown. At early waiting times crosspeaks indicate that the corresponding diagonal peaks are electronically coupled. At later waiting times the appearance of crosspeaks indicates energy transfer between the different electronic transitions. (Reprinted from Anna, J.M. et al., *BioScience*, 64, 15, 2014.)

waiting times before dynamic processes occur, the schematic $t_2 = 0$ fs 2D spectrum, the crosspeaks indicate that the corresponding diagonal peaks **a** and **b** arise from transitions with common ingredients—common molecular orbitals or a pair of exciton states. The 2DES experiment uses the properties of short laser pulse photo-excitation to probe the properties of the electronic states in light-harvesting complexes. The femtosecond pulse contains a spectrum of colors that are all "in phase." Therefore, it can photo-excite absorption bands in phase. This gives the crosspeaks in the 2D spectrum additional properties when the absorption bands have a common character. From these crosspeaks, information such as the magnitude of coupling and the relative orientation of the excitonic states can be extracted (Cho 2008, 2009, Golonzka and Tokmakoff 2001, Hamm and Zanni 2011, Jonas 2003, Schlau-Cohen et al. 2011). These crosspeaks also have a well-characterized waiting time dependence: their amplitudes are modulated as a result of a superposition (i.e., quantum coherence) created between the excitonic states **a** and **b** (Cho 2008, 2009, Hamm and Zanni 2011, Jonas 2003, Mukamel 2000). As t_2 increases, the amplitude of these crosspeaks will oscillate at a frequency given by the difference in energy between the two strongly coupled states, $((E_a - E_b)/\hbar)$. By monitoring these oscillations, which are typically rapidly damped, we can determine the lifetime of the coherence.

From the schematic spectra at later waiting times, we see the appearance of crosspeaks among weakly coupled excitonic states. These crosspeaks indicate that energy is transferred between the corresponding diagonal peaks. From the schematic 2D spectra at $t_2 > 0$ fs, the appearance of crosspeaks below the diagonal indicates that energy is transferred downhill from state **c** to states **a** and **b**, while the appearance of crosspeaks above the diagonal indicates uphill energy transfer from states **a** and **b** to state **c**. Monitoring the time-dependent amplitudes of these crosspeaks thereby allows detailed information on energy flow pathways to be determined.

19.4 EXAMPLES OF SYSTEMS MEASURED WITH 2DES

These aspects of 2DES were applied to explore a wide range of photosynthetic complexes (Anna et al. 2012, Brixner et al. 2005, Calhoun et al. 2009, Cho 2008, Collini et al. 2010, Dostál et al. 2012, Engel et al. 2007, Ferretti et al. 2014, 2016, Fuller et al. 2014, Harel and Engel 2012, Lewis and Ogilvie 2012, Myers et al. 2010, Ostroumov et al. 2013, Panitchayangkoon et al. 2010, Schlau-Cohen et al. 2009, 2011, 2012,b, Turner et al. 2012, Zigmantas et al. 2006), including electronic energy transfer in reaction centers (Fuller et al. 2014, Myers et al. 2010, Romero et al. 2014, Schlau-Cohen et al. 2012), the chlorosome (Dostál et al. 2012), trimeric PSI (Anna et al. 2012), LH2 (Harel and Engel 2012, Ostroumov et al. 2013), LH3 (Zigmantas et al. 2006), LHCII (Calhoun et al. 2009, Schlau-Cohen et al. 2009), FMO (Brixner et al. 2005; Engel et al. 2007, Panitchayangkoon et al. 2010), and phycobiliproteins (Collini et al. 2010, Turner et al. 2012). Here we briefly summarize our recent results on two different light-harvesting complexes: FMO (Engel et al. 2007, Panitchayangkoon et al. 2010) and PC645 (Collini et al. 2010, Turner et al. 2012), illustrating how electronic coherences (superpositions between different electronic states) are observed by 2DES.

19.4.1 Fenna–Matthews–Olson complex

The FMO is among the most well-studied models of energy transfer. Found in many green sulfur bacteria, this complex acts as a spacer between the light-harvesting antenna and the reaction center. The complex is effectively a little "excitonic wire" that conducts excitations from the chlorosome to the reaction center while allowing sufficient exposure to the protoplasm to ensure that reductants diffuse in to replenish electrons to the reaction center (Blankenship 2002, van Amerongen et al. 2000).

The FMO complex is an ideal model system to study energy transfer (Brixner et al. 2005, Engel et al. 2007, Ishizaki and Fleming 2009a, Mueh et al. 2007, Savikhin et al. 1997b). It contains eight bacteriochlorophyll a molecules, of which seven are strongly coupled and strongly bound (Tronrud et al. 2009). (The eighth bacteriochlorophyll is solvent exposed and would press up against the chlorosome antenna in nature.) The x-ray crystal structure of the complex is well resolved (Camara-Artigas et al. 2003, Li et al. 1997). The complex itself is robust and water soluble. The complex is asymmetric, allowing definitive identification of

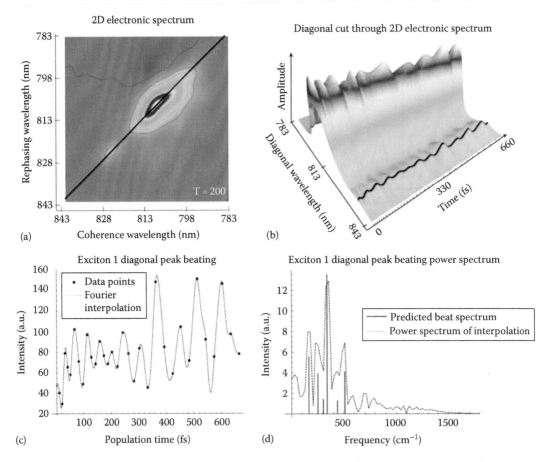

Figure 19.4 Cuts along the diagonal of a time series of 2D spectra of FMO at 77K **(a)** show clear evidence of non-exponential decay **(b)**. The beating pattern at the location of exciton 1 in the spectrum **(c)** is consistent with electronic coherence among excitons **(d)**. This evidence suggested that coherence among excited states persists for at least 660 fs while estimates of the dephasing time from homogeneous linewidth (observable in **(a)**) indicate that dephasing occurs between the ground and excited state within 70 fs. (Reprinted from Engel, G. et al., *Nature*, 446, 782, 2007.)

all excited states. Thus, the seven strongly coupled bacteriochlorophylls provide a wonderful system for both spectroscopic and theoretical studies.

In 2007, using 2DES, we reported the first direct observations of wavelike energy transfer based on long-lived coherent beating among electronic states in FMO (Engel et al. 2007). Similar signals had been observed using pump-probe anisotropy previously (Savikhin et al. 1997a,b), but the importance of these signals to energy transfer was not broadly discussed at that time. In our experiments, we observed coherent oscillations among excited states for over 660 fs at 77 K (Figure 19.4). This result was surprising because we expected to see the coherence among excited states dephase on the same time scale as excited state coherences

with the ground-state dephase (70 fs). What made these results so interesting is that they pointed to a very unusual and structured relationship between the system and its bath. The energy gap between excited states and the ground state appeared to fluctuate, but the energy gaps among excited states did not. Of course, for energy transfer, only the gaps between excited states matter. However, for spectroscopy (the tool used to understand energy transfer), the gaps between the ground and excited states are typically measured.

In subsequent experiments in 2010, we observed that this same phenomenon extended all the way to physiological temperatures (277 K) (Panitchayangkoon et al. 2010), that is, the energy gaps among excited states dephase approximately

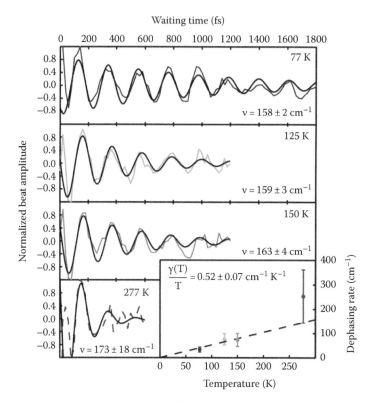

Figure 19.5 Beating patterns evident in cross peaks of FMO spectra at a variety of temperatures show clear signs of decaying quantum beats at a frequency consistent with the electronic spacing between excitons 1 and 3. At physiological temperature (277 K), the dephasing rate is approximately 250 cm⁻¹ showing that coherence persists for hundreds of femtoseconds—the same time scales as energy transfer within the complex. (Reprinted from Panitchayangkoon, G. et al., *Proc. Natl. Acad. Sci. USA*, 107, 12766, 2010.)

20 times slower than the energy gaps between the ground and excited states. Thus, coherences among excited states persist to about 300 fs—long enough to matter for energy transfer steps on the same time scale (Figure 19.5). Microscopically, this means that fluctuations in the environment surrounding the chromophores affect all the excited states in a similar, correlated fashion. This observation radically changes the picture of energy transfer from an incoherent hopping picture to one that includes a coherent component.

These observations have spawned many theoretical works examining the dynamics in FMO and attempting to understand the system-bath coupling in detail. It is this coupling that governs all the dynamics, and the coherences provide new mechanisms to understand how the bright states interact with the bath surrounding them and how that bath affects the relationships between the states. The coherences in FMO appear to be predominantly electronic in nature (Fransted et al. 2012), that is, they involve the states that shuttle the electronic excitations to the reaction center. The coherences prove to be robust to isotopic substitution that would scramble vibrational coherences (Hayes et al. 2011). However, some vibronic contribution must exist—in fact, it is completely reasonable to define the "system" to include those coupled bath motions to give a vibronic system.

19.4.2 Cryptophyte algae

In 2010, we reported oscillations found at ambient temperature by performing 2DES experiments on cryptophyte algae antenna complexes (Collini et al. 2010). The linear absorption spectrum along with the structure of one of the cryptophyte (*Chroomonas* sp.) antenna complexes, known as PC645, is displayed in Figure 19.6a and b. There are eight chromophores contributing to the linear electronic absorption spectrum, and the electronic transition frequencies associated with the system are indicated as sticks in the spectrum. The 2D spectra at three different waiting times are displayed in Figure 19.6d through f, with the crosspeak between the excitonically coupled states indicated with an arrow. The waiting time–dependent amplitude of the crosspeak is plotted in Figure 19.6c where it can be seen that oscillations persist for hundreds of femtoseconds.

Identifying the electronic coherences has been a challenge because molecular spectroscopy involves electronic excitations as well as excitation of vibrational energy levels of the molecules. Owing to the "vibronic" nature of the spectroscopy of molecules, upon excitation with a short optical laser pulse (broad spectral bandwidth) vibrational wave packets, coherent superpositions between vibrational levels, in addition to possible electronic superpositions, can be prepared (Bitto and Huber 1992, Heller 1981, Jonas and Fleming 1995, Zewail 2000). Oftentimes, these vibrational frequencies are quite similar in magnitude to the difference in energy between the excitonic states, and one of the challenges in 2DES is assigning the waiting time–dependent oscillations to vibrational and electronic coherences. It turns out that more detailed analysis of 2DES can discriminate electronic and vibrational coherences (Butkus et al. 2012, Novoderezhkin et al. 2015, Turner et al. 2011). On that basis, it was concluded that a long-lived electronic coherence (having a dephasing time of 170 fs) contributes

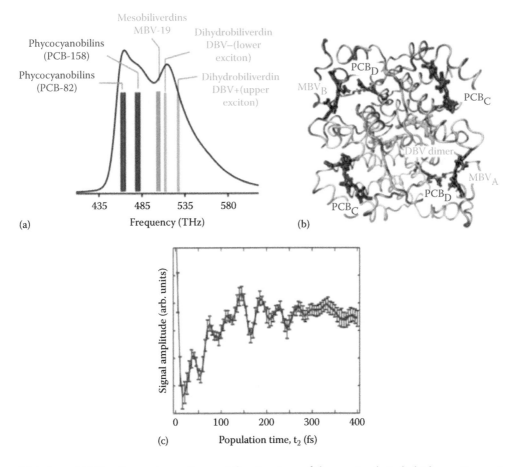

Figure 19.6 **(a and b)** The linear absorption and the structure of the cryptophyte light-harvesting antenna complex PC645 from *Chroomonas* sp. Strain CCMP270 are shown. **(c)** The waiting time dependent amplitude of the crosspeak (indicated with an arrow in the 2D electronic spectra) is shown. The electronic coherence, determined from analyzing different contributions tp the 2DES spectra, was found to dephase with a time constant of 170 fs. *(Continued)*

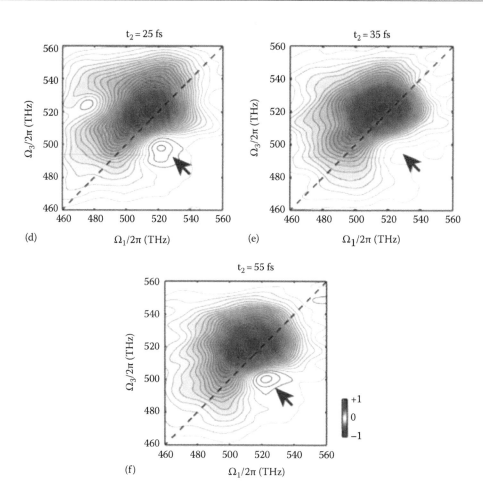

Figure 19.6 (*Continued*) **(d–f)** 2D electronic spectra at three different waiting times recorded at ambient temperature. The appearance and disappearance of the crosspeak is indicated with an arrow. (Adapted from Collini, E. et al., *Nature*, 463, U644, 2010; Turner, D.B. et al., *J. Phys. Chem. Lett.*, 2, 1904, 2011.)

to the oscillatory amplitude of the crosspeak in PC645. While 170 fs is incredibly short on the time scale of biology, it is comparable to the time scale of the most rapid energy transfer processes.

19.5 CONCLUSION

In this chapter we discussed the measurement of coherences in light-harvesting complexes. But, it can be difficult to appreciate why these properties observed via 2DES matter. Are these measured properties that are influenced by the incoming coherent light source relevant in nature? The main point is that the experiments expose unforeseen properties of the electronic structure of the antenna complex, specifically those effects that require theoretical models beyond Förster theory. Biophysicists

are interested in these mechanistic aspects of the process of electronic energy transfer and how it causes excitation to move among the chromophores. Understanding how it works in detail will help to reveal differences between the various light-harvesting complexes found in nature. For example, there is a striking difference between the arrangements of molecules in the LH2 complex from purple bacteria compared to the LHCII complex from higher plants (Figure 19.2). Why? Does one structure function more efficiently than the other? Answering such questions will, in turn, yield guidance for the design of artificial systems for energy conversion, sensing, or even processing of excitation energy.

It is possible that interplay between coherence and appropriate vibrational levels of the chromophores could be beneficial for rapid energy jumps (Chin et al. 2013, Kolli et al. 2012, Novoderezhkin et al. 2015, Pullerits et al. 2013, Tiwari et al. 2013). One qualitative illustration of the possible role of the protein matrix in ultrafast energy and electron transfer can be considered as follows. In a disordered energy landscape, an excitation could move around from one site to another, but since the roughness of the energy landscape has a molecular length scale, the localized excitation might easily be trapped and be unsuccessful in reaching the reaction center. Suppose that one specific protein vibrational mode is resonant with the difference between the electronic energy levels, a mode/deformation that steers the excitation toward the reaction center. Then the roughness of the energy landscape would be overcome. It would be like kicking a football on a rough playground and at the moment you kick the ball a channel opens that moves coherently with the ball to "guide" it toward its goal. The interaction between the ball (the excitation) and the playground (the protein) must be "coherent" in the sense that they travel together, and this would of course dramatically increase the chance of scoring a goal (be trapped by the reaction center). In this way, the long-lived vibrational coherences observed in photosynthetic light-harvesting systems can dramatically affect the efficiency by which reaction centers trap the excitations of the surrounding 200 chlorophylls. The challenge is to discover these vibrational modes and how nature has achieved to select them and suppress others.

This paper leads to the key question: Coherence is detected in femtosecond laser experiments used to examine photoinduced processes in various light-harvesting complexes, but is it exploited in biological function? Owing to the strong interactions between chromophores and the environment, as well as the fact the antenna complexes are inherently complex (there are many pathways through space the excitation can traverse on its way to the reaction center), quantum coherent effects should be lost over long length and time scales, yet they can have a significant influence initiating ultrafast processes. The role of coherence appears to be the way it interplays with incoherent transport rather than dominating how excitation flows (Chang and Cheng 2012, Hoyer et al. 2012). These kinds of arguments indicate that we should not think of quantum coherence as providing a smoother and more efficient passage of excitation energy through the antenna complex. Instead,

we should enquire how coherence on short length and time scales might seed some kind of property or function of the system that is not itself quantum in nature. We have in mind here *emergent phenomena* (Anders and Wiesner 2011). Researchers acquainted with that field will appreciate how difficult it will be to unravel biological functions to expose the roles played by quantum effects—and we see this as one of the next big challenges. The difficulty in pinpointing such an effect lies in the immense complexity of biological systems.

REFERENCES

Anders J, Wiesner K. 2011. Increasing complexity with quantum physics. Chaos: An Interdisciplinary. *Journal of Nonlinear Science* 21: 037102–037109.

Andrews DL. 1989. A unified theory of radiative and radiationless molecular-energy transfer. *Chemical Physics* 135: 195–201.

Anna JM. 2014. A little coherence in photosynthetic light harvesting. *Bioscience* 64(1): 14–25 by permission of Oxford University Press.

Anna JM, Ostroumov EE, Maghlaoui K, Barber J, Scholes GD. 2012. Two-dimensional electronic spectroscopy reveals ultrafast downhill energy transfer in photosystem I trimers of the cyanobacterium *Thermosynechococcus elongatus*. *The Journal of Physical Chemistry Letters* 3(24): 3677–3684.

Arnett DC, Moser CC, Dutton PL, Scherer NF. 1999. The first events in photosynthesis: Electronic coupling and energy transfer dynamics in the photosynthetic reaction center from *Rhodobacter sphaeroides*. *The Journal of Physical Chemistry B* 103: 2014–2032.

Bitto H, Huber JR. 1992. Molecular quantum beats—High-resolution spectroscopy in the time domain. *Accounts of Chemical Research* 25: 65–71.

Blankenship RE. 2002. *Molecular Mechanisms of Photosynthesis*. Malden, MA: Blackwell Science Ltd.

Brixner T, Stenger J, Vaswani H, Cho M, Blankenship R, Fleming G. 2005. Two-dimensional spectroscopy of electronic couplings in photosynthesis. *Nature* 434: 625–628.

Butkus V, Zigmantas D, Valkūnas L, Abramavičius D. 2012. Vibrational vs. electronic coherences in 2D spectrum of molecular systems. *Chemical Physics Letters* 545: 40–43.

Calhoun TR, Ginsberg NS, Schlau-Cohen GS, Cheng YC, Ballottari M, Bassi R, Fleming GR. 2009. Quantum coherence enabled determination of the energy landscape in light-harvesting complex II. *Journal of Physical Chemistry B* 113: 16291–16295.

Camara-Artigas A, Blankenship RE, Allen JP. 2003. The structure of the FMO protein from Chlorobium tepidum at 2.2 angstrom resolution. *Photosynthesis Research* 75: 49–55.

Caruso F, Chin AW, Datta A, Huelga SF, Plenio MB. 2009. Highly efficient energy excitation transfer in light-harvesting complexes: The fundamental role of noise-assisted transport. *Journal of Chemical Physics* 131: 105106-1–105106-15.

Chachisvilis M, Pullerits T, Jones MR, Hunter CN, Sundström V. 1994. Vibrational dynamics in the light-harvesting complexes of the photosynthetic bacterium *Rhodobacter sphaeroides*. *Chemical Physics Letters* 224: 345–351.

Chang H-T, Cheng Y-C. 2012. Coherent versus incoherent excitation energy transfer in molecular systems. *The Journal of Chemical Physics* 137: 165103.

Cheng Y-C, Fleming GR. 2009. Dynamics of light harvesting in photosynthesis. *Annual Review of Physical Chemistry* 60: 241–262.

Chin AW, Prior J, Rosenbach R, Caycedo-Soler F, Huelga SF, Plenio MB. 2013. The role of non-equilibrium vibrational structures in electronic coherence and recoherence in pigment-protein complexes. *Nature Physics* 9: 113–118.

Cho M. 2008. Coherent two-dimensional optical spectroscopy. *Chemical Reviews* 108: 1331–1418.

Cho M. 2009. *Two-Dimensional Optical Spectroscopy*. Boca Raton, FL: CRC Press.

Collini E, Wong CY, Wilk KE, Curmi PMG, Brumer P, Scholes GD. 2010. Coherently wired light-harvesting in photosynthetic marine algae at ambient temperature. *Nature* 463: U644–U669.

Dostál J, Mančal T, Augulis R, Vácha F, Pšenčik J, Zigmantas D. 2012. Two-dimensional electronic spectroscopy reveals ultrafast energy diffusion in chlorosomes. *Journal of the American Chemical Society* 134: 11611–11617.

Engel G, Calhoun T, Read E, Ahn T, Mancal T, Cheng Y, Blankenship R, Fleming G. 2007. Evidence for wavelike energy transfer through quantum coherence in photosynthetic systems. *Nature* 446: 782–786.

Fassioli F, Dinshaw R, Arpin PC, Scholes GD. 2014. Photosynthetic light harvesting: Excitons and coherence. *Journal of the Royal Society Interface* 11: 20130901.

Ferretti M, Hendrikx R, Romero E, Southall J, Cogdell RJ, Novoderezhkin VI, Scholes GD, Van Grondelle R. 2016. Dark states in the light-harvesting complex 2 revealed by two-dimensional electronic spectroscopy. *Scientific Reports* 6: 20834.

Ferretti M, Novoderezhkin VI, Romero E, Augulis R, Pandit A, Zigmantas D, Rv G. 2014. The nature of coherences in the B820 bacteriochlorophyll dimer revealed by two-dimensional electronic spectroscopy. *Physical Chemistry Chemical Physics* 16: 9930–9939.

Fleming GR, van Grondelle R. 1997. Femtosecond spectroscopy of photosynthetic light-harvesting systems. *Current Opinion in Structural Biology* 7: 738–748.

Förster T. 1965. Delocalized excitation and excitation transfer, pp. 93–137, in *Modern Quantum Chemistry. Part III*, Sinanoglu O, ed. New York: Academic Press.

Fransted KA, Caram JR, Hayes D, Engel GS. 2012. Two-dimensional electronic spectroscopy of bacteriochlorophyll a in solution: Elucidating the coherence dynamics of the Fenna-Matthews-Olson complex using its chromophore as a control. *The Journal of Chemical Physics* 137: 125101-1–125101-8.

Frauenfelder H, Wolynes PG. 1985. Rate theories and puzzles of hemeprotein kinetics. *Science* 229: 337–345.

Fuller FD, Pan J, Gelzinis A, Butkus V, Senlik SS, Wilcox DE, Yocum CF, Valkunas L, Abramavicius D, Ogilvie JP. 2014. Vibronic coherence in oxygenic photosynthesis. *Nature Chemistry* 6: 706–711.

Golonzka O, Tokmakoff A. 2001. Polarization-selective third-order spectroscopy of coupled vibronic states. *Journal of Chemical Physics* 115: 297–309.

Hamm P, Zanni M. 2011. *Concepts and Methods of 2D Infrared Spectroscopy*. New York: Cambridge University Press.

Harel E, Engel GS. 2012. Quantum coherence spectroscopy reveals complex dynamics in bacterial light-harvesting complex 2 (LH2). *Proceedings of the National Academy of Sciences of the United States of America* 109: 706–711.

Hayes D, Wen J, Panitchayangkoon G, Blankenship RE, Engel GS. 2011. Robustness of electronic coherence in the Fenna-Matthews-Olson complex to vibronic and structural modifications. *Faraday Discussions* 150: 459–469.

Heller EJ. 1981. The semi-classical way to molecular-spectroscopy. *Accounts of Chemical Research* 14: 368–375.

Hildner R, Brinks D, Nieder JB, Cogdell RJ, van Hulst NF. 2013. Quantum coherent energy transfer over varying pathways in single light-harvesting complexes. *Science* 340: 1448–1451.

Hoyer S, Ishizaki A, Whaley KB. 2012. Spatial propagation of excitonic coherence enables ratcheted energy transfer. *Physical Review E* 86: 041911.

Hoyer S, Sarovar M, Whaley KB. 2010. Limits of quantum speedup in photosynthetic light harvesting. *New Journal of Physics* 12: 065041.

Hu XC, Ritz T, Damjanovic A, Schulten K. 1997. Pigment organization and transfer of electronic excitation in the photosynthetic unit of purple bacteria. *Journal of Physical Chemistry B* 101: 3854–3871.

Ishizaki A, Fleming GR. 2009a. Theoretical examination of quantum coherence in a photosynthetic system at physiological temperature. *Proceedings of the National Academy of Sciences of the United States of America* 106: 17255–17260.

Ishizaki A, Fleming GR. 2009b. Unified treatment of quantum coherent and incoherent hopping dynamics in electronic energy transfer: Reduced hierarchy equation approach. *The Journal of Chemical Physics* 130: 234111-1–234111-10.

Ishizaki A, Tanimura Y. 2005. Quantum dynamics of system strongly coupled to low-temperature colored noise bath: Reduced hierarchy equations approach. *Journal of the Physical Society of Japan* 74: 3131–3134.

Jonas D. 2003. Two-dimensional femtosecond spectroscopy. *Annual Review of Physical Chemistry* 54: 425–463.

Jonas D, Fleming GR. 1995. *Vibrationally Abrupt Pulses in Pump-Probe Spectroscopy*. Oxford, U.K.: Blackwell Science.

Kolli A, O'Reilly EJ, Scholes GD, Olaya-Castro A. 2012. The fundamental role of quantized vibrations in coherent light harvesting by cryptophyte algae. *The Journal of Chemical Physics* 137: 174109–174115.

Krueger BP, Scholes GD, Fleming GR. 1998. Calculation of couplings and energy transfer pathways between the pigments of LH2 by the ab initio transition density cube method. *Journal of Physical Chemistry B* 102: 5378–5386.

Lee H, Cheng YC, Fleming GR. 2007. Coherence dynamics in photosynthesis: Protein protection of excitonic coherence. *Science* 316: 1462–1465.

Lewis KLM, Ogilvie JP. 2012. Probing photosynthetic energy and charge transfer with two-dimensional electronic spectroscopy. *Journal of Physical Chemistry Letters* 3: 503–510.

Li YF, Zhou WL, Blankenship RE, Allen JP. 1997. Crystal structure of the bacteriochlorophyll a protein from *Chlorobium tepidum*. *Journal of Molecular Biology* 271: 456–471.

Martin JL, Vos MH. 1992. Femtosecond biology. *Annual Review of Biophysics and Biomolecular Structure* 21: 199–222.

Miller WH. 2012. Perspective: Quantum or classical coherence? *The Journal of Chemical Physics* 136: 210901.

Mohseni M, Rebentrost P, Lloyd S, Aspuru-Guzik A. 2008. Environment-assisted quantum walks in photosynthetic energy transfer. *Journal of Chemical Physics* 129: 174106-1–174106-9.

Mohseni M, Shabani A, Lloyd S, Rabitz H. 2014. Energy-scales convergence for optimal and robust quantum transport in photosynthetic complexes. *The Journal of Chemical Physics* 140: 035102-1–035102-12.

Mueh F, Madjet ME-A, Adolphs J, Abdurahman A, Rabenstein B, Ishikita H, Knapp E-W, Renger T. 2007. Alpha-helices direct excitation energy flow in the Fenna-Matthews-Olson protein. *Proceedings of the National Academy of Sciences of the United States of America* 104: 16862–16867.

Mukamel S. 2000. Multidimensional femtosecond correlation spectroscopies of electronic and vibrational excitations. *Annual Review of Physical Chemistry* 51: 691–729.

Myers JA, Lewis KLM, Fuller FD, Tekavec PF, Yocum CF, Ogilvie JP. 2010. Two-dimensional electronic spectroscopy of the D1-D2-cyt b559 photosystem II reaction center complex. *Journal of Physical Chemistry Letters* 1: 2774–2780.

Novoderezhkin VI, van Grondelle R. 2010. Physical origins and models of energy transfer in photosynthetic light-harvesting. *Physical Chemistry Chemical Physics* 12: 7352–7365.

Novoderezhkin VI, Romero E, van Grondelle R. 2015. How exciton-vibrational coherences control charge separation in the photosystem II reaction center. *Physical Chemistry Chemical Physics* 17(46): 30828–30841.

Ogilvie JP, Kubarych KJ. 2009. Multidimensional electronic and vibrational spectroscopy: An ultrafast probe of molecular relaxation and reaction dynamics, pp. 249–321, in *Advances in Atomic, Molecular, and Optical Physics*, vol. 57, Arimondo E, Berman PR, Lin CC, eds., Academic Press, London, U.K.

Ostroumov EE, Mulvaney RM, Cogdell RJ, Scholes GD. 2013. Broadband 2D electronic spectroscopy reveals a carotenoid dark state in purple bacteria. *Science* 340: 52–56.

Panitchayangkoon G, Hayes D, Fransted KA, Caram JR, Harel E, Wen JZ, Blankenship RE, Engel GS. 2010. Long-lived quantum coherence in photosynthetic complexes at physiological temperature. *Proceedings of the National Academy of Sciences of the United States of America* 107: 12766–12770.

Prince SM, Papiz MZ, Freer AA, McDermott G, Hawthornthwaite-Lawless AM, Cogdell RJ, Isaacs NW. 1997. Apoprotein structure in the LH2 complex from Rhodopseudomonas acidophila strain 10050: Modular assembly and protein pigment interactions. *Journal of Molecular Biology* 268: 412–423.

Pullerits T, Zigmantas D, Sundström V. 2013. Beatings in electronic 2D spectroscopy suggest another role of vibrations in photosynthetic light harvesting. *Proceedings of the National Academy of Sciences of the United States of America* 110: 1148–1149.

Redfield A. 1957. On the theory of relaxation processes. *IBM Journal of Research Development* 1: 19–31.

Redfield AG. 1965. The theory of relaxation processes. *Advanced Magnetic Resonance* 1: 1–32.

Renger T. 2009. Theory of excitation energy transfer: From structure to function. *Photosynthesis Research* 102: 471–485.

Romero E, Augulis R, Novoderezhkin VI, Ferretti M, Thieme J, Zigmantas D, Van Grondelle R. 2014. Quantum coherence in photosynthesis for efficient solar-energy conversion. *Nature Physics* 10(9): 676–682.

Savikhin S, Buck DR, Struve WS. 1997a. Oscillating anisotropies in a bacteriochlorophyll protein: Evidence for quantum beating between exciton levels. *Chemical Physics* 223: 303–312.

Savikhin S, Buck DR, Struve WS. 1997b. Pump-probe anisotropies of Fenna-Matthews-Olson protein trimers from *Chlorobium tepidum*: A diagnostic for exciton localization? *Biophysical Journal* 73: 2090–2096.

Schlau-Cohen GS, Calhoun TR, Ginsberg NS, Read EL, Ballottari M, Bassi R, van Grondelle R, Fleming GR. 2009. Pathways of energy flow in lhcii from two-dimensional electronic spectroscopy. *Journal of Physical Chemistry B* 113: 15352–15363.

Schlau-Cohen GS, Dawlaty JM, Fleming GR. 2012. Ultrafast multidimensional spectroscopy: Principles and applications to photosynthetic systems. *IEEE Journal of Selected Topics in Quantum Electronics* 18: 283–295.

Schlau-Cohen GS, De Re E, Cogdell RJ, Fleming GR. 2012. Determination of excited-state energies and dynamics in the b band of the bacterial reaction center with 2D electronic spectroscopy. *Journal of Physical Chemistry Letters* 3: 2487–2492.

Schlau-Cohen GS, Ishizaki A, Fleming GR. 2011. Two-dimensional electronic spectroscopy and photosynthesis: Fundamentals and applications to photosynthetic light-harvesting. *Chemical Physics* 386: 1–22.

Scholes GD. 2003. Long-range resonance energy transfer in molecular systems. *Annual Review of Physical Chemistry* 54: 57–87.

Scholes GD. 2010. Quantum-coherent electronic energy transfer: Did nature think of it first? *Journal of Physical Chemistry Letters* 1: 2–8.

Scholes GD, Mirkovic T, Turner DB, Fassioli F, Buchleitner A. 2012. Solar light harvesting by energy transfer: From ecology to coherence. *Energy & Environmental Science* 5: 9374–9393.

Scholes GD, Rumbles G. 2006. Excitons in nanoscale systems. *Nature Materials* 5: 683–696.

Schrödinger E. 1944. *What Is Life?* Cambridge, U.K.: Cambridge University Press.

Spano FC. 2010. The spectral signatures of Frenkel polarons in H- and J-aggregates. *Accounts of Chemical Research* 43: 429–439.

Standfuss R, van Scheltinga ACT, Lamborghini M, Kuhlbrandt W. 2005. Mechanisms of photoprotection and nonphotochemical quenching in pea light-harvesting complex at 2.5 A resolution. *EMBO Journal* 24: 919–928.

Stock G, Thoss M. 1997. Semiclassical description of nonadiabatic quantum dynamics. *Physical Review Letters* 78: 578.

Sundström V. 2000. Light in elementary biological reactions. *Progress in Quantum Electronics* 24: 187–238.

Tanimura Y. 2014. Reduced hierarchical equations of motion in real and imaginary time: Correlated initial states and thermodynamic quantities. *The Journal of Chemical Physics* 141: 044114-1–044114-13.

Tanimura Y, Kubo R. 1989. Time evolution of a quantum system in contact with a nearly Gaussian-Markoffian noise bath. *Journal of the Physical Society of Japan* 58: 101–114.

Tiwari V, Peters WK, Jonas DM. 2013. Electronic resonance with anticorrelated pigment vibrations drives photosynthetic energy transfer outside the adiabatic framework. *Proceedings of the National Academy of Sciences of the United States of America* 110: 1203–1208.

Tronrud DE, Wen J, Gay L, Blankenship RE. 2009. The structural basis for the difference in absorbance spectra for the FMO antenna protein from various green sulfur bacteria. *Photosynthesis Research* 100: 79–87.

Turner DB, Dinshaw R, Lee KK, Belsley MS, Wilk KE, Curmi PMG, Scholes GD. 2012. Quantitative investigations of quantum coherence for a light-harvesting protein at conditions simulating photosynthesis. *Physical Chemistry Chemical Physics* 14: 4857–4874.

Turner DB, Wilk KE, Curmi PMG, Scholes GD. 2011. Comparison of electronic and vibrational coherence measured by two-dimensional electronic spectroscopy. *Journal of Physical Chemistry Letters* 2: 1904–1911.

van Amerongen H, Valkūnas L, van Grondelle R. 2000. *Photosynthetic Excitons*. Singapore: World Scientific Publishing Co. Pte. Ltd.

van Grondelle R, Novoderezhkin VI. 2006. Energy transfer in photosynthesis: Experimental insights and quantitative models. *Physical Chemistry Chemical Physics* 8: 793–807.

van Grondelle R, Dekker JP, Gillbro T, Sundström V. 1994. Energy-transfer and trapping in photosynthesis. *Biochimica Et Biophysica Acta-Bioenergetics* 1187: 1–65.

Vos MH, Lambry JC, Robles SJ, Youvan DC, Breton J, Martin JL. 1991. Direct observation of vibrational coherence in bacterial reaction centers using femtosecond absorption-spectroscopy. *Proceedings of the National Academy of Sciences of the United States of America* 88: 8885–8889.

Voth GA, Hochstrasser RM. 1996. Transition state dynamics and relaxation processes in solutions: A Frontier of physical chemistry. *Journal of Physical Chemistry* 100: 13034–13049.

Yang M, Damjanovic A, Vaswani HM, Fleming GR. 2003. Energy transfer in photosystem I of cyanobacteria *Synechococcus elongatus*: Model study with structure-based semi-empirical Hamiltonian and experimental spectral density. *Biophysical Journal* 85: 140–158.

Zewail AH. 2000. Femtochemistry: Atomic-scale dynamics of the chemical bond. *Journal of Physical Chemistry A* 104: 5660–5694.

Zigmantas D, Read EL, Mancal T, Brixner T, Gardiner AT, Cogdell RJ, Fleming GR. 2006. Two-dimensional electronic spectroscopy of the B800-B820 light-harvesting complex. *Proceedings of the National Academy of Sciences of the United States of America* 103: 12672–12677.

20

Systems biophysics: Global and target analysis of light harvesting and photochemical quenching *in vivo*

IVO VAN STOKKUM

ABBREVIATIONS

Chl	Chlorophyll
DAS	Decay-Associated Spectrum
EAS	Evolution-Associated Spectrum
EET	excitation energy transfer
FWHM	full width at half maximum
IRF	instrument response function
LHC	light-harvesting complex
MA	magic angle
$n\lambda$	number of wavelengths
nt	number of time points
PS	photosystem
RC	reaction center
SAS	Species-Associated Spectrum
SVD	singular value decomposition
TCSPT	time-correlated single photon timing

20.1 SUMMARY

Systems biophysics can describe the input to a photosynthetic system and thus contribute a building block to the systems biology of photosynthesis. Systems biophysics mathematically describes the processes during the light reactions of photosynthesis: light harvesting, energy transfer, photochemical quenching (charge separation), nonphotochemical quenching, and state transitions. The aim of *systems biophysics* is to develop mathematical models that describe the functioning of complex photosynthetic systems. Such models are based upon measured time-resolved absorption and emission spectra, which are two-dimensional data sets. Theoretical methods and software have been developed to identify the model and estimate the biophysical parameters that describe all data. These methods are termed global and target analysis. Key ingredients are compartmental, spectral, and thermodynamic models. The methods will be demonstrated with case studies of cyanobacterial light harvesting and photochemical quenching *in vivo*.

20.2 INTRODUCTION

Systems theory methods are widely used in biophysics to describe the functioning of systems with the help of mathematical models. Light absorbed by a light-harvesting complex provides the input to a photosynthetic system, for example, the thylakoid membrane. The final outputs of the thylakoid membrane are energized molecules: reduction equivalents (NADPH) and ATP. This chapter will be limited to light harvesting, energy transfer, and photochemical quenching (charge separation) in photosystems (van Grondelle et al. 1994; Croce and van Amerongen 2013; van Amerongen and Croce 2013).

Cyanobacteria are model systems for studies of photosynthesis. They can be mutated, which allows us to study the properties of subsystems *in vivo*. In particular, the BE and PAL mutant of *Synechocystis* (Ajlani and Vernotte 1998; Krumova et al. 2010; Tian et al. 2013) can be considered elementary photosynthetic systems. They lack the phycobilisome antenna. The BE mutant only contains photosystem I (PSI) complexes, whereas the PAL mutant contains both PSI and photosystem II (PSII) complexes.

Time-resolved spectroscopy has proven to be extremely useful in photosynthesis research in the past decades. Both absorption and emission spectroscopy have been employed in a variety of wavelength ranges. When the superposition principle applies, the system can be considered linear. After excitation by a short laser flash, the (excited state) dynamics of the system can be studied by measuring a time-resolved spectrum. In the language of linear systems theory, the time-resolved spectrum can be considered the response of the system, which is a convolution of the impulse response (the response after a unit impulse of negligible width) and the instrument response function (IRF) (van Stokkum et al. 2004).

This chapter will be limited to measurements of chlorophyll *a* (Chl *a*) emission from PSI or PSII at magic angle. For analysis of polarized light experiments where anisotropy comes into play and for transient absorption experiments the reader is referred to van Stokkum et al. (2004). Time-resolved emission spectra can be measured with low excitation intensities, thus avoiding nonlinear effects like annihilation, which can be problematic with transient absorption experiments (Muller et al. 1996).

20.3 FIRST DATA: THE TIME-RESOLVED EMISSION SPECTRUM OF BE CELLS

To monitor the fluorescence, a synchroscan streak camera in combination with a spectrograph is employed. It can simultaneously record the temporal dynamics and wavelength of fluorescence representable as an image with time and wavelength along the axes (van Stokkum et al. 2008). The time-resolved spectrum of BE cells at room temperature after 400 nm excitation is depicted in Figure 20.1. The emission evolves both temporally and spectrally: it rises due to the finite rise time of the instrument; the spectral shape evolves due to the equilibration between Chl pigments with different properties; and finally, it decays due to photochemical quenching (charge separation, trapping) in the reaction center (RC) (Gobets et al. 2001).

20.3.1 Usage of the singular value decomposition

The matrix structure of the streak data enables the usage of matrix decomposition techniques, in particular, the singular value decomposition (SVD) (Shrager 1986; Henry and Hofrichter 1992; Hendler and Shrager 1994). Formally, the data matrix can be decomposed as (Golub and Van Loan 1996)

$$\psi(t, \lambda) = \sum_{l=1}^{m} u_l(t) s_l w_l(\lambda)$$

where

u_l and w_l are the left and right singular vectors
s_l is the sorted singular value
m is the minimum of the number of rows and columns of the data matrix

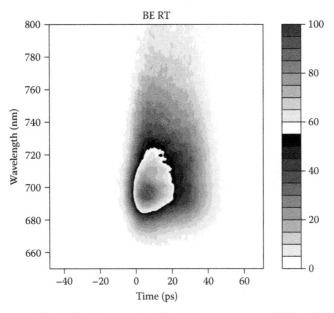

Figure 20.1 Filled contour plot of the time-resolved emission spectrum of BE cells (containing only PSI) at room temperature after 400 nm excitation. Time step $\approx$ 0.8 ps, wavelength step $\approx$ 2 nm.

The singular vectors are orthogonal and provide an optimal least squares approximation of the matrix. From the SVD, the rank of the data matrix can be estimated, as judged from the singular values and singular vector pairs significantly different from noise. This rank corresponds to the number of spectrally and temporally independent components n_{comp}. When the data matrix has not been corrected for dispersion, this is no longer strictly true, and then the singular values provide only an indication of n_{comp}. Furthermore, the SVD of the matrix of residuals is useful to diagnose shortcomings of the model used, or systematic errors in the data.

Figure 20.2a through c depicts the SVD of the PSI data, where three singular values and singular vector pairs are significantly different from noise. These first three singular values account for 96.5% of the variance of the data matrix. The left and right singular vectors are both linear combinations of the true concentration profiles and SAS and are hard to interpret. The first pair (squares) represents a kind of average. The SVD of the residual matrix will be discussed later.

20.4 GLOBAL AND TARGET ANALYSIS

The aim of global data analysis in general is to obtain a model-based description of the *full* data set in terms of a model containing a small number of precisely estimated parameters. Our main assumption here is that the time and wavelength properties of the system of interest are *separable*, which means that the spectra of species or states are independent of time. Analogously, the dynamics of species or states are assumed to be wavelength independent. When applicable, a wavelength dependence of the IRF location can be described parametrically (see Section 20.4.1). Thus, the adjective *global* (Beechem et al. 1985; Beechem 1989) indicates that the kinetic and spectral parameters of the model apply to the *full* data set. The adjective *target* (Arcioni and Zannoni 1984) refers to the targeted model.

A description of the basic ingredient of kinetic models, the exponential decay, will be given first, followed by a description of how to use these ingredients for global and target analysis (see, e.g., the reviews by Holzwarth (1996) and van Stokkum et al. (2004)) of the full data. The parameter estimation is generally based on nonlinear least squares. Because of the wealth of conditionally linear parameters, the variable projection algorithm is crucial (Golub and Pereyra 1973, 2003; Golub and LeVeque 1979; Nagle 1991; Mullen and van Stokkum 2009). For further details on parameter estimation techniques, the reader is also referred to the previously cited reviews and references cited therein and to van Stokkum (2005). Software

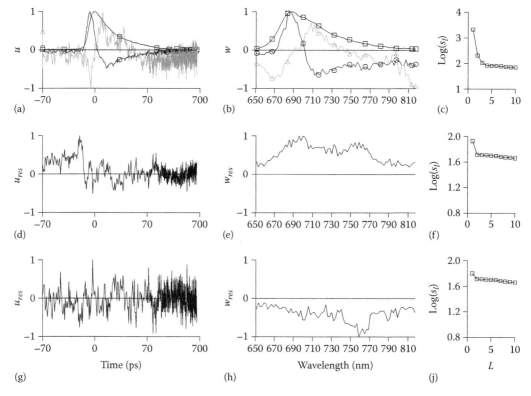

Figure 20.2 SVD of the PSI data matrix (a–c) and matrix of residuals (d–i). (a) First 3 (order squares, circles, triangles) left singular vectors u_l, (b) first 3 right singular vectors w_l, (c) first 10 singular values s_l on a logarithmic scale. Panels (d–f) represent the matrix of residuals from a fit using a simple Gaussian IRF: the first left singular vector $u_{res,1}$, the first right singular vector $w_{res,1}$, and the first 10 singular values $s_{res,l}$ on a logarithmic scale. Panels (g–i) represent the matrix of residuals from a fit using a triple Gaussian IRF. Further explanation in text.

issues are discussed in Snellenburg et al. (2012); van Stokkum and Bal (2006); Mullen and van Stokkum (2007).

20.4.1 Modeling an exponential decay

Here, an expression is derived for describing the contribution of an exponentially decaying component to the streak image. The IRF $i(t)$ can usually adequately be modeled with a Gaussian with parameters μ and Δ for, respectively, location and full width at half maximum (FWHM):

$$i(t) = \frac{1}{\tilde{\Delta}\sqrt{2\pi}} \exp\left(-\log(2)\left(2(t-\mu)/\Delta\right)^2\right)$$

where $\tilde{\Delta} = \Delta / \left(2\sqrt{2\log(2)}\right)$. The adequacy of the Gaussian approximation of the IRF shape is depicted in Figure 20.3a of van Stokkum et al. (2008).

The convolution (indicated by an $*$) of this IRF with an exponential decay (with rate k) yields an analytical expression, which facilitates the estimation of the IRF parameters μ and Δ:

$$c(t, k, \mu, \Delta) = \exp(-kt) * i(t)$$

$$= \frac{1}{2}\exp(-kt)\exp\left(k\left(\mu + \frac{k\tilde{\Delta}^2}{2}\right)\right)$$

$$\times \left\{1 + \mathrm{erf}\left(\frac{t - \left(\mu + k\tilde{\Delta}^2\right)}{\sqrt{2}\tilde{\Delta}}\right)\right\}$$

The periodicity of the synchroscan results in the detection of the fluorescence that remains after multiples of half the synchroscan period T (typically $T \approx 13$ ns). Therefore, if lifetimes longer than ≈ 1 ns occur in a sample, the said expression

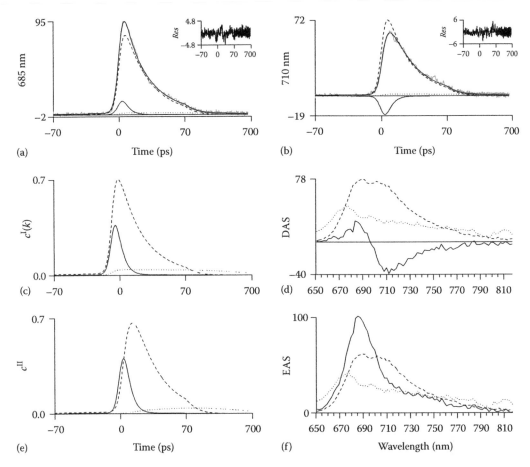

Figure 20.3 Results from global analysis of BE mutant (PSI) data depicted in Figure 20.1. Note that in a–c and e, the time axis is linear from –70 to +70 ps relative to the maximum of the IRF and logarithmic thereafter. Insets in a and b show residuals. (a) Data (in grey) and fit (in black) of 685 nm emission trace showing multiexponential decay. Contributions of the three exponential decays with different lifetimes (shown in c) multiplied by their amplitudes at 685 or 710 nm (shown in d) are indicated by line type in panels a and b. (b) Data (in grey) and fit (in black) of 710 nm emission trace showing multiexponential rise (visible as a contribution of $c^I(k_1)$ with negative amplitude) and decay. (c) Exponential decays $c^I(k_i)$. Estimated lifetimes (in panels c and e): 5.8 ps (solid), 26.4 ps (dashed), 0.59 ns (dotted). Note that in panels c and e, the dotted concentration profile was divided by 25. (d) Estimated Decay-Associated Spectra (DAS). Note that in panels d and f, the dotted spectrum was multiplied by 25 to increase its visibility. (e) Evolutionary concentration profiles c_l^{II} (assuming a sequential kinetic scheme with increasing lifetimes). (f) Estimated Evolution-Associated Spectra (EAS).

should be extended with a summation over the signal contributions that result from forward and backward sweeps:

$$c(t, k, T) = \sum_{n=0}^{\infty} e^{-kTn} \left\{ e^{-k(t-\mu+T)} + e^{-k(T/2-t-\mu)} \right\}$$

$$= \frac{\left\{ e^{-k(t-\mu+T)} + e^{-k(T/2-t-\mu)} \right\}}{\left(1 - e^{-kT} \right)}$$

Note that it is assumed here that time zero of the time base corresponds to the zero crossing of the sweep and that the convolution with the IRF is no longer necessary at times longer than $T/2$. Adding the previous expressions provides the full model function for an exponential decay recorded with a synchroscan streak camera and will henceforth be denoted by $c^I(k)$:

$$c^I(k) \equiv c(t, k, \mu, \Delta, T) = c(t, k, \mu, \Delta) + c(t, k, T)$$

Because fluorescence samples are relatively dilute, elastic scattering or Raman scattering of the excitation light by water (or of other solvents) can complicate the measurement if they occur within the analyzed wavelength interval. Such contributions can be modeled with an extra component with a time course identical to the IRF $i(t)$. Usually, it is possible to restrict the contribution of scattering to a limited wavelength region.

If the streak image has not been corrected for the instrumental curvature, the wavelength dependence of the IRF location μ can be modeled with a polynomial (usually a parabola is adequate). Sometimes, the IRF shape is better described by a superposition of two or even three Gaussians (with common location μ), leading to a superposition description of the exponential decay (van Stokkum, 2005). We will return to this later.

The IRF description given earlier for the synchroscan streak camera is fairly complicated. With time-resolved absorption spectroscopy, the Gaussian approximation of the IRF shape describes the pump-probe overlap rather well. With time-correlated single photon timing (TCSPT) the IRF can be measured via scatter (at the excitation wavelength) or with a fast-decaying reference compound (at the detection wavelength). In practice, the TCSPT IRF can be well approximated by a superposition of three or four Gaussians (with different locations μ).

20.4.2 The superposition principle

The basis of global analysis is the superposition principle, which states that the measured data $\psi(t, \lambda)$ result from a superposition of the spectral properties $\varepsilon_l(\lambda)$ of the components present in the system of interest weighted by their concentration $c_l(t)$:

$$\psi(t, \lambda) = \sum_{l=1}^{n_{comp}} c_l(t) \varepsilon_l(\lambda)$$

The $c_l(t)$ of all n_{comp} components are described by a compartmental model that consists of first-order differential equations, with sums of exponential decays as solution. We will consider three types of compartmental models: (1) a model with components decaying monoexponentially in parallel, which yields Decay-Associated Spectra (DAS); (2) a sequential model with increasing lifetimes, also called an unbranched unidirectional model,

giving rise to Evolution-Associated Spectra (EAS); and (3) a full compartmental scheme, which may include possible branchings and equilibria, yielding Species-Associated Spectra (SAS). The latter is most often referred to as target analysis, where the target is the proposed kinetic scheme, including possible spectral assumptions.

20.4.3 Components decaying in parallel: DAS

With components decaying in parallel, the model reads

$$\psi(t, \lambda) = \sum_{l=1}^{n_{comp}} c^l(k_l) DAS_l(\lambda)$$

The DAS thus represent the estimated amplitudes of the defined exponential decays $c^l(k_l)$.

Three components are sufficient to describe the data. The adequacy of the fit is judged from the SVD of the first left and right singular vectors of the matrix of residuals. With a simple Gaussian IRF shape, the residuals (shown in Figure 20.2d through f) show some structure, especially at times before the maximum of the IRF, and at all wavelengths where there is signal. This indicates that the IRF shape is more complicated. Therefore, we applied a triple Gaussian IRF shape with FWHM 9 ps (61% of the area), 15 ps (29%), and 204 ps (10%). We attribute these small fractions of wider IRF to shortcomings of the synchroscan. The trends are no longer present in Figure 20.2g and h. The rms error of the fit decreased from 1.154 to 1.136, which is just over 1% of the peak in Figure 20.1.

The estimated lifetimes (reciprocals of estimated rate constants k_l) are 5.8 ± 0.2 ps, 26.4 ± 0.2 ps, and 0.59 ± 0.02 ns.

The estimated DAS from the PSI data are shown in Figure 20.3d. Several observations can be made: the 5.8 ps DAS (solid) is nearly conservative, that is, the positive and negative areas are almost equal. It represents decay of more blue and rise of more red emission and can be interpreted as energy transfer from Bulk to Red Chl a, that is, Chl a that absorbs at wavelengths longer than the primary electron donor P700. Thus, *in vivo*, the excited state equilibration time constant of PSI is 5.8 ps (Krumova et al. 2010; Chukhutsina et al. 2013). The 26.4 ps DAS represents the trapping spectrum. The long-lived (0.59 ns) DAS (dotted) was multiplied by 25 to increase its visibility. It has a maximum near 670 nm and resembles the shape of a Chl a fluorescence spectrum. Most probably it is not connected to PSI. It could be the single Chl a present in cytochrome b6f (Peterman et al. 1998). Henceforth, it will be referred to as "free" Chl a. Note that in a steady-state spectrum its contribution will be as large as that of PSI, because of its long lifetime. Clearly, the first two DAS do not represent pure species, and they are interpreted as linear combinations (with positive and negative contributions) of true species spectra.

Note that the ultrafast ($\ll 1$ ps) rise due to the relaxation from the initially excited Soret state (higher excited state, of which the ≈ 460 nm emission is outside the detection range) to the Q_y emission (lowest excited state) cannot be resolved in this experiment due to the limited time resolution (the main FWHM Δ of the IRF was 9 ps).

20.4.4 Sequentially decaying components: EAS

The sequential model reads

$$\psi(t,\lambda) = \sum_{l=1}^{n_{comp}} c_l^{II} EAS_l(\lambda)$$

where each concentration is a linear combination of the exponential decays (Nagle et al. 1982), $c_l^{II} = \sum_{j=1}^{l} b_{jl} c^I(k_l)$, and the amplitudes b_{jl} are given by $b_{11} = 1$ and for $j \leq l$:

$$b_{jl} = \frac{\prod_{m=1}^{l-1} k_m}{\prod_{n=1,n\neq j}^{l}(k_n - k_j)}$$

Examples of c_l^{II} are depicted in Figure 20.3e, whereas the estimated EAS are shown in Figure 20.3f. With increasing lifetimes and thus decreasing rates k_j, the first EAS (equal to the sum of DAS) corresponds to the spectrum at time zero with an ideal infinitely small IRF, $i(t) = \delta(t)$. The first EAS (solid line in Figure 20.3f) represents the sum of the spectra of all excitations that have arrived from the Soret region and is dominated by Bulk Chl a. The second EAS, which is formed with a time constant of 5.8 ps and decays with a time constant of 26.4 ps,

contains more Red Chl a emission and represents a Boltzmann equilibrium of all PSI Chl a pigments. In this interpretation, we neglect the small contribution of the long-lived (0.59 ns) "free Chl a." When we collate the concentration profiles of the parallel and sequential model in matrices C^I and C^{II}, with nt rows and n_{comp} columns, respectively, these obey the matrix relation $C^{II} = C^I B$, with the elements of the $n_{comp} \times n_{comp}$ upper triangular B matrix defined earlier. Likewise, we form $n\lambda \times n_{comp}$ matrices DAS and EAS that contain the DAS_l and EAS_l of the components. In matrix form, the superposition model for the $nt \times n\lambda$ data matrix Ψ is expressed as $\Psi = C^I \cdot DAS^T$ (parallel) or $\Psi = C^{II} \cdot EAS^T$ (sequential). In global analysis (without any spectral constraints, see Section 20.4.5), the parallel and sequential model result in *exactly* the same residuals and *exactly* the same estimated lifetimes. The estimated concentration profiles obey $C^{II} = C^I B$, and therefore the estimated DAS and EAS are related as $C^{II} \cdot EAS^T = C^I B \cdot EAS^T = C^I \cdot DAS^T$, which leads to $B \cdot EAS^T = DAS^T$ or when transposed, $EAS \cdot B^T = DAS$ or $EAS = DAS \cdot B^{-T}$. The lower triangular matrix B^T can be inverted, yielding $b_{11}^{-1} = 1 = b_{11}^{-T}$ and

$$b_{jl}^{-1} = \prod_{n=1}^{j-1} \frac{(k_n - k_l)}{k_n} = b_{lj}^{-T}$$

For the three components of the global analysis, we can compute (inserting $k_1 = 0.173$, $k_2 = 0.038$, $k_3 = 0.0016$)

$$B = \begin{bmatrix} 1 & -k_1/(k_1-k_2) & k_1/(k_1-k_2)\cdot k_2/(k_1-k_3) \\ 0 & k_1/(k_1-k_2) & k_1/(k_1-k_2)\cdot k_2/(k_3-k_2) \\ 0 & 0 & k_1/(k_1-k_3)\cdot k_2/(k_2-k_3) \end{bmatrix}$$

$$= \begin{bmatrix} 1 & -1.28 & 0.28 \\ 0 & 1.28 & -1.34 \\ 0 & 0 & 1.06 \end{bmatrix}$$

$$B^{-T} = \begin{bmatrix} 1 & 0 & 0 \\ 1 & (k_1-k_2)/k_1 & 0 \\ 1 & (k_1-k_3)/k_1 & (k_1-k_3)/k_1\cdot(k_2-k_3)/k_2 \end{bmatrix}$$

$$= \begin{bmatrix} 1 & 0 & 0 \\ 1 & 0.78 & 0 \\ 1 & 0.99 & 0.95 \end{bmatrix}$$

This leads to the following relations for the DAS_l and EAS_l of the components:

$$EAS_1 = DAS_1 + DAS_2 + DAS_3$$

$$EAS_2 = 0.78DAS_2 + 0.99DAS_3$$

$$EAS_3 = 0.95DAS_3$$

$$DAS_1 = EAS_1 - 1.28EAS_2 + 0.28EAS_3$$

$$DAS_2 = 1.28EAS_2 - 1.34EAS_3$$

20.4.5 Target model: SAS

When neither of these two simple models is applicable, a full kinetic scheme may be appropriate. This is called a compartmental model. The problem with most kinetic schemes is that while the kinetics are described by more than n_{comp} microscopic rate constants (due to branchings or equilibria), the data only allows for the estimation of n_{comp} decay rates (or lifetimes). Then additional information is required to estimate the microscopic rates (Nagle 1991), which can be spectral constraints (zero contribution of SAS at certain wavelengths) or spectral relations. This is detailed in van Stokkum et al. (2004).

Now the model reads

$$\psi(t, \lambda) = \sum_{l=1}^{n_{comp}} c_l^{III} SAS_l(\lambda)$$

Disregarding the small contribution of the long-lived (0.59 ns) "free Chl a," we can describe the PSI dynamics by the following two coupled differential equations (omitting the superscript III):

$$\frac{d}{dt}\begin{bmatrix} c_1(t) \\ c_2(t) \end{bmatrix}$$

$$= \begin{bmatrix} -(k_{01} - k_{21}) & k_{12} \\ k_{21} & -(k_{02} - k_{12}) \end{bmatrix}\begin{bmatrix} c_1(t) \\ c_2(t) \end{bmatrix} + \begin{bmatrix} x_1 \\ x_2 \end{bmatrix}i(t)$$

where the indices 1 and 2 refer to Bulk and Red Chl a, respectively. It is known that a PSI monomer contains about six Red Chl a and 90 Bulk Chl a; thus, the inputs to the system are $x_1 = 90/96$ and $x_2 = 6/96$. There are four microscopic rate constants, k_{21} and k_{12} representing the equilibration, k_{01} representing the photochemical quenching (trapping), and

k_{02} representing the natural lifetime (k_f) of a Chl a. Typically, $k_f \approx 1/(2ns)$, but this means that still three microscopic rate constants need to be estimated, whereas only two lifetimes and two SAS_l can be estimated. Thus, spectral constraints imposed on the two SAS_l are needed. When modeling emission data, a natural spectral constraint is nonnegativity of the SAS_l (Mullen and van Stokkum 2009). However, this constraint is still not enough. A strong spectral constraint is $SAS_2 = 0$ in an appropriate wavelength range (typically below 685 nm for Red Chl a). This strong constraint is enough to estimate the equilibrium between the two states (Muller et al. 1992; Gobets et al. 2001). A mild spectral constraint is to assume that the area under the SAS_l is equal for states of equal oscillator strength like Bulk and Red Chl a (Snellenburg et al. 2013). This mild constraint is also enough to estimate the equilibrium between the two states and is especially useful when a strong zero constraint cannot be applied. Formally, when the ratio between the oscillator strengths of two Chl species with SAS_i and SAS_j is assumed to be α, a penalty can be imposed, which is added to the least squares criterion of the fit:

$$penalty_{ij} = weight_{ij} \int_{\lambda min}^{\lambda max} |SAS_i(\lambda) - \alpha SAS_j(\lambda)| d\lambda$$

Without a priori knowledge, we assume that all Chl species possess the same oscillator strength, that is, $\alpha = 1$. Here, $weight_{ij}$ is used to tune the importance of this area constraint in the least squares fitting process.

20.4.5.1 SOLVING THE COMPARTMENTAL MODEL

The general compartmental model with p compartments can be abbreviated as $\frac{d}{dt}c(t) = Kc(t) + j(t)$ with $c(t) = [c_1(t) \ldots c_p(t)]^T$, $j(t) = i(t)x; x = [x_1 \ldots x_p]^T$ is a vector that describes the amount of excitation of each compartment. In this example, the concentration vector is $c(t) = [c_1(t) \quad c_2(t)]^T$ and the input vector is $j(t) = i(t)[x_1 \quad x_2]^T$. The transfer matrix K contains off-diagonal elements k_{ij}, representing the microscopic rate constant from compartment j to compartment i. k_{0j} indicates decay from compartment j to the outside of the system described.

The diagonal elements contain the total decay rates of each compartment. To solve a compartmental model, we employ the eigendecomposition of K given by $U\Lambda U^{-1}$ with $\Lambda = diag(-\kappa_1, -\kappa_2, \ldots, -\kappa_p)$, the diagonal matrix containing the eigenvalues $-\kappa_l$ of K, and U, the matrix with the eigenvectors as columns (check: $KU = U\Lambda U^{-1}U = U\Lambda$, thus each eigenvector is scaled by its eigenvalue). Thus, we have $e^{Kt} = Ue^{\Lambda t}U^{-1}$ $\left(\text{since } \exp(Kt) = \sum_{n=0}^{\infty}(Kt)^n/n! \text{ and} \right.$ $K^n = U\Lambda U^{-1} \ldots U\Lambda U^{-1} = U\Lambda^n U^{-1}$, thus $\sum_{n=0}^{\infty}(Kt)^n/n! =$ $\left. U\left(\sum_{n=0}^{\infty}(Kt)^n/n!\right)U^{-1} = U\exp(\Lambda t)U^{-1}\right)$ and

$$c^{\mathrm{III}}(t) = e^{Kt} * j(t)$$

$$= U\mathrm{diag}(U^{-1}x)\begin{bmatrix} e^{-\kappa_1 t} * i(t) \\ \ldots \\ e^{-\kappa_p t} * i(t) \end{bmatrix}$$

$$= U\mathrm{diag}(U^{-1}x)\begin{bmatrix} c^{\mathrm{I}}(\kappa_1) \\ \ldots \\ c^{\mathrm{I}}(\kappa_p) \end{bmatrix}$$

Thus, the solution of the general compartmental model is a linear combination of $c^{\mathrm{I}}(\kappa_l)$. The concentration of each compartment is a linear combination of at most p exponential decays (convolved with the IRF), with coefficients that depend upon the microscopic rate constants that describe the transitions between all the compartments.

20.4.5.2 DETAILED BALANCE

When two compartments, i and j, are at equilibrium, we have

$$k_{ij}c_j(t) = k_{ji}c_i(t)$$

and we can compute the Gibbs free energy difference between compartments i and j as

$$\Delta G_{ij} = k_B T \ln\left(\frac{k_{ji}}{k_{ij}}\right)$$

For a closed system at equilibrium, the sum of the Gibbs free energy differences of a cycle equals zero.

Thus, for example, with three compartments, the product of the clockwise rate constants $1\to2\to3\to1$ is equal to the product of the counterclockwise rate constants $1\leftarrow2\leftarrow3\leftarrow1$:

$$k_{21}k_{32}k_{13} = k_{12}k_{31}k_{23}$$

This is called a detailed balance condition. It must be fulfilled and reduces the number of unknown microscopic rate constants (Nagle 1991).

20.4.5.3 TARGET ANALYSIS OF THE FIRST DATA

The kinetic scheme used for the target analysis of the BE PSI data is depicted in Figure 20.4a. As explained earlier, three microscopic rates, the forward and backward rate constants between the Bulk (B) and Red (R) Chl a compartment and the rate of photochemical quenching from B, were estimated with the help of the mild spectral constraint that the area under the SAS_l is equal for states B and R. The shapes of the estimated SAS shown in Figure 20.4c are realistic. Very small amplitudes of SAS_R are present below 690 nm.

The free energy difference between two compartments depends upon the energy difference between the two emission maxima and upon the relative number of pigments of the compartments ($k_B T \ln(N_1/N_2)$). We estimate the energetic contribution from the difference between the average emission wavelengths of the B and R SAS, 703.6 and 722.7 nm, for B* and R*, respectively. This corresponds to 47 meV. The observed free energy difference is $\Delta G_{12} = k_B T\ln(k_{21}/k_{12}) = k_B T\ln(61/126) = -18$ meV. The entropic free energy difference thus equals $47 + 18 = 65$ meV, which is equal to $k_B T\ln(N_1/N_2)$, with N_1 and N_2, the number of pigments in the B and R compartments. Using $k_B T = 25$ meV, we thus find $N_1/N_2 = 13.1$, which is consistent with N_1 and N_2 comprising ≈ 92 and ≈ 7 pigments, respectively. This is very close to the assumption that the PSI monomer contains about six Red Chl a and 90 Bulk Chl a.

The eigenvalues of the K matrix correspond to lifetimes of 4.7 and 28 ps and 0.68 ns. These lifetimes differ slightly from those estimated in the global analysis (5.6 and 26.4 ps and a very small contribution of 0.59 ns). The difference is attributed to the limited signal to noise ratio of the data. Apparently, the relative precision of the estimated lifetimes is typically about 10%. The quality of the fit (indicated by rms errors and

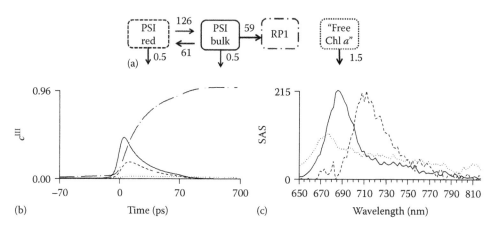

Figure 20.4 **(a)** Kinetic scheme used for the target analysis of the BE PSI data depicted in Figure 20.1. After excitation in the Soret band, three compartments are populated: Bulk Chl a (B), Red Chl a (R), and a small fraction of "free Chl a" (F). The first two compartments equilibrate, and excitations are trapped from B. Rates in 1/ns. **(b)** Concentration profiles c_l^{III}; note that the time axis is linear from −70 to +70 ps relative to the maximum of the IRF and logarithmic thereafter. **(c)** Species-Associated Spectra (SAS). Key in a, b, and c: Bulk Chl a (solid), Red Chl a (dashed), "free Chl a" (dotted). In addition, panel b depicts the rise of the charge-separated state (radical pair) RP1 (reaching 0.96) in dot-dash.

structure of the residual matrix) of the global analysis and of the target analysis are practically the same.

Finally, the amplitude matrix A is presented in Table 20.1, which expresses the relation between the concentration profiles c_l^{III} and the exponential decays $c^I(\kappa_l)$, $C^{III} = A \cdot C^I$.

The interpretation of the amplitude matrix is as follows (neglecting the IRF):

$$c_B = 0.392\exp(-t/4.7) + 0.508\exp(-t/28),$$

$$c_R = -0.282\exp(-t/4.7) + 0.342\exp(-t/28),$$

$$c_{RP1} = -0.110\exp(-t/4.7) - 0.838\exp(-t/28) + 0.948.$$

The total PSI excitation is 0.96, and the RP1 yield is 0.948/0.96 = 99%. No attempt is made to resolve the equilibration within the Bulk compartment to and from the RC. Several models are discussed in van Stokkum et al. (2013); Muller et al. (2010).

20.5 SECOND DATA: THE TIME-RESOLVED EMISSION SPECTRUM OF PAL CELLS

The time-resolved spectrum of PAL cells at room temperature after 400 nm excitation is depicted in Figure 20.5. The PSIIRC is in the open state (a) or in the closed state due to the presence of DCMU (b). Note that PAL cells contain both PSI and PSII, and thus a target model must be a superposition of target models for both PSI and PSII.

Again, the emission evolves both temporally and spectrally: it rises due to the finite rise time of the instrument, the spectral shape evolves due to equilibration between Chl pigments with different properties, and finally it decays due to photochemical quenching (charge separation, trapping) in the RCs (PSIRC and PSIIRC).

Table 20.1 Amplitude matrix for the different lifetimes of the BE PSI data

Compartment	Red	Bulk	RP1	F
Excitation	0.06	0.90	0	0.024
4.7 ps	−0.282	0.392	−0.110	
28 ps	0.342	0.508	−0.838	
Long lived			0.948	
0.68 ns				0.024

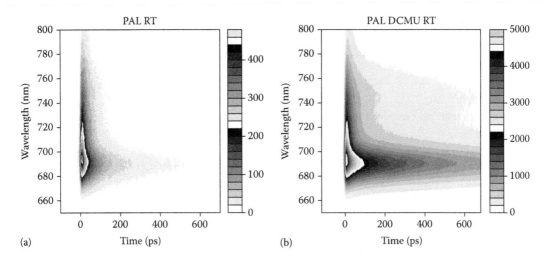

Figure 20.5 Filled contour plot of the time-resolved emission spectrum of PAL cells (that contain both PSI and PSII) at room temperature after 400 nm excitation. Time step ≈ 0.8 ps, wavelength step ≈ 2 nm. **(a)** PSIIRC in the open state, **(b)** PSIIRC in the closed state due to the presence of DCMU.

20.5.1 Global analysis of PAL with PSIIRC in the open state

Four components are sufficient to describe the data. The estimated lifetimes (reciprocals of estimated rate constants k_l) are 4.6 ± 0.2, 16.5 ± 0.3, 45.8 ± 0.5, and 382 ± 3 ps. Judging the DAS is less straightforward than with BE. The fastest lifetime (4.6 ps) again corresponds to equilibration in PSI, where the rise of the Red Chl a is evidenced by the negative amplitudes above 690 nm. The other three DAS are all positive and demonstrate photochemical quenching on three time scales. Apparently, both PSI and PSII show quenching on time scales of 16 and 46 ps. The final lifetime of 382 ps can safely be assigned to PSII. The EAS confirm this rough picture. However, target analysis is mandatory to disentangle the contributions of PSI and PSII. From the SVD of the data (not shown), it is clear that three temporally and spectrally different components are present. The fourth pair of left and right singular vectors represents noise. Thus, it can be concluded that in these data, the PSII dynamics are describable with a single spectrum decaying multiexponentially (Figure 20.6).

20.5.2 Simultaneous target analysis of both PAL data sets

Although a target analysis of each individual PAL data set is feasible, it is more instructive to present here a simultaneous analysis in order to demonstrate that both PAL data sets can be described by a single model that differs only in the parameters describing the open or closed condition of the PSIIRC (Figure 20.7).

Now, there are three K matrices. The eigenvalues of the PSI K matrix correspond to lifetimes of 3.7 and 26.6 ps. These lifetimes are slightly shorter than those estimated in the BE target analysis. The estimated PSI SAS of the BE and PAL mutants are practically identical, apart from a small shift due to wavelength calibration uncertainty. Hence, the PSI results of the PAL mutant will not be discussed further. The relative contributions of PSI and PSII can be estimated. Assuming that PSI is in trimeric form and PSII in dimeric form, they contain $3 \times 96 = 288$ and $2 \times 35 = 70$ Chl a, respectively. We then estimate a stoichiometry of 1.74 or 1.83 PSII dimer per PSI trimer for the samples with PSIIRC in the open or closed condition, respectively. When expressed as the ratio of PSIRC to PSIIRC, this corresponds to 0.86 or 0.82, which is in agreement with the values reported by Stadnichuk et al. (2009) of 0.7 ± 0.2 and 0.76 (Krumova et al. 2010; Tian et al. 2013).

Finally, the amplitude matrices A describing the PSII dynamics with PSIIRC in the open or closed condition are presented in Tables 20.2 and 20.3, respectively. The interpretation of the amplitude matrix is as follows (neglecting the IRF):

$$c_{B2,open} = 0.226 \exp(-t/55) + 0.071 \exp(-t/417),$$
$$c_{B2,closed} = 0.108 \exp(-t/123) + 0.200 \exp(-t/877).$$

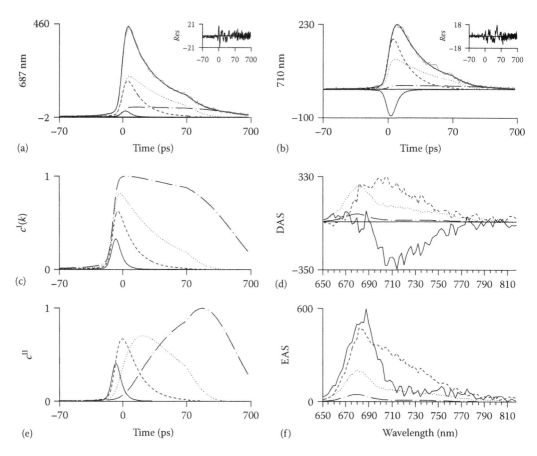

Figure 20.6 Results from global analysis of PAL mutant data with PSIIRC in the open state depicted in Figure 20.5a. Note that in a–c and e, the time axis is linear from −70 to +70 ps relative to the maximum of the IRF and logarithmic thereafter. Insets in a and b show residuals. **(a)** Data (in grey) and fit (in black) of 687 nm emission trace showing multiexponential decay. Contributions of the four exponential decays with different lifetimes (shown in c) multiplied by their amplitudes at 687 or 710 nm (shown in d) are indicated by line type in panels a and b. **(b)** Data (in grey) and fit (in black) of 710 nm emission trace showing multiexponential rise (visible as a contribution of $c^l(k_1)$ with negative amplitude) and decay. **(c)** Exponential decays $c^l(k_l)$. Estimated lifetimes (in panels c and e): 4.6 ps (solid), 16.5 ps (dashed), 45.8 ps (dotted), and 382 ps (dot-dashed). **(d)** Estimated Decay-Associated Spectra (DAS). **(e)** Evolutionary concentration profiles c_l^{II} (assuming a sequential kinetic scheme with increasing lifetimes). **(f)** Estimated Evolution-Associated Spectra (EAS).

Effectively, in PSII cores two lifetimes are sufficient to describe photochemical quenching *in vivo* in the PAL mutant. *In vitro*, two additional short lifetimes (<10 ps) with small contributions can be resolved from time-resolved emission spectra of PSII cores of *Thermosynechococcus elongatus* with open PSIIRC (Miloslavina et al. 2006; van der Weij–de Wit et al. 2011). The *in vitro* PSII core emission from van der Weij–de Wit et al. (2011) can be well fitted with the simple two-compartment scheme of bulk and radical pair, resulting in $c_{B2,open}^{in\,vitro} = 0.834 \exp(-t/42) + 0.156 \exp(-t/173)$.

The photochemical quenching parameters of PSII are collated in Table 20.4.

Thus, the main differences with open RC are that *in vivo* photochemical quenching is a bit slower than *in vitro* and that the free energy difference between B2 and RP2 is smaller. The transmembrane potential could be the cause thereof. *In vivo* with closed RC, the negative charge on Q_A^- lifts the RP2 to a higher free energy than that of B2.

Since it is not possible to resolve the small contributions of the faster equilibration lifetimes (<10 ps), no attempt is made to resolve the equilibration

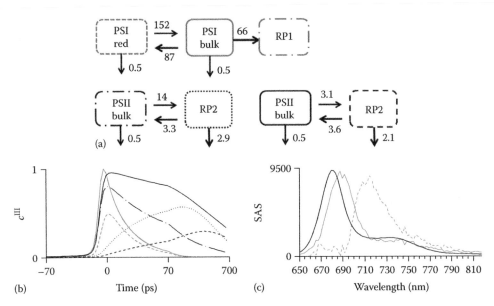

Figure 20.7 **(a)** Kinetic scheme used for the target analysis of the PAL data depicted in Figure 20.5. After excitation in the Soret band, three compartments are populated: Bulk Chl a (B1), Red Chl a (R1) of PSI, and Bulk Chl a (B2) of PSII. B2 equilibrates with a nonradiative radical pair (RP2) compartment, thus effectively describing the biexponential decay of the PSII emission. Excitations are trapped from B1 or B2. Rates in 1/ns. **(b)** Concentration profiles c_i^{III}; note that the time axis is linear from –70 to +70 ps relative to the maximum of the IRF and logarithmic thereafter. **(c)** Species-Associated Spectra (SAS). Key in a, b, and c: PSI in grey with B1 (solid), R1 (dashed), PSII in black with (in the closed state) B2 (solid), RP2 (dashed), and (in the open state) B2 (dot-dashed), RP2 (dotted).

Table 20.2 Amplitude matrix for the different lifetimes of the PAL data with open PSIIRC describing PSII

Compartment	B2	RP2
Excitation	0.297	0
55 ps	0.226	−0.262
417 ps	0.071	0.262

Table 20.3 Amplitude matrix for the different lifetimes of the PAL data with closed PSIIRC describing PSII

Compartment	B2	RP2
Excitation	0.308	0
123 ps	0.108	−0.137
877 ps	0.200	0.137

Table 20.4 Estimated photochemical quenching parameters of PSII, rates in 1/ns, $\Delta G_{B2,RP2}$ in meV

Sample	$k_{B2\rightarrow RP2}$	$k_{RP2\rightarrow B2}$	$k_{RP2\rightarrow}$	$\Delta G_{B2,RP2}$
In vitro, open	20	2.1	6.5	57
In vivo, open	14	3.3	2.9	37
In vivo, closed	3.1	3.6	2.1	−4

within the B2 compartment between the PSIIRC and the CP43 and CP47 antenna complexes, which is still disputed (Barter et al. 2001; Miloslavina et al. 2006; Raszewski and Renger 2008; van der Weij–de Wit et al. 2011; Tian et al. 2013).

20.6 CONCLUSIONS

Systems biophysics can resolve *in vivo* processes. Combined with knowledge from the literature, simultaneous analysis of different experiments can describe the functioning of complex photosynthetic systems. The high time resolution experiments of *in vitro* systems, preferably with selective excitation of, for example, the RC (Groot et al. 2005; Holzwarth et al. 2006; Muller et al. 2010) can reveal details of ultrafast processes that are obscured by the relatively slow dynamics of energy transfer to the RC. Cyanobacteria are promising model systems for a full description of the intact thylakoid membrane with the help of target analysis.

ACKNOWLEDGMENTS

The data analyzed have been expertly measured by Lijin Tian under the supervision of Herbert van Amerongen in the Laboratory of Biophysics of Wageningen University. Joris Snellenburg is thanked for critically reading the text. This project was partly carried out within the research program of BioSolar Cells, cofinanced by the Dutch Ministry of Economic Affairs, Agriculture and Innovation. Financial support of the European Research Council is acknowledged (Advanced Grant proposal 267333 PHOTPROT to Rienk van Grondelle).

REFERENCES

Ajlani G, Vernotte C (1998) Construction and characterization of a phycobiliprotein-less mutant of *Synechocystis* sp. PCC 6803. *Plant Molecular Biology* 37 (3):577–580.

Arcioni A, Zannoni C (1984) Intensity deconvolution in fluorescence depolarization studies of liquids, liquid-crystals and membranes. *Chemical Physics* 88 (1):113–128.

Barter LMC, Bianchietti M, Jeans C, Schilstra MJ, Hankamer B, Diner BA, Barber J, Durrant JR, Klug DR (2001) Relationship between excitation energy transfer, trapping, and antenna size in photosystem II. *Biochemistry* 40 (13):4026–4034.

Beechem JM (1989) A 2nd generation global analysis program for the recovery of complex inhomogeneous fluorescence decay kinetics. *Chemistry and Physics of Lipids* 50 (3–4):237–251.

Beechem JM, Ameloot M, Brand L (1985) Global and target analysis of complex decay phenomena. *Analytical Instrumentation* 14 (3–4):379–402.

Chukhutsina V, Tian L, Ajlani G, van Amerongen H (2013) Time-resolved fluorescence of photosystem I in vivo: Global and target analysis. In: Kuang T, Lu C, Zhang L (eds.) *Photosynthesis Research for Food, Fuel and Future: 15th International Conference on Photosynthesis*, Beijing, China,

Symposium 16 Photoprotection, Photoinhibition and Dynamics. Springer, Heidelberg, Germany, pp. 465–468.

Croce R, van Amerongen H (2013) Light-harvesting in photosystem I. *Photosynthesis Research* 116 (2–3):153–166.

Gobets B, van Stokkum IHM, Rogner M, Kruip J, Schlodder E, Karapetyan NV, Dekker JP, van Grondelle R (2001) Time-resolved fluorescence emission measurements of photosystem I particles of various cyanobacteria: A unified compartmental model. *Biophysical Journal* 81 (1):407–424.

Golub G, Pereyra V (2003) Separable nonlinear least squares: The variable projection method and its applications. *Inverse Problems* 19 (2):R1–R26.

Golub GH, LeVeque RJ (1979) Extensions and uses of the variable projection algorithm for solving nonlinear least squares problems. *Proceedings of the 1979 Army Numerical Analysis and Computers Conference*, White Sands Missile Range, NM, ARO Report 79-3.

Golub GH, Pereyra V (1973) Differentiation of pseudo-inverses and nonlinear least-squares problems whose variables separate. *SIAM Journal on Numerical Analysis* 10 (2):413–432.

Golub GH, Van Loan CF (1996) *Matrix Computations*. The Johns Hopkins University Press, Baltimore, MD.

Groot ML, Pawlowicz NP, van Wilderen L, Breton J, van Stokkum IHM, van Grondelle R (2005) Initial electron donor and acceptor in isolated Photosystem II reaction centers identified with femtosecond mid-IR spectroscopy. *Proceedings of the National Academy of Sciences of the United States of America* 102 (37):13087–13092.

Hendler RW, Shrager RI (1994) Deconvolutions based on singular-value decomposition and the pseudoinverse—A guide for beginners. *Journal of Biochemical and Biophysical Methods* 28 (1):1–33.

Henry ER, Hofrichter J (1992) Singular value decomposition—Application to analysis of experimental-data. *Methods in Enzymology* 210:129–192.

Holzwarth AR (1996) Data analysis of time-resolved measurements. In: Amesz J, Hoff AJ (eds.) *Biophysical Techniques in Photosynthesis*. Kluwer, Dordrecht, the Netherlands, pp. 75–92.

Holzwarth AR, Muller MG, Reus M, Nowaczyk M, Sander J, Rogner M (2006) Kinetics and mechanism of electron transfer in intact photosystem II and in the isolated reaction center: Pheophytin is the primary electron acceptor. *Proceedings of the National Academy of Sciences of the United States of America* 103 (18):6895–6900.

Krumova SB, Laptenok SP, Borst JW, Ughy B, Gombos Z, Ajlani G, van Amerongen H (2010) Monitoring photosynthesis in individual cells of *Synechocystis* sp. PCC 6803 on a picosecond timescale. *Biophysical Journal* 99 (6):2006–2015.

Miloslavina Y, Szczepaniak M, Muller MG, Sander J, Nowaczyk M, Rogner M, Holzwarth AR (2006) Charge separation kinetics in intact photosystem II core particles is trap-limited. A picosecond fluorescence study. *Biochemistry* 45 (7):2436–2442.

Mullen KM, van Stokkum IHM (2007) TIMP: An R package for modeling multi-way spectroscopic measurements. *Journal of Statistical Software* 18 (3):1–46.

Mullen KM, van Stokkum IHM (2009) The variable projection algorithm in time-resolved spectroscopy, microscopy and mass spectrometry applications. *Numerical Algorithms* 51 (3):319–340.

Muller MG, Griebenow K, Holzwarth AR (1992) Primary processes in isolated bacterial reaction centers from *Rhodobacter-sphaeroides* studied by picosecond fluorescence kinetics. *Chemical Physics Letters* 199 (5):465–469.

Muller MG, Hucke M, Reus M, Holzwarth AR (1996) Annihilation processes in the isolated D1-D2-cyt-b559 reaction center complex of photosystem II. An intensity-dependence study of femtosecond transient absorption. *Journal of Physical Chemistry* 100 (22):9537–9544.

Muller MG, Slavov C, Luthra R, Redding KE, Holzwarth AR (2010) Independent initiation of primary electron transfer in the two branches of the photosystem I reaction center. *Proceedings of the National Academy of Sciences of the United States of America* 107 (9):4123–4128.

Nagle JF (1991) Solving complex photocycle kinetics—Theory and direct method. *Biophysical Journal* 59 (2):476–487.

Nagle JF, Parodi LA, Lozier RH (1982) Procedure for testing kinetic-models of the photocycle of Bacteriorhodopsin. *Biophysical Journal* 38 (2):161–174.

Peterman EJG, Wenk SO, Pullerits T, Palsson LO, van Grondelle R, Dekker JP, Rogner M, van Amerongen H (1998) Fluorescence and absorption spectroscopy of the weakly fluorescent chlorophyll a in cytochrome b(6)f of *Synechocystis* PCC6803. *Biophysical Journal* 75 (1):389–398.

Raszewski G, Renger T (2008) Light harvesting in photosystem II core complexes is limited by the transfer to the trap: Can the core complex turn into a photoprotective mode? *Journal of the American Chemical Society* 130 (13):4431–4446.

Shrager RI (1986) Chemical-transitions measured by spectra and resolved using singular value decomposition. *Chemometrics and Intelligent Laboratory Systems* 1 (1):59–70.

Snellenburg JJ, Dekker JP, van Grondelle R, van Stokkum IHM (2013) Functional compartmental modeling of the photosystems in the thylakoid membrane at 77 K. *The Journal of Physical Chemistry B* 117 (38):11363–11371.

Snellenburg JJ, Laptenok SP, Seger R, Mullen KM, van Stokkum IHM (2012) Glotaran: A Java-based graphical user interface for the R-package TIMP. *Journal of Statistical Software* 49(3):1–22.

Stadnichuk IN, Lukashev EP, Elanskaya IV (2009) Fluorescence changes accompanying short-term light adaptations in photosystem I and photosystem II of the cyanobacterium *Synechocystis* sp PCC 6803 and phycobiliprotein-impaired mutants: State 1/State 2 transitions and carotenoid-induced quenching of phycobilisomes. *Photosynthesis Research* 99 (3):227–241.

Tian L, Farooq S, van Amerongen H (2013) Probing the picosecond kinetics of the photosystem II core complex in vivo. *Physical Chemistry Chemical Physics* 15 (9):3146–3154.

van Amerongen H, Croce R (2013) Light harvesting in photosystem II. *Photosynthesis Research* 116 (2–3):251–263.

van der Weij–de Wit CD, Dekker JP, van Grondelle R, van Stokkum IHM (2011) Charge separation is virtually irreversible in photosystem II core complexes with oxidized primary quinone acceptor. *The Journal of Physical Chemistry A* 115 (16):3947–3956.

van Grondelle R, Dekker JP, Gillbro T, Sundström V (1994) Energy-transfer and trapping in photosynthesis. *Biochimica et Biophysica Acta* 1187 (1):1–65.

van Stokkum IHM (2005) Global and target analysis of time-resolved spectra. Lecture Notes for the Troisième Cycle de la Physique en Suisse Romande. Department of Physics and Astronomy, Vrije Universiteit, Amsterdam, the Netherlands.

van Stokkum IHM, Bal HE (2006) A problem solving environment for interactive modelling of multiway data. *Concurrency and Computation: Practice and Experience* 18:263–269.

van Stokkum IHM, Desquilbet TE, van der Weij-de Wit CD, Snellenburg JJ, van Grondelle R, Thomas J-C, Dekker JP, Robert B (2013) Energy transfer and trapping in red-chlorophyll-free photosystem I from *Synechococcus* WH 7803. *The Journal of Physical Chemistry B* 117 (38):11176–11183.

van Stokkum IHM, Larsen DS, van Grondelle R (2004) Global and target analysis of time-resolved spectra. *Biochimica et Biophysica Acta* 1657:82–104.

van Stokkum IHM, van Oort B, van Mourik F, Gobets B, van Amerongen H (2008) (Sub)-Picosecond spectral evolution of fluorescence studied with a synchroscan streak-camera system and target analysis. In: Aartsma TJ, Matysik J (eds.) *Biophysical Techniques in Photosynthesis*, vol. II. Advances in Photosynthesis and Respiration, vol. 26. Springer, Dordrecht, the Netherlands, pp. 223–240.

PART 5

Artificial and Natural Photosynthesis

21

Light harvesting, photoregulation, and photoprotection in selected artificial photosynthetic systems

KATHERINE WONGCARTER, MANUEL J. LLANSOLA-PORTOLES,
GERDENIS KODIS, DEVENS GUST, ANA L. MOORE, AND
THOMAS A. MOORE

21.1 INTRODUCTION TO ARTIFICIAL PHOTOSYNTHESIS

Artificial photosynthesis (AP) includes the design and synthesis of molecular systems that manifest selected aspects of the natural photosynthetic process [1–8]. The essential photochemistry of these synthetic systems follows from that of their natural counterparts: absorption of sunlight, energy transfer, regulation of energy flow, protection from photodamage, and conversion of excited states to redox potential. Also, AP includes catalysis and, ultimately, self-repair and replication. Following the example from nature, the goal of AP is to assemble molecular systems into larger-scale constructs capable of storing a fraction of the energy carried by sunlight in energy-rich compounds.

AP is a diverse field that is in its infancy and already involves contributions from disciplines ranging from physics through chemistry and material science to synthetic biology. AP includes a future in which photosynthesis is reengineered for higher efficiency using the design principles developed in simpler artificial systems and new engineered forms of life where heterotrophic organisms are converted to autotrophs by interfacing them to sources of electrons and electromotive force [9,10]. These hybrid constructs will have retained the essential features of living organisms including repair, replication, and self-assembly, but their metabolism will be driven by solar photovoltaic cells [11].

AP research today is largely focused on the understanding and mimicry of the steps (processes) employed by photosynthesis to produce an energy-rich fuel. These processes are linked to photosynthetic membranes where antenna pigments convert solar photons to excited states, regulate the flow of collected excitation energy to reaction centers (RCs), and suppress the undesirable production of reactive oxygen species (ROS) such as molecular singlet oxygen. RCs use the excited states of specialized chlorophyll pigments to generate high-potential and low-potential redox species; electron flow down the resulting redox gradient is used to pump protons across the photosynthetic membrane, thereby generating proton motive force. In algae, photosynthetic cyanobacteria, and higher plants, the electrons and hydrogen ions from water oxidation are used to reduce carbon dioxide. AP systems designed today use these same sources of electrons and carbon that nature uses. For the foreseeable future, this is certainly appropriate because the imperative to develop carbon-neutral energy and to draw down atmospheric CO_2 will require a source of electrons that can only be met by water oxidation.

One of many approaches to the assembly of AP model compounds is to use synthetic chromophores that are related to the natural pigments (bacteriochlorophylls, chlorophylls, carotenoids, etc.), but to use covalent bonds in place of the proteins that are used in natural systems to hold and organize the pigments. In the pigment–protein complexes of photosynthetic membranes, the electronic interactions, for example, between carotenoids and chlorophylls, that give rise to energy and electron transfer are in part a consequence of the proximity of their π electron systems, which is a function of the distance and geometry imposed on the pigments by the secondary, tertiary, and quaternary protein structure. In general, these may be thought of as through-space interactions, although protein moieties may play a part. In our model systems, interchromophore distance and geometry are controlled by the linkage bonds joining the carotenoid and cyclic tetrapyrrole moieties. There are two roles for the linker: in cases where the electronic structure of the linkage participates in the coupling, the interaction

is referred to as through bond and in cases where the linkage does not participate to a measureable extent, the coupling is referred to as through space. These two interactions are not mutually exclusive, and hence AP models can be designed to involve both through-space and through-bond electronic couplings.

In our experience, regardless of whether the coupling is a consequence of through-space or through-bond interactions, in a wide variety of photophysical and photochemical measurements, our model systems have demonstrated the same essential features of electronic coupling required for the energy and electron transfer as are observed in the natural systems. In this chapter, we review a selection of the AP constructs developed in our laboratories that mimic many of the photochemical steps that occur in the natural photosynthetic antenna systems. Reviews of our work in AP reaction centers, proton pumps, water oxidation, and proton reduction have been published elsewhere [1–8,12].

21.2 ANTENNA FUNCTION

In nature, chlorophylls and bacteriochlorophylls are prevalent throughout the photosynthetic antenna, but they only absorb light strongly in the blue and red regions of the visible solar spectrum. By incorporating other pigments that collect photons from the remaining part of the available wavelengths of sunlight, plants, algae, and photosynthetic bacteria are able to collect more energy (Figure 21.1). Carotenoids play a key role in absorbing the green and yellow photons of the solar spectrum (400–550 nm region) [13]. An efficient energy transport mechanism is needed in order to pass the energy from the carotenoids and other pigments to the reaction center, where the excitation energy is transformed into chemical potential in the form of a charge-separated state [14,15].

In this section, we will describe a series of artificial antennas that exhibit very low excitation energy transfer (EET) efficiency and others where the efficiency reaches unity. The artificial antenna compounds are presented in chronological order to demonstrate the evolution of synthetic antenna design over time. The measured EET efficiency as a function of the electronic coupling between the chromophores guided the synthetic work.

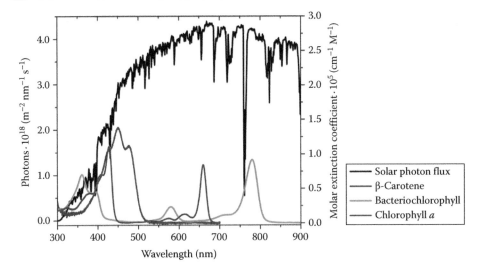

Figure 21.1 Solar irradiance spectrum (left axis) and absorption spectra of some photosynthetic pigments (right axis).

21.2.1 Carotenoid-based antennas

The first artificial antenna molecules synthesized by our research group were carotenoporphyrins [16,17]. These systems consisted of a porphyrin covalently attached to carotenoids with 6 or 10 carbon–carbon double bonds (db) conjugated with a phenyl ring (compounds 1 and 2 in Figure 21.2). Carotenoids with different conjugation lengths, or number of db, were used to tune the wavelength of absorption. The absorption spectra of 1 and 2 are essentially identical to the sum of the spectra of related unlinked carotenoids and porphyrins, indicating that there is no strong electronic coupling between the chromophores. The ester linkage constrains the edges of the carotenoid and porphyrin π-orbitals to be close to one another and provides weak electronic coupling. Using steady-state fluorescence excitation spectroscopy, it was found that in 1, the efficiency of EET from the carotenoid to the porphyrin is 80%. However, this short carotenoid absorbs light in about the same spectral region as the porphyrin and hence does not add significantly to the spectral range used. The longer carotenoid employed in 2 collects light from an underutilized region of the solar spectrum, but none of the light absorbed by this carotenoid led to fluorescence emission from the porphyrin in dyad 2, indicating that EET was vanishingly small. This result suggests that a greater degree of electronic

interaction between the energy donor and acceptor moieties is required.

A constitutional isomer of 2, dyad 4 (Figure 21.2), wherein the carotenoid is attached to an *ortho* position of the *meso*-aryl ring of the porphyrin was synthesized in order to achieve a closer interaction of the two chromophores while maintaining the same ester linkage. NMR studies of 4 demonstrated that as a result of the *ortho* attachment of the carotenoid to the porphyrin *meso*-aryl group, the carotenoid folds back across the porphyrin so that the π-systems are essentially in van der Waals contact. Molecular mechanics calculations determined that the separation of the two π-stacked systems is ~4 Å and the fluorescence measurements showed 25% EET efficiency.

Dyads 2 and 4, illustrate the use of through-bond and through-space coupling that can be accessed in models. The electronic coupling provided by the linkage, as reflected by the number of bonds and the hybridization of the atoms, is essentially the same in both cases and is insufficient to mediate singlet energy transfer (antenna function) from the carotenoid to the porphyrin and fast triplet–triplet (T–T) energy transfer. However, by adopting a folded conformation, dyad 4 acquires through-space interactions between the π systems resulting in both singlet–singlet (S–S) energy transfer and ultrafast T–T transfer (*vide infra*). This clearly demonstrates that through-space electronic interactions can be

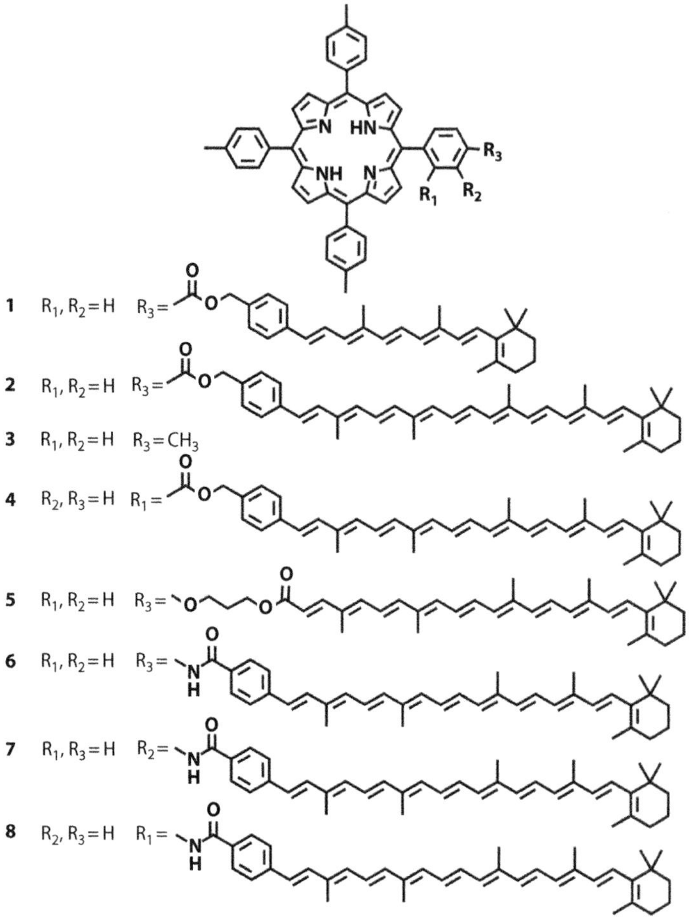

Figure 21.2 Molecular structures of carotenoporphyrin dyads and *meso*-tetra(*para*-tolyl)porphyrin (TTP), **3**.

mimicked in covalently linked dyads, provided a conformation can be acquired that gives the necessary orbital overlap.

Dyad variations of the *para*- and *ortho*-positioned carotenoids with longer (propyl) linkers to the porphyrin (e.g., **5**) and to pyropheophorbide (a derivative of natural chlorophyll) were also investigated for their EET capability in low- and high-viscosity environments [18–22]. The overall results corroborate the previous observation that significant EET requires relatively strong electronic interaction between the chromophores such as that provided by van der Waals contact. Later, it was found that increased electronic interactions between the carotenoid and cyclic tetrapyrrole moieties can also be achieved by involving the linkage itself [23].

The potential involvement of the linkage in the electronic interaction of carotenoids and cyclic tetrapyrroles was investigated in a study of carotenoporphyrin dyads **6–8** (Figure 21.2). These dyads consist of identical carotenoid and porphyrin moieties, which are joined by a relatively rigid amide group, but the point of attachment of the carotenoid to the porphyrin *meso*-aryl group was varied. The S–S energy transfer quantum yields, measured by fluorescence, were 0.17, 0.10, and 0.13 for **8**, **7**, and **6**, respectively. A lower energy transfer quantum yield and therefore a slower singlet energy transfer rate was observed for the *meta* isomer, **7**, than for the *para* isomer, **6**, despite the fact that molecular mechanics calculations showed that the separation of the two chromophores is smaller for **7**. These results were consistent with energy transfer mediated by the linkage bonds rather than by a through-space mechanism.

Qualitatively, this superexchange, or through-bond, hypothesis was evaluated using simple Hückel molecular orbital theory, assuming that the highest occupied (HOMO) and lowest unoccupied (LUMO) molecular orbitals of the linker were involved in the interaction. These results showed that for both HOMO and LUMO, the coefficient of the wavefunction is greater at the *ortho* and *para* positions than at the *meta* position.

One of the big challenges in building artificial photosynthetic antennas with carotenoids is the unique photophysics of linear polyenes. Carotenoids typically have very high extinction coefficients in the 450–500 nm spectral region (on the order of 10^5 M^{-1} cm^{-1}, see Figure 21.1). This strongly allowed transition is from the ground state, $S_0(1^1A_g^-)$, to the second excited singlet state, $S_2(1^1B_u^+)$ (group theory symbols for C_{2h} symmetry). As expected for an upper excited singlet state, the S_2 lifetime is extremely short (<300 fs). The transition from the ground state, $S_0(1^1A_g^-)$, to the lowest-lying excited singlet state of carotenoids, $S_1(2^1A_g^-)$, is single photon forbidden, and the state is populated by relaxation from S_2. The lifetime of the S_1 state is on the order of 1–100 ps for most carotenoids and is inversely dependent on the length of the polyene. There are other dark excited states found in carotenoids, such as the $1^1B_u^-$ and $3^1A_g^-$ states that have been theoretically predicted and the S^* and S^+ that are readily seen in transient absorption experiments [14,15,24]. In general, the unusual characteristics of the carotenoid excited states (see Chapter 3) impose strict constraints on the electronic coupling and thermodynamics necessary for the energy and electron transfer processes between them and cyclic tetrapyrroles, which are crucial to photosynthesis and AP.

Even though the carotenoid S_2 state lifetime is extremely short, we were able to achieve efficient EET in dyad **9** (Figure 21.3) by careful attention to electronic coupling and thermodynamic parameters [25]. Measured by fluorescence upconversion spectroscopy, the S_2 lifetime of a model for the carotenoid in dyad **9**, a phenylacetamido substituted carotenoid with 10 db, was found to be 150 fs. In dyad **9**, the carotenoid moiety was found to have a considerably shorter lifetime, ~40 fs. If this quenching is attributed entirely to EET from the carotenoid S_2 state to the purpurin moiety, the quantum yield of this process would be 0.73 from

$$\Phi = 1 - \frac{\tau_q}{\tau_f} \qquad (21.1)$$

where τ_q and τ_f are the S_2 lifetimes of the carotenoid moiety of the dyad and the carotenoid model alone, respectively. Indeed, steady-state fluorescence excitation measurements revealed that the quantum yield was in agreement with that calculated from the kinetics. The rate constant for EET in this system (~2 × 10^{13} s^{-1}) is of the order of that observed in natural light-harvesting antennas [13]. Further evidence for EET from S_2 comes from the observation that the lifetime of the S_1 state of the carotenoid (7.8 ps) was the same in both a carotenoid model and the dyad. Moreover, the 10 db (double bond) carotenoid S_1 state energy is below that of the lowest singlet excited state, Q_y, of the purpurin or the lowest singlet excited state, Q_x, of porphyrins (Platt's nomenclature for cyclic tetrapyrroles

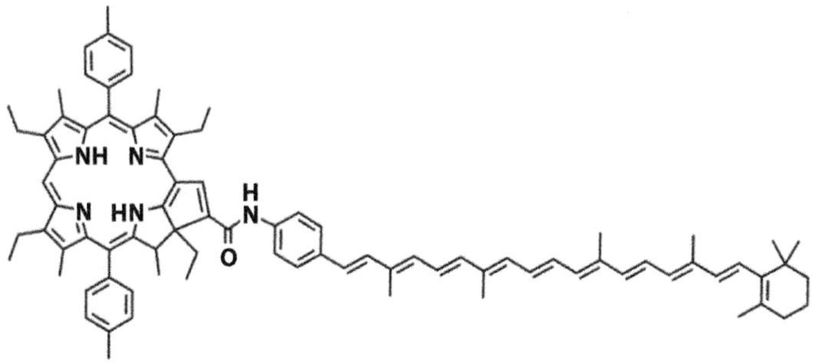

Figure 21.3 Molecular structure of dyad **9**, a 10 db carotenoid linked through a phenylamide group to the fused cyclopentadiene ring of a purpurin.

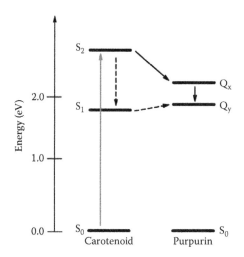

Figure 21.4 Singlet energy transfer pathways from the carotenoid S_2 state (solid arrow) or involving the carotenoid S_1 state route (dashed arrows, not observed) to the purpurin in dyad **9**. The grey arrow indicates the population of S_2 by direct excitation.

of D_{2h} symmetry) [26]. Therefore, EET is only possible from the S_2 state of the carotenoid. We attribute efficient EET from S_2 to the electronic coupling provided by the amide group and the favorable thermodynamics achieved by overlap between the purpurin Q_x absorption bands and the carotenoid S_2 state emission band (Figure 21.4).

Triad **10** (Figure 21.5) was designed to address the question of energy transfer from carotenoids to cyclic tetrapyrroles as a function of the relative orientations of the pigments [27]. Inspiration for this series comes from the relative orientation of carotenoid to bacteriochlorophyll in the B800/B850 LH II of *Rhodopseudomonas acidophila*, where the carotenoids are approximately perpendicular to the plane of the macrocycle in B800 and approximately parallel the plane in B850 [28]. In triad **10**, two 9 db carotenoids are linked on opposite sides to a silicon phthalocyanine (Pc) through an ester linkage at the center of the macrocycle and extend axially at an angle of approximately 90° to the plane of the cyclic tetrapyrrole (Figure 21.5). It was found that efficient EET was achieved by using both the S_2 and S_1 states. To determine in detail how the carotenoid S_1 and S_2 excited states are involved in the flow of energy to Pc, the lifetimes of these excited states were measured in a model carotenoid and in triad **10**. By fluorescence upconversion spectroscopy, it was

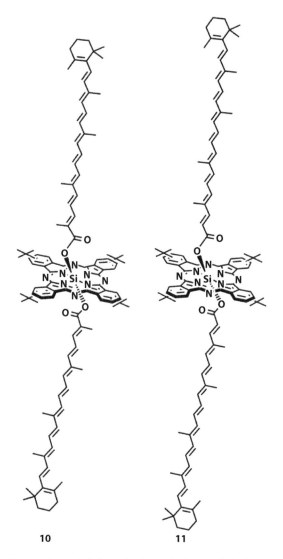

10 **11**

Figure 21.5 Axially substituted silicon phthalocyanine triads with 9 db, **10**, or 10 db, **11**, carotenoids.

demonstrated that the S_2 state of the carotenoid was quenched from 82 fs in the model carotenoid to 53 fs in **10**, resulting in an EET efficiency of 35% from the S_2 state. The contribution of S_1 was investigated by transient absorption measurements. The observed difference in the S_1 state lifetime of a carotenoid in triad **10** (2.6 ps) and the carotenoid model (24.4 ps), taking into account that the quantum yield of formation of the S_1 state was only 65%, gives the quantum yield of energy transfer from S_1 as ~58%. Thus, the overall quantum yield of EET from carotenoid to Pc from both the S_2 (35%) and S_1 (58%) states was 93%, in excellent agreement with

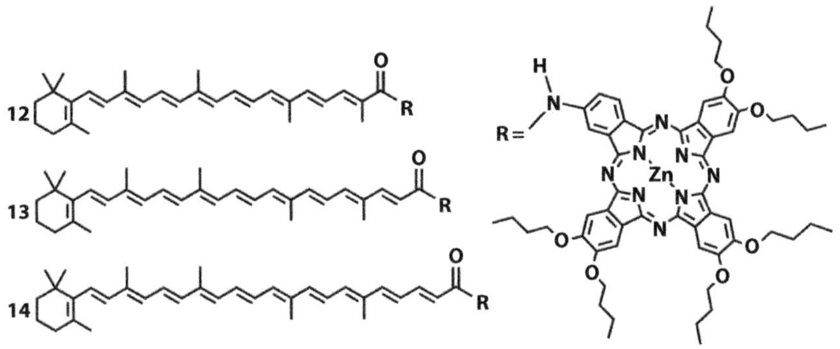

Figure 21.6 Three carotenophthalocyanine dyads where the two chromophores are connected via an amide linkage. Dyad **12** has 9 db, **13** has 10 db, and **14** has 11 db.

the value estimated from steady-state fluorescence excitation measurements [27].

As illustrated earlier, we note that steady-state fluorescence excitation spectroscopy can only give the total EET efficiency. Time-resolved methods are required to resolve the contributions from the individual states. As we shall demonstrate in the following text, these contributions are a sensitive function of the electronic structure of the carotenoid moiety. Extending the carotenoid conjugation by one double bond from 9 to 10 in triad **11** (Figure 21.5) reduced the overall EET efficiency from ~90% to 30%. This change is a result of the S_1 level in the longer carotenoid moving below that of the lowest singlet state of Pc making energy transfer from the S_1 state of carotenoid to the Pc thermodynamically unfavorable. Thus, in triad **11**, EET to Pc occurs only from the carotenoid S_2 state. Transient absorption studies on triads **10** and **11** also provided evidence for the formation of the S^* state from S_2 in both of the triads but not in the isolated carotenoid models. The S^* state had a lifetime of ~26 ps in both triads [29]. Interestingly, in triad **10**, the decay of S^* was associated with an increase in the Pc ground-state bleaching near 690 nm, indicating that S^* transfers energy to Pc in this case. The situation in triad **11** was different in that the Pc ground-state bleaching and stimulated emission signals did not rise concomitantly with the decay of S^* (26 ps). Thus, the efficient antenna function of the 9 db carotenoid in **10** is associated not only with the S_1 and S_2 states; all three states S_2, S_1, and S^* contribute singlet electronic energy to Pc. This illustrates the importance of using multiple pathways for the EET process when using carotenoids in the construction of highly efficient artificial photosynthetic antennas.

Similar dependence of the EET from carotenoids to Pc on carotenoid length was observed in another set of dyads (Figure 21.6). In contrast to the case discussed where the carotenoids were axially linked to the Pc, in dyads **12**, **13**, and **14**, with carotenoids having 9, 10, and 11 db, respectively, the carotenoids were attached to the periphery of the macrocycle of a zinc Pc via an amide linkage [30]. This orientation is reminiscent of the relationship between the carotenoid and the bacteriochlorophyll in B850 [28]. Ultrafast transient absorption spectroscopy and global analysis [31] of the data were used to evaluate the energy transfer and excited state deactivation pathways following excitation of the allowed carotenoid S_2 state. The S_2 state rapidly relaxed to the S^* and S_1 states. In all systems, a pathway of energy deactivation within the carotenoid manifold in which the S^* state acts as an intermediate state in the S_2 to S_1 internal conversion on a subpicosecond time scale was detected. In dyad **12**, similar to the case with triad 10, all carotenoid singlet excited states S_2, S^*, and S_1 contribute to S–S energy transfer to Pc, making the process very efficient (>90%), while for dyads **13** and **14**, the S_1 energy transfer channel is precluded by thermodynamic considerations and only S_2 was capable of transferring energy to Pc. Finding S^* in our dyads and triads is particularly gratifying because the same state is seen in the natural systems; this is a further illustration that the essential features of the electronic coupling and spectroscopy in native antenna pigment–protein complexes is recapitulated in these models [32].

The fluorescence excitation spectra of carotenophthalocyanines **12** and **13** indicated a weak but measureable dependence of the energy transfer efficiency on excitation wavelength, suggesting an increase of energy transfer with excitation at the blue edge of the absorption band (to the higher vibronic states) of the carotenoid. This phenomenon was more evident in a series of carotenophthalocyanines where the two chromophores were tightly coupled through an arylamine moiety (dyads 26–29, Figure 21.11). In addition to fluorescence excitation measurements, ultrafast transient absorption measurements also indicated that the population of the carotenoid hot S_2 state results in enhanced energy transfer to the Pc. It was hypothesized that the higher energy transfer efficiency with excess vibrational excitation results either from a decreased internal conversion rate to lower-lying optically forbidden states or from enhanced coupling between higher vibrational S_2 states and Pc [33].

The spectroscopic results obtained for these model systems in conjunction with the crystallographic data, which shows very close intermolecular contacts in natural antenna pigment-protein complexes, suggests that classical Förster theory is not adequate to explain the mechanism of S–S energy transfer between longer carotenoids and cyclic tetrapyrroles. First, at the short interchromophore distances necessary for EET, the Förster dipole approximation breaks down. At these distances, coulombic coupling using transition monopoles is appropriate [34,35]. Furthermore, at the distances required for efficient energy and electron transfer, there is clearly orbital overlap in the natural systems; this level of electronic coupling in synthetic models is provided by the linkage or a combination of the linkage and a conformation that brings the π systems into van der Waals contact (vide supra). Orbital overlap introduces contributions to the electronic coupling that arise from electron exchange interactions. Interestingly, both of these mechanisms may promote T–T as well as S–S energy transfer (see the last paragraph of this chapter for evolutionary implications of this electronic coupling).

21.2.2 Tetrapyrrole antennas

EET between (bacterio)chlorophyll molecules within antennas and funneling of the energy to reaction centers is ubiquitous in light-harvesting complexes (LHCs) of natural photosynthetic systems. In collaboration with J. Lindsey, we constructed antenna molecules consisting of four covalently linked zinc tetraarylporphyrins, for example, tetrad 15, and linked them to free base porphyrin-fullerene artificial reaction centers to give hexads 16 and 17 (Figure 21.7) [36,37].

Time-resolved absorption and emission studies showed that excitation of any peripheral zinc porphyrin moiety was followed by EET to the central zinc porphyrin with a time constant of ~50 ps. The excitation energy was then transferred to the free base porphyrin of 16 in 240 ps, which decayed by electron transfer to the fullerene with a time constant of 3 ps. The final charge-separated state had a lifetime of ~1 ns and was generated with a quantum yield of 0.69 based on light absorbed by the zinc porphyrin antenna.

We note that EET rate between the zinc porphyrin in the antenna of 16 and the free base porphyrin was relatively slow (240 ps)$^{-1}$ compared with EET among the zinc porphyrins (50 ps)$^{-1}$. Examination of the frontier molecular orbitals of the porphyrins involved in the linkage provides one feasible explanation. The symmetry of the HOMOs of porphyrins are either a_{2u}, which has lobes at the meso positions, or a_{1u}, which has nodes at the meso positions. Indeed, there are examples of meso-linked arrays of porphyrins with a_{2u} HOMOs, which undergo rapid energy transfer, whereas similar arrays of porphyrins with a_{1u} HOMOs undergo much slower energy transfer [38–40]. The presence of the eight β-alkyl substituents of the free base porphyrin of 16 results in the HOMO having a_{1u} symmetry and therefore diminished electronic coupling with the linker. Additionally, in molecules with β-alkyl substituents, steric effects result in an increased dihedral angle between the planes of the porphyrin ring and the meso-aryl ring. These factors undoubtedly reduced the electronic coupling and thereby reduced through bond–mediated contributions to EET. In hexad 17, we addressed these issues by changes in the structure, which changed the symmetry of the HOMO and relaxed the steric-induced dihedral angle. The resulting EET from the antenna to the free base porphyrin occurs in 32 ps, an eightfold increase. Increasing the rate of this process by addressing the factors that control the symmetry of the HOMO and steric effects and thereby the electronic coupling that mediates EET is an example of using theory to guide design to optimize function. This strategy resulted in an increase in the yield of the final charge-separated

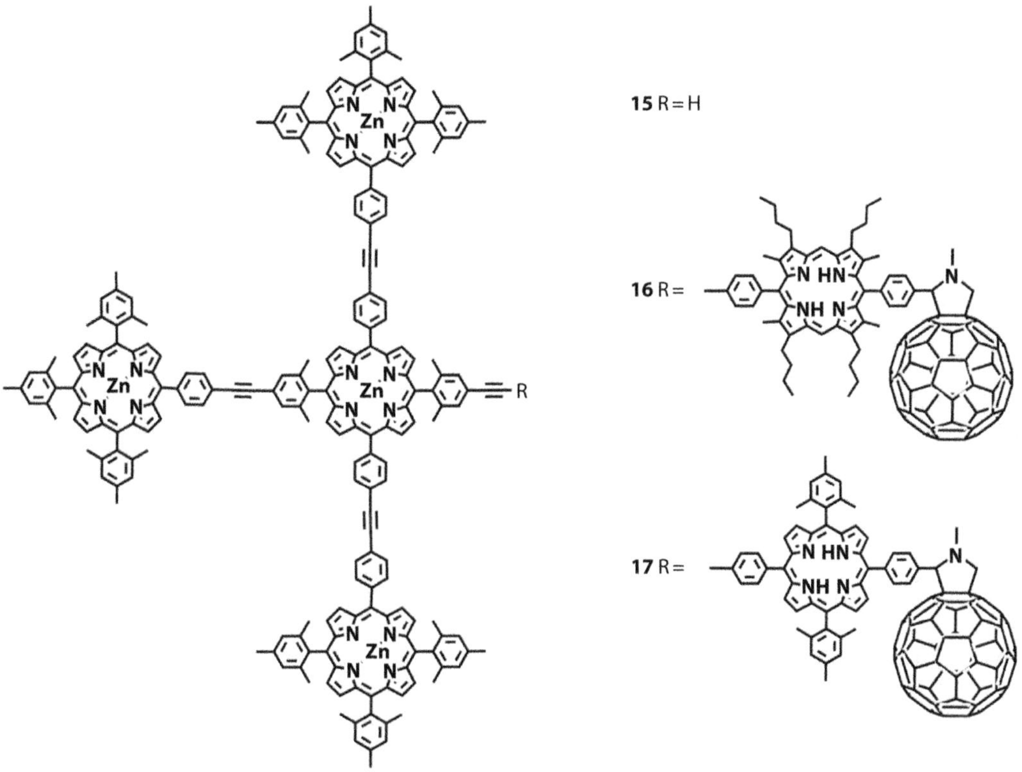

Figure 21.7 Structures of a tetrad antenna **15** and two different hexad antenna-reaction center models, one with an octalkyl-*meso*-diphenylporphyrin-C_{60} reaction center, **16**, and another with a *meso*-tetraaryl porphyrin-C_{60} reaction center, **17**.

state from 69% in **16** to 90% in **17** [37]. Following these and related principles, it was possible to design antennas consisting of different multiple chromophores. With these systems, efficient light harvesting throughout the visible spectrum was achieved and the resulting charge-separated state stored more than 50% of the energy of the lowest excited singlet state of the cyclic tetrapyrrole [41–44].

The results described earlier demonstrate that it is possible to design and prepare synthetic antenna–reaction center complexes that mimic basic photochemical functions of natural photosynthetic light-harvesting antennas and reaction centers in simple, structurally well-defined constructs.

21.3 PHOTOREGULATION

Antennas are essential for collecting the photons that drive photosynthesis, but there are times when the antennas overdo their job, particularly under conditions of high light intensity. It is hypothesized

that in the evolution of photosynthesis there was, and continues to be, a selective advantage for the organism to make optimum use of low-light conditions to drive the relatively slow downstream metabolic reactions of legacy biochemistry [45]. Low light requires large antennas, resulting in a situation where during much of the day, light in excess of that required to drive downstream metabolism is absorbed. If excess light energy is not dealt with properly, it would cause oxidative damage to the organism (see Section 21.4). As photosynthesis changed the earth's atmosphere from reductive to oxidative *ca.* 2.6–2 billion years ago, photosynthetic organisms developed myriad ways to prevent oxidative damage that occurs under excess light. One of these is known as nonphotochemical quenching (NPQ) [46]. NPQ is a biological switch that results in the deactivation of the singlet excited states of chlorophylls in the antenna system. There are several triggers of NPQ including, in green plants, luminal acidification beyond the normal range required to maintain proton motive force. It is thought that essentially all oxygenic photosynthetic organisms have developed some form of NPQ [47].

Discovering the precise mechanisms of NPQ is an active research field (see Chapter 24). NPQ operates on several time scales from short times required for sensing and protein conformational changes to longer time scales for enzyme activation to even longer time scales for gene expression. Our models seek to mimic the rapid events known as the qE component of NPQ that results in quenching chlorophyll excited states and thereby interrupting EET to reaction centers. There is now general agreement that qE occurs following changes in the protein structure in pigment–protein complexes. These changes induce quenching of chlorophyll singlet excited states by interaction with specific carotenoids. In certain organisms under high light intensity, it has been demonstrated that zeaxanthin, a carotenoid having 11 db, is formed by de-epoxi-dase enzymes in the xanthophyll cycle; it becomes the predominant carotenoid. It has been proposed that its longer conjugated system allows it to perform NPQ by some of the mechanisms discussed in Sections 21.3.1–21.3.3 [46,48,49].

While NPQ is a necessary regulatory system in photosynthetic organisms, it is also one of the important factors responsible for the low overall efficiency of photosynthesis [9,45]. Anticipating that ways will be discovered to accelerate the rate-limiting downstream dark reactions, there will be a need to recalibrate NPQ to allow the photosynthetic process to proceed more efficiently. But, before that can be done, it is imperative to understand the detailed mechanisms of NPQ. Studies with simple artificial light-harvesting antennas have helped in this regard, and some examples of the systems and photophysical studies are briefly described in the following section.

21.3.1 Quenching of excited states of cyclic tetrapyrroles by electron transfer

Carotenoids have a paradoxical role in photosynthesis: they participate in EET to chlorophylls but when there is excess energy carotenoids are also responsible for quenching the excited states of chlorophylls. Since the synthesis of our first carotenoporphyrin [16], we observed quenching of the porphyrin fluorescence and speculated that electron transfer to the excited porphyrin from the carotenoid would

be a reasonable explanation. As the synthesis of artificial photosynthetic antennas progressed, the attachment of the carotenoids, the nature of the porphyrins, and the number of conjugated db in the carotenoid were systematically changed. With caro-tenoporphyrins 6–8 (Figure 21.2), the degree of porphyrin fluorescence quenching depends on the position of carotenoid attachment to the meso-aryl group of the porphyrin [23]. In addition to steady-state porphyrin fluorescence intensity measurements, the fluorescence lifetime was used to assess the quenching. The porphyrin model for dyads 6–8 has an 11 ns excited singlet state lifetime, while the ortho, meta, and para isomers have shorter lifetimes (Table 21.1). This effect was observed with the dyads dissolved in toluene. When the solvent polarity was increased, lifetimes got even shorter (Table 21.1). This quenching process was interpreted as a possible intramolecular electron transfer from the carotenoid to the porphyrin excited state via coupling through the linkage bonds. The ortho and para isomers were more strongly quenched and, as discussed previously for EET, the ortho- and para- linked isomers have better electronic coupling [23]. Moreover, as will be shown in Section 21.4, the same pattern is observed for T–T energy transfer (see Figure 21.18).

No spectroscopic evidence for the formation of the radical pair species ($C^{\cdot+}$-$P^{\cdot-}$) was seen in 6–8. In order to increase the probability of observing electron transfer, we designed dyad 18, which featured a powerfully oxidizing porphyrin. We reasoned that electron transfer would be in the Marcus normal region so that if more driving force for

Table 21.1 Fluorescence lifetimes for carotenoporphyrins in nonpolar solvent (toluene) and polar solvent (acetonitrile)[a]

Molecule	Toluene (ns)	Acetonitrile (ns)
Model porphyrin 3	11.0	11.0
Ortho amide dyad 6	5.2	1.0[b]
Meta amide dyad 7	7.2	5.8
Para amide dyad 8	5.9	3.9

[a] All lifetimes were calculated from single exponential fits to the fluorescence decays.

[b] In the case of 6 in acetonitrile, three exponential components were required. The major (80%) component is reported.

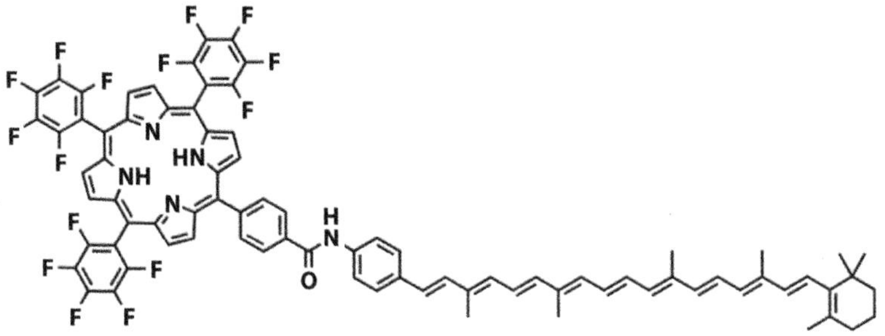

Figure 21.8 Molecular structure of dyad **18**; the fluorinated porphyrin is covalently attached to a carotenoid containing 10 db through an *N*-phenylamide linkage.

photoinduced electron transfer was available, the process would be faster and the transient concentration of the charge-separated species ($C^{\bullet+}$-$P^{\bullet-}$) would build up to a detectable level. In **18**, three of the tolyl groups of the porphyrin were replaced by pentafluorophenyl groups and the carotenoid in **18** was a better electron donor than that in **6–8**. The driving force for the process C-*P → $C^{\bullet+}$-$P^{\bullet-}$ is estimated to be ~400 meV greater in **18** than in **6–8** [50] (Figure 21.8).

The porphyrin fluorescence in **18** was strongly quenched, and using transient absorption spectroscopy, we detected the formation and decay of the carotenoid cation radical, which has strong absorption at around 975 nm in butyronitrile. This construct demonstrated what has come to be called inverted kinetics in which the transient species decays faster than it forms. In this case, the cation radical forms in ~50 ps and decays to the ground state in ~3 ps. We concluded that the necessary driving force for the 50 ps formation of the charge-separated species is provided by the favorable combination of increased oxidation strength of the fluorinated porphyrin with a carotenoid that is easier to oxidize. This work established electron transfer quenching as a viable mechanism for cyclic tetrapyrrole quenching by carotenoids.

One consequence of inverted kinetics is that the concentration of the intermediate does not build up, making it difficult (and sometimes impossible) to detect. In the conclusion of this paper [50] we stated, "…the observation of porphyrin fluorescence quenching by electron transfer from a carotenoid provides a mechanism for similar energy-dissipating electron-transfer quenching *in vivo* and warns that because of the kinetics it could be very difficult to detect the transient species in pigment-bearing proteins. A rapid charge recombination process would be advantageous because it would preclude deleterious reactions of the radical ions in the photosynthetic membranes. Through energy dissipation, carotenoids could play a role in controlling photosynthetic energy flow." Twelve years later in the laboratory of G. R. Fleming, it was possible to detect the formation of the carotenoid radical cation in intact leaves under conditions necessary to induce NPQ [51].

Once the electron transfer mechanism was confirmed to operate in a carotenoporphyrin dyad, the length of the carotenoid conjugated system was studied in relation to its quenching ability. This study was relevant to NPQ because, as mentioned before, during the xanthophyll cycle it is the longer carotenoid (zeaxanthin with 11 db) that is presumably involved in the quenching of the chlorophyll excited state and not the shorter carotenoids (antheraxanthin and violaxanthin with 10 and 9 db, respectively). A series of carotenoporphyrins were synthesized with carotenoids ranging from 5 to 11 double bonds (Figure 21.9) and their porphyrin fluorescence quenching properties evaluated [52]. Steady-state fluorescence studies showed that shorter carotenoids, between 5 and 7 double bonds, do not quench the porphyrin fluorescence while longer carotenoids (8–11 double bonds) are able to quench the porphyrin fluorescence (Figure 21.9).

Electrochemical studies of the same set of carotenoporphyrin dyads and models for their carotenoid moieties resulted in data that are summarized in Figure 21.9c [53]. Carotenoids with more than eight double bonds can form charge-separated states ($C^{\bullet+}$-$P^{\bullet-}$) that are lower in energy than the singlet excited state of a tetra-p-tolylporphyrin (TTP). If the formation of $C^{\bullet+}$-$P^{\bullet-}$ takes place, it would explain the fluorescence quenching of the porphyrin. Although we did not detect the radical

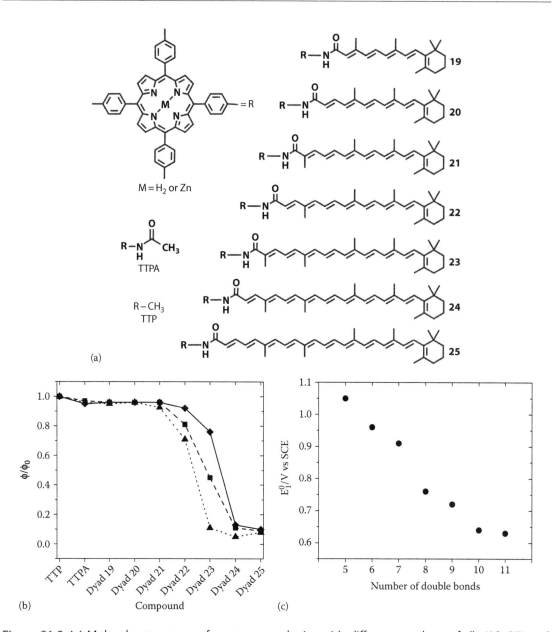

Figure 21.9 **(a)** Molecular structures of carotenoporphyrins with different numbers of db (**19–25**) and 5-(4-acetamidophenyl)-10,15,20-tris(4-methylphenyl)porphyrin (TTPA); **(b)** relative quantum yields of the porphyrin fluorescence of the carotenoporphyrins relative to free base or zinc tetra tolyl porphyrins (TTP): (◆) free base porphyrins and carotenoporphyrins in toluene, (■) free base porphyrins and carotenoporphyrins in acetonitrile, and (▲) zinc porphyrins and carotenoporphyrins in toluene; **(c)** correlation between the first oxidation potential of the carotenoid and the number of db of the carotenoid.

pair species directly, the solvent and redox potential dependence are good evidence that the quenching was by electron transfer to the excited state of the porphyrin. This work established the feasibility of NPQ based on controlling the chlorophyll excited state quenching by adjusting the number of conjugated db in the carotenoid. As shown in Figure 21.9b, a change of one db changes the quenching by >5-fold; in the natural system, the change from zeaxanthin to violaxanthin is two db.

The radical pair species characteristic of electron transfer quenching was subsequently observed in polar solvents in triads **10** and **11**. The sensitivity of this quenching to the dielectric properties of the surrounding matrix provides another mechanism in the natural system for a protein conformational change to induce quenching. We also note that sensitivity to solvent dielectric is not necessarily specific to electron transfer quenching (see Section 21.3.2).

21.3.2 Quenching of the excited state of cyclic tetrapyrroles by energy transfer to the carotenoid S_1 state

We initiated the study of carotenophthalocyanine dyads, expecting to further characterize electron transfer quenching. Phthalocyanine tetrapyrroles with their large extinction coefficients of the Q bands more closely resemble chlorophylls than porphyrins do, and we were interested in exploring this feature for artificial reaction centers. Carotenophthalocyanines **12**, **13**, and **14** with amide groups linking carotenoid moieties with 9, 10, or 11 db, respectively, (see Figure 21.6) to the periphery of a zinc Pc were studied by transient absorption spectroscopy with excitation at 680 nm into the lowest Q band of the Pc [54]. The Pc fluorescence lifetimes indicated that carotenoids with 10 and 11 db quenched the singlet excited state of Pc, while the 9 db carotenoid did not. In dyad **13** (10 db), the Pc singlet lifetime is reduced by a factor of 10, to 300 ps with respect to **12** (9 db, 3 ns), and in dyad **14** (11 db), the carotenoid becomes an even stronger quencher with a fluorescence lifetime of 56 ps.

As we mentioned earlier, the switch from non-quenching to quenching upon an increase in the conjugation length of the carotenoid by one double bond was a remarkable observation and supports the hypothesis that the activation of NPQ in the violaxanthin cycle occurs by increasing the conjugation of the carotenoid. It only remained to show that the quenching mechanism was electron transfer as in the case of dyad **18**. However, repeated and careful time-resolved measurements failed to detect the characteristic spectroscopic signature of the carotenoid radical cation expected from electron transfer quenching. Turning to target analysis, we used a reasonable kinetic scheme and were able to resolve the "pure" spectra of transient species and discovered the characteristic spectrum of the S_1 carotenoid species [31]. The spectral and kinetic

parameters clearly showed that the carotenoid S_1 state is the quencher of the singlet excited state of Pc in dyads **13** and **14**. In other words, this is energy transfer quenching. This work provided the model for detecting the carotenoid S_1 state in these systems, and similar techniques were subsequently used in a study of preparations isolated from LHCII, the major antenna of photosystem II (PSII) of plants. In this natural system, clear evidence of the S_1 state of one of the carotenoids, lutein 1, acting as the quencher of the excited state of chlorophylls was found just a year later [55]. Recently, in a cyanobacteria protein (HliD), which is an ancestor of the whole LHC superfamily, the Polívka group measured the S–S energy transfer rate between Chl-*a* and β-carotene. Two rate constants were determined: 2.1 and 30 ps depending on which Chl-*a* was quenched [56,57].

Figure 21.10 shows the energy level diagram summarizing the proposed Pc Q state quenching mechanism observed in dyads **13** and **14**. Dyad **12** is not able to quench the singlet excited state of Pc because the S_1 state of the carotenoid with 9 db is above the Q state of Pc. On the other hand, the S_1 state of the 9 db carotenoid of dyad **12** acts as energy donor and is responsible for ~20% of the EET from

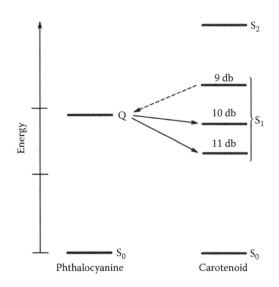

Figure 21.10 Energy diagram showing the relative energy levels of phthalocyanine (zinc Pc) and carotenoids with 9, 10, and 11 db. Solid arrows indicate energy transfer from the Pc Q state to the carotenoid S_1 states, and the dashed arrow indicates energy transfer in the reverse sense from the carotenoid S_1 state to the Pc Q state.

the carotenoid to the Pc [30]. This set of dyads provides a clear example of the "molecular gear shift" mechanism proposed by Frank and coworkers in 1994 to explain the quenching effect of zeaxanthin (11 db carotenoid) and the antenna function of violaxanthin (9 db carotenoid) [58].

It also was observed that quenching of Pc Q state occurs to a larger extent in more polar solvents such as acetone or DMSO. We interpreted this observation as evidence for involvement of a carbonyl carotenoid intramolecular charge transfer (ICT) state in the energy transfer mechanism. ICT involvement could be real or virtual; its energy is a function of solvent dielectric, and as the dielectric increases, it moves the ICT state closer to resonance between the S_1 and Q states. ICT states in carotenoids have been reported by Frank in a number of natural carotenoids containing carbonyl groups [15,59,60]. This finding suggests that the "molecular gear shift" mechanism may be easily controlled by the environment surrounding carotenoids.

21.3.3 Quenching of the excited state of cyclic tetrapyrroles by excitonic coupling with the S_1 state of carotenoids

Dyads 12, 13, and 14 employed an amide linkage, and therefore a carbonyl group was part of the structure. In order to determine whether the carbonyl group was essential to the energy transfer, we designed a series of dyads linked by an arylamino group (Figure 21.11) [61]. The fact that such changes can easily be made in artificial systems highlights the flexibility of working with such systems.

Transient absorption measurements of these dyads in toluene revealed completely different behavior from that observed with carotenophthalocyanines 12–14. Even though the quenching of the Pc singlet state was a function of the number of db, in dyads 28 and 29, the spectral signature of the carotenoid S_1 state was present in the transient spectrum immediately after the excitation of the Pc with a ~100 fs laser pulse [61]. In an attempt to resolve the rise time of the S_1 spectrum, dyad 28 was studied with ~30 fs time resolution and the S_1 spectrum was fully developed on that time scale [62]. This observation indicates that on the 30 fs time scale, the quenching of the Q state of Pc could not be described by a sequential ^{1}Pc to carotenoid S_1 energy transfer scheme as shown with the dyads 12–14 of Figure 21.6. This suggests that the quenching of the Q state of Pc was mediated through the formation of an exciton state, which is described as a composite or shared state between the Pc and the carotenoid. Such an exciton state would be a consequence of the strong electronic coupling between the chromophores in these systems. A schematic diagram showing formation and decay of the exciton state is depicted in Figure 21.12. Importantly, such a state can be populated from either the Pc Q state or the carotenoid S_1 state [61,63,64]. Selective two-photon excitation fluorescence experiments have provided evidence for the existence of such states in the natural systems [63–65].

Dyads 26–29 dissolved in THF showed the characteristic spectroscopic signatures of carotenoid

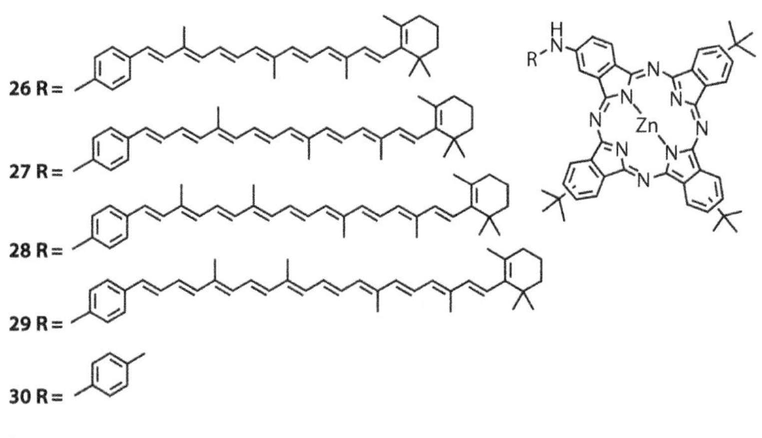

Figure 21.11 Molecular structures of carotenophthalocyanines with 8–11 db carotenoids with phenylamino linkages (26–29) to a zinc Pc and a model zinc Pc (30).

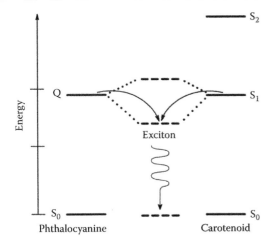

Figure 21.12 Energy diagram showing the exciton state formed in carotenophthalocyanine dyads, which has contributions from the Pc Q and carotenoid S_1 states. After occupying the excitonic state (curved arrows), the excited energy is rapidly dissipated as heat and the system returns to the ground state.

radical cation species, indicating that photoinduced electron transfer contributes to the quenching for these dyads in more polar solvents.

In summary, on the basis of our work on various carotenophthalocyanine dyads, we have identified three mechanisms of cyclic tetrapyrrole singlet excited state quenching by carotenoids in artificial systems:

1. Electron transfer from the carotenoid to the tetrapyrrole and fast recombination
2. Energy transfer from the lowest excited Q state of the tetrapyrrole to form the carotenoid S_1, followed by fast internal conversion to the carotenoid ground state
3. Excitonic state formation between the lowest Q state of the tetrapyrrole and carotenoid S_1 state, followed by rapid internal conversion to the ground state

The dominant quenching mechanism is a function of the molecular structure and dielectric environment. The study of these simple synthetic systems provides the basis for a detailed understanding of the interactions between carotenoids and cyclic tetrapyrroles and plays a role in advancing our understanding of the control mechanisms that operate in natural photosynthetic systems.

21.3.4 Self-regulating artificial photosynthetic systems

We have designed and synthesized an artificial reaction center complex, **31**, with an integral control element that demonstrates how the electron transfer process can be downregulated as a function of light intensity [66]. Molecular pentad **31** consists of two light-absorbing antenna moieties, a porphyrin primary electron donor, a fullerene electron acceptor, and a photochrome control unit that functions as a light intensity sensor. At low light intensities, the photochrome exists in a closed, dihydroindolizine form (DHI) that has no effect upon the photochemistry of the rest of the molecule. EET from the antenna dyes to the porphyrin is essentially 100% efficient. The porphyrin excited singlet state is quenched from 11.4 to 2.0 ns to form the $P^{\bullet+}$-$C_{60}^{\bullet-}$ charge-separated state with a quantum yield of 82%. Increasing the actinic light intensity drives photoisomerization of the control unit from the DHI to the open form, betaine (BT). The BT quenches the porphyrin excited state to 33 ps by EET. The BT reverts thermally to the DHI form. The fraction of pentads with the control unit in the BT form is a reversible function of the actinic light intensity and temperature. This rapid quenching of the porphyrin primary electron donor effectively competes with photoinduced electron transfer to fullerene to dramatically lower the yield of charge-separation involving the fullerene electron acceptor. Thus, this self-regulating molecule adapts its function as an artificial reaction center to the light level and mimics the regulatory mechanism observed in cyanobacteria promoted by the photoactive orange carotenoid protein [67,68] (Figure 21.13).

Another self-regulating system, antenna **32**, consisting of five zinc porphyrin moieties and a pH-sensitive control unit (Figure 21.14) was prepared and studied [69]. When dissolved in an organic solvent, it absorbs light and exchanges singlet excitation energy among the porphyrins with a time constant of ~180 ps. The porphyrin excited states decay by the normal photophysical processes in ~2 ns, which is the fluorescence lifetime of a monomeric zinc porphyrin. Although not coupled to an artificial reaction center at this time, this excited state lifetime is certainly long enough to yield highly efficient charge separation. The control feature is demonstrated when acetic acid is added and the pH-sensitive control unit is protonated and

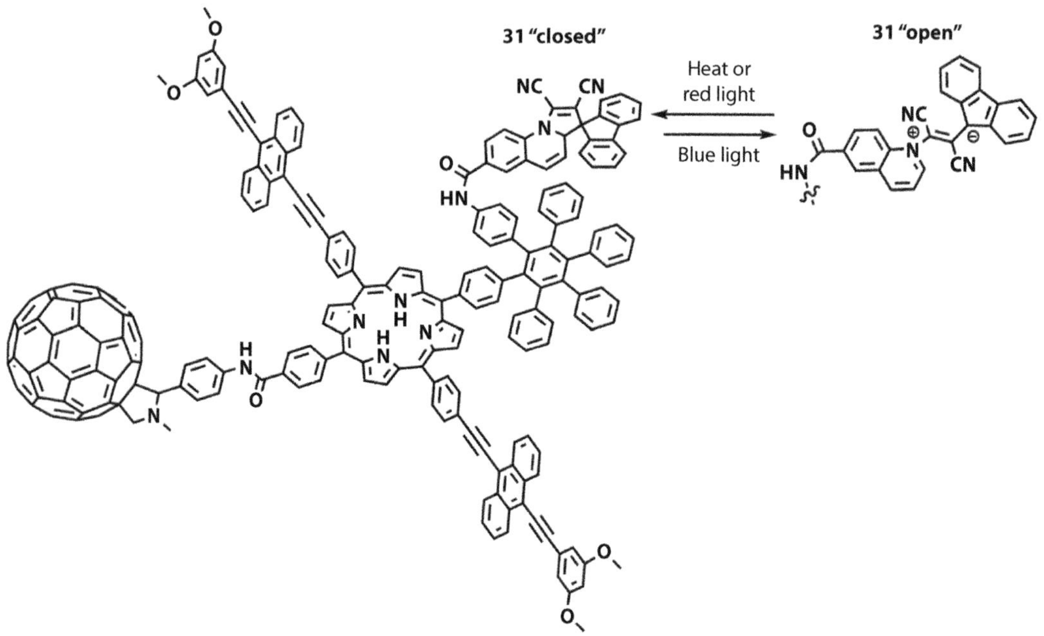

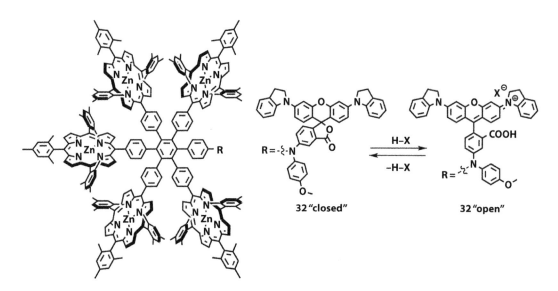

Figure 21.13 Structure of pentad **31** featuring a porphyrin as the primary electron donor linked to a fullerene electron acceptor and two bis(phenylethynyl)anthracene (BPEA) light-absorbing antennas. A photochromic control unit that can exist in two isomeric forms is an integral part of the structure.

Figure 21.14 Structure of the hexad **32** featuring an integral pH sensor in either its closed or open form. HX is an acid that converts the pH sensor from closed to open.

converted to an open form that rapidly quenches the first excited singlet states of the antenna system to <40 ps, rendering them kinetically incompetent to form charge-separated states. The quenching process was ascribed to S–S energy transfer from the porphyrins to the pH-sensitive dye, and the transfer rates were consistent with those calculated using Förster theory. The resulting excited dye rapidly decays to the ground state, converting the excitation energy to heat. This system mimics the role of the antenna in NPQ triggered by low pH in the lumen of the thylakoid membrane. The quenching of the excited states of porphyrins in **31** and **32** is perfectly reversible and demonstrates that incorporation of self-regulation in response to external stimuli is possible in artificial photosynthetic constructs.

21.4 PHOTOPROTECTION

Molecular oxygen formed as by-product of photosynthesis is an essential component of the biosphere. As a terminal electron acceptor in aerobic metabolism, it provides us with abundant energy. However, its reactivity is treacherous. The bad side comes from the facile sensitization of oxygen to yield its lowest excited state known as singlet oxygen ($^1O_2$). $^1O_2$ is an important member of the ROS family; it can damage lipids, proteins, cofactors, and other cellular components and is a serious threat to living cells. In photosynthetic membranes, in addition to the small probability of intersystem crossing from chlorophyll singlet excited states under normal photosynthetic operation, conditions of excess light absorption can lead to radical pair recombination at reaction centers and the formation of triplet chlorophyll. In the oxygen-rich environment of the photosynthetic membrane or in anaerobic photosynthetic bacteria and archaea, which adventitiously encounter an oxygenic environment, these triplet species can sensitize the formation of $^1O_2$. Control mechanisms such as NPQ are partially effective at dissipating excess energy at the chlorophyll singlet excited state level and thereby controlling chlorophyll triplet formation. But, because intersystem crossing and radical pair recombination do occur to a small extent, photosynthetic membranes have a second line of defense against $^1O_2$ involving carotenoid pigments as quenchers of chlorophyll triplet species and $^1O_2$ itself.

Figure 21.15 illustrates the photoprotective processes in photosynthetic membranes that suppress $^1O_2$ sensitization. The lowest excited triplet energy level of carotenoids (having >8 db) is below that of chlorophyll and the $^1O_2$ energy level, which sets the stage for two distinct protective mechanisms:

1. The carotenoid ground-state species can quench the chlorophyll triplet state by energy transfer (step 1), precluding sensitization by the triplet chlorophyll to form $^1O_2$ (step 2). The triplet carotenoid then decays harmlessly, yielding heat and the ground-state carotenoid (step 3).
2. The carotenoid can quench singlet oxygen by energy transfer (step 4), which returns the $^1O_2$ to its ground state; the resulting triplet carotenoid decays harmlessly to its ground state (step 3). Additionally, carotenoids can act as chemical quenchers of $^1O_2$, β-carotene endoperoxide being a major oxidation product in PSII reaction centers [70]. This is considered a minor pathway in comparison with the energy transfer mechanism.

Mechanism 1 actually prevents $^1O_2$ from forming and would appear to be the more effective of the two. It does require that T–T transfer be much faster than diffusion-limited $^1O_2$ sensitization. The rate of this energy transfer process is controlled by thermodynamics and the electronic coupling between the pigments, which depends upon the extent of orbital overlap [71]. In mechanism 2,

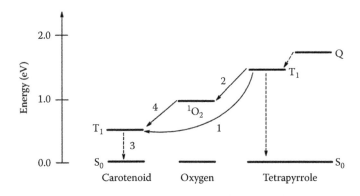

Figure 21.15 An energy level diagram with energy levels of a generic carotenoid and cyclic tetrapyrrole interacting with singlet oxygen ($^1O_2$). Photoprotective processes are represented by numbered arrows: dashed arrows are intersystem crossing events and solid arrows refer to energy transfers.

the cell attempts to trap any $^1O_2$ that does form before it can damage cellular components. This is more risky and would depend on the carotenoid intercepting the trajectory of the $^1O_2$ before it finds a target. In any case, PSII is particularly vulnerable because the slow steps of water oxidation may lead to radical pair recombination at P680 and therefore triplet formation. These triplets cannot be protected by mechanism 1 because a carotenoid in van der Waals contact with P680 would be oxidized faster than the Mn cluster and would preclude water oxidation. This leaves only mechanism 2, which must be only partly effective because PSII frequently succumbs to oxidative damage at high light levels. Indeed, our studies of AP models indicated that mechanism 2 is not highly effective (*vide infra*) [72]. We also predicted that photoprotection in photosynthetic membranes would require van der Waals contact between carotenoid and (bacterio)chlorophyll pigments [16,17,73,74]. The crystal structure of the reaction center of *Rhodobacter sphaeroides (R. sphaeroides)* confirmed our prediction [75].

The effectiveness of mechanism 1 depends critically on the relative rates of the T–T energy transfer that quenches the chlorophyll triplet species and the diffusion-limited rate of singlet oxygen sensitization. This is illustrated in Figure 21.16 in which sensitizers having different T–T energy transfer rates to the attached carotenoid pigment were

tested as $^1O_2$ sensitizers [76]. As a control, TTP with no carotenoid present and therefore no photoprotection was used; it can be seen (Figure 21.16, [□]) that $^1O_2$ was produced at a high rate. β-Carotene added to the TTP solution (Δ) partially quenched singlet oxygen and protected the $^1O_2$ indicator dye, diphenylisobenzofuran (BPDH), by some combination of mechanisms 1 and 2 but at the concentrations used, primarily by mechanism 2. Dyad **2** (●) (Figure 21.2), having porphyrin to carotenoid T–T energy transfer with a 1.5 μs time constant, showed better photoprotection of the BPDH than β-carotene in solution, but some $^1O_2$ was produced. The amount produced is qualitatively consistent with diffusion-limited collisions between oxygen and the porphyrin during its triplet state lifetime of 1.5 μs (note that spin statistics limit the collisions producing $^1O_2$ to 1/9 of the Debye diffusion-limited collisions). Qualitatively, mechanism 2 played a minor role in quenching $^1O_2$ even though the carotenoid pigment was nearby. Consistent with the multitude of expected trajectories of $^1O_2$ following the sensitization collision with the triplet porphyrin moiety, we can assume that most of the trajectories of the nascent $^1O_2$ were not intercepted by the attached carotenoid. In trace (■), an isomer of dyad **2**, dyad **4** (Figure 21.2), in which T–T energy transfer is ultrafast, demonstrated no $^1O_2$ sensitization, that is, complete photoprotection of the DPBH. We assign the photoprotection of DPBH to mechanism 1.

This work exploring the relative efficiencies of $^1O_2$ sensitization as a function of the T–T energy transfer rates and therefore the electronic coupling between cyclic tetrapyrrole and carotenoid has taken on new significance. It has been reported that in the cases looked at to date, T–T energy transfer rates are slower in anaerobic bacteria than in oxygenic photosynthesis [71,77]. Assuming that photoprotection is a fitness parameter in evolutionary selection, this would suggest that in the pigment–protein complexes binding chlorophyll and carotenoid pigments, the electronic coupling necessary for adequate photoprotection is driven by the ambient levels of molecular oxygen, O_2. If this is the case, then oxygenic photosynthesis today may need improved photoprotection when reengineered to provide improved efficiencies. A 10-fold increase in photosynthetic efficiency would mean a 10-fold increase in O_2 flux and therefore a greater challenge for quenching whole arrays of chlorophyll triplet states.

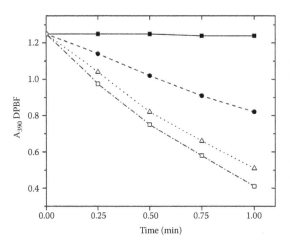

Figure 21.16 Change in absorbance measured at 390 nm due to the bleaching of diphenylisobenzofuran (DPBF), a singlet oxygen indicator, in the presence of (■) *ortho*-ester carotenoporphyrin **4**, (●) *para*-ester carotenoporphyrin **2**, (□) TTP **3**, or (Δ) mixture of TTP and β-carotene in toluene.

As mentioned earlier in connection with $^1O_2$ suppression, T–T energy transfer in the reaction center of *R. sphaeroides*, which was found to depend on temperature [78,79], has been explored using our model systems. In this case, the atomic level resolution of the reaction center crystal structure provided a clue. In fact, the carotenoid is located in van der Waals contact with an accessory bacteriochlorophyll, which is near the reaction center primary donor (the so-called special pair of bacteriochlorophylls) but on the "inactive branch" with respect to electron flow processes. The second lowest triplet level in the system is located on the special pair (the carotenoid triplet is the lowest excited state). A triplet formed on the special pair by radical pair recombination must migrate to the accessory bacteriochlorophyll, an endergonic process, before spontaneously migrating to the carotenoid pigment.

In order to mimic this triplet energy transfer relay [72] and to take advantage of the subtle variations in structure afforded by models to fully explore the kinetic and thermodynamic parameters controlling $^1O_2$ suppression, we designed triad **33** and its derivative, **34**, as triplet relay constructs in which the energy landscape of the triplet levels includes an endergonic step (Figure 21.17). Triad **33** features a free base porphyrin and triad **34** a zinc porphyrin. The pyropheophorbide lowest triplet excited state energy level is lower than that of either porphyrin; it is a mimic of the special pair and the porphyrins are mimics of the accessory bacteriochlorophyll in the *R. sphaeroides* reaction center. Following excitation of the triad, singlet excitation is localized on the pyropheophorbide where intersystem crossing forms the triplet state and activated T–T energy transfer occurs to populate the carotenoid triplet level. Because the zinc porphyrin has a higher triplet excited state energy level than the free base, energy transfer to the carotenoid in triad **34** is slower than in **33**.

Generally, we observed the same thermally activated flow of triplet energy in these triads that was observed in the natural system. In all cases, at 77 K there was no evidence of T–T energy transfer. At room temperature, $^1O_2$ sensitization was suppressed in triad **33** but not prevented. In the case of **34**, due to slow T–T energy transfer to the carotenoid, the quantum yield of $^1O_2$ formation was not suppressed by the attached carotenoid pigment. We conclude that even though the carotenoid is near the site of $^1O_2$ formation, it is not effective at $^1O_2$ suppression, implying that mechanism 2 is not effective.

It is worth noting that the temperature dependence establishes that the accessory bacteriochlorophyll in the natural system and the porphyrins in

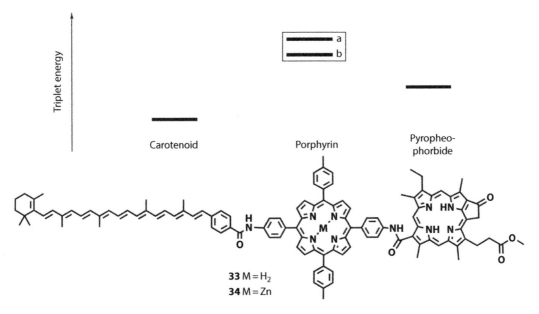

Figure 21.17 Energy diagram for triad **34**, which contains a zinc porphyrin with a higher porphyrin triplet energy level (a) than triad **33** (b).

triads **33** and **34** are real, not virtual, intermediates in the triplet energy transfer process.

Interestingly, the necessary conformation for electronic coupling need not be the most stable one, provided sufficient thermally activated conformations are available for the dyad to transiently acquire the required orbital overlap. In 1982, we reported a carotenoporphyrin (dyad **5**, Figure 21.2) [18] in which the linkage between the chromophores was electronically insulating but sufficiently flexible to allow the dyad to transiently "fold" and undergo electron exchange–mediated T–T energy transfer. Freezing these dyads to restrict thermal motion completely abolished the T–T energy transfer. But, because the carotenoid singlet states are very short lived, the thermal motion required for electronic coupling is not sufficiently fast to mediate S–S energy transfer; the efficiency of S–S transfer from the carotenoid to the porphyrin was vanishingly low.

As mentioned earlier in connection with singlet energy transfer and photoregulation, the role of electron exchange interactions in mediating energy and electron transfer is further illustrated by the series of *ortho*, *meta*, and *para* amide-linked dyads **6**, **7**, and **8** (Figure 21.2) [23]. In each case, the number of bonds and the hybridization of the atoms in the bonds are the same. As shown in Figure 21.18 and Table 21.1, the photophysics of the three isomers depend in the same way on the attachment point of the linkage to the aryl group. This dependence is adequately explained by electronic coupling terms that depend upon the overlap integrals that arise in electron exchange interactions, which

are a sensitive function of the *ortho*, *meta*, and *para* positions on the aromatic ring, as explained earlier.

Mimicking the photophysics of nonadiabatic processes in photosynthesis involves managing the thermodynamic parameters controlling the process as well as the electronic coupling. The thermodynamic parameters in EET and T–T energy transfer are quantitatively characterized by using spectral overlap to evaluate the density of states term in the Fermi Golden Rule. Here, we present one example where the thermodynamics of electron transfer, specifically the reorganization energy, played a pivotal role in biomimicry. In the reaction centers of photosynthetic purple bacteria, radical pair recombination to yield a triplet state localized on the special pair was one of the most intriguing and characteristic electron transfer processes observed in the early events of photosynthesis. The role of radical pair recombination was touched on in the introduction of Section 21.4. Both transient absorption and magnetic resonance techniques have been used to characterize the exchange coupling and dynamics that lead to charge recombination involving triplet excited states [80–82]. In a series of papers, we reported the requisite conditions in artificial reaction centers for observing such recombination [83,84]. We found that observing recombination to a triplet energy surface in these systems depends crucially on the reorganization energy. When the internal reorganization energy was low, implemented in our system by using a C_{60} moiety as the electron acceptor and a porphyrin as primary donor in carotenoporphyrin-C_{60} triad **35** (Figure 21.19), the solvent dielectric

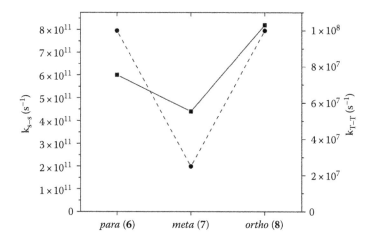

Figure 21.18 Singlet–singlet (k_{S-S}, ■) and triplet–triplet energy transfer rates (k_{T-T}, ●) measured for three carotenoporphyrin amide isomers. See also Table 21.1.

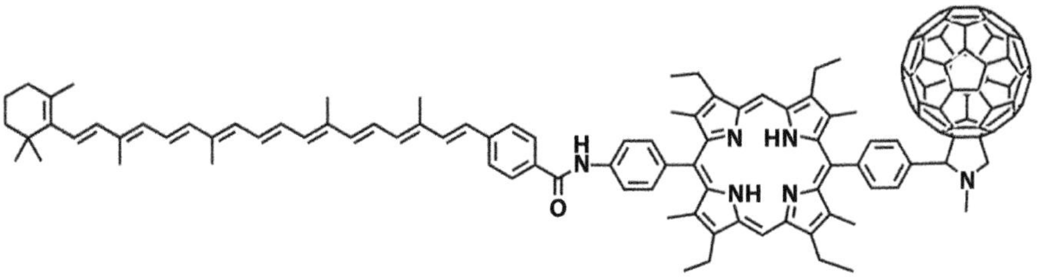

Figure 21.19 An AP reaction center: carotenoporphyrin–C_{60} triad **35**. It undergoes photoinduced charge separation that recombines to the carotenoid triplet state.

could be varied to modulate the yield of recombination to the triplet species from vanishingly small to almost 100%. Unexpectedly, understanding in detail the photophysics, dynamics, and thermodynamics of recombination in these triads could play an important role in unlocking some of the secrets of navigation in migrating birds! [85].

21.5 CONCLUSIONS

In conclusion, the fundamental photochemical interactions among chromophores that are used in photosynthesis for light gathering and photoprotection can be realized and studied in small, covalently linked model systems. The study of these simpler systems contributes to the understanding of energy flow in the much more complex natural systems. In addition, the principles embodied in the synthetic molecules could be used to engineer photosynthesis-like functions into artificial photosynthetic solar energy conversion devices.

We also hypothesize that the photoprotective role of carotenoids is an example of evolutionary exaptation. We imagine that in the beginning, the only photophysical properties important in nature's involvement with carotenoids were the highly allowed absorption to the S_2 state and the highly forbidden nature of the S_1 state. The strong absorption of the S_2 state for light gathering was advantageous for the nascent solar-powered organisms. In fact, even shorter carotenoids were finding their way into photoprocesses as revealed by the ubiquitous distribution of rhodopsins in the biosphere. Even in the early anoxic atmosphere, photosynthesis began the process of freeing organisms from the requirement of a redox gradient in the earth's crust to drive their metabolism. Solar-powered organisms surely thrived. Because carotenoid S_2 and S_1 states are very short lived, making use of those states in light harvesting required the strong electronic coupling

to the chlorophyll pigment system as described in this review. Evolutionary pressure drove this coupling to the point where light harvesting acquired the necessary efficiency for a particular organism to be reproductively successful in its environmental niche.

Serendipitously, the carotenoid pigments came with another photophysical property, a triplet level below $^1O_2$ [86]. Because the evolution of photosynthesis occurred in an anoxic atmosphere, there was no selective pressure for $^1O_2$ suppression. However, the electronic coupling necessary to make the combination of carotenoid S_2 and S_1 states successful at singlet energy transfer to the chlorophyll system assured the electronic coupling necessary for T–T energy transfer and photoprotection. Thus, the stage was set for the evolution of water oxidation and O_2 production. In other words, photoprotection was in place before it was needed; it preexisted. Water oxidation and O_2 production were a huge success for photosynthetic organisms and, ironically, resulted in the extinction of what must have been a wonderful cornucopia of anaerobic life forms.

ACKNOWLEDGMENT

Much of the work reported in this chapter was supported by the Office of Basic Energy Sciences, Division of Chemical Sciences, Geosciences, and Energy Biosciences, Department of Energy, under Contract DE-FG02-03ER15393.

REFERENCES

1. Sherman, B. D., M. D. Vaughn, J. J. Bergkamp, D. Gust, A. L. Moore, and T. A. Moore. 2014. Evolution of reaction center mimics to systems capable of generating solar fuel. *Photosynth. Res.* 120:59–70.

2. Concepcion, J. J., R. L. House, J. M. Papanikolas, and T. J. Meyer. 2012. Chemical approaches to artificial photosynthesis. *Proc. Natl. Acad. Sci. U. S. A.* 109:15560–15564.

3. Hambourger, M., G. F. Moore, D. M. Kramer, D. Gust, A. L. Moore, and T. A. Moore. 2009. Biology and technology for photochemical fuel production. *Chem. Soc. Rev.* 38:25–35.

4. Gust, D., T. A. Moore, and A. L. Moore. 2000. Mimicking photosynthetic solar energy transduction. *Acc. Chem. Res.* 34:40–48.

5. Gust, D., T. A. Moore, and A. L. Moore. 2009. Solar fuels via artificial photosynthesis. *Acc. Chem. Res.* 42:1890–1898.

6. Llansola-Portoles, M. J., D. Gust, T. A. Moore, and A. L. Moore. 2017. Artificial photosynthetic antennas and reaction centers. *C. R. Chim.* 20:296–313.

7. Llansola-Portoles, M. J., R. E. Palacios, D. Gust, T. A. Moore, and A. L. Moore. 2015. Artificial photosynthesis: From molecular to hybrid nanoconstructs. In *From Molecules to Materials—Pathways to Artificial Photosynthesis*. E. Rozhkova and K. Ariga eds. Springer International Publishing, Cham, Switzerland.

8. Gust, D. 2016. Chapter one—An illustrative history of artificial photosynthesis. In *Advances in Botanical Research*. R. Bruno ed. Academic Press Ltd-Elsevier Science Ltd., London, England, pp. 1–42.

9. Blankenship, R. E., D. M. Tiede, J. Barber, G. W. Brudvig, G. Fleming, M. Ghirardi, M. R. Gunner et al. 2011. Comparing photosynthetic and photovoltaic efficiencies and recognizing the potential for improvement. *Science* 332:805–809.

10. Moore, T. A., A. L. Moore, and D. Gust. 2013. Artificial photosynthesis combines biology with technology for sustainable energy transformation. *AIP Conf. Proc.* 1519:68–72.

11. Lovley, D. R., T. Ueki, T. Zhang, N. S. Malvankar, P. M. Shrestha, K. A. Flanagan, M. Aklujkar et al. 2011. Geobacter: The microbe electric's physiology, ecology, and practical applications. In *Advanced Microbial Physiology*. K. P. Robert ed. Academic Press, London, U.K., pp. 1–100.

12. Gust, D., T. A. Moore, and A. L. Moore. 2012. Realizing artificial photosynthesis. *Faraday Discuss.* 155:9–26.

13. Croce, R. and H. van Amerongen. 2014. Natural strategies for photosynthetic light harvesting. *Nat. Chem. Biol.* 10:492–501.

14. Polívka, T. and H. A. Frank. 2010. Molecular factors controlling photosynthetic light harvesting by carotenoids. *Acc. Chem. Res.* 43:1125–1134.

15. Polívka, T. and V. Sundström. 2004. Ultrafast dynamics of carotenoid excited states–from solution to natural and artificial systems. *Chem. Rev. (Washington, DC)* 104:2021–2072.

16. Dirks, G., A. L. Moore, T. A. Moore, and D. Gust. 1980. Light absorption and energy transfer in polyene-porphyrin esters. *Photochem. Photobiol.* 32:277–280.

17. Moore, A. L., G. Dirks, D. Gust, and T. A. Moore. 1980. Energy transfer from carotenoid polyenes to porphyrins: A light-harvesting antenna. *Photochem. Photobiol.* 32:691–695.

18. Moore, A. L., A. Joy, R. Tom, D. Gust, T. A. Moore, R. V. Bensasson, and E. J. Land. 1982. Photoprotection by carotenoids during photosynthesis: Motional dependence of intramolecular energy transfer. *Science* 216:982–984.

19. Liddell, P. A., G. A. Nemeth, W. R. Lehman, A. M. Joy, A. L. Moore, R. V. Bensasson, T. A. Moore, and D. Gust. 1982. Mimicry of carotenoid function in photosynthesis: Synthesis and photophysical properties of a carotenopyropheophorbide. *Photochem. Photobiol.* 36:641–645.

20. Gust, D., T. A. Moore, R. V. Bensasson, P. Mathis, E. J. Land, C. Chachaty, A. L. Moore, P. A. Liddell, and G. A. Nemeth. 1985. Stereodynamics of intramolecular triplet energy transfer in carotenoporphyrins. *J. Am. Chem. Soc.* 107:3631–3640.

21. Frank, H. A., B. W. Chadwick, J. Jin Oh, D. Gust, T. A. Moore, P. A. Liddell, A. L. Moore, L. R. Makings, and R. J. Cogdell. 1987. Triplet-triplet energy transfer in B800–850 light-harvesting complexes of photosynthetic bacteria and synthetic carotenoporphyrin molecules investigated by electron spin resonance. *Biochim. Biophys. Acta Bioenerg.* 892:253–263.

22. Moore, T. A., D. Gust, and A. L. Moore. 1989. The function of carotenoid pigments in photosynthesis and their possible involvement in the evolution of higher plants. In *Carotenoids*. N. Krinsky, M. Mathews-Roth, and R. Taylor eds. Springer, New York, pp. 223–228.

23. Gust, D., T. A. Moore, A. L. Moore, C. Devadoss, P. A. Liddell, R. Hermant, R. A. Nieman, L. J. Demanche, J. M. DeGraziano, and I. Gouni. 1992. Triplet and singlet energy transfer in carotene-porphyrin dyads: Role of the linkage bonds. *J. Am. Chem. Soc.* 114:3590–3603.

24. Polívka, T. and V. Sundström. 2009. Dark excited states of carotenoids: Consensus and controversy. *Chem. Phys. Lett.* 477:1–11.

25. Macpherson, A. N., P. A. Liddell, D. Kuciauskas, D. Tatman, T. Gillbro, D. Gust, T. A. Moore, and A. L. Moore. 2002. Ultrafast energy transfer from a carotenoid to a chlorin in a simple artificial photosynthetic antenna. *J. Phys. Chem. B* 106:9424–9433.

26. Gouterman, M. 1961. Spectra of porphyrins. *J. Mol. Spectrosc.* 6:138–163.

27. Mariño-Ochoa, E., R. Palacios, G. Kodis, A. N. Macpherson, T. Gillbro, D. Gust, T. A. Moore, and A. L. Moore. 2002. High-efficiency energy transfer from carotenoids to a phthalocyanine in an artificial photosynthetic antenna. *Photochem. Photobiol.* 76:116–121.

28. Cogdell, R. J., A. Gall, and J. Köhler. 2006. The architecture and function of the light-harvesting apparatus of purple bacteria: From single molecules to in vivo membranes. *Q. Rev. Biophys.* 39:227–324.

29. Kodis, G., C. Herrero, R. Palacios, E. Mariño-Ochoa, S. Gould, L. de la Garza, R. van Grondelle et al. 2004. Light harvesting and photoprotective functions of carotenoids in compact artificial photosynthetic antenna designs. *J. Phys. Chem. B* 108:414–425.

30. Berera, R., I. H. M. van Stokkum, G. Kodis, A. E. Keirstead, S. Pillai, C. Herrero, R. E. Palacios et al. 2007. Energy transfer, excited-state deactivation, and exciplex formation in artificial caroteno-phthalocyanine light-harvesting antennas. *J. Phys. Chem. B* 111:6868–6877.

31. van Stokkum, I. H. M., D. S. Larsen, and R. van Grondelle. 2004. Global and target analysis of time-resolved spectra. *Biochim. Biophys. Acta Bioenerg.* 1657:82–104.

32. Papagiannakis, E., J. T. M. Kennis, I. H. M. van Stokkum, R. J. Cogdell, and R. van Grondelle. 2002. An alternative carotenoid-to-bacteriochlorophyll energy transfer pathway in photosynthetic light harvesting. *Proc. Natl. Acad. Sci. U. S. A.* 99:6017–6022.

33. Kloz, M., S. Pillai, G. Kodis, D. Gust, T. A. Moore, A. L. Moore, R. v. Grondelle, and J. T. M. Kennis. 2012. New light-harvesting roles of hot and forbidden carotenoid states in artificial photosynthetic constructs. *Chem. Sci.* 3:2052–2061.

34. Scholes, G. D., X. J. Jordanides, and G. R. Fleming. 2001. Adapting the Förster theory of energy transfer for modeling dynamics in aggregated molecular assemblies. *J. Phys. Chem. B* 105:1640–1651.

35. Krueger, B. P., G. D. Scholes, and G. R. Fleming. 1998. Calculation of couplings and energy-transfer pathways between the pigments of LH2 by the *ab initio* transition density cube method. *J. Phys. Chem. B* 102:5378–5386.

36. Kuciauskas, D., P. A. Liddell, S. Lin, T. E. Johnson, S. J. Weghorn, J. S. Lindsey, A. L. Moore, T. A. Moore, and D. Gust. 1999. An artificial photosynthetic antenna-reaction center complex. *J. Am. Chem. Soc.* 121:8604–8614.

37. Kodis, G., P. A. Liddell, L. de la Garza, P. C. Clausen, J. S. Lindsey, A. L. Moore, T. A. Moore, and D. Gust. 2002. Efficient energy transfer and electron transfer in an artificial photosynthetic antenna–reaction center complex. *J. Phys. Chem. A* 106:2036–2048.

38. Strachan, J.-P., S. Gentemann, J. Seth, W. A. Kalsbeck, J. S. Lindsey, D. Holten, and D. F. Bocian. 1997. Effects of orbital ordering on electronic communication in multiporphyrin arrays. *J. Am. Chem. Soc.* 119:11191–11201.

39. Yang, S. I., J. Seth, T. Balasubramanian, D. Kim, J. S. Lindsey, D. Holten, and D. F. Bocian. 1999. Interplay of orbital tuning and linker location in controlling electronic communication in porphyrin arrays. *J. Am. Chem. Soc.* 121:4008–4018.

40. Yang, S. I., J. Seth, J.-P. Strachan, S. Gentemann, D. Kim, D. Holten, J. S. Lindsey, and D. F. Bocian. 1999. Ground and excited state electronic properties of halogenated tetraarylporphyrins. Tuning the building blocks for porphyrin-based photonic devices. *J. Porphyrins Phthalocyanines* 3:117–147.

41. Kodis, G., Y. Terazono, P. A. Liddell, J. Andréasson, V. Garg, M. Hambourger, T. A. Moore, A. L. Moore, and D. Gust. 2006. Energy and photoinduced electron transfer

in a wheel-shaped artificial photosynthetic antenna-reaction center complex. *J. Am. Chem. Soc.* 128:1818–1827.

42. Terazono, Y., G. Kodis, P. A. Liddell, V. Garg, T. A. Moore, A. L. Moore, and D. Gust. 2009. Multiantenna artificial photosynthetic reaction center complex. *J. Phys. Chem. B* 113:7147–7155.

43. Terazono, Y., T. A. Moore, A. L. Moore, and D. Gust. 2012. Light harvesting, excitation energy/electron transfer, and photoregulation in artificial photosynthetic systems. In *Multiporphyrin Arrays: Fundamentals and Applications*. D. Kim ed. Pan Stanford Publishing, Singapore.

44. Garg, V., G. Kodis, P. A. Liddell, Y. Terazono, T. A. Moore, A. L. Moore, and D. Gust. 2013. Artificial photosynthetic reaction center with a coumarin-based antenna system. *J. Phys. Chem. B* 117:11299–11308.

45. Gust, D., D. Kramer, A. Moore, T. A. Moore, and W. Vermaas. 2008. Engineered and artificial photosynthesis: Human ingenuity enters the game. *MRS Bull.* 33:383–387.

46. Demmig-Adams, B. and W. W. Adams. 2002. Antioxidants in photosynthesis and human nutrition. *Science* 298:2149–2153.

47. Niyogi, K. K. 1999. Photoprotection revisited: Genetic and molecular approaches. *Annu. Rev. Plant Physiol. Plant Mol. Biol.* 50:333–359.

48. Havaux, M. and K. K. Niyogi. 1999. The violaxanthin cycle protects plants from photooxidative damage by more than one mechanism. *Proc. Natl. Acad. Sci. U. S. A.* 96:8762–8767.

49. Baroli, I., A. D. Do, T. Yamane, and K. K. Niyogi. 2003. Zeaxanthin accumulation in the absence of a functional xanthophyll cycle protects chlamydomonas reinhardtii from photooxidative stress. *Plant Cell* 15:992–1008.

50. Hermant, R. M., P. A. Liddell, S. Lin, R. G. Alden, H. K. Kang, A. L. Moore, T. A. Moore, and D. Gust. 1993. Mimicking carotenoid quenching of chlorophyll fluorescence. *J. Am. Chem. Soc.* 115:2080–2081.

51. Holt, N. E., D. Zigmantas, L. Valkunas, X.-P. Li, K. K. Niyogi, and G. R. Fleming. 2005. Carotenoid cation formation and the regulation of photosynthetic light harvesting. *Science* 307:433–436.

52. Cardoso, S. L., D. Nicodem, T. A. Moore, A. L. Moore, and D. Gust. 1996. Synthesis and fluorescence quenching studies of a series of carotenoporphyrins with carotenoids of various lengths. *J. Braz. Chem. Soc.* 7:19–30.

53. Fungo, F., L. Otero, E. Durantini, W. J. Thompson, J. J. Silber, T. A. Moore, A. L. Moore, D. Gust, and L. Sereno. 2003. Correlation of fluorescence quenching in carotenoporphyrin dyads with the energy of intramolecular charge transfer states. Effect of the number of conjugated double bonds of the carotenoid moiety. *Phys. Chem. Chem. Phys.* 5:469–475.

54. Berera, R., C. Herrero, I. H. M. van Stokkum, M. Vengris, G. Kodis, R. E. Palacios, H. van Amerongen et al. 2006. A simple artificial light-harvesting dyad as a model for excess energy dissipation in oxygenic photosynthesis. *Proc. Natl. Acad. Sci. U. S. A.* 103:5343–5348.

55. Ruban, A. V., R. Berera, C. Ilioaia, I. H. M. van Stokkum, J. T. M. Kennis, A. A. Pascal, H. van Amerongen, B. Robert, P. Horton, and R. van Grondelle. 2007. Identification of a mechanism of photoprotective energy dissipation in higher plants. *Nature* 450:575–578.

56. Llansola-Portoles, M. J., R. Sobotka, E. Kish, M. K. Shukla, A. A. Pascal, T. Polívka, and B. Robert. 2017. Twisting a β-carotene, an adaptive trick from nature for dissipating energy during photoprotection. *J. Biol. Chem.* 292:1396–1403.

57. Staleva, H., J. Komenda, M. K. Shukla, V. Šlouf, R. Kaňa, T. Polívka, and R. Sobotka. 2015. Mechanism of photoprotection in the cyanobacterial ancestor of plant antenna proteins. *Nat. Chem. Biol.* 11:287–291.

58. Frank, H., A. Cua, V. Chynwat, A. Young, D. Gosztola, and M. Wasielewski. 1994. Photophysics of the carotenoids associated with the xanthophyll cycle in photosynthesis. *Photosynth. Res.* 41:389–395.

59. Bautista, J. A., R. E. Connors, B. B. Raju, R. G. Hiller, F. P. Sharples, D. Gosztola, M. R. Wasielewski, and H. A. Frank. 1999. Excited state properties of peridinin: Observation of a solvent dependence of the lowest excited singlet state lifetime and spectral behavior unique among carotenoids. *J. Phys. Chem. B* 103:8751–8758.

60. Frank, H. A., J. A. Bautista, J. Josue, Z. Pendon, R. G. Hiller, F. P. Sharples, D. Gosztola, and M. R. Wasielewski. 2000. Effect of the solvent environment on the spectroscopic properties and dynamics of the lowest excited states of carotenoids. *J. Phys. Chem. B* 104:4569–4577.

61. Kloz, M., S. Pillai, G. Kodis, D. Gust, T. A. Moore, A. L. Moore, R. van Grondelle, and J. T. M. Kennis. 2011. Carotenoid photoprotection in artificial photosynthetic antennas. *J. Am. Chem. Soc.* 133:7007–7015.

62. Maiuri, M., J. J. Snellenburg, I. H. M. van Stokkum, S. Pillai, K. WongCarter, D. Gust, T. A. Moore et al. 2013. Ultrafast energy transfer and excited state coupling in an artificial photosynthetic antenna. *J. Phys. Chem. B* 117:14183–14190.

63. Liao, P.-N., S. Pillai, D. Gust, T. A. Moore, A. L. Moore, and P. J. Walla. 2011. Two-photon study on the electronic interactions between the first excited singlet states in carotenoid–tetrapyrrole dyads. *J. Phys. Chem. A* 115:4082–4091.

64. Liao, P.-N., S. Pillai, M. Kloz, D. Gust, A. Moore, T. Moore, J. M. Kennis, R. Grondelle, and P. Walla. 2012. On the role of excitonic interactions in carotenoid–phthalocyanine dyads and implications for photosynthetic regulation. *Photosynth. Res.* 111:237–243.

65. Bode, S., C. C. Quentmeier, P.-N. Liao, N. Hafi, T. Barros, L. Wilk, F. Bittner, and P. J. Walla. 2009. On the regulation of photosynthesis by excitonic interactions between carotenoids and chlorophylls. *Proc. Natl. Acad. Sci. U. S. A.* 106:12311–12316.

66. Straight, S. D., G. Kodis, Y. Terazono, M. Hambourger, T. A. Moore, A. L. Moore, and D. Gust. 2008. Self-regulation of photoinduced electron transfer by a molecular nonlinear transducer. *Nat. Nanotechnol.* 3:280–283.

67. Tian, L., I. H. M. van Stokkum, R. B. M. Koehorst, A. Jongerius, D. Kirilovsky, and H. van Amerongen. 2011. Site, rate, and mechanism of photoprotective quenching in cyanobacteria. *J. Am. Chem. Soc.* 133:18304–18311.

68. Wilson, A., C. Punginelli, A. Gall, C. Bonetti, M. Alexandre, J.-M. Routaboul, C. A. Kerfeld et al. 2008. A photoactive carotenoid protein acting as light intensity sensor. *Proc. Natl. Acad. Sci. U. S. A.* 105:12075–12080.

69. Terazono, Y., G. Kodis, K. Bhushan, J. Zaks, C. Madden, A. L. Moore, T. A. Moore, G. R. Fleming, and D. Gust. 2011. Mimicking the role of the antenna in photosynthetic photoprotection. *J. Am. Chem. Soc.* 133:2916–2922.

70. Ramel, F., S. Birtic, S. Cuiné, C. Triantaphylidès, J.-L. Ravanat, and M. Havaux. 2012. Chemical quenching of singlet oxygen by carotenoids in plants. *Plant Physiol.* 158:1267–1278.

71. Ho, J., E. Kish, D. Méndez-Hernández, K. WongCarter, S. Pillai, G. Kodis, D. Gust et al. 2017. Triplet-triplet energy transfer in artificial photosynthetic antennas. *Proc. Natl. Acad. Sci. U. S. A.* 114 (28):E5513–E5521.

72. Gust, D., T. A. Moore, A. L. Moore, A. A. Krasnovsky, P. A. Liddell, D. Nicodem, J. M. DeGraziano, P. Kerrigan, L. R. Makings, and P. J. Pessiki. 1993. Mimicking the photosynthetic triplet energy-transfer relay. *J. Am. Chem. Soc.* 115:5684–5691.

73. Schenck, C. C., P. Mathis, M. Lutz, D. Gust, and T. A. Moore. 1983. Triplet state of bacteriochlorophyll and carotenoid in bacterial reaction centers. *Biophys. J.* 41:113a–160a.

74. Liddell, P. A., D. Barrett, L. R. Makings, P. J. Pessiki, D. Gust, and T. A. Moore. 1986. Charge separation and energy transfer in carotenopyropheophorbide-quinone triads. *J. Am. Chem. Soc.* 108:5350–5352.

75. Yeates, T. O., H. Komiya, A. Chirino, D. C. Rees, J. P. Allen, and G. Feher. 1988. Structure of the reaction center from *Rhodobacter sphaeroides* R-26 and 2.4.1: Protein-cofactor (bacteriochlorophyll, bacteriopheophytin, and carotenoid) interactions. *Proc. Natl. Acad. Sci. U. S. A.* 85:7993–7997.

76. Bensasson, R. V., E. J. Land, A. L. Moore, R. L. Crouch, G. Dirks, T. A. Moore, and D. Gust. 1981. Mimicry of antenna and photo-protective carotenoid functions by a synthetic carotenoporphyrin. *Nature* 290:329–332.

77. Gall, A., R. Berera, M. T. A. Alexandre, A. A. Pascal, L. Bordes, M. M. Mendes-Pinto, S. Andrianambinintsoa et al. 2011. Molecular adaptation of photoprotection: Triplet states in light-harvesting proteins. *Biophys. J.* 101:934–942.

78. Frank, H. A., J. Machnicki, and R. Friesner. 1983. Energy transfer between the primary donor bacteriochlorophyll and carotenoids in *Rhodopseudomonas sphaeroides*. *Photochem. Photobiol.* 38:451–455.

79. Schenck, C. C., P. Mathis, and M. Lutz. 1984. Triplet formation and triplet decay in reaction centers from the photosynthetic bacterium *Rhodopseudomonas sphaeroides*. *Photochem. Photobiol.* 39:407–417.

80. Blankenship, R. E. 2014. *Molecular Mechanisms of Photosynthesis*. Blackwell Science, Oxford, U.K.

81. Blankenship, R. E. 1981. Chemically induced magnetic polarization in photosynthetic systems. *Acc. Chem. Res.* 14:163–170.

82. Thurnauer, M. C., J. J. Katz, and J. R. Norris. 1975. The triplet state in bacterial photosynthesis: Possible mechanisms of the primary photo-act. *Proc. Natl. Acad. Sci. U. S. A.* 72:3270–3274.

83. Kuciauskas, D., P. A. Liddell, S. Lin, S. G. Stone, A. L. Moore, T. A. Moore, and D. Gust. 2000. Photoinduced electron transfer in carotenoporphyrin–fullerene triads: Temperature and solvent effects. *J. Phys. Chem. B* 104:4307–4321.

84. Carbonera, D., M. Di Valentin, C. Corvaja, G. Agostini, G. Giacometti, P. A. Liddell, D. Kuciauskas, A. L. Moore, T. A. Moore, and D. Gust. 1998. EPR investigation of photoinduced radical pair formation and decay to a triplet state in a carotene–porphyrin–fullerene triad. *J. Am. Chem. Soc.* 120:4398–4405.

85. Maeda, K., K. B. Henbest, F. Cintolesi, I. Kuprov, C. T. Rodgers, P. A. Liddell, D. Gust, C. R. Timmel, and P. J. Hore. 2008. Chemical compass model of avian magnetoreception. *Nature* 453:387–390.

86. Moore, T. A., A. L. Moore, and D. Gust. 2002. The design and synthesis of artificial photosynthetic antennas, reaction centres and membranes. *Philos. Trans. R. Soc. Lond. B* 357:1481–1498.

22

Light to useful charge in nanostructured organic and hybrid solar cells

TÖNU PULLERITS AND VILLY SUNDSTRÖM

22.1 INTRODUCTION

Many different types of nanostructured thin-film materials and combination of materials are considered in the search for efficient, cheap, and environmentally benign solar cells. The introduction of dye-sensitized solar cells[1] in the early 1990s was a major step, and intense development work in many different directions have led to solar cells with power conversion efficiencies approaching 15%.[2] Here we will describe some of our recent work aiming at correlating dye–semiconductor electron transfer, sensitizer binding geometry, and solar cell efficiency. The ultimate aim of this work is to develop methods to produce a sensitizer/semiconductor material with predictable light-harvesting and electron transfer properties.

Only a few years ago, this field took an exciting new direction when organometal halide perovskites (OMHPs) were introduced as sensitizers,[3–5] and within this very short time, solar cells based on such perovskites are already over a remarkable power conversion of 22%.[6] Despite this very fast progress in material and device performance, many questions regarding the fundamental properties underlying this astonishing progress remain unanswered. Here we will summarize some of our very recent results on exciton and charge carrier dynamics in these novel materials.

Yet another direction within the dye sensitized solar cell (DSC) field is to replace the often-used Ru-sensitizers with earth-abundant and benign materials. Fully organic dyes is one approach that has been successfully explored. Another strategy to achieve this goal is to replace the Ru-based dyes with analogs based on earth-abundant metals, for example, iron (Fe). Until very recently this work was not rewarded with much success. By replacing the conventionally used polypyridyl ligands in transition metal complexes with N-heterocyclic carbenes (NHCs) in Fe-complexes, we could increase the triplet metal to ligand charge transfer (^{3}MLCT) state lifetime by a factor of hundred as compared to conventional Fe-polypyridyl complexes.[7] With this increase of the MLCT lifetime, Fe-based sensitizers are becoming of interest for DSC. We briefly summarize this work.

Semiconductor nanocrystals, so-called quantum dots (QDs), have emerged as a viable alternative to replace the dye molecules in DSC to form QD-sensitized solar cells (QDSC).[8] In QDs a continuum of states in the conduction and valence bands of a semiconductor transforms into sets of discrete energy levels due to quantum confinement of carriers in the tiny volume of a "dot." The level spacing qualitatively follows the particle-in-a-box principle. In this way, the absorption spectrum of the QD can be easily tuned by changing the size of the dots. The term "rainbow solar cell" was coined[9] to artistically visualize the ease and beauty of covering the whole solar spectrum by such simple means. Perhaps even more promising is the ability of QDs to generate more than one electron-hole pair from a single high-energy photon.[10] Such multiple exciton generation (MEG) allows to break the Shockley-Queisser thermodynamic limit of single-junction solar cells.[11] In the following, we will give a brief account of the QDSC and summarize some of our own work. Besides the QDSC, QDs can be used in various other solar cell configurations. For example, the successive ionic-layer adsorption and reaction (SILAR)[12] method enables putting QDs directly on the electron-accepting electrode without involving any linker molecule. Various strategies based on thin films of colloidal QDs are also being developed.[13]

Both small molecules and polymer materials are considered for fully organic solar cells. The bulk heterojunction (BHJ) concept[14,15] has been successful in combining efficient light harvesting and charge collection through the nanostructure of the material. A typical BHJ material consists of a conjugated polymer for light harvesting and hole transport, while a fullerene acts as an electron acceptor and electron transport material. All aspects of the charge generation process have been extensively studied.[16-26] How free mobile charges that can be extracted from the material are formed appears to be a key process for the operation of a polymer solar cell. Using a combination of ultrafast methods, we have addressed this issue and can provide insights into the factors controlling charge separation and formation of mobile charges.

22.2 FROM DYE-SENSITIZED TO ORGANOMETAL HALIDE PEROVSKITE SOLAR CELLS

22.2.1 Harvesting the photons— Electron injection from sensitizer to semiconductor

Initial work on excited state and electron dynamics in dye-sensitized solar cell materials focused on the dye-to-semiconductor-electron-injection process. A large body of work identified this process as decisive for efficient light harvesting[27-36] and conversion of the light energy to energy-rich electrons. For efficient utilization of absorbed photons and excited state energy of the sensitizer, electron injection into the semiconductor has to be significantly faster than the sum of all other excited-state deactivation processes. For many of the sensitizers used, nanosecond and longer excited-state lifetimes are not unusual, implying that injection times on the few picosecond time scale are sufficient for close to 100% quantum efficiency injection. Our early work on RuN3[28-30,37]-sensitized nanostructured TiO_2 electrodes illustrates this point. The strong visible absorption of metal–polypyridyl molecules is due to singlet and triplet excited states. For the Ru-based molecules discussed here, the lowest excited state is a ^{3}MLCT state, and there is a singlet MLCT (^{1}MLCT) state at higher energy responsible for the main absorption band. Light absorption into this band therefore generates the excited ^{1}MLCT state, but within a very short time (~100 fs), the molecule has relaxed into the lowest ^{3}MLCT state, because of efficient intersystem crossing (ISC) caused by the heavy metal atom. For efficient electron injection and energy conversion in the sensitized semiconductor system, the energy of the lowest ^{3}MLCT state has to be above the conduction band edge of the semiconductor. The resulting scenario is illustrated in Figure 22.1,

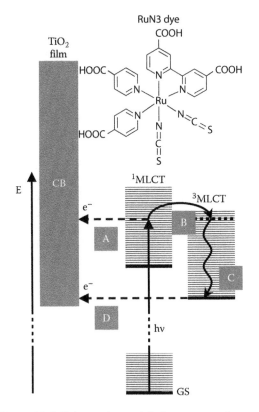

Figure 22.1 Schematic model of two-state electron injection and structure of RuN3. Following MLCT excitation (at 530 nm) of the RuN3-sensitized TiO_2 film, an electron is promoted from a mixed ruthenium NCS state to an excited p* state of the dcbpy-ligand and injected into the conduction band (CB) of the semiconductor. GS: ground state of RuN3. Channel A: electron injection from the nonthermalized, ^{1}MLCT excited state. Channel B and C: intersystem crossing (ISC) followed by internal vibrational relaxation in the ^{3}MLCT excited state. Channel D: electron injection from the thermalized, ^{3}MLCT excited state.

showing the valence and conduction bands of the semiconductor, as well as the ground and excited states of the RuN3 sensitizer. Our work[28,30,33,34,38] shows that following light absorption to the ^{1}MLCT state, ~60% of the molecules inject electrons directly from this state into the semiconductor conduction band with a characteristic time constant of 50 fs. Upon excitation to higher-lying vibrational states of the ^{1}MLCT state, even faster injection occurs (~20 fs), in competition with vibrational energy relaxation and redistribution. The residual ~40% of the excited sensitizers relax to the triplet state, from which they inject

electrons much more slowly on the 1–100 ps time scale.[37,39] Many other Ru-sensitizers,[27,40–42] but also porphyrins,[41,43] phtalocyanines, and organic dyes,[44–47] have demonstrated fast and efficient injection. Thus, to achieve efficient electron injection from a sensitizer dye to a metal oxide (MO) nanostructured film appears to be a relatively manageable task.

22.2.2 Electron–cation recombination

Electron injection from dye to semiconductor is just the first step in a series of processes that eventually lead to a photocurrent and voltage in an external circuit. For the electrons to be extracted in high yield, recombination with the holes on the oxidized dye has to be much slower than the re-reduction of the oxidized sensitizer by the redox couple of the hole transport material (HTM). The rate of this process depends on the nature of the HTM and if it is a liquid- or solid-state material. Electron–hole recombination times on the hundreds of nanosecond and slower time scale are generally sufficient for efficient utilization of the light-generated charges. The realization that electron–hole recombination is a process directly related to solar cell efficiency—every recombined electron is a lost electron and lost photocurrent—has motivated work to understand the factors controlling the process. We have performed a systematic study of how a sensitizer binding to the semiconductor surface controls the electron transfer processes in general and electron–cation (EC) recombination in particular.

For Ru-polypyridyl dyes (e.g., RuN3, the black dye) resulting in the most efficient solar cells, EC recombination has been shown to be very slow (microsecond time scale)[48–50] and slower than regeneration of the oxidized sensitizer by the redox mediators,[51] and therefore not a limiting factor for the efficiency of a solar cell based on these dyes. This fact, established for some Ru-polypyridyl sensitizers often seems to have been extrapolated to suggest that this is also the case for other dyes,[43] leading to a picture where variations in solar cell efficiency have been directly correlated to the efficiency and rate of electron injection.[43] Below, we will show that EC recombination can be a more important process for controlling solar cell efficiency.[41] This will also lead to a correlation of electron injection and recombination rates with the binding geometry of the sensitizer, and hence the steps toward the goal of designing nanostructured dye-sensitizer materials with predictable electron transfer properties. Many sensitizer molecules are developed with simple binding models guiding the design of the molecules. Detailed studies of the electron transfer dynamics (see Figures 22.2 through 22.6) show that considerably more advanced modeling of the dye–semiconductor interactions is required in order to predict the optimal geometry correctly. We will use some of our recent results for Zn-porphyrin/TiO$_2$ electrodes to illustrate this. A combination of ultrafast spectroscopy, surface-sensitive vibrational sum frequency generation (SFG), and measurements of solar cell power conversion efficiency leads to a picture where we can correlate binding geometry to electron transfer dynamics, which in its turn controls the conversion efficiency of the solar cell.

By using the series of Zn-porphyrins (Zn-P) depicted in Figure 22.2, we could vary several molecular properties of importance for dye–semiconductor binding.[27,41] The electron transfer dynamics were monitored by transient absorption (TA) spectra and kinetics.[41,42] The time evolution of the TA spectrum of a Zn-porphyrin/TiO$_2$ electrode can be described as a nonexponential process involving only two species, the singlet excited state and the Zn-porphyrin radical cation formed as a result of electron injection from the Zn-P to TiO$_2$. The excited state of the

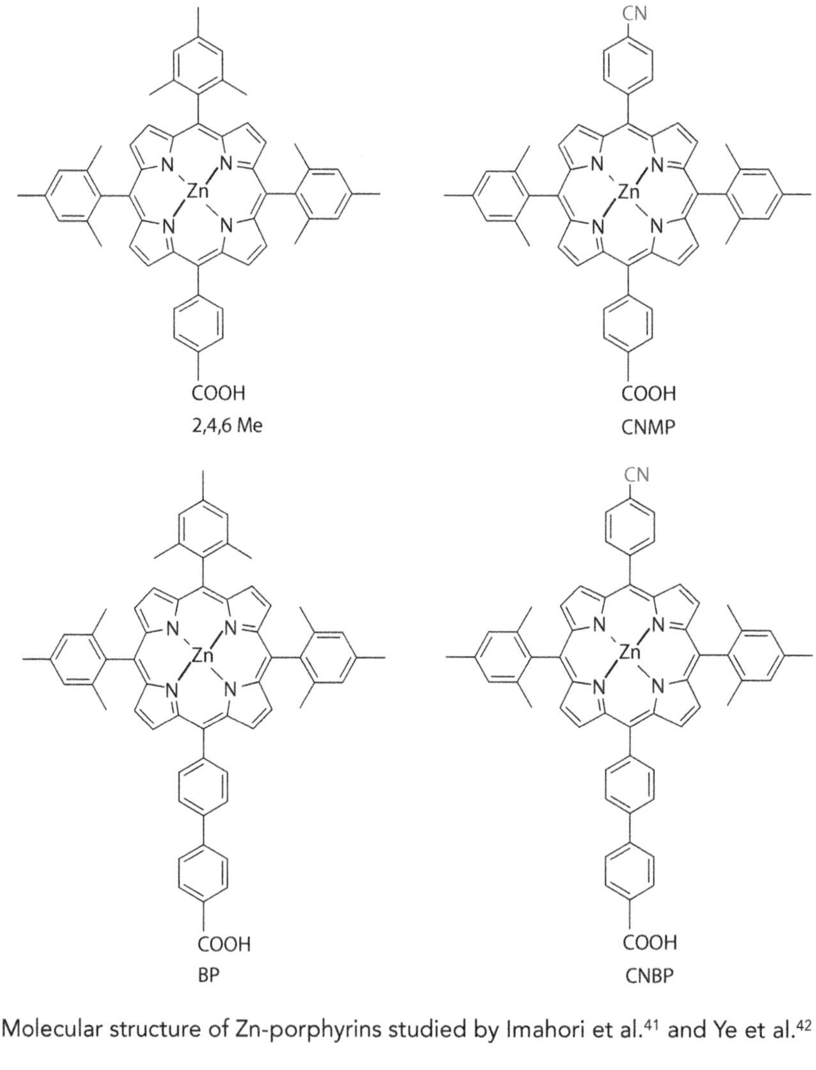

Figure 22.2 Molecular structure of Zn-porphyrins studied by Imahori et al.[41] and Ye et al.[42]

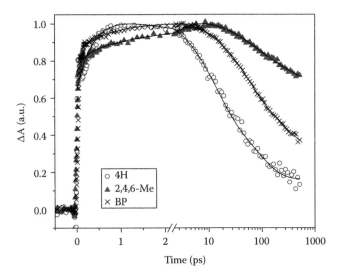

Figure 22.3 TA kinetics of 2,4,6-Me (full, red triangles), BP (black cross), and 4H (open, blue circles) sensitized TiO$_2$ in t-BuOH/acetonitrile for 1 h. Kinetics are normalized to their maxima and black solid lines represent fits of the kinetics.

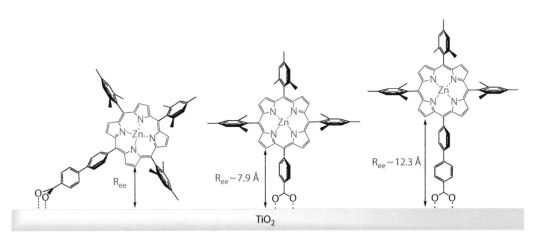

Figure 22.4 Binding model for Zn-porphyrins to TiO$_2$. The edge-to-edge distance (R$_{ee}$) is decreased upon tilting the dye molecule.

sensitizer is formed within the time resolution (<100 fs) of the experiment and then transforms with multi-exponential kinetics to the oxidized sensitizer, which then decays back to the ground state by charge recombination with conduction band electrons. This recombination can be described for most dyes by two lifetimes, one on the tens to hundreds of picoseconds time scale and another much slower, >50 ns.

From Marcus theory of electron transfer,[52–54] and its modifications for interfacial electron transfer,[55,56] it is expected that the electron transfer rate should have a strong (exponential) distance dependence. If electron transfer between the porphyrin core and the semiconductor occurs via the connecting spacer, as often envisaged, making this spacer longer should slow down the transfer. With the help of the sensitizers 2,4,6-Me and BP, we can test this expectation; introduction of the extra phenyl moiety in the biphenyl spacer of TiO$_2$/BP relative to TiO$_2$/2,4,6-Me would result in approximately 1.5 times longer distance between the porphyrin core and the TiO$_2$ surface and therefore result in a considerably smaller electronic

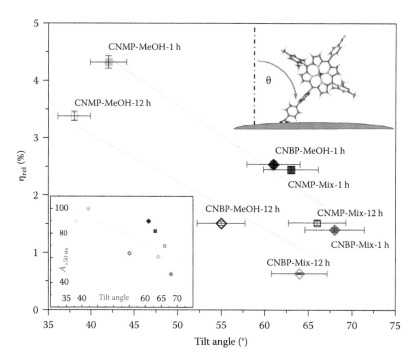

Figure 22.5 Solar cell efficiency normalized for surface coverage (η_{rel}) as a function of tilt angle. Closed and open symbols correspond to the sensitization time of 1 and 12 h, respectively. The inset shows the correlation between tilt angle and amplitude of >50 ns charge recombination, A > 50 ns [i.e., concentration of long-lived electrons in the TiO$_2$ conduction band (%)]. The color and symbol coding is the same as in main figure.

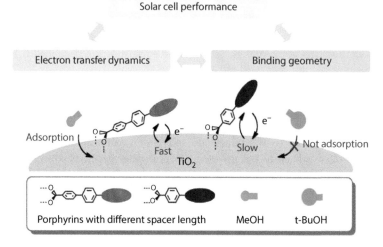

Figure 22.6 Schematic illustration of factors controlling the sensitizer–TiO$_2$ binding geometry for Zn-porphyrin sensitizers of the type studied here.[41,42] In addition to sensitizer structure and spacer length, the nature of the solvent influences the binding geometry.

coupling and much slower electron injection and recombination. Figure 22.3 shows the TA kinetics of the two molecules attached to the TiO_2 film over a time window of 500 ps. Already a superficial glance at the two kinetic curves shows that the charge recombination process (decay of the curves) does not meet this expectation—the BP sensitizer with the longer connecting spacer has a much higher amplitude of fast recombination components and a lower amplitude of long-lived signal (32% for BP vs 64% for 2,4,6-Me) that accounts for long-lived charges that can be harvested as photocurrent in a solar cell. Not only the amplitude of the >50 ns decay component is higher for 2,4,6,-Me/TiO_2 than for BP/TiO_2, but the lifetimes of the faster decays are shorter for BP/TiO_2. Also the electron injection (rise of the kinetics) is faster for BP/TiO_2.

This shows that both electron injection and recombination are overall faster for the sensitizer with the longer connecting spacer. For recombination, this appears in two ways—a lower amplitude of long-lived (>50 ns) charges and faster recombination on the tens and hundreds of picoseconds time scale for the Zn-porphyrin with the longer connecting spacer (BP). For the 4H molecule, lacking the methyl groups on the phenyl substituents on the porphyrin core, the effect is even more pronounced—there is almost no slow injection and most of the recombination is complete after 500 ps. Obviously, electron transfer does not occur as could be anticipated via the spacer connecting the porphyrin core to the TiO_2 surface. Instead, we suggest a picture where the single carboxyl anchoring group allows a flexible binding geometry; for some of the porphyrins, depending on structural factors such as length of the spacer group and bulkiness of the porphyrin core, a fraction of the molecules is bound at an angle to the semiconductor surface and electron transfer occurs through space rather than through the linker group connecting the porphyrin core to the anchoring COOH group[41] (Figure 22.4).

A variation of the tilt angle as a result of a modified porphyrin molecule structure leads to a change of the distance between the porphyrin core and semiconductor surface, which in its turn will lead to a change in the through-space electron transfer rate.

Owing to the expected exponential distance dependence of electron transfer, only a modest change of distance (and thus angle) will have a dramatic impact on the transfer rate. As an example,

let us consider the situation when an upright (perpendicular) orientation of the porphyrin results in a distance of 12 Å between the porphyrin molecule and the TiO_2 surface and a recombination time of ~10 ns (similar to the measured slow part of the recombination). A decrease of the porphyrin–TiO_2 distance to 8 Å, which corresponds to a tilt angle of 42° (relative to the surface normal), results in a dramatic shortening of the recombination time to ~50 ps, assuming all other factors in the rate equation are constant. This simple picture is just to illustrate the idea, without consideration of the shape and size of the porphyrin core, which definitely will add complexity. Thus, we see that a modest variation in binding geometry can give rise to the very large variation in recombination rates that we observed from the studied Zn-porphyrins.

Electron transfer rates are, of course, not an unambiguous measure of binding geometry; for that an experimental method providing more direct structural information is required. To this end we have used vibrational SFG spectroscopy on the Zn-porphyrins of Figure 22.2 labeled with a CN infrared active chromophore.[42] The IR transition dipole moment of the CN-group is along the symmetry axis of the Zn-porphyrin molecules; SFG will therefore give the orientation of the porphyrin relative to the semiconductor surface (the tilt angle). The inset of Figure 22.5 shows that there is a direct correlation between tilt angle of the Zn-P molecule and amplitude of long-lived ($A_{>50 ns}$) conduction band electrons that can contribute to photocurrent—a smaller tilt angle leads to a higher amplitude of long-lived electrons. By comparing tilt angles obtained from the SFG measurements with solar cell power conversion (η_{rel}, Figure 22.5), we can take this correlation one step further. Thus, we can see that a smaller tilt angle leads to a higher cell efficiency, meaning that solar cell power conversion efficiency is directly related to the amplitude of slow EC recombination, that is, the concentration of long-lived electrons in the conduction band. The results suggest a method to characterize the structure of the sensitizer/semiconductor interface and thus pave the way toward providing DSC materials with predictable electron transfer properties. Despite the very simple model, a strong correlation between tilt angle, electron transfer rate, and solar cell efficiency was confirmed. The binding geometry model and factors controlling the binding are schematically illustrated in Figure 22.6.

22.2.3 Electron–cation electrostatic interaction in dye-sensitized solar cells

As described earlier, optical spectroscopy can provide detailed information about excited states and intermediates participating in the light-to-charge conversion process of DSC materials. As can be easily appreciated, electron transport within and between the semiconductor nanostructures is an important process for the functioning of a solar cell. Optical techniques unfortunately do not provide much information about charge transport. More information is obtained from transient far-infrared conductivity spectra and kinetics measured by time-resolved terahertz (THz) spectroscopy (TRTS).[57] This technique allows noncontact characterization of photoconductivity with sub-picosecond time resolution.[58,59] The amplitude of the photoconductivity is then a direct measure of the population of injected charge carriers and carrier mobility. In addition, from the shape of the transient conductivity spectrum, it is possible to infer the mechanisms of the charge transport or to distinguish between the response of free charge carriers and localized excitations.[60] The strong interaction of THz radiation with free charge carriers in semiconductors makes the TRTS an ideal tool for the investigation of charge carrier dynamics in semiconductors and DSC.

The generally accepted picture of DSC is that mobile electrons appear in concert with injection. Our results show that charge injection and formation of mobile charges are not necessarily

directly connected and that charge transport in the active solar cell material can be different from that in nonsensitized semiconductors. This is related to strong electrostatic interaction between injected electrons and dye cations at the surface of the semiconductor nanoparticle. This interaction in addition may give rise to fast EC recombination, decreasing the efficiency of a solar cell. To illustrate this, we take advantage of combining results of visible and THz pump–probe spectroscopy. Probing in the visible range typically provides information about the electronic state of the dye molecules, that is, the formation of oxidized dye molecules (cations) upon electron injection is monitored. Probing in the THz spectral range is sensitive to the transport of mobile charge carriers (injected electrons). We investigated in detail the dynamics in ZnO and TiO$_2$ nanoparticle thin films sensitized by Zn(II)-5-(3,5-dicarboxyphenyl) phenyl-10,15,20-triphenylporphyrin (ZnTPP-Ipa) and RuN3 dyes.[57,61]

Measurement of the THz kinetics after photoexcitation of the sensitizer reflects the population of mobile electrons injected into the semiconductor. For the Zn-porphyrin molecule, ZnTPP-Ipa, attached to a nanocrystalline ZnO film, the rise in the population of injected mobile electrons is slow and occurs on the tens to hundreds of picoseconds time scale (Figure 22.7). After reaching its maximum value, it does not decay for at least 10 ns, which implies that recombination of injected electrons is slow. This behavior is in sharp contrast to the optical TA dynamics of oxidized dye (cation) (not shown); the formation is essentially completed in 5 ps[41] and the decay

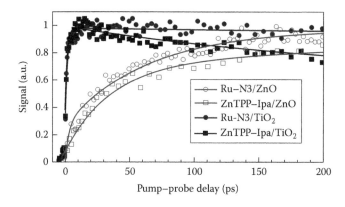

Figure 22.7 Evolution of transient THz conductivity (normalized to unity). The lines serve only to guide the eye.

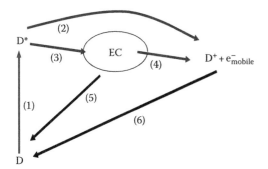

Figure 22.8 Scheme of the processes in dye-sensitized ZnO and TiO$_2$. (1) Dye excitation, (2) direct electron injection, (3) formation of an EC complex, (4) dissociation of EC complex, (5) recombination of EC complex, and (6) charge recombination. D = dye molecule.

exactly matches the slow THz rise of mobile electrons in Figure 22.7. This behavior can be understood by the kinetic scheme depicted in Figure 22.8, involving an intermediate state between the excited state of the sensitizer and cation with mobile electrons in the conduction band of ZnO. Thus, photoexcitation of the sensitizer (1) leads to an EC complex within ~5 ps (3) in which the electron is strongly bound to the cation and therefore does not contribute to the THz signal. The bound EC state can either recombine (5) or dissociate into a mobile electron in ZnO and cation (4). Very similar results were obtained for RuN3/ZnO (Figure 22.7), showing that this behavior is a property of ZnO, rather than the sensitizer. These results point to the key role of the EC complex in controlling charge recombination and, therefore, possibly solar cell efficiency. The injection/recombination dynamics observed for sensitized nanocrystalline TiO$_2$ films is fundamentally different from that observed for sensitized ZnO. The THz and TA kinetics of both dyes (Figure 22.7 and Reference 61) show that both cations and mobile conduction band electrons appear on the same ultrafast time scale (100 fs–10 ps). This suggests that charge injection into TiO$_2$ is a direct process [(2) in Figure 22.8], not involving the recombination-promoting EC complex. The reason for this difference in the formation of mobile charge carriers in ZnO and TiO$_2$ is the difference in screening of the electrostatic interaction in the two MOs, due to the much higher dielectric permittivity of TiO$_2$.[61]

22.2.4 Quantum dot–sensitized solar cells

Photoinduced electron injection from QDs to MO acceptors is a key process in QDSC. Various studies using QD-related signals have suggested a broad range of time scales for the process from sub-picosecond,[62] tens of picoseconds[63] to nanosecond[64] time scales depending on the measurement method and the QD attachment method. If only the QD signal is measured, the question remains whether the kinetics is truly caused by the electron injection or by some possible trapping process at the QD–MO interface. Combined measurement of the QD TA signal recovery and the appearance of the transient THz signal from the injected electrons in the MO acceptor[65] have provided a direct proof that the electron injection in CdSe QDs attached to ZnO nanorods via a 2-mercaptopropionic acid (2MPA) linker occurs in 3–12 ps, depending on the QD size, as shown in Figure 22.9. We also obtained evidence that the electrons are injected via an intermediate charge transfer state.

Electron transfer through the bridge is usually seen as a tunneling process where the coupling and thereby the injection rate depend exponentially on the linker length.[66] In a recent study, the electron injection from the CdSe QD to ZnO was studied through three different linkers: mercaptoacetic acid (MAA), 2MPA, and 3-meraptopropionic acid (3MPA).[67] Despite the same length of the MAA and 2MPA, the electron injection is quite different for these two linkers. The 3MPA is the longest of the three, and indeed, its injection is slowest too. However, even this case does not quantitatively follow the expected exponential distance dependence. Instead it was shown that the orbital topology of the linker molecule at the attachment points controls the injection rate.

The tunable transition energy of QDs is very useful for covering the whole solar spectrum. However, the most promising in novel solar cell applications is the so-called multiple exciton generation (MEG) in QDs. Via this effect one can "harvest" the excess energy of the absorbed photon, which otherwise would be converted to heat. While the practical potential of the effect in drastically improving the efficiency of the solar cells is apparent, the reported yields of the process are contradictory[68,69] and understanding of the basic principles is still obscure. The process is commonly

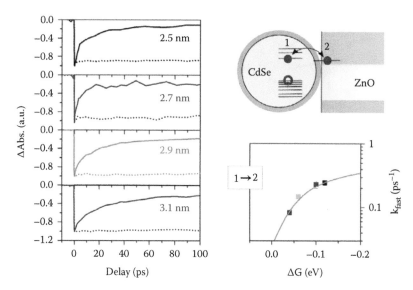

Figure 22.9 Left-hand side: normalized TA kinetics of pure QDs (dotted lines) and sensitized ZnO nanowires (NW) (solid lines) with different QD sizes (measured in spectral position of the sample maximum signal). Top right-hand side: scheme of reversible initial electron transfer in sensitized NW (1, electron in CdSe QD; 2, electron localized near ZnO surface). Bottom right: dependence of fast TA decay rate (solid squares) described by Marcus theory (gray line), λ_{exc} = 470 nm, λ_{probe} (2.5 nm) = 506 nm, λ_{probe} (2.7 nm) = 516 nm, λ_{probe} (2.9 nm) = 526 nm, λ_{probe} (3.1 nm) = 540 nm.

described in terms of impact ionization where multiple excitons are generated in a sequential process.[70] The observed dynamics is very fast—multiple excitons are formed within the time resolution of the experiments carried out so far (<50 fs). This has called for alternative explanations via processes where partially coherent intermediates and dephasing are involved.[71] In a computational approach, an additional stage, multiple exciton fission to independent excitons was proposed as the final step before charges can be injected.[72] The details of all these processes depend on the mixing of the electronic states due to the Coulomb interaction, relaxation through nonadiabatic couplings, and dephasing. Our own recent work has proposed a refined, consistent analysis procedure for using TA spectroscopy in MEG quantification.[73] We found that many earlier studies use unjustified approximations assuming that the state with two electron–hole pairs (biexciton) gives double the signal compared to the one electron–hole pair. We have shown that this assumption generally does not hold,[74] which can partially explain existing controversial reports.

Multiple excitons in a QD recombine via the Auger process to a single exciton.[75] The rate of the process depends on the size of the QD, and the phenomenon resembles a so-called excitation annihilation in multichromophoric systems.[76] In order to make use of MEG, the electron injection from the QD to an acceptor system must be considerably faster than the Auger relaxation (annihilation) process. Our own studies of electron injection from CdSe QDs to a ZnO acceptor show that the process can be as fast as 10 ps, certainly a promising result. In another study, we investigated the transport of the second electron in the case of doubly excited QDs.[77] The injection does slow down compared to the first electron, but even the second electron is injected sufficiently fast to escape the Auger process. Figure 22.10 depicts a schematic representation of the possible relaxation pathways in a doubly excited QD. Multiple exciton collection has been observed in the case of a specially prepared TiO_2 surface,[78] clearly showing that QDSCs where MEG is implemented are feasible. For quite a long time, the clear bottleneck in QDSC performance has been the hole-scavenging efficiency.[79] Recent development in ligand exchange for band alignment of QDs[80] has opened new promising avenues for QDSCs, and the approach has very recently led to solar cells with an efficiency above 8%.[81]

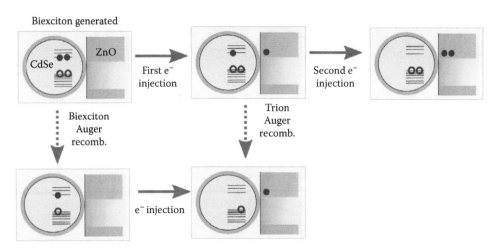

Figure 22.10 Electron injection and Auger recombination processes present in doubly excited QDs attached to ZnO.

22.2.5 Earth-abundant sensitizers for dye-sensitized solar cells

Most studies of dye-sensitized solar cells have been performed on ruthenium(II) complexes that have favorable excited-state properties for solar cells, for example, long excited-state lifetimes, efficient electron-injection-to-metal-oxide semiconductors, and generally slow electron–hole recombination.[27,82–84] Using iron instead of ruthenium in such complexes would be an important step to promote light-harvesting applications on a large scale, as iron is earth abundant, inexpensive, and environmentally benign. However, its intense MLCT absorption has been considered unexploitable in energy conversion applications,[85] due to the low-lying metal-centered (MC) quintet (Q) high-spin state that typically deactivates the ^{1,3}MLCT manifolds on a sub-picosecond time scale.[86] Studies of the prototype [Fe(bpy)$_3$]$^{2+}$ (bpy = 2,2-bipyridine) complex have revealed an excited-state decay mechanism that involves ultrafast ISC from the first populated ^{1}MLCT state to the ^{3}MLCT state, followed by an ultrafast (sub-picosecond) cascade of ISCs to the ^{3}MC, and then to the ^{5}MC state.[87,88] Thus, destabilizing these MC states should result in a longer-lived MLCT state. The ^{3}MC state in Ru-bis(tridentate)[89] and tris(bidentate)[90] polypyridine complexes has been effectively destabilized through the introduction of a strongly σ-donating ligand in the first coordination sphere, likely resulting in an increased t_{2g}/e_g-like orbital splitting. We recently successfully used this approach by

synthesizing a Fe-NHC complex, [Fe(CNC)$_2$](PF$_6$)$_2$ (CNC = 2,6-bis(3-methylimidazole-1-ylidine)pyridine) (Figure 22.11), and ultrafast TA measurements (Figure 22.12) showed a 100-fold extended excited-state lifetime of 9 ps as compared to previously known Fe^{II}-polypyridyl complexes. The spectral evolution in Figure 22.12 was tentatively assigned as <100 fs relaxation from a spectrally very broad (500–700 nm) ^{1}MLCT state to the ^{3}MLCT state characterized by a TA band at ~530 nm decaying with a ~9 ps time constant. No spectral signatures of a ^{5}MC state, and therefore no significant population of such a state, was observed. Recovery of the ground state was observed to occur, within experimental error, with the same 9 ps time constant as the decay of the ^{3}MLCT state. This suggested a <2 ps lifetime of the metal-centered excited states, dramatically different from much longer, several hundred picoseconds, lifetimes known for Fe^{II}-polypyridyl complexes.[7,27,82]

In order to better understand the origin of the 9 ps ^{3}MLCT lifetime, why it is so much longer than for conventional polypyridyl Fe^{II}-complexes, and how to further improve Fe^{II}-complexes for light harvesting, the measured excited-state dynamics were rationalized with the help of DFT and TD-DFT calculations. The calculations showed that the exceptionally long excited-state lifetime for this type of Fe-complex (~9 ps) is achieved through a significant destabilization of both triplet and quintet metal-centered scavenger states compared to other Fe^{II}-complexes. In addition, a shallow ^{3}MLCT potential energy surface with

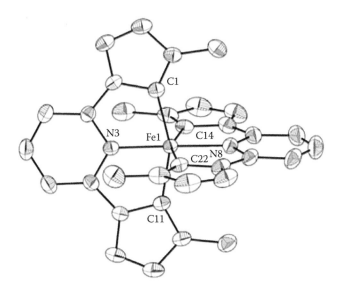

Figure 22.11 Structure of [Fe(CNC)$_2$]$^{2+}$ (CNC = 2,6-bis(3-methylimidazole-1-ylidine)pyridine).

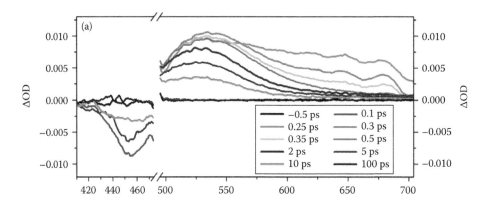

Figure 22.12 TA spectra measured at different times of the Fe-NHC molecule in Figure 22.11.

a low-energy transition path from the ^{3}MLCT to ^{3}MC and facile crossing from the ^{3}MC state to the ground state were identified as key features for the excited-state deactivation. The pathway of excited-state relaxation is indicated by the arrows in the potential energy curve diagram of Figure 22.13. It is encouraging to see that these potential energy curves nicely explain both the several-picoseconds lifetime of the ^{3}MLCT state, as well as the unusually short lifetime of the ^{3}MC state through its near crossing with the ground state. The ^{5}MC state is not populated at all because of its high energy. We believe that the combination of experimental work and calculations as illustrated here for [Fe(CNC)$_2$]$^{2+}$ yields an improved level of understanding of the excited-state

dynamics and provides solid ground to further explore how to extend the ^{3}MLCT-state lifetime of Fe^{II}-complexes, in order to achieve competitive Fe-photosensitizers for large-scale production of dye-sensitized solar cells.

22.2.6 Organometal halide perovskite solar cells

Organometal halide perovskite (OMHP)-based solar cells have recently been reported to be highly efficient, giving an overall power conversion efficiency of up to and exceeding 15%.[5,27,91–93] However, much of the fundamental photophysical properties underlying this performance have remained unknown. For instance, it is not known whether the

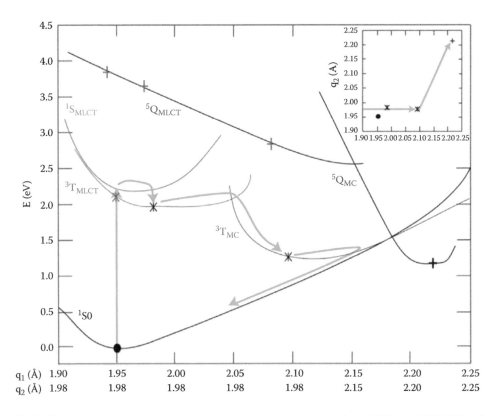

Figure 22.13 Schematic excitation and deactivation pathways based on DFT- and TD-DFT-calculated potential energy curves. (From Fredin, L.A. et al., *J. Phys. Chem. Lett.*, 5, 2066, 2014.)

reported exceptionally long diffusion lengths are related to molecular excitons, that is, tightly bound electron–hole pairs, or to highly mobile charge carriers. Hence, the nature of the initial photophysical product is not known. The exact role of the metal oxide TiO_2, as electron acceptor and charge transport layer in an OMHP-based photovoltaic cell, is also not understood. Moreover, the influence of introducing OMHP in the open voids of sintered isolating nanoparticles, for example, Al_2O_3 on the optoelectronic properties, has not yet been addressed. We recently applied a combination of photoluminescence (PL), transient absorption (TA), time-resolved terahertz spectroscopy (TRTS), and time-resolved microwave conductivity (TRMC) measurements[94] to determine the time scales of generation and recombination of charge carriers as well as their transport properties in solution-processed $CH_3NH_3PbI_3$ perovskite materials.

Time-resolved THz conductivity measurements provide information about the formation of mobile charges.[57,95] The early time THz kinetics of a thin film

of $CH_3NH_3PbI_3$, mesoscopic $CH_3NH_3PbI_3/Al_2O_3$, and $CH_3NH_3PbI_3/TiO_2$ are shown in Figure 22.14a. For $CH_3NH_3PbI_3$ and $CH_3NH_3PbI_3/Al_2O_3$, an instrument-limited rise (A = 70%) assigned to the generation of electron–hole pairs is followed by a second (~2 ps, A = 30%) rise before reaching its maximum. Upon photoexcitation, electron–hole pairs are initially Coulombically bound and require an activation energy on the order of the thermal energy, kT, to dissociate into mobile charges. This dissociation could take a few picoseconds and is in this case manifested as the ~2 ps rise in the THz kinetics of $CH_3NH_3PbI_3$ and $CH_3NH_3PbI_3/Al_2O_3$. For $CH_3NH_3PbI_3/TiO_2$, the THz transient rises faster than in the other two materials and grows with a response-limited rise time. We previously showed that in the presence of a MO with a high electron affinity, electron injection from a photoexcited sensitizer to the oxide can be ultrafast, for example, in CdSe quantum dot–sensitized ZnO[65] and dye-sensitized TiO_2.[28] We therefore assign the response-limited rise in THz photoconductivity

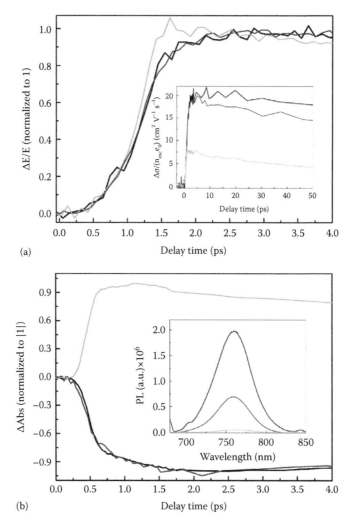

Figure 22.14 Early time dynamics of a thin film of $CH_3NH_3PbI_3$ (black), mesoscopic $CH_3NH_3PbI_3/Al_2O_3$ (red), and $CH_3NH_3PbI_3/TiO_2$ (green). **(a)** THz kinetics (λ_{pump} = 400 nm, I_{exc} = 1.7 × 10^{13} ph cm^{-2} per pulse) normalized to |1|. Inset is the THz photoconductivity kinetics for the first 50 ps. **(b)** TA kinetics (λ_{pump} = 603 nm, λ_{probe} = 970 nm, I_{exc} = 6.0 × 10^{14} ph cm^{-2} per pulse). Inset is the corresponding PL spectra (λ_{pump} = 550 nm).

for $CH_3NH_3PbI_3/TiO_2$ as sub-picosecond electron injection from the OMHP to TiO2. Moreover, the transient THz photoconductivity kinetics (Figure 22.14a inset),[94] normalized with respect to excitation density, shows that the charge mobility of $CH_3NH_3PbI_3/TiO_2$ is ~7.5 cm^2 V^{-1} s^{-1}, approximately 3–4 times lower than in $CH_3NH_3PbI_3$ and the $CH_3NH_3PbI_3/Al_2O_3$ (i.e., ~20 cm^2 V^{-1} s^{-1}), supporting the proposal that electron injection occurs in $CH_3NH_3PbI_3/TiO_2$. The THz photoconductivity kinetics also shows that the mobility remains high for many tens of picoseconds. THz measurements at lower intensity and comparison with optical TA measurements (which measure the concentration of charges, in contrast to THz measurements that measure the product of concentration and mobility) show that mobility remains high and unchanged for at least a nanosecond. In other words, the decay of THz and TA signal observed at higher excitation densities is a result of nongeminate electron–hole recombination.

Optical TA kinetics of $CH_3NH_3PbI_3$ and $CH_3NH_3PbI_3/Al_2O_3$ (Figure 22.14b)[94] are characterized by an instrument-limited negative response followed by an approximately 2 ps of further decrease. The time scale of this two-component signal is consistent with the two-step rise in THz kinetics, confirming that charges are not instantaneously created. Furthermore, the negative signal, in a wavelength region lacking ground-state absorption, is a characteristic of stimulated emission, which agrees with the steady-state PL spectra (Figure 22.14b inset). In contrast, a single-component ultrafast rise with a positive sign characterizes the kinetics of $CH_3NH_3PbI_3/TiO_2$, consistent with the time scale of the appearance of electrons in TiO_2[5] and identical to the rise time of the THz kinetics (Figure 22.14a). Although not complete, strong quenching of the PL in $CH_3NH_3PbI_3/TiO_2$ (Figure 22.14b inset) is additional evidence for electron injection.

Slow electron–hole recombination and persistent high mobility are essential features for an efficient solar cell. By using the TRMC technique, we could monitor the changes in photoconductance up to a hundred microseconds at very low excitation density (carrier density), minimizing the contribution of nongeminate recombination and allowing characterization of even very slow recombination dynamics. Figure 22.15 shows TRMC kinetics of $CH_3NH_3PbI_3/Al_2O_3$ measured over two orders of excitation fluence (5.9×10^9–6×10^{11} ph cm^{-2} per pulse) and up to 100 μs. All transients exhibit a nonexponential decay with a fast (~100 ns) initial part—faster at higher intensity—and a slow microsecond tail with increasing lifetime and amplitude for lower intensity, a characteristic behavior for nongeminate charge recombination.[20,21,96] The high mobility of electrons and holes in the OMHP phase allows rapid diffusion away from their locus of generation, effectively retarding geminate recombination. The consequence of this is intensity-dependent TRMC kinetics dominated by nongeminate electron–hole recombination. Intensity-dependent TRMC kinetics down to the lowest intensity measured shows that geminate recombination must be very slow, at least tens of microseconds (Figure 22.15). However, since the TRMC response is a product of carrier concentration and mobility, just as in TRTS, it is not possible to extract from a single transient the precise contribution of charge carrier recombination and carrier relaxation to the observed TRMC decay. Nevertheless, at the lowest excitation intensity (5.9×10^9 ph cm^{-2} per pulse), the decay of the TRMC kinetics occurs on a 1–10 μs time scale and longer, showing that geminate and nongeminate recombination as well as charge carrier relaxation are microsecond-domain processes. From the obtained mobilities and carrier lifetimes, we can estimate carrier diffusion lengths exceeding 1 μm, favorable for efficient charge extraction from a solar cell device.

Unbalanced electron and hole mobilities, which are orders of magnitude different in BHJ solar cells, result in space charge-limited photocurrents lowering the PCE.[97] Therefore it is important to assess the

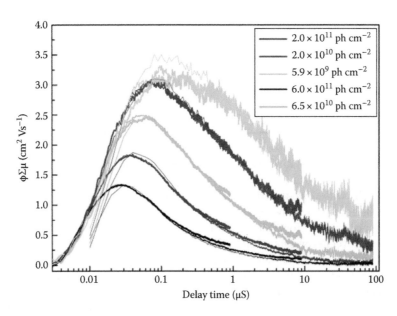

Figure 22.15 TRMC kinetics of $CH_3NH_3PbI_3/Al_2O_3$ at different excitation densities measured up to 100 μs.

mobilities of both electrons and holes. By comparing THz conductivities for several materials with known electron or hole mobilities, these quantities can be derived. Thus, we could conclude that holes in the OMHP phase have a mobility of ~7.5 cm^2 V^{-1} s^{-1} and electrons ~12.5 cm^2 V^{-1} s^{-1}.[94] The $12.5/7.5 \approx 2$ ratio of electron and hole mobilities in the OMHP phase is in agreement with the recent calculations of the relative effective masses of electrons and holes[98] and shows that electron and hole mobilities are almost balanced—key information for understanding the OMHP-only or OMHP/Al_2O_3 solar cell function. Further TRMC and PL measurements were used to investigate the temperature dependence of the carrier generation, mobility, and recombination in $(CH_3NH_3)PbI_3$. At temperatures maintaining the tetragonal crystal phase of the perovskite, an exciton-binding energy of about 32 meV was found, leading to a temperature-dependent yield ($\varphi = 0.84$ at 300 K) of highly mobile (6.2 cm^2 V^{-1} s^{-1} at 300 K) charge carriers. At higher laser intensities, second-order recombination with a rate constant of $\gamma = 13 \times 10^{-10}$ cm^3 s^{-1} becomes apparent. Reducing the temperature results in increasing charge carrier mobilities following a $T^{-1.6}$ dependence, which we attribute to a reduction in phonon scattering. Despite the fact that carrier mobility increases with decreasing temperature, γ diminishes with a factor of six upon lowering the temperature to 165 K, implying that the charge recombination in $(CH_3NH_3)PbI_3$ is temperature activated. The results underline the importance of the perovskite crystal structure, the exciton binding energy, and the activation energy for recombination as key factors in optimizing new perovskite materials.

22.3 POLYMER SOLAR CELLS

The power conversion efficiency of organic polymer/fullerene solar cells has dramatically increased over the last 10 years, from ~2% to now over 10%,[99] and thus comparable with dye-sensitized solar cells,[1,2] but they still have some way to go before they catch up with more conventional silicon thin-film technology and novel perovskite[4,5] solar cells. This progress is a result of material development and improved understanding of how factors like material morphology and carrier dynamics correlate to device efficiency. Nevertheless, there are many gaps in our understanding of how light

energy absorbed by the active solar cell material is converted into useful photocurrent. For an efficient solar cell, the energy of light has to be converted into free mobile charges that can be extracted from the active material into the electrodes of the solar cell. This might sound like a trivial statement, but considering that positive and negative charges are formed at a short distance in a low-dielectric constant medium by the absorbed light—and therefore supposedly strongly interacting—immediately poses the question: "How are these charges separated escaping their mutual electrostatic interaction?" Here we will highlight how this may happen with the help of some recent results, but first we briefly describe the various steps from light absorption to free charges.

There is generally a good understanding and more or less consensus of what are the main steps of free charge formation, as illustrated in Figure 22.16. Polymer excitons are formed through light absorption (step 1), and they break up into a closely bound charge pair (also called CT state by many authors) on the ultrafast sub-picosecond time scale (step 2). This charge pair then dissociates into eventually free noninteracting mobile charges (step 3). During this dissociation process, charges may recombine geminately (step 4), and free charges formed through separate light absorption events may also recombine in a nongeminate process (step 5), on a time scale determined by the carrier density. If there is significant agreement on the presence of these steps, there is considerably less agreement on the details of how they occur and which are the underlying mechanisms. Adding to the complexity is an increasing awareness that different materials (different polymer/fullerene blends) and different material processing and handling may lead to different dynamics. Whether the studied material is a thin film deposited on a substrate or part of a functioning device may also influence the dynamics. We will briefly highlight some of the most important issues and different views for each of the reaction steps shown in Figure 22.16.

Step 1—Exciton generation: The polymer exciton formed by light absorption can be thought of as a localized excited state or a delocalized (extended) exciton state, depending on the character of the light-absorbing unit (spectroscopic unit). For a homopolymer with identical repeat

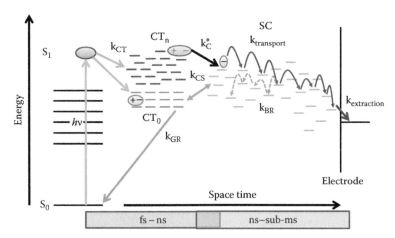

Figure 22.16 Schematic summary of light-induced processes in a polymer/fullerene solar cell, from light absorption by the polymer, over charge separation and transport, to charge extraction at the counter electrode. Notation: CT = charge transfer states, or charge pair states; CS = charge separated states; k_{GR} = geminate recombination; k_{BR} = bimolecular (nongeminate) recombination; k_{CS} = charge separation.

units, the exciton picture is probably plausible, but for a heteropolymer with chemically different units in the chain, the localized picture appears more appropriate. A delocalized exciton would probably localize on an ultrafast time scale due to the combined action of static and dynamic disorder, as has been observed in molecular aggregates and photosynthetic antenna pigments.[100] The role of hot and delocalized excitons has been discussed in the context of ultrafast charge generation.[17]

Step 2—Charge generation: Already the early time-resolved studies of polymer/fullerene blends observed ultrafast (<100 fs) generation of charges, and many later studies have confirmed this result.[20,101,102] With a picture of polymer/fullerene morphology consisting of 10-nm or larger domains of neat polymer and fullerene, it becomes a challenge to explain the ultrafast charge generation. A localized polymer exciton would need tens of picoseconds to diffuse from the interior of a polymer domain to the polymer/fullerene interface where charge generation occurs. Large exciton delocalization could possibly resolve this dilemma, but charge generation would in any case have to compete with exciton localization, which most likely occurs on the ~10 fs time scale.[100,103–105] Other more intriguing,[106] and disputed,[107] explanations to the ultrafast charge generation have also been proposed. For an amorphous polymer lacking large crystalline domains, ultrafast charge generation could be explained by a more or less molecular mixing of polymer and fullerene.[20,102]

Step 3—Charge separation: This is where much of the discussion and controversies stand at the moment. The nature of the CP (or CT) state[108] has been debated, and some authors view it as a strongly coupled electron–hole pair,[21,102] whereas others consider it as being a unique state and attach specific significance to it as a gateway for the formation of free mobile charges.[16,109,110] The dissociation of the CP (CT) state into free charges is a key issue addressed recently by several authors—is it ultrafast or through slow diffusion of charges? When do free independent charges appear and how is this related to recombination? We will highlight this issue using some of our own recent results.

Steps 4 and 5—Recombination: Initial reports on electron–hole recombination as mainly controlled by a geminate or nongeminate process appeared as contradictory. However, now with results becoming available for many more polymer/fullerene blends, it is increasingly evident that this process is strongly material dependent.[20,21,96,111–114] In some materials the geminate recombination is quite fast (nanosecond time scale),[20,21] meaning that experimental conditions can easily be achieved where the nongeminate recombination process is much slower than the geminate one (i.e., low-excitation light intensity), and therefore not contributing substantially to the charge decay. In other materials, excitation-intensity-dependent recombination kinetics is observed down to the lowest excitation intensity attainable in a time-resolved measurement, implying that geminate recombination is very slow.[96,115] Understanding how energetic and structural properties of different polymer/fullerene blends control the relative importance of the two recombination processes is obviously important for optimizing solar cell performance. Recombination dynamics, particularly on the slow micro- and millisecond time scales, appears to be very sensitive to the microscopic morphology of the material. Thus, it is not unusual that the dynamics of a polymer/fullerene thin film deposited on a substrate is different from that of the same film incorporated in a solar cell device under open-circuit conditions.[116] The consequence of all this is of course that great care has to be exercised when comparing results obtained under different conditions. We will now discuss the charge separation step in some more detail by briefly summarizing some of our recent results addressing this process.

22.3.1 Direct visualization of the charge separation process

The separation of photogenerated bound charges can be visualized with the help of the time-resolved electric field–induced second harmonic (TREFISH) method.[117,118] The intensity of the TREFISH signal measures the electric field inside the polymer/fullerene blend, which changes when photogenerated charge carriers drift in a charged-plate capacitor-like solar cell. Polarization of molecular charge densities and conductivity currents following excitation may be evaluated from the field dynamics, which allows us to monitor the field-induced carrier displacement, that is, charge separation caused by the drift of the electron and hole in opposite directions. From the drift dynamics, carrier mobility averaged over electrons and holes is also obtained. The field-induced charge pair displacements, extracted directly from the measured time-dependent TREFISH signal is shown in Figure 22.17 for a poly(3-hexylthiophene) (P3HT)/[6,6]-phenyl-C61butyric acid methyl ester (PCBM) (P3HT/PCBM) solar cell. The charge displacement distances increase on a tens of picosecond time scale.

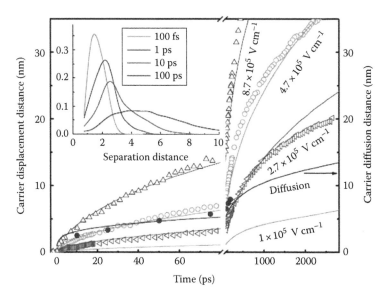

Figure 22.17 Time-dependent experimental (symbols) and calculated (black lines) average charge-pair separation distances along electric fields of different strength. The electric field was created by the applied external voltage and by different work functions of used electrodes of about 0.7 eV. The solid blue line shows the calculated absolute carrier diffusion distance at zero electric field. It shows that diffusion, rather than drift, is the dominant component at t < 100 ps at electric fields equivalent to an operating solar cell (1 V). The insert shows the evolution of the distribution of the absolute carrier separation distances at zero electric field.

Interestingly, we observe only very small (<1 nm) carrier displacement distances on a sub-picosecond time scale. We attribute this modest initial separation to the field-induced polarization of molecular electron and hole charge densities, similar to exciton polarization observed in neat polymers.[117] It has been established that charge transfer from the polymer to the fullerene acceptor in BHJ blends occurs on a time scale as short as 0.1 ps.[20,101,102,116,119] The small charge-pair displacement on a sub-picosecond time scale shows that the applied electric field has no, or only weak, influence on the spatial separation of these nascent bound pair states. The absence of a major ultrafast TREFISH response therefore leads us to conclude that no long-range charge separation occurs on an ultrafast sub-picosecond time scale.

Motion of the charge carriers in an applied electric field may be characterized by a time-dependent effective mobility. We can evaluate the instantaneous effective carrier mobility, averaged over electrons and holes, as the time derivative of the carrier displacement kinetics in Figure 22.17. The effective mobility decreases gradually on the picosecond to nanosecond time scale and is independent of the electric field strength within experimental accuracy. The latter suggests that the charge-pair separation is not electric field dependent in this material system since any field dependence of carrier separation would lead to field dependence of the mobility at short times. The initial mobility is at least two orders of magnitude higher than the long-time mobility[120] and is comparable to that obtained in TRTS measurements on P3HT/PCBM[121] and other polymer/PCBM blends.[60,122–124]

Separation of charges in an electric field has two contributions, field-induced drift (as measured in the TREFISH experiments) and diffusion, and we would like to know their respective contributions to the charge separation. The time-dependent diffusion coefficient D(t) can be obtained from the measured carrier mobility μ through the Einstein relation, $D = \mu k_B T/q$, where q is the electron charge, k_B and T are the Boltzmann constant, and temperature, respectively. From the time-dependent D(t), we can obtain distributions of the diffusion distances r as a function of time and hence estimate the average diffusion distance as the peak of this distribution function (Figure 22.17 inset). These results suggest that charges separate rapidly through diffusion—already at 10 ps, the peak

of the displacement distance distribution is at ~3 nm and a significant fraction of the charges have reached a distance of >5 nm; at 100 ps the majority of the charges have separated to distances exceeding 5 nm by diffusion. We can also apply Monte Carlo simulations of the carrier motion to compare with the observed time-dependent carrier mobility. The simulations enable us to determine the diffusion contribution to the charge carrier separation by calculating absolute carrier separation distances at zero-applied field. The results in Figure 22.17 nicely agree with those of the Einstein diffusion calculations and iterate that the carrier separation during the initial several picoseconds is dominated by diffusion, which drags charge carriers apart by approximately 3 nm. The drift starts to contribute more substantially to the carrier separation during the slower (t > 10 ps) phase, but diffusion still strongly dominates at typical electric fields in solar cells (~1 × 10⁷ V m⁻¹), even on a sub-nanosecond time scale when the Coulomb electron–hole attraction is effectively overcome (~7 to 8 nm). The carrier motion slows down rapidly with time as carriers relax into deeper energetic traps within the DOS. Extended polymer segments are also expected to assist the initial charge separation process by allowing carriers to find the fastest transport pathways through the material. The charge separation process is schematically illustrated in Figure 22.18.

22.3.2 Carrier mobility from time-resolved THz conductivity

From what was discussed, it is clear that charge carrier mobility is an important property characterizing the ability of charges to escape from their mutual Coulombic interaction and form free mobile charges that can be extracted as photocurrent. Time-resolved THz conductivity is a noncontact technique to measure charge carrier mobilities with temporal resolution in the sub-picosecond time domain. THz conductivity dynamics have been studied for many polymer/PCBM thin films, and despite the different chemical nature of the materials very similar, if not identical, dynamics were observed[124]; photogenerated species are created almost instantaneously after photoexcitation followed by an almost complete picosecond decay of the conductivity signal. This very fast decay was variably assigned to exciton–exciton annihilation, interfacial charge transfer, and trapping

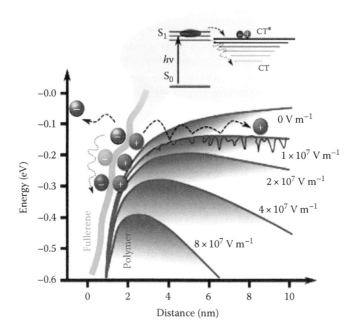

Figure 22.18 Charge carrier separation model. Pink lines show potential surfaces of charge carriers in the vicinity of the counter charge along the applied electric field of different strength calculated by using the point charge approximation. Dotted line shows hole motion and the blue line illustrates the potential energy landscape of the disordered polymer. The green line shows the polymer and PCBM interface. Following photoexcitation of the polymer, the manifold of CT states with the excess energy (see insert) is populated. Hot CT states, at a particular site, relax toward the lowest CT states in competition with overcoming the CT-state binding energy and exploring the CT density of states (at other sites). During the charge separation monitored by the TREFISH measurement, the hole on the polymer samples the energy landscape of the polymer and its diffusion coefficient remains high as long as the hole is not trapped in low-energy sites. Successful charge separation occurs when diffusion-driven overcoming the CT binding energy leads to free charges, before the hole loses mobility in a deep trap.

and cooling of highly mobile charges. At the same time, optical TA measurements on many polymer/fullerene materials (see, e.g., References 20,21,96) have demonstrated carrier lifetimes on the nanoseconds and longer time scale. High excitation densities (generally much higher than in optical TA measurements) are routinely used in THz measurements to ensure that enough charge carriers are photogenerated to extract a decent signal. Under such excitation conditions, TA measurements have demonstrated extensive nongeminate charge recombination that could severely shorten the carrier lifetime.[20]

By pushing the sensitivity limits of the THz technique, we could obtain ultrafast THz dynamics of carriers in a polymer/fullerene thin film generated by excitation light pulses varying in intensity over two orders.[124] Two

polymers TQ1 (2,3-bis-(3-octyloxyphenyl)quinoxaline-5,8-diyl-*alt*-thiophene-2,5-diyl) and APFO-3 (poly([2,7-(9,9-dioctyl-fluorene)-alt-5,5-(4′,7′-di-2-thienyl-2′,1′,3-benzothiadiazole)])) blended with the PCBM electron acceptor were studied.[124] Both showed excitation fluence dependence of the early time, 1–10 ps decay. At lower pump power ($\sim 10^{12}$–10^{13} ph^{-1} cm^{-2} pulse^{-1}), the decay of transient photoconductivity, that is, the polaron pair concentration mobility product, was estimated to be in the hundreds of picoseconds, while at high fluence and thus high charge density ($\sim 10^{14}$–10^{15} ph^{-1} cm^{-2} pulse^{-1}), the conductivity decays in several picoseconds due to polaron pair annihilation (Figure 22.19). Thus, these results show that carrier cooling and trapping in these materials occur on the hundreds of picosecond and nanosecond time scale, much slower than

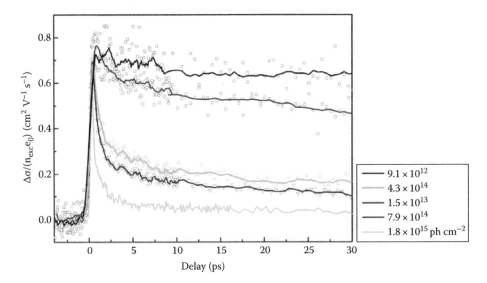

Figure 22.19 Evolution of transient conductivity of TQ1/PCBM at different excitation densities. Lines are smoothed data for clarity of the plot except for highest excitation fluence.

previously reported.[57,60,122,125–127] This implies that the photogenerated charges maintain a very high mobility ($\sim$0.1 cm^2 V^{-1} s^{-1}) for hundreds of picoseconds, which of course has far-reaching implications for how photogenerated charges separate and form free mobile charges that can be extracted as photocurrent in a functioning solar cell.

The results summarized earlier suggest that efficient charge separation is driven by the rapid diffusion of electron and holes on the time scale of tens and hundreds of picoseconds, that is, on times that are still short as compared to geminate electron–hole recombination in polymer/fullerene materials of efficient solar cells. In other words, charge separation on the 10–100 ps time scale is sufficient to outcompete charge recombination and efficiently form free mobile charges. Experimental work by Friend and coworkers[119,128] and theoretical work by Deibel et al.[129] have proposed a picture where either the electrons in the fullerene or holes in the polymer phase initially are delocalized and that this helps breaking up the bound charge pair states. We suggest that this view can be combined with the diffusion picture proposed earlier: somewhat delocalized electrons or holes reduces the CP-state binding energy to a level ($\sim$0.2 eV) where the initially fast diffusion of the charges can take them to a distance ($\sim$10 nm) where Coulomb interaction is effectively broken in a time which is much shorter ($\sim$100 ps) than the geminate recombination time.[57,60,118,122]

ACKNOWLEDGMENTS

We would like to thank all our coworkers and collaborators at the Division of Chemical Physics, Lund University, for their contributions to this work and for many fruitful discussions. The Swedish Research Council, the Knut and Alice Wallenberg Foundation, the Swedish Energy Administration, and the European Research Council are acknowledged for financial support.

REFERENCES

1. Oregan, B.; Gratzel, M. A. Low-cost, high-efficiency solar-cell based on dye-sensitized colloidal TiO$_2$ films. *Nature* 1991, *353* (6346), 737–740.
2. Mathew, S.; Yella, A.; Gao, P.; Humphry-Baker, R.; Curchod, B. F. E.; Ashari-Astani, N.; Tavernelli, I.; Rothlisberger, U.; Nazeeruddin, M. K.; Grätzel, M. Dye-sensitized solar cells with 13% efficiency achieved through the molecular engineering of porphyrin sensitizers. *Nature Chemistry* 2014, *6*, 242–247.
3. Kojima, A.; Teshima, K.; Shirai, Y.; Miyasaka, T. Organometal halide perovskites as visible-light sensitizers for photovoltaic cells. *Journal of the American Chemical Society* 2009, *131* (17), 6050–6051.

4. Kim, H. S.; Lee, C. R.; Im, J. H.; Lee, K. B.; Moehl, T.; Marchioro, A.; Moon, S. J. et al. Lead iodide perovskite sensitized all-solid-state submicron thin film mesoscopic solar cell with efficiency exceeding 9%. *Scientific Reports* 2012, *2*, 592 (1–7).

5. Lee, M. M.; Teuscher, J.; Miyasaka, T.; Murakami, T. N.; Snaith, H. J. Efficient hybrid solar cells based on meso-superstructured organometal halide perovskites. *Science* 2012, *338* (6107), 643–647.

6. Snaith, H. J. Perovskites: The emergence of a new era for low-cost, high-efficiency solar cells. *Journal of Physical Chemistry Letters* 2013, *4* (21), 3623–3630.

7. Liu, Y.; Harlang, T.; Canton, S. E.; Chabera, P.; Suarez-Alcantara, K.; Fleckhaus, A.; Vithanage, D. A. et al. Towards longer-lived metal-to-ligand charge transfer states of iron(II) complexes: An N-heterocyclic carbene approach. *Chemical Communications* 2013, *49* (57), 6412–6414.

8. Kamat, P. V. Quantum dot solar cells. The next big thing in photovoltaics. *Journal of Physical Chemistry Letters* 2013, *4* (6), 908–918.

9. Kongkanand, A.; Tvrdy, K.; Takechi, K.; Kuno, M.; Kamat, P. V. Quantum dot solar cells. Tuning photoresponse through size and shape control of CdSe-TiO$_2$ architecture. *Journal of the American Chemical Society* 2008, *130* (12), 4007–4015.

10. Nozik, A.; Beard, M.; Luther, J.; Law, M.; Ellingson, R.; Johnson, J. Semiconductor quantum dots and quantum dot arrays and applications of multiple exciton generation to third-generation photovoltaic solar cells. *Chemical Reviews* 2010, *110* (11), 6873–6890.

11. Shockley, W.; Queisser, H. J. Detailed balance limit of efficiency of P-N junction solar cells. *Journal of Applied Physics* 1961, *32* (3), 510.

12. Lee, H.; Wang, M.; Chen, P.; Gamelin, D. R.; Zakeeruddin, S. M.; Graetzel, M.; Nazeeruddin, M. Efficient CdSe quantum dot-sensitized solar cells prepared by an improved successive ionic layer adsorption and reaction process. *Nano Letters* 2009, *9* (12), 4221–4227.

13. Sargent, E. H. Colloidal quantum dot solar cells. *Nature Photonics* 2012, *6* (3), 133–135.

14. Halls, J. J. M.; Walsh, C. A.; Greenham, N. C.; Marseglia, E. A.; Friend, R. H.; Moratti, S. C.; Holmes, A. B. Efficient photodiodes from interpenetrating polymer networks. *Nature* 1995, *376* (6540), 498–500.

15. Yu, G.; Gao, J.; Hummelen, J. C.; Wudl, F.; Heeger, A. J. Polymer photovoltaic cells—Enhanced efficiencies via a network of internal donor-acceptor heterojunctions. *Science* 1995, *270* (5243), 1789–1791.

16. Veldman, D.; Ipek, O.; Meskers, S. C.; Sweelssen, J.; Koetse, M. M.; Veenstra, S. C.; Kroon, J. M.; van Bavel, S. S.; Loos, J.; Janssen, R. A. Compositional and electric field dependence of the dissociation of charge transfer excitons in alternating polyfluorene copolymer/fullerene blends. *Journal of the American Chemical Society* 2008, *130* (24), 7721–7735.

17. Grancini, G.; Maiuri, M.; Fazzi, D.; Petrozza, A.; Egelhaaf, H. J.; Brida, D.; Cerullo, G.; Lanzani, G. Hot exciton dissociation in polymer solar cells. *Nature Materials* 2013, *12* (1), 29–33.

18. Bakulin, A. A.; Dimitrov, S. D.; Rao, A.; Chow, P. C.; Nielsen, C. B.; Schroeder, B. C.; McCulloch, I.; Bakker, H. J.; Durrant, J. R.; Friend, R. H. Charge-transfer state dynamics following hole and electron transfer in organic photovoltaic devices. *Journal of Physical Chemistry Letters* 2013, *4* (1), 209–215.

19. Fishchuk, I.; Kadashchuk, A.; Genoe, J.; Ullah, M.; Sitter, H.; Singh, T.; Sariciftci, N.; Baessler, H. Temperature dependence of the charge carrier mobility in disordered organic semiconductors at large carrier concentrations. *Physical Review B* 2010, *81*, 045202 (1–12).

20. De, S.; Pascher, T.; Maiti, M.; Jespersen, K. G.; Kesti, T.; Zhang, F. L.; Inganas, O.; Yartsev, A.; Sundstrom, V. Geminate charge recombination in alternating polyfluorene Copolymer/Fullerene blends. *Journal of the American Chemical Society* 2007, *129* (27), 8466–8472.

21. Pal, S. K.; Kesti, T.; Maiti, M.; Zhang, F. L.; Inganas, O.; Hellstrom, S.; Andersson, M. R. et al. Geminate charge recombination in polymer/fullerene bulk heterojunction films and implications for solar cell function. *Journal of the American Chemical Society* 2010, *132* (35), 12440–12451.

22. Cabanillas-Gonzalez, J.; Grancini, G.; Lanzani, G. Pump-probe spectroscopy in organic semiconductors: Monitoring fundamental processes of relevance in optoelectronics. *Advanced Materials* 2011, *23* (46), 5468–5485.

23. Grancini, G.; Polli, D.; Fazzi, D.; Cabanillas-Gonzalez, J.; Cerullo, G.; Lanzani, G. Transient absorption imaging of P3HT:PCBM photovoltaic blend: Evidence for interfacial charge transfer state. *Journal of Physical Chemistry Letters* 2011, *2* (9), 1099–1105.

24. Howard, I. A.; Mauer, R.; Meister, M.; Laquai, F. Effect of morphology on ultrafast free carrier generation in polythiophene: Fullerene organic solar cells. *Journal of the American Chemical Society* 2010, *132* (42), 14866–14876.

25. Howard, I. A.; Laquai, F. Optical probes of charge generation and recombination in bulk heterojunction organic solar cells. *Macromolecular Chemistry and Physics* 2010, *211* (19), 2063–2070.

26. Howard, I. A.; Meister, M.; Baumeier, B.; Wonneberger, H.; Pschirer, N.; Sens, R.; Bruder, I. et al. Two channels of charge generation in perylene monoimide solid-state dye-sensitized solar cells. *Advanced Energy Materials* 2014, *4*, 1300640 (1–9).

27. Ardo, S.; Meyer, G. J. Photodriven heterogeneous charge transfer with transition-metal compounds anchored to TiO$_2$ semiconductor surfaces. *Chemical Society Reviews* 2009, *38* (1), 115–164.

28. Benko, G.; Kallioinen, J.; Korppi-Tommola, J. E. I.; Yartsev, A. P.; Sundstrom, V. Photoinduced ultrafast dye-to-semiconductor electron injection from nonthermalized and thermalized donor states. *Journal of the American Chemical Society* 2002, *124* (3), 489–493.

29. Benko, G.; Myllyperkio, P.; Pan, J.; Yartsev, A. P.; Sundstrom, V. Photoinduced electron injection from Ru(dcbPY)(2)(NCS)(2) to SnO$_2$ and TiO$_2$ nanocrystalline films. *Journal of the American Chemical Society* 2003, *125* (5), 1118–1119.

30. Benko, G.; Kallioinen, J.; Myllyperkio, P.; Trif, F.; Korppi-Tommola, J. E. I.; Yartsev, A. P.; Sundstrom, V. Interligand electron transfer determines triplet excited state electron injection in RuN3-sensitized TiO$_2$ films. *Journal of Physical Chemistry B* 2004, *108* (9), 2862–2867.

31. Furube, A.; Katoh, R.; Yoshihara, T.; Hara, K.; Murata, S.; Arakawa, H.; Tachiya, M. Ultrafast direct and indirect electron-injection processes in a photoexcited dye-sensitized nanocrystalline zinc oxide film: The importance of exciplex intermediates at the surface. *Journal of Physical Chemistry B* 2004, *108* (33), 12583–12592.

32. Katoh, R.; Furube, A.; Barzykin, A. V.; Arakawa, H.; Tachiya, M. Kinetics and mechanism of electron injection and charge recombination in dye-sensitized nanocrystalline semiconductors. *Coordination Chemistry Reviews* 2004, *248* (13–14), 1195–1213.

33. Katoh, R.; Furube, A.; Fuke, N.; Fukui, A.; Koide, N. Ultrafast relaxation as a possible limiting factor of electron injection efficiency in black dye sensitized nanocrystalline TiO$_2$ films. *Journal of Physical Chemistry C* 2012, *116* (42), 22301–22306.

34. Katoh, R.; Yaguchi, K.; Furube, A. Effect of dye concentration on electron injection efficiency in nanocrystalline TiO$_2$ films sensitized with N719 dye. *Chemical Physics Letters* 2011, *511* (4–6), 336–339.

35. Thorsmolle, V. K.; Wenger, B.; Teuscher, J.; Bauer, C.; Moser, J. E. Dynamics of photoinduced interfacial electron transfer and charge transport in dye-sensitized mesoscopic semiconductors. *Chimia* 2007, *61* (10), 631–634.

36. Wenger, B.; Gratzel, M.; Moser, J. E. Rationale for kinetic heterogeneity of ultrafast light-induced electron transfer from Ru(II) complex sensitizers to nanocrystalline TiO$_2$. *Journal of the American Chemical Society* 2005, *127* (35), 12150–12151.

37. Kallioinen, J.; Benko, G.; Sundstrom, V.; Korppi-Tommola, J. E. I.; Yartsev, A. P. Electron transfer from the singlet and triplet excited states of Ru(dcbpy)(2)(NCS)(2) into nanocrystalline TiO$_2$ thin films. *Journal of Physical Chemistry B* 2002, *106* (17), 4396–4404.

38. Kallioinen, J.; Benko, G.; Myllyperkio, P.; Khriachtchev, L.; Skarman, B.; Wallenberg, R.; Tuomikoski, M.; Korppi-Tommola, J.; Sundstrom, V.; Yartsev, A. P. Photoinduced ultrafast dynamics of Ru(dcbpy)(2)(NCS)(2)-sensitized nanocrystalline TiO$_2$ films: The influence of sample preparation and experimental conditions. *Journal of Physical Chemistry B* 2004, *108* (20), 6365–6373.

39. Myllyperkio, P.; Benko, G.; Korppi-Tommola, J.; Yartsev, A. P.; Sundstrom, V. A study of electron transfer in Ru(dcbpy)(2)(NCS)(2) sensitized nanocrystalline TiO2 and SnO$_2$ films induced by red-wing excitation. *Physical Chemistry Chemical Physics* 2008, *10* (7), 996–1002.

40. Katoh, R.; Furube, A.; Kasuya, M.; Fuke, N.; Koide, N.; Han, L. Photoinduced electron injection in black dye sensitized nanocrystalline TiO$_2$ films. *Journal of Materials Chemistry* 2007, *17* (30), 3190–3196.

41. Imahori, H.; Kang, S.; Hayashi, H.; Haruta, M.; Kurata, H.; Isoda, S.; Canton, S. E. et al. Photoinduced charge carrier dynamics of Zn-porphyrin-TiO$_2$ electrodes: The key role of charge recombination for solar cell performance. *Journal of Physical Chemistry A* 2011, *115* (16), 3679–3690.

42. Ye, S.; Kathiravan, A.; Hayashi, H.; Tong, Y.; Infahsaeng, Y.; Chabera, P.; Pascher, T. et al. Role of adsorption structures of Zn-porphyrin on TiO$_2$ in dye-sensitized solar cells studied by sum frequency generation vibrational spectroscopy and ultrafast spectroscopy. *Journal of Physical Chemistry C* 2013, *117* (12), 6066–6080.

43. Imahori, H.; Umeyama, T.; Ito, S. Large pi-aromatic molecules as potential sensitizers for highly efficient dye-sensitized solar cells. *Accounts of Chemical Research* 2009, *42*, 1809–1818.

44. Zhao, J.; Yang, X.; Cheng, M.; Li, S.; Sun, L. New organic dyes with a phenanthrenequinone derivative as the pi-conjugated bridge for dye-sensitized solar cells. *Journal of Physical Chemistry C* 2013, *117* (25), 12936–12941.

45. Zhao, J.; Yang, X.; Cheng, M.; Li, S.; Wang, X.; Sun, L. Highly efficient iso-quinoline cationic organic dyes without vinyl groups for dye-sensitized solar cells. *Journal of Materials Chemistry A* 2013, *1* (7), 2441–2446.

46. Hao, Y.; Yang, X.; Cong, J.; Hagfeldt, A.; Sun, L. Engineering of highly efficient tetrahydroquinoline sensitizers for dye-sensitized solar cells. *Tetrahedron* 2012, *68* (2), 552–558.

47. Jiang, X.; Karlsson, K. M.; Gabrielsson, E.; Johansson, E. M.; Quintana, M.; Karlsson, M.; Sun, L.; Boschloo, G.; Hagfeldt, A. Highly efficient solid-state dye-sensitized solar cells based on triphenylamine dyes. *Advanced Functional Materials* 2011, *21* (15), 2944–2952.

48. Haque, S. A.; Handa, S.; Peter, K.; Palomares, E.; Thelakkat, M.; Durrant, J. R. Supermolecular control of charge transfer in dye-sensitized nanocrystalline TiO$_2$ films: Towards a quantitative structure-function relationship. *Angewandte Chemie-International Edition* 2005, *44* (35), 5740–5744.

49. Moser, J. E.; Gratzel, M. Observation of temperature independent heterogeneous electron-transfer reactions in the inverted Marcus region. *Chemical Physics* 1993, *176* (2–3), 493–500.

50. Oregan, B.; Moser, J.; Anderson, M.; Gratzel, M. Vectorial electron injection into transparent semiconductor membranes and electric-field effects on the dynamics of light-induced charge separation. *Journal of Physical Chemistry* 1990, *94* (24), 8720–8726.

51. Fitzmaurice, D. J.; Frei, H. Transient near-infrared spectroscopy of visible-light sensitized oxidation of I- at colloidal TiO$_2$. *Langmuir* 1991, *7* (6), 1129–1137.

52. Marcus, R. A.; Sutin, N. Electron transfers in chemistry and biology. *Biochimica et Biophysica Acta* 1985, *811* (3), 265–322.

53. Marcus, R. A. Theory of oxidation-reduction reactions involving electron transfer .1. *Journal of Chemical Physics* 1956, *24* (5), 966–978.

54. Marcus, R. A. Chemical + electrochemical electron-transfer theory. *Annual Review of Physical Chemistry* 1964, *15*, 155–196.

55. Gerischer, H. Electrochemical techniques for study of photosensitization. *Photochemistry and Photobiology* 1972, *16* (4), 243–260.

56. Gerischer, H. Charge transfer processes at semiconductor-electrolyte interfaces in connection with problems of catalysis. *Surface Science* 1969, *18* (1), 97–122.

57. Nemec, H.; Kuzel, P.; Sundstrom, V. Charge transport in nanostructured materials for solar energy conversion studied by time-resolved terahertz spectroscopy. *Journal of Photochemistry and Photobiology A-Chemistry* 2010, *215* (2–3), 123–139.

58. Beard, M. C.; Turner, G. M.; Schmuttenmaer, C. A. Progress towards two-dimensional biomedical imaging with THz spectroscopy. *Physics in Medicine and Biology* 2002, *47* (21), 3841–3846.

59. Turner, G. M.; Beard, M. C.; Schmuttenmaer, C. A. Carrier localization in nanocrystalline TiO$_2$. *Abstracts of Papers of the American Chemical Society* 2002, *224*, 271.

60. Nemec, H.; Nienhuys, H. K.; Zhang, F.; Inganas, O.; Yartsev, A.; Sundstrom, V. Charge carrier dynamics in alternating polyfluorene copolymer: Fullerene blends probed by terahertz spectroscopy. *Journal of Physical Chemistry C* 2008, *112* (16), 6558–6563.

61. Nemec, H.; Rochford, J.; Taratula, O.; Galoppini, E.; Kuzel, P.; Polivka, T.; Yartsev, A.; Sundstrom, V. Influence of the electron-cation interaction on electron mobility in dye-sensitized ZnO and TiO(2) nanocrystals: A study using ultrafast terahertz spectroscopy. *Physical Review Letters* 2010, *104* (19), 197401.

62. Tisdale, W.; Zhu, X. Artificial atoms on semiconductor surfaces. *Proceedings of the National Academy of Sciences of the United States of America* 2011, *108* (3), 965–970.

63. Tvrdy, K.; Frantsuzov, P. A.; Kamat, P. V. Photoinduced electron transfer from semiconductor quantum dots to metal oxide nanoparticles. *Proceedings of the National Academy of Sciences of the United States of America* 2011, *108* (1), 29–34.

64. Dibbell, R. S.; Youker, D. G.; Watson, D. F. Excited-state electron transfer from CdS quantum dots to TiO$_2$ nanoparticles via molecular linkers with phenylene bridges. *Journal of Physical Chemistry C* 2009, *113* (43), 18643–18651.

65. Zidek, K.; Zheng, K.; Ponseca, C. S.; Messing, M. E.; Wallenberg, L.; Chabera, P.; Abdellah, M.; Sundstrom, V.; Pullerits, T. Electron transfer in quantum-dot-sensitized ZnO nanowires: Ultrafast time-resolved absorption and terahertz study. *Journal of the American Chemical Society* 2012, *134* (29), 12110–12117.

66. Barbara, P. F.; Meyer, T. J.; Ratner, M. A. Contemporary issues in electron transfer research. *Journal of Physical Chemistry* 1996, *100* (31), 13148–13168.

67. Hansen, T.; Zidek, K.; Zheng, K.; Abdellah, M.; Chabera, P.; Persson, P.; Pullerits, T. Orbital topology controlling charge injection in quantum-dot-sensitized solar cells. *Journal of Physical Chemistry Letters* 2014, *5* (7), 1157–1162.

68. Nair, G.; Geyer, S. M.; Chang, L. Y.; Bawendi, M. G. Carrier multiplication yields in PbS and PbSe nanocrystals measured by transient photoluminescence. *Physical Review B* 2008, *78*, 125325 (1–10).

69. Trinh, M.; Houtepen, A. J.; Schins, J. M.; Hanrath, T.; Piris, J.; Knulst, W.; Goossens, A. P.; Siebbeles, L. D. In spite of recent doubts carrier multiplication does occur in PbSe nanocrystals. *Nano Letters* 2008, *8* (6), 1713–1718.

70. Fu, Y.; Zhou, Y.; Su, H.; Boey, F.; Agren, H. Impact ionization and auger recombination rates in semiconductor quantum dots. *Journal of Physical Chemistry C* 2010, *114* (9), 3743–3747.

71. Schaller, R. D.; Agranovich, V. M.; Klimov, V. I. High-efficiency carrier multiplication through direct photogeneration of multi-excitons via virtual single-exciton states. *Nature Physics* 2005, *1* (3), 189–194.

72. Madrid, A. B.; Hyeon-Deuk, K.; Habenicht, B. F.; Prezhdo, O. V. Phonon-induced dephasing of excitons in semiconductor quantum dots: Multiple exciton generation, fission, and luminescence. *ACS Nano* 2009, *3* (9), 2487–2494.

73. Karki, K. J.; Ma, F.; Zheng, K.; Zidek, K.; Mousa, A.; Abdellah, M. A.; Messing, M. E.; Wallenberg, L.; Yartsev, A.; Pullerits, T. Multiple exciton generation in nano-crystals revisited: Consistent calculation of the yield based on pump-probe spectroscopy. *Scientific Reports* 2013, *3*, 2287 (1–5).

74. Lenngren, N.; Garting, T.; Zheng, K.; Abdellah, M.; Lascoux, N.; Ma, F.; Yartsev, A.; Zidek, K.; Pullerits, T. Multiexciton absorption cross sections of CdSe quantum dots determined by ultrafast spectroscopy. *Journal of Physical Chemistry Letters* 2013, *4* (19), 3330–3336.

75. Klimov, V. I.; Mikhailovsky, A. A.; McBranch, D. W.; Leatherdale, C. A.; Bawendi, M. G. Quantization of multiparticle Auger rates in semiconductor quantum dots. *Science* 2000, *287* (5455), 1011–1013.

76. Bruggemann, B.; Herek, J. L.; Sundstrom, V.; Pullerits, T.; May, V. Microscopic theory of exciton annihilation: Application to the LH2 antenna system. *Journal of Physical Chemistry B* 2001, *105* (46), 11391–11394.

77. Zidek, K.; Zheng, K.; Abdellah, M.; Lenngren, N.; Chabera, P.; Pullerits, T. Ultrafast dynamics of multiple exciton harvesting in the CdSe-ZnO system: Electron injection versus auger recombination. *Nano Letters* 2012, *12* (12), 6393–6399.

78. Sambur, J. B.; Novet, T.; Parkinson, B. Multiple exciton collection in a sensitized photovoltaic system. *Science* 2010, *330* (6000), 63–66.

79. Kamat, P. V.; Christians, J. A.; Radich, J. G. Quantum dot solar cells: Hole transfer as a limiting factor in boosting the photoconversion efficiency. *Langmuir* 2014, *30* (20), 5716–5725.

80. Brown, P. R.; Kim, D.; Lunt, R. R.; Zhao, N.; Bawendi, M. G.; Grossman, J. C.; Bulovic, V. Energy level modification in lead sulfide quantum dot thin films through ligand exchange. *ACS Nano* 2014, *8*, 5863–5872.

81. Chuang, C. H. M.; Brown, P. R.; Bulovic, V.; Bawendi, M. G. Improved performance and stability in quantum dot solar cells through band alignment engineering. *Nature Materials* 2014, *13* (8), 796–801.

82. Gratzel, M. Recent advances in sensitized mesoscopic solar cells. *Accounts of Chemical Research* 2009, *42* (11), 1788–1798.

83. Lobello, M. G.; Wu, K. L.; Reddy, M. A.; Marotta, G.; Graetzel, M.; Nazeeruddin, M. K.; Chi, Y.; Chandrasekharam, M.; Vitillaro, G.; De Angelis, F. Engineering of Ru(II) dyes for interfacial and light-harvesting optimization. *Dalton Transactions* 2014, *43* (7), 2726–2732.

84. Graetzel, M.; Janssen, R. A.; Mitzi, D. B.; Sargent, E. H. Materials interface engineering for solution-processed photovoltaics. *Nature* 2012, *488* (7411), 304–312.

85. Meyer, T. J. Photochemistry of metal coordination-complexes—Metal to ligand charge-transfer excited-states. *Pure and Applied Chemistry* 1986, *58* (9), 1193–1206.

86. Cho, H.; Strader, M. L.; Hong, K.; Jamula, L.; Gullikson, E. M.; Kim, T. K.; de Groot, F. M.; McCusker, J. K.; Schoenlein, R. W.; Huse, N. Ligand-field symmetry effects in Fe(II) polypyridyl compounds probed by transient X-ray absorption spectroscopy. *Faraday Discussions* 2012, *157*, 463–474.

87. Sousa, C.; de Graaf, C.; Rudavskyi, A.; Broer, R.; Tatchen, J.; Etinski, M.; Marian, C. M. Ultrafast deactivation mechanism of the excited singlet in the light-induced spin crossover of [Fe(2,2'-bipyridine)(3)](2+). *Chemistry: A European Journal* 2013, *19* (51), 17541–17551.

88. Zhang, W.; Onso-Mori, R.; Bergmann, U.; Bressler, C.; Chollet, M.; Galler, A.; Gawelda, W. et al. Tracking excited-state charge and spin dynamics in iron coordination complexes. *Nature* 2014, *509* (7500), 345–348.

89. Bomben, P. G.; Robson, K. C.; Koivisto, B. D.; Berlinguette, C. P. Cyclometalated ruthenium chromophores for the dye-sensitized solar cell. *Coordination Chemistry Reviews* 2012, *256* (15–16), 1438–1450.

90. Bessho, T.; Yoneda, E.; Yum, J. H.; Guglielmi, M.; Tavernelli, I.; Imai, H.; Rothlisberger, U.; Nazeeruddin, M. K.; Graetzel, M. New paradigm in molecular engineering of sensitizers for solar cell applications. *Journal of the American Chemical Society* 2009, *131* (16), 5930–5934.

91. Fredin, L. A.; Papai, M.; Rozsalyi, E.; Vanko, G.; Wärnmark, K.; Sundström, V.; Persson, P. Exceptional excited-state lifetime of an iron(II)-N-heterocyclic carbene complex explained. *Journal of Physical Chemistry Letters* 2014, *5*, 2066–2071.

92. Heo, J. H.; Im, S. H.; Noh, J. H.; Mandal, T. N.; Lim, C. S.; Chang, J. A.; Lee, Y. H. et al. Efficient inorganic-organic hybrid heterojunction solar cells containing perovskite compound and polymeric hole conductors. *Nature Photonics* 2013, *7* (6), 487–492.

93. Burschka, J.; Pellet, N.; Moon, S. J.; Humphry-Baker, R.; Gao, P.; Nazeeruddin, M. K.; Graetzel, M. Sequential deposition as a route to high-performance perovskite-sensitized solar cells. *Nature* 2013, *499* (7458), 316–319.

94. Ponseca, C. S.; Savenije, T. J.; Abdellah, M.; Zheng, K.; Yartsev, A.; Pascher, T.; Harlang, T. et al. Organometal halide perovskite solar cell materials rationalized: Ultrafast charge generation, high and microsecond-long balanced mobilities, and slow recombination. *Journal of the American Chemical Society* 2014, *136* (14), 5189–5192.

95. Beard, M. C.; Turner, G. M.; Schmuttenmaer, C. A. Terahertz spectroscopy. *Journal of Physical Chemistry B* 2002, *106* (29), 7146–7159.

96. Vithanage, D. A.; Wang, E.; Wang, Z.; Ma, F.; Inganas, O.; Andersson, M. R.; Yartsev, A.; Sundstrom, V.; Pascher, T. Charge carrier

dynamics of polymer: Fullerene blends: From geminate to non-geminate recombination. *Advanced Energy Materials* 2014, *8*, 1301706.

97. Blom, P. W.; Mihailetchi, V. D.; Koster, L.; Markov, D. E. Device physics of polymer: Fullerene bulk heterojunction solar cells. *Advanced Materials* 2007, *19*(12), 1551–1566.

98. Umari, P.; Mosconi, E.; De Angelis, F. Relativistic GW calculations on $CH_3NH_3PbI_3$ and $CH_3NH_3SnI_3$ perovskites for solar cell applications. *Scientific Reports* 2014, *4*, 4467 (1–7).

99. You, J.; Dou, L.; Yoshimura, K.; Kato, T.; Ohya, K.; Moriarty, T.; Emery, K. et al. A polymer tandem solar cell with 10.6% power conversion efficiency. *Nature Communications* 2013, *4*, 1446 (1–10).

100. van Amerongen, H.; Valkunas, L.; van Grondelle, R. *Photosynthetic Excitons*, World Scientific, Singapore, 2010.

101. Brabec, C. J.; Zerza, G.; Cerullo, G.; De Silvestri, S.; Luzzati, S.; Hummelen, J. C.; Sariciftci, S. Tracing photoinduced electron transfer process in conjugated polymer/fullerene bulk heterojunctions in real time. *Chemical Physics Letters* 2001, *340* (3–4), 232–236.

102. De, S.; Kesti, T.; Maiti, M.; Zhang, F.; Inganas, O.; Yartsev, A.; Pascher, T.; Sundstrom, V. Exciton dynamics in alternating polyfluorene/fullerene blends. *Chemical Physics* 2008, *350* (1–3), 14–22.

103. Scheblykin, I. G.; Yartsev, A.; Pullerits, T.; Gulbinas, V.; Sundstroem, V. Excited state and charge photogeneration dynamics in conjugated polymers. *Journal of Physical Chemistry B* 2007, *111* (23), 6303–6321.

104. Beenken, W. J. D.; Pullerits, T. Spectroscopic units in conjugated polymers: A quantum chemically founded concept? *Journal of Physical Chemistry B* 2004, *108* (20), 6164–6169.

105. Yang, X. J.; Dykstra, T. E.; Scholes, G. D. Photon-echo studies of collective absorption and dynamic localization of excitation in conjugated polymers and oligomers. *Physical Review B* 2005, *71*, 045203 (1–15).

106. Kaake, L. G.; Moses, D.; Heeger, A. J. Coherence and uncertainty in nanostructured organic photovoltaics. *Journal of Physical Chemistry Letters* 2013, *4* (14), 2264–2268.

107. Mukamel, S. Comment on coherence and uncertainty in nanostructured organic photovoltaics. *Journal of Physical Chemistry A* 2013, *117* (40), 10563–10564.

108. Bredas, J. L.; Norton, J. E.; Cornil, J.; Coropceanu, V. Molecular understanding of organic solar cells: The challenges. *Accounts of Chemical Research* 2009, *42* (11), 1691–1699.

109. Tvingstedt, K.; Vandewal, K.; Gadisa, A.; Zhang, F.; Manca, J.; Inganas, O. Electroluminescence from charge transfer states in polymer solar cells. *Journal of the American Chemical Society* 2009, *131* (33), 11819–11824.

110. Vandewal, K.; Tvingstedt, K.; Gadisa, A.; Inganas, O.; Manca, J. V. Relating the open-circuit voltage to interface molecular properties of donor: Acceptor bulk heterojunction solar cells. *Physical Review B* 2010, *81*, 125204 (1–8).

111. Nogueira, A. F.; Montanari, I.; Nelson, J.; Durrant, J. R.; Winder, C.; Sariciftci, N. S. Charge recombination in conjugated polymer/fullerene blended films studied by transient absorption Spectroscopy. *Journal of Physical Chemistry B* 2003, *107* (7), 1567–1573.

112. Nelson, J.; Choulis, S. A.; Durrant, J. R. Charge recombination in polymer/fullerene photovoltaic devices. *Thin Solid Films* 2004, *451*, 508–514.

113. Offermans, T.; Meskers, S. C. J.; Janssen, R. A. J. Charge recombination in a poly(para-phenylene vinylene)-fullerene derivative composite film studied by transient, nonresonant, hole-burning spectroscopy. *Journal of Chemical Physics* 2003, *119* (20), 10924–10929.

114. Savenije, T. J.; Kroeze, J. E.; Wienk, M. M.; Kroon, J. M.; Warman, J. M. Mobility and decay kinetics of charge carriers in photoexcited PCBM/PPV blends. *Physical Review B* 2004, *69*, 155205 (1–11).

115. Andersson, L.; Melianas, A.; Infahasaeng, Y.; Tang, Z.; Yartsev, A.; Inganas, O.; Sundstrom, V. Unified study of recombination in polymer: Fullerene solar cells using transient absorption and charge-extraction measurements. *Journal of Physical Chemistry Letters* 2013, *4* (12), 2069–2072.

116. Infahsaeng, Y. PhD thesis. Light induced charge generation, recombination, and transport in organic solar cells. Lund, Sweden, 2014.

117. Devizis, A.; Serbenta, A.; Meerholz, K.; Hertel, D.; Gulbinas, V. Ultrafast dynamics of carrier mobility in a conjugated polymer probed at molecular and microscopic length scales. *Physical Review Letters* 2009, *103*, 027404 (1–4).

118. Vithanage, D.; Devizis, A.; Abramavicius, V.; Infahsaeng, Y.; Abramavicius, D.; MacKenzie, R.; Keivanidis, P. et al. Visualizing charge separation in bulk heterojunction organic solar cells. *Nature Communications* 2013, *4*, 2334 (1–6).

119. Bakulin, A. A.; Rao, A.; Pavelyev, V. G.; van Loosdrecht, P. H.; Pshenichnikov, M. S.; Niedzialek, D.; Cornil, J.; Beljonne, D.; Friend, R. H. The role of driving energy and delocalized states for charge separation in organic semiconductors. *Science* 2012, *335* (6074), 1340–1344.

120. Mihailetchi, V. D.; Xie, H. X.; de Boer, B.; Koster, L. J. A.; Blom, P. W. M. Charge transport and photocurrent generation in poly (3-hexylthiophene): Methanofullerene bulk-heterojunction solar cells. *Advanced Functional Materials* 2006, *16* (5), 699–708.

121. Parkinson, P.; Lloyd-Hughes, J.; Johnston, M.; Herz, L. Efficient generation of charges via below-gap photoexcitation of polymer-fullerene blend films investigated by terahertz spectroscopy. *Physical Review B* 2008, *78*, 115321 (1–11).

122. Nemec, H.; Nienhuys, H. K.; Perzon, E.; Zhang, F. L.; Inganas, O.; Kuzel, P.; Sundstrom, V. Ultrafast conductivity in a low-band-gap polyphenylene and fullerene blend studied by terahertz spectroscopy. *Physical Review B* 2009, *79*, 245326 (1–8).

123. Ponseca, C. S.; Nemec, H.; Vukmirovic, N.; Fusco, S.; Wang, E.; Andersson, M. R.; Chabera, P.; Yartsev, A.; Sundstrom, V. Electron and hole contributions to the terahertz photoconductivity of a conjugated polymer: Fullerene blend identified. *Journal of Physical Chemistry Letters* 2012, *3* (17), 2442–2446.

124. Ponseca, C. S.; Yartsev, A.; Wang, E.; Andersson, M. R.; Vithanage, D.; Sundstrom, V. Ultrafast terahertz photoconductivity of bulk heterojunction materials reveals high carrier mobility up to nanosecond time scale. *Journal of the American Chemical Society* 2012, *134* (29), 11836–11839.

125. Grzegorczyk, W. J.; Savenije, T. J.; Dykstra, T. E.; Piris, J.; Schins, J. M.; Siebbeles, L. D. Temperature-independent charge carrier photogeneration in P3HT-PCBM blends with different morphology. *Journal of Physical Chemistry C* 2010, *114* (11), 5182–5186.

126. Ferguson, A. J.; Kopidakis, N.; Shaheen, S. E.; Rumbles, G. Dark carriers, trapping, and activation control of carrier recombination in neat P3HT and P3HT: PCBM blends. *Journal of Physical Chemistry C* 2011, *115* (46), 23134–23148.

127. Ulbricht, R.; Hendry, E.; Shan, J.; Heinz, T. F.; Bonn, M. Carrier dynamics in semiconductors studied with time-resolved terahertz spectroscopy. *Reviews of Modern Physics* 2011, *83* (2), 543–586.

128. Gelinas, S.; Rao, A.; Kumar, A.; Smith, S. L.; Chin, A. W.; Clark, J.; van der Poll, T. S.; Bazan, G. C.; Friend, R. H. Ultrafast long-range charge separation in organic semiconductor photovoltaic diodes. *Science* 2014, *343* (6170), 512–516.

129. Deibel, C.; Strobel, T.; Dyakonov, V. Origin of the efficient polaron-pair dissociation in polymer-fullerene blends. *Physical Review Letters* 2009, *103*, 036402 (1–4).

23

Chlorophyll fluorescence as a tool for describing the operation and regulation of photosynthesis *in vivo*

JEREMY HARBINSON

23.1 INTRODUCTION

In very broad terms, photosynthesis begins with the absorption of light (the photosynthetically active kind) and ends with assimilation—the fixation of CO_2—a process that is fundamental to trophic networks in the biosphere. It is a relatively straightforward task to measure the photosynthetically active irradiance (PAR) incident on a leaf (or any other photosynthetic system) and even to measure the amount of that irradiance that is absorbed. The measurement of assimilation is, likewise, readily easily accomplished using infrared gas analysis systems. However, in between the absorption of light and the act of assimilation, there are the intermediate processes of photosynthesis: photochemistry, electron transport and energy transduction, metabolism, and gaseous diffusion processes. The rate of assimilation will ultimately be determined by the limiting activity of these processes. There are also complex regulatory networks that coordinate the activities of the many subprocesses of photosynthesis whose combined activity is necessary for assimilation. To understand the relationship between light absorption and assimilation, it is necessary to understand operation of the intermediate processes, how they interact with each other, and how they individually or collectively limit the overall efficiency of assimilation.

Destructive *in vitro* techniques can provide snapshots of the photosynthetic proteome and metabolome, and though these can provide copious amounts of data, their destructive character limits both their usage and usefulness. Much of the operational state of photosynthesis cannot be captured by *in vitro* techniques (e.g., conductances for gaseous diffusion), and many important aspects of the operational state of photosynthesis are simply too fugitive to be captured by existing *in vitro* methods (e.g., quantum efficiencies of photosystems I [PSI] and II [PSII], transthylakoid ion gradients). Destructive techniques therefore fail to capture the operational state of the photosynthetic system in the detail required to allow the correlations between the state or activity of different components of the photosynthetic system to be described. An additional major limitation of destructive approaches is that they make it impossible to describe how the state of the photosynthetic system changes in a specific leaf in response to changing environmental factors, such as irradiance, CO_2 concentration, and temperature. These factors are expected to affect the operation of

photosynthesis in specific ways, provoking changes in the regulatory state of the system. Correlations between the state of the photosynthetic system and activities of different component processes of photosynthesis, measured *in vivo*, often in response to changing environmental factors form the basis of our understanding of how photosynthesis is regulated and limited *in vivo*. Nonetheless, destructive techniques have proven to be and continue to be essential to our understanding of the composition, organization, and acclimation of the photosynthetic apparatus across many levels of integration. Without this basic knowledge of the organization of photosynthesis, nondestructive techniques of the kind described in this chapter could not exist as the powerful tools they have proven to be.

The value of nondestructive approaches to measuring different aspects of photosynthesis has led to a large number of techniques being co-opted as tools in photosynthesis research. These include gas analysis for CO_2 and water vapor (Coombs et al. 1985; Ball 1987) and O_2 (Delieu and Walker 1981; Walker and Walker 1987; Oja and Laisk 2000; Laisk et al. 2012), online analysis of the isotopic content of CO_2 and O_2 using mass-spectroscopic or tuned diode laser techniques (Barbour et al. 2007; Pons et al. 2009; Driever and Baker 2011), photoacoustic techniques (Bults et al. 1982; Kanstad et al. 1983; Poulet et al. 1983), photothermal radiometry (Bults et al. 1982; Kanstad et al. 1983, Malkin et al. 1991), fluorescence of NADP/NADPH (Duysens and Sweep 1957; Cerovic et al. 1993), and chlorophyll fluorescence (the subject of this chapter). In some cases, the specific implementation of a technique that is, in principle, nondestructive requires the use of a sample of tissue, such as a leaf disc, which in the longer term will die (the length of time a sample, such as a leaf disc, can be used depends upon the species). Of these techniques, some are infrequently encountered either because they require specialized, and possibly expensive, equipment (such as mass spectroscopy), while other techniques (such as infrared gas analysis for CO_2 and water vapor) are in routine use.

In this chapter, we will describe the principles and use of a major, widely used nondestructive method for measuring the operation and regulation of PSII not only in leaves and similar photosynthetic plant tissues, but also in *in vitro* samples: steady-state chlorophyll fluorescence. This class of methods (I use the term class because there are, overall, a range of different strategies for making these measurements [e.g., nonimaging and imaging techniques]) is based upon more or less continuous measurement of chlorophyll fluorescence and, in particular, the relative yield of chlorophyll fluorescence. It is technically quite distinct from fluorescence lifetime techniques, though the techniques do (and should) give results that agree with each other. Chlorophyll fluorescence provides information about the operation and regulation of PSII, in particular the light use efficiency of linear electron transport and the nonphotochemical dissipation processes that result in the nonphotochemical quenching (NPQ) of chlorophyll fluorescence. It is a relatively simple method to apply, and the equipment required is relatively inexpensive with the result that it has become widely used (and sometimes abused) in not only photosynthesis research but as a general tool in plant environmental physiology. This chapter begins by introducing fluorescence in general, followed by an introduction to chlorophyll fluorescence in solution and *in vitro*, then progress to an outline of the most widely used parameters that can be derived from fluorescence and the problems of measuring fluorescence in leaves. There are other similar descriptions of the application of chlorophyll fluorescence (Maxwell and Johnson 2000; Baker et al. 2007; Baker 2008), which will provide a similar perspective on this topic, albeit with different emphasis.

23.2 FLUORESCENCE: GENERAL PRINCIPLES

Fluorescence, in general, is about the absorption of photons, the formation of excited states, and the decay or dissipation of those excited states by the emission of a photon or some other path. To understand the phenomenon of fluorescence and its measurement, it is necessary to understand something of the physics of fluorescence, which is what will be covered in this section before moving on to describe how chlorophyll fluorescence is used in practice. The following description of the principles of light absorption and fluorescent emission processes is a very simplified one, but it should make clear the basic nature of excitation and fluorescence.

Photons are the basic packets of energy that comprise electromagnetic radiation, such as light, ultraviolet radiation, infrared radiation, etc. The absorption of a photon by a molecule or atom (from now on I will refer only to molecules) will increase the energy content of that molecule, resulting in the formation of an excited state (Figure 23.1a). It is the nature of molecules that their electrons can only have certain specific energies, and these energies correspond to "orbitals," which are the volumes of space surrounding the nuclei of a molecule in which an electron with a specific energy is likely to be found. Every orbital is quantum mechanically unique, though different orbitals can have the same

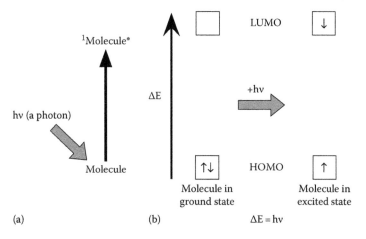

Figure 23.1 (a) The excitation of a molecule in the ground state ("molecule") by the absorption of photon to a form a singlet excited state ("^{1}molecule*"); "h" is Planck's constant (6.626×10^{-34}) and "ν" is the frequency of the photon. (b) The change in the electronic configuration of a molecule in the ground state, where the highest-energy electrons are in the highest occupied molecular orbital (HOMO), following the absorption of a photon with an energy (E) of $h\nu$, which lifts an electron from the HOMO to the lowest unoccupied molecular orbital (LUMO) with no change of spin.

energy level. Each orbital can contain two electrons, which are distinct from each other as they have a different "spin"—spin is a quantum mechanical property of electrons and there are two possible spin states for an electron. The electrons of a molecule fill the energy levels or orbitals from lowest energy to highest, and the highest energy electrons in a molecule occupy the so-called highest occupied molecular orbital or "HOMO." In most molecules, the HOMO contains two paired electrons, each with different spins. This paired spin configuration is the singlet, or S, state and the ground state is the S0 state. All orbitals with a higher energy than the HOMO are normally empty—they contain no electrons, and one of these, the lowest unoccupied molecular orbital or "LUMO," is often particularly important in fluorescence. The higher energy content of the excited state in comparison to the unexcited, or ground state, of the molecule is the result of the energy of the photon being used to lift an electron from the HOMO (or even from a lower-energy orbital) to a higher, empty orbital (Figure 23.1b). As these orbitals have distinct energies, only photons with certain energies can successfully lift an electron from, for example, the HOMO to a higher energy level, and it is this energy specificity that gives rise to the specific light absorption spectrum of molecules. Excitation does not change the spin of the excited electron, so just ground state of the molecule is nearly always a singlet state then so are the excited state. Each spectral absorption band corresponds to a different singlet state, with the lowest energy excited state being numbered S1, the next most energetic S2 etc. Just as the absorption of a photon is accompanied by excitation, relaxation from an excited state can be accompanied by the emission of a photon, giving rise to the phenomenon of fluorescence (Figure 23.2). In most molecules, however, fluorescence is observed only from the S1 → S0 transition (Kasha's rule) because in most cases states higher than S1 relax to the S1 state within about 10^{-14} s with the extra energy being released as heat. The predominant role of the S1 → S0 transition as a source of fluorescence (though note that there are exceptions to this) means that a fluorescence emission spectrum is largely the mirror image of the S0 → S1 absorption spectrum.

The process of fluorescent emission (also called radiative decay) occurs at a rate that depends on the energy difference of HOMO and the LUMO and the strength of the absorption band corresponding to

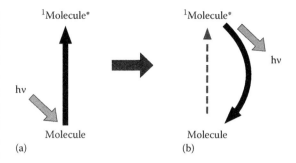

Figure 23.2 The relationship between excitation **(a)** in which an excited molecule is formed by absorbing a photon and **(b)** fluorescence, in which an excited molecule relaxes (and loses energy) to the ground state by the emission of a photon.

excitation from the HOMO to the LUMO. In any physical context (e.g., in a particular solvent or bound to a specific matrix, such as a protein), these are fixed properties of the molecule, so the intrinsic rate of fluorescence decay is fixed. It is also a first-order process and has a first-order rate constant, k_f. The fixed nature of k_f means that the amount of fluorescence and the actual rate of decay of fluorescence are powerful tools with which to follow the fate of excited states in molecules such as chlorophyll; any changes in either the lifetime of fluorescence or the amount of fluorescence imply that some other process is quenching, or dissipating, excited states that would otherwise relax to the ground state via fluorescence. The fixed value of k_f means that fluorescence acts as an internal standard against which the impact of other potentially variable processes can be measured—this is an extremely important concept that underlies the application of chlorophyll fluorescence to photosynthesis. The impact of other decay or relaxation processes on fluorescence can be quantified by means of the quantum yield (sometimes just shortened to yield) of fluorescence, Φ_F. If all excited states relax to the ground state via fluorescence decay then the yield of fluorescence will be 1, but if an alternative decay path is present (e.g., photochemistry in the case of photosynthesis), then the yield of fluorescence will be reduced (Figure 23.3). As the alternative quenching or dissipation process has its own first-order rate constant k_{alt}, then its impact on fluorescence yield can be quantified simply (Figure 23.4). The sensitivity of fluorescence yield to the activity of alternative decay processes (also known as quenching processes) allows changes in chlorophyll yield to be used to calculate changes in the activity of the alternative decay processes as illustrated in Figure 23.4; calculations of this

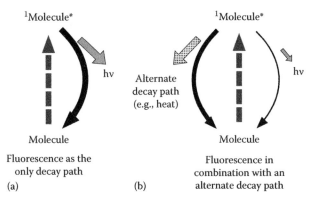

Figure 23.3 The qualitative impact on fluorescence of an alternate decay, or relaxation, path for an excited state leading the ground state: if fluorescence is the only path **(a)**, then the amount of fluorescence is maximal, while if an alternate decay path is present **(b)**, then some of the population of molecules in the excited state will relax via that alternate path, reducing the number of the molecules that can relax via the fluorescence pathway, reducing the intensity, or amount, of fluorescence photons emitted.

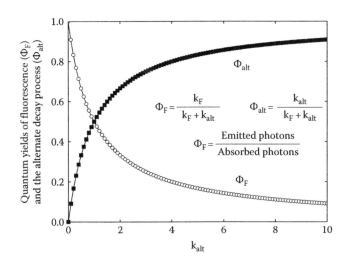

Figure 23.4 The quantitative impact on the yield fluorescence of an alternate decay, or relaxation, path for an excited state. The yield of fluorescence (Φ_F) is the ratio of the number of photons emitted as fluorescence and the number of photons absorbed to form an excited state. When the rate constant for the alternate decay path (k_{alt}) is zero, the yield of fluorescence is 1; in the example shown, the rate constant for fluorescence, k_f, is set to 1. As k_{alt} increases, the yield of fluorescence decreases and the yield of the alternate decay path increases according the equation for Φ_F.

kind are used extensively in the derivation of physiologically meaningful parameters from chlorophyll fluorescence changes.

23.3 CHLOROPHYLL *a* AND FLUORESCENCE

Though the focus of this and subsequent sections will be on chlorophyll *a* (Chl *a*), the comments will

also apply qualitatively to chlorophyll *b*. The main use of Chl *a* fluorescence is to probe photosynthesis *in* or *ex vivo* and under conditions where the Chl *a* is bound to pigment-binding proteins and will have spectral and related properties that differ from Chl *a* in solution both in terms of absorbance and fluorescence. The properties of Chl *a in vivo* are derived from those *in vitro*, so it is helpful to describe the *in vitro* properties of Chl *a* with respect to the absorbance of light and fluorescence before moving on to the same phenomena *in vitro*. Chl *a* is a complex molecule with complex absorbance features extending from a strong absorbance peak in the red part of the spectrum through to the UV region, with other minor absorbance peaks at 530, 575, and 615 nm and strong absorption peaks in the blue and UV regions; in total, to accurately simulate or fit the absorption spectrum of Chl *a* in the range 600–700 nm, it is necessary to include contributions from nine or more Gaussian absorbance bands (Fragata et al. 1988; Trissl 2003) (Figure 23.5 [see also Chapter 1]). These data are all for free Chl *a* in solution (e.g., dissolved in ethoxyethane [ether])—*in vivo* the absorbance bands are red-shifted, so the absorption peak at 660 nm in ethoxyethane solution moves to longer wavelengths *in vivo* (Figure 23.5; see also Chapter 4); most textbooks show the absorption spectrum of chlorophylls a and b in solution even though this is not the *in vivo* absorption spectrum. Compared to the simple portrayal of molecular excitation and relaxation via fluorescence shown in Section 23.2 and Figures 23.1 through 23.3, the situation with a real-world molecule like Chl *a* is more complex owing to coupling of electronic and vibrational transitions and the occurrence of mixed excitations (Fragata et al. 1988; Parusel and Grimme 2000; Reimers et al. 2013). Despite this extra complexity, the excitation of Chl *a* still results in the formation of S1 or higher excited states, followed by the relaxation to the S1

excited state. The absorbance peak at 660 nm in solution, which is shifted to longer wavelengths *in vivo*, corresponds to the S1 state, and excitation to any higher S state will relax to the S1 state within 10^{-14} s with the extra energy being lost as heat and no detectable fluorescence is observed from these higher S states. It is from the S1 state (the lowest energy-excited state) that chlorophyll fluorescence occurs and from which other biologically important photo-physical processes, such as triplet state formation or nonphotochemical dissipation, and photochemistry, occur. This S1 excited state of Chl *a* will be abbreviated to ^{1}chl* throughout this chapter. As fluorescence is emitted from the S1 state, the basic spectrum of Chl *a* fluorescence is a mirror image of the S1 absorption spectrum (Figure 23.5); the fluorescence spectrum has two clear peaks with the strongest close to 666 nm due to the S1(0)-to-S0(0) transition, while the weaker, longer wavelength peak at 720 nm is due to the S1(0)-to-S0(1) transition (the numbers in brackets refer to vibrational levels associated with the S states) (Papageorgiou 2004). An implication of this excitation and fluorescence behavior is that a range of wavelengths can excite Chl *a* to any one of several S states, but because of internal conversion, whatever the energy of the S state initially formed, it will relax to the S1 state from which fluorescence and other processes will occur. At least as far as the Chl *a* molecule is concerned, the fluorescence spectrum is therefore independent of

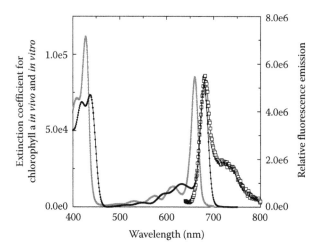

Figure 23.5 The wavelength dependency of the extinction coefficient for chlorophyll *a* in solution (ethoxyethane as a solvent) (grey symbols) and for chlorophyll *a in vivo* (black symbols) and the spectrum of total (i.e., from PSI and PSII) relative chlorophyll fluorescence emission from a white rose petal (open symbols).

the excitation spectrum and any wavelength that is absorbed by Chl *a* can be, in practice, used to excite fluorescence from ^{1}chl*.

23.3.1 Nonfluorescent decay of ^{1}chl* in solution

The ^{1}chl* formed in solution does not decay to ground state solely via fluorescence, and it is possible to simply calculate, as shown in Figures 23.3 and 23.4, the activity of the alternate decay paths. The radiative lifetime (lifetime is $1/k_f$) of Chl *a* fluorescence is 20 ns ($k_f = 5 \times 10^7$ s^{-1}) (Connolly et al. 1982), while the measured lifetime is 5.5–6 ns ($k_{f(apparent)} = 1.67 – 1.8 \times 10^8$ s^{-1}) (Kaplanová et al. 1981; Connolly et al. 1982). This difference in lifetimes implies that ^{1}chl* has other decay paths in addition to fluorescence. Taking a value for $k_{f(apparent)}$ of 1.8×10^8 s^{-1}, the value for k_{alt} (Figure 23.4) would be 1×10^8 s^{-1} and the yield of fluorescence would be 0.33. This simple use of rate constants is a model for the more complex use of rate constants for different ^{1}chl* decay paths to analyze the fate of ^{1}chl* *in vivo*. The major alternate (i.e., nonfluorescent) decay path for ^{1}chl* relaxation in solution is intersystem crossing that leads to triplet state formation, which is an excited state where the spins of electrons in the HOMO and the LUMO (one electron in each) are the same. This should be contrasted with the singlet state where all electron spins are paired. The yield of Chl *a* triplet formation (^{3}chl*) in ethoxyethanol is 0.64. These triplet states are potentially long-lived (lifetime for ^{3}chl* of 2 ms [Krasnovsky and Kovalev 2014]) and give rise to phosphorescence. Molecular oxygen (O_2) is unusual in that it has a triplet ground state, and if thermodynamically possible, two triplet molecules can interact to form two singlet molecules. In the case of O_2 and triplet Chl *a*, the reaction is exergonic (about −32 kJ mol^{-1}), resulting in singlet O_2 ($^1O_2$) and ground-state (S0) Chl *a*. This quenching of triplet Chl *a* by O_2 in solution is very efficient; Krasnovsky (1979) measured the same quantum yields for triplet Chl *a* formation and $^1O_2$ formation. Importantly, *in vivo* triplet Chl *a* are quenched via the transfer of the triplet state from Chl *a* to a carotenoid (the triplet valve; Wolff and Witt 1969; Kramer and Mathis 1980).

23.4 CHLOROPHYLL *a IN VIVO*

In the viridiplantae the chlorophylls present (always Chl *a* and normally Chl *b*—some mutants lack Chl *b*) are to only a negligible extent found dissolved free in the thylakoid membranes (Markwell et al. 1979). This implies that *in vivo*, the chlorophylls are almost exclusively found bound, along with carotenoids, to proteins where their photoactivity can be controlled and harnessed. Other taxonomic groups of oxygenic photosynthetic organisms will contain at least Chl *a* and carotenoids and, depending on the taxon, possibly other photosynthetic pigments (e.g., chlorophylls *c*, *d*, and *f* and phycobilins, phycoerythrins, and phycocyanins) (see also Chapter 8).

The combination of Chl *a* with other photosynthetic pigments changes the action spectrum for ^{1}chl* formation because the absorbance spectrum of these pigments differs from that of Chl *a*. For embryophytes, the impact of these other pigments (in this case these pigments are chlorophyll *b* and carotenoids [predominantly lutein, neoxanthin, α-carotene, β-carotene and violaxanthin [e.g., Matsubara et al. 2009]]) on the action spectrum for photosynthesis results in increases in activity around 450–500 and 630–650 nm (Boichenko 1998). In the viridiplantae, the absorbance spectra of the various photosynthetic pigments overlap, so though it is not possible to selectively excite just one pigment type, limited preferential excitation can be achieved (e.g., Ballottari et al. 2007). Irrespective of whether light is absorbed by a chlorophyll, a carotenoid, or a phycobilin, the excited state is rapidly transferred to Chl *a* to form ^{1}chl*, so fluorescence arises mainly from Chl *a*. The absorbance maxima in the red of Chl *a* and Chl *b*, for example, are separated by about 20 nm, corresponding to an energy difference of about 8 kJ mol^{-1}. At temperatures permissive for physiology, kT is about 2.5 kJ mol^{-1}, so approximately 5% of fluorescence could come from Chl *b*. While Chl *b* fluorescence can be distinguished from that of Chl *a in vitro*, well enough that fluorescence-based assays work well for measuring high Chl *a*/*b* ratios *in vitro* (Boardman and Thorne 1971; Talbot and Sauer 1997), I know of no studies, that have quantified the amount of Chl *b* fluorescence *in vivo*, and it is generally assumed that all chlorophyll fluorescence *in vivo* is from Chl *a*. It is also, therefore, assumed that the presence of photosynthetic pigments other than Chl *a* does not change the fluorescence emitter, which remains ^{1}chl*, so the spectrum is therefore derived from ^{1}chl* in some form or another (see succeeding text). The presence of other photosynthetic pigments does, however, change the excitation spectrum of fluorescence.

As has already been noted, the binding of Chl *a* to proteins results in a red shift of both the absorption and fluorescence emission spectra, but in addition to this spectral shift, there are other ways that chlorophyll *in vivo* differs to that in solution. The proteins to which Chl *a*, Chl *b*, and carotenoids are bound assemble into supercomplexes, PSI and PSII. Within these photosystems, ^{1}chl* can be quenched (i.e., converted to the ground state) by mechanisms, such as photochemistry and inducible thermal dissipation pathways (such as q_E) that are not found in solution. In addition to photosynthetic pigments, other nonphotosynthetic pigments may be present *in vivo*, and these nonphotosynthetic pigments can also absorb light in the photosynthetically active part of the spectrum (PAR; nominally this is 400–700 nm), competing with photosynthetic pigments for this quantum flux. Finally, *in vivo* and especially in leaves, the chlorophylls and other photosynthetic pigments are present at considerable concentrations, resulting in high absorbances for not only incident PAR but also some wavelengths of fluorescence emission, which can result in considerable reabsorption of fluorescence at these wavelengths (Figure 23.6). Each of these factors has an influence on the properties of chlorophyll fluorescence, and it is because of these influences that chlorophyll fluorescence has proven to be such a useful tool.

23.5 BASICS OF FLUORESCENCE FROM Chl *a* IN BIOLOGICAL SYSTEMS; PHOTOCHEMISTRY AND QUENCHING

Fluorescence from PSI and PSII has different spectra and average intensities and responds differently to the state of the reaction center and thus the presence or absence of photochemical dissipation of ^{1}chl*. In both PSI and PSII, ^{1}chl* is the driving force for photochemistry (for reviews of this process see, for example, Nelson and Yocum (2006), Jansson et al. (2007) and van Amerongen and Croce (2013) and see Chapters 9 and 15), so the photochemical use of ^{1}chl* therefore competes with chlorophyll fluorescence and other processes that dissipate, or quench, ^{1}chl*, ultimately regenerating ground-state Chl *a*. The quantum efficiencies of PSI and PSII for electron transport (the end result of photochemical quenching of ^{1}chl*) in both photosystems is high. Surprisingly, consensus values based on a body of robust experimental data for the quantum efficiencies for photosynthetic electron transport by either photosystem do not exist. The efficiency of PSI is normally considered to be close to one (Nelson and Yocum 2006), while that of PSII is considered to be lower, at about 0.9 (Roelofs et al. 1992; Pfundel 1998; Wientjes et al. 2013). Given these high yields for photochemistry, it would be expected (albeit simplistically) that the presence or absence of photochemistry would have a major effect on the yield of fluorescence.

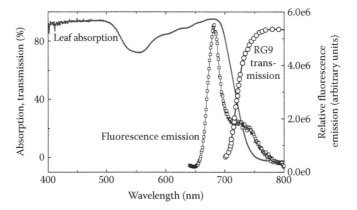

Figure 23.6 The wavelength dependency of light absorption by a tomato leaf (on a percentage basis) compared to the total (i.e., from PSI and PSII) relative chlorophyll fluorescence emission spectrum from a white rose petal. The percentage absorption of an RG9 filter (typical of the colored glass filters used to measure chlorophyll fluorescence) is also shown. Note the strong overlap between the fluorescence emission spectrum and leaf absorption spectrum and the extent of the overlap between the response of the RG9 filter and both the fluorescence emission spectrum and the leaf absorption spectrum.

23.5.1 Impact of photochemistry on PSII fluorescence

In PSII, charge separation, the point at which the photophysical phenomenon of excitation becomes the chemical phenomenon of oxidation and reduction, begins with oxidation of a specific chlorophyll (chl_{D1}) and reduction of phaeophytin, followed by the rapid transfer of the "hole" to another chlorophyll P_{D1} forming P_{680}^+, which is then reduced by Y_z, (Telfer 2005) and the transfer of the electron to Q_A (see Chapter 9). Interruption of any of these processes could in principle block charge separation and close the trap. *In vivo* (and very often *in vitro*) oxidized P_{680}^+ is rapidly reduced, and no phaeophytin accumulation occurs, so closing of the PSII reaction center occurs via Q_A reduction. This reduction of Q_A results in a fluorescence yield increase by a factor of about 5 from the F_o level to the F_m level (Figures 23.9 and 23.10). The F_o level corresponds to completely open PSII reaction centers with all Q_A in the nonreduced state and therefore with maximum efficiency of photochemical dissipation of $^1chl^*$, while the F_m level corresponds to complete reduction of the Q_A pool and zero photochemical dissipation of $^1chl^*$ (Joliot and Joliot 1964, 1979; Krause et al. 1982; Barber et al. 1989; Lazar 2003) (i.e., PSII photochemistry no longer competes with chlorophyll fluorescence as a route for the dissipation of $^1chl^*$). Note that, as will be explained in more detail below, the F_o and F_m fluorescence levels are not due solely to PSII fluorescence alone.

23.5.2 Impact of photochemistry on PSI fluorescence

While the yield of chlorophyll fluorescence in PSII is strongly dependent on the openness of the PSII reaction center, PSI fluorescence shows only minor changes upon changes in the state of the PSI reaction center. Relative to the size of the fluorescence, yield changes arising from PSII, those arising from PSI are so small that fluorescence yield change phenomena can normally be interpreted solely in terms of PSII-related processes. Nonetheless, while changes in PSI fluorescence are normally ignored because of their small size, there are changes in the yield of PSI fluorescence due to trap closure, and it is worth documenting what is known about these changes. The reaction center of PSI can close for two reasons. First, P700 (the primary electron donor in

the reaction center of PSI) can oxidize, forming P700⁺, which can accumulate in PSI resulting in a loss of quantum efficiency for electron transport (Harbinson and Woodward 1987; Harbinson et al. 1989; Genty and Harbinson 1996). Second, electrons can accumulate on the acceptor side of PSI, blocking the oxidation of P700 and therefore charge separation (Harbinson and Hedley 1993).

Oxidized P_{700} has often been considered to be as equally good a quencher of $^1chl^*$ in PSI as P_{700} (e.g., Schlodder et al. 2011); the lifetime of PSI fluorescence is the same when P_{700} is oxidized or nonoxidized (e.g., Nuijs et al. 1986). More detailed measurements have, however, shown that P_{700}^+ is a slightly less good quencher than is P_{700} and fluorescence yields increase by about 15% (Trissl 1997; Byrdin et al. 2000) or 4% (Wientjes and Croce 2012) if P700 is oxidized. Using a modelling approach, Lazàr concluded that the contribution of PSI fluorescence to the total fluorescence signal changes during the fluorescence rise from F_o to F_m could be 8%–15% due to P_{700}^+ formation–induced changes in the PSI fluorescence yield (Lazár 2013). Notably, PSI fluorescence yield changes have largely been described in respect of PSI trap closure due to P_{700} oxidation and not due to overreduction of the PSI acceptor pool even though this can also occur, for example, during photosynthetic induction or when the PSI acceptor pool is not regenerated for other reasons (Klughammer and Schreiber 1991; Harbinson and Hedley 1993; Klughammer and Schreiber 1994). During photosynthetic induction, maximal overreduction of the PSI acceptor pool occurs when chlorophyll fluorescence is at the F_m state and the changes in fluorescence yield due to PSII are at their greatest (i.e., small changes due to PSI fluorescence would be easy to miss). Yield changes of PSI fluorescence as a result of trap-closure due to overreduction have been described by Ikegami (1976) and Trissl (1997). The results of Trissl are easier to interpret quantitatively, and they show an increase in PSI fluorescence yield comparable to that due to P_{700}^+ formation (about 10%–15%) and which was attributed to recombination between P_{700}^+ and either A_0^- (A_0 is the primary electron acceptor of PSI) or A_1^- (A_1 is the secondary electron acceptor of PSI). *In vivo* conditions that would favor overreduction of the PSI acceptor pool block the steady-state accumulation of P_{700}^+ (Harbinson and Hedley 1993), but P_{700}^+ will still be formed transiently before being reduced by a back-reaction from the PSI reaction center acceptors (Brettel and Leibl 2001). Which back-reaction will actually occur will depend on which PSI acceptors can photoaccumulate when electron donation to P_{700}^+ from plastocyanin continues while electron transport from PSI is blocked. Electron transport in the PSI reaction center proceeds from P700 to ferredoxin via A_0, A_1, F_x, F_A, and F_B. If forward electron transfer along this chain is blocked, then the electron can re-reduce P_{700}^+ but only if P_{700}^+ has not in the meantime been reduced by plastocyanin. It is therefore a matter of the balance of the kinetics of P_{700}^+ reduction by plastocyanin or by a PSI acceptor. Donation from plastocyanin to P_{700}^+ is biphasic with half-times of 10 and approximately 100 μs (Drepper et al. 1996). Any back-reaction slower than this will not be able to outcompete forward donation into P_{700}^+, and that acceptor will photoaccumulate, while if the back-reaction is faster than the forward donation, then the back-reaction will predominate. When the F_x, F_A, and F_B pools are reduced back-reactions from A_0^- and A_1^- have submicrosecond kinetics (Brettel and Leibl 2001), so it is likely that the back-reaction from A_1^- to P_{700}^+ will be that which predominantly regenerates P_{700} when the forward electron transport path is blocked, and this reaction forms P_{700} and not excited P_{700} (i.e., ^{1}chl* has been quenched)(Rutherford et al. 2012). This still remains, however, to be proven to be the case *in vivo*.

It is still clear that for PSI, the reaction center is largely a quencher irrespective of whether the reaction center is open or closed, with only small increases (10%–15%) in fluorescence yield occurring upon closure. The small size of these changes compared to those that occur in PSII mean that changes in PSI fluorescence can be nearly always ignored when analyzing fluorescence changes from leaves or similar photosynthetic tissues. The fact that PSI shows only minor yield change upon trap closure also implies that PSI does have nonphotochemical dissipation of ^{1}chl*, whether that is due to quenching by P_{700}^+ or quenching by a back-reaction between P_{700}^+ and A_1^- (or possibly A_0^-). Finally, although changes in fluorescence yield are predominantly due to fluorescence originating from PSII, it is important to remember that there remains a largely constant chlorophyll fluorescence emission from PSI contributing to the overall fluorescence signal.

23.5.3 PSII and PSI fluorescence emission spectra and the relative contribution of PSI fluorescence to the total fluorescence signal

The relative contributions of PSI and PSII fluorescence to the total chlorophyll emission from leaves (etc.) are best understood with reference to their spectra. The pure fluorescence emission spectrum (i.e., that unaffected by reabsorption effects) of PSI at room temperature has a peak at 730 nm and a shoulder at about 690 nm (most clearly seen in Figure 23.7) (Franck et al. 2002), while that of PSII has a strong peak at 683 nm and a shoulder

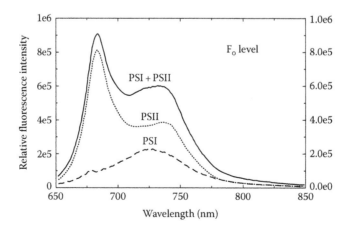

Figure 23.7 The calculated relative emission spectra for chlorophyll fluorescence from PSI, PSII, and PSI and PSII combined, calculated or measured at the F_o fluorescence level. (Data from Franck, F. et al., *Biochim. et Biophys. Acta*, 1556, 239, 2002.)

or peak at 740 nm (Figures 23.7 and 23.8) (Barber et al. 1989; Franck et al. 2002). The relative contributions of PSI and PSII fluorescence to the total fluorescence signal depends on the wavelength and the state of the PSII reaction center and (though to a much smaller extent) on the state of the PSI reaction center. Importantly, in contrast to the intensity of fluorescence, the shapes of the fluorescence emission spectra are not noticeably affected by the state of reaction centers. Owing to the small change in the yield of PSI fluorescence upon moving from the F_o to F_m states, the relative contribution of PSI fluorescence is greater and easier to analyze in the F_o state (Figure 23.7). Shorter measuring wavelengths below 700 nm more strongly favor PSII fluorescence, while longer wavelengths favor PSI fluorescence (Franck et al. 2002), so at wavelengths above 775 nm, the relative contributions of PSI and PSII fluorescence are similar; around 725 nm, the contribution of PSI fluorescence is about one-third that of the total fluorescence signal, while at 680 nm, the fluorescence signal is 90% from PSII (Franck et al. 2002). In the F_m state (Figure 23.8), the contribution of PSI fluorescence is much smaller because of the weak variable fluorescence from PSI, but the wavelength bias is the same; below 700 nm, the impact of PSI fluorescence is very small, while at around 725 nm, it is still about 83% from PSII (Franck et al. 2002). The relative total fluorescence emission spectra measured at the F_o and F_m states is shown in Figure 23.9—note that owing to reabsorption of fluorescence by chlorophyll, the spectra shown

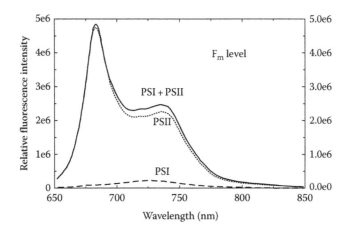

Figure 23.8 The calculated relative emission spectra for chlorophyll fluorescence from PSI, PSII, and PSI and PSII combined, calculated or measured at the F_m fluorescence level. (Data from Franck, F. et al., *Biochim. et Biophys. Acta*, 1556, 239, 2002.)

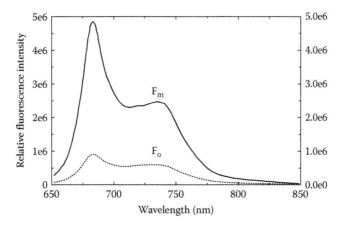

Figure 23.9 A comparison of the relative combined (PSI and PSII) fluorescence emission spectra at the F_o and F_m levels. (Data from Franck, F. et al., *Biochim. et Biophys. Acta*, 1556, 239, 2002.)

in Figure 23.9 will not often be measured *in vivo*, though the emission spectrum measured from a white rose petal (Figures 23.5 and 23.6) comes close. Chlorophyll fluorescence is often measured at wavelengths above 700 nm using a long-pass optical filter, such as a Schott RG9 (Figure 23.6). When fluorescence is measured using this filter, the contribution of PSI fluorescence to the F_o fluorescence level (that measured when the Q_A pool of PSII is oxidized) has been estimated to be about 30% of the total fluorescence signal (Pfundel 1998), which is both a considerable contribution and consistent with the spectra shown in Figure 23.7.

23.6 PHYSIOLOGY OF FLUORESCENCE FROM PSII AND ITS USE IN THE MEASUREMENT OF PHOTOSYNTHESIS I MEASUREMENTS ON DARK-ADAPTED MATERIAL

23.6.1 Variable fluorescence from PSII, Q_A redox state, and NPQ of ¹chl* and its terminology

In contrast to PSI, the yield of fluorescence from PSII is strongly dependent on the openness of the PSII traps (or reaction centers). The capacity for photochemical dissipation in PSII is dependent on whether Q_A is oxidized or reduced; this has already been introduced in broad terms as a major factor influencing the yield of PSII chlorophyll fluorescence. In addition, PSII fluorescence yield is significantly influenced by a range of other factors that collectively give rise to the phenomenon of NPQ of PSII fluorescence (Horton et al. 2005; Horton and Ruban 2005; Demmig-Adams and Adams 2006) (see also Chapter 11). NPQ is rapidly activated (e.g., Quick and Horton 1984; Schreiber et al. 1986; Li et al. 2000) in the light and relaxes in the dark with half-times of 10s or longer (Quick and Stitt 1989; Walters and Horton 1991), so it is an inducible process whose activity can be varied and which is under close physiological control (e.g., Bilger and Björkman 1990, 1991; Walters and Horton 1991; Demmig-Adams et al. 1999).

NPQ also creates some terminology problems because the term "quenching" is used in photochemistry to describe two distinct but linked phenomena;

it can be used to describe a decrease in fluorescence (or fluorescence yield), or it can be used to describe the dissipation of excited states via a nonfluorescent pathway (as illustrated in Figures 23.3 and 23.4). This is a potentially confusing situation because "quenching" can refer to both a mechanism of fluorescence change (the dissipation of excited states) or an effect (the decrease of fluorescence yield). To avoid this, I will restrict the use of NPQ to changes in fluorescence yield and use the term NPD to include all nonphotochemical and nonfluorescent dissipation of ¹chl* in PSII; NPQ is an effect of NPD on fluorescence. Another complication is that in physiological terms, the processes that give rise to NPQ are removed by dark adaptation treatment of leaves, chloroplasts, etc., with the implication that in this dark-adapted state, there is no NPQ (or NPD). In terms of the way that "NPQ" is used, physiologically this is correct, but in terms of photophysics, this implication is incorrect. We have already seen that ¹chl* in solution does not relax to ground state solely via fluorescent decay but also by conversion to triplet states, a process that also occurs *in vivo* (Wolff and Witt 1969; Kramer and Mathis 1980). The formation of chlorophyll triplet states leads to a nonradiative decay process, so it is therefore a NPD process but one which is active in the dark-adapted state where NPQ is often considered to be absent. Other nonphotochemical processes also seem to be active in dissipating ¹chl* in dark-adapted PSII because the lifetime of fluorescence in isolated LHCII is longer than that in intact thylakoid membranes with PSII in a state where photochemistry is blocked and fluorescence lifetime is maximum (e.g., Belgio et al. 2012), contributing to NPD in this state. It is necessary to distinguish between basal NPD processes present in dark-adapted material in which (physiological) NPQ should be absent and those inducible NPD processes that give rise to NPQ in the light; these will be termed NPD_b and NPD_i, respectively. The distinction between NPD_b and NPD_i is not clear and simple, especially in leaves where PSII has been photoinhibited or subject to slowly reversible downregulation. However, as it is possible to estimate the quantum yields for nonphotochemical dissipation by PSII in terms of basal and inducible nonphotochemical dissipation processes and a distinction is routinely made between the impact of basal and inducible dissipation of ¹chl* in calculations of fluorescence and other yields, it is necessary to consider which processes are NPD_i and which are NPD_b.

23.6.2 Measurement of relative fluorescence yields F_o and F_m

The changes in chlorophyll fluorescence relative to the F_o level are so strongly dominated by changes in the state of PSII that the contribution of PSI is normally ignored (see preceding text). Fluorescence changes are therefore interpreted in terms of changes in photochemical dissipation (quenching) and nonphotochemical dissipation of $^1chl^*$ in PSII. Photochemical dissipation in PSII is the starting point of linear electron transport, while nonphotochemical dissipation is important in understanding damage to, and protection of, PSII (see also Chapter 11). The importance of these processes for the operation and regulation of PSII and the fact that fluorescence changes mirror them, has driven the development of chlorophyll fluorescence as tool for measuring photosynthesis. This development began with the pioneering work of Butler and coworkers (Butler and Kitajima 1975) and has continued ever since. Butler developed the basic terminology used to describe the changes in chlorophyll fluorescence yield from dark-adapted photosynthetic material (Butler and Kitajima 1975), a terminology that has been expanded to include phenomena in light-adapted material (e.g., Krause et al. 1982; Dietz et al. 1985; Kooten and Snel 1990; Baker 2008) (Table 23.1). Butler developed his approach to a considerably more complex level than the simple representation used here.

The basic fluorescence states of a dark-adapted leaf (or similar) subjected to a saturating light pulse (i.e., an irradiance sufficient to completely reduce all the Q_A in the subject) is shown in Figure 23.10. An important feature of these results, which are typical of those obtained using a modulated fluorescence measuring system, is that they show changes in the relative chlorophyll fluorescence yield. The ability of these instruments to measure the relative fluorescence yield, though this is usually just referred to as a fluorescence yield, is a consequence of the operation of these devices. A weak modulated measuring beam is applied to the subject (a leaf for example), and the fluorescence (also modulated) produced by this measuring beam is picked up by a detector, usually a photodiode filtered by a long-pass optical filter, such as an RG9 (Figure 23.6), to prevent any modulated nonfluorescent radiation reaching the detector. Fiber optics might be used to convey the excitation light to the leaf or bring fluorescence to the detector. The detector will receive not only modulated fluorescence produced by the pulsed measuring beam but also other nonmodulated fluorescent and scattered actinic radiation, but these other signals are rejected by signal processing electronics or software to leave only a signal corresponding to the fluorescence produced by the weak measuring beam. As the intensity of the measuring beam is either constant or adjusted in a carefully controlled way, any change in the intensity of the fluorescence produced by this beam can be interpreted as a change in the relative fluorescence yield. Normally, the modulated measuring beam is produced by a light-emitting diode (LED), usually filtered by a short or band-pass optical filter that blocks any long wavelengths that would be detected by the fluorescence detector. LEDs are easy to use in this application because they are small, have a reasonably pure spectral output to begin with, and can be easily switched on and off. Their output, however, is temperature sensitive, so it is important to be sure that any changes in the apparent fluorescence yield are not due to warming or cooling of the LED producing the measuring light pulses. The interpretation of the fluorescence produced by the measuring beam as a relative fluorescence yield also implicitly assumes that the absorbance of the leaf or similar (i.e., the excitation input that forms $^1chl^*$) does not change during the course of measurement. This assumption will not be met if there is chloroplast movement (Brugnoli and Björkman 1992), which occurs in most plants in response to high irradiances, particularly when blue light is present in the spectrum (Haupt and Scheuerlein 1990). There is also the complication that arises if light absorbed by photosynthetic pigments associated with PSII in the dark-adapted state transfer excitation to PSI—the so-called state transition. This will be discussed more in Section 23.7 (see also Chapter 4).

In a dark-adapted leaf, chloroplast suspension, etc., applying a measuring beam should produce an increase in fluorescence yield to the F_o level—this rise should be rapid and limited by the response time of the instrument and recording system and should have no slow phase (slow being several seconds). How long the duration of dark adaptation should be depends on the experimental question, but at least 20 min is normal. A slow phase after switching on the measuring light would indicate that the measuring beam intensity is too high and is producing an actinic effect on the sample.

Table 23.1 The routinely measured fluorescence parameters, the conditions under which they are measured, and their relationship to the rate constants for fluorescence, photochemistry, and the basal and inducible thermal dissipation processes

Fluorescence parameter	Actinic irradiance on or off?	Saturation pulse on or off?	QA reduction state	Constitutive (or basal) nonphotochemical quenching state	Inducible nonphotochemical quenching state	Fluorescence yield equation
F_o	Off	Off	Fully oxidized	Present	Absent	$F_o = k_f/\left(k_f + k_d + k_{P([QA]_{rel}=1)}\right)$
F_m	Off	On	Fully reduced	Present	Absent	$F_m = k_f/\left(k_f + k_d\right)$
F'	On	Off	Usually partly reduced	Present	Usually present	$F' = k_f/\left(k_f + k_d + k_{P(0<[QA]_{rel}<1)} + k_{Dvar}\right)$
F_m'	On	On	Fully reduced	Present	Usually present	$F_m' = k_f/\left(k_f + k_d + k_{Dvar}\right)$
F_o'	Off	Off	Fully oxidized	Present	Usually present	$F_o' = k_f/\left(k_f + k_d + k_{P([QA]_{rel}=1)} + k_{Dvar}\right)$

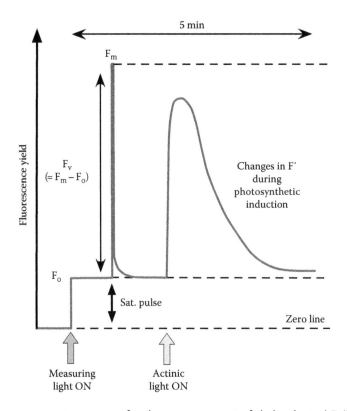

Figure 23.10 The measurement sequence for the measurement of dark-adapted F_v/F_m and the changes in relative fluorescence yield and the major basic fluorescence parameters (F_o, F_m, F_v and F') that occur during this measurement and during the response of fluorescence to the application of continuous actinic irradiance. The application of a weak fluorescence excitation irradiance (the measuring light) to the subject, such as a leaf, allows measurement of the F_o level (a short, weak far-red irradiance is often added during the measurement of F_o to ensure that all Q_A is oxidized (not illustrated). The application of saturating light pulse (typical duration: 1–2 s) reduces all Q_A in the subject and produces a fluorescence rise to F_m—the difference between F_m and F_o is F_v. At the end of the saturating pulse, the fluorescence decays quickly to the F_o level as the reduced Q_A is reoxidized. Applying actinic (i.e., an irradiance sufficiently intense to produce significant photosynthetic activity) produces an increase, a, in fluorescence yield as the Q_A pool is (partially) reduced following by a decrease in yield as photochemical and NPQ increase—the process known as photosynthetic induction. The yield of fluorescence under continuous actinic irradiance (F') normally declines to a level close to F_o (it may even be lower than F_o).

Typical average intensities (the beam is pulsed) of a measuring beam for F_o are less than 1 μmol m^{-2} s^{-1}, with an irradiance closer to 0.1 μmol m^{-2} s^{-1} being typical. To ensure that all Q_A is oxidized, a short pulse of far-red (>700–720 nm) radiation, which will preferentially excite PSI, should be applied to the leaf. Note that some dark reduction of the Q_A pool is normal and an increase in chlorophyll fluorescence in the dark is routinely used as an indicator of NADH:ubiquinone oxidoreductase (NDH) activity (Burrows et al. 1998). The F_o fluorescence state is the minimum fluorescence yield that can be obtained from a dark-adapted sample and corresponds to a physiological state where the photochemical light use efficiency of PSII is maximum (i.e., all PSII traps are open) and with only the competing dissipation processes of NPD$_b$ and fluorescence active. Applying a saturating light pulse to rapidly reduce all Q_A produces a rise in the fluorescence yield from F_o to F_m—"saturating" in this context means sufficiently strong to reduce all the Q_A in the sample, which for a leaf, or similar, would require an irradiance of about 3000 μmol m^{-2} s^{-1}. This pulse is normally applied for about 1 s, though a shorter duration would still be sufficient (e.g., Lazar 2003); the 1 s option offers a margin of safety, though repeated application of such a pulse can result in photodamage. The rise of fluorescence from F_o to F_m has complex kinetics, which have been the subject of much analysis (Lazar 2003; Zhu et al. 2005) but will not be further discussed here because they provide only limited information about the performance of PSII in the light. The F_m state, with all Q_A reduced, corresponds to a physiological state where the photochemical light use efficiency by PSII is zero—all traps are closed—and all dissipation of $^1chl^*$ in PSII is now via only fluorescence and NPD_b. The most useful measure of PSII function that can be obtained from the F_o and F_m measurements is the F_v/F_m value, where $F_v = F_m - F_o$ and which is commonly interpreted as the maximum light use efficiency of PSII electron transport—which is not wholly true (see following text).

23.6.3 Conversion of F_o and F_m fluorescence yields into a measure of light use efficiency, F_v/F_m

The ability to routinely measure the relative fluorescence yields corresponding to two limiting physiological states (F_o and F_m) of PSII is impressive, but on their own, these yields have limited use. Moving from these fluorescence yields to a yield of PSII photochemistry, something much more useful, requires a calculation based on a framework for analyzing chlorophyll introduced by Butler (Butler and Kitajima 1975). In this framework, all processes that dissipate $^1chl^*$ in PSII are treated as being simple first-order kinetic processes with associated first-order rate constants k (cf Figure 23.4). This framework is very widely used in developing methods to obtain physiologically useful values from basic fluorescence data (Butler and Kitajima 1975; Hendrickson et al. 2004; Kramer et al. 2004), though note that this approach is a substantial simplification of the physiology of $^1chl^*$ dissipation in PSII (e.g., Lavergne and Trissl 1995; Croce and van Amerongen 2011; van Amerongen and Croce 2013) The rate constants for different processes are distinguished by different subscripts: k_f is the rate constant for fluorescence; k_D is the rate constant for those inducible dissipation processes that give rise to NPQ (NPD_i); k_d is the rate constant for

dissipation via triplet formation and other basal dissipation processes (NPD_b); and k_p is the rate constant for photochemistry by PSII. Of these, only k_D and k_p are treated as variable, with the value of k_p being determined by the redox state of Q_A, so k_p is often written as $k_{p[QA]}$ or similar to indicate this. The yield (often abbreviated as Φ or F if for fluorescence yield; "Y" should not be used) of any dissipation process is given as the ratio of its rate constant divided by the sum of the rate constants for those processes acting to dissipate $^1chl^*$ (Figure 23.4). So the yield of fluorescence at F_o (Φ_{Fo}) is given as the ratio $k_f/(k_f + k_d + k_{p([QA]rel = 1)})$; $[Q_A]_{rel}$ is the relative oxidation state of the Q_A pool, the "1" indicates that it is fully oxidized), while for F_m it is $k_f/(k_f + k_d)$. The fluorescence yield equations for all the commonly measured fluorescence values (F_o, F_m, F', F_m', and F_o'), along with the physiological states corresponding to those values and the state of the measuring system, are listed in Table 23.1.

All fluorescence yield equations have, of course, k_f as the numerator, while a yield for (e.g.) photochemistry would have k_p as the numerator. By rearranging F_o and F_m (Figure 23.11), it is possible to generate an equation for the yield of PSII photochemistry (Φ_{PSII}) in the dark-adapted state, which of course should be $k_{p([QA]rel=1)}/(k_f + k_d + k_{p([QA]rel=1)})$. While I will make distinctions between different redox states of Q_A in k_p and indicate that k_D is variable (k_{Dvar}), these are often not always indicated explicitly elsewhere but nonetheless should normally be taken as implicit. The rearrangement shown in Figure 23.11 shows how a measure of the maximum quantum yield for PSII electron transport in the absence of NPD_b (there is no k_{Dvar} in the equation), which is a basic property of PSII, can be obtained from a simple combination of two easily measured fluorescence values, F_o and F_m. A similar approach has been developed for other parameters that describe the state and regulation of PSII not in the dark-adapted state but in the presence of actinic light when there is NPQ and NPD_i is therefore active (Genty et al. 1989; Kramer et al. 2004; Baker et al. 2007; Baker 2008).

With regard to the F_v/F_m value, the presence of PSI fluorescence is one source of error. The framework upon which the F_v/F_m calculation is based is for PSII fluorescence and takes no account of the PSI fluorescence that is present at an almost constant

$$1.\ F_o = k_f/(k_f + k_d + k_{p([Q_A]_{rel}=1)}) \quad F_m = k_f/(k_f + k_d)$$

$$2.\ \Phi_{PSII_{dark\ adapted}} = \frac{(F_m - F_o)}{F_m}$$

$$3.\ \Phi_{PSII_{dark\ adapted}} = \frac{\left(k_f/(k_f + k_d) - k_f/(k_f + k_d + k_{p([Q_A]_{rel}=1)})\right)}{k_f/(k_f + k_d)}$$

$$4.\ \Phi_{PSII_{dark\ adapted}} = \frac{k_f k_{p([Q_A]_{rel}=1)}}{(k_f + k_d)(k_f + k_d + k_{p([Q_A]_{rel}=1)})}\frac{(k_f + k_d)}{k_f}$$

$$5.\ \Phi_{PSII_{dark\ adapted}} = \frac{k_{p([Q_A]_{rel}=1)}}{(k_f + k_d + k_{p([Q_A]_{rel}=1)})}$$

Figure 23.11 The fluorescence yield equations of F_o and F_m (Equation 1) in terms of the first-order rate constants that represent dissipation, or quenching, of ^{1}chl* in PSII by chlorophyll fluorescence (k_f), basal nonphotochemical and nonfluorescence processes (NPD_b; k_d) and photochemistry (k_p, here qualified as $[Q_A]_{rel} = 1$ to indicate that under the measurement conditions, the Q_A pool is fully oxidized). Equations 2 through 5 show the derivation, the yield equation for photochemistry ($\Phi_{PSII\ (dark\ adapted)}$ - more commonly called the dark-adapted F_v/F_m) from the fluorescence yield equations for the Fm and Fo parameters—this type of rearrangement has often been used to derive physiologically useful parameters from the basic measurable fluorescence parameters F_o, F_m, F_v, F_o', F_m', and F' (Figures 23.10 and 23.12 and Table 23.1).

intensity. This PSI fluorescence contributes more to the level of the F_o signal than to that of the F_m, and its presence makes the F_v/F_m value lower than would be so if all fluorescence were from PSII. Pfündel (1998) has tried to quantitatively compensate for the error in F_v/F_m due to PSI and estimated that for a C3 leaf, the F_v/F_m in the absence of PSI fluorescence should be 0.88, while the measured values for C3 leaves were 0.835–0.84 (the values for C4 leaves were lower 0.768–0.789, owing to the higher amounts of PSI in these leaves). Another possible error arises because it is assumed in the derivation of F_v/F_m that all rate constants behave independently of each other, so when k_p decreases to zero, k_d does not change, and this might not be true. In the absence of photochemistry and NPDi, the mean chlorophyll fluorescence lifetime is 2 ns (Genty et al. 1992; Belgio et al. 2012), while in solution, it is 5.5–6 ns (Kaplanová et al. 1981; Connolly et al. 1982), which indicates that *in vivo*, there is an extra quenching process. Lavergne and Trissl (Lavergne and Trissl 1995) concluded that this was due to an increase in k_d (i.e., NPD_b) in the PSII reaction center upon Q_A reduction (i.e., k_d depends on k_p). If this interpretation is correct, then then there would be another error in the F_v/F_m calculation because this extra quenching in the F_m state would make F_m lower than expected. A recent paper by Belgio et al. (Belgio et al. 2012) argued, however, that the low fluorescence lifetime at F_m was due to

the aggregation of chlorophylls in LHCII (at least) and was not, therefore, due to reaction center processes. If the conclusion of Belgio et al. were to be found to be true, then this possible extra error in F_v/F_m could be discounted.

Values for F_v/F_m in healthy, unstressed leaves are quite conservative among groups of plants. One extensive study of C3 leaves found values of 0.832 ± 0.004 (Björkman and Demmig 1987)—the exact value of F_v/F_m will vary from instrument to instrument, depending on filters, background fluorescence, and other errors. The constancy of F_v/F_m coupled with its ease of measurement and the sensitivity of F_v/F_m to stress, senescence, and other factors has made the value of F_v/F_m a very widely used measure of plant stress and postharvest product deterioration (DeEll and Toivonen 2003). The simplicity of the measurement is often in contrast to the complexity of the mechanisms by which stresses and developmental changes produce the decline in F_v/F_m from its maximum value. So it might be easy to measure a loss of F_v/F_m but hard to fully explain why it has declined. The usefulness of F_v/F_m is enhanced by its broader physiological meaning—its value correlated with that of the quantum yield of photosynthetic oxygen evolution (Björkman and Demmig 1987), so it is not only theoretically a measure of the quantum efficiency of PSII electron transport but has been shown to be empirically correlated with a leaf-level effect of photosynthetic activity.

23.7 PHYSIOLOGY OF FLUORESCENCE FROM PSII AND ITS USE IN THE MEASUREMENT OF PHOTOSYNTHESIS; II MEASUREMENTS ON ILLUMINATED PHOTOSYNTHETIC TISSUES

In illuminated leaves and other photosynthetic tissues, the physiological state is different and more complex. Photosynthetic responses to irradiance are complex and depend on numerous other factors, such as temperature, stomatal conductance, nutrient supply, developmental stage, and plant history in terms of not only the history of that photosynthetic tissue or plant but also in the time frame of evolution. The photosynthetic properties of leaves and other photosynthetic tissue are therefore very diverse in detail, but within this diversity, there are common factors that need to be accounted for if chlorophyll fluorescence is to be used to measure the operation and regulation of PSII and photosynthesis. The operation of photosynthesis is all about transforming the free energy of light into chemical, metabolic free energy, and this requires the formation of free energy gradients that are used to drive the metabolic fluxes that are basic to assimilation and growth. Light drives photosynthetic electron transport in PSII and does so with an efficiency—the operating efficiency (Φ_{PSII}, or F_q'/F_m' in the terminology of Baker 2008)—that varies depending on the basic capacity and regulation of the electron transport chain and metabolic demand (e.g., Genty and Harbinson 1996).

The modulation of the efficiency of PSII electron transport occurs in the first instance by the closing of PSII traps via reduction of some part of the Q_A pool, but in addition, PSII efficiency is also modulated by the activation, or formation, of a range of processes that dissipate $^1chl^*$ in PSII without producing photochemistry in PSII (Krause and Jahns 2004). These are the processes that give rise to NPQ and are collectively referred to here as NPD_i. These NPD_i processes mostly act by thermally dissipating $^1chl^*$ in PSII, but one mechanism, the state transition, acts to at least some extent by transferring part of the LHC of PSII to PSI. This results in some $^1chl^*$, which in the absence of the state transition could

drive PSII photochemistry and produce fluorescence from PSII, being delivered to PSI where they drive PSI photochemistry and produce PSI fluorescence. This regulated transfer of some LHCII from PSII to PSI therefore reduces PSII fluorescence and fluorescence overall. In illuminated leaves and other photosynthetic tissues, various processes need to be accounted for:

- The measurement of Φ_{PSII} needs to account for the combined effects of trap closure and NPD_i.
- The activity of NPD_i needs to be measured and understood in terms of the different processes that contribute to it.
- The impact of NPD_i and NPD_b on the inefficiency of PSII electron transport needs to be assessed.

The steady-state fluorescence yield measured from illuminated material is usually termed F', F_s, or F_t—the use of a "'" in the fluorescence terminology used here implies that the yield is measured in illuminated material or material that has been illuminated but is not yet fully dark adapted (this also applies to F_o' and F_m') (Figure 23.12). The value of F' is the result of competition between fluorescent decay of $^1chl^*$ in PSII on the one hand and on the other hand, dissipation of $^1chl^*$ via NPD_b mechanisms (represented by the rate constant k_d), photochemistry ($k_{p[QA]}$), and the processes that give rise to NPD_i (k_{Dvar}) (Table 23.1). Using chlorophyll fluorescence measurements, it is possible to calculate yields for all these dissipation processes and measures that describe the regulatory state of PSII, and thus it is possible to comprehensively analyze the operation and regulation of PSII.

23.7.1 Quantifying the impact of NPD_i on chlorophyll fluorescence: NPQ and $F_m/F_m' - 1$

When a saturating light pulse is applied to an illuminated leaf (or any other photosynthetic tissue), the fluorescence yield rises from the steady-state fluorescence level F' to a value, F_m', that is maximal and which can be light saturated (Markgraf and Berry 1990; Loriaux et al. 2013). Despite being a maximum, this maximum is local and applies only for that tissue under the conditions of measurement

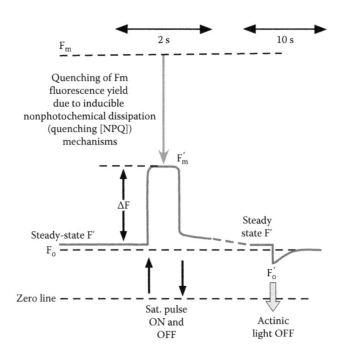

Figure 23.12 The measurement sequence for Φ_{PSII} in actinic irradiance, which is sometimes termed the operating efficiency of PSII to distinguish it from the dark-adapted measurement (Figure 23.10), and the changes in relative fluorescence yield and the major basic fluorescence parameters (F', F'_m, ΔF, and F'_o) that occur during this measurement. The fluorescence yield in steady-state irradiance (F') is normally close to the F_o (it can be anywhere from slightly above to slightly below F_o). The application of a saturating light pulse (typical duration 1–2 s) produces an increase in fluorescence yield to the F'_m level—this is lower than F_m because of NPQ, but the extent of the decrease will depend upon the regulation of PSII in the subject being measured (e.g., a leaf). The difference between F' and F'_m is $\Delta F'$ (or just ΔF). Removing the actinic irradiance causes the fluorescence yield to decrease from F' to F'_o ($F'_v = F'_m - F'_o$; not shown). To ensure complete oxidation of Q_A during F'_o, a weak far-red light is often superimposed upon the measuring beam (not shown).

and is always lower than the dark-adapted F_m (Figure 23.12). As with F_m, the measurement of F'_m requires that all Q_A is reduced, and the use of "saturating" within the term "saturating light pulse" means that the intensity is sufficient to reduce Q_A completely. In the case of light-adapted leaves, this can be a difficult requirement to achieve. While irradiances of 6000–8000 µmol m^{-2} s^{-1} are often assumed to be saturating, these may not be fully saturating. Extrapolation of the apparent F'_m measured at subsaturating irradiances to an infinite irradiance may be necessary to obtain an accurate F'_m (Markgraf and Berry 1990; Loriaux et al. 2013). It is assumed that during the rise from F' to F'_m, no change in NPD$_i$ occurs (this can be tested by measuring the stability of F'_m [Loriaux et al. 2013]). The dissipation action of NPD$_i$, represented by the rate constant k_{Dvar} in the denominator of the fluorescence yield equation for F'_m (Table 23.1) results in

F'_m being lower than F_m (Figure 23.12). The impact of NPD$_i$ on the F_m fluorescence level is commonly quantified using the Stern–Volmer quenching parameter: $F_m/F'_m - 1$ (Table 23.2, Equation 1 [Krause and Jahns 2004]). This quantification does not distinguish between the different mechanisms that contribute to NPD$_i$ nor is it a measure of the quantum yield of dissipation via NPD$_i$ (there is no $k_{p[QA]}$ in either the fluorescence yield equation for either F_m or F'_m, so the impact of photochemistry is excluded). The value of the Stern–Volmer parameter is that it is proportional to the concentration of a putative quencher, so it has proven to be a useful way to analyze the possible mechanisms of NPQ (e.g., Noctor et al. 1991; Ruban et al. 1993; Bilger and Björkman 1994—in these papers NPQ is identified as q_E, which is often the major component of NPQ), and the value of the parameter is equivalent to the ratio of k_{Dvar} and the sum of k_f and k_d

Table 23.2 The most commonly encountered fluorescence-derived parameters, their purpose, and their relationship to the routinely measured fluorescence parameters and the rate constants for photochemistry, fluorescence, and the basal and inducible thermal dissipation processes

	Equation	Description
1	$NPQ = \left(\dfrac{F_m}{F'_m}\right) - 1 = \left(\dfrac{F_m - F'_m}{F'_m - F'_o}\right) \times \dfrac{F'_v}{F'_m}$	The relationship between the Stern-Volmer parameter for NPQ and F'_v/F'_m
2	$NPQ = \left(\dfrac{F_m}{F'_m}\right) - 1 = \left(\dfrac{k_D}{(k_f + k_d)}\right) = \dfrac{\Phi_{NPQ}}{\Phi_{NO}}$	The relationship between the Stern-Volmer parameter for NPQ and the first-order rate constants for all dissipation processes for ^{1}chl* in PSII, and the yields for (1) inducible nonphotochemical dissipation processes in PSII (Φ_{NPQ}) and (2) the basal nonphotochemical dissipation processes in PSII (Φ_{NO})
3	$q_Q = q_P = \left(\dfrac{F'_m - F'}{F'_m - F'_o}\right)$	The calculation of q_P (or q_Q) from F'_m, F' and F'_o
4	$q_L = \left(\dfrac{F'_m - F'}{F'_m - F'_o}\right) \times \dfrac{F'_o}{F'}$	The calculation of qL (relative QA redox state) from F'_o, F' and F'_o
5	$\dfrac{F'_v}{F'_m} = \left(\dfrac{F'_m - F'_o}{F'_m}\right) = \dfrac{k_{P_{([QA]_{rel}=1)}}}{\left(k_f + k_d + k_{P_{([QA]_{rel}=1)}} + k_{Dvar}\right)}$	The calculation of F'_v/F'_m from F'_m and F'_o and its relationship to the first-order rate constants for dissipation of ^{1}chl* in PSII
6	$\Phi_{PSII} = \dfrac{F'_v}{F'_m} \times q_P = \dfrac{k_{P_{(0\leq[QA]_{rel}\leq1)}}}{\left(k_f + k_d + k_{P_{(0\leq[QA]_{rel}\leq1)}} + k_{Dvar}\right)}$	The relationship between Φ_{PSII} and F'_v/F'_m and q_P and its definition in terms of the rate constants for dissipation of ^{1}chl* in PSII
7	$\Phi_{PSII} = \dfrac{F'_v}{F'_m} \times q_P = \dfrac{F'_m - F'}{F'_m} = \dfrac{\Delta F}{F'_m}$	The relationship between Φ_{PSII} and F'_m and F', ΔF and F'_m and F'
8	$J_{e_{\bar{PSII}}} = Irradiance \times 0.84 \times fraction_{PSII} \times \Phi_{PSII}$	The commonly used equation for *approximately* converting Φ_{PSII} into a rate of electron transport
9	$\Phi_{NP}^c = \Phi_{f,D} = \Phi_{NO} = \dfrac{F'}{F'_m} = \dfrac{k_f + k_d}{\left(k_f + k_d + k_{P_{(1\leq[QA]_{rel}\leq0)}} + k_{Dvar}\right)}$	The three forms of the parameter used to describe the yield of dissipation of ^{1}chl* in PSII by the basal nonphotochemical dissipation processes in terms of F_m and F', and the rate constants for all dissipation processes
10	$\Phi_{NP}^r = \Phi_{NPQ} = \dfrac{F'}{F'_m} - \dfrac{F'}{F_m} = \dfrac{k_{Dvar}}{\left(k_f + k_d + k_{P_{(1\leq[QA]_{rel}\leq0)}} + k_{Dvar}\right)}$	The two forms of the parameter used to describe the yield of dissipation of ^{1}chl* in PSII by the inducible nonphotochemical dissipation processes in terms of F_m, F'_m and F', and the rate constants for all dissipation processes

(the basal nonphotochemical dissipation pathways) (Table 23.2, Equation 2).

The measure $F_m / F'_m - 1$, while giving a means of describing the extent of NPQ in a useful way, does not offer a means for determining which of the numerous possible mechanisms that contribute to NPD_i are actually active and which therefore give rise to NPQ. These mechanisms are diverse, and their interrelationships are sometimes uncertain, so analyzing which mechanisms are contributing to NPD_i is not that easy. Using chlorophyll fluorescence alone (more is possible if fluorescence is combined with a complex approach based on mutants and inhibitors), it is possible to distinguish between different mechanisms (which may each be groups of mechanisms) by measuring the rate relaxation of NPQ (i.e., the increase in F'_m in the dark; Horton and Hague 1988; Quick and Stitt 1989; Walters and Horton 1991; Müller et al. 2001; Krause and Jahns 2004). The relaxation kinetics can often be split into three components: fast relaxing processes with a half-time of around 30–60 s that are normally attributed to energy-state quenching (the NPQ due to this component is termed q_E); processes relaxing with a half-time of 5–10 min that are usually attributed to state transitions (this component is termed q_T) but could also include chloroplast movements; and slowly relaxing processes that decay with a half-time of over 30 min, and sometimes days, that are attributed to photodamage to PSII (Aro et al. 1993) or slowly reversible downregulation of PSII (Demmig-Adams and Adams 2006) (this component is q_I). Maxwell and Johnson (Maxwell and Johnson 2000) give equations that allow the total NPQ to split into q_E, q_T, and q_I components based on relaxation kinetics of NPQ (and see also Krause and Jahns 2004). Some terminology issues are created by the slow reversibility of the q_I component of NPQ and therefore the NPD processes that give rise to this, when taken together with the typical dark adaptation time of either less than 60 min (20 or 30 min is more usual) or sometimes overnight (in some field studies the dark-adapted F_v/F_m is measured predawn). If the q_I component of NPQ takes hours or days to fully relax, then even an overnight dark adaptation may not eliminate it—there will still be some NPD_i even after dark adaptation (i.e., some of what was k_{Dvar} will now be k_d). The solution to this is a pragmatic one: as far as fluorescence analysis goes, whatever

NPD processes are left after dark adaptation are normally just included in the basal NPD (i.e., accounted for by k_d) rather than by creating a new rate constant or adding k_{Dvar} to the equations for F_o and F_m. This, of course, does not change the specific physiological character of the persistent, inducible quenching processes, just the way they are dealt with in fluorescence and other yield equations.

The mechanisms that are thought to contribute to each of these phases are still the subject of considerable debate, and it is not possible to usefully discuss this rather rich topic here, but for more information see Krause and Jahns (2004), Demmig-Adams et al. (2014), and Chapter 11. One distinction that needs to be drawn, however, is between q_E and q_I on the one hand and q_T on the other. Both q_E and q_I are thermal dissipation mechanisms that quench $^1chl^*$ in PSII once the $^1chl^*$ have been formed. On the other hand, while the mechanisms and consequences of the state transitions that give rise to q_T are not completely clear, at least in some cases, a fraction of the PSII light-harvesting complexes (LHCII) transfer excitation energy to PSI via a mechanism that involves phosphorylation of that LHCII in response to overexcitation of PSII relative to PSI (Allen 1992; Samson and Bruce 1995; Vener et al. 1998; Kouril et al. 2005), see also Chapter 4. In other cases, a decrease in PSII cross section has been observed with no increase in PSI cross section (Haworth and Melis 1983) nor any increase in light use efficiency for assimilation (Andrews et al. 1993). Whether or not some of the excitation that was channeled to PSII in the dark-adapted state is transferred to PSI, the loss of F'_m associated with the state transition implies that the excitation is no longer transferred to PSII once the state transition has occurred—state transitions decrease (and appear to increase quenching of) F_m by preventing formation of $^1chl^*$ in PSII in first place. In other words, the mechanisms that give rise to q_E and q_I are demand-side mechanisms (they increase dissipation of $^1chl^*$ in PSII once they have been formed), while q_T is a supply-side mechanism (it acts by forming less $^1chl^*$ in PSII in the first place). However, as the intensity of chlorophyll fluorescence emission from PSII is $k_f \cdot [^1chl^*]_{PSII}$, increasing a demand-side process or decreasing a supply process will have the same effect—a loss of PSII fluorescence. Consistent with state transitions acting as a supply-side process, they can be measured not only via a decrease in F_m but also of F_o (Samson and Bruce 1995), and decreases in F_o associated with a state transition are not associated with a noticeable loss in F_v'/F_m', a measure of the efficiency of an open PSII reaction center in illuminated leaves that is analogous to the dark-adapted F_v/F_m parameter (Hogewoning et al. 2012). If the ^{1}chl* that would be formed in PSII are in PSI, then there would, in principle, be an increase in PSI fluorescence and an increase in F_o or F_o', which should to a limited extent counteract the loss of F_o or F_o' due to the effect of the state transition on PSII, but no analysis of room-temperature fluorescence has addressed this point. In contrast, losses of F_m' due to the mechanisms that give rise to the qE component of NPQ do result in a loss of F_v'/F_m'. The fluorescence yield equations (Table 23.1) deal only with demand-side processes—processes that consume ^{1}chl* in PSII—so state transitions, which change the rate of formation rather than the dissipation of ^{1}chl* in PSII, are not accounted for by an analysis of PSII dissipation.

23.7.2 Quantifying the efficiency of PSII in steady-state illumination: F_v'/F_m'

Upon applying a saturating irradiance to an illuminated leaf (or some other photosynthetic tissue), the relative fluorescence yield will rise from F' to F_m' (Figure 23.12) and if the actinic irradiance that has given rise to F' is removed (i.e., there is darkness) will decrease to F_o' (Figure 23.12). This fluorescence yield is analogous to F_o, except that the denominator of the fluorescence yield equation for F_o' also contains the k_{Dvar} term, consistent with it being measured in material where NPD_i processes could be active. Just as F_v/F_m gives a measure of the quantum yield for electron transport by PSII in the dark-adapted state (i.e., no NPD_i is present and when all traps are open), there is an analogous parameter for F_v'/F_m' measured in the light when NPD_i might be active— this is termed F_v'/F_m'. The derivation of this analog for F_v/F_m in the light-adapted state is the same as that for F_v/F_m (Figure 23.11), except that while for F_v/F_m, the denominator of the fluorescence yield equations for F_o and F_m is $k_f + k_{p([QA]rel=1)} + k_{d;}$ in the case of F_o' and F_m' (Figure 23.12), it is $k_f + k_{p([QA]rel=1)} + k_d + k_{Dvar}$ (Tables 23.1 and 23.2, Equation 5).

23.7.3 Quantifying the efficiency of PSII in steady-state illumination: q_P

The F' yield lies between the lower and upper yield limits set by F_o' and F_m'. Based on *in vitro* studies of the fluorescence response of illuminated chloroplasts and Chlorella cells to DCMU (an inhibitor Q_A reoxidation) (Krause et al. 1982) and the response of fluorescence of illuminated leaves to laser pulses (Bradbury and Baker 1984), it was clear that the level of F' between F_o' and F_m' was determined by the extent of Q_A reduction (Dietz et al. 1985). At F_m' all Q_A is reduced and at F_o' all Q_A is oxidized—the level of F' is therefore determined by the openness of PSII traps. The level of F' relative to F_o' and F_m' is quantified by the q_P parameter (sometimes termed q_Q) (Table 23.2, Equation 3); if F' is the same as F_o', then q_P is 1. q_P has been interpreted as being linearly related to the relative Q_A oxidation state, but while this is semiquantitatively true, the relationship between Q_A redox state and q_P is nonlinear because of interconnectivity between PSII units (Paillotin 1976; Butler 1980; Lavergne and Trissl 1995). The result of this is that when q_P is close to 1, it is relatively insensitive to changes in Q_A redox state, while when it is close to 0, it is relatively more sensitive to changes in Q_A redox state. A parameter, q_L, developed by Kramer and coworkers (Kramer et al. 2004) attempts to provide an estimate of actual Q_A redox state from fluorescence parameters (Table 23.2, Equation 4), compensating for the nonlinearity between Q_A redox state and q_P. Despite its nonlinear dependency on Q_A redox state, q_P is still an important parameter because it is a measure of the impact that closure of PSII traps has on the quantum yield for PSII photochemistry.

If q_P is not a linear measure of Q_A oxidation, what exactly is it? Baker introduced the term "PSII efficiency factor" to describe the impact of q_P on Φ_{PSII}, and this defines well its role in relation to F_v'/F_m' and Φ_{PSII} in Equation 6 (Table 23.2). More subtle is the basis of those physical processes, which are reflected in the value q_P and which reduce the quantum efficiency of PSII electron transport below the value of F_v'/F_m'. q_P is dependent on Q_A redox state, but ^{1}chl* migrating through the PSII pigment bed are not significantly quenched just by encountering a PSII reaction center with a reduced Q_A (Lavergne and Trissl 1995), and the migration continues. In this migration, the ^{1}chl* are, however, exposed to nonphotochemical and fluorescence dissipation in the PSII pigment bed. q_P is therefore probably best seen as a factor, which adjusts F_v'/F_m' for the increase in the time required for photochemical quenching to occur due to a fraction of PSII reaction centers being closed by Q_A reduction. If all PSII are open, then the time to quenching is minimum; ^{1}chl* are exposed to pigment bed quenching (via k_f, k_d, and k_{Dvar}) to the least extent and Φ_{PSII} equals F_v'/F_m'. As Q_A becomes more reduced, the time to find an open trap increases and the ^{1}chl* are exposed to the quenching environment of the pigment bed for longer and the losses of ^{1}chl* due to fluorescence, NPD$_t$, and NPD$_b$ increase, reducing Φ_{PSII} by a factor, q_P, relative to the maximum, F_v'/F_m'.

23.7.4 Quantifying the efficiency of PSII in steady-state illumination: The operating efficiency of PSII, Φ_{PSII}

The recognition that F_v'/F_m' was a measure of the efficiency for electron transport by open PSII traps and that the q_P parameter was a measure of the impact of trap closure on the efficiency of PSII electron transport led Genty and coworkers to show that the product of F_v'/F_m' and q_P would give the quantum efficiency of PSII electron transport (Φ_{PSII}; Table 23.2, Equation 6) in a state where NPDi could be active (i.e., there was NPQ) and not all PSII traps were open (Genty et al. 1989). The measurement of both F_v'/F_m' and q_P requires that F_0' be known and the product of F_v'/F_m' and q_P simplifies to $\left(F_m'-F'\right)/F_m'$ (Table 23.2, Figure 23.7; $\Delta F/F_m'$ [Figure 23.12]) (Genty et al. 1989), so Φ_{PSII} can be measured without the dark interval required for the measurement of F_0', which is extremely useful, especially for measurements in the field. In the event that F_0' cannot be measured but it is necessary to split Φ_{PSII} into F_v'/F_m' and q_P components, F_0' can be estimated using an equation developed by Oxborough and Baker (Oxborough and Baker 1997), though note that this equation cannot compensate for changes in F_0 due to q_T, only for those due to q_E and q_I components (remember the fluorescence yield equation for F_0' contains k_{Dvar} [Table 23.1]).

The light use efficiency for electron transport by PSII is also a measure for the light use efficiency for linear electron transport (except when plastid terminal oxidase diverts electrons from the plastoquinone pool to O_2 [Cornic and Miginiac-Maslow 1985]—an apparently rare phenomenon). The dominant sink for the products of linear electron transport in mature leaves is photosynthetic carbon reduction and photosynthetic carbon oxidation (photorespiration), and gross assimilation is the result of these two opposing processes (Farquhar et al. 1980). Gross assimilation is easy to measure, and in the absence of photorespiration, assimilation is easy to interpret as an electron transport rate. Under conditions where photorespiration is eliminated, the relationship between Φ_{PSII} estimated from chlorophyll fluorescence and the light use efficiency of gross assimilation (or rates of electron transport estimated from Φ_{PSII} and rates of gross assimilation) have been found to be linear (e.g., Genty et al. 1989; Genty et al. 1990; Harbinson et al. 1990; Peterson 1991, 1994; Loriaux et al. 2013), and though the relationship is often less linear in normal air where photorespiration is active, it is still often predominantly linear (e.g., Genty et al. 1990; Harbinson et al. 1990; Peterson 1994). The same reservations are expressed about the accuracy of the fluorescence-derived estimate of the operating light use efficiency for electron transport by PSII as have been voiced for that of F_v/F_m—the presence of PSI fluorescence in the F_0' fluorescence is certainly an error, and there might also be an error in the F_m' measurement (Section 23.6.3). For these reasons, it is not appropriate to treat a fluorescence-derived estimate of Φ_{PSII} as an accurate measure of the actual Φ_{PSII}, but the linearity between the fluorescence-derived Φ_{PSII} and gross assimilation implies that Φ_{PSII} calculated from fluorescence is a linear index of the light use efficiency of PSII electron transport. In addition, obtaining an accurate F_m' requires a higher irradiance than has been commonly used; irradiances of 6000 μmol m^{-2} s^{-1}, or even less, have in the past been used for this purpose, and these will not always saturate Q_A reduction (Markgraf and Berry 1990; Loriaux et al. 2013). In the absence of an F_m', irradiance response is not possible to predict the extent to which an irradiance will be subsaturating, but it seems that the higher the light-saturated assimilation rate of the leaf (the higher the light-saturated electron transport rate) and the higher the actinic irradiance, the higher will be the irradiance needed to produce the same extent of Q_A reduction. As a result, at least some fluorescence-based estimates of Φ_{PSII} will be underestimates owing the underestimate of F_m', and care ought to be taken to ensure the extent to which the available saturating light pulse is actually saturating, especially if the accuracy of the value of the fluorescence-derived parameter (such as Φ_{PSII}) is critical.

A conversion of fluorescence-derived estimates of Φ_{PSII} to rates of electron transport is often done using an equation such as Equation 8 (Table 23.2). To correct for incomplete light absorption by the leaf, irradiance is scaled by 0.84 (or a specific measurement of light absorption), and to correct for light partitioning between PSI and PSII, a scaling factor is used that represents the fraction of irradiance absorbed by PSII (often set to 0.5). Finally, the estimated irradiance input to PSII is multiplied by the light use efficiency for PSII electron transport. An approach of this kind is at best semiquantitative. The light absorption for a typical green leaf has been measured to be around 0.84, but this was for filtered radiation from a quartz-halogen light source (Björkman and Demmig 1987); this will not be appropriate for hairy, scaly, chlorotic leaves, etc., and even for normal green leaves, the estimate of absorptance will depend on the spectrum of the irradiance, so the value of 0.84 would not necessarily be appropriate for another spectrum. Even on an absorbed light basis, not all photons are equally efficient at driving photosynthesis; wavelengths less than around 560 nm have a lower quantum yield than photons in the range 560–690 nm (McCree 1972; Inada 1976; Hogewoning et al. 2012), and this extra inefficiency factor is not included in equations of the kind shown in Equation 8. Some methods have been proposed for estimating irradiance partitioning between PSI and PSII (e.g., Laisk et al. 2006), but these also involve assumptions and have not been well validated outside the original publications. Finally, the value of Φ_{PSII} obtained from fluorescence is in error because of the presence of fluorescence from PSI and possibly changes in k_d in PSII due to Q_A reduction. If it is desired to convert a measure of Φ_{PSII} into an electron transport rate, it is necessary to calibrate the product of Φ_{PSII} and irradiance against a measure of electron transport, such as the carbon dioxide fixation rate in the absence of photorespiration. Though better than using Equation 8, this approach does not take into account oxygen reduction (the Mehler reaction) (Driever and Baker 2011) by photosynthetic

electron transport nor other metabolic uses of electron transport, such as nitrite reduction.

23.7.5 Quantifying the inefficiency of PSII in steady-state illumination

The light use efficiency of PSII electron transport based on chlorophyll fluorescence (and ignoring any errors implicit in this method) has a maximum value of 0.83. With increasing irradiance, it decreases, and by implication, the yield of nonphotochemical processes (including fluorescence, NPD_b, and NPD_i) must have increased—after all, the $^1chl^*$ have to be dissipated by something. Equations for yield of basal dissipation (termed Φ_{NP}^c, $\Phi_{f,D}$, or Φ_{NO}; Table 23.2, Equation 9) due to fluorescence and NPD_b that will be active in the dark-adapted and light-adapted states, and the yield of inducible dissipation (Φ_{NP}^r, Φ_{NPQ} Table 23.2, Equation 10) due to NPD_i, which will only be present in the light-adapted state, have been derived (Cailly et al. 1996; Hendrickson et al. 2004; Kramer et al. 2004). The Stern–Volmer parameter for NPQ (Table 23.2, Equation 2) does not provide a measure of the yield of inducible dissipation (it has no k_p term in the denominator), but the ratio Φ_{NPQ}/Φ_{NO} is the Stern–Volmer parameter for NPQ (cf Equations 2, 9, and 10, Table 23.2). As with other fluorescence-derived parameters, the errors due to the presence of PSI fluorescence in the F_o' signal, and that of F_m' due to increased reaction center quenching when Q_A is reduced, need to be borne in mind.

23.8 COMPLICATIONS OF FLUORESCENCE MEASUREMENTS FROM LEAVES

The chlorophyll fluorescence emission spectra shown in Figure 23.9 are rarely measured from leaves or similar. The fluorescence emission spectrum overlaps strongly with the leaf absorption spectrum (Figure 23.6), and as a result, the shorter wavelengths of the pure emission spectra shown in Figures 23.7 through 23.9 are subject to considerable distortion due to this reabsorption. The peak at around 680 nm, in particular, is strongly reabsorbed (Figure 23.13) (e.g., Peterson et al. 2014), but reabsorption effects extend through much of the emission spectrum (e.g., Gitelson et al. 1998). The effect of

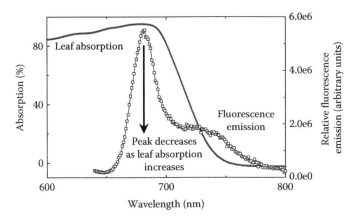

Figure 23.13 Detail of the chlorophyll fluorescence emission spectrum from a white rose petal (so very low reabsorption of fluorescence) and the leaf absorption spectrum showing the major change in the chlorophyll fluorescence spectrum that occurs as the absorption of the leaf increases, leading to greater reabsorption of the fluorescence.

leaf reabsorption due to the presence of chlorophyll (and especially Chl a given the wavelength range of the fluorescence emission spectrum) is so strong and reproducible that the ratio of the fluorescence emission at 700 and 735 nm can be used to accurately estimate leaf chlorophyll content (Gitelson et al. 1999). The sensitivity of the fluorescence emission spectrum on leaf chlorophyll content highlights the importance of not using leaf emission spectra for other purposes, such as PSII/PSI ratios—these other uses of the emission spectrum, which can be reliably used in more controlled conditions *in vitro*, cannot be used reliably in leaves. The spectrum shown in Figure 23.6, while very similar to the pure fluorescence spectra shown in Figures 23.7 through 23.9, was in fact measured from a white rose petal in which reabsorption effects would be negligible. A consequence of reabsorption is that while the basic fluorescence peak at 680 nm would seem to offer the strongest fluorescence signal, the strong reabsorption of this peak means that in this spectral region, the emission may be lower than in the longer wavelength region around 720–730 nm (Peterson et al. 2014). These gross effects of reabsorption on the fluorescence spectrum overlie a more complex relationship between leaf light absorption and fluorescence that has a considerable impact on the use and interpretation of fluorescence data.

Leaves contain high concentrations of photosynthetic pigments (deciduous and tropical tree leaves have up to the order of 1 mmol total chlorophyll per square meter, about 900 mg chlorophyll per square meter) (Gitelson et al. 2003; Asner et al. 2008) and the optical properties of leaves, particularly because of light scattering by the leaf structure, are complex. Overall, across the range of what is normally considered to be PAR (400–700 nm), leaves are good absorbers, and absorbance across the range of PAR (400–700 nm) is not strongly affected by total chlorophyll content over the normal range of leaf chlorophyll contents found *in vivo* (Björkman and Demmig 1987; Evans 1993; de Groot et al. 2003). The absorptance is maximum in the blue (400–500 nm) and red (670–690 nm) spectral regions, with a minimum at about 560 nm (Figure 23.6). The depth of the 560 nm minimum relative to the maxima in the blue and the red depends on leaf chlorophyll content (Björkman and Demmig 1987; Merzlyak et al. 2009). This implies that the apparent extinction coefficient of leaves is higher in the red and blue parts of the spectrum compared to green. The apparent extinction coefficient is a combination of the actual extinction coefficient of the photosynthetic pigments and light scattering (Merzlyak et al. 2009), with relative contribution of scattering increasing as the pure absorbance (i.e., the absorbance of the extracted pigments) of the photosynthetic pigments decreases. The result is that light penetration into leaves (and by implication, the ability of fluorescence to exit leaves) is strongly dependent on wavelength in a way that is not simply dependent on leaf light absorbance (Cui et al. 1991; Merzlyak and Gitelson 1995; Merzlyak et al. 2009). Another problem of leaf optics is the presence of nonphotosynthetic pigments. In some cases, these are conspicuous anthocyanins and carotenoids formed as a result of stress or senescence but in some cases also in unstressed healthy leaves in garden forms of plants and in some wild types (e.g., *Setcreasea purpurea*). Additionally, leaves have less conspicuous nonphotosynthetic pigments that absorb strongly at wavelengths less than 400 nm but also still absorb in the region of 400–500 nm or longer (Cerovic et al. 2002).

The consequences of these observations for the excitation and measurement of chlorophyll fluorescence in leaves is significant. The wavelength of the excitation light will determine how deeply into the leaf this light will penetrate and profile of the gradient of excitation that is produced. As fluorescence is measured at wavelengths of 680 nm or (usually) higher, the excitation wavelengths used in practice are less than 680 nm, and commercially the choice is often guided by the availability of high-output power LEDs, so spectra centered on 660, 620–640, and 450–460 nm are commonly used. Wavelengths less than about 400 nm are strongly absorbed by nonphotosynthetic pigments—a phenomenon that allows chlorophyll fluorescence to be used to measure the presence of these pigments in plants (Cerovic et al. 2002). Within the range of 400–680 nm, other wavelengths can easily be used as excitation radiation with a little engineering knowledge. The choice of excitation wavelength has implications for which region within the depth profile of a leaf is being measured. Radiation around 560 nm will penetrate relatively deeply into the leaf, wavelengths in the range 660 nm less deeply (Cui et al. 1991; Sun et al. 1998; Oguchi et al. 2011). Fluorescence will be emitted along the excitation intensity gradient created as the excitation radiation penetrates and is absorbed by the leaf. This fluorescence will then experience wavelength-dependent reabsorption as it moves through the leaf to the upper or lower surfaces toward the fluorescence detector. Shorter fluorescence wavelengths are more strongly absorbed than longer. If fluorescence is emitted from deep inside the leaf (or from tissue on the other side of the leaf from the detector), the shorter wavelengths will be attenuated while longer wavelengths will be less strongly absorbed. The choice of excitation and measurement wavelengths will therefore have a strong effect on which parts of a leaf are actually being measured by fluorescence and result in bias in the measurements. Green excitation

light and far-red measurement wavelengths will tend to measure more evenly across the depth of the leaf, while blue excitation and shorter fluorescence measurement wavelengths will result in fluorescence being detected from a tissue close to the surface of the leaf (a fact that has already been exploited [Malkin et al. 1981]). In many fluorescence measurement systems, an RG9 filter (or similar) is used to select a range of fluorescent wavelengths. The transmission of this filter does overlap with the light absorption spectrum of a typical leaf, so even though this filter admits only far-red fluorescence for detection, there is likely to be a bias in the measurement if the detector is placed on the same side of the leaf as the excitation source or on the other side. Using an RG9-filtered detector to measure fluorescence from a leaf and using either strongly absorbed 660 nm or more deeply penetrating 560 nm radiation to measure, for example, leaf responses to irradiance, results in the 660 nm excited fluorescence reporting on tissues experiencing a higher irradiance than the 560 nm excited fluorescence.

ACKNOWLEDGMENT

JH acknowledges support from the Dutch national program "Biosolar Cells."

REFERENCES

Allen, J. F. (1992), How does protein phosphorylation regulate photosynthesis? *Trends Biochemical Science 17*, 12–17.

Andrews, J. R., Bredenkamp, G. J., and Baker, N. R. (1993), Evaluation of the role of State transitions in determining the efficiency of light utilisation for CO_2 assimilation in leaves, *Photosynthesis Research 38*, 15–26.

Aro, E. M., Virgin, I., and Andersson, B. (1993), Photoinhibition of photosystem II. Inactivation, protein damage and turnover, *Biochimica et Biophysica Acta 1143*, 113–134.

Asner, G. P., Jones, M. O., Martin, R. E., Knapp, D. E., and Hughes, R. F. (2008), Remote sensing of native and invasive species in Hawaiian forests, *Remote Sensing of Environment 112*, 1912–1926.

Baker, N. R. (2008), Chlorophyll fluorescence: A probe of photosynthesis *in vivo*, *Annual Review of Plant Biology 59*, 89–113.

Baker, N. R., Harbinson, J., and Kramer, D. M. (2007), Determining the limitations and regulation of photosynthetic energy transduction in leaves, *Plant, Cell & Environment 30*, 1107–1125.

Ball, J. T. (1987), Calculations related to gas exchange. In *Stomatal Function*, pp. 445–476. Eds. E. Zeiger, G. D. Farquhar, and I. R. Cowan. Stanford University Press Stanford, Stanford, CA.

Ballottari, M., Dall'Osto, L., Morosinotto, T., and Bassi, R. (2007), Contrasting behavior of higher plant photosystem I and II antenna systems during acclimation, *Journal of Biological Chemistry 282*, 8947–8958.

Barber, J., Malkin, S., Telfer, A., and Schreiber, U. (1989), The origin of chlorophyll fluorescence *in vivo* and its quenching by the photosystem II reaction centre [and Discussion], *Philosophical Transactions of the Royal Society of London B Biological Sciences 323*, 227–239.

Barbour, M. M., Farquhar, G. D., Hanson, D. T., Bickford, C. P., Powers, H., and McDowell, N. G. (2007), A new measurement technique reveals temporal variation in $\delta^{18}O$ of leaf-respired CO_2, *Plant, Cell & Environment 30*, 456–468.

Belgio, E., Johnson, M. P., Jurić, S., and Ruban, A. V. (2012), Higher plant photosystem II light-harvesting antenna, not the reaction center, determines the excited-state lifetime—Both the maximum and the nonphotochemically quenched, *Biophysical Journal 102*, 2761–2771.

Bilger, W. and Björkman, O. (1990), Role of the xanthophyll cycle in photoprotection elucidated by measurements of light-induced absorbance changes, fluorescence and photosynthesis in leaves of *Hedera canariensis*, *Photosynthesis Research 25*, 173–185.

Bilger, W. and Björkman, O. (1991), Temperature dependence of violaxanthin de-epoxidation and non-photochemical fluorescence quenching in intact leaves of *Gossypium hirsutum* L. and *Malva parviflora* L, *Planta 184*, 226–234.

Bilger, W. and Björkman, O. (1994), Relationships among violaxanthin deepoxidation, thylakoid membrane conformation, and nonphotochemical chlorophyll fluorescence quenching in leaves of cotton (*Gossypium hirsutum* L.), *Planta 193*, 238–246.

Björkman, O. and Demmig, B. (1987), Photon yield of O_2 evolution and chlorophyll fluorescence characteristics at 77 K among vascular plants of diverse origins, *Planta 170*, 489–504.

Boardman, N. K. and Thorne, S. W. (1971), Sensitive fluorescence method for the determination of chlorophyll *a*/chlorophyll *b* ratios, *Biochimica et Biophysica Acta 253*, 222–231.

Boichenko, V. (1998), Action spectra and functional antenna sizes of photosystems I and II in relation to the thylakoid membrane organization and pigment composition, *Photosynthesis Research 58*, 163–174.

Bradbury, M. and Baker, N. R. (1984), A quantitative determination of photochemical and non-photochemical quenching during the slow phase of the chlorophyll fluorescence induction curve of bean leaves, *Biochimica et Biophysica Acta 765*, 275–281.

Brettel, K. and Leibl, W. (2001), Electron transfer in photosystem I, *Biochimica et Biophysica Acta 1507*, 100–114.

Brugnoli, E. and Björkman, O. (1992), Chloroplast movements in leaves: Influence on chlorophyll fluorescence and measurements of light-induced absorbance changes related to ΔpH and zeaxanthin formation, *Photosynthesis Research 32*, 23–35.

Bults, G., Horwitz, B. A., Malkin, S., and Cahen, D. (1982), Photoacoustic measurements of photosynthetic activities in whole leaves. Photochemistry and gas exchange, *Biochimica et Biophysica Acta 679*, 452–465.

Burrows, P. A., Sazanov, L. A., Svab, Z., Maliga, P., and Nixon, P. J. (1998), Identification of a functional respiratory complex in chloroplasts through analysis of tobacco mutants containing disrupted plastid ndh genes, *EMBO Journal 17*, 868–876.

Butler, B. and Kitajima, K. (1975), Fluorescence quenching in photosystem II of chloroplasts, *Biochimica et Biophysica Acta 376*, 116–125.

Butler, W. L. (1980), Energy transfer between photosystem II units in a connected package model of the photochemical apparatus of photosynthesis, *Proceedings of the National Academy of Sciences 77*, 4697–4701.

Byrdin, M., Rimke, I., Schlodder, E., Stehlik, D., and Roelofs, T. A. (2000), Decay kinetics and quantum yields of fluorescence in photosystem I from *Synechococcus elongatus* with P700 in the reduced and oxidized state: Are the kinetics of excited state decay trap-limited or transfer-limited? *Biophysical Journal 79*, 992–1007.

Cailly, A. L., Rizza, F., Genty, B., and Harbinson, J. (1996), Fate of excitation at PS II in leaves. The non-photochemical side, *Plant Physiology and Biochemistry*, Special Issue 86.

Cerovic, Z. G., Bergher, M., Goulas, Y., Tosti, S., and Moya, I. (1993), Simultaneous measurement of changes in red and blue fluorescence in illuminated isolated chloroplasts and leaf pieces: The contribution of NADPH to the blue fluorescence signal, *Photosynthesis Research 36*, 193–204.

Cerovic, Z. G., Ounis, A., Cartelat, A., Latouche, G., Goulas, Y., Meyer, S., and Moya, I. (2002), The use of chlorophyll fluorescence excitation spectra for the non-destructive in situ assessment of UV-absorbing compounds in leaves, *Plant, Cell & Environment 25*, 1663–1676.

Connolly, J. S., Janzen, A. F., and Samuel, E. B. (1982), Fluorescence lifetimes of chlorophyll *a*: Solvent, concentration and oxygen dependence, *Photochemistry and Photobiology 36*, 559–563.

Coombs, J., Hall, D. O., and Long, S. P. (1985), *Techniques in Bioproductivity and Photosynthesis*. Pergamon Press, Oxford, U.K.

Cornic, G. and Miginiac-Maslow, M. (1985), Photoinhibition of photosynthesis in broken chloroplasts as a function of electron transfer rates during light treatment, *Plant Physiology 78*, 724–729.

Croce, R. and van Amerongen, H. (2011), Light-harvesting and structural organization of photosystem II: From individual complexes to thylakoid membrane, *Journal of Photochemistry and Photobiology B: Biology 104*, 142–153.

Cui, M., Vogelmann, T. C., and Smith, W. K. (1991), Chlorophyll and light gradients in sun and shade leaves of *Spinacia oleracea*, *Plant, Cell & Environment 14*, 493–500.

de Groot, C. C., van den Boogaard, R., Marcelis, L. F., Harbinson, J., and Lambers, H. (2003), Contrasting effects of N and P deprivation on the regulation of photosynthesis in tomato plants in relation to feedback limitation, *Journal of Experimental Botany 54*, 1957–1967.

DeEll, J. R. and Toivonen, P. M. A. (2003), *Practical Applications of Chlorophyll Fluorescence in Plant Biology*. Kluwer Academic Publishers, Boston, MA.

Delieu, T. and Walker, D. A. (1981), Polarographic measurement of photosynthetic oxygen evolution by leaf discs, *New Phytologist 89*, 165–178.

Demmig-Adams, B. and Adams, W. W. (2006), Photoprotection in an ecological context: The remarkable complexity of thermal energy dissipation, *New Phytologist 172*, 11–21.

Demmig-Adams, B., Adams, W. W., Ebbert, V., and Logan, B. A. (1999), Ecophysiology of the xanthophyll cycle. In *The Photochemistry of Carotenoids*, pp. 245–269. Eds. H. Frank, A. Young, G. Britton, and R. Cogdell. Springer, Dordrecht, the Netherlands.

Demmig-Adams, B., Garab, G., Adams, W., and Govindjee. (2014), *Non-Photochemical Quenching and Energy Dissipation in Plants, Algae and Cyanobacteria*. Springer, Dordrecht, the Netherlands.

Dietz, K. -J., Schreiber, U., and Heber, U. (1985), The relationship between the redox state of Q_A and photosynthesis in leaves at various carbon-dioxide, oxygen and light regimes, *Planta 166*, 219–226.

Drepper, F., Hippler, M., Nitschke, W., and Haehnel, W. (1996), Binding dynamics and electron transfer between plastocyanin and photosystem I, *Biochemistry 35*, 1282–1295.

Driever, S. M. and Baker, N. R. (2011), The water-water cycle in leaves is not a major alternative electron sink for dissipation of excess excitation energy when CO_2 assimilation is restricted, *Plant, Cell & Environment 34*, 837–846.

Duysens, L. N. M. and Sweep, G. (1957), Fluorescence spectrophotometry of pyridine nucleotide in photosynthesizing cells, *Biochimica et Biophysica Acta 25*, 13–16.

Evans, J. R. (1993), Photosynthetic acclimation and nitrogen partitioning within a lucerne canopy. II. Stability through time and comparison with a theoretical optimum, *Australian Journal of Plant Physiology 20*, 69–82.

Farquhar, G. D., Caemmerer, S., and Berry, J. A. (1980), A biochemical model of photosynthetic CO_2 assimilation in leaves of C_3 species, *Planta 149*, 78–90.

Fragata, M., Norden, B., and Kurucsev, T. (1988), Linear dichroism (250–700 nm) of chlorophyll *a* and pheophytin *a* oriented in a lamellar phase of glycerylmonooctanoate/H_2O. Characterization of electronic transitions, *Photochemistry and Photobiology 47*, 133–143.

Franck, F., Juneau, P., and Popovic, R. (2002), Resolution of the photosystem I and photosystem II contributions to chlorophyll fluorescence of intact leaves at room temperature, *Biochimica et Biophysica Acta 1556*, 239–246.

Genty, B., Briantais, J. -M., and Baker, N. R. (1989), The relationship between the quantum yield of photosynthetic electron transport and quenching of chlorophyll fluorescence, *Biochimica et Biophysica Acta 990*, 87–92.

Genty, B., Goulas, Y., Dimon, B., Peltier, J. M., and Moya, I. (1992), Modulation efficiency of primary conversion in leaves, mechanisms involved at PSII. In *Research in Photosynthesis*, pp. 603–610. Ed. N. Murata. Kluwer Academic Publishers, Dordrecht, the Netherlands.

Genty, B. and Harbinson, J. (1996), Regulation of light utilization for photosynthetic electron transport. In *Photosynthesis and the Environment*, pp. 67–99. Ed. N. R. Baker. Kluwer Academic Publishers, Dordrecht, the Netherlands.

Genty, B., Harbinson, J., and Baker, N. R. (1990), Relative quantum efficiencies of the two photosystems of leaves in photorespiratory and nonrespiratory conditions, *Plant Physiology and Biochemistry (Paris) 28*, 1–10.

Gitelson, A. A., Buschmann, C., and Lichtenthaler, H. K. (1998), Leaf chlorophyll fluorescence corrected for re-absorption by means of absorption and reflectance measurements, *Journal of Plant Physiology 152*, 283–296.

Gitelson, A. A., Buschmann, C., and Lichtenthaler, H. K. (1999), The chlorophyll fluorescence ratio F735/F700 as an accurate measure of the chlorophyll content in plants, *Remote Sensing of Environment 69*, 296–302.

Gitelson, A. A., Gritz, Y., and Merzlyak, M. N. (2003), Relationships between leaf chlorophyll content and spectral reflectance and algorithms for non-destructive chlorophyll assessment in higher plant leaves, *Journal of Plant Physiology 160*, 271–282.

Harbinson, J., Genty, B., and Baker, N. R. (1989), Relationship between the quantum efficiencies of photosystems I and II in pea leaves, *Plant Physiology 90*, 1029–1034.

Harbinson, J., Genty, B., and Baker, N. R. (1990), The relationship between CO_2 assimilation and electron transport in leaves, *Photosynthesis Research 25*, 213–224.

Harbinson, J. and Hedley, C. L. (1993), Changes in P-700 oxidation during the early stages of the induction of photosynthesis, *Plant Physiology 103*, 649–660.

Harbinson, J. and Woodward, F. I. (1987), The use of light induced absorbance changes at 820 nm to monitor the oxidation state of P700 in leaves, *Plant Cell and Environment 10*, 131–140.

Haupt, W. and Scheuerlein, R. (1990), Chloroplast movement, *Plant, Cell & Environment 13*, 595–614.

Haworth, P and Melis, A. (1983), Phosphorylation of chloroplast membrane proteins does not increase the absorption cross-section of photosystem I, *FEBS Letters 160*, 277–280.

Hendrickson, L., Furbank, R. T., and Chow, W. S. (2004), A simple alternative approach to assessing the fate of absorbed light energy using chlorophyll fluorescence, *Photosynthesis Research 82*, 73–81.

Hogewoning, S. W., Wientjes, E., Douwstra, P., Trouwborst, G., van Ieperen, W., Croce, R., and Harbinson, J. (2012), Photosynthetic quantum yield dynamics: From photosystems to leaves, *The Plant Cell 24*, 1921–1935.

Horton, P. and Hague, A. (1988), Studies on the induction of chlorophyll fluorescence in isolated barley protoplasts. IV. Resolution of non-photochemical quenching , *Biochimica et Biophysica Acta 932*, 107–115.

Horton, P. and Ruban, A. (2005), Molecular design of the photosystem II light-harvesting antenna: Photosynthesis and photoprotection, *Journal of Experimental Botany 56*, 365–373.

Horton, P., Wentworth, M., and Ruban, A. (2005), Control of the light harvesting function of chloroplast membranes: The LHCII-aggregation model for non-photochemical quenching, *FEBS Letters 579*, 4201–4206.

Ikegami, I. (1976), Fluorescence changes related in the primary photochemical reaction in the P700-enriched particles isolated from spinach chloroplasts, *Biochimica et Biophysica Acta 449*, 245–258.

Inada. (1976), Action spectra for photosynthesis in higher plants, *Plant and Cell Physiology 17*, 355–365.

Jansson, S., Leister, D., Robinson, C., and Scheller, H. (2007), Structure, function and regulation of plant photosystem I, *Biochimica et Biophysica Acta 1767*, 335–352.

Joliot, A. and Joliot, P. (1964), Etude cinetique de la reaction photochimique liberant l'oxygene au cours de la photosynthese, *Comptes Rendus de l'Académie des Sciences, Paris, 258*, 4622–4625.

Joliot, P. and Joliot, A. (1979), Comparative study of the fluorescence yield and of the C550 absorption change at room temperature, *Biochimica et Biophysica Acta 546*, 93–105.

Kanstad, S. O., Cahen, D., and Malkin, S. (1983), Simultaneous detection of photosynthetic energy storage and oxygen evolution in leaves by photothermal radiometry and photoacoustics, *Biochimica et Biophysica Acta 722*, 182–189.

Kaplanová, M., Kaplanova, M., and Čermák, K. (1981), Effect of reabsorption on the concentration dependence of fluorescence lifetimes of chlorophyll a, *Journal of Photochemistry 15*, 313–319.

Klughammer, C. and Schreiber, U. (1991), Analysis of light-induced absorbance changes in the near-infrared region. I. Characterization of various components in isolated chloroplasts, *Zeitschrift Fur Naturforschung C 46*, 233–244.

Klughammer, C. and Schreiber, U. (1994), An improved method, using saturating light pulses, for the determination of photosystem I quantum yield via P-700$^+$-absorbance changes at 830 nm, *Planta 192*, 261–268.

Kooten, O. and Snel, J. F. (1990), The use of chlorophyll fluorescence nomenclature in plant stress physiology, *Photosynthesis Research 25*, 147–150.

Kouril, R., Zygadlo, A., Arteni, A. A., de Wit, C. D., Dekker, J. P., Jensen, P. E., Scheller, H. V., and Boekema, E. J. (2005), Structural characterization of a complex of photosystem I and light-harvesting complex II of *Arabidopsis thaliana*, *Biochemistry 44*, 10935–10940.

Kramer, D. M., Johnson, G., Kiirats, O., and Edwards, G. E. (2004), New fluorescence parameters for the determination of Q_A redox state and excitation energy fluxes, *Photosynthesis Research 79*, 209–218.

Kramer, H. and Mathis, P. (1980), Quantum yield and rate of formation of the carotenoid triplet state in photosynthetic structures, *Biochimica et Biophysica 593*, 319–329.

Krasnovsky, A. A. (1979), Photoluminescence of singlet oxygen in pigment solutions, *Photochemistry and Photobiology 29*, 29–36.

Krasnovsky, A. A. and Kovalev, Y. V. (2014), Spectral and kinetic parameters of phosphorescence of triplet chlorophyll a in the photosynthetic apparatus of plants, *Biochemistry* (Moscow) *79*, 349–361.

Krause, G. H. and Jahns, P. (2004), Non-photochemical energy dissipation determined by chlorophyll fluorescence quenching: Characterization and function. In *Chlorophyll a Fluorescence: A Signature of Photosynthesis*, pp. 463–495. Eds. G. C. Papageorgiou and Govindjee, Springer, Dordercht, the Netherlands.

Krause, G. H., Vernotte, C., and Briantais, J. M. (1982), Photoinduced quenching of chlorophyll fluorescence in intact chloroplasts and algae. Resolution into two components, *Biochimica et Biophysica Acta 679*, 116–124.

Laisk, A., Eichelmann, H., Oja, V., Rasulov, B., and Rämma, H. (2006), Photosystem II cycle and alternative electron flow in leaves, *Plant and Cell Physiology 47*, 972–983.

Laisk, A., Oja, V., and Eichelmann, H. (2012), Oxygen evolution and chlorophyll fluorescence from multiple turnover light pulses: Charge recombination in photosystem II in sunflower leaves, *Photosynthesis Research 113*, 145–155.

Lavergne, J. and Trissl, H. (1995), Theory of fluorescence induction in photosystem II: Derivation of analytical expressions in a model including exciton-radical-pair equilibrium and restricted energy transfer between photosynthetic units, *Biophysical Journal 68*, 2474–2492.

Lazar, D. (2003), Chlorophyll a fluorescence rise induced by high light illumination of dark-adapted plant tissue studied by means of a model of photosystem II and considering photosystem II heterogeneity, *Journal of Theoretical Biology 220*, 469–503.

Lazár, D. (2013), Simulations show that a small part of variable chlorophyll a fluorescence originates in photosystem I and contributes to overall fluorescence rise, *Journal of Theoretical Biology 335*, 249–264.

Li, X. P., Bjorkman, O., Shih, C., Grossman, A. R., Rosenquist, M., Jansson, S., and Niyogi, K. K. (2000), A pigment-binding protein essential for regulation of photosynthetic light harvesting, *Nature 403*, 391–395.

Loriaux, S. D., Avenson, T. J., Welles, J. M., McDermitt, D. K., Eckles, R. D., Riensche, B., and Genty, B. (2013), Closing in on maximum yield of chlorophyll fluorescence using a single multiphase flash of sub-saturating intensity, *Plant, Cell & Environment 36*, 1755–1770.

Malkin, S., Armond, P. A., Mooney, H. A., and Fork, D. C. (1981), Photosystem II photosynthetic unit sizes from fluorescence induction in leaves: Correlation to photosynthetic capacity, *Plant Physiology 67*, 570–579.

Malkin, S., Schreiber, U., Jansen, M., Canaani, O., Shalgi, E., and Cahen, D. (1991), The use of photothermal radiometry in assessing leaf photosynthesis: I. General properties and correlation of energy storage to P700 redox state, *Photosynthesis Research 29*, 87–96.

Markgraf, T. and Berry, J. (1990), Measurement of photochemical and non-photochemical quenching: Correction for turnover of PS2 during steady-state photosynthesis. In *Current Research in Photosynthesis*, pp. 279–282. Ed. Baltscheffsky, M. Kluwer, Springer, Dordrecht, the Netherlands.

Markwell J. P., Thornber J. P., and Boggs R. T. (1979), Higher plant chloroplasts: Evidence that all the chlorophyll exists as chlorophyll—Protein complexes. *Proceedings of the National Academy of Science 76*, 1233–1235.

Matsubara, S., Krause, G. H., Aranda, J., Virgo, A., Beisel, K. G., Jahns, P., and Winter, K. (2009), Sun-shade patterns of leaf carotenoid composition in 86 species of neotropical forest plants, *Functional Plant Biology 36*, 20.

Maxwell, K. and Johnson, G. N. (2000), Chlorophyll fluorescence—A practical guide, *Journal of Experimental Botany 51*, 659–668.

McCree, K. J. (1972), The action spectrum, absorptance and quantum yield of photosynthesis in crop plants, *Agricultural Meteorology 9*, 191–216.

Merzlyak, M. N., Chivkunova, O. B., Zhigalova, T. V., and Naqvi, K. R. (2009), Light absorption by isolated chloroplasts and leaves: Effects of scattering and packing, *Photosynthesis Research 102*, 31–41.

Merzlyak, M. N. and Gitelson, A. (1995), Why and what for the leaves are yellow in autumn? On the interpretation of optical spectra of senescing leaves (*Acer platanoides* L.), *Journal of Plant Physiology 145*, 315–320.

Müller, P., Li, X. P., and Niyogi, K. K. (2001), Non-photochemical quenching. A response to excess light energy, *Plant Physiology 125*, 1558–1566.

Nelson, N. and Yocum, C. (2006), Structure and function of photosystems I and II, *Annual Reviews of Plant Biology 57*, 521–565.

Noctor, G., Rees, D., Young, A., and Horton, P. (1991), The relationship between zeaxanthin, energy-dependent quenching of chlorophyll fluorescence, and trans-thylakoid pH gradient in isolated chloroplasts, *Biochimica et Biophysica Acta 1057*, 320–330.

Nuijs, A. M., Shuvalov, V. A., van Gorkom, H. J., Plijter, J. J., and Duysens, L. N. M. (1986), Picosecond absorbance difference spectroscopy on the primary reactions and the antenna-excited states in Photosystem I particles, *Biochimica et Biophysica Acta 850*, 310–318.

Oguchi, R., Douwstra, P., Fujita, T., Chow, W. S., and Terashima, I. (2011), Intra-leaf gradients of photoinhibition induced by different color lights: Implications for the dual mechanisms of photoinhibition and for the application of conventional chlorophyll fluorometers, *The New Phytologist 191*, 146–159.

Oja, V. and Laisk, A. (2000), Oxygen yield from single turnover flashes in leaves: Non-photochemical excitation quenching and the number of active PSII, *Biochimica et Biophysica Acta 1460*, 291–301.

Oxborough, K. and Baker, N. R. (1997), Resolving chlorophyll *a* fluorescence images of photosynthetic efficiency into photochemical and non-photochemical components—Calculation of qP and F_v'/F_m' without measuring F_o, *Photosynthesis Research 54*, 135–142.

Paillotin, G. (1976), Movement of excitations in the photosynthetic domains of photosystem II, *Journal of Theoretical Biology 58*, 237–252.

Papageorgiou, G. C. (2004), Fluorescence of photosynthetic pigments *in vitro* and *in vivo*. In *Chlorophyll a Fluorescence: A Signature of Photosynthesis*, pp. 43–63. Eds. G. C. Papageorgiou and Govindjee. Springer, Dordrecht, the Netherlands.

Parusel, A. B. and Grimme, S. (2000), A theoretical study of the excited states of chlorophyll a and pheophytin a, *The Journal of Physical Chemistry B 104*, 5395–5398.

Peterson, R. B. (1991), Effects of O_2 and CO_2 concentrations on quantum yields of photosystem-I and photosystem-II in tobacco leaf tissue, *Plant Physiology 97*, 1388–1394.

Peterson, R. B. (1994), Regulation of electron transport in photosystems I and II in C3, C3-C4, and C4 species of *Panicum* in response to changing Irradiance and O_2 levels, *Plant Physiology 105*, 349–356.

Peterson, R. B., Oja, V., Eichelmann, H., Bichele, I., DallOsto, L., and Laisk, A. (2014), Fluorescence F_o of photosystems II and I in developing C3 and C4 leaves, and implications on regulation of excitation balance, *Photosynthesis Research 122*, 1–16.

Pfundel, E. (1998), Estimating the contribution of photosystem I to total leaf chlorophyll fluorescence, *Photosynthesis Research 56*, 185–195.

Pons, T. L., Flexas, J., von Caemmerer, S., Evans, J. R., Genty, B., Ribas-Carbo, M., and Brugnoli, E. (2009), Estimating mesophyll conductance to CO_2: Methodology, potential errors, and recommendations, *Journal of Experimental Botany 60*, 2217–2234.

Poulet, P., Cahen, D., and Malkin, S. (1983), Photoacoustic detection of photosynthetic oxygen evolution from leaves. Quantitative analysis by phase and amplitude measurements, *Biochimica et Biophysica Acta 724*, 433–446.

Quick, W. P. and Horton, P. (1984), Studies on the induction of chlorophyll fluorescence in barley protoplasts. II. Resolution of fluorescence quenching by redox state and the transthylakoid pH gradient, *Proceedings of the Royal Society of London. Series B Biological Sciences 220*, 371–382.

Quick, W. P. and Stitt, M. (1989), An examination of factors contributing to non-photochemical quenching of chlorophyll fluorescence in barley leaves, *Biochimica et Biophysica Acta 977*, 287–296.

Reimers, J. R., Cai, Z. L., Kobayashi, R., Rätsep, M., Freiberg, A., and Krausz, E. (2013), Assignment of the Q-bands of the chlorophylls: Coherence loss via Qx–Qy mixing, *Scientific Reports 3*, 2761.

Roelofs, T. A., Lee, C. H., and Holzwarth, A. R. (1992), Global target analysis of picosecond chlorophyll fluorescence kinetics from pea chloroplasts: A new approach to the characterization of the primary processes in photosystem II α-and β-units, *Biophysical Journal 61*, 1147–1163.

Ruban, A. V., Young, A. J., and Horton, P. (1993), Induction of nonphotochemical energy dissipation and absorbance changes in leaves (evidence for changes in the state of the light-harvesting system of photosystem II *in vivo*), *Plant Physiology 102*, 741–750.

Rutherford, A. W., Osyczka, A., and Rappaport, F. (2012), Back-reactions, short-circuits, leaks and other energy wasteful reactions in biological electron transfer: Redox tuning to survive life in O_2, *FEBS Letters 586*, 603–616.

Samson, G. and Bruce, D. (1995), Complementary changes in absorption cross-sections of photosystems I and II due to phosphorylation and Mg^{2+}—Depletion in spinach thylakoids, *Biochimica et Biophysica Acta 1232*, 21–26.

Schlodder, E., Hussels, M., Cetin, M., Karapetyan, N. V., and Brecht, M. (2011), Fluorescence of the various red antenna states in photosystem I complexes from cyanobacteria is affected differently by the redox state of P700, *Biochimica et Biophysica Acta 1807*, 1423–1431.

Schreiber, U., Schliwa, U., and Bilger, W. (1986), Continuous recording of photochemical and non-photochemical chlorophyll fluorescence quenching with a new type of modulation fluorometer, *Photosynthesis Research 10*, 51–62.

Sun, J., Nishio, J. N., and Vogelmann, T. C. (1998), Green light drives CO_2 fixation deep within leaves, *Plant and Cell Physiology 39*, 1020–1026.

Talbot, M. F. J. and Sauer, K. (1997), Spectrofluorimetric method for the determination of large chlorophyll a/b ratios, *Photosynthesis Research 53*, 73–79.

Telfer, A. (2005), Too much light? How β-carotene protects the photosystem II reaction center, *Photochemical and Photobiological Science 4*, 950–956.

Trissl, H. W. (1997), Determination of the quenching efficiency of the oxidized primary donor of photosystem I, P700+: Implications for the trapping mechanism, *Photosynthesis Research 54*, 237–240.

Trissl, H. W. (2003), Modeling the excitation energy capture in thylakoid membranes. In *Photosynthesis in Algae*, pp. 245–276. Eds. W. Larkum, S. E. Douglas, and J. A. Raven. Springer, Dordrecht, the Netherlands.

van Amerongen, H. and Croce, R. (2013), Light harvesting in photosystem II, *Photosynthesis Research 116*, 251–263.

Vener, A. V., Ohad, I., and Andersson, B. (1998), Protein phosphorylation and redox sensing in chloroplast thylakoids, *Current Opinion in Plant Biology 1*, 217–223.

Walker, D. and Walker, R. (1987), *The Use of the Oxygen Electrode and Fluorescence Probes in Simple Measurements of Photosynthesis*. Oxygraphics, Sheffield, U.K.

Walters, R. G. and Horton, P. (1991), Resolution of components of non-photochemical chlorophyll fluorescence quenching in barley leaves, *Photosynthesis Research 27*, 121–133.

Wientjes, E. and Croce, R. (2012), PMS: Photosystem I electron donor or fluorescence quencher, *Photosynthesis Research 111*, 185–191.

Wientjes, E., van Amerongen, H., and Croce, R. (2013), Quantum yield of charge separation in photosystem II: Functional effect of changes in the antenna size upon light acclimation, *Journal of Physical Chemistry B 117*, 11200–11208.

Wolff, C. and Witt, H. T. (1969), On metastable states of carotenoids in primary events of photosynthesis. Registration by repetitive ultra-short-flash photometry, *Zeitschrift fur Naturforschung [B] 24*, 1031–1037.

Zhu, X. G., Govindjee, Baker, N. R., deSturler, E., Ort, D. O., and Long, S. P. (2005), Chlorophyll a fluorescence induction kinetics in leaves predicted from a model describing each discrete step of excitation energy and electron transfer associated with photosystem II, *Planta 223*, 114–133.

24

Harvesting sunlight with cyanobacteria and algae for sustainable production in a bio-based economy

PASCAL VAN ALPHEN AND KLAAS J. HELLINGWERF

ABBREVIATIONS

APC	Allophycocyanin
LHC	Light-harvesting complex
NPQ	Nonphotochemical quenching
PAR	Photosynthetically active radiation
PBR	Photobioreactor
PBS	Phycobilisome
PC	Phycocyanin
PUFA	polyunsaturated fatty acid
ROS	Reactive oxygen species
RuBisCO	Ribulose-1,5-bisphosphate carboxylase oxygenase
Tla	Truncated light-harvesting antenna

24.1 INTRODUCTION: LIMITS AND OPPORTUNITIES IN OXYGENIC PHOTOSYNTHESIS

Oxygenic photosynthesis in cyanobacteria, algae, and plants provides directly or indirectly (via fossil reserves) the vast majority of the energy for the living world. Cyanobacteria and microalgae are responsible for about half of this conversion of free energy [1]. These organisms convert solar energy into chemical energy by the conversion of CO_2 into organic carbon compounds, evolving oxygen—the essential substrate of respiration in chemoheterotrophs—in parallel. If all solar energy that reaches the earth's surface were used with 100% efficiency, it would take only 1 hour to satisfy the energy demand of the world's population for a year [2]. Even though clearly not all of this radiation can be harvested in oxygenic photosynthesis, it is important to analyze the efficiency of light energy conversion in (oxygenic) photosynthesis in detail, in order to identify molecular targets for improvement. Over the years, the results of a wide range of measurements and estimates of the "solar-to-biomass" free-energy conversion efficiency have been published. The most-cited values range from 4.6% to 6% for C_3 and C_4 plants, respectively [2], and 8% to 10% for cyanobacteria and algae [3,4]. Which of the latter two single-celled (micro) organisms actually is most efficient in converting radiation free energy into the free energy of biomass is difficult to decide on the basis of the available data, though it is clear that the prokaryotic representatives among them have a distinctly lower maintenance energy demand for growth. These efficiency numbers were derived from lab-scale experiments. The highest reported solar-to-biomass conversion yield reported for algal productivity at a larger scale so far is only 3%, which still is significantly higher than for a plant such as switchgrass (i.e., 0.2%) [4].

Free-energy losses in photosynthesis occur at various points in the light-energy transduction pathway, such as the limited spectral width of photosynthetically active radiation (PAR, equal to ≈45% of the incident radiation from the sun), as well as losses due to nonphotochemical quenching (NPQ) and those due to futile cycles in downstream metabolism, such as some forms of "photorespiration." Hence, it is often observed that oxygenic photosynthesis requires significantly more than the theoretical eight photons [5,6] per molecule of fixed CO_2. Actual numbers in fact range from 9 to 17 [7–11].

Compared to land plants, cyanobacteria and microalgae have little nonphotosynthetic overhead (i.e., roots, stems, etc.). Furthermore, they can grow at high densities and growth rates and are capable of continued production throughout the year. Together with their high intrinsic photosynthetic efficiency, this makes them promising candidates for becoming the preferred production host for biofuels and fine chemicals in a sustainable bio-based society.

Currently, algal cultivation is mainly done in open ponds, which can be subdivided into natural waters (lakes, lagoons, ponds) and artificial ponds or containers [12,13]. The most commonly used systems include large shallow ponds and raceway ponds, either paddle stirred or unstirred, with a water depth of 15–30 cm [14]. Even though these are relatively simple to build and operate, they are only suited for growth of (consortia of) algae that can cope with predators and invaders, or with the extreme conditions required to suppress the former, as such open systems will inevitably be infected. In the absence of efficient mixing and hence slow mass transfer of CO_2 and with poor light utilization, this will lead to low productivity. Beyond these open (pond) systems, also many closed photobioreactors (PBRs) have been developed, which are less prone to infection and often more efficient in terms of mass transfer and light utilization [12,13]. However, these closed systems are more difficult to scale up to mass-production capacity without reducing the illumination surface and/or mixing efficiency and require significant capital investment.

Efficient product formation in large-scale culturing facilities for cyanobacteria and microalgae generally requires dense cultures with very efficient light harvesting. These are characteristics for which the organisms are not selected in their natural environment. As light intensity in nature varies depending on the time of the day, the weather, the season, and other environmental factors, such organisms are adapted to maximize light harvesting under varying and suboptimal conditions so that they deprive their competitors of light. However, photosynthesis only initially responds linearly to an increase in light intensity: It already saturates at moderate (sunlight) intensities. In fact, on a sunny day at sea level and moderate latitude, the photosynthetic capacity in cyanobacteria and microalgae is saturated most of the time [4]. This leads to dissipation of excess energy into heat by photoprotective mechanisms to avoid production of an excess of reactive oxygen species (ROS) in the organisms involved.

The cellular capacity for light harvesting is based on antenna size, which is defined as the number of light-harvesting molecules (chlorophyll a and b, phycobilisomes (PBS), carotenoids, etc.) that are attached to and serve the two photosystems, and the abundance of these photosystems. Green algae contain approximately the same number of each photosystem, whereas in cyanobacteria, photosystem I (PSI) is ≈5-fold more abundant than photosystem II (PSII) [15–17]. Both photosystems, particularly PSI, contain several dozen chlorophyll molecules as intrinsic antenna pigments [18,19]. In addition to those intrinsic antennae, mobile antennae exist—notably PBS in cyanobacteria and the Light-Harvesting Complex (LHC) in green algae—that can adjust exciton transfer to the two separate photosystems, via so-called "state transitions," so that equal excitation rates can be obtained in the two photosystems, which is a prerequisite for balanced phototrophic growth. Both changes in light color and light intensity may necessitate such a redistribution, which usually takes several minutes in LHC-containing organisms [20,21] and tens of seconds in cyanobacteria [22–24].

In dense (mass) cultures, light penetration is poor, which will lead to a large part of the culture being in light intensities below the compensation point, that is, the light intensity at which photosynthesis equals respiration. This in turn will lower the overall efficiency of such a production system to less than the theoretical maximum. Additionally, mixing of these cultures causes cells to experience light/dark cycles as they shuttle between the dark deeper layers and high light intensities near the surface of the reactor, which can have positive as well as adverse effects on biomass yield, depending, among others, on the incident light intensity and the geometry and mixing characteristics of the reactor [3,25,26]. The highest productivity per volume is achieved with a biomass concentration and PBR configuration that leaves light intensity above the compensation point at the far end of the PBR in order to minimize losses from respiration. In addition, incident intensities should not be much higher than the saturation intensity. The former point has experimentally been confirmed with mass cultures of *Arthrospira platensis* [27] and computationally rationalized [1] (Figure 24.1). In outdoor culturing, nevertheless, incident light intensity is primarily dictated by latitude and climate. Further adjustment can only be made

through the geometrical design of the PBR [28] and via genetic engineering of the cell's antenna complexes. The latter aspect will be discussed in Part 3.

24.2 LARGE-SCALE CULTURING OF OXYPHOTOTROPHS FOR A BIO-BASED ECONOMY

Direct exploitation of the free energy from sunlight through natural oxygenic photosynthesis for use in a bio-based economy requires the use of large shallow PBRs. The distributed nature of sunlight implies that it is not of much use to have a light path in the reactor of more than a few centimeters at moderate cell densities. In its simplest form, this approach aims at producing biomass, which subsequently can be processed, for example, via anaerobic digestion, into methane or even into crude oil via pyrolysis or other chemical conversion processes [29]. With optimal areal productivity of the algae, this approach has been brought close to economic feasibility and is expected to reach this point within the next 10–15 years [30]. However, as biomass may contain specific components (polyunsaturated fatty acids [PUFAs], carotenoids, etc.) with a much higher value than the average (liquid) energy carrier, a bio-refinery approach in which such valuable products are first separated may significantly enhance the profitability of these culturing systems [31].

The bio-refining approach can be extended to the subcellular fraction that contains the

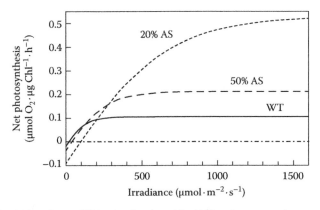

Figure 24.1 Photosynthesis–Irradiance (PI) curve simulated for WT and antenna size mutants with 20% and 50% of WT antenna levels, respectively. The dash-dotted line at 0 $\mu mol\ O_2 \cdot \mu g\ Chl^{-1} \cdot h^{-1}$ indicates the compensation point. Though the compensation point is reached at a higher light intensity than WT, antenna size mutants have a much higher saturation level and allow for better light penetration, leading to overall higher productivity in dense cultures. (Reproduced from Formighieri, C. et al., *J. Biotechnol.*, 162, 115, 2012.)

polysaccharides or the triglycerides. These can then be converted into fermentation products such as ethanol, butanol, etc., and into biodiesel after transesterification with methanol, respectively [30,32–34]. For this latter approach, (eukaryotic) microalgae may be preferable over cyanobacteria in view of their larger cellular storage capacity. This also holds for the production of products with even higher added value that are stored in the cytoplasmic compartment like, for example, pharmaceutical proteins (note that some polysaccharides synthesized by cyanobacteria may form an exception as they are stored in the extracellular compartment). Directed engineering to increase the cellular content of the refined product in the algae may further boost this approach, be it that this engineering is complicated by the presence of multiple subcellular compartments, several of which have their own dedicated genetic system. Hence, complicated regulatory circuits will need to be optimized, similar to the "crisscross" regulation in sporulation [35] as, for example, in terpene synthesis [36].

For maximal overall efficiency, however, it is beneficial to engineer a shortcut in phototrophic metabolism, which then allows such engineered organisms to directly channel fixed carbon into a fermentation/fuel product or into a chemical commodity [37–39]. This strategy can be applied to both microalgae [40] and cyanobacteria but has been much more successful in the latter class of organisms. This may be due to the ease with which the prokaryotic phototrophs can be engineered genetically, but also an increased permeability of their cell envelope may play a role in this. This "shortcut" strategy—also referred to as the "Photanol approach," see Hellingwerf and Teixeira de Mattos [38]—is particularly attractive if the final product of metabolism can leave the cells, either through cloning of a dedicated export protein (as for sucrose [41]) or via passive leakage (as for lactic acid [42] and ethanol [37,43]). Several large-scale initiatives are underway that are based on a cyanobacterial cell factory for a small molecule. Their economic competitiveness is currently under investigation [44].

24.2.1 Cyanobacterial cell factories

Use of cyanobacteria for the production of bio-fuel and chemical commodities has primarily been focused on *Synechocystis* and *Synechococcus*, with smaller efforts directed toward *Anabaena* and *Cyanothece* (a nitrogen-fixing strain). A proper molecular genetic toolbox is available for all four of these species, be it that knowledge about (regulatable) promoters and regulatory translation signal is still limited. Various metabolic pathways have been introduced into these organisms, like for the production of ethanol [37,38], lactic acid [42,45], sucrose [41], ethylene [46], 2,3-butanediol [47,48], isoprene [49], isobutyraldehyde [50], isobutanol [51], 1-butanol [39], fatty acids [52], and alkanes [53].

After the expansion of cyanobacterial metabolism with a heterologous metabolic pathway and the subsequent optimization of the pathway itself [39], optimization of product formation can be pursued by streamlining cyanobacterial CO_2 fixation and intermediary metabolism. For sucrose production, this was not even necessary: A salt stress makes the cells channel 90% of its fixed CO_2 into this sugar and even increases the overall rate of CO_2 fixation [41]. However, most of the products listed above are derived from pyruvate, of which the level in wild-type cyanobacteria is rather low [54]. To increase the rate and level of formation of these products, the capacity of the exogenous catabolic/fermentation pathway needs to be optimized [43,46,55]. Furthermore, competing pathways that catabolize pyruvate need to be eliminated (and pyruvate formation stimulated). This strategy has successfully been applied, for example, in the production of ethanol and lactic acid [56]. The choice of enzymes that can contribute to increased pyruvate formation extends all the way to Ribulose-1,5-bisphosphate carboxylase oxygenase (RuBisCO) [50], which catalyzes the initial reaction of the Calvin cycle, be it that engineering of an "electron sink" by itself already leads to increased RuBisCO activity, provided the heterologous metabolic pathway has sufficient capacity [41,55].

The developing field of synthetic-systems biology clearly has the potential to help resolve all the complicated mechanisms that lead to (sink) regulation in cyanobacterial photosynthesis. Eventually, the point will be reached in which nearly all (i.e., >95%) carbon fixed by an engineered cyanobacterium is channeled into product and only a small fraction into the synthesis of new cells. Such cells then would truly be "cell factories." Ideally, the remaining growth rate will equal the death rate of these cells in a production environment. Metabolic control theory can then provide the tools to predict the optimal expression level of the heterologous product-forming metabolic pathway to prevent that this pathway would become a metabolic burden [42].

24.2.2 Development of microalgal cell factories

Because of their specific and complex genetic and spatial organization, eukaryotic algae are attractive for cell factory applications for which a large intracellular storage capacity is required. As they are able to either directly synthesize (e.g., in the chloroplast) or temporarily store (i.e., in the vacuole) products within cellular subcompartments, this increases the flexibility and efficiency of design-driven synthetic biology strategies. A further advantage is their ease of harvesting due to their large cell size (>10 μm).

Chlamydomonas reinhardtii is generally considered as the best genetically accessible green alga. Both its chloroplast genome [57,58] and its nuclear genome can be targeted selectively for knockout mutagenesis and gene insertion. Heterologous gene expression, combined with cytosolic protein production is well established in this organism [59], be it that their efficiency could further be improved. Of course, *C. reinhardtii*, in contrast to the cyanobacteria, brings all the advantages of a eukaryotic production system in terms of the presence of systems for posttranslational modification. Significantly, also efficient protein secretion has recently been engineered into this organism [60,61]. Neutral lipid metabolism in microalgae is complex, compartmentalized, and not yet fully understood [62], which makes it difficult to define unambiguous targets for molecular engineering [63]. Nevertheless, in parallel to the approach of synthetic biology, also new physiological strategies are being developed to enhance lipid production in microalgae [64–67].

Besides *C. reinhardtii*, for only a few additional green algae (e.g., *Nannochloropsis gaditana* and *Ostreococcus tauri*) and one diatom (i.e., *Phaeodactylum tricornutum*), a toolbox for genetic engineering has been developed [68–70]. *C. reinhardtii* strains with improved phenotypes of interest have already been constructed [71,72]. The well-established protocols for transformation of *C. reinhardtii* with heterologous genes have also been the driver of the application of this organism in the expression of edible therapeutics in its chloroplasts [73].

Another microalga for which many researchers eagerly await increased accessibility of its genome(s) is *Botryococcus braunii*. This green alga can produce large amounts of the triterpene botryococcene, in such a way that the organism is able to deposit a large part of this "biocrude" product in the extracellular matrix. This has great advantages for downstream processing of this fuel product. Molecular techniques are now rapidly providing a view on the mechanisms and pathway by which this product is made and deposited [74,75], but the race is on to see whether genetic engineering to enhance its productivity will take place in this green alga itself or in a relative that now already is genetically accessible and in which the relevant genes may be heterologously expressed.

24.3 IMPROVING LIGHT UTILIZATION BY TRUNCATING ANTENNA SIZE

As discussed previously, a moderation of the changes in light intensity to which cells that grow in a well-mixed PBR are exposed may optimize their photosynthetic efficiency. A promising approach to achieve this is to reduce their antenna size in such a way that it keeps the PSII/PSI excitation ratio in balance. This brings the advantages that (1) cells will be less prone to quench excitation energy nonproductively and (2) the contents of the PBR will become more transparent. The latter aspect will make light gradients in the reactor shallower, and therefore put less stringent demands on cellular adaptation mechanisms to changes in light intensity, and may simplify the engineering of PBRs in which cells can carry out photosynthesis with high efficiency.

Though initially it turned out to be difficult to generate truly useful mutants with a truncated light-harvesting antenna (Tla), indeed there are now examples of mutants in which a significantly reduced antenna size leads to a higher biomass yield in dense cultures. In green algae, random mutagenesis has led to the identification of various mutants with reduced antenna size [76–78], from which rationally designed mutants were derived [79–81]. One particular Tla mutant that has extensively been studied is Tla1 in the green alga *Chlamydomonas reinhardtii*. It was originally identified after random mutagenesis but has also been used in knockdown and overexpression experiments [82]. The *tla1* gene is thought to be a regulator of chlorophyll antenna size as its knockdown significantly reduces the antenna size of both photosystems and chlorophyll *b* synthesis. Furthermore, its overexpression causes a slight increase in antenna size and chlorophyll *b* synthesis. Other related mutants like Tla2 and Tla3 have a similar phenotype in which the peripheral antenna complexes are strongly reduced in abundance, as well as the PSI and PSII content [76,77]. The genes *tla2* and *tla3* are postulated to encode chaperone proteins CpSRP43 and CpFTSY, which assist the translocation of LHCs to the thylakoid membrane. In their absence, the membrane proteins aggregate and are subsequently degraded, possibly leading to feedback inhibition in light-harvesting protein and chlorophyll synthesis, thereby also indirectly causing the reduction of PSI and PSII content observed in these mutants. All Tla mutants display a reduction in the number of auxiliary chlorophyll-containing proteins rather than in the number of chlorophyll molecules associated with the core photosystem proteins, which contain only a fraction of the total chlorophyll content of the cell. This is similar to the high-light phenotype [83,84]. However, these auxiliary complexes have other functions beyond light harvesting, making a complete knockout undesirable [20,80]. In both green algae and in cyanobacteria, antenna mutants are also shown to have an extensive impact on cell morphology, particularly on the thylakoid membranes [15,82,85]. A challenge for the future is to find mutants that selectively show a truncation of the fixed antennae rather than of the mobile ones and hence a reduction of the number of light-harvesting proteins rather than their absence, thereby preserving state transitions and thylakoid morphology. Whereas a change in morphology might not be so harmful under mass culture conditions, there is no escape from (quickly) changing light conditions in sunlight-driven reactors because of the inherent fluid dynamics and weather conditions. Such fluctuations in light intensity may be especially problematic for mutants impaired in light adaptation.

Though the issue of light attenuation and PBR design clearly leaves much room for further improvement in solar-to-biomass conversion efficiency, electron transport rate is an area in which production in cyanobacteria can be optimized. In cyanobacteria, the PSII/PSI ratio is much lower (1:5–6)

than usually found in algae and plants (where this ratio is around 1). Interestingly, despite the stark difference in ratio, cyclic electron transport only "significantly contributes (80%) to overall electron transport under mixotrophic growth conditions, compared to photoautotrophic growth (when it is below 10%) [86]." Cyclic electron transport has been shown to be required for photosynthesis to adjust the NADPH/ATP ratio generated for assimilation of CO_2 and other processes in the cell [87], as well as for the detection of excess of light energy, through the acidification of the lumen [88,89]. In order to maximize electron transport toward NADPH, the ratio should be 1:1, assuming equal photon capture of both systems. Fortunately, the PSII/PSI ratio is tunable by changes in light quality/quantity and knocking out selected genes, among which there are genes that directly affect the antenna size of the PBS [86,90,91].

Similar to truncation mutants in green algae, in cyanobacteria selected, truncation mutants do not affect the light-harvesting ability of the two photosystems themselves but rather that of the external antenna [92,93]. PBS deletion mutants in *Synechocystis* sp. PCC 6803 were shown to also affect the PSII/PSI ratio, which is increased to plant levels [15,86]. Contrary to the Tla2 and Tla3 mutants in *C. reinhardtii*, the research of Bernát et al. [86] shows that deletion of the PBS (i.e., the PAL mutant) causes the PSII/PSI ratio to increase significantly due to a reduction of the amount of PSI while simultaneously the amount of PSII increases. Linear electron transport in this mutant was increased up to 5.5-fold compared to wild type. Considering that the rate of linear electron flow in cyanobacteria appears to be low, this is a welcome additional effect, though the growth rate of the PAL mutant was slower than that of the corresponding wild type.

An antenna mutant called the "Olive" mutant, which leaves the core structure of the PBS largely intact but lacks phycocyanin (PC), also shows an increased rate of linear electron transport, as well as a significantly faster growth rate (>2×) under the selected growth conditions [86]. To date, few studies used cyanobacteria with a modified PBS antenna size in a (small-scale) mass cultivation system, and this yielded mixed results [94–96]. The mutants used had only one layer of PC attached to the core of the PBS (i.e., the CB mutant), just the allophycocyanin (APC) core of the PBS (CK and Δ*cpc*), or no PBS at all (PAL). Figure 24.2 (adapted from [97])

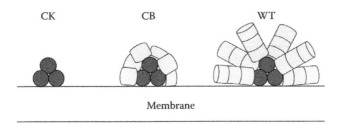

Figure 24.2 Structure of the phycobilisome (PBS) of Synechocystis sp. PCC6803 (WT) and mutants (CK and CB). The PBS consists of a tricylindrical core, from which radiate six rods composed of three phycocyanin (PC) hexamers. In the CB mutant, only one layer of PC hexamers remains and no PC remains in the CK mutant. The PAL mutant lacks all components of the PBS. (Adapted from Tian, L. et al., *Biophys. J.*, 102, 1692, 2012.)

schematically depicts the core, the rods, and the PBS mutants. For additional details, the reader is referred to [98] and elsewhere in this book. All mutants did show higher productivity at high culture density, in agreement with predictions based on green algae research, but grew slower than the wild type. The fact that growth rates during exponential growth were lower compared to wild type, contrary to the "Olive" mutant discussed earlier (which is similar to CK), indicates that antenna mutants are not easily comparable and that multiple aspects of the selected growth conditions are likely to be important. The results of Collins et al. show that in the case of the CB mutant, significant amounts of PC were found in the cytosol, indicating that despite truncation of the PBS, significant overhead is created. Additionally, the morphology of the thylakoid membranes was disturbed in both the CK and PAL mutants, further suggesting that at this stage a clear-cut comparison to wild type is hard to make.

Nevertheless, algal Tla mutants have been shown to have increased productivity on a per-Chl basis as a result of less wasteful dissipation of energy and increased light penetration into deeper layers of the culture [78,80,84,85]. The apparently conflicting results point toward stringent requirements for the optimization of these two properties, that is, antenna size and rate of linear electron transport, for actual improvements. These truncations can therefore only be an improvement for solar energy conversion if a strain complies with the following characteristics [4,86]:

- The absolute amount of PSII per cell or chloroplast is not reduced.
- The functional PSII and PSI antenna size is smaller than in wild type.

- The quantum yield of photosynthesis, that is, the number of photons required per CO_2 fixed, is unaffected.
- The light-saturated rate of photosynthesis, as measured in $O_2 \cdot [Chl]^{-1} \cdot s^{-1}$, should be inversely proportional to the measured antenna size.
- The absolute rate of linear electron transport is not decreased.
- The phenotype does not create respiratory "drag" by useless synthesis of proteins.

Additionally, it has been suggested by Formighieri et al. [1] that respiration increases in low-chlorophyll mutants as they can grow at higher densities, and thus the compensation point is reached at a higher light intensity. This would then require the light intensity to be significantly higher before a net gain is observed (Figure 24.1). Because NPQ partly relies on pigments such as carotenoids that are also involved in light harvesting, care must be taken to not hamper the photoprotective mechanism in an effort to reduce the antenna size [99].

In this chapter, we have summarized multiple ways in which photosynthetic organisms can be genetically adjusted to increase photosynthetic efficiency. In most cases, this is akin to irreversible photoacclimatization, which leads to reduced antenna size, either via chlorophyll content or via PBS composition. Regulation of expression and assembly of these systems is still poorly understood, which has led to the identification of several ineffective mutants that superficially appear to have the desired traits yet don't necessarily improve productivity in mass cultures. Therefore, it is both of practical and of fundamental value to further study these aspects in order to develop strategies to domesticate algae as we have done with crops over thousands of years [100]. Fortunately, our tools have improved.

24.4 LARGE-SCALE CULTURING IN 2D AND 3D PBRs

The use of large-scale cyanobacterial PBRs for direct product formation has several advantages; not the least of which are (1) the avoidance of the need to use arable land, (2) its closed minerals cycle, and (3) its low water requirement [32,101]. This approach may already compete favorably with traditional agriculture as is apparent from the study of Ducat et al. [41] on sucrose production by an engineered cyanobacterium, which led these authors to conclude that "Sucrose production in engineered *S. elongatus* compares favorably with sugarcane and other agricultural crops." An important aspect of this approach, which so far received relatively little attention, is the downstream processing for product recovery. Depending on the specific product selected for large-scale production, use can be made of selected physicochemical properties like volatility and water miscibility [102,103]. As an alternative, two-stage approaches can be selected, like in Günther et al. [40]: Glycolate is produced by excessive photorespiration in green algae, which then leaks out into the external medium. In a second reactor/phase, this product is subsequently converted into methane, which is then recovered [40]. Likewise, it would be possible to convert the lactic acid produced by cyanobacteria into an organic solvent through fermentation by lactic acid bacteria.

Nevertheless, in a future, sustainable society that puts efforts into closing the global carbon cycle, additional forms of application of cyanobacterial and algal mass cultures can be foreseen. In that society, a large part of the energy needs will (have to) be covered by the generation of renewable electricity, with the use of solar panels, wind turbines, and tidal and hydroelectric plants. As the daily rate of generation of renewable electricity will not be constant, new innovative ways will have to be developed to use "peak electricity." Because a closed carbon cycle also implies that fossil long-chain hydrocarbons will no longer be available, this provides an opportunity for engineered cyanobacteria and algae to show that they can use their natural capacity for CO_2 fixation as an alternative source.

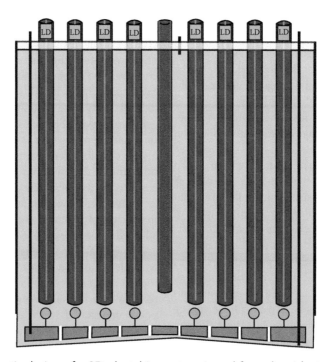

Figure 24.3 A schematic design of a 3D photobioreactor viewed from the side. Indicated are LED (red) and sensor rods (dark grey) submerged in the culture volume and bubbling vents (light grey). Critical in its design is efficient illumination of the culture volume in order to minimize respiration and photoinhibition as well as efficient gassing for proper mixing, to provide CO_2, and to remove O_2.

In the slightly more distant future, this may even lead to an entirely new form of agriculture: Current best photovoltaic cells have a solar power to electricity conversion efficiency of 44.5%; one may therefore anticipate that within a decade panels with 50% efficiency will be available [104].

Likewise, the efficiency of light emission by LEDs shows a trend of improvement that make one predict that in the same period, 70% efficiency may be achievable [105]. Therefore, the free energy of sunlight should be transportable to a 3D (i.e., volumetric) PBR with 35% power efficiency. If we let these LEDs emit light with a wavelength of ~690 nm, textbook schemes suggest that the energy of this monochromatic light should be convertible to NADPH and ATP and hence to Calvin cycle intermediates with approximately 30% free energy conversion efficiency [106]. Hence, this type of photosynthesis should have an overall efficiency of $0.5 \times 0.7 \times 0.3$ ($\times 100\%$) = 10.5%. This is considerably higher than any type of plant photosynthesis and is not hampered by factors such as insects, drought, hail, etc., factors that bring the actual crop yield significantly below the theoretical maximum. A schematic design of such a 3D PBR is shown in Figure 24.3.

REFERENCES

1. Formighieri, C., F. Franck, and R. Bassi. 2012. Regulation of the pigment optical density of an algal cell: Filling the gap between photosynthetic productivity in the laboratory and in mass culture. *J. Biotechnol.* 162, 115–123.

2. Zhu, X., S. P. Long, and D. R. Ort. 2008. What is the maximum efficiency with which photosynthesis can convert solar energy into biomass? *Curr. Opin. Biotechnol.* 19, 153–159.

3. Janssen, M., M. Janssen, M. de Winter, J. Tramper, L. R. Mur, J. Snel, and R. H. Wijffels. 2000. Efficiency of light utilization of *Chlamydomonas reinhardtii* under medium-duration light/dark cycles. *J. Biotechnol.* 78, 123–137.

4. Melis, A. 2009. Solar energy conversion efficiencies in photosynthesis: Minimizing the chlorophyll antennae to maximize efficiency. *Plant Sci.* 177, 272–280.

5. Bolton, J. R. and D. O. Hall. 1991. The maximum efficiency of photosynthesis. *Photochem. Photobiol.* 53, 545–548.

6. Yin, X. and P. C. Struik. 2010. Modelling the crop: From system dynamics to systems biology. *J. Exp. Bot.* 61, 2171–2183.

7. Björkman, O. and B. Demmig. 1987. Photon yield of O_2 evolution and chlorophyll fluorescence characteristics at 77 K among vascular plants of diverse origins. *Planta* 170, 489–504.

8. Evans, J. R. 2013. Improving photosynthesis. *Plant Physiol.* 162, 1780–1793.

9. Evans, J. 1987. The dependence of quantum yield on wavelength and growth irradiance. *Funct. Plant Biol.* 14, 69–79.

10. Ley, A. C. and D. C. Mauzerall. 1982. Absolute absorption cross-sections for photosystem II and the minimum quantum requirement for photosynthesis in *Chlorella vulgaris. Biochim. Biophys. Acta (BBA)—Bioenergetics* 680, 95–106.

11. Schuurmans, R. M., P. van Alphen, J. M. Schuurmans, H. C. Matthijs, and K. J. Hellingwerf. 2015. Comparison of the photosynthetic yield of cyanobacteria and green algae: Different methods give different answers. *PloS One* 10, e0139061.

12. Ugwu, C. U. and H. Aoyagi. 2012. Designs, operation and applications. *Biotechnology* 11, 127–132.

13. Ugwu, C., H. Aoyagi, and H. Uchiyama. 2008. Photobioreactors for mass cultivation of algae. *Bioresour. Technol.* 99, 4021–4028.

14. Pulz, O. 2001. Photobioreactors: Production systems for phototrophic microorganisms. *Appl. Microbiol. Biotechnol.* 57, 287–293.

15. Collins, A. M., M. Liberton, H. D. Jones, O. F. Garcia, H. B. Pakrasi, and J. A. Timlin. 2012. Photosynthetic pigment localization and thylakoid membrane morphology are altered in *Synechocystis* 6803 phycobilisome mutants. *Plant Physiol.* 158, 1600–1609.

16. Fujita, Y., A. Murakami, and K. Aizawa. 1995. The accumulation of protochlorophyllide in cells of *Synechocystis* PCC 6714 with a low PSI/PSII stoichiometry. *Plant Cell Physiol.* 36, 575–582.

17. Moal, G. and B. Lagoutte. 2012. Photo-induced electron transfer from photosystem I to NADP: Characterization and tentative simulation of the in vivo environment. *Biochim. et Biophys. Acta (BBA)-Bioenergetics* 1817, 1635–1645.

18. Glick, R. E. and A. Melis. 1988. Minimum photosynthetic unit size in system I and system II of barley chloroplasts. *Biochim. et Biophys. Acta (BBA)-Bioenergetics* 934, 151–155.

19. Rögner, M., P. J. Nixon, and B. A. Diner. 1990. Purification and characterization of photosystem I and photosystem II core complexes from wild-type and phycocyanin-deficient strains of the cyanobacterium *Synechocystis* PCC 6803. *J. Biol. Chem.* 265, 6189–6196.

20. Allen, J. F. 2003. State transitions—A question of balance. *Science* 299, 1530–1532.

21. Wollman, F. 2001. State transitions reveal the dynamics and flexibility of the photosynthetic apparatus. *EMBO J.* 20, 3623–3630.

22. Acuña, A. M., R. Kaňa, M. Gwizdala, J. J. Snellenburg, P. van Alphen, B. van Oort, D. Kirilovsky, R. van Grondelle, and I. H. van Stokkum. 2016. A method to decompose spectral changes in *Synechocystis* PCC 6803 during light-induced state transitions. *Photosynth. Res.* 130, 237–249.

23. Kaňa, R., E. Kotabová, O. Komárek, B. Šedivá, G. C. Papageorgiou, and O. Prášil. 2012. The slow S to M fluorescence rise in cyanobacteria is due to a state 2 to state 1 transition. *Biochim. et Biophys. Acta (BBA)-Bioenergetics* 1817, 1237–1247.

24. Mullineaux, C. W. and D. Emlyn-Jones. 2005. State transitions: An example of acclimation to low-light stress. *J. Exp. Bot.* 56, 389–393.

25. Matthijs, H. C., H. Balke, U. M. Van Hes, B. Kroon, L. R. Mur, and R. A. Binot. 1996. Application of light-emitting diodes in bioreactors: Flashing light effects and energy economy in algal culture (*Chlorella pyrenoidosa*). *Biotechnol. Bioeng.* 50, 98–107.

26. Tredici, M. R. and G. C. Zittelli. 1998. Efficiency of sunlight utilization: Tubular versus flat photobioreactors. *Biotechnol. Bioeng.* 57, 187–197.

27. Hu, Q., H. Guterman, and A. Richmond. 1996. A flat inclined modular photobioreactor for outdoor mass cultivation of photoautotrophs. *Biotechnol. Bioeng.* 51, 51–60.

28. Granata, T. 2016. Dependency of microalgal production on biomass and the relationship to yield and bioreactor scale-up for biofuels: A statistical analysis of 60 years of algal bioreactor data. *BioEnergy Res.* 10, 267–287.

29. Czernik, S. and A. Bridgwater. 2004. Overview of applications of biomass fast pyrolysis oil. *Energy Fuels* 18, 590–598.

30. Wijffels, R. H. and M. J. Barbosa. 2010. An outlook on microalgal biofuels. *Science (Washington)* 329, 796–799.

31. Vanthoor-Koopmans, M., R. H. Wijffels, M. J. Barbosa, and M. H. Eppink. 2013. Biorefinery of microalgae for food and fuel. *Bioresour. Technol.* 135, 142–149.

32. Anemaet, I. G., M. Bekker, and K. J. Hellingwerf. 2010. Algal photosynthesis as the primary driver for a sustainable development in energy, feed, and food production. *Marine Biotechnol.* 12, 619–629.

33. Hannon, M., J. Gimpel, M. Tran, B. Rasala, and S. Mayfield. 2010. Biofuels from algae: Challenges and potential. *Biofuels.* 1, 763–784.

34. Singh, A., P. S. Nigam, and J. D. Murphy. 2011. Renewable fuels from algae: An answer to debatable land based fuels. *Bioresour. Technol.* 102, 10–16.

35. Losick, R. and P. Stragier. 1992. Crisscross regulation of cell-type-specific gene expression during development in *B. subtilis. Nature.* 355, 601–604.

36. Okada, K. 2011. The biosynthesis of isoprenoids and the mechanisms regulating it in plants. *Biosci. Biotechnol. Biochem.* 75, 1219–1225.

37. Deng, M. and J. R. Coleman. 1999. Ethanol synthesis by genetic engineering in cyanobacteria. *Appl. Environ. Microbiol.* 65, 523–528.

38. Hellingwerf, K. and M. Teixeira de Mattos. 2009. Alternative routes to biofuels: Light-driven biofuel formation from CO_2 and water based on the 'photanol'approach. *J. Biotechnol.* 142, 87–90.

39. Lan, E. I. and J. C. Liao. 2012. ATP drives direct photosynthetic production of 1-butanol in cyanobacteria. *Proc. Natl. Acad. Sci.* 109, 6018–6023.

40. Günther, A., T. Jakob, R. Goss, S. König, D. Spindler, N. Räbiger, S. John, S. Heithoff, M. Fresewinkel, and C. Posten. 2012. Methane production from glycolate excreting algae as a new concept in the production of biofuels. *Bioresour. Technol.* 121, 454–457.

41. Ducat, D. C., J. A. Avelar-Rivas, J. C. Way, and P. A. Silver. 2012. Rerouting carbon flux to enhance photosynthetic productivity. *Appl. Environ. Microbiol.* 78, 2660–2668.

42. Angermayr, S. A., M. Paszota, and K. J. Hellingwerf. 2012. Engineering a cyanobacterial cell factory for production of lactic acid. *Appl. Environ. Microbiol.* 78, 7098–7106.

43. Gao, Z., H. Zhao, Z. Li, X. Tan, and X. Lu. 2012. Photosynthetic production of ethanol from carbon dioxide in genetically engineered cyanobacteria. *Energy Environ. Sci.* 5, 9857–9865.

44. Algenol Biofuels. 2017. www.algenol.com.

45. Niederholtmeyer, H., B. T. Wolfstädter, D. F. Savage, P. A. Silver, and J. C. Way. 2010. Engineering cyanobacteria to synthesize and export hydrophilic products. *Appl. Environ. Microbiol.* 76, 3462–3466.

46. Takahama, K., M. Matsuoka, K. Nagahama, and T. Ogawa. 2003. Construction and analysis of a recombinant cyanobacterium expressing a chromosomally inserted gene for an ethylene-forming enzyme at the *psbAI* locus. *J. Biosci. Bioeng.* 95, 302–305.

47. Oliver, J. W., I. M. Machado, H. Yoneda, and S. Atsumi. 2013. Cyanobacterial conversion of carbon dioxide to 2, 3-butanediol. *Proc. Natl. Acad. Sci.* 110, 1249–1254.

48. Savakis, P. E., S. A. Angermayr, and K. J. Hellingwerf. 2013. Synthesis of 2, 3-butanediol by *Synechocystis* sp. PCC6803 via heterologous expression of a catabolic pathway from lactic acid-and enterobacteria. *Metab. Eng.* 20, 121–130.

49. Lindberg, P., S. Park, and A. Melis. 2010. Engineering a platform for photosynthetic isoprene production in cyanobacteria, using *Synechocystis* as the model organism. *Metab. Eng.* 12, 70–79.

50. Atsumi, S., W. Higashide, and J. C. Liao. 2009. Direct photosynthetic recycling of carbon dioxide to isobutyraldehyde. *Nat. Biotechnol.* 27, 1177–1180.

51. Lan, E. I. and J. C. Liao. 2012. Microbial synthesis of n-butanol, isobutanol, and other higher alcohols from diverse resources. *Bioresour. Technol.* 135, 339–349.

52. Liu, X., S. Fallon, J. Sheng, and R. Curtiss. 2011. CO_2-limitation-inducible green recovery of fatty acids from cyanobacterial biomass. *Proc. Natl. Acad. Sci.* 108, 6905–6908.

53. Wang, W., X. Liu, and X. Lu. 2013. Engineering cyanobacteria to improve photosynthetic production of alka(e)nes. *Biotechnol. Biofuels* 6, 1–9.

54. Takahashi, H., H. Uchimiya, and Y. Hihara. 2008. Difference in metabolite levels between photoautotrophic and photomixotrophic cultures of *Synechocystis* sp. PCC 6803 examined by capillary electrophoresis electrospray ionization mass spectrometry. *J. Exp. Bot.* 59, 3009–3018.

55. Angermayr, S. A. and K. J. Hellingwerf. 2013. On the use of metabolic control analysis in the optimization of cyanobacterial biosolar cell factories. *J. Phys. Chem. B.* 117, 11169–11175.

56. Angermayr, S. A., A. D. Van der Woude, D. Correddu, A. Vreugdenhil, V. Verrone, and K. J. Hellingwerf. 2014. Exploring metabolic engineering design principles for the photosynthetic production of lactic acid by *Synechocystis* sp. PCC6803. *Biotechnol. Biofuels* 7, 1.

57. Bateman, J. and S. Purton. 2000. Tools for chloroplast transformation in *Chlamydomonas*: Expression vectors and a new dominant selectable marker. *Mol. General Gene. MGG* 263, 404–410.

58. Franklin, S. E. and S. P. Mayfield. 2004. Prospects for molecular farming in the green alga *Chlamydomonas reinhardtii*. *Curr. Opin. Plant Biol.* 7, 159–165.

59. Coragliotti, A. T., M. V. Beligni, S. E. Franklin, and S. P. Mayfield. 2011. Molecular factors affecting the accumulation of recombinant proteins in the *Chlamydomonas reinhardtii* chloroplast. *Mol. Biotechnol.* 48, 60–75.

60. Lauersen, K. J., H. Berger, J. H. Mussgnug, and O. Kruse. 2012. Efficient recombinant protein production and secretion from nuclear transgenes in *Chlamydomonas reinhardtii*. *J. Biotechnol.* 167, 101–110.

61. Rasala, B. A., P. A. Lee, Z. Shen, S. P. Briggs, M. Mendez, and S. P. Mayfield. 2012. Robust expression and secretion of Xylanase1 in

Chlamydomonas reinhardtii by fusion to a selection gene and processing with the FMDV 2A peptide. *PloS One* 7, e43349.

62. La Russa, M., C. Bogen, A. Uhmeyer, A. Doebbe, E. Filippone, O. Kruse, and J. H. Mussgnug. 2012. Functional analysis of three type-2 DGAT homologue genes for triacylglycerol production in the green microalga *Chlamydomonas reinhardtii*. *J. Biotechnol.* 162, 13–20.

63. Yu, W., W. Ansari, N. G. Schoepp, M. J. Hannon, S. P. Mayfield, and M. D. Burkart. 2011. Modifications of the metabolic pathways of lipid and triacylglycerol production in microalgae. *Microb. Cell Fact.* 10, 91.

64. Gouveia, L. and A. C. Oliveira. 2009. Microalgae as a raw material for biofuels production. *J. Ind. Microbiol. Biotechnol.* 36, 269–274.

65. Griffiths, M. J., R. P. van Hille, and S. T. Harrison. 2012. Lipid productivity, settling potential and fatty acid profile of 11 microalgal species grown under nitrogen replete and limited conditions. *J. Appl. Phycol.* 24, 989–1001.

66. Hu, Q., M. Sommerfeld, E. Jarvis, M. Ghirardi, M. Posewitz, M. Seibert, and A. Darzins. 2008. Microalgal triacylglycerols as feedstocks for biofuel production: Perspectives and advances. *Plant J.* 54, 621–639.

67. Klok, A. J., D. E. Martens, R. H. Wijffels, and P. P. Lamers. 2013. Simultaneous growth and neutral lipid accumulation in microalgae. *Bioresour. Technol.* 134, 233–243.

68. Derelle, E., C. Ferraz, S. Rombauts, P. Rouzé, A. Z. Worden, S. Robbens, F. Partensky, S. Degroeve, S. Echeynié, and R. Cooke. 2006. Genome analysis of the smallest free-living eukaryote *Ostreococcus tauri* unveils many unique features. *Proc. Natl. Acad. Sci.* 103, 11647–11652.

69. Merchant, S. S., S. E. Prochnik, O. Vallon, E. H. Harris, S. J. Karpowicz, G. B. Witman, A. Terry, A. Salamov, L. K. Fritz-Laylin, and L. Maréchal-Drouard. 2007. The *Chlamydomonas* genome reveals the evolution of key animal and plant functions. *Science* 318, 245–250.

70. Radakovits, R., R. E. Jinkerson, S. I. Fuerstenberg, H. Tae, R. E. Settlage, J. L. Boore, and M. C. Posewitz. 2012. Draft genome sequence and genetic transformation of the oleaginous alga *Nannochloropis gaditana*. *Nat. Commun.* 3, 686.

71. Molnar, A., A. Bassett, E. Thuenemann, F. Schwach, S. Karkare, S. Ossowski, D. Weigel, and D. Baulcombe. 2009. Highly specific gene silencing by artificial microRNAs in the unicellular alga *Chlamydomonas reinhardtii*. *Plant J.* 58, 165–174.

72. Neupert, J., N. Shao, Y. Lu, and R. Bock. 2012. Genetic transformation of the model green alga *Chlamydomonas reinhardtii*. In: J. Dunwell and A. Wetten (eds), *Transgenic Plants Methods in Molecular Biology*, Humana Press, New York, pp. 35–47.

73. Mayfield, S. P., A. L. Manuell, S. Chen, J. Wu, M. Tran, D. Siefker, M. Muto, and J. Marin-Navarro. 2007. *Chlamydomonas reinhardtii* chloroplasts as protein factories. *Curr. Opin. Biotechnol.* 18, 126–133.

74. Baba, M., M. Loki, N. Nakajima, Y. Shiraiwa, and M. M. Watanabe. 2012. Transcriptome analysis of an oil-rich race A strain of *Botryococcus braunii* (BOT-88-2) by *de novo* assembly of pyrosequencing cDNA reads. *Bioresour. Technol.* 109, 282–286.

75. Molnár, I., D. Lopez, J. H. Wisecaver, T. P. Devarenne, T. L. Weiss, M. Pellegrini, and J. D. Hackett. 2012. Bio-crude transcriptomics: Gene discovery and metabolic network reconstruction for the biosynthesis of the terpenome of the hydrocarbon oil-producing green alga, *Botryococcus braunii* race B (showa). *BMC Genomics* 13, 576.

76. Kirst, H., J. G. García-Cerdán, A. Zurbriggen, and A. Melis. 2012. Assembly of the light-harvesting chlorophyll antenna in the green alga *Chlamydomonas reinhardtii* requires expression of the TLA2-CpFTSY gene. *Plant Physiol.* 158, 930–945.

77. Kirst, H., J. G. Garcia-Cerdan, A. Zurbriggen, T. Ruehle, and A. Melis. 2012. Truncated photosystem chlorophyll antenna size in the green microalga *Chlamydomonas reinhardtii* upon deletion of the TLA3-CpSRP43 gene. *Plant Physiol.* 160, 2251–2260.

78. Polle, J. E., S. Kanakagiri, and A. Melis. 2003. Tla1, a DNA insertional transformant of the green alga *Chlamydomonas reinhardtii* with a truncated light-harvesting chlorophyll antenna size. *Planta* 217, 49–59.

79. Beckmann, J., F. Lehr, G. Finazzi, B. Hankamer, C. Posten, L. Wobbe, and O. Kruse. 2009.

Improvement of light to biomass conversion by de-regulation of light-harvesting protein translation in *Chlamydomonas reinhardtii*. *J. Biotechnol.* 142, 70–77.

80. Oey, M., I. L. Ross, E. Stephens, J. Steinbeck, J. Wolf, K. A. Radzun, J. Kügler, A. K. Ringsmuth, O. Kruse, and B. Hankamer. 2013. RNAi knock-down of LHCBM1, 2 and 3 increases photosynthetic H_2 production efficiency of the green alga *Chlamydomonas reinhardtii*. *PloS One* 8, e61375.

81. Perrine, Z., S. Negi, and R. T. Sayre. 2012. Optimization of photosynthetic light energy utilization by microalgae. *Algal Res.* 1, 134–142.

82. Mitra, M., H. Kirst, D. Dewez, and A. Melis. 2012. Modulation of the light-harvesting chlorophyll antenna size in *Chlamydomonas reinhardtii* by TLA1 gene over-expression and RNA interference. *Philos. Trans. R. Soc. B: Biol. Sci.* 367, 3430–3443.

83. Masuda, T., A. Tanaka, and A. Melis. 2003. Chlorophyll antenna size adjustments by irradiance in *Dunaliella salina* involve coordinate regulation of chlorophyll a oxygenase (CAO) and *lhcb* gene expression. *Plant Mol. Biol.* 51, 757–771.

84. Melis, A., J. Neidhardt, I. Baroli, and J. R. Benemann. 1998. Maximizing photosynthetic productivity and light utilization in microalgae by minimizing the light-harvesting chlorophyll antenna size of the photosystems. In: O. R., Zaborsky, J. R. Benemann, T. Matsunaga, J. Miyake, and A. San Pietro (eds), *BioHydrogen*, Springer, Boston, MA, pp. 41–52.

85. Mussgnug, J. H., S. Thomas-Hall, J. Rupprecht, A. Foo, V. Klassen, A. McDowall, P. M. Schenk, O. Kruse, and B. Hankamer. 2007. Engineering photosynthetic light capture: Impacts on improved solar energy to biomass conversion. *Plant Biotechnol. J.* 5, 802–814.

86. Bernát, G., N. Waschewski, and M. Rögner. 2009. Towards efficient hydrogen production: The impact of antenna size and external factors on electron transport dynamics in *Synechocystis* PCC 6803. *Photosynth. Res.* 99, 205–216.

87. Kramer, D. M. and J. R. Evans. 2011. The importance of energy balance in improving photosynthetic productivity. *Plant Physiol.* 155, 70–78.

88. Kramer, D. M., T. J. Avenson, and G. E. Edwards. 2004. Dynamic flexibility in the light reactions of photosynthesis governed by both electron and proton transfer reactions. *Trends Plant Sci.* 9, 349–357.

89. Munekage, Y., M. Hashimoto, C. Miyake, K. Tomizawa, T. Endo, M. Tasaka, and T. Shikanai. 2004. Cyclic electron flow around photosystem I is essential for photosynthesis. *Nature* 429, 579–582.

90. Abe, S., A. Murakami, K. Ohki, Y. Aruga, and Y. Fujita. 1994. Changes in stoichiometry among PSI, PSII and cyt b6-f complexes in response to chromatic light for cell growth observed with the red alga *Porphyra yezoensis*. *Plant Cell Physiol.* 35, 901–906.

91. Kondo, K., Y. Ochiai, M. Katayama, and M. Ikeuchi. 2007. The membrane-associated CpcG2-phycobilisome in *Synechocystis*: A new photosystem I antenna. *Plant Physiol.* 144, 1200–1210.

92. Elmorjani, K., J. Thomas, and P. Sebban. 1986. Phycobilisomes of wild type and pigment mutants of the cyanobacterium *Synechocystis* PCC 6803. *Arch. Microbiol.* 146, 186–191.

93. Nakajima, Y. and R. Ueda. 1999. Improvement of microalgal photosynthetic productivity by reducing the content of light harvesting pigment. *J. Appl. Phycol.* 11, 195–201.

94. Kirst, H., C. Formighieri, and A. Melis. 2014. Maximizing photosynthetic efficiency and culture productivity in cyanobacteria upon minimizing the phycobilisome light-harvesting antenna size. *Biochim. et Biophys. Acta (BBA)-Bioenergetics.* 1837, 1653–1664.

95. Lea-Smith, D. J., P. Bombelli, J. S. Dennis, S. A. Scott, A. G. Smith, and C. J. Howe. 2014. Phycobilisome-deficient strains of *Synechocystis* sp. PCC 6803 have reduced size and require carbon-limiting conditions to exhibit enhanced productivity. *Plant Physiol.* 165, 705–714.

96. Page, L. E., M. Liberton, and H. B. Pakrasi. 2012. Reduction of photoautotrophic productivity in the cyanobacterium *Synechocystis* sp. strain PCC 6803 by phycobilisome antenna truncation. *Appl. Environ. Microbiol.* 78, 6349–6351.

97. Tian, L., M. Gwizdala, I. H. van Stokkum, R. B. Koehorst, D. Kirilovsky, and H. van Amerongen. 2012. Picosecond kinetics of light harvesting and photoprotective quenching in wild-type and mutant phyco-bilisomes isolated from the cyanobacterium *Synechocystis* PCC 6803. *Biophys. J.* 102, 1692–1700.

98. Arteni, A. A., G. Ajlani, and E. J. Boekema. 2009. Structural organisation of phyco-bilisomes from *Synechocystis* sp. strain PCC6803 and their interaction with the membrane. *Biochim. Et Biophys. Acta (BBA)-Bioenergetics* 1787, 272–279.

99. Havaux, M., L. Dall'Osto, and R. Bassi. 2007. Zeaxanthin has enhanced antioxidant capac-ity with respect to all other xanthophylls in *Arabidopsis* leaves and functions inde-pendent of binding to PSII antennae. *Plant Physiol.* 145, 1506–1520.

100. Doebley, J. F., B. S. Gaut, and B. D. Smith. 2006. The molecular genetics of crop domestication. *Cell* 127, 1309–1321.

101. Angermayr, S. A., K. J. Hellingwerf, P. Lindblad, T. de Mattos, and M. Joost. 2009. Energy biotechnology with cyanobacteria. *Curr. Opin. Biotechnol.* 20, 257–263.

102. Bentley, F. K. and A. Melis. 2012. Diffusion-based process for carbon dioxide uptake and isoprene emission in gaseous/aqueous two-phase photo-bioreactors by photosynthetic microorganisms. *Biotechnol. Bioeng.* 109, 100–109.

103. Varman, A. M., Y. Xiao, H. B. Pakrasi, and Y. J. Tang. 2013. Metabolic engineering of *Synechocystis* sp. strain PCC 6803 for isobu-tanol production. *Appl. Environ. Microbiol.* 79, 908–914.

104. Atwater, H. A. and A. Polman. 2010. Plasmonics for improved photovoltaic devices. *Nat. Mater.* 9, 205–213.

105. Piprek, J. and Z. S. Li. 2013. Origin of InGaN light-emitting diode efficiency improvements using chirped AlGaN multi-quantum barriers. *Appl. Phys. Lett.* 102, 023510–023510-4.

106. Berg, J., J. Tymoczko, and L. Stryer. 2002. *Biochemistry.* W. H. Freeman, New York.

Index

Printed and bound by CPI Group (UK) Ltd, Croydon, CR0 4YY

01/11/2024

01782604-0015